Trace Elements in Man and Animals 6

Trace Elements in Man and Animals 6

Edited by
Lucille S. Hurley,
Carl L. Keen,
Bo Lönnerdal,
and Robert B. Rucker
University of California, Davis
Davis, California

PLENUM PRESS • NEW YORK AND LONDON

Library of Congress Cataloging in Publication Data

International Symposium on Trace Elements in Man and Animals (6th: 1987: Pacific
Grove, Calif.)
    Trace elements in man and animals 6 / edited by Lucille S. Hurley, . . . [et al.].
        p.   cm.
    "Proceedings of the Sixth International Symposium on Trace Elements in Man and
Animals, held May 31–June 5, 1987, in Pacific Grove, California"—T.p. verso.
    Includes bibliographical references and index.
    ISBN 0-306-43004-5
    1. Trace elements in the body—Congresses. 2. Trace elements—Physiological
effect—Congresses. 3. Trace elements in nutrition—Congresses. I. Hurley, Lucille S.
II. Title.
QP534.I575   1987   88-22554
599′.019′214—dc19   CIP

Proceedings of the Sixth International Symposium on
Trace Elements in Man and Animals, held May 31–June 5, 1987,
in Pacific Grove, California

© 1988 Plenum Press, New York
A Division of Plenum Publishing Corporation
233 Spring Street, New York, N.Y. 10013

Printed in the United States of America

PREFACE

This book is the published proceedings of the Sixth International Symposium on Trace Element Metabolism in Man and Animals. The Symposium was held at the Asilomar Conference Center in Pacific Grove, California, U.S.A. from May 31 through June 5, 1987. The decision to hold TEMA-6 at Asilomar was made at TEMA-5 in 1985. The International Guidance Committee decided to hold the meeting in California in part to recognize the significant contributions made to the field of trace element metabolism by Professor Lucille S. Hurley. As such, she was the obvious choice as chair of the local organizing committee. One of the principal goals of Professor Hurley was that TEMA-6 serve as a forum for discussing the use and application of newer methodologies, such as molecular biology, computer modelling and stable isotopes, in studies of trace element metabolism. Based on the comments which the local organizing committee has received, this goal was achieved.

The Symposium was attended by 275 scientists from 32 countries covering 6 continents. Twenty-five speakers were chosen for our plenary sessions. One hundred contributions were presented verbally and an additional 180 were presented as posters. Particularly gratifying to the Organizing Committee was the high attendance at both the evening sessions as well as those on the last day of the meeting, suggesting that, despite the pleasant surroundings of the Carmel Valley, the topics presented were of interest to all participants. We would like to express our sincere thanks to all the participants and session chairs for their valuable contributions. We are also very grateful to the members of the International Guidance Committee for their assistance in arranging this meeting, and to the graduate students in attendance from the University of California, Davis, who assisted in various ways in keeping the meeting running smoothly. Moreover, the editors also thank Ms. Barbara Brandon and Ms. Carol Biancalana for their excellent secretarial and managerial assistance, without which this book would not have been published in the timely fashion that it has been.

In August of 1988 we received the first galleys for this book, which were proofed by Professor Hurley, who expressed at the time her satisfaction with the Proceedings. It was a tremendous shock to us when, a few weeks later, Professor Hurley died from complications arising from surgery. The members of the local organizing committee would like to hope that, in some small way, the Proceedings of this meeting serve as a memorial to Professor Hurley and the outstanding scientific as well as personal contributions that she made to the field of trace element metabolism.

C. L. Keen, B. Lönnerdal and R. B. Rucker, Davis, 1988.

CONTENTS

TRACE ELEMENTS AND IMMUNE FUNCTION

TRACE ELEMENTS AND ENDOCRINE/EXOCRINE FUNCTION: Cu, Ca, I

MECHANISMS OF TRACE ELEMENT TRANSPORT

EPIDEMIOLOGICAL STUDIES OF TRACE ELEMENT
NUTRITION IN MAN AND ANIMALS

TRACE ELEMENTS AND FREE RADICALS

TRACE ELEMENTS IN CALCIFIED TISSUES AND MATRIX BIOLOGY

MOLECULAR BIOLOGY OF TRACE ELEMENT FUNCTION AND METABOLISM

TRACE ELEMENT INTERACTIONS 3

TRACE ELEMENTS AND ENDOCRINE/EXOCRINE FUNCTION

TRACE ELEMENTS AND IMMUNE FUNCTION/BACTERIOLOGY

TRACE ELEMENT TRANSPORT

TRACE ELEMENT EPIDEMIOLOGY

TRACE ELEMENT INTERACTIONS

THE GLOBAL SELENIUM AGENDA

Orville A. Levander

U.S. Department of Agriculture
Beltsville Human Nutrition Research Center
Beltsville, Maryland   USA   20705

Rapid increases have taken place in our awareness of the importance of selenium in human health and disease during the last decade. In 1979, two reports, one from New Zealand and the other from the People's Republic of China, described conditions related to human selenium deficiency. The New Zealand report concerned a total parenteral nutrition patient whose muscle pain and tenderness responded to selenium treatment (1). This patient was considered to be the first clinical case supporting the essential role of selenium in human nutrition. The report from China presented evidence that Keshan disease (see below) is associated with low selenium status (2). This experiment of nature is the result of the selenium-impoverished soils located in certain regions of China. On the other hand, another Chinese paper that appeared in 1983 described an outbreak of endemic human selenosis due to the consumption of foods that contained toxic amounts of selenium (3). The selenium in that episode leached from selenium-rich coal into the soil and was then available for uptake by food crops.

Such dramatic differences in the amount of selenium occurring naturally in the food supply have posed problems for public health officials in various countries. The purpose of this review is to document the worldwide differences in dietary selenium intakes, to discuss briefly the impact of low and high selenium intakes on human health, and to present our current knowledge regarding possible dietary standards for selenium.

WORLDWIDE DIFFERENCES IN SELENIUM INTAKE

Adult dietary selenium intakes as low as 7 ug/day have been measured in Keshan disease areas of China. On the other hand, estimated adult dietary selenium intakes in the human selenosis area of China reached 38,000 ug/day. These figures are the extreme values that have been reported for dietary selenium intakes around the world (4). Dietary intakes in New Zealand and Finland, for example, are low but range between 28 and 30 ug/day, still comfortably above that seen in the Keshan disease areas. Intakes approaching 600 ug/day were recently observed in South Dakota, U.S.A., but no evidence of selenium poisoning was found in that population (5).

1

The amount of selenium in the diet, of course, is only one factor that determines an individual's selenium status since the nutritional bioavailability (and hence presumably also the toxicological potency) of the selenium in foods varies considerably. Thus, the selenium in mushrooms was only 4% as effective as that in Brazil nuts for increasing hepatic glutathione peroxidase activity in rats depleted of selenium (6).

SELENIUM INTAKE AND HUMAN HEALTH

Two endemic diseases associated with poor selenium status have been extensively studied in China: Keshan disease, a cardiomyopathy that affects primarily infants, children, and women of child-bearing age, and Kashin-Beck disease, a juvenile osteoarthropathy (7). The names are confusing but Keshan is a county in northeastern China where the cardiomyopathy occurs and Kashin and Beck were two Russian scientists who conducted detailed investigations of the osteoarthropathy in eastern Siberia at the turn of the century. The relationship between selenium deficiency and Keshan disease now seems fairly clear but the etiology of Kashin-Beck disease is much less certain. Millions of persons are at risk with regard to these diseases and both diseases are considered national priority health problems in China.

Animal models have provided clues that severe selenium deficiency might under some conditions contribute to the development of heart disease of the cardiovascular type by modifying the metabolism of arachidonic acid. In rats fed selenium-deficient diets, the biosynthesis of prostacyclin was reduced whereas the production of thromboxane was stimulated (8). Such an alteration in the prostacyclin/ thromboxane ratio could lead to an increased proaggregatory tendency in vivo. Despite this plausible mechanistic hypothesis, the epidemiological evidence linking low selenium status and cardiovascular disease is less than satisfactory. Of five prospective epidemiological studies conducted in Scandinavia, one found an inverse association between the risk of death from ischemic heart disease and poor selenium status, two found no association, and two were equivocal (9).

The most significant unresolved problem in current research on human selenium nutrition in North America is whether or not selenium intakes at levels higher than those needed to prevent nutritional deficiency diseases of the classical type are of any benefit against cancer. The concept that selenium may play a role in cancer chemoprevention is based on both epidemiological studies and experiments with laboratory animals. Some individual-based prospective epidemiological studies are in agreement with an inverse relationship between selenium status and cancer risk in humans but others are not (4). Many animal studies have now been published which indicate that under certain conditions relatively high dietary levels of selenium can protect against a variety of experimentally induced tumors in rats and mice (10). On the other hand, selenium has been shown to increase the incidence of experimental cancer in animals under some conditions and has led some investigators to caution against the use of selenium compounds in humans for cancer prevention (11).

From the above, the rationale for the use of selenium as a cancer chemoprevention agent seems rather mixed, but some limited selenium intervention trials in humans are already underway and more extensive studies are being planned for the future. For that reason, it is important to learn as much as possible about the metabolism and possible hazards of elevated intakes of selenium in people. Two large-scale field trials are currently underway in high-selenium areas of the US and China in an attempt to define with greater precision the dietary intakes of selenium that are well-tolerated by humans.

Self-dosing with superpotent "health food" supplements that contained 182 times more selenium than was stated on the label caused acute selenium intoxication in thirteen persons in the U.S.A. in 1984 (12). Signs and symptoms included nausea, abdominal pain, diarrhea, hair and nail changes, peripheral neuropathy, fatigue, and irritability. Although this outbreak of human selenium poisoning was because of faulty manufacturing practice, the incident nonetheless emphasizes the toxic potential of such products.

Ecological selenium contamination apparently is a problem in certain parts of California, U.S.A. because of agricultural wastewaters that run off irrigated soils of high selenium content (13). Waterfowl nesting on ponds receiving such drainwater exhibit reproductive difficulties resembling signs of selenium toxicity in domestic birds. Fear of possible adverse human health effects due to selenium overexposure led the California State Department of Health Services to post health advisories limiting the consumption of game birds taken from these contaminated waters.

DIETARY SELENIUM STANDARDS

The first attempt to define a dietary standard for selenium was that of the US National Research Council in 1980 which suggested an estimated safe and adequate daily dietary intake of 50 to 200 ug for adults with correspondingly lower intakes for infants and children (14). This estimate was based primarily on animal experiments since few human data were available at that time.

Since 1980, several studies have appeared that have clarified considerably our understanding of human selenium requirements. For example, dietary surveys conducted in Keshan and non-Keshan disease areas of China indicated that the disease did not exist in those areas where the selenium intakes were at least 13 and 19 ug/day in female and male adults, respectively (15). These intakes presumably represent minimum selenium requirements.

The time-honored metabolic balance approach has now been employed by a number of investigators to estimate human selenium requirements. However, an international comparison reveals great variation in the "balance requirement" obtained in different countries. For example, Chinese men living in a Keshan-disease area whose typical dietary selenium intake averaged 8.8 ug/day needed only 7.4 ug of selenium per day to maintain balance (16). On the other hand, North American men who normally consumed about 90 ug/day required 80 ug of selenium per day to achieve balance (17). That is, the "balance requirement" appears to be determined primarily by a person's usual dietary selenium consumption (although body size may also play a role). Thus, the balance technique is not particularly useful in delineating human nutritional requirements for selenium.

Our best estimate of human selenium requirement derives from the elegant work of Yang and associates in Beijing who repleted Chinese men of low selenium status (usual dietary intake 10 ug/day) with graded doses of selenomethionine and followed changes in plasma glutathione peroxidase activity (15). This enzymatic activity plateaued in all subjects receiving a total of 40 ug of selenium or more daily (diet plus supplement), so this intake was considered to satisfy the physiological selenium requirement of Chinese men of this size (about 60 kg).

One could now formulate a rational dietary selenium standard based on Yang's results. To translate Yang's Chinese data on requirements into a dietary recommendation appropriate for the West, it would be necessary to incorporate a safety factor for individual variation as well as an adjustment for larger body size. If a coefficient of variation of 15% is arbitrarily assumed and if the standard North American male and female are presumed to weigh 79 and 62 kg, it is possible to calculate a dietary selenium recommendation rounded to 70 and 55 ug/day for adult men and women, respectively.

The above calculations of course do not take into account any hypothetical anti-cancer effects of elevated pharmacological selenium intakes.

THE GLOBAL SELENIUM AGENDA

An agenda is defined as a "plan of things to be done" and indeed some countries have already taken steps to ensure that their populations are protected against poor selenium status. In low-selenium areas of China, for example, children are given selenium pills to protect against the dread Keshan disease. In Finland, selenium is now being routinely incorporated into all the main fertilizers so that the national food supply will provide an average dietary selenium intake between 50 and 100 ug per day (18). On the other hand, no official program of selenium supplementation or fortification exists in New Zealand, another country with low-selenium soils. Moreover, a review prepared by the US National Institutes of Health concluded that present evidence does not favor dietary selenium supplementation for either life extension or cancer prevention (19). Clearly, authorities in every country will need to formulate their own policies regarding the necessity for or desirability of national selenium intervention programs.

REFERENCES

1. Van Rij, A.M., Thomson, C.D., McKenzie, J.M & Robinson, M.F. (1979) Selenium deficiency in total parenteral nutrition. Am. J. Clin. Nutr. 32: 2076-2085.
2. Keshan Disease Research Group (1979) Epidemiologic studies on the etiologic relationship of selenium and Keshan disease. Chin. Med. J. 92: 477-482.
3. Yang, G., Wang, S., Zhou, R. & Sun, S. (1983) Endemic selenium intoxication of humans in China. Am. J. Clin. Nutr. 37: 872-881.
4. Levander, O.A. (1987) A global view of human selenium nutrition. Ann. Rev. Nutr. 7: 227-250.
5. Longnecker, M.P., Taylor, P.R., Levander, O.A., Howe, S.M., Veillon, C., et al. (1987) Tissue selenium (Se) levels and indices of Se exposure in a seleniferous area. Federation Proc. 46: 587.
6. Chansler, M.W., Mutanen, M., Morris, V.C. & Levander, O.A. (1986) Nutritional bioavailability to rats of selenium in Brazil nuts and mushrooms. Nutr. Res. 6: 1419-1428.
7. Yang, G. (1987) Research on selenium-related problems in human health in China. In: Selenium in Biology and Medicine, Part A (Combs, G.F., Spallholz, J.E., Levander, O.A. & Oldfield, J.E., eds.) pp. 9-32, Van Nostrand Reinhold, New York.
8. Schoene, N.W., Morris, V.C. & Levander, O.A. (1986) Altered arachidonic acid metabolism in platelets and aortas from selenium-deficient rats. Nutr. Res. 6: 75-83.

9. Salonen, J.T. & Huttunen, J.K. (1986) Selenium in cardiovascular disease. Ann. Clin. Res. 6: 75-83.
10. Ip, C. (1986) The chemopreventive role of selenium in carcinogenesis. J. Am. Coll. Toxicol. 5: 7-20.
11. Birt, D.F. (1986) Update on the effects of vitamins A, C, and E and selenium on carcinogenesis. Proc. Soc. Exp. Biol. Med. 183: 311-320.
12. Helzlsouer, K., Jacobs, R., & Morris, S. (1985) Acute selenium intoxication in the United States. Federation Proc. 44: 1670.
13. Ohlendorf, H.M., Hoffman, D.J., Saiki, M.K., & Aldrich, T.W. (1986) Embryonic mortality and abnormalities of aquatic birds: apparent impacts of selenium from irrigation drainwater. Sci. Total Environ. 52: 49-63.
14. National Research Council (1980) Recommended Dietary Allowances, 9th ed. National Academy of Sciences, Washington.
15. Yang, G.Q., Zhu, L.Z., Liu, S.J., Gu, L.Z., Qian, P.C., et al. (1987) Human selenium requirements in China. In: Selenium in Biology and Medicine (Combs, G.F., Spallholz, J.E., Levander, O.A., & Oldfield, J.E., eds.), pp. 589-607, Van Nostrand Reinhold, New York.
16. Luo, X., Wei, H., Yang, C., Xing, J., Qiao, C., et al. (1985) Selenium intake and metabolic balance of 10 men from a low- selenium area of China. Am. J. Clin. Nutr. 42: 439-448.
17. Levander, O.A. & Morris, V.C. (1984) Dietary selenium levels needed to maintain balance in North American adults consuming self-selected diets. Am. J. Clin. Nutr. 39: 809-815.
18. Koivistoinen, P. & Huttunen, J.K. (1986) Selenium in food and nutrition in Finland. An overview on research and action. Ann. Clin. Res. 18: 13-17.
19. Schneider, E.L. & Reed, J.D. (1985) Life extension. N. Engl. J. Med. 312: 1159-1168.

THE PREDICAMENT - HOW SHOULD SELENOPROTEINS BE HYDROLYZED AND BIOAVAILABILITY STUDIES WITH SELENIUM CONDUCTED?

P. D. Whanger

Department of Agricultural Chemistry
Oregon State University
Corvallis, OR 97330

The majority of the selenium (Se) bioavailability studies have been conducted with selenite as the standard.[1,2] However, using Se accumulation as the basis, this could lead to erroneous conclusions if the test samples contain selenomethionine (SeMet) as the form of Se. This is demonstrated by a study with rats where Se (0.2 to 4.0 ppm) as either selenite or SeMet was fed to rats (see Figure).

At least two conclusions can be drawn from this work. Blood Se levels may not always reflect the tissue levels of Se, particularly in the muscle. The magnitude of difference in deposition of Se in tissues between selenite and SeMet becomes more pronounced as the dietary Se levels are increased. The average fold increase of both blood and liver Se was 1.2, 1.8, 2.7 and 3.7 in rats fed SeMet as compared to selenite. In contrast, the fold increase of Se in muscle due to SeMet versus selenite was 2.7, 9, 17 and 26 respectively for rats fed 0.2, 1.0, 2.0 and 4.0 ppm Se. Using the slope-ratio analyses procedure[2] for muscle Se, SeMet would be calculated to be 73 times more available than selenite. This factor could account for values reported in the literature where Se in the test substance was calculated to be several times more available than selenite. Therefore, the same chemical forms of Se should be used as the standard

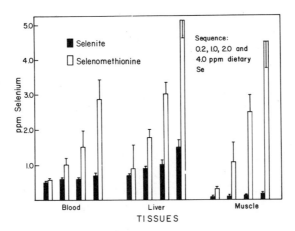

Fig. 1. Effects of feeding various levels of selenium as either selenite or selenomethionine on tissue levels of selenium in rats.

as present in the test substance. Interestingly, the glutathione peroxidase (GPx) activity did not differ significantly between tissues from rats fed SeMet versus selenite even though the Se content varied markedly.[3] This is an example where GPx activity is not always correlated with Se content.

To determine the chemical forms of Se in test materials, they must be hydrolyzed. Methods in which both SeMet and selenocysteine (CySe) could be determined on the same hydrolyzate were evaluated. Three hydrolysis methods [6N HCl, proteolytic enzymes (pronase plus prolidase) and thioethane-sulfonic acid (TESA)] were tested with hemoglobin, Se enriched yeast, high Se wheat and liver. Hydrolysis with TESA gave acceptable recovery of SeMet but this procedure destroyed CySe. Hydrolysis with the proteolytic enzymes usually gave good recovery but that for CySe in yeast sometimes resulted in poor recovery. Hydrolysis with HCl gave the most satisfactory recovery of both SeMet and CySe with more of the test materials. It should be emphasized that hydrolysis with HCl must be done under an absolutely oxygen-free atmosphere.

The hydrolyzates were chromatographed on an amino acid analyzer column (Dionex DC6A resin) with 0.1 M sodium citrate buffers.[4] Standard CySe and tissue hydrolyzates were chromatographed with added cysteine (CyS) so that it would elute as the mixed selenodisulfide of cysteine (CySSeCy). This is necessary to clearly separate CySe from SeMet. The elution times (in minutes) for some selenocompounds on this column were respectively 11, 13, 54, 71, 74, 75, 76 and 76 for selenate, selenite, selenodiglutathione, selenodisulfide of cysteine (CySSeCy), selenotrisulfide of cysteine (CySSeSCY), SeMet and selenodimercapoethanol. Therefore, since the last 4 compounds elute near the same position, it is prudent to confirm the identity of the compound suspected to be SeMet. One way is to determine whether this compound can be converted to Se-adenosylselenomethionine when incubated with S-adenosyl methionine synthetase and ATP. Another approach is to determine whether treatment with methyl cyanogen bromide yields methyl selenocyanate. For other selenoamino acids like CySe, the carboxylmethyl derivative has been commonly made to confirm its identity.[6] However, this cannot be done when both SeMet and CySe are to be determined in one preparation because the carboxylation process will destroy SeMet.

In conclusion, it is suggested that the same chemical forms of Se be used in the standards as in the test materials when bioavailability studies are conducted with Se. When both SeMet and CySe are to be determined, hydrolysis with 6N HCl under an oxygen-free atmosphere appears to be adequate with most selenoproteins. It is always prudent, however, to confirm the identity of the selenocompound by a second method, especially when new selenoproteins are to be analyzed.

ACKNOWLEDGEMENTS

Supported by NIH grant NS 07413.

REFERENCES

1. O.A. Levander, Selenium in Biology and Medicine, Third International Symposium, Beijing, PRC, G.F. Combs, J.E. Spallholz, O.A. Levander and J.E. Oldfield, Eds., Avi, Van Nostrand Reinhold Co., New York, N.Y., p. 403.
2. A.R. Alexander, P.D. Whanger and L.T. Miller, J. Nutr. 113:196 (1983).
3. P.D. Whanger and J.A. Butler. Biol. Trace Element Res. (subm. 1987).
4. M.A. Beilstein and P.D. Whanger, J. Nutr. 116:1711 (1986).
5. M.G.N. Hartmanis and T.C. Stadman. Proc. Natl. Acad. Sci. 79:4912 (1982).
6. J.W. Forstrom, J.J. Zakowski and A.L. Tappel. Biochemistry 17:2639 (1978).

BIOAVAILABILITY OF SELENIUM IN WHEAT AND MUSHROOMS AS ASSESSED BY A SHORT-TERM AND A LONG-TERM EXPERIMENT

Maija Vainio, Marja Mutanen and Hannu Mykkänen

Department of Nutrition
University of Helsinki
00710 Helsinki, Finland

INTRODUCTION

Assessment of selenium bioavailability in foods requires a lengthy feeding period before a measurable change in body selenium status can be observed. The present study was designed to find out whether bioavailability of selenium to humans can be measured in a short-term experiment, that is within 1-3 hrs after giving a large single dose of selenium-containing food. Alterations in plasma and erythrocyte selenium in the short-term experiment were compared to changes in plasma and erythrocyte selenium of subjects receiving supplemental selenium for 4 weeks.

METHODS

In the short-term experiment 3 subjects received a single dose of 300 µg wheat-Se as buns made of Se-rich wheat flour (6.8 µg Se/g flour). The other 3 subjects received a single dose of 200 µg mushroom-Se as bread made of mushrooms containing 12.6 µg Se/g dried mushrooms, Boletus edulis. Plasma and erythrocyte Se were measured at 0, 1 and 3 hours after the dose using the fluorometric method.

In the long-term experiment 4 subjects received daily 150 µg wheat-Se or 100 µg mushroom-Se for 4 weeks. Plasma and erythrocyte Se were measured at 0, 4 and 8 weeks after starting the supplementation.

RESULTS AND CONCLUSIONS

Neither wheat-Se nor mushroom-Se altered significantly (P 0.05) the plasma and erythrocyte Se in the short-term experiment (Figure 1). In the long-term experiment plasma Se increased in the group receiving wheat-Se (P 0.01), while erythrocyte Se increased in both groups during the 4 weeks of supplementation (Figure 2).

Alterations in plasma Se were similar in the short-term and long-term experiments, but the increase within 3 hours after dosing was small even with wheat-Se, which is generally considered a highly bioavailable source of selenium. Therefore this short-term experiment cannot be used in assessing the bioavailability of selenium in foods.

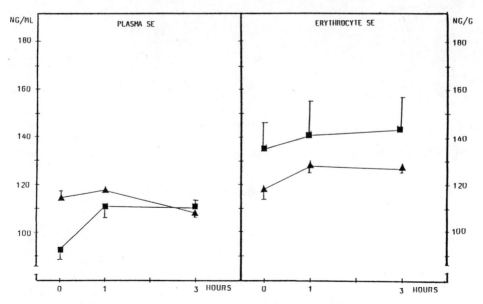

FIGURE 1. Effect of a single dose of 300 µg wheat-Se (■) or 200 µg mushroom-Se (▲) on plasma and erythrocyte Se concentrations.

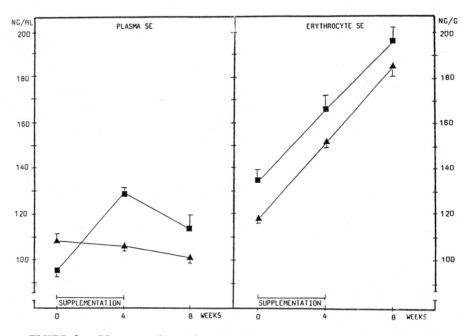

FIGURE 2. Plasma and erythrocyte Se concentrations in subjects supplementing daily 150 µg wheat-Se (■) or 100 µg mushroom-Se (▲) for 4 weeks.

SELENIUM STATUS DURING RECOVERY FROM MALNUTRITION:

EFFECT OF SELENIUM SUPPLEMENTATION

C. Murphy, B. Golden, D. Ramdath, and M. Golden

Wellcome Trace Element Research Group, Tropical Metabolism
Research Unit, University of the West Indies, Kingston 7,
Jamaica

Erythrocyte glutathione peroxidase activity (E-GPx) is an established measure of selenium status. However, plasma glutathione peroxidase activity (P-GPx) responds more rapidly to selenium supplementation in rats[1] and children.[2] Low plasma selenium concentrations (P-Se) were found in Guatemalan children with kwashiorkor.[3] In Jamaica, we found that E-GPx was particularly low in malnourished children with hepatomegaly and oedema: such children tended to develop heart failure.[4] As hepatic dysfunction and cardiomyopathy are both recognised features of selenium deficiency, we surmised that our children were suffering clinically from selenium deficiency. Therefore, this study was designed to assess more definitively, selenium status of Jamaican children before and after recovery from severe malnutrition and the effect of selenium supplementation.

SUBJECTS AND METHODS

Expt 1. Forty-one children, 2 to 27 months old, with marasmus or oedematous malnutrition were studied on admission to hospital, and 4 to 8 weeks later when they had reached > 90% expected weight for height by the Boston Standards.[5] During recovery, they were fed ad libitum a cow's milk based formula (Pelargon, Nestlé) enriched with corn oil. E-GPx, P-GPx and P-Se were measured on 2.5 ml blood samples using, for GPx, a modification of Paglia and Valentine's method[6] and for Se, graphite furnace atomic absorption spectrophotometry (Perkin Elmer 5000). Controls comprised healthy children of similar ages.

Expt 2. Twenty-four malnourished children with low E-GPx on admission (mean 17 U/g Hb vs 36 U/g Hb in controls) were also studied longitudinally 8 times – during recovery. For the first 3 weeks of recovery, seventeen of these were supplemented with 20 μg Se/Kg/day as sodium selenate solution. Otherwise, their management was similar to that of the 7 unsupplemented children. Only P-GPx and P-Se were measured in this series.

RESULTS AND DISCUSSION

For Expt 1, results are shown as means (SEM) in Table 1.

Table 1. Selenium status (E-GPx, P-GPx, P-Se) of children studied.

Groups	E-GPx U/g Hb	P-GPx U/l	P-Se ug/l
Control	36 (2)	140 (9)	86 (4)
Marasmus	24 (4)	142 (15)	76 (12)
Oedematous	21 (3)	98 (8)	53 (5)
Recovered	20 (2)	88 (7)	41 (4)

Student's t tests were performed to test differences, which were assumed significant when $p < 0.05$.

E-GPx was similar in marasmic and oedematous children, and significantly lower than E-GPx in controls. P-GPx and P-Se values were similar in control and marasmic children but both were significantly lower in oedematous children. All 3 values, E-GPx, P-GPx and P-Se remained significantly low after recovery.

In Expt 2, selenium supplementation resulted in a rapid increase in P-Se from a mean value of 45 µg/1 on Day 0, to 148 µg/1 on Day 6. By discharge, 1 to 4 weeks after supplementation ceased, P-Se remained high at 118 ug/1. P-GPx behaved similarly. In the supplemented children, P-GPx rose from 116 U/1 on Day 0 to 167 U/1 on Day 6. By discharge, it was still significantly higher (147 U/1) than in the unsupplemented children (94 U/1).

Thus, many malnourished children had evidence of chronic selenium deficiency. This did not resolve during recovery on our high energy formula. Oedema was associated with low P-Se and P-GPx. These values also remained low during recovery unless the children were supplemented with selenium. In this case, they rose rapidly and dramatically and remained within the control range after supplementation ceased.

ACKNOWLEDGEMENT

This work was supported by the Wellcome Trust.

REFERENCES

1. C. K. Chow and A. L. Tappel, Responses of glutathione peroxidase to dietary selenium in rats, J. Nutr. 104:444-451 (1974).
2. H. J. Cohen, M. E. Chovaniec, D. Mistretta, and S. S. Baker, Selenium repletion and glutathione peroxidase – differential effects on plasma and red blood cell enzyme activity, Am. J. Clin. Nutr. 41:735-747 (1985).
3. R. F. Burk, W. N. Pearson, R. P. Wood, and F. Viteri, Blood selenium levels and in vitro red cell uptake of 75 Se in kwashiorkor, Am. J. Clin. Nutr. 20:723-733 (1967).
4. F. Bennett, M. H. N. Golden, and B. E. Golden, Red cell glutathione peroxidase concentrations in Jamaican children with malnutrition, West Ind. Med. J. 32(suppl):28 (1983).
5. V. C. Vaughan and R. J. McKay, "Nelson: Textbook of Pediatrics," 10th Ed., W. B. Saunders Co., Philadelphia (1975).
6. D. E. Paglia and W. N. Valentine, Studies on the quantitative and qualitative characterization of erythrocyte glutathione peroxidase, J. Lab. Clin. Med. 70:158-169 (1967).

SELENIUM AND GLUTATHIONE PEROXIDASE IN MOTHERS EXPERIENCING

SUDDEN INFANT DEATH SYNDROME

Jane Valentine, Bahram Faraji and Kathy Akashi

School of Public Health
University of California
Los Angeles, CA 90024

INTRODUCTION

The relationship of selenium to the sudden infant death syndrome (SIDS) has been evaluated only summarily. Findings in the animal studies done by Money[1] led him to suggest that selenium deficiency could be a causative factor in reported human cases. A study by Rhead et al.[2] failed to support this contention when whole blood selenium levels were evaluated in 12 SIDS cases and compared to 4 control infants. Rhead's report has been cited as evidence refuting selenium's role in SIDS. We feel, nonetheless, that this small number of observations weakens the rebuttal of Money's hypothesis.

Lombeck et al.[3] have shown the lowest serum selenium concentrations to exist in infants between 1 and 4 months of age. The activity of glutathione peroxidase (GSH-Px) in erythrocytes was also decreased. We thus felt it worthwhile to further evaluate selenium's role in SIDS.

METHOD

Mothers experiencing SID within two years of our evaluation were chosen as subjects, and their selenium and GSH-Px values were compared to those of the youngest live birth of each. A neighborhood mother who had never experienced SID and her child were selected for controls.

GSH-Px was determined by the procedure of Thomson et al.,[4] with the modification of 0.5 ml of blood being washed with 3 ml isotonic saline using centrifugation and diluting the final sample to 1 ml in saline. Hydrogen peroxide, 0.1 ml, 1.1 mM, was used to initiate the reaction; NADPH to NADP conversion was measured on a Coleman Hitachi (Model 124) spectrophotometer.

Selenium in whole blood was determined using a modified method of Kang and Valentine,[5] with 6N HCl used to dilute nitric/perchloric acid digests. Hydrogen selenide gas, generated with sodium borohydride, was analyzed on a Perkin-Elmer (Model 603) atomic absorption spectrophotometer.

RESULTS

Blood selenium and GSH-Px are shown in the table. Selenium levels of the SID mothers were found to be lower than those of control mothers; the difference was not statistically significant ($p > 0.05$). Blood selenium

Table 1. Selenium Concentrations and GSH-Px Activities in Whole Blood of Mothers and Children Studied.

Subjects/Ages	N	Se, µg/100 ml Mean ± SD	N	GSH-Px, U/ml Mean ± SD
Adults				
SID experience/				
20-30 years	19	6.6 ± 2.4	20	2.3 ± 0.45
Controls/				
19-34 years	8	7.2 ± 2.7	8	2.41 ± 0.76
Children				
SID siblings/				
1-3 years	5	4.1 ± 1.2	5	2.03 ± 0.22
0.3-0.8 years		Not determined	4	1.55 ± 0.27
0.3-2.9 years*	5	4.1 ± 1.2	9	1.82 ± 0.34
Controls/				
1-3 years	6	5.7 ± 2.4	6	2.23 ± 0.54
0.4-5.2 years*	7	6.1 ± 2.4	8	2.16 ± 0.47

* Total age group. [Note: Statistical comparisons performed using Student's T-test.]

levels for both groups were within normal range previously reported for our laboratory[6] and by Welz and Melcher,[7] but lower than reported by Rhead et al.[2] and Butler and Whanger.[8] No statistically significant difference was found for GSH-Px when SID and control mothers were compared, though levels in the former were lower than the control. SID siblings had selenium concentrations lower than those of their mothers ($p < 0.05$). SID siblings also had lower selenium concentrations than the control children; the difference had only borderline statistical significance ($p < 0.10$). No difference in blood selenium was found in comparing control children with their mothers' values. GSH-Px in SID siblings was found to be lower than that of control children when all ages were compared and when subgroups aged 1-3 years were compared; these differences were not statistically significant ($p > 0.05$). SID siblings were found to have lower GSH-Px than their mothers ($p < 0.01$) only when all ages of siblings were included in the comparison or when just infants under one year were included.

REFERENCES

1. D. F. L. Money, Vitamin E and selenium deficiencies and their possible aetiological role in the sudden death in infants syndrome. N. Z. Med. J., 71:32 (1970).
2. W. J. Rhead, E. E. Cary, W. H. Allaway, S. L. Saltzstein, and G. N. Schrauzer, The vitamin E and selenium status of infants and the sudden infant death syndrome, Bioinorg. Chem., 1:289 (1977).
3. J. Lombeck, K. Kasperek, H. D. Harbisch, L. E. Feinendegen, and H. J. Bremer, The selenium state of healthy children, Eur. J. Pediatr., 125:81 (1977).
4. C. D. Thomson, H. M. Rea, V. M. Doesburg, and M. F. Robinson, Selenium concentrations and glutathione peroxidase activities in whole blood of New Zealand residents, Br. J. Nutr., 37:457 (1977).
5. H. K. Kang and J. L. Valentine, Acid interference in the determination of arsenic by atomic absorption spectrometry, Anal. Chem., 49:1829 (1977).
6. J. L. Valentine, L. S. Reisbord, H. K. Kang, and M. D. Schluchter, Effects on human health of exposure to selenium in drinking water, in: "Selenium in Biology and Medicine," G. F. Combs et al., ed., Van Nostrand Rheinhold Co., New York (1987).
7. B. Welz and M. Melcher, Accuracy of selenium determination in human body fluids using hydride-generation atomic absorption spectrometry, in: "Selenium in Biology and Medicine," (ibidem).
8. J. A. Butler and P. D. Whanger, Dietary selenium requirements of pregnant women and their infants, in: "Selenium in Biology and Medicine," (ibidem).

SELENIUM RESPONSIVE CONDITIONS AND THE CONCENTRATIONS

OF SELENIUM IN SKELETAL MUSCLE IN YOUNG SHEEP

D.W. Peter, D.J. Buscall and P. Young

CSIRO Division of Animal Production, Private Bag, PO Wembley,
WA 6014 Australia

INTRODUCTION

The occurrence of selenium (Se) responsive conditions in weaner sheep
in southern Western Australia cannot be accurately predicted from pasture,
blood or liver Se concentrations[1,2]. This paper now briefly describes
several studies of seasonal changes in Se concentration in skeletal muscle
of grazing weaner sheep to evalute its potential for predicting Se
responsive conditions.

MATERIALS AND METHODS

Trials were carried out on small plots in a low Se area at Bakers Hill,
Western Australia using Merino weaner wethers. In two of the trials (1 and
3) half the animals in each plot were given a high density intraruminal Se
pellet (Permasel, ICI Aust. Ltd) in late spring while in the third trial (2)
half the animals grazed plots fertilised in early winter with 17 g Se/ha as
sodium selenite. In trials 1 and 3 sheep in half the plots were fed 250 g/d
of lupin grain from early summer until late autumn (nil or +lupins). In
addition, sheep in trial 3 had free access to a mineral lick either
containing or not containing sulphur (gypsum; nil or +S). The sheep were
weighed regularly and shorn in October when greasy fleece weights (GFW) were
recorded. Samples of plasma, liver (two trials only) and semitendinosus
muscle were obtained from 4-10 sheep/ treatment group prior to Se
supplementation and at varying intervals thereafter. Se analysis was
according to the method of Wakinson.[3]

RESULTS AND DISCUSSION

Se supplementation resulted in positive, and at times significant
($P<0.05$), improvements in liveweight in all but one of the treatment groups
in the three trials. However, the time of maximum response varied between
trials. GFW was also increased by Se supplementation in all but two of all
treatment groups, although the increase was only statistically significant
($P<0.01$) in the +lupin group in trial 1. Negative responses occurred in
liveweight (non-significant) and GFW ($P<0.05$) in the +lupin, nil S group in
trial 3. (See Table 1 for examples.)

Table 1. Examples of seasonal changes in tissue Se concentrations and of production responses to Se supplementation

Trial/Season[a]		Se concentration[b] and production response[c]				
		A	B	C	D	E
1. Nil lupins	Plasma	9±1	10±1	15±1	10±2	13±1
	Liver	222±14	195±20	252±22	134±8	-
	Muscle	108±19	76±8	72±4	67±4	-
	Δ L.Wt	0	0.1	1.6	1.7	0.7 (0.10)
1. +lupins	Plasma	8±1	14±2	21±3	13±1	12±1
	Liver	213±11	238±6	50±8	67±4	-
	Muscle	119±9	99±8	50±8	67±4	-
	Δ L.Wt	0	0.9	1.4	2.0	1.0 (0.47)
2. (Se by fertiliser at C)	Plasma	8±0	(17)	19±2	18±2	-
	Liver	198±12	289±17	294±17	22±6	-
	Muscle	98±9	121±6	133±8	76±5	-
	Δ L.Wt	-	0	0.3	1.1	1.9 (0.15)
3. +lupins, nil S	Plasma	9±1	21±3	28±2	14±1	11±1
	Muscle	85±6	136±10	116±12	85±4	-
	Δ L.Wt	0.3	0	0.4	0.4	1.3 (-0.30)

[a] A = late spring/early summer; B = early-mid autumn; C = late autumn/early winter (break of season); D = mid-late winter; E = mid-spring (shearing).

[b] Se concentrations in plasma, liver and muscle were increased three to fivefold by Se treatment. Concentration data shown and referred to in the text are for nil Se sheep only.

[c] Plasma [Se] in µg/L; muscle and liver [Se] in µg/kg DM. Difference (kg) in liveweight (L.Wt) and GFW (in parenthesis at E) of + and nil Se treatment groups.

Concentrations of Se in plasma and liver increased during summer/autumn (dry pasture) and decreased during winter/spring (green pasture). Changes in muscle Se concentrations were less regular, varying both within and between trials, and were generally unrelated to changes in plasma and liver. Using all data collected prior to shearing in mid-spring there was a significant negative correlation ($P<0.05$) between muscle Se concentration of nil Se sheep and liveweight response to Se supplementation.

It is concluded that measurements of muscle Se concentration in young sheep during late spring may assist in predicting likely liveweight responses to Se treatment over the summer/autumn period although subsequent changes in concentration may modify the expected response. These results suggest that a production response in weaner sheep is likely if the Se concentration in muscle is 100 µg/kg DM or less.

REFERENCES

1. B.J. Gabbedy, H. Masters and E.B. Boddington, White muscle disease of sheep and associated tissue selenium levels in Western Australia. Aust. Vet. J. 53:482 (1977).

2. R.A. Hunter, D.W. Peter, M.P. Quinn and B.D. Siebert, Intake of selenium and other nutrients in relation to selenium status and productivity of grazing sheep, Aust. J. Agric. Res. 33:637 (1982).

3. J.H. Watkinson, Semi-automated fluorimetric determination of nanogram quantities of selenium in biological material, Anal. Chim. Acta. 105:319 (1979).

SELENIUM STATUS IMPROVED AMONG ELDERLY

DUE TO SELENIUM-ENRICHMENT OF FERTILIZERS IN FINLAND

M. Tolonen*, M. Halme*, S. Sarna*, T. Westermarck**,
U.-R. Nordberg**, and W. Bayer***

Dept. Publ. Hlth., University of Helsinki, Finland*
Helsinki Central Inst. for Mentally Retarded, Kirkkonummi**
Lab Dr. Bayer, Stuttgart, Germany***

INTRODUCTION

Since summer 1984 all commercial fertilizers in Finland are enriched with Se in order to increase the notoriously low Se intake. As a result, Finnish foodstuffs contain more Se since late 1985. The Se intake from an average 10 MJ diet has increased from 30 to 90-100 ug.[1]

As part of a clinical trial (Westermarck et al. in this volume), we studied the Se status in a group of 25 elderly at an old people's home (the placebo group). Our study coincided with the increase of Se in the Finnish foodstuffs, i.e. November 1985 through November 1986.

SUBJECTS AND METHODS

The study group comprised 21 women and 4 men living at Tapanila old people's home in the city of Lahti. Their mean age was 80 years (66-94). All of them ate their meals together at the nursing home.

Whole blood Se (B-Se) was analysed in Nov. 1985 and at 3, 8 and 12 month intervals. Blood samples were collected in heparinized tubes after a night's fasting. Se determinations were made using hydride AAS at dr. Bayer's laboratory. For quality control, five samples were also analysed in Oulu Regional Institute of Occupational Health, using hydride AAS, as well. The results were within five per cent range in the two laboratories.

Erythrocyte glutathione peroxidase (E-GSHPx) activity was measured at 0, 3 and 12 months according to Beutler et al.[2] at dr. Westermarck's laboratory. The same laboratory determined serum lipid peroxides by means of thiobarbituric acid reaction.[3]

RESULTS AND DISCUSSION

Initial mean B-Se (ug/l) was 85 $\pm$ 17.3, at 3 months 119 $\pm$ 19.7, at 8 months 161 $\pm$ 19.2 and at 12 months 162 $\pm$ 19.6. The corresponding values for E-GSHPx were initially 20 $\pm$ 5.7 U/gHb, at 3 months 24 $\pm$ 5.8 and at 12 months 28 $\pm$ 5.6. Figure 1. illustrates the increases of B-Se and E-GSHPx. Serum lipid peroxides (umol/l) were initially 2.7 $\pm$ 0.69, at 3 months 2.8 $\pm$ 0.70 and at 12 months 2.8 $\pm$ 0.75.

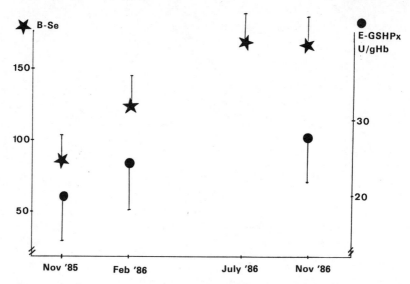

Fig. 1. Mean and SD values for whole blood selenium (B-Se) and erythrocyte glutathione peroxidase (E-GSHPx) activity in Finnish elderly 1985-1986 (N=25)

Initially, the B-Se levels were low, at the same level as in adults in 1977.[4] During the study period, the Se status improved considerably; the mean B-Se (162 ug/l) was sufficient to saturate the E-GSHPx, as recently shown by us.[2] It is therefore noteworthy, that even this significant increase in the Se status seemed not to reduce the relatively high serum lipid peroxides of the elderly. Yet, a significant decrease was noted, when the B-Se exceeded 200 ug/l in another group of elderly, who took Se as dietary supplements (see Westermarck et al. in this volume). Our data suggest that Se may exert an antioxidative function unrelated to GSHPx. Further studies are needed on the optimal Se intake and balance in man, particularly in regard to aging.

REFERENCES

1. Working Group on Selenium, First Annual Report, Ministry of Agriculture and Forestry, Helsinki 1986 (in Finnish)
2. E. Beutler, K. D. Blume, J. C. Kaplan, G. W. Löhr, B. Ramot, and W. N. Valentine, Recommended methods for red cell enzyme analysis, Br. J. Haematol. 35, 331 (1977)
3. K. Yagi, Assay for serum lipid peroxide level and its clinical signicance in: "Lipid Peroxides in Biology and Medicine". Academic Press New York (1982)
4. T. Westermarck, Selenium content of tissues in Finnish infants and adults with various diseases, and studies on the effects of selenium supplementation in neuronal ceroid lipofuscinosis patients, Acta pharmacol. toxicol. 41, 121 (1977)
5. M. Tolonen, T. Westermarck, M. Halme, S. Sarna, Selen-Gehalt des Vollblutes und Erythrozyten-Glutathionperoxidase in Finnland - Eine Korrelationstudie, Vita. Min. Spur. 1, 21 (1986)

APPARENT PROTEOLYTIC MODIFICATION OF GLUTATHIONE PEROXIDASE[*]

Roger A. Sunde, Simon A.B. Knight, Troy W. Geyman,
and Jacqueline K. Evenson

Department of Nutrition and Food Science
University of Arizona, Tucson AZ 85721 USA

Little is known about the posttranslational processing of glutathione peroxidase (GSH-Px)(GSH:H_2O_2oxidoreductase, EC 1.11.1.9). The difference of 10 amino acids in the reported position of selenocysteine in bovine erythrocyte GSH-Px, as determined by crystallographic analysis[1] as compared to chemical sequence analysis,[2] suggests that a polypeptide may have been cleaved from the crystallized GSH-Px. In a study of rat liver mitochondrial GSH-Px, Voight and Autor[3] reported that the initial GSH-Px gene product may be as much as 5 kDa larger than the mature enzyme. These two reports thus suggest that newly synthesized GSH-Px may be processed to a smaller molecular weight form during maturation. We have been studying various aspects of GSH-Px synthesis using immunoblotting with anti-GSH-Px antibodies to measure levels of GSH-Px protein, and using SDS-gel electrophoresis to follow ^{75}Se incorporation into selenoproteins. In the course of these studies, we have observed that GSH-Px can be detected in several forms that differ in their apparent molecular weight. The purpose of this report is to further describe these different molecular weight forms of GSH-Px.

Methods

Male weanling rats (Holtzman Co., Madison, WI) were fed a 30% torula yeast-based diet[4] supplemented with 0 or 0.2 ppm Se as Na_2SeO_3. GSH-Px, purified from rat liver,[5] was used to elicit production of antibodies in rabbits, and the immunoglobin G (IgG) fraction was purified and then used for immunoblotting against rat liver cytosol or purified GSH-Px.[4] To determine the ^{75}Se incorporation into GSH-Px, rats were injected iv with 50 uCi [^{75}Se]selenite (100-450 mCi/mg Se, ICN Radiochemicals, Irvine, CA) and killed 3 or 72 h later. Liver cytosol (1500 ug protein) from these ^{75}Se-injected rats was subjected to SDS-gel electrophoresis on 7.5-20% acrylamide gradient 3 mm slab gels. The fixed sample lanes were then cut into 2 mm slices and counted to determine the ^{75}Se incorporation into selenoproteins.[6]

Results

When purifed GSH-Px was subjected to SDS-electrophoresis, we often observed a Coomassie blue-stained doublet of subunits with molecular weights of 20 kDa and 23 kDa as calculated from molecular weight standards run on the same gel.[4] Immunoblot analysis showed that both species in the doublet were reactive with anti-GSH-Px IgG. Immunoblots of Se-adequate liver cytosol (47 ug protein) also showed a reactive band corresponding to the 23 kDa subunit;

[*]Supported by the University of Arizona Experiment Station and NIH (AM 32942)

no reactive band at 20 kDa was observed but a new reactive band corresponding to 25 kDa was present in Se-adequate rat liver cytosol. No reactive GSH-Px subunit bands were detected for Se-deficient cytosol.[4]

In further experiments, Se-adequate and Se-deficient rats were injected with ^{75}Se, killed 3 or 72 h later, and the liver cytosols subjected to both the immumoblot and SDS-PAGE procedures. SDS-PAGE (1500 ug protein) and ^{75}Se counting showed that the 23 kDa GSH-Px subunit was the only major ^{75}Se-labeled protein at 3 and 72 h in Se-adequate liver cytosol; at 72 h GSH-Px contained 50% more ^{75}Se than at 3 h (not shown). In Se-deficient cytosol, GSH-Px contained only 1/6 as much ^{75}Se as in Se-adequate cytosol. Immunoblotting (50 ug protein) showed a major reactive band against 23 kDa GSH-Px and a minor reactive band at 25 kDa in Se-adequate cytosol, but no reactivity against GSH-Px in Se-deficient cytosol.

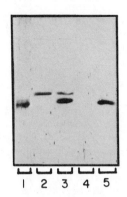

Fig. 1. Immunoblot of purified GSH-Px (lanes 1,3,5), +Se cytosol (lanes 2,3), and -Se cytosol (lanes 4,5).

To directly determine whether the difference in subunit size of purified GSH-Px as compared to GSH-Px in liver cytosol was due to a difference in mobility or in molecular weight, purified GSH-Px, Se-adequate and Se-deficient cytosol were mixed and analyzed. These experiments used a second anti-GSH-Px antibody which was specific for GSH-Px but which also reacted with a 80 kDa polypeptide. (This 80 kDa species was not labeled by ^{75}Se, nor was its presence altered by Se status). The immunoblot (Fig. 1) clearly showed that purified GSH-Px (20 kDa) and GSH-Px in Se-adequate liver cytosol (23 kDa) are distinct forms. Again no reactive GSH-Px subunits were detected in Se-deficient cytosol. To determine where the change in molecular weight occurred, fractions from each step of purification were immunoblotted. This experiment (not shown) demonstrated that the reduction to 20 kDa occurred during dialysis of liver supernatant prior to ion exchange chromatography.

Discussion

These experiments have shown that GSH-Px purified using our procedure has an apparent molecular weight about 3 kDa lower than the major 23 kDa GSH-Px subunit observed in Se-adequate rat liver cytosol. The continued activity of this purified GSH-Px[5] clearly shows that this modification does not result in the loss of enzyme activity. A 25 kDa immunoreactive species was also observed in Se-adequate cytosol, but no reactive GSH-Px subunits were detected in Se-deficient cytosol. While these experiments have not shown that the apparent different GSH-Px subunits are the result of proteolysis, proteolytic loss of polypeptide fragments is the likely cause. The presence of the 23 and 25 kDa forms of GSH-Px in liver cytosol suggests that proteolytic modification of GSH-Px may also occur in vivo.

References

1. O. Epp, R. Ladenstein, and A. Wendel, Eur. J. Biochem. 133: 51 (1983).
2. W. A. Gunzler, G. J. Steffens, A. Grossmann, S.-M. A. Kim, F. Otting, A. Wendel, and L. Flohe, Hoppe-Seyler's Z. Physiol. Chem. 365:195 (1984).
3. J. M. Voight, and A. P. Autor, Fed. Proc. 42:806 (1983).
4. S. A. B. Knight, and R. A. Sunde, J. Nutr. 117:   (1987).
5. R. A. Sunde, and J. K. Evenson, J. Biol. Chem. 262:933 (1987).
6. J. K. Evenson, and R. A. Sunde, Fed. Proc., 45: 371 (1986).

LONG-RANGE ELECTRON TRANSFER IN RUTHENIUM-MODIFIED PROTEINS

Charles M. Lieber, Jennifer L. Karas, Stephen L. Mayo,
Andrew W. Axup, Michael Albin, R.J. Crutchley, W.R. Ellis, Jr.,
and Harry B. Gray*

Arthur Amos Noyes Laboratory
California Institute of Technology
Pasadena, CA 91125

INTRODUCTION

Long-range-electron transfer (ET) reactions are important mechanistic steps in many biological processes.[1] Current research in this area has focused on elucidating the factors that control these ET processes in donor-acceptor (D-A) systems.[2-9] A general two-site model consists of a donor (D) and acceptor (A) that are fixed in a rigid structure (e.g., protein matrix) and separated by a through-space distance, d. Donor to acceptor ET pathways through chemical bonds are also possible, but generally are longer than the through-space distance.

A widely accepted classical description of the ET rate (k_{ET}) for such systems has been derived by Marcus:[2] $k_{ET} = \Gamma exp(-\Delta G*/RT)$. This expression is composed of two major parts; an electronic term, Γ, and a free-energy term, $\Delta G*$ (eqn. 1a and 1b, respectively).

1a. $\Gamma = 10^{13} exp[-\beta(d-3)]$

1b. $\Delta G* = (\Delta G^\circ + \lambda)^2/4\lambda$

The electronic part, eqn. 1a, consists of a frequency factor (barrier crossing rate) of 10^{13} s^{-1}, and an exponential decay. The exponential decay reflects the decrease in D-A electronic coupling with distance, and the rate of this decay is given by the parameter β. The activation free-energy factor $\Delta G*$ (eqn. 1b) is dependent on the overall reaction free energy (or driving force), ΔG°, and the reorganization energy, λ. With this theoretical framework as a guide, we have investigated the dependence of the ET rate on ΔG° and d, as described below.

EXPERIMENTAL SYSTEM

The approach that we employ to prepare our D-A models involves co-valently linking a redox-active ruthenium complex to a surface histidine of a crystallographically characterized metalloprotein.[5-9] Using this strategy it is possible to prepare D-A systems in which the ET distance and inter-vening medium are known. In addition, we can tune the reaction free energy through systematic changes in the ruthenium complex.

Sperm whale myoglobin has four surface histidine residues that lie between 13 and 22 Å from the heme center. These four histidine residues have been modified with several different ruthenium reagents using procedures developed in our laboratory.[6-8] For example, pentaammineruthenium (a_5Ru) has been attached to each of the histidine residues of Mb. Native Mb is reacted with excess $a_5Ru^{II}(OH_2)$ to form singly and multiply labeled species. The singly modified derivatives are separated from unreacted and multiply labeled Mb using preparative IEF, and the four isomerically pure Ru(His-X)Mb species are isolated using cation-exchange chromatography. The specific labeling sites have been identified by peptide mapping and proton NMR studies. Spectroscopic (UV-VIS, CD, EPR) and crystallographic results[10] demonstrate that the native and ruthenium-modified structures are virtually the same. Therefore, we can prepare ruthenium-modified Mb with one Ru per molecule attached to a specific histidine residue (any one of the four) at a known distance. In the following sections we will discuss the results we have obtained in studying ET in these systems.

DISTANCE DEPENDENCE OF k_{ET}

Our initial studies[6a,c] with $a_5Ru(48)MbFe$ (Fe denotes the native heme center) indicated that a large driving force would be required to study ET to the three long-distance sites (histidines 12, 81, 116). The driving force in the heme system is only 20 mV.

To prepare a high driving force system, we have removed the heme in Mb and replaced it with a photoactive porphyrin, Zn(mesoporphyrin IX) (ZnP). The triplet excited state of ZnP (ZnP*) is long-lived and highly reducing, properties that are ideal for our distance dependence studies. The ET rates in these systems can be determined by monitoring the change in absorbance of the ground-state Soret, as outlined below.

Pulsed laser excitation generates ZnP*, which subsequently returns to the ground state via nonradiative decay and ET pathways. Since $k_b > k_{ET}$, the observed decay rate is equal to $k_D + k_{ET}$. The experimental results for the forward ET process to each of the four histidine sites are summarized in Table 1.[9] The driving force for this reaction is 0.8 V.

Table 1. $a_5Ru^{3+}(His-X)MbZnP* \longrightarrow a_5Ru^{2+}(His-X)MbZnP^+$

His-X	k_{ET} (s^{-1}) at 25°C
48	70,000
12	100
81	90
116	90

The exponential damping factor (eqn. 1a) can be determined from a plot of $\ln(k_{ET})$ vs. d: $\ln(k_{ET}) = -\beta(d-3) + constant$; however, the actual value of d requires some discussion. The through-space ET distance can be evaluated by several methods including measurement of the edge-edge (E-E) and metal-metal (M-M) distances. The E-E distance corresponds to that measured from the edge of the porphyrin ring to the edge of the histidine imidazole to which the ruthenium is bound. The actual distances, both E-E and M-M, used in the following analysis have been determined from a rigid body conformational search of the ruthenium-modified proteins; the uncertainty in distance was taken to be the variation found at 6.5 kcal above the calculated potential minimum (Table 2).

Table 2. ET Distances in Ruthenium-Modified Myoglobin[a,b]

Site (His-X)	M-M (Å)	E-E (Å)
48	16.6-23.9 (17.1)	11.8-16.6 (12.7)
81	24.1-26.6 (25.1)	18.8-19.3 (19.3)
116	26.8-27.8 (27.7)	19.8-20.4 (20.1)
12	27.8-30.5 (29.3)	21.5-22.3 (22.0)

a) see ref. 9. b) distance at potential minimum in parenthesis.

The limits on β determined from least-squares analyses of the E - E and M-M $\ln(k_{ET})$ vs. distance plots using our experimental data (Table 1) and the results for a_5Ru-modified Zn-substituted cytochrome c are 0.87-0.99 and 0.75-0.90 Å^{-1}, respectively.[9] These two ranges overlap somewhat, indicating that the method of defining the site to site distance is not that crucial in our system. Since there is some evidence that the ET rate determined for a_5Ru(His-12)Mb is enhanced via electronic interactions with the Trp-14 residue that lies in the ET pathway, we have reanalyzed our data without the His-12 results. The limits on β obtained from least squares analyses of the E-E and M-M plots are 0.99-1.12 and 0.84-1.09 Å^{-1}.[11] The larger β determined without the His-12 data is indicative of a weaker D-A electronic coupling. At this time, however, it is unclear whether or not the ET rate in a_5Ru(His-12)Mb is enhanced by the medium.

In summary, the unbiased range of β determined using all of our experimental points and E-E distances is 0.87-0.99 Å^{-1}. Although both E-E and M-M plots have been used to evaluate β, we favor the E-E measurement because the ZnP excited state is a delocalized porphyrin state and because the ruthenium and imidazole wavefunctions interact strongly.[12] Additional work on the distance dependence (i.e., more points) and the influence of the medium (e.g., Trp-14 question) is needed to explore these issues more fully. Studies employing site-directed mutagenesis to prepare systems that can address these points are currently in progress in our laboratory (i.e., we plan to introduce histidines in order to probe a range of distances and the role of specific residues in the ET pathway).

FREE ENERGY AND REORGANIZATION ENERGY EFFECTS

In this section we report our investigations probing the dependence of k_{ET} on driving force that have allowed an evaluation of the magnitude of the reorganization energy in the myoglobin system. Our experimental approach involves changing the redox potential of either D or A to vary $\Delta G°$, using substituted ruthenium complexes of the general form a_4LRu- (L = ammine (a), pyridine (py), etc.), and/or different metal-porphyrin centers.

Using flash photolysis we have measured the Fe^{2+} to Ru^{3+} ET rates in myoglobin that has been modified at histidine-48 with a_5Ru and $a_4$pyRu: Ru^{3+}-(48)MbFe$^{2+} \rightarrow Ru^{2+}$(48)MbFe^{3+}.[6a] The increase in rate, 0.04 to 2.5 s^{-1}, observed when changing from the a_5Ru^{3+} acceptor ($\Delta E° = 0.02$ V) to a_4pyRu^{3+} ($\Delta E° = 0.24$ V) agrees with that predicted by Marcus theory. The limits on the reorganization energy determined from these two points are $\lambda = $ 1-2.5 eV. A better estimate of the reorganization energy can be made using higher driving force systems. That is, the change in ET rates for a well-defined step in driving force (i.e., a_5Ru to $a_4$pyRu) at an overall $\Delta E°$ near 1 V is predicted to be small for $\lambda = 1$ eV and significantly larger for $\lambda = 2.5$ eV.

To address this problem we have replaced the heme in the above two ruthenium-modified derivatives with Pd(mesoporphyrin IX) (PdP).[6b] PdP possesses a highly reducing excited state (PdP*) and therefore can be used to study ET in ruthenium-modified (His-48)Mb but at much larger $\Delta E°$'s. The ET rates, $Ru^{3+}(48)MbPdP^* \longrightarrow Ru^{2+}(48)MbPdP^+$, in the a_5Ru ($\Delta E° = 0.7$ V) and a_4pyRu ($\Delta E° = 0.92$ V) systems were determined by monitoring the quenching of the PdP* emission following pulsed laser excitation. These results and those determined for the heme system are summarized in Table 3.

Table 3. Electron-Transfer Rates in Ru(48)Mb[a]

Reaction	$\Delta E°$ (V)	k (s^{-1})
$a_5Ru^{III}MbFe^{II} \longrightarrow a_5Ru^{II}MbFe^{III}$	0.020	0.040
$a_4pyRu^{III}MbFe^{II} \longrightarrow a_4pyRu^{II}MbFe^{III}$	0.240	2.5
$a_5Ru^{III}MbPdP^* \longrightarrow a_5Ru^{II}MbPdP^+$	0.70	9100
$a_4pyRu^{III}MbPdP^* \longrightarrow a_4pyRu^{II}MbPdP^+$	0.92	90,000

a) references 6a and 6b.

The experimental limits on the reorganization energy for the myoglobin system, which were calculated using Marcus theory, are $\lambda = 2.05-2.45$ eV.[6b] The relatively large reorganization energy is similar to those reported recently for [ruthenium-modified] zinc-cytochrome c[13] and [Zn,Fe]-hybrid hemoglobin.[14] We believe that the large λ in these ET systems is due mainly to reorganization of the aqueous medium around the solvent accessible redox sites.

There are several important conclusions to be drawn from this work. Since the electron-transfer rate is strongly dependent on the magnitude of λ, we suggest that long-range ET reactions may be controlled through variations in the reorganization energy. The potentially large degree of control available through variations of λ is illustrated by the calculated values of k_{ET} set out below (Table 4). The calculations were made using our experimentally determined value of β (0.91 Å^{-1}).

Table 4. Calculated ET Rates as a Function of d, $\Delta E°$, and λ

d(Å)	Electron-Transfer Rate: k (s^{-1})									
	0.5		1.0		1.5		2.0		2.5	= λ
	0.2	0.5	0.2	0.5	0.2	0.5	0.2	0.5	0.2	0.5 = $\Delta E°$
5	3E11	2E12	3E9	1E11	3E7	3E9	3E5	3E7	2E3	3E5
10	3E9	2E10	4E7	2E9	3E5	3E7	3E3	3E5	2E1	3E3
15	3E7	2E8	4E5	2E7	3E3	3E5	3E1	3E3	2E-1	3E1
20	3E5	2E6	4E3	2E5	4E1	3E3	3E-1	4E1	3E-3	4E-1
25	4E3	2E4	4E1	2E3	4E-1	3E1	3E-3	4E-1	3E-5	4E-3
30	4E1	2E2	4E-1	2E1	4E-3	3E-1	3E-5	4E-3	3E-7	4E-5

It may be possible by judicious use of Table 4 to extract ET rates from estimated distances or ET distances from rates in systems that are not completely characterized. In addition, the strikingly large variations in k_{ET} at constant d and $\Delta G°$ indicate that specificity in biological ET reactions could be achieved through changes in the reorganization energy. The reorganization energy for systems containing redox centers buried in a low dielectric medium (e.g., a lipid bilayer) should be considerably smaller than those for which the D/A is exposed to an aqueous environment. We are currently investigating this matter through the synthesis of systems in which the redox-active metal complex has a substantially lower λ than the ruthenium complexes used in this study.

Acknowledgment. Our research on biological ET is supported by NIH (DK-19038) and NSF (CHE85-18793; CHE85-09637). Fellowships: NIH (CML) and AT&T (SLM). Contribution No.7615 (Noyes Laboratory).

References

1. (a) Y. Hatefi, <u>Annu. Rev. Biochem.</u> 54:1015 (1985). (b) B.P.S.N. Dixit and J. M. Vanderkooi, <u>Curr. Top. Bioenerg.</u> 13:159 (1984).
2. (a) R. A. Marcus and N. Sutin, <u>Biochim. Biophys. Acta</u> 811:265 (1985). (b) D. DeVault, "Quantum-Mechanical Tunneling in Biological Systems," 2nd Ed., Cambridge University Press (1984).
3. G. McLendon, T. Guarr, M. McGuire, K. Simolo, S. Strauch, and K. Taylor, <u>Coord. Chem. Rev.</u> 64:113 (1985).
4. S.E. Peterson-Kennedy, J.L. McGourty, P.S. Ho, C.J. Sutoris, N. Liang, H. Zemel, N.V. Blough, E. Margoliash, and B.M. Hoffman, <u>Coord. Chem. Rev.</u> 64:125 (1985).
5. (a) S.L. Mayo, W.R. Ellis, R.J. Crutchley, and H.B. Gray, <u>Science</u> 233:948 (1986). (b) H.B. Gray, <u>Chem. Soc. Rev.</u> 15:17 (1986).
6. (a) C.M. Lieber, J.L. Karas, and H.B. Gray, <u>J. Am. Chem. Soc.</u> in press. (b) J.L. Karas, C.M. Lieber, and H.B. Gray, submitted for publication. (c) R.J. Crutchley, W.R. Ellis, and H.B. Gray, <u>J. Am. Chem. Soc.</u> 107:5002 (1986). (d) R.J. Crutchley, W.R. Ellis, and H.B. Gray, in "Frontiers in Bioinorganic Chemistry;" A.V. Xavier, Ed., VCH Verlagsgesellschaft, Weinheim (1986).
7. (a) K.M. Yocom, J.B. Shelton, J.R. Shelton, W.A. Schroeder, G. Worosila, S.S. Isied, E. Bordignon, and H.B. Gray, <u>Proc. Nat. Acad. Sci. (USA)</u> 79:7052 (1982). (b) D.G. Nocera, J.R. Winkler, K.M. Yocom, E. Bordignon, and H.B. Gray, <u>J. Am. Chem. Soc.</u> 106:5145 (1984). (c) S.S. Isied, C. Kuehn, and G. Worosila, <u>J. Am. Chem. Soc.</u> 106:1722 (1984).
8. (a) N.M. Kostic, R. Margalit, C.-M. Che, and H.B. Gray, <u>J. Am. Chem. Soc.</u> 105:7765 (1983). (b) R. Margalit, N.M. Kostic, C.-M. Che, D. Blair, H.-J. Chiang, I.Pecht, J.B. Shelton, J.R. Shelton, W.A. Schroeder, and H.B. Gray, <u>Proc. Nat. Acad. Sci.(USA)</u> 81:6554 (1984).
9. A.W. Axup, M. Albin, S.L. Mayo, R.J. Crutchley, and H.B. Gray, <u>J. Am. Chem. Soc.</u>, submitted for publication.
10. J. Mottonen, D. Ringe, and G. Petsko, unpublished results.
11. S. Mayo, unpublished results.
12. K. Krogh-Jespersen, J.D. Westbrook, J.A. Potenza, and H.J. Schugar, <u>J. Am. Chem. Soc.</u>, submitted for publication.
13. H. Elias, M.H. Chou, and J.R. Winkler, <u>J. Am. Chem. Soc.</u>, submitted for publication.
14. S.E. Peterson-Kennedy, J.L. McGourty, J.A. Kalweit, and B.M. Hoffman, <u>J. Am. Chem. Soc.</u> 108:1739 (1986).

FACTORS CONTROLLING THE RATES OF ELECTRON TRANSFER IN PROTEINS

David Whitford[*], David W. Concar, Yuan Gao, Gary J. Pielak and Robert J.P. Williams

Department of Biochemistry[*] and Inorganic Chemistry Laboratory,
University of Oxford, South Parks Road, Oxford OX1 3QR, U.K.

INTRODUCTION

A prerequisite of electron transfer between two metal centres is effective electronic coupling. For reactions involving metalloproteins the coupling of the redox centres is generally brought about by the formation of a protein-protein complex, the geometric and dynamic properties of which reflect the nature of the protein surface interaction. Two types of intracomplex electron transfer reactivity can be distinguished. In the first type effective electronic coupling and, hence, electron transfer occurs as a direct consequence of protein-protein binding. If any reorganisation in complex structure is required to optimise the coupling it is not rate limiting. It is probable that the reactions of cytochrome c with cytochrome b_5 and cytochrome c with cytochrome c peroxidase fall into this category. The second type is characterised by a relatively large activation barrier to effective redox coupling within the electron transfer complex. In this case protein-protein binding in itself does not give rise to a mutual orientation or combination of protein states which are favourable to electron transfer. The kinetic barrier may be related to unfavourable electronic spin-states or a relatively large redox centre separation in such a way that a change in protein conformation or orientation is required before electron transfer can occur. Intracomplex reactions of this type may be susceptible to catalysis or control by additional molecular species which facilitate the necessary conformational or electronic changes. Controlled electron transfer is likely to be an important aspect of the reactivities of many biological systems including cytochrome oxidase, nitrogenase and cytochrome P-450. The purpose of this article is to summarise the progress that has been made in relating protein structure to the observed first order electron transfer rates of both "simple" and "controlled" reactions.

THE RELATIONSHIP BETWEEN THE KINETIC PARAMETERS OF ELECTRON TRANSFER AND PROTEIN STRUCTURE

Insight into the dependence of the observed first order electron transfer rates, k_{et}, on protein structure can be gained by considering equations of the type:

$$k_{et} = \left[A \exp(-E_r/kT) \right] \left[k_o \exp(-\beta r) \right] \left[\exp\left(\frac{-(\Delta G + \lambda)^2}{4\lambda kT} \right) \right]$$

The activation energy, E_r, is related to the reorganisation in the protein complex required to optimise electronic coupling and for most simple electron transfer processes this can be assumed to be small. The second term reflects the degree of electronic coupling present and is dependent on the redox site separation distance, r, together with the donor-acceptor orientation and intervening medium parameters (k_o and β). The third term arises from the nuclear motion which accompanies charge transfer. λ is a relaxation energy term related to the shift in nuclear coordinates during the reaction and its contribution to the activation energy is modulated by the thermo-dynamic driving force of the reaction, ΔG[1]. It is important to consider how a protein matrix can influence these parameters if specificty and control in biological electron transfer are to be understood. These parameters are described below:

(1) The redox centre separation distance and intervening medium

Distances of above 20Å have been shown to result in slow electron transfer[2], so that despite some previous estimates[3] it is unlikely that a pair of small spherical proteins with metal atoms at their centres would approach metal/metal distances in excess of 20Å. Additionally the redox centres of small proteins and even large membrane bound complexes, appear to be considerably off centre giving a preferred electron transfer distance and orientation. The protein-solvent medium is likely to have some influence on the effectiveness of through-space electronic coupling.

(2) The reactive surfaces of redox partners

The rate of electron transfer is enhanced if the lifetime of the complex is such that dissociation occurs rapidly after electron transfer. The selectivity of protein surfaces necessary for facile electron transfer is achieved either by the interaction of complementary charge groups, as in the case of cytochrome c with cytochrome b_5[4] or cytochrome c peroxidase[5], or alternatively by specific ion binding to charged surface groups.

(3) The redox potentials of the protein sites

It is through the ΔG term that a protein exerts a thermodynamic influence on the rate of electron transfer. The redox potentials of metal centres are controlled through a multitude of effects including the first coordination sphere, the outer sphere groups, the exposure to solvent and the presence of buried or surface charges[6].

(4) Relaxation processes

In a discussion of the influence of relaxation on electron transfer rates the cooperative nature of protein structure must be considered. For relatively rapid electron transfer the degree of protein relaxation must be minimised. In cytochrome c, for example, the redox state conformational change is known to be small from X-ray crystallographic and nmr studies[7,8]. The relaxation energy has been estimated at less than 10kJ mol^{-1}.

A conformational change may of course alter the effects of any of these physical constraints in order to enhance or reduce the rate of electron transfer. Allosteric processes may control electron transfer by transmitting the effects of substrate binding or a distant spin state change to the interior or surface of the protein. The interplay between relatively rigid and more mobile regions of the protein matrix may be particularly important in achieving this. Allosteric events,

30

particularly spin-state changes, may act in the cytochrome P-450 set of redox centres.

Proteins may fold to give significant distortion of the inner sphere geometry away from that expected for a given combination of metal and ligand-environment. This effect has implications for relaxation, as well as influencing the redox potential. It is perhaps best illustrated by type 1 "blue" copper proteins such as plastocyanin or azurin. It was postulated as early as 1961[9] that the copper was constrained in a centre close to tetrahedral by the protein fold. This is generally agreed to be the case although the long Cu-Met S bond (~3.00Å) has been a contentious issue. Equally, the recent suggestion of an additional Cu-O bond derived from a polypeptide chain carbonyl group, and the presence of trigonal bipyramidal geometry[10], does not detract from the general idea that the protein redox centre is arranged to minimise reorganisation from one state to another. Most model complexes of copper with uni- or bidentate ligands do not mimic proteins because only the latter possess this particular rigid frame.

In addition to the "blue" copper proteins, the other most simple electron transfer molecules are cytochromes c[11]. Once again these proteins are tightly folded and rigid, despite the high content of α-helices. The tightness of the fold, which contains the partially exposed heme edge, through which electron transfer occurs, is due to three factors: (1) cross-linking through thioether bridges, (2) cross-linking via the iron atom and (3) the large planar hydrophobic platform provided by the porphyrin and an evolutionary invariant tryptophan residue. The very low mobility of most of the protein structure has been demonstrated by the slow flip rates of several aromatic and aliphatic residues[12,13]. This low internal mobility coupled with minimal changes in bond lengths and angles on going from low spin Fe(II) to low spin Fe(III) provides a centre which requires little reorganisation during electron transfer[7]. Other cytochromes such as the membrane bound components of respiratory and photo-synthetic chains may differ considerably in this respect (see the end of this article) but the family of cytochromes c, including the multiheme proteins provide a rich pool with which to study the dependence of electron transfer on protein interiors.

The other major electron transfer centres in biology are the flavins and the iron-sulphur centres (Fe-S). Both these centres are found in β-sheet proteins which in the case of the Fe-S proteins are strongly cross linked to the metal and in that of the flavins are stabilised by the hydrophobic platform of the molecule. Once again relaxation of the redox centre is likely to be small in these proteins where the metal is buried deep within the protein, perhaps 10-20Å. Electron transfer from one iron centre to another is rapid ($>10^3$ s^{-1}) and it is much faster than in mixed oxidation model solid state devices where the relaxation is usually large[14].

THE USE OF NMR TO INVESTIGATE PROTEIN ELECTRON TRANSFER

Many studies of the kinetics of first order electron transfer reactions are aimed at understanding how effective coupling can occur at relatively large separation distances of between 10 and 25Å in the absence of an obvious conjugated pathway. However, the effect of the proteinaceous framework on electronic coupling is not easy to determine since the distance and medium parameters must be systematically changed without seriously affecting the reorganisation energy terms which are sensitive functions of protein structure. In this laboratory, distance, medium and reorganisation effects are being investigated and correlated for a series of site-specifically modified variants of cytochrome c. A relationship between structure and mechanism for a number of reaction types is beginning to emerge from structural, dynamic and kinetic data obtained by detailed nmr studies.

The study of electron transfer at the atomic level is hampered by the large size of many of the proteins present in respiratory and photosynthetic systems. Our approach to this problem has involved using nmr to study electron transfer processes particularly in small, freely mobile, proteins such as cytochrome c. The spectrum of horse heart cytochrome c has been assigned in sufficient detail that electron transfer can be described in terms of (i) conformational changes at the iron centre, (ii) the internal mobility of residues and (iii) most importantly the surface of the protein. These nmr methods involve using the heme as a paramagnetic probe of the core of the molecule and the use of indirect external relaxation reagents such as chromium hexacyanide to define binding surfaces.

Systematic alteration of the protein structure has involved site directed mutagenesis and chemical modification. Phe-87 in yeast iso-1-cytochrome c has been a target residue for site-directed mutagenesis as a result of its proximity to the electron transfer region and its evolutionary invariance (Phe-87 of iso-1-cytochrome c corresponds to Phe-82 of mammalian cytochromes c). The nmr study shows that the variant cytochromes differ only slightly in structure (G.J. Pielak et al., unpublished observations). Substitution of Phe-87 by glycine or serine markedly effects the redox potential although other bulky or hydrophobic amino acids such as tyrosine or leucine are without major effects[15]. Furthermore, the reaction of the reduced variants with the π-cation radical of Zn-substituted cytochrome c peroxidase in a preformed 1:1 complex has a large thermodynamic driving force and showed that substitution of a Phe-87 with non-aromatic amino acids decreased the unimolecular rate[16]. These data agree with model studies showing the involvement of aromatic groups in the pathway of electron transfer at potentials in excess of 1 volt[17].

Nmr studies have also yielded information on controlled electron transfer where a reorientation in the configuration of the protein partners is required. In the catalysis of cytochrome c electron self-exchange by sodium hexametaphosphate (HMP) the proteins form an electron transfer active dimer which is effectively the only species in solution. The electron transfer rate can be determined by exchange induced line broadening experiments and is high ($\sim 10^3 s^{-1}$) at room temperature but it also has an unexpectedly high activation energy ($\sim 70 kJmol^{-1}$) (D.W. Concar et al., unpublished results). This implies there is a rearrangement within the dimer to form a geometry favourable to charge transfer. In view of the observed rate of electron transfer the true rate for self exchange in the most favourable protein orientation is likely to be very fast. The action of hexametaphosphate provides a useful analogy to the likely surface phenomena in vivo.

Additionally, cytochrome c site-specifically modified at lysines 72 or 13 using 4-chloro 3,5-dinitrobenzoic acid (CDNB), has been investigated in both anion-catalysed electron self-exchange (unimolecular) and bimolecular reactions. High resolution nmr confirmed there are no major alterations in the tertiary structure of these derivatives. Consequently differences in the reactivity of CDNP-K13 and K72 in both bimolecular and unimolecular processes could be ascribed to the local site of modification. Significantly, lysine modification results in changes in the association constant and the unimolecular rate. In self-exchange reactions the K72 derivative has an observed rate 10 times that of the K13 isomer and over 100 times that of the native protein[18]. The disparity of these rates indicates the presence of reorientational changes within the reactive complex although it is presently unclear whether such adjustments control the electron transfer.

The implications of these observations for the reactions of cytochrome c in biology are clear. If the most favourable binding configuration of cytochrome c with its donor/acceptor partner involves optimal electronic coupling then the observed activation energy corresponds to that of the electron transfer event itself and the reaction may be rapid. Cytochrome c can bind non-specifically to anionic surfaces and it is therefore possible that optimal coupling with a redox partner may require a rearrangment of the donor/acceptor complex prior to rapid charge transfer. For example, the reaction of cytochrome $\underline{c}$ with cytochrome $\underline{c}$ oxidase is modulated by anions which include cardiolipin and ATP[19],[20]. At present the effect of these anions on the unimolecular rate is unknown. Cytochrome $\underline{c}$ appears to act as a conformational effector of the oxidase so that unimolecular events between the Cu and Fe centre may also be influenced.

A different example of a high energy of activation for electron transfer is observed in the tetraheme protein cytochrome $\underline{c}_{554}$[21]. One step of a redox transition had an activation energy of about $80kJ$ $mole^{-1}$. This was attributed to an electron transfer step which was dependent upon a conformational change associated with a high-spin to low spin switch of at least one of the hemes.

Perhaps the ultimate example of electron transfer linked to a conformational change is that coupled to vectorial proton trans-location across membranes during energy conservation[22]. Here a proton is forced to move against a concentration gradient as the electron drops through a potential gradient of ~0.4 volts. It is likely that cytochrome $\underline{c}$ oxidase and the cytochrome $\underline{b}-\underline{c}(f)$ complexes of mitochondria, chloroplasts and bacteria are typical examples of redox linked proton pumps. The electron transfer distance in cytochrome oxidase is unlikely to be greater than 20Å and it should be noted that the increased metal composition described recently for this complex[23] would tend to limit redox centre separation. For cytochromes $\underline{a}$ and $\underline{b}$ there is no rigid cross linking of the heme to the protein but the protein in all probability, remains, helical[24]. Proteins of predominantly helical composition are able to undergo helix/helix translational/rotational motion. One can imagine that in these proteins, as for hemoglobin and calmodulin, changes in conformation or charge could cause helix motion which is coupled to the gating and pumping of protons.

REFERENCES

1) R.A. Marcus and N. Sutin, Biochim. Biophys. Acta 811 265-322 (1985).
2) S.L. Mayo, W.R. Ellis, R.J. Crutchley and H.B. Gray Science 233 948-952 (1986) and references therein.
3) P.L. Dutton, J.S. Leigh, R.C. Prince and D.M. Tiede, in "Tunnelling in Biological Systems" (eds B. Chance, D.C. DeVault, H. Frauenfelder, R.A. Marcus, J.R. Schrieffer, and N. Sutin) pp 319-352 Academic Press New York (1979).
4) T.L. Poulos and J. Kraut, J. Biol. Chem. 255 10322-10330 (1980).
5) F.R. Salemme, J. Mol. Biol. 102 563-568 (1976).
6) G.R. Moore, G.W. Pettigrew and N.K. Rogers, Proc. Natl. Acad. Sci. USA 83 4998-4999 (1986).
7) T. Takano and R.E. Dickinson, J. Mol. Biol. 153 79-94 (1980).
8) G. Williams, G.R. Moore, R. Porteous, M.N. Robinson, N. Soffe and R.J.P. Williams, J. Mol. Biol. 183 409-428 (1985).
9) B.R. James and R.J.P. Williams, J. Chem. Soc. 2007 (1961).
10) G.E. Norris, B.F. Anderson and E.N. Baker, J. Amer. Chem. Soc. 108 2784-2785 (1986).
11) T.E. Meyer and M.D. Kamen, Adv. Protein Chem. 35 105-212 (1982).
12) G.R. Moore and R.J.P. Williams, Eur. J. Biochem. 103 533-541 (1980).
13) A.P. Boswell, G.R. Moore, R.J.P. Williams, J.C.W. Chien and L.C. Dickinson, J. Inorg. Biochem. 13 347-352 (1980).
14) D. Culpin, P. Day, P.R. Edwards and R.J.P. Williams, J. Chem. Soc. 1155-1163, (1968A).
15) G.J. Pielak, A.G. Mauk and M. Smith, Nature 313 152-154.
16) N. Liang, G.J. Pielak, A.G. Mauk, M. Smith and B.M. Hoffman, Proc. Natl. Acad. Sci. USA 84 1249-1252 (1987).

17) P.S. Burns, J.F. Harrod, R.J.P. Williams and P.E. Wright,Biochim. Biophys. Acta, 428 261-268 (1976).

18) D.W. Concar, H.A.O. Hill, G.R. Moore, D. Whitford and R.J.P. Williams, FEBS Lett. 206 15-19 (1986).

19) R.A. Capaldi, Biochim. Biophys. Acta 694 291-306 (1982).

20) F.J. Huther and B. Kadenbach, FEBS Lett. 207 89-94 (1986).

21) A.A. DiSpirito, C. Balny and A.B. Hooper, Eur. J. Biochem. 162 299-304 (1987).

22) R.J.P. Williams, J. Theor. Biol. 1 1-13 (1961).

23) G.C.M. Steffens, R. Biewald and G. Buse. Eur. J. Biochem. 164 295-300 (1987).

24) M. Saraste, FEBS Lett. 166 367-372 (1984).

KINETIC LABILITY OF ZN BOUND TO METALLOTHIONEIN (MT) IN

EHRLICH CELLS

D. Petering, S. Krezoski, J. Villalobos, and C. F. Shaw III

Department of Chemistry,
University of Wisconsin-Milwaukee
Milwaukee, Wisconsin 53201

INTRODUCTION

The functions of Zn-Mt in cells have not been well defined. One hypothesis views Mt as a passive store of Zn, which releases metal upon rate-limiting biodegradation of its protein structure. Chemical studies indicate that Zn-Mt is unusually reactive in ligand substitution and metal exchange reactions[1]. Not only do small ligands compete sucessfully for Mt-bound Zn, but apo-carbonic anhydrase is also effective[1]. This study explores whether the metal clusters of Zn-Mt interact directly with other ligands in cells, such as apo-Zn proteins, to carry out ligand exchange reactions.

RESULTS AND DISCUSSION

It has been shown in rat kidney and Ehrlich tumor cells that during Zn-deficiency Mt is the one obvious pool of Zn which is lost[2,3]. Zinc deficiency establishes a condition in which there is a metabolic demand for the metal which cannot be met by extracellular Zn. In rapidly dividing Ehrlich cells, there is a continual need to synthesize new macromolecules including Zn metalloproteins. Thus, one expects in Zn deficiency that any storage pools of Zn will be used to constitute such proteins until these sites are depleted of available zinc. Experiments were designed to observe the redistribution of intracellular Zn in Ehrlich cells after they were placed in Zn-depleted media. The objective was to see whether the behavior of Zn-Mt suggested that it was undergoing ligand substitution reactions.

Two procedures were used to deplete the growth medium of Zn. Chelex 100 resin equilibrated with Eagles minimal essential medium plus Earles salts was added directly to the cell suspension to chelate Zn or Chelex was incubated with and then separated from the complete medium, which also contained 2.5% fetal calf serum, prior to the experiment. Then at time zero cells were added to Zn-depleted media. Incubations of cells in either medium for 24 hrs halted proliferation but left them >90% impermeable to Trypan Blue. When transferred back to complete growth media, these cells divided at normal rates. The time-dependent distribution of cellular Zn was measured by Sephadex G-75 chromatography of sonicated cell supernatant. To follow the rate of biodegradation of high molecular weight and Mt proteins, cells were prelabelled with 35-S

cystine. Later, during the experiment, supernatant was treated with 2.5%
trichloroacetic acid to precipitate protein. Only labelled Mt remained
in solution, which was separated from low molecular weight species by
Sephadex G-25 chromatography and then counted.

Over a 20 hr period little if any Zn is lost from high molecular
weight protein. In contrast, Zn is lost from Mt with a first order rate
constant of 0.6-1.0 hr^{-1}. Mt-bound 35-S turns over much slower whether
or not cold cystine is included in the medium. Depending on conditions,
first order rate constants of 0.08-0.17 hr^{-1} were calculated. The rest
of the protein is degraded in a slower rate process. Total Mt protein
declines more slowly than Mt label because protein continues to be
synthesized even in the absence of zinc. The loss of Zn from Mt is
virtually complete before there is any inhibition of DNA synthesis.

There are three ways Zn may leave Zn-Mt. It may be lost during
rate-limiting protein degradation; it may dissociate from Mt and then
become bound to another, thermodynamically favorable binding site; or it
can undergo direct ligand substitution reactions. During short term Zn-
deficiency, loss of Zn from Mt in Ehrlich cells is not rate limited by
biodegradation of protein. Initial, full dissociation of Zn from Zn-Mt
is unlikely. Ligand substitution chemistry remains as an explanation for
the rapid redistribution of Zn from Zn-Mt under the conditions of this
experiment.

Once Metallothionein has been depleted of Zn, cells can be
transferred back into Zn-sufficient, complete medium. These cells
contain apo-Mt, a normal binding site for Zn. According to preliminary
results, Zn accumulates in Mt over the next 3-4 hours to restore the
normal Zn distribution. Because this experiment was done under
conditions of cellular demand for Zn and because the source of metal is
the labile, high molecular weight pool in fetal calf serum, a
physiologically relevant transport-distribution process is being
observed.

ACKNOWLEDGEMENT

Supported by grants from NIH and the Milwaukee Foundation.

REFERENCES

1. T.-Y. Li, A. Kraker, C.F. Shaw III, and D. H. Petering, Ligand
    substitution reactions of metallothioneins with EDTA and apo-
    carbonic anhydrase. Proc. Natl. Acad. Sci. U.S.A., 77:6338 (1980).
2. D. H. Petering, J. Loftsgaarden, J. Schneider, and B. Fowler,
    Metabolism of cadmium, zinc and copper in the rat kidney: the role
    of metallothionein and other binding sites. Environ. Health
    Perspectives, 54:73 (1984).
3. A. J. Kraker, G. Krakower, C. F. Shaw III, and D. H. Petering, Zinc
    metabolism in Ehrlich cells: properties and role of a
    metallothionein-like zinc-binding protein. Cancer Res, accepted for
    publication.

THE Cu-THIOLATE CLUSTER IN MUTANTS OF YEAST METALLOTHIONEIN

Dennis R. Winge

Departments of Medicine and Biochemistry
University of Utah Medical Center
Salt Lake City, Utah 84132

The class of proteins known as metallothionein is presently only defined structurally since the physiological function is unresolved. These proteins are characterized by being low molecular weight (<10,000 daltons) polypeptides enriched in cysteinyl residues and containing numerous metal ions coordinated within polynuclear metal-thiolate clusters. The structure of mammalian Cd,Zn-metallothionein has been resolved by both x-ray analysis and NMR spectroscopy (1,2). The molecule exists as a two domain protein with the seven metal ions coordinated tetrahedrally in two polynuclear clusters with the 20 cysteines serving as terminal and bridging ligands.

Information concerning Cu-metallothionein is incomplete but equally important in that the protein may well function in some aspect of Zn and Cu metabolism in addition to its clear capacity in metal detoxification. It is clear that mammalian metallothionein binds Cu(I) in a distinct configuration and stoichiometry from Cd(II) or Zn(II). Twelve Cu ions can bind maximally to mammalian protein with a distribution of 6 ions per cluster, although Cu binding is usually restricted to the amino-terminal β domain to yield hybrid Cu,Zn-molecules (3,4). Within the β domain the Cu ions appear to bind with a trigonal geometry unlike the tetrahedral coordination seen with Zn- and Cd,Zn-metallothioneins (5). Further structural characterization of the mammalian Cu-protein has been hampered by instability problems.

Saccharomyces cerevisiae cells produce a Cu-metallothionein that is relatively stable and amenable to study. The Cu-binding protein encoded by the CUP1 locus is classified as metallothionein in having the mentioned properties and exhibiting homology to the mammalian protein (6,7). The yeast MT gene encodes a 61 residue polypeptide although the isolated protein lacks the amino-terminal 8 residues (7,8). The 53 residue molecule contains 12 cysteinyl residues and is capable of binding either 8 Cu(I) or 4 Cd(II) ions (7). Extended x-ray absorption fine structure analysis of yeast CuMT indicates Cu binding to cysteinyl thiolates with apparent trigonal geometry as is the case with mammalian Cu-metallothionein. Cu binding to the apoprotein is highly interactive so no intermediate forms can be identified. Addition of 1 mol eq Cu(I) leads to saturation binding to 1/8 of the molecules. The highly interactive binding is suggestive but not proof that one large polynuclear cluster exists in the protein. Cu_8S_{12} structural units appear to be a stable arrangement of Cu(I) complexes with dianionic sulfure chelates (9,10).

Mutant of the CUP1 Cu-protein have been prepared in which peptide segments from the carboxyl-terminus are deleted or in which pairs of cysteines have been mutated to seryl residues (11). A Cu-containing 48 residue polypeptide with 2 of the 5 deleted residues being cysteines has been isolated. Despite the reduction in the number of cysteines, the maximal binding stoichiometry is still 8 mol eq. Not only is the binding stoichiometry unaffected but other properties such as the binding stability, sensitivity to oxygen, absorption of ultraviolet light and luminescence are equivalent to those of the wild-type Cu-MT. A second carboxyl-terminal truncated MT is a 35 residue polypeptide lacking 4 of the 12 cysteines but 7-8 Cu ions are still associated with the protein. This molecule is much less stable than wild-type MT and the luminescence of the Cu cluster is highly quenched suggesting that the cluster is accessible to solvent. Four double mutants of MT have also been isolated. In each of these mutants pairs of cysteines have been converted to serines, the numbers of the mutants specify the sequence positions of the residues altered. Mutant 49,50-MT is analogous to Cu-T48 in that it is indistinguishable from wild-type MT in binding stoichiometry, cluster stability, oxygen sensitivity, and luminescence. Mutants 9,11-MT and 24,26-MT are much less stable Cu-proteins and possess Cu clusters readily accessible to solvent. Although the stability of the clusters is affected by these mutations the binding stoichiometry is unaffected. Yeast harboring genes for 9,11-MT or 24,26-MT are much more sensitive to the Cu concentration in the growth media relative to 49,50-MT or wild-type MT. The implication of the mutation analysis is that not all cysteines are equivalent within the cluster. If cysteines 49 and 50 participate in the cluster they can be replaced without affecting the integrity of the cluster.

REFERENCES

1. W.F. Furey, A.H. Robbins, L.L. Clancy, E.R. Winge, and D.R. Stout, Science 231:704 (1986).
2. G Wagner, D. Neuhaus, E. Worgotter, M. Vasak, J.H.R. Kagi, and K. Wuthrich, J. Mol. Biol. 187:131 (1986).
3. K.B. Nielson, C.L. Atkin, and D.R. Winge, J. Biol. Chem. 260:5342 (1985).
4. K.B. Nielson, D.R. Winge, J. Biol. Chem. 260:8698 (1985).
5. G.N. George, D.R. Winge, C.D. Stout, and S.P. Cramer, J. Inorganic Biochem. 27:213 (1986).
6. M. Karin, R. Najarian, A. Haslinger, P. Valenzuela, J. Welch and S. Fogel, Proc. Natl. Acad. Sci USA 81:337 (1984).
7. D.R. Winge, K.B. Nielson, W.R. Gray, and D.H. Hamer, J. Biol. Chem. 260:14464 (1985).
8. C.F. Wright, K. McKenney, D.H. Hamer, J. Byrd, and D.R. Winge. J. Biol Chem, in press (1987).
9. F.J. Hollander, D. Coucouvanis, J. Amer. Chem. Soc. 96:5646 (1974).
10. L.E. McCandlish, E.C. Bissell, D. Coucouvanis, J.P. Fackler, and K. Knox, J. Ameri. Chem. Soc. 90:7357 (1986)
11. C.F. Wright, D.H. Hamer, and K. McKenney, Nuc. Acid. Res. 14:8489 (1987).

ACCUMULATION OF HEPATIC Zn_7 AND Zn/Cu METALLOTHIONEIN IN COPPER-LOADED

CHICKS: ISOLATION AND CHARACTERIZATION

Charles C. McCormick and Lih-Y. Lin

Department of Poultry & Avian Sci. and Division of
Nutritional Sciences, Cornell University, Ithaca, NY,
14853, USA

INTRODUCTION

The induction of metallothionein (MT) by copper has historically been
a controversial area of research. Initial work in the mid 1970's
suggested that parenteral copper caused induction of a cysteine-rich
protein (copper chelatin) distinct from MT. However, this protein is now
considered to be an artifact.[1]

Copper given parenterally is a relatively weak inducer of MT
synthesis in rat or mouse liver when compared to that of zinc or cadmium.[2]
In addition, feeding copper can result in little or no accumulation of
hepatic MT despite massive changes in hepatic copper.[3] Although these
findings cast doubt on the "physiological" significance of MT induction by
copper, there is little doubt that copper binds avidly to MT and that the
nature of copper binding is unique relative to that of zinc and/or
cadmium.[4]

Studies in our laboratory have focused on the nature of
copper-induced MT accumulation in chick tissue. This species normally
possesses a single form of MT but as shown for the first time in this
communication, possesses two forms of MT under conditions of copper
loading; forms which differ only in metal composition.

METHODS AND MATERIALS

Four-week-old male chicks were given (i.p.) 2 mg Cu/kg as $CuSO_4$ for 3
consecutive days. Twenty-four hours after the last injection, chicks were
killed and livers excised and immediately stored at $-20^{\circ}C$. A portion of
liver was combined (1:2 w/v) in ice-cold buffer (0.25 M sucrose/0.01M
Tris-HCl/0.1% NaN_3/5 mM dithiothreitol, pH 8.6) and homogenized using a
Potter-Elvehjem glass homogenizer held in ice. The homogenate was
centrifuged at 100,000 x g for 1 hour at $4^{\circ}C$. Fifteen ml of cytosol were
applied to column (5 x 50 cm) of Sephadex G-75 equilibrated in 2 mM
Tris-acetate (pH 7.4) buffer and eluted with the same buffer. The Zn and
Cu content of each fraction were measured and those fractions in the VeMT
were pooled and applied directly to a column (0.9 x 30 cm) containing DEAE
Sephadex A25 equilibrated in 2 mM Tris-acetate (pH 7.4). The material was
eluted with a linear gradient of 2- 80 mM Tris-acetate (pH 7.4, total 200
ml). All steps of purification were performed at $4^{\circ}C$.

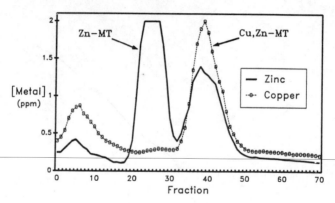

Figure 1. Anion exchange chromatography of G75 fractions (VeMT) obtained from hepatic cytosol of chicks given parenteral copper.

RESULTS AND DISCUSSION

The ratio of copper and zinc in the pooled G75 fractions in the VeMT was approximately 0.95. Subsequent anion exchange chromatography of these pooled fractions is shown in Figure 1. The recovery of copper and zinc from this and several preparations was 30% and 60%, respectively. The ion exchange chromatography showed two major peaks, one containing zinc only and the other containing both zinc and copper. Amino acid compositional analysis of both peaks revealed similar amino acid profiles, each indicative of MT, i.e., 30% Cys, 0% Phe, 0% Ile, and 0% Tyr.

Identical procedures for samples from rats given parenteral copper showed a G75 VeMT copper/zinc ratio of 3.7 and ion exchange recovery of both zinc and copper at 60%. In contrast, rat samples contained two peaks (MT1 and MT2) both containing copper and zinc.

Metal analysis of purified Cu,Zn-MT from chick liver indicated a metal/MT molar ratio of 6.0 and 4.4 for Cu and Zn, respectively. The ratio for Zn-MT was 6.92. These results are consistent with metal stoichiometries of the beta and alpha domains for mammalian MTs.[5]

We believe this to be the first report of two forms of MT (present in vivo) differing only in metal composition. We propose that the relative proportions of copper and zinc in chick tissue and cooperative binding of copper to MT[5] are responsible for the accumulation of two forms of MT in chick liver subsequent to parenteral copper. Preliminary data from in vitro copper substitution studies support this suggestion.

REFERENCES

1) I. Bremner, Involvement of metallothionein in the hepatic metabolism of copper. J. Nutr. 117:19-29 (1987)
2) D.M. Durnam and R.D. Palmiter, Transcriptional regulation of the mouse metallothionein-I gene by heavy metals. J.Biol.Chem. 256:5712-5716 (1981)
3) I. Bremner, R.K. Mehra, J.N. Morrison and A.M. Wood, Effects of dietary copper supplementation of rats on the occurrence of metallothionein-I in liver and its secretion into blood, bile and urine. Biochem. J. 235:735-739 (1986)
4) K.B. Nielson, C.L. Atkin and D.R. Winge, Distinct metal-binding configurations in metallothionein. J. Biol. Chem. 260:5342-5350 (1985)
5) K.B. Nielson and D.R. Winge, Preferential binding of copper to the beta domain of metallothionein. J. Biol. Chem. 259:4941-4946 (1984)

FACTORS INFLUENCING THE ACCUMULATION OF

METALLOTHIONEIN IN RAT BLOOD CELLS

James N. Morrison, Anne M. Wood and Ian Bremner

Rowett Research Institute
Bucksburn
Aberdeen,   AB2 9SB, U.K.

INTRODUCTION

It has been shown by radioimmunoassay that small amounts of metal-lothionein (MT) are present in blood plasma and cells of normal rats. Concentrations in plasma generally increase after induction of tissue MT synthesis but little is known of the effects of nutritional and physiological factors on the occurrence of MT-I in blood cells or on its distribution among sub-populations of cells. As part of an investigation into the use of MT assays for the assessment of trace element status, we have examined the effects of zinc deficiency, of zinc and cadmium injection, of endotoxaemia and of induced reticulocytosis on the occurrence of MT-I in blood cells of rats.

MATERIALS AND METHODS

Male Hooded Lister rats (Rowett strain), usually aged 5-6 weeks, were used. They were given either a semi-synthetic diet containing 3, 6 or 12 mg zinc/kg or stock colony diet ad libitum. Groups of at least 5 rats were injected as appropriate with zinc (3 mg/kg body weight, i.p.), cadmium (1.0 mg/kg,s.c.), copper (3 mg/kg,i.p.), endotoxin (1 mg/kg,i.p.) and phenyl-hydrazine (100 mg/kg,s.c.). Rats were killed at various intervals thereafter by exsanguination under ether anaesthesia. Concentractions of MT-I in lysed blood cells was measured by radioimmunoassay (Mehra & Bremner, 1983). In some experiments blood cells were fractionated by density gradient centrifugation on Percoll.

RESULTS AND DISCUSSION

When rats were given semi-purified diets containing only 3 or 6 mg zinc/kg blood cell MT-I levels decreased rapidly. After 3 days they were reduced to only 27 and 46% of those in control rats given 12 mg zinc/kg diet. After 7 days concentrations in the rats given 3, 6 and 12 mg zinc/kg were 3.5 ± 1.7, 14.2 ± 1.8 and 45.3 ± 5.0 ng/ml blood respectively. No further reduction in concentrations occurred in the zinc-deficient animals over the next 7 days, although levels in the control rats showed an age-dependent reduction to 30 ng/ml.

Table 1. Changes in Blood Cell MT-I Concentrations in Rats

Days	Zinc	Cadmium	Endotoxin	Phenylhydrazine
		Treatment		
1	580	161	60	281
2	190		62	
3	250		100	289
4	160		155	
7		11500		625
22		1570		

Details of the injections are given in the text. Concentrations are given as % of values in control rats, which were about 25 ng/ml blood.

Injection of rats with zinc and cadmium caused rapid increases in blood cell MT-I levels but injection of endotoxin had little effect (Table 1). Maximum MT-I concentrations were attained 1 day after zinc injection, whereupon values decreased over the next 3 days to just above control levels. Concentrations in cadmium-injected rats did not increase as rapidly as in zinc-injected rats, but the final values were very much higher and remained elevated for a longer period. Even after 22 days MT-I levels in the blood cells were still 15 times greater than those in control animals. Efficient mechanisms clearly exist for the rapid elimination of MT-I from the blood cells of zinc-injected and zinc-deficient rats but whether this involves degradation of the protein, its leakage from the cell or simply loss of metal from the protein is unknown. However loss of cadmium-MT from the cells occurs much less readily, possibly because this metalloform is more resistant to degradation.

Rapid appearance and disappearance of MT from blood cells of zinc-deficient and zinc-injected rats would be easier to understand if the protein were associated with leucocytes rather than erythrocytes. However when blood cells from normal and zinc-injected rats were fractionated on Percoll, no significant amount of MT-I was associated with the white cells. Four sub-populations of red cells were separated. The lightest fraction, which contained a high proportion of reticulocytes, only accounted for about 2% of the total erythrocytes but nevertheless contained 50% of the blood cell MT-I. In the zinc-injected rats MT-I concentrations in the four sub-populations of cells (lightest to heaviest) were 336, 4, 4 and 8 pg/million cells. The importance of reticulocytes in the binding of red cell MT-I was confirmed in rats which were injected with phenylhydrazine to induce reticulocytosis. As can be seen in Table 1, there were major increases in blood cell MT-I concentrations in these animals after 7 days.

These results indicate that MT is probably synthesised in immature erythroblasts in response to injection of certain metals but not in response to acute endotoxin shock. As the cells mature the protein appears in the circulation but, with the exception of cadmium-dosed animals, only remains there for a relatively short period. Concentrations of blood cell MT decrease substantially in zinc-deficient animals.

REFERENCES

Mehra, R.K., and Bremner, I., 1983, Development of a radioimmunoassay for rat liver metallothionein-I and its application to the analysis of rat plasma and kidneys, Biochem. J. 213:459-465.

REGULATION OF PANCREATIC EXOCRINE FUNCTION BY MANGANESE

Murray Korc and Patsy M. Brannon

Departments of Internal Medicine
and Nutrition and Food Science
University of Arizona
Tucson, AZ

INTRODUCTION

The pancreas is a heterogenous organ consisting of clusters of
endocrine cells that are dispersed throughout the exocrine tissue. The
endocrine cells secrete insulin, glucagon, somatostatin, and a variety of
other peptides into the blood. The exocrine tissue consists of acinar
cells that synthesize digestive enzymes and duct cells that produce
bicarbonate rich fluid. Under normal physiological conditions numerous
factors interact to enable the endocrine and exocrine pancreas to respond
to a meal in a highly coordinated manner that contributes to the digestion
and subsequent assimilation of ingested food. This regulation involves an
interplay between neurotransmitters, nutrients, gastrointestinal hormones,
and islet cell hormones acting through a variety of second messengers.
Because of the relative richness of the pancreas in manganese[1], it has
been suggested that this divalent cation may also participate in the
regulation of pancreatic function. This hypothesis is supported by several
types of observations. Thus, manganese is taken up from the systemic
circulation by the pancreas, and the concentration of manganese in the
pancreatic duct is greater than in the blood[2]. Second-
generation manganese-deficient animals may exhibit ultrastructural damage
or complete atrophy of the pancreatic acinar cell[3,4]. These animals also
exhibit beta cell dysfunction that is manifested by decreased insulin
secretion and synthesis, and enhanced insulin degradation[5].

Effects of Manganese on the Pancreatic Acinar Cell

Numerous studies have demonstrated that manganese exerts direct
effects on the exocrine pancreas. Manganese stimulates amylase secretion
in guinea pig[6], rat[7], and mouse[8] pancreas, and hyperpolarizes the
pancreatic acinar cell[8]. Manganese also modulates acinar cell protein
synthesis[9], enhancing synthesis at low concentrations (0.03 mM), and
inhibiting synthesis at high concentrations (1 mM). The stimulatory effect
of manganese occurs after a lag period of 30 min, and, in normal rat acini,
is abolished in the presence of extracellular calcium[9]. Pancreatic
secretagogues such as cholecystokinin-octapeptide (CCK_8) and carbachol
also regulate protein synthesis, enhancing synthesis over a concentration
range that is maximal for enzyme secretion, and inhibiting synthesis at
higher concentrations[10]. However, manganese does not act via either the
cholecystokinin or acetylcholine receptor[11].

The mechanisms underlying the actions of manganese in the pancreas are not well understood. Nonetheless, manganese regulates the activity of protein kinases and calmodulin in several tissues[12,13], modulates protein phosphorylation in pancreatic acini (unpublished observations), and enhances calcium efflux in these cells[14,15]. In contrast to the action of CCK$_8$, the effect of manganese on calcium efflux is slow in onset[15]. Manganese may act, therefore, on cellular calcium stores that are distinct from those mobilized following phospholipid hydrolysis by CCK$_8$. These observations suggest that manganese may directly modulate the activity of protein kinases, or may act indirectly by altering cellular calcium pools.

In addition to its direct actions, manganese blocks the stimulatory effects of CCK$_8$ on both pancreatic enzyme secretion[6,8] and protein synthesis[16]. During short incubation periods manganese does not alter the basal rates of calcium influx or efflux, but blocks the action of CCK$_8$ on calcium influx[16]. It is therefore possible that manganese antagonizes the stimulatory effects of CCK$_8$ by interfering with CCK$_8$-induced calcium influx. Alternatively, manganese may exert this effect by acting on some distal intracellular processes.

Manganese Action in the Diabetic Rat Pancreas

Both the direct and antagonistic effects of manganese in the exocrine pancreas occur at relatively high concentrations of the divalent cation. Further, many of these effects are observed only in the absence of extracellular calcium. The physiological relevance of manganese in the regulation of pancreatic exocrine function is not, therefore, readily apparent. Nonetheless, very low concentrations of manganese (0.01 mM) stimulate pancreatic protein synthesis in acini from diabetic rats that are incubated in the presence of extracellular calcium[9]. The magnitude of the response is also greatly enhanced in these acini. The increased responsiveness of acini from diabetic rats is not due to in vitro repletion of cellular manganese because manganese levels are not greater in the pancreas of diabetic rats than in normal rats[9]. Further, magnesium, cobalt, and nickel do not mimic the stimulatory effect of manganese[13]. These findings emphasize the specificity of manganese action in the exocrine pancreas by comparison to other metal cations.

Omission of calcium from incubation medium markedly attenuates the stimulatory effect of manganese on protein synthesis in diabetic rat acini[11]. Further, replacement of calcium by either barium or strontium abolishes the action of manganese[13]. Similarly, lanthanum, an agent that blocks calcium influx, completely blocks the action of manganese[11]. In contrast, both CCK$_8$ and insulin exert a full stimulatory effect on protein synthesis in pancreatic acini from diabetic rats when calcium is omitted from the incubation medium (Figure 1), and lanthanum does not block insulin-mediated stimulation of protein synthesis[16,17]. It is unlikely, therefore, that either calcium omission or lanthanum non-specifically interfere with stimulation of protein synthesis. These observations suggest that manganese enhances protein synthesis in diabetic rat acini via a mechanism that is dependent either on calcium influx or on the presence of a tightly bound pool of calcium on the outer aspect of the cell membrane. This calcium pool also contributes to the CCK$_8$-mediated rise in cytosolic free calcium[18]. Both the requirement for calcium and the enhanced sensitivity to manganese indicate that, in the face of insulin deficiency, manganese may participate in the regulation of pancreatic protein synthesis in vivo. Inasmuch as pancreatic protein synthesis is directed mainly toward the production of pancreatic digestive enzymes, these observations suggest that manganese may modulate the synthesis of these enzymes.

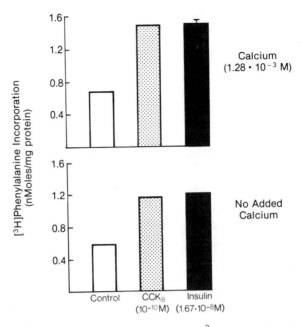

Figure 1. Effects of CCK_8 and insulin on ^{3}H-phenylalanine incorporation into protein in diabetic rat acini. CCK_8 (0.1nM) and insulin (16.7 nM) were present for 40 and 105 min, respectively, prior to the addition of ^{3}H-phenylalanine for 15 min. Values are means ± SE of 3 experiments (only the largest SE is shown). The stimulatory effects were similar for both agonists, were highly significant, and were not attenuated when cells were incubated in medium containing no added calcium.

The ability of manganese to greatly enhance pancreatic protein synthesis in diabetic rat acini in the presence of calcium may be due to altered cellular calcium homeostasis in these cells. By comparison to normal rat acini, diabetic rat acini exhibit a decrease in the basal levels of cytosolic free calcium, an enhanced sensitivity to the actions of manganese on calcium efflux, and an attenuated rise in free calcium levels following CCK receptor activation[15]. These observations suggest that insulin may regulate the efficiency of signal transduction systems that lead to calcium mobilization in the pancreatic acinar cell.

Effects of Dietary Manganese Deficiency on the Exocrine Pancreas

Manganese is an essential cofactor for RNA synthesis in vitro[19], and is especially abundant in nuclei[20]. Further, manganese deficiency has been shown to cause swelling in the endoplasmic reticulum[3]. It is therefore possible that manganese participates in the transcriptional and translational regulation of gene expression, and interacts with hormones and nutrients to regulate pancreatic digestive enzyme synthesis. To evaluate this possibility, the effects of manganese deficiency on pancreatic enzyme content were determined[21]. Rats were fed for 10 weeks a high carbohydrate diet containing 20 percent casein protein and 65 percent cornstarch[22]. The diet was supplemented in the control group with 39.6 ppm manganese, and in the deficient group with 0.5 ppm manganese. During the 10 week period, amylase and lipase activities did

Table 1. Effect of CCK₈ on amylase release
in manganese-deficient rats.

Addition		Amylase Release (percent above control)
CCK₈,	0.01 nM	253 ± 31
	0.1 nM	722 ± 72
	1.0 nM	545 ± 66
	10 nM	287 ± 30

Sprague-Dawley rats were placed on a manganese-deficient
diet for 8 weeks. Pancreatic acini were then prepared
and incubated for 30 min at 37°C in the presence of the
indicated additions. Data are means ± SE from 7 separate
experiments, revealing a characteristically biphasic
stimulatory effect on amylase release.

not vary significantly in the control group; chymotrypsin activity
decreased significantly from weeks 2 through 4, and trypsin activity
increased significantly from weeks 1 through 6. In the manganese-deficient
group pancreatic amylase activity increased gradually by comparison to the
control group. This increase achieved significance at week 8. Lipase and
proteolytic enzyme activities were not altered in these animals. The
increase in amylase content was not due to decreased secretion of this
enzyme inasmuch as basal amylase release, as well as the magnitude of the
secretory response to CCK₈ (Table 1), were comparable in acini prepared
from manganese-deficient and control rats.

When rats are rendered manganese deficient and then placed on a high
carbohydrate, high protein, or high fat diet, they exhibit increases in the
content of pancreatic amylase, trypsinogen, and lipase, respectively[23].
However, in each dietary group pancreatic amylase content is greater in the
manganese-deficient rats than in the corresponding control rats. Further,
lipase content is significantly greater in manganese-deficient rats that
are on a high fat diet than in the corresponding control rats. In
contrast, the levels of proteolytic enzymes are not altered by manganese
deficiency. These findings indicate that manganese may play a role in the
regulation of pancreatic content of amylase, and in the dietary adaptation
of lipase to high fat diets. Taken together with the in vitro studies, the
dietary observations support the hypothesis that manganese participates in
the regulation of pancreatic exocrine function.

Acknowledgement

This study was supported by National Institutes of Health Research Grant
AM-32561.

REFERENCES

1. C. L. Keen, B. Lonnerdal, and L. S. Hurley, Manganese, in:
    Biochemistry of the Essential Ultratrace Elements, E. Frieden, ed.,
    Plenum Pub. Co., New York (1984).
2. W. T. Burnett, R. R. Bigelow, A. W. Kimball, and C. W. Sheppard,
    Radiomanganese studies on the mouse, rat and pancreatic fistula dog,
    Am. J. Physiol. 168:620 (1952).

3. L. T. Bell, and L. S. Hurley, Ultrastructural effects of manganese deficiency in liver, heart, kidney, and pancreas of mice, Lab. Invest. 29:723 (1973).
4. R. E. Schrader, and G.J. Everson, Pancreatic pathology in manganese-deficient guinea pigs, J. Nutr. 94:269 (1968).
5. D. L. Baly, D. L. Curry, C. L. Keen, and L. S. Hurley, Dynamics of insulin and glucagon release in rats: influence of dietary manganese, Endocrinology 116:1734 (1985).
6. S. Abdelmoumene, and J. D. Gardner, Effect of extracellular manganese on amylase release from dispersed pancreatic acini, Am. J. Physiol. 241:G359 (1981).
7. T. Kanno, and O. Nishimura, Stimulus-secretion coupling in pancreatic acinar cells: inhibitory effects of calcium removal and manganese addition on pancreozymin-induced amylase release, J. Physiol. (London) 257:309 (1976).
8. O. H. Petersen, and N. Ueda, Pancreatic acinar cells: the role of calcium in stimulus-secretion coupling, J. Physiol. (London) 254:583 1976.
9. M. Korc, Manganese action on pancreatic protein synthesis in normal and diabetic rats, Am. J. Physiol. 245:G628 (1983).
10. M. Korc, A. C. Bailey, and J.A. Williams. Regulation of protein synthesis in isolated rat pancreatic acini by cholecystokinin, Am. J. Physiol. 241:G116 (1981).
11. M. Korc, Manganese action on protein synthesis in diabetic rat pancreas: evidence for a possible physiological role, J. Nutr. 2119 (1984).
12. C. S. Rubin, J. Erlichman, and O. M. Rosen, Cyclic adenosine 3',5'-monophosphate-dependent protein kinase of human erythrocyte membranes, J. Biol. Chem. 247:6135 (1972).
13. Y. M. Lin, Y. P. Liu, and W. Y. Cheung, Cyclic 3':5'-nucleotide phosphodiesterase purification, characterization, and active form of the protein activator from bovine brain, J. Biol. Chem. 249:4943 (1974).
14. B. E. Argent, R. M. Case, and F. C. Hirst, The effects of manganese, cobalt and calcium on amylase secretion and calcium homeostasis in rat pancreas, J. Physiol. (London) 323:353 (1982).
15. M. Korc, and M. H. Schoni, Quin-2 and manganese define multiple alterations in cellular calcium homeostasis in diabetic rat pancreas, Diabetes, In Press.
16. M. Korc, Effect of lanthanum on pancreatic protein synthesis in streptozotocin-diabetic rats. Am. J. Physiol. 244:G321 (1983).
17. M. Korc, Regulation of pancreatic protein synthesis by cholecystokinin and calcium. Am. J. Physiol. 243: G69-75, 1982.
18. M. Korc, and M. H. Schoni, Modulation of cytosolic free calcium levels by extracellular phosphate and lanthanum, Proc. Natl. Acad. Sci. USA 84:1282 (1987).
19. Y. Nagamine, D. Mizuno, and S. Natori, Differences in the effects of manganese and magnesium on initiation and elongation in the RNA polymerase I reaction, Biochim. Biophys. Acta, 519:440, (1978).
20. R. E. Thiers, and B. L. Vallee, Distribution of metals in subcellular fractions of rat liver, J. Biol. Chem. 226:911 (1957).
21. P. M. Brannon, V. P. Collins and M. Korc, Alterations of pancreatic digestive enzyme content in the manganese-deficient rat, J. Nutr. 117:305, (1987).
22. P. M. Brannon, A. S. Demarest, J. Sabb, and M. Korc, Dietary modulation of epidermal growth factor action in cultured pancreatic acinar cells of the rat, J. Nutr. 116:1306 (1986).
23. L. Werner, M. Korc, and P. M. Brannon, Effects of manganese deficiency on dietary adaptation of the pancreas, Fed. Proc. 45:368 (1986).

EFFECT OF MANGANESE DEFICIENCY ON GLUCOSE TRANSPORT AND INSULIN BINDING

IN RAT ADIPOCYTES

Deborah L. Baly

Department of Nutrition
Rutgers University
New Brunswick, N.J.

Manganese deficiency results in altered carbohydrate metabolism; both at the level of pancreatic insulin biosynthesis (1) and gluconeogenesis (2). The effects of manganese deficiency on peripheral actions of insulin, however, have not been examined. Evidence of a role for manganese in mediating peripheral actions of insulin has come from several studies. It is well recognized that the insulin receptor is a hormone-dependent kinase that is stimulated by both Mg^{2+} and Mn^{2+} (3). Ueda and coworkers (4) have also demonstrated that Mn^{2+} mildly enhances the extracellular binding of insulin to its receptor, facilitates the physiological actions of insulin and mimics the action of the hormone. Although the role of insulin receptor phosphorylation in the mechanism of insulin action is still disputed, the authors postulate that the insulinomimetic action of Mn^{2+} may be related to its ability to cause the phosphorylation of the β-subunit of the insulin receptor and thus "turn on" insulin-related events. Alternatively, Mn^{2+} may be acting at some step distal to the insulin receptor, such as adenylate cyclase linked processes. In the present studies, we have begun to examine the effect of Mn deficiency on these processes.

Isolated adipose cells were prepared by collagenase digestion of the epididymal fat pad. For the studies on glucose transport, adipose cells were incubated in the presence or absence of insulin (0-7 nM). After a 30 min incubation period, $[^{14}C]$-3-0-methyl glucose (30MG) transport was measured (5). Insulin binding studies were carried out with tracer amounts of ^{125}I insulin (2 x 20^{-10} M) and varying amounts of cold insulin. Cells were incubated for 45 min at 20°C. At the end of the incubation, aliquots were removed and centrifuged through oil. The radioactivity in the cells (bound) and infranate (free) were measured. Nonspecific binding was determined in the presence of 1 μM unlabeled insulin. Less than 5% of the ^{125}I-labeled insulin was degraded at the end of the incubation period as assessed by solubility in 10% trichloroacetic acid. Binding data were analyzed by the method of Scatchard (6). Adipose cell size was determined by the osmic acid fixation (7).

Epididymal fat pad weight and adipose cell size were not affected by manganese deficiency. Both basal and insulin-stimulated 30MG transport were significantly lower in Mn- rats compared to control (Table 1). However, the fold-stimulation above basal values by insulin was not different between control and Mn- rats. Manganese deficient rats had lower glucose

Table 1. Effect of Mn Deficiency on Glucose Transport and Insulin Binding

Group	Glucose Transport		Insulin Receptor	
	Basal	Insulin Stimulated	Ro	K_D
	fmole/min/cell		Sites/Cell	nM
Control	0.218±.030	4.79±.27	124,346	3.2
Mn Deficient	0.107±.013*	2.99±.19*	91,668*	2.0*

* Significantly different from control level ($p<.05$).

transport rates than controls at all insulin concentrations examined, averaging 50% of control values (data not shown). In addition, no difference in the half maximal stimulation of 30MG uptake was observed (195 pM insulin for both Mn- and control rats). The insulin binding studies demonstrated that cold insulin could maximally displace tracer in both control and Mn- rats. Although the insulin dose response curves were parallel, the Mn- rats demonstrated a leftward shift in the binding curve, indicative of a higher affinity receptor. Kinetic analysis of the data revealed that Mn- rats had fewer ($p<.05$) insulin receptors and a lower K_D for insulin binding (Table 1). This lower K_D in Mn- rats results primarily from changes in affinity of the high capacity binding site.

These data demonstrate that Mn deficiency has a pronounced effect on glucose transport in the adipose cell. The finding of both decreased basal and insulin-stimulated glucose uptake suggests that the defect may be at the level of the transport protein: such as a decrease in the number of glucose carriers available for transport or an altered affinity of the carrier for glucose. The first possibility seems likely, since changes in transporter number have been shown to accompany the altered transport capacity observed in several disease states such as obesity (8) and diabetes (9).

Other mechanisms that might explain the impaired ability of insulin to stimulate glucose transport include lesions in the insulin receptor or post receptor defects. A defect at the level of the receptor is unlikely since maximal displacement of tracer by cold insulin was observed in both Mn- and control rats. Although a 30% decrease in insulin receptor number was observed in Mn- rats, this should not compromise transport, as only 5% receptor occupancy is required for full stimulation of glucose transport. In addition, insulin receptor in Mn- rats had a higher affinity for its ligand.

In summary, these data suggest that Mn deficiency affects glucose transport in the isolated adipose cell. The apparent defect lies distal to the insulin receptor. These alterations may contribute to the altered glucose homeostasis observed in Mn- animals.

REFERENCES

1. D. L. Baly, D. L. Curry, C. L. Keen, and L. S. Hurley, Endocrinology 116:1734 (1985).
2. D. L. Baly, C. L. Keen, and L. S. Hurley, J. Nutr. 115:872 (1985).
3. M. P. Czech, Ann. Rev. Biochem. 46:359 (1977).
4. M. Ueda, F. W. Robinson, M. M. Smith, and T. Kono, J. Biol. Chem. 259:9520 (1984).
5. J. Gliemann and R. R. Whitesell, Diabetologia 13:396 (1977).
6. G. Scatchard, Ann. N.Y. Acad. Sci., 52:660 (1949).
7. J. Hirsch and E. Gallian, J. Lipid Res. 9:110 (1968).
8. L. J. Wardzala, M. Crettaz, E. D. Horton, B. Jeanrenaud, and E. S. Horton, Am. J. Physiol. 243:E418 (1982).
9. E. Karnieli, P. J. Hissin, I.A. Simpson, L. B. Salans, and S. W. Cushman, J. Clin. Invest. 68:811 (1981).

ZINC ABSORPTION IN PANCREATIC INSUFFICIENCY

W.S. Watson[1], G. McLauchlan[2], T.D.B. Lyon[3], I. Pattie[4] and
G.P. Crean[2]

Department of Clinical Physics[1] and Gastro-Intestinal Centre[2],
Southern General Hospital, Department of Biochemistry, Royal
Infirmary[3] and Department of Biochemistry, Gartnavel General
Hospital[4], Glasgow, Scotland, U.K.

INTRODUCTION

The zinc tolerance test has been used to demonstrate impaired handling
of orally administered zinc in patients with pancreatic insufficiency
(Boosalis et al, 1983). This test requires a non-physiological oral load of
zinc (> 300 umol) and does not provide an absolute measure of zinc
absorption. In order to obtain more quantitative information, we have
employed the radiotracer zinc-65 and whole-body counting techniques to study
zinc absorption and turnover in patients with pancreatic insufficiency.

MATERIALS AND METHODS

Six patients (5M, 1F) with a history of chronic pancreatitis and
clinical evidence of pancreatic insufficiency (PI) were studied. All six
were taking pancreatic extract for steatorrhoea.

Zinc absorption was estimated from zinc-65 whole-body retention 14 days
after oral ingestion of a zinc-65 test dose containing 75 umol stable zinc.
Absorption was measured three times in each subject as follows (1) baseline
absorption, (2) absorption plus pancreatic extract (one Pancrex V capsule,
Pabyrn, UK) and (3) absorption plus 0.5g citric acid. The total activity of
zinc-65 administered to each PI subject was 150 kiloBecquerels. In 4/6
subjects, long term zinc-65 retention was measured between 80 and 100 days
after the completion of the absorption studies. Total body potassium (TBK),
a measure of lean body mass, was estimated from the whole-body activity of
the naturally-occurring potassium radioisotope, potassium-40.

The normal range for zinc absorption was estimated in nine healthy
controls (4M, 5F), while the long term retention values for the PI subjects
were compared with the values obtained for a group of 10 healthy controls
(7M, 3F) studied out to 365 days.

Plasma zinc and 24 hour urine zinc were also measured.

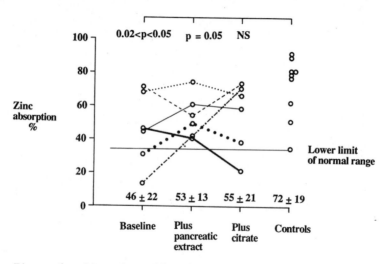

Figure 1. Zinc absorption in pancreatic insufficiency.

RESULTS AND CONCLUSIONS

Plasma zinc was marginally low in 2/6 PI patients, while 24 hour urine zinc was normal in all six. The absorption results are shown in Figure 1. Baseline absorption for the group was significantly less than normal (46% v 72%, p < 0.05) with 2/6 subjects below the lower limit of normal at 34%. Absorption in both these subjects was within the normal range when pancreatic extract or citrate was given with the zinc-65 test dose. However, the group mean absorption when pancreatic extract was given was still significantly lower than normal (53% v 72%, p = 0.05).

The long term retention values at 80 to 110 days in the four PI subjects measured were within the normal range; these four included the two subjects who malabsorbed zinc. This suggests that the fractional turnover of zinc, ie endogenous zinc loss per day/total body zinc (TBZn) was normal in these subjects. Total body potassium levels ranged from 75% to 102% of the predicted normal values with 4/6 PI subjects having TBK values on or below the lower limit of normal at 84% predicted. In a previous study of zinc metabolism in controls and Crohn's disease patients (Mitchell et al, 1985), we showed that there was a positive correlation between TBZn and TBK with no significant difference in TBZn/TBK ratio between controls and Crohn's disease patients. Therefore, the low TBK results obtained in 4/6 PI subjects in the present study suggest that TBZn was also low in these subjects.

The reduction in lean body mass in these subjects, as evidenced by low TBK, may indicate an adaptive response to low zinc absorption in an attempt to maintain normal lean tissue zinc concentrations.

REFERENCES

Boosalis MG, Evans GW and McClain CJ, 1983, Impaired handling of orally administered zinc in pancreatic insufficiency, Am J Clin Nutr, 37:268.
Mitchell KG, Watson WS, Lyon TDB, Bethel M and Crean GP, 1985, Zinc metabolism in Crohn's disease, in: "Trace Elements in Man and Animals- TEMA 5", C F Mills, I Bremner and J K Chesters, eds, Commonwealth Agricultural Bureaux, Slough, UK.

IDENTIFICATION OF ABNORMAL SITES OF ZN TRANSPORT AND

METABOLISM IN PATIENTS WITH SENSORY AND ENDOCRINE DISORDERS

M.E. Wastney, D.M. Foster and R.I. Henkin

Georgetown University Medical Center,
Washington D.C. 20007 and University of
Washington, Seattle, WA 98195, USA

INTRODUCTION

Abnormal Zn metabolism is associated with several clinical disorders. To determine the role of Zn in these disorders it is necessary to first define the sites of abnormal Zn metabolism. Once these sites are defined the mechanisms involved may be determined and the metabolism of Zn then related to the disorder. To define sites where Zn metabolism is abnormal it is necessary to compare data from normal and abnormal states. A useful technique for comparing kinetic data is by mathematical modeling.

Modeling Kinetic Data

Modeling is a powerful tool for analysing kinetic data since data from many tissues may be analysed simultaneously. A large number of parameters may be determined from these data including absorption, distribution, tissue uptake and release as well as secretion, excretion and the amount of a compound in tissues and whole body. Modeling can also be used to define sites of abnormal metabolism.

USE OF MODELING TO DEFINE SITES OF ABNORMAL METABOLISM

A model consists of a set of mathematical equations which, when solved, predict the behavior of a substance in a system. Thus, when the model for Zn metabolism (Wastney et al., 1986) is solved the calculated values fit the observed kinetic data obtained from each tissue.

When metabolism is abnormal one or more parameters of the system are perturbed to new values. The problem is, given a set of data from an abnormal state, to determine which parameters of metabolism changed to generate these data. The parameters which change in an abnormal state can be determined by systematically changing parameter values of the model until the model solution fits the abnormal data. These parameters then represent sites of abnormal metabolism.

The combination of the fewest parameters necessary to fit the abnormal data are selected. This is the maxim of 'necessary and sufficient' changes in parameters to fit data from perturbed states (Berman, 1963).

The present studies were undertaken to determine the sites of abnormal Zn metabolism in two human disorders. One is an endocrine disorder, adrenal corticosteroid insufficiency and the second a sensory disorder, taste and smell dysfunction.

Sites of Abnormal Zn Metabolism in an Endocrine Disorder

Adrenal corticosteroid insufficiency is characterized by high Zn in serum and low Zn in urine. Two patients were studied, one had Addison's disease and other had an adrenalectomy (Henkin et al., 1984). The subjects were studied for two 7 day periods, once while on treatment (20 mg/d prednisone) and then while off treatment, without carbohydrate-active steroids. At the start of each period patients were given ^{69m}Zn by intravenous injection and tracer was measured in plasma, red blood cells (RBC), urine, feces and over liver and thigh regions for 7 days.

Kinetic data from all tissues were fitted using SAAM (Berman and Weiss, 1978) and a model for Zn metabolism (see Henkin et. al., 1984). Data from the untreated state were fitted by changes in two parameters representing low RBC and liver uptake of Zn. In each patient Zn uptake by both tissues increased with treatment.

Sites of Abnormal Zn Metabolism in a Sensory Disorder

Taste and smell dysfunction in humans is characterized by low Zn in saliva and red blood cells (RBC). Patients (n=10) and normal volunteers (n=32) were studied for 250 d following oral administration of ^{65}Zn. Tracer was measured in plasma, RBC, urine, feces and over whole body, liver and thigh. Data from patients were compared to normals using a model for Zn metabolism (Wastney et al., 1986). Patient data were fitted by changes in four parameters. In patients absorption and RBC uptake of Zn were low while fecal excretion of Zn and the fraction of body Zn, that turned over in 50 d, observed in liver were high compared to normals.

SUMMARY AND CONCLUSION

Two sites of abnormal Zn metabolism were defined in patients with untreated corticosteroid insufficiency and four sites of abnormal Zn metabolism were defined in patients with taste and smell dysfunction.

Modeling kinetic data from normal and abnormal states can define sites of abnormal metabolism and the degree of abnormality at each site.

REFERENCES

Berman, M., 1963, A postulate to aid in model building, J. Theoret. Biol. 4:229.
Berman, M., and Weiss, M.F., 1978, SAAM Manual. Washington, DC: U.S. Printing Office, [DHEW Publication No. (NIH)78-180].
Henkin, R. I., Foster, D. M., Aamodt, R. L., and Berman, M. 1984, Zinc metabolism in adrenal cortical insufficiency: Efffects of carbohydrate-active steroids, Metab., 33:491
Wastney, M. E., Aamodt, R. L., Rumble, W. F., and Henkin, R. I., 1986, Kinetic analysis of zinc metabolism and its regulation in normal humans, Am. J. Physiol. 251(Regulatory Integrative Comp. Physiol. 20):R398.

SELENIUM AND SELENOPROTEINS IN TISSUES WITH

ENDOCRINE FUNCTIONS

D. Behne, H. Gessner, H. Hilmert, and S. Scheid

Hahn-Meitner-Institut Berlin
Glienicker Str. 100
D-1000 Berlin 39, West Germany

INTRODUCTION

In studies on rats it has been shown that regulatory mechanisms exist, with the help of which the organism strives
to maintain the Se level in the testis (Behne et al., 1982).
It can be assumed that the regulation mainly serves to ensure
the supply of sufficient amounts of the element to the spermatozoa. However, for the adrenals similar results were later
obtained (Behne and Höfer-Bosse, 1984), and in the testis of
severely Se-depleted rats a decreased testosterone secretion
was found (Behne et al., 1986). As these findings suggest
that Se is also involved in endocrine functions, several experiments on rats were carried out in order to investigate
the Se metabolism and the selenoproteins in the different endocrine organs.

MATERIALS AND METHODS

Wistar rats were kept for three generations on either a
low Se diet (2 µg Se/kg) or the same diet with 300 µg Se/kg
added as selenite. 30 µCi of ^{75}Se-selenite (0.13 µg Se) were
injected twice with an interval of 4 days into Se-deficient
rats or control animals of the 2nd generation. The animals
were killed after 6 weeks and the ^{75}Se activity in the tissues was determined.

In the Se-deficient animals, which received the ^{75}Se
dose five times at weekly intervals, the ^{75}Se-containing proteins in tissue homogenates were separated by means of SDS-
polyacrylamide gel electrophoresis (SDS-PAGE) (Laemmli, 1970)
and identified by autoradiography. Their relative molecular
masses were determined by marker proteins labeled with ^{125}J.

In a third experiment with rats of the 3rd generation
^{75}Se was given either to deficient animals or to animals during Se repletion. The ^{75}Se activity in the separated selenoproteins was measured using a Ge(Li) well-type detector.

Table 1. ^{75}Se-retention in tissues of Se-deficient male rats

Tissue	RF [1]	Tissue	RF [1]
Brain	51.9	Heart	1.9
Thyroid	46.9	Muscle	1.8
Pituitary	42.0	Plasma	1.7
Testes	22.5	Liver	0.9
Adrenals [2]	21.9	Erythrocytes	0.6
Ovaries [2]	22.6		
Corpora lutea [2]	42.7		

[1] Retention factor: Ratio of the specific ^{75}Se activities in depleted rats to that in control animals
[2] Tissues from similarly treated female rats

RESULTS AND DISCUSSION

The retention factors calculated from the results of the first experiment (Table 1) show that in severely depleted rats, after administration of a very small amount of Se, the retention of the element varied greatly in the different organs. In tissues such as the heart, muscle and liver, in which in Se deficiency or a combined Se and vitamin E deficiency state lesions occur, the retention in the deficient rats was only slightly different from that in the animals which had been fed sufficient amounts of the element. However, in the brain and in several organs with endocrine functions the depleted animals retained about 20 to 50 times more of the Se dose administered than the controls. The results show that in Se deficiency there is a priority transport of the element to certain target tissues. As it can be assumed that Se is supplied above all to the sites in which it is most needed, these findings indicate important functions of the element in endocrine organs and in the brain.

In order to find out to what extent other Se compounds besides glutathione peroxidase (GSH-Px) should be included in the study of the functions of Se in these organs, the Se-containing proteins were investigated. Analytical SDS-PAGE was chosen for the protein fractionation because of its high resolution. With this method only very small protein amounts can be separated. However, due to the greatly increased retention of ^{75}Se in the depleted animals the specific activity in the ^{75}Se-containing proteins was sufficiently high to allow them to be detected by autoradiography.

In this way a total of 12 Se-containing proteins or protein subunits was found. Their relative molecular masses are listed in Table 2. In other studies, using chromatographic methods, fewer Se-containing proteins were found (e.g. Hawkes et al., 1985). This difference is most probably due to the higher resolution of the separation method used here.

As with the distribution of ^{75}Se among the tissues the distribution of the tracer at a molecular level depended to a large extent on the Se status of the animals. The data obtained in the third experiment show that, compared with the

Table 2. Se-containing proteins in rat tissues [1]

Protein	M_r [2]	Protein	M_r
1 +++ [3]	14400 ± 200	7 +	33300 ± 1300
2 +++	15600 ± 300	8 ++	55500 ± 1300
3 +++	18000 ± 300	9 +++	59900 ± 2200
4 +++	19700 ± 400	10 +++	64900 ± 1200
5 +++	23700 ± 700	11 +	70100 ± 1300
6 +	27800 ± 400	12 ++	75400 ± 2300

[1] Adrenals, brain, corpora lutea, heart, kidneys, liver, lungs, muscle, ovaries, pituitary, prostate, spleen, testes, thyroid
[2] Relative molecular mass
[3] Detected in a few (+), nearly all (++) or all (+++) of the tissues investigated

Se-repleted rats, in all the tissues of the Se-deficient animals a much higher percentage of the ^{75}Se activity was retained in other selenoproteins than in the GSH-Px. In the adrenal cortex, for instance, the percentage of the tissue ^{75}Se activity contained in the GSH-Px increased from 9 % in the deficient male rats to 48 % in the repleted controls, and in the liver from 7 % to 56 %.

From the findings of this study it can be concluded that in Se deficiency there is a priority supply of the element to certain specific target organs, and at a molecular level to Se-containing proteins other than the GSH-Px. The results indicate important biological functions of these selenoproteins especially in the brain and several endocrine organs.

ACKNOWLEDGEMENTS

This study was supported by the Deutsche Forschungsgemeinschaft (Grant Be 977/1-2).

REFERENCES

Behne, D., Höfer, T., von Berswordt-Wallrabe, R., and Elger, W., 1982, J. Nutr. 112:1682.

Behne, D. & Höfer-Bosse, T., 1984, J. Nutr. 114:1289.

Behne, D., Höfer-Bosse, T., and Elger, W., 1986, in: "5. Spurenelement-Symposium," M. Anke, W. Baumann, H. Bräunlich, C. Brückner, B. Groppel, eds., Karl-Marx-Universität Leipzig, Friedrich-Schiller-Universität Jena.

Hawkes, W. C., Wilhelmsen, E. C., and Tappel, A. L., 1985, J. Inorg. Biochem. 23:77.

Laemmli, U. K., 1970, Nature 227:680.

THE EFFECT OF SELENIUM DEFICIENCY ON PLASMA THYROID HORMONE CONCENTRATIONS AND ON HEPATIC TRI-IODOTHYRONINE PRODUCTION

John R. Arthur, Fergus Nicol and *Geoffrey J. Beckett

Rowett Research Institute   *Clinical Chemistry Dept.
Bucksburn   The Royal Infirmary
Aberdeen,   AB2 9SB, U.K.   Edinburgh,   EH3 9YW, U.K.

INTRODUCTION

Although many effects of Se deficiency in animals can be explained by decreases in the cytosolic activity of the seleno-enzyme glutathione peroxidase (Se-GSHPx) which participates in the antioxidant systems of the cell, evidence is accumulating that Se may have other essential biochemical functions (Reiter and Wendel, 1984; Arthur et al., 1986; Hill et al., 1987). Severe selenium deficiency can induce characteristic changes in many hepatic enzyme activities in rats and mice and increase hepatic glutathione synthesis and release in rats. The changes in enzyme activity can, however, be reversed by administration of very small doses of selenium which do not produce detectable changes in cytosolic Se-GSHPx activity (Reiter and Wendel, 1984; Hill et al., 1987). Total hepatic glutathione S-transferase (GST) activity is increased by severe selenium deficiency in the rat (Hill et al., 1987) and since hypothyroidism can have a similar effect on GST activity (Arias et al., 1976) this study was initiated to determine whether Se deficiency could influence thyroid hormone metabolism in the rat.

METHODS

Experiments were performed with weanling male Hooded Lister rats of the Rowett Institute strain offered selenium-deficient diets ad libitum (5 µg Se/kg diet, Abdel-Rahim et al., 1986) in which Torula yeast was replaced by a mixture of purified amino acids. Control rats received the same diet supplemented with 0.1 mg Se/kg as Na_2SeO_3. Se-GSHPx and GST activities were measured using previously described methods (Abdel-Rahim et al., 1986; Hill et al., 1987). Thyroid hormones, thyroxine (T4) and, the more metabolically active, 3,5,3′ triiodothyronine (T3) were determined in plasma and in liver homogenates using double antibody radio-immunoassays.

RESULTS AND DISCUSSION

Four weeks after the start of the experiments hepatic Se-GSHPx activity had fallen to 1% of control values and hepatic GST activity was significantly increased in the Se-deficient rats. This occurred without the Se deficiency affecting the body weight or thyroid weight of the rats. Plasma T4 concentrations were increased by 80% and plasma T3 concentrations were decreased by 20% in the Se-deficient group compared to Se-supplemented controls. Most circulating T3 is derived from deiodination of T4 in liver and other organs. Thus the changes in circulating hormones are consistent with the decreased T3 formation in liver homogenates from Se-deficient rats (Table 1).

Table 1. Plasma Thyroid Hormone Concentrations, Hepatic Enzyme Activities, Thyroid Gland and Body Weights in Se-Deficient and Se-Supplemented Rats

Variable	Group		
	+Se	-Se	-Se (10µg Se)
Se-GSHPx (mU/mg protein)	1550 ± 128	14 ± 2[a]	11 ± 1[a]
GST (µmol CDNB/ min.mg protein)	0.37 ± 0.03	0.47 ± 0.02[b]	0.37 ± 0.02
T4 (nmol/l)	76.7 ± 5.2	137.0 ± 6.1[a]	131.7 ± 1.73[a]
T3 (nmol/l)	1.70 ± 0.12	1.43 ± 0.09	1.45 ± 0.12
Body wt. (g)	201 ± 7	199 ± 4	187 ± 11
Thyroid wt. (mg)	13.9 ± 1.5	11.1 ± 0.9	nr.
T3 production (fmol/min.mg protein in liver homogenate)	14.5 ± 2.5	2.0 ± 0.6[a]	nr.

Significantly different from +Se group, [a]$p<0.001$, [b]$p<0.02$, nr. not recorded; results are means ± SEMs for 6 rats.

Injection of Se-deficient rats with 10µg Se/kg body wt. (as Na_2SeO_3) 5 days before they were killed reversed the induction of GST activity without affecting hepatic Se-GSHPx activity or plasma T4 or T3 concentrations (-Se (10µg Se) group, Table 1).

Thus Se deficiency can change the concentrations of circulating thyroid hormones in the rat probably by decreasing conversion of T4 and T3 in liver and other tissues. However this is not closely related to the effects of Se deficiency on hepatic GST activity. In view of the importance of normal thyroid status to the maintenance of cell metabolism, alterations in the circulating concentrations of thyroid hormones caused by inadequate Se intake may have adverse effects on the rat and other animals.

REFERENCES

Abdel-Rahim, A. G., Arthur, J. R., and Mills, C. F., 1986, Effects of dietary copper, cadmium, iron, molybdenum and manganese on selenium utilisation by the rat, J. Nutr., 116:403.
Arias, I. M., Fleischer, G., Kirsch, R., Mishkin, S. and Gatmaintan, Z., 1976, On the structure and function of ligandin, in "Glutathione: metabolism and function", I. M. Arias and W. Jackoby, eds., Kroc. Found. Ser. Vol 6, pp 175.
Arthur, J. R., Boyne, R., Morrice, P. C. and Nicol, F., 1986, Selenium and neutrophil function in mice, Proc. Nutr. Soc., 45:63A.
Hill, K. E., Burk, R. F. and Lane, J. M., 1987, Effect of selenium depletion and repletion on plasma glutathione and glutathione dependent enzymes in the rat, J. Nutr., 117:99.
Reiter, R. and Wendel, A., 1984, Selenium and drug metabolism-II Independence of glutathione peroxidase and reversibility of hepatic enzyme modulations in selenium deficient mice, Biochem. Pharmacol., 33:1923.

PYRIDOXAL ISONICOTINOYL HYDRAZONE (PIH) AND ITS ANALOGUES: A NEW GROUP

OF EFFECTIVE CHELATORS

P. Ponka[a,b], H.M. Schulman[a] and J. Edward[c]

[a]Lady Davis Institute for Medical Research, Jewish
  General Hospital and Departments of [b]Physiology and
[c]Chemistry, McGill University, Montreal, Quebec, Canada

INTRODUCTION

Iron, which is involved in many metabolic processes including transport and storage of oxygen and oxidation-reduction reactions is an essential element for all living cells with the one possible exception lactic acid bacilli. However, organisms are not equipped with active excretory systems for iron and, therefore, if excessive amounts of iron get to the organism they accumulate without being excreted. This results in an iron overload which is a common finding in patients with refractory anemias such as thalassemia major, some other hemoglobinopathies and certain other anemias[1-4]. The overload is a consequence of increased iron absorption, but primarily of long-term transfusion therapy. Excess iron accumulates in liver, heart and endocrine glands causing serious dysfunction and death. Excess iron is removed from such patients by administration of chelating agents, primarily desferrioxamine. However, since desferrioxamine is inadequate and expensive, new iron chelating agents are needed[5].

One iron chelator that seems to be especially promising is pyridoxal isonicotinoyl hydrazone (PIH), first identified by Ponka et al[6]. PIH is tolerated in animals and, in contrast to desferrioxamine, it is also effective when given orally. The discovery of PIH and its potential to mobilize iron from cells in vitro[6-10] and from animals[11,12] prompted us to synthesize other compounds with iron binding structures similar to PIH. So far we have synthesized 62 various acylhydrazones which were prepared by condensation of aromatic aldehydes with acid hydrazides[13]. The purpose of this study was to compare the relative effectiveness of some of these compounds in terms of their capacity to mobilize ^{59}Fe from reticulocytes loaded with ^{59}Fe in their nonheme iron pool[7,8]. The reason for selecting reticulocytes is that this system is relatively simple, allowing testing of a large number of drugs, highly reproducible and involves aspects of membrane transport[8].

MATERIALS AND METHODS

Chelating Agents

Schiff base condensation products of aromatic aldehydes (pyridoxal salicylaldehyde, 2-hydroxy-1-naphtylaldehyde), with a series of various acid hydrazides (see Table 1) were prepared using standard procedures[14]. The compounds formed were characterized with regard to their molecular weights, melting points and infra-red spectra as will be published elsewhere[13]. Each chelator and its iron complex was assayed for its lipophilicity by extraction into the organic solvent ethylacetate.

Table 1. List of Hydrazones Examined

Hydrazone	Pyridoxal	Salicylaldehyde	2-Hydroxy-1-Naphtylaldehyde
Benzoyl	101	201	301
p-Hydroxybenzoyl	102	202	302
p-Methylbenzoyl	103	203	303
p-Nitrobenzoyl	104	204	304
p-Aminobenzoyl	105	205	305
p-t-Butylbenzoyl	106	206	306
p-Methoxybenzoyl	107	207	307
m-Chlorobenzoyl	108	208	308
m-Fluorobenzoyl	109	209	309
m-Bromobenzoyl	110	210	310
Isonicotinoyl	111	-	-

Reticulocytes

Reticulocytes were obtained from chronically bled rabbits and purified, prepared and incubated as in previous experiments. Reticulocytes with a high level of non-heme ^{59}Fe were prepared by incubation of cells with 1 mM succinylacetone[15] to inhibit heme synthesis. After 15 min ^{59}Fe-labelled transferrin was added and the incubation continued for a further 60 min. Incubation of reticulocytes with ^{59}Fe-transferrin and heme synthesis inhibitors was previously shown[15,16] to label their non-heme iron pool with ^{59}Fe and such cells are referred to as "^{59}Fe-reticulocytes".

Washed ^{59}Fe-reticulocytes (25-30 μl) were incubated (60 min) in buffered salt solution (250 μl, final volume) with various hydrazones as specified in the results section. Succinylacetone (1 mM) was present in all incubations to prevent the utilization of non-heme ^{59}Fe for heme synthesis during reincubation period. ^{59}Fe was measured both in washed reticulocytes and in the medium, and the percentage of ^{59}Fe mobilized from the reticulocytes was calculated[6]. Each value presented is the mean of duplicate determinations, the variation between duplicates never exceeding 5%.

RESULTS AND DISCUSSION

Effect of Various Hydrazones on the Mobilization of ^{59}Fe from the ^{59}Fe-Labelled Reticulocytes

Figure 1 shows that some of the compounds (No. 101, 103, 107-110, 301) were more effective than PIH in promoting ^{59}Fe release from reticulocytes. However, several compounds were less effective (102, 104, 105,

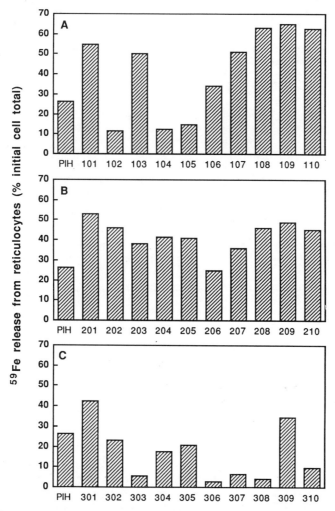

Fig. 1. ^{59}Fe-Reticulocytes were prepared as described in Materials and Methods and incubated without (control) or with indicated substances (50 μmol/L, final concentration) for 60 min following which ^{59}Fe-radioactivities in the medium and reticulocytes were determined and percentages of ^{59}Fe released from the reticulocytes calculated. Control ^{59}Fe release was always less than 1%/h.

303, 306-308, 310) and the rest of the hydrazones stimulated ^{59}Fe release from reticulocytes with efficiencies similar to those of PIH.

In screening potential iron-chelating drugs one must think in terms of several assays in order to develop a clear picture of potential efficacy. It is apparent that the reticulocyte-testing system, which has been used previously in our laboratory[6-8], can only screen for chelating agents that can cross the cell membrane and deplete cellular iron stores. ^{59}Fe-reticulocyte assay system is useful, namely for characterizing the mechanism of action of iron-chelating agents including investigations of membrane transport phenomena[8]. ^{59}Fe-labelled macrophages and hepatocyte screen systems may be more useful in evaluating the effectiveness of iron-chelating drugs since they present in vitro analogues of cells which become heavily iron overloaded in patients with transfusion-requiring refractory anemias. Experiments are in progress to evaluate the effectiveness of available acylhydrazones to mobilize iron from macrophages and hepatocytes.

The most active PIH analogues in reticulocyte screening system were pyridoxal benzoyl hydrazone, pyridoxal m-fluorobenzoyl hydrazone and salicylaldehyde benzoyl hydrazone. Although their high activity may be related to their high lipophilicity, this is not the only factor explaining their efficiency. Some compounds, which were also highly lipophilic, were much less effective in mobilizing cellular iron. Therefore, further characteristics of the compounds such as their affinity for iron, should be determined to reveal structure-function relationship of the chelating agents.

ACKNOWLEDGEMENTS

This research was supported by the Medical Research Council of Canada. We would like to thank Ms. Cheryl Partridge for technical assistance, Mrs. Sandy Fraiberg for typing the manuscript and Ms. Christine Lalonde for the illustrations.

REFERENCES

1. A. W. Nienhuis, Moderator, Thalassemia major: molecular and clinical aspects, Ann. Intern. Med. 91:883 (1979)
2. A. Jacobs, The pathology of iron overload, in: "Iron in Biochemistry and Medicine II", A. Jacobs and M. Worwood, eds., Academic Press, New York, pp. 427-459 (1980).
3. T. H. Bothwell and R. W. Charlton, A general approach to the problems of iron deficiency and iron overload in the population at large, Seminars Hematol. 19:54 (1982).
4. A. V. Hoffbrand, Transfusion siderosis and chelation therapy, in: "Iron in Biochemistry and Medicine II", A. Jacobs and M. Worwood, eds., Academic Press, New York, pp. 499-527 (1980).
5. D. J. Weatherall, M. J. Pippard, and S. F. Callender, Editorial retrospective. Iron loading in thalassemia - five years with the pump, N. Engl. J. Med. 308:456 (1983).
6. P. Ponka, J. Borova, J. Neuwirt, and O. Fuchs, Mobilization of iron from reticulocytes: Identification of pyridoxal isonicotinoyl hydrazone as a new iron chelating agent, FEBS Lett. 97:317 (1979).

7. P. Ponka, J. Borova, J. Neuwirt, O. Fuchs, and E. Necas, A study of intracellular iron metabolism using pyridoxal isonicotinoyl hydrazone and other synthetic chelating agents, Biochim. Biophys. Acta 586:278 (1979).

8. A. Huang and P. Ponka, A study of the mechanism of action of pyridoxal isonicotinoyl hydrazone at the cellular level using reticulocytes loaded with non-heme ^{59}Fe, Biochim. Biophys. Acta 757:306 (1983).

9. E. H. Morgan, Chelator-mediated iron efflux from reticulocytes. Biochim. Biophys. Acta 733:39 (1983).

10. L. Baker, M. L. Vitolo, and J. Webb, Hepatocytes in culture as an in vitro screen for iron chelators: Evaluation of analogues of PIH, in: "Structure and Function of Iron Storage Proteins", I. Urushizaki, P. Aisen, I. Listowsky, and J. W. Drysdale, eds., Elsevier, pp. 457-460 (1983).

11. M. Cikrt, P. Ponka, E. Necas, and J. Neuwirt, Biliary iron excretion in rats following pyridoxal isonicotinoyl hydrazone, Brit. J. Haematol. 45:275 (1980).

12. C. Hershko, C. Avramovici-Grisaru, G. Link, L. Gelfand, and S. Sarel, Mechanism of in vivo iron chelation by pyridoxal isonicotinoyl hydrazone and other imino derivatives of pyridoxal, J. Lab. Clin. Med. 98:99 (1981).

13. P. Ponka, M. Gauthier, F. L. Chubb, and Edward, T. J., Synthesis and characterization of iron-chelating acylhydrazones derived from pyridoxal isonicotinoyl hydrazone (PIH), J. Chem. Eng. Data, submitted for publication.

14. F. Wild, in: "Characterization of Organic Compounds, 2nd ed.", Cambridge University Press, p. 110 (1958).

15. P. Ponka, A. Wilczynska, and H. M. Schulman, Iron utilization in rabbit reticulocytes: A study using succinylacetone as an inhibitor of heme synthesis, Biochim. Biophys. Acta 720:96 (1982).

16. P. Ponka and J. Neuwirt, The use of reticulocytes with high non-haem iron pool for studies of regulation of haem synthesis, Brit. J. Haematol. 19:593 (1970).

IRON TOXICITY AND CHELATING THERAPY

C. Hershko, G. Link, A. Pinson, S. Sarel, S. Grisaru, Y.
Hasin, and R.W. Grady

Dept. Medicine, Shaare Zedek Med. Center, Dept. Nutrition,
Biochemistry and Cardiology, Hebrew University Hadassah
Med. School, Hebrew Univ. School of Pharmacy, Jerusalem,
and Dept. Pediatrics, Cornell University, New York

INTRODUCTION

Iron is one of the most common elements on Earth and yet, because
of its low solubility in nature, iron deficiency is one of the most
common forms of nutritional deficiency. Because of the biological
importance of iron as a catalyzer of one-electron redox reactions,
evolution has provided us with efficient mechanisms for the acquisiti-
on, transfer and storage of iron, but not with mechanisms for the
excretion of excess iron. The clinical consequences of iron
accumulation in hemochromatosis are manifested in abnormal function of
a number of vital organs, the most important of which is the heart. In
the following, I would like to focus on four aspects of iron toxicity
and chelation: (a) Studies conducted in an in vitro model of iron
chelation and toxicity in rat myocardial cell cultures; (b) Evidence
for the clinical effectiveness of iron chelating therapy in transfusional iron overload; (c) The potential usefulness of iron chelators
in diseases unrelated to iron overload, and; (d) The development of
new orally active iron chelators.

a. MYOCARDIAL CELL CULTURES

Research on iron toxicity has been greatly hindered by the
failure to reproduce hemochromatosis in an experimental animal model.
Because the most critical manifestation of iron overload is damage to
the heart, we have decided to study the harmful effects of iron
loading in cultured rat myocardial cells (1). A unique feature of
these cultures is the ability to differentiate into spontaneously
contracting cells, providing an opportunity to study simultaneously
the biochemical and functional effects of iron toxicity in cardiac
cells. These cells are identified by their characteristic elongated
shape, the presence of typical myofibrils and Z bands on electron
microscopy, and above all, their spontaneous and autonomous
contractility in situ, in tissue culture plates. Exposure of cultured
myocardial cells to increased concentrations of iron supplied in the
form of ferric ammonium citrate is followed by the endocytosis of
visible, 10-25 A iron aggregates into membrane-bound structures, and
subsequently their incorporation into cellular ferritin. Dose-response

curves showed that optimal iron uptake occurs at an iron concentration of 20 µg/ml and this concentration has been employed in all subsequent studies. The rate of iron uptake was greatly affected by the composition of culture medium: Iron uptake was highest in serum-free medium, and lowest with transferrin iron, indicating the protective role of transferrin in preventing the cardiac uptake of non-transferrin-bound low molecular weight iron complexes.

The effect of iron on cell contraction was studied in perfusion chambers at constant temperatures and pH. Increasing the iron content of the perfusion medium resulted in a rapid decrease in the amplitude of cell motion, and a slight reduction in rates of contraction. This could be reversed by simple washout, or by the addition of desferrioxamine (DF) to the perfusate. In order to determine the mechanism of damage to cell contractility, a series of antidotes with a positive inotropic effect have been empoyed in the continued presence of iron: calcium, ouabain and adrenaline had only a slight effect or no effect at all on cell motion. In contrast, caffeine corrected entirely the negative inotropic effect of iron. There are 3 known effects of caffeine promoting positive inotropy: (a) it activates sarcolemmal calcium channels via cyclic AMP; (b) it promotes the release of calcium from sarcoplasmic reticulum, and; (c) it increases the sensitivity of myofilaments to calcium. Because of its late timing and failure to respond to ouabain and adrenaline, reduced contractility is not likely to be the product of a sarcolemmal membrane effect of iron. By exclusion, the toxic effect of iron is most probably exerted either on the sarcoplasmic reticulum (SR) or the myofilaments. As SR in neonatal myocytes is poorly developed, the most likely explanation for the reduced contractility of cultured myocardial cells exposed to iron is reduced sensitivity of cardiac myofilaments to calcium.

One of the most important and specific effects of iron toxicity is promotion of membrane lipid peroxidation resulting in the accumulation of the lipid peroxidation product malonyl-dialdehyde (MDA). This is the consequence of the catalytic effect of ferrous iron in the Haber-Weiss reaction converting superoxide and hydrogen peroxide into the highly reactive and toxic free hydroxyl radicals (2). Iron-induced lipid peroxidation is accelerated in vitro by ascorbic acid (reducing ferric to ferrous iron) and inhibited by α-tocopherol, (an antioxidant interfering with the chain reaction of membrane lipid peroxidation). Iron loading resulted in a 15-fold increase in myocardial MDA concentrations, in direct proportion to iron uptake. Ascorbate treatment resulted in a 7-fold increase in lipid peroxidation relative to iron uptake, whereas α-tocopherol treatment resulted in a striking inhibition of lipid peroxidation. Hypoxia enhanced iron uptake but had only a marginal effect on lipid peroxidation. The interaction of ascorbate, α-tocopherol and the iron chelating effect of DF is of particular clinical relevance: As expected, DF treatment resulted in a reduction in cellular iron content as well as in MDA concentrations. Ascorbate had no effect on iron mobilization but prevented the beneficial effect of DF on MDA production. Conversely, α-tocopherol enhanced the inhibitory effect of DF on MDA production. The accelerated rate of myocardial iron deposition in hypoxia may be regarded as a further argument in favour of maintaining near-normal hemoglobin levels in thalassemia major and other iron-loading anemias. The demonstration of increased myocardial lipid peroxidation following ascorbate therapy underlines the controversy regarding ascorbate supplementation in thalassemia. Finally, the protective effect of α-tocopherol against iron-induced lipid peroxidation in myocardial cells underlines the need to explore the therapeutic potential of this compound in transfusional hemosiderosis.

b. CLINICAL EFFECTS OF IRON CHELATION

The pioneering studies of Barry et al (3) demonstrating the ability of long-term DF treatment to arrest hepatic siderosis in thalassemia, stimulated the adoption of chelating programs by the great majority of centers responsible for the care of thalassemic children. Ideally, such long-term and expensive chelating programs costing 4000 to 6000 dollars per patient year should be justified by the clearcut demonstration of improved survival in prospective, controlled clinical trials. Such studies, however, are not available and it is highly unlikely that they will be forthcoming in the future, since witholding DF treatment from thalassemic patients for the sake of research may no longer be ethically acceptable.

Continued follow up of the original group reported by Barry et al showed that 6 of 10 control patients died compared with 1 of 9 in the treated group. This difference, however, did not reach statistical significance because of the limited number of observations and a slightly higher age of the control patients. In a retrospective analysis of survival in 92 thalassemic patients in Britain (4) a significant positive correlation was found between dose of DF received on the preceding 5 years and mortality during the following year. However, 2 of the 13 deaths in the untreated thalassemics with an observed/expected mortality ratio of 1.5 were unrelated to iron overload. Probably the most impressive and up to date data on mortality and morbidity in non-compliant vs. compliant patients are those reported by Wolfe et al in 1985 (5). Only 1 of 17 compliant thalassemic patients developed cardiac disease and died, compared to 12 of 19 non-compliant patients who developed heart failure and of whom 7 have died. It should be remembered, however, that the age of non-compliant patients was significantly higher than of compliant patients and further follow up is needed to demonstrate improved survival in compliant patients.

The most convincing evidence for the beneficial effect of DF on hemosiderotic heart disease is the reversal of established myocardiopathy in some far-advanced cases. Earlier experience in hereditary hemochromatosis has shown that the myocardiopathy of iron overload is potentially curable by effective iron mobilization through phlebotomy. However, in transfusional hemosiderosis, the course of established myocardial disease is uniformly fatal and, until recently has been considered to be non-responsive to iron chelating therapy. Several recent reports indicate that such patients may still be responsive to aggressive chelating treatment. Marcus et al (6) described the reversal of established symptomatic myocardial disease in 3 of 5 patients by continuous high-dose (85-200 mg/kg/d) i.v. DF therapy at the cost of severe reversible retinal toxicity. No such toxicity has been described in 3 patients on high-dose s.c. and i.v. DF treatment reported by Hyman et al (7) who survived 2, 7 and 8 years after the onset of congestive heart failure. Recognition of early myocardiopathy is now possible with the introduction of exercise radionuclide angiography. The report by Freeman et al (8) in asymptomatic thalassemic patients has shown that ventricular response during exercise may be improved in about one third of subjects after one year of high-dose s.c. DF therapy.

c. DF IN CONDITIONS UNRELATED TO IRON OVERLOAD

Interest in the therapeutic potential of iron chelators in biological processes not related to iron overload has been stimulated by the important work of Halliwell and Gutteridge on the role of iron and iron chelation in oxygen toxicity (2). Activation of the

respiratory burst in granulocytes and macrophages is associated with a
sharp increase in the production of superoxide and peroxide,
terminating in the production of free hydroxyl radicals through the
iron-catalysed Haber-Weiss reaction. This reaction is inhibited by the
high-affinity binding of ferric iron to DF. In experimental animal
models, interference with free hydroxyl radical formation by DF
treatment was shown to be effective in suppressing arthritis,
complement- and neutrophil-dependent pulmonary vasculitis,
immune-complex-induced vasculitis, chronic islet cell allograft
rejection, and bleomycin or paraquat toxicity (9,10). Increased
production of the highly toxic free hydroxyl radicals is a common
feature of all of these conditions, and this may be prevented by the
effective inactivation of ferric iron through DF chelation. Initial
attempts to treat rheumatoid arthritis patients with DF yielded rather
disappointing results (11). However, inadequate methods of DF delivery
may have been partly responsible for these ea: y therapeutic failures.
DF is rapidly cleared from the circulation, and bolus injections are
unable to provide a steady supply of the drug required to inactivate a
rapidly exchanging low-molecular weight chelatable iron pool, such as
that existing in the synovial tissues of patients with rheumatoid
arthritis. Similarly, bolus injections of DF are unable to modify
paraquat toxicity whereas continuous DF infusion is highly effective.

Ribonucleotide reductase is a rate-controlling enzyme in DNA
synthesis. This enzyme contains a thyrosine free-radical structure
essential to its activity, which requires the presence of iron and
oxygen (12). In vitro DF treatment of human lymphocytes results in a
reversible S-phase inhibition of their proliferation through
ribonucleotide reductase inactivation. Similarly, in vitro growth of
Plasmodium falciparum and in vivo proliferation of P vinckei can be
inhibited by DF treatment. In a preliminary report DF was shown to
achieve temporary control of drug-resistant acute leukemia (13).
Whether or not these antiproliferative effects of DF may be of
clinical usefulness in the management of human malaria or of malignant
lymphoproliferative disease, remains to be seen.

Finally, DF treatment was shown to enhance the activity of
uroporphyrinogen decarboxylase in porphyria cutanea tarda with the
same efficiency as phlebotomy, which is currently regarded as the
treatment of choice for this inherited metabolic disease (14). This
observation is of immediate practical value for patients for whom
coexistent refractory anemia or advanced liver disease may preclude
the use of phlebotomy.

Modification of disease by preventing the formation of free
radicals, the powerful final effectors of tissue damage resulting from
the activation of granulocytes and macrophages participating in the
inflammatory response, is an exciting new concept in pharmacologic
intervention. Although much more experimental work is required, this
new approach may have far-going implications in the management of such
diverse conditions as immune complex disease, adult respiratory
distress syndrome, and allotransplantation.

d. DEVELOPMENT OF NEW CHELATORS

In spite of an increasing body of evidence supporting the
clinical efficacy of DF, a number of serious limitations remain,
underlining the need for the development of other, improved iron
chelating medications. These limitations include (a) the need for
parenteral administration caused by the inefficient intestinal
absorption of DF; (b) its almost prohibitive price, and; (c) its short
duration of action caused by a combination of rapid clearance and
enzymatic breakdown. New, inexpensive, and orally effective iron
chelators would not only make long-term chelating therapy more

acceptable, but would increase its efficiency by a continuous supply of circulating drug by virtue of their slow intestinal absorption. A number of orally effective new iron chelating compounds have been identified in recent years, all of which are superior to DF in their in vivo chelating efficiency. The most prominent among these are the derivatives of pyridoxal isonicotinoyl hydrazone, the phenolic EDTA derivatives, and the α-ketohydroxypyridones (15-17). Common features of this new generation of iron chelating drugs are their improved intestinal absorption; increased lipid solubility; preferential interaction with parenchymal, and therefore more toxic iron stores, and; an 8 to 15 fold increase in in vitro iron chelating efficiency compared with DF. However, experience gained in the management of thalassemic patients and in experimental animal models of immune-complex-induced vasculitis, islet cell allografts, and paraquat toxicity has shown, that it is not only chelating power, but the uninterrupted supply of a chelating drug and its ability to penetrate the critical extra- or intracellular compartments, that determines its ability to modify disease.

The resources required for the development of a new and promising iron chelating compound into a commercially available medication approved for clinical use after extensive toxicity testing, are beyond the reach of any single group of academic investigators. It is only through the intervention of governmental agencies or the pharmaceutic industry that such compounds may be made available for clinical use in the foreseeable future. Thus far, drug industry has shown limited interest in the development of new iron chelating drugs, perhaps on the assumption that clinical application of such drugs may be limited to thalassemic patients. However, recognition of the potential of iron chelators to modify disease severity in a large number of clinical conditions in which free hydroxyl radicals may be instrumental in the pathogenesis of disease, may result in a reassessment of this hesitant attitude.

ACKNOWLEDGEMENTS

Supported by grant HL 34062-01 of the National Heart, Lung and Blood Institute and grant 032-4133 of the Fund for Basic Research of the Israel Academy of Sciences to CH, and the Benador Foundation for Heart Research to AP.

REFERENCES

1. Link G, Pinson A, Hershko C. J Lab Clin Med 106: 147, 1985.
2. Halliwell B, Gutteridge ML. Biochem J 219: 1, 1984.
3. Barry M, Flynn DM, Letsky EA et al. Brit Med J 2: 16, 1974.
4. Modell B, Letsky EA, Flynn DM et al. Brit Med J 284: 1081, 1982.
5. Wolfe L, Olivieri N, Sallan D et al. New Engl J Med 312: 1600,1985
6. Marcus RE, Davies SC, Bantock HM et al. Lancet 1: 392, 1984.
7. Hyman CB, Agness CL, Rodriguez R et al. Ann N Y Acad Sci 445: 293, 1985.
8. Freeman AP, Giles RW, Berdoukas VA et al. Ann Intern Med 99: 450, 1983.
9. Ward PA, Till GO, Kunkel R et al. J Clin Invest 72: 789, 1983.
10. Bradley B, Prowse SJ, Bauling P et al. Diabetes 35: 550, 1986.
11. Polson RJ, Jawad ASM, Bomford A et al. Q J Med 61: 1153, 1986.
12. Reichard P, Ehrenberg A. Science 221: 514, 1983.
13. Estrow Z, Tawa A, Wang XH et al. Blood 69: 757, 1987.
14. Rocchi E, Gilbertini P, Cassanelli M et al. Brit J Derm 114: 621, 1986.
15. Hershko C, Avramovici GS, Link G et al. J Lab Clin Med 98: 99, 1981.
16. Hershko C, Grady RW, Link G. J Lab Clin Med 103: 337, 1984.
17. Kontoghiorghes GJ. Molec Pharm 30: 670, 1987.

TOXIC AND ESSENTIAL METAL LOSS DURING CHELATION THERAPY OF LEAD-POISONED

CHILDREN

David J. Thomas and J. Julian Chisolm, Jr.

Department of Pediatrics, University of Nebraska Medical
Center, Omaha, NE and The Kennedy Intstitute and Department
of Pediatrics, School of Medicine, The Johns Hopkins
University, Baltimore, MD

Pb poisoning in children is a serious public health problem in the
United States. Chelating agents are commonly used to reduce the body burden
of Pb. Calcium disodium ethylenediamine tetraacetate (Ca disodium EDTA)
and 2,3-dimercapto-1-propanol (British antilewisite, BAL) are commonly used
to treat Pb poisoning in children (Piomelli et al., 1984). A number of new
chelating agents are under evaluation for use in the treatment of Pb
poisoning. 2,3-Dimercaptopropane-1-sulfonate (DMPS) is a water soluble
analog of BAL which has been used to treat Pb, methyl Hg and inorganic Hg
poisoning (Aposhian, 1983). In the work reported here, the decorporation
of Pb and the essential metals, Zn and Cu, has been examined in Pb-poisoned
children treated with Ca disodium EDTA or with DMPS.

The source population for these studies was children attending the
Lead Poisoning Diagnostic, Treatment and Follow-up Clinic of the Kennedy
Institute in Baltimore. Children recruited for these studies were 30 to 72
months old, were toilet trained and had no intercurrent disease. All
studies were performed in the Pediatric Clinical Research Unit of the Johns
Hopkins Hospital and treatment protocols and consent forms were approved by
human studies committees of Hopkins Hospital and the Kennedy Institute.
The dosage level for Ca disodium EDTA was 500 mg per m^2 surface area given
im every 12 hours for a total of 10 doses. DMPS was given po at dosage
levels of 200 or 400 mg per m^2 per day given in four equal doses each day
for five days. Blood Pb (PbB) and plasma Zn (ZnP) and Cu (CuP) concentrations were determined before, during and after courses of treatment with
these chelating agents. Twenty four hour urine collections were obtained
before, during and after chelation therapy and concentrations of Pb, Zn
and Cu in urine were determined.

Pretreatment PbB averaged 55.7 ug per 100 ml of whole blood in 10
children treated with Ca disodium EDTA. Within 72 hours after the onset
of treatment, PbB fell to about 70% of its pretreatment value. ZnP
declined to about 60% of its pretreatment value by 120 hours after the
beginning of therapy; CuP was unaffected by chelation therapy. After
cessation of Ca disodium EDTA treatment, ZnP returned quickly to its pretreatment value. PbB rose slowly after treatment. Ca disodium EDTA
treatment increased the daily loss of Pb in urine about 21-fold over that
seen before chelation therapy. Urinary loss of Zn during treatment
increased about 17-fold over its pretreatment value. Urinary loss of Cu

did not rise above its endogenous level. A significant correlation was found between the pretreatment PbB and the chelatable Pb burden as estimated either by compartmental analysis or by the method of Araki and Ushio (1982). However, no relationship was found between pretreatment ZnP and the loss of Zn in urine during chelation therapy.

Pretreatment PbB in 12 children treated with DMPS averaged 52 ug per 100 ml of whole blood. Both dosage levels of DMPS reduced PbB. After 96 hours of treatment, PbB in the lower dosage level group was 76% of its pretreatment value; in the higher dosage group, it was 68% of pretreatment value. Little rebound in PbB occurred after the cessation of DMPS treatment. ZnP and CuP were unaffected by DMPS treatment. Cumulative urinary loss of either Pb or Zn during DMPS treatment was two to six times that which would occur in the absence of chelation therapy. Cumulative loss of Cu in urine during a five day course of DMPS ranged from four to 25 times that which would occur in the absence of DMPS treatment. A trend of increasing urinary loss of Cu with increasing cumulative dosage of DMPS was noted.

Taken together, these data indicate that treatment with either Ca disodium EDTA or DMPS reduced PbB and increased the urinary loss of this toxic metal. However, these agents also increased the loss of essential metals. With Ca disodium EDTA treatment, ZnP declined markedly and urinary Zn loss rose. With DMPS treatment, urinary loss of Cu increased but CuP was unaffected. These data reflect in part the nonselectivity of these agents as metal chelators. Further, they suggest that essential metal loss may be an unrecognized hazard of treatment with these agents.

REFERENCES

Aposhian, H.V., 1983, DMSA and DMPS- Water soluble antidotes for heavy metal poisoning, Annu. Rev. Pharamacol. Toxicol., 23:193.
Araki, S., and Ushio, K., 1982, Assessment of the body burden of chelatable lead: A model and its application to lead workers, Br. J. Ind. Med., 39: 157.
Piomelli, S., Rosen, J.F., Chisolm, J.J. Jr., and Graef, J.W., 1984, Management of childhood lead poisoning, J. Pediatr., 105:523.

USE OF METAL CHELATING AGENTS TO MODULATE EHRLICH CELL GROWTH MULTIPLE

METAL-LIGAND INTERACTIONS

D. H. Petering, S. Krezoski, D. Lehn, D. Stone, and H.
Loomans

Department of Chemistry,
University of Wisconsin-Milwaukee
Milwaukee, WI 53201

INTRODUCTION

Metal chelating agents are used to simulate metal-deficient
conditions in culture or to modulate the content and distribution of
specific intracellular metals. In general, multidentate ligands can form
stable complexes with a variety of metal ions. They may compete
selectively with biological metal complexes for metal. Depending on
their solubility characteristics, they will also differ in their access
to intracellular metal. Because of these complications, studies have
been conducted to determine the effects of a series of commonly used
ligands on metal distribution in Ehrlich cells.

RESULTS AND DISCUSSION

The following ligands were used to halt cell growth in culture:
Chelex 100 resin, bearing iminodiacetate chelating groups, EDTA, EGTA
(ethylene glycol bis(2-aminoethylether)-N,N-tetraacetic acid), NTA
(nitrilo-triacetate), and $\underline{o}$-phen (1,10-phenanthroline). The culture
medium contained Eagles minimal essental medium plus Earles salts and
2.5% fetal calf serum. Zinc, iron, and copper distributions in
supernatant of sonicated cells were determined using Sephadex G-75
chromatography. Ehrlich cells contain high molecular weight Zn, Fe
(ferritin), and Cu (high M_r, >30,000 daltons), and lower M_r Zn and Cu
bound to metallothionein (Mt). The effects of ligand concentrations
throughout the dose-response curve were examined. Results are reported
for levels which cause growth inhibition.

As controls for these studies, mice were made Zn or Fe deficient and
injected with tumor cells. After about two weeks, cells were harvested
and their intracellular metal distributions determined. In Zinc
deficient cells only Mt lost Zn within the error of the measurement. In
Fe deficiency supernatant ferritin iron was depressed.

Chelex resin (10 g/100mL) interacts only with extracellular metals.
It has no effect on Fe or Cu content but does deplete the growth medium
of Zn and Ca. Sephadex chromatography showed that the only cellular
site to lose Zn, Cu, or Fe was metallothionein. The same result was
obtained when EDTA (4-6 nmol/10[7] cells)[2] was used[1]. As with Chelex,
little if any EDTA enters Ehrlich cells[2]. EGTA also inhibits growth,

75

but at much higher concentrations than EDTA. Although one might suggest that this is an effect on calcium because EGTA is commonly used to modulate Ca levels, in fact, CaEGTA exhibits the same concentration dependence for growth inhibition as EGTA. Thus, chelation of Ca does not seem to be involved in the effects of this agent.

When cells are incubated with NTA (12-25 nmol/10^7 cells), both high M_r and Mt Zn lowered. There is no observable effect on Cu. One experiment showed that ferritin Fe had not changed in the presence of NTA. Since neither extracellular chelating agent--Chelex and EDTA-- affects high M_r Zn, it is suggested that NTA gets into these cells and directly reacts with this pool of Zn to extract metal.

Finally, o-phen (25-50 nmol/10^7 cells) depresses both pools of Zn, lowers ferritin Fe and, redistributes Cu. This ligand readily enter cells and has multiple effects once inside. Like NTA it probably reacts directly with high M_r Zn. The effect on ferritin is interesting because it requires a valence change--Fe(III) to Fe(II)-- as well as ligand substitution. The finding that Cu is lost from both pools suggests that a Cu(I) species of o-phen is an intermediate in the redistribution process. When metals are added back to see which can restore growth, several metal ions which bind well to 1,10-phenanthroline are effective[3]. When Zn is used, the reversal of growth inhibition has the appearance of a titration of total ligand[2].

Considering the role of metals in the growth inhibitory properties of these ligands, it is evident that multiple sites and metals may need examination. In addition, one needs to investigate the toxicity of metal complexes, which form when ligands encounter cells. For example, Cu(I) (o-phen)$_2$ is known to cause the strand scission of DNA.

ACKNOWLEDGEMENT

Supported by NIH grant CA-22184.

REFERENCES

1. D. H. Petering, S. Krezoski, J. Villalobos, C. F. Shaw III, and J. D. Otvos, Cadmium-zinc interactions in the Ehrlich cell: metallothionein and other sites, in Metallothionein. Proceedings of the Second International Meeting on Metallothionein and Other Low Molecular Weight Metal-Binding Proteins, J. H. R. Kagi and Y. Kojima, eds., Birkhauser Verlag, Basal, 1986, in press.
2. C. Krishnamurti, L. A. Saryan, and D. H. Petering, Effects of ethylenediaminetetraacetic acid and 1,10-phenanthroline on cell proliferation and DNA synthesis of Ehrlich ascites cells. Cancer Res., 40:4092 (1980).
3. N. A. Burger, E. S. Johnson, and A. M. Skinner, Orthophenanthroline inhibition of DNA synthesis in mammalian cells. Expl. Cell Res., 66:145 (1975).

EFFECTS OF CHELATING COMPOUNDS FORMED ON FOOD PROCESSING ON ZINC

METABOLISM IN THE RAT

Diane E. Furniss, Jacques Vuichoud, Paul-André Finot
and Richard F. Hurrell

NESTEC LTD., Nestlé Research Centre, Vers-chez-les-Blanc
CH-1800 Vevey, Switzerland

INTRODUCTION

Lysinoalanine (LAL) and Maillard reaction products (MRP's) formed in foods during processing chelate Zn in vitro. When fed to rats these compounds induce hyperzincuria[1]. In the following study we have examined the effect of free and protein-bound LAL, and of MRP's from heated casein-glucose or heated casein-lactose mixtures on Zn metabolism in the rat.

MATERIALS AND METHODS

Test products. Free LAL was prepared by the method of Okuda and Zahn[2]. Protein-bound LAL was prepared by heating casein for 3h at 90°C in 0.1M NaOH followed by freeze-drying. MRP's were formed by mixing 4kg casein with 5.7kg of glucose or lactose monohydrate in 20L distilled water, spray-drying and then further heating in sealed containers (refer to Table 1).

Experimental design. The test diets were fed ad libitum to weanling rats for 21d. A Zn balance study was performed over days 9-14. At 21d animals were sacrificed and blood, liver, kidney and femurs were removed for Zn analysis.

Analytical methods. Total lysine and lysinoalanine were determined by ion-exchange chromatography. Lysine as fructose-lysine or lactose-lysine, and lysine destruction were determined by the furosine method[3]. Zn was analysed by flame atomic absorption spectrophotometry.

Diet Composition. All diets contained 20% protein and 20mg/kg Zn. In the test diets containing the alkaline-treated and the Maillardised caseins 10% of the protein was provided by unheated casein. All nutrients were provided in adequate amounts.

RESULTS (Table 1)

Feeding free LAL (0.76g/kg diet) increased urinary Zn 11-fold and also significantly increased faecal Zn ($p < 0.05$). Protein-bound LAL (2.84g/kg diet) had no effect on urinary Zn loss but faecal Zn was increased from 45% to 60% of the intake.

Table 1. The Influence of Feeding Lysinoalanine (LAL) and Maillard Reaction Products (MRP's) on ZN.

Group	Test Product. (g/kg diet)		Zn Excretion (% intake, 9-14 d)				Femur Zn (21 d) (μg/g dry weight)	
			Urine		Faeces			
			Mean	SE	Mean	SE	Mean	SE
	LAL							
Control	-		0.6	0.1	44.8	2.6	210	7
LAL (Free)	0.76		6.5**	0.3	54.3*	3.9	193	5
LAL (Protein -bound)	2.84		0.6	0.1	60.5**	3.7	189*	5
	MRP's [a]							
Casein- glucose	Early	Advanced						
Unheated	-		0.7	0.1	37.9	2.2	206	4
SD[b]	3.69	0.82	3.7**	0.3	40.7	2.6	208	5
SD, heated 3d at 60°	4.73	2.41	4.8**	0.4	34.8	1.3	207	5
Casein- lactose								
Unheated	-	-	1.2	0.2	34.6	4.3	218	8
SD, heated 1d at 50°	3.39	0.51	1.5	0.4	34.0	1.4	222	7
SD, heated 3d at 60°	3.89	3.12	2.2*	0.3	44.6	4.3	212	5

[a]Measured as lysine combined as the early MRP's, fructose-lysine or lactose-lysine; or as lysine combined as the advanced MRP's (3)
[b]Spray-dried. * p <0.05 **p <0.01

Urinary Zn was increased 5-fold and 7-fold respectively on feeding the first and second levels of MRP's in the heated casein-glucose. Urinary Zn was not increased by the first level of MRP's in heated casein-lactose (p>0.05), and increased only 2-fold on feeding the higher level of MRP's. Faecal Zn was not changed on feeding the heated casein-glucose diets, but was higher when the second level of casein-lactose MRP's were fed.

There were no significant differences in Zn concentrations of serum, liver, and kidney in the test groups relative to their respective controls in either study (p>0.05). Femur Zn concentrations were reduced by 10% on feeding LAL in either form but were not influenced by the MRP's.

DISCUSSION

LAL does not occur in the free form in processed foods but is protein-bound. Heat-sterilised milk contains up to 1.lg LAL/kg protein, and some food ingredients may contain up to 50g LAL/kg protein[4]. Levels of MRP's comparable to those fed in the rat study can occur in processed foods[4].

The large urinary Zn losses suggest that free LAL and MRP, particularly those formed from casein-glucose can form Zn chelates which are absorbed by the rat and excreted in the urine. Faecal Zn excretion was most probably increased by the formation of chelates which were not absorbed. Both protein-bound LAL and MRP's of the advanced type are known to be poorly absorbed. At the levels fed, LAL, but not the MRP's, lowered Zn retention and femur Zn.

In infant formulas the MRP's formed in the reaction of milk protein with lactose are mainly of the "early" type such as were present in the least-heated casein-lactose which did not influence Zn metabolism. Nevertheless, when evaluating food as a source of Zn the presence of both LAL and MRP's should be considered.

REFERENCES

1.   D.E. Furniss, R.F. Hurrell, D. de Weck, and P.A. Finot. In: Trace
     Element Metabolism in Man and Animals - TEMA-5. C.F. Mills,
     I. Bremner and J.K. Chester, eds, Commonwealth Agricultural Bureaux
     (1985).
2.   T. Okuda and H. Zahn. Chem. Ber. 98:1164 (1965).
3.   E. Bujard and P.A. Finot. Ann. Nutr. Alim. 32:291 (1978).
4.   R.F. Hurrell. In: "Development in Food Proteins - 3",
     B.J.F. Hudson, ed., Elsevier Applied Science Publishers, London
     (1984).

CHANGES IN ZINC METABOLISM DURING THE USE OF EDTA

H. Spencer, S. K. Agrawal, S. J. Sontag and D. Osis

Metabolic Research, Veterans Administration Hospital
Hines, Illinois   60141

INTRODUCTION

Ethylenediaminetetraacetic acid, EDTA, a chelating agent with a high stability constant for lead, is commonly used for the diagnosis and treatment of lead poisoning[1,2]; but EDTA has also a very high stability constant for zinc[3] and is therefore expected to bind and remove this essential trace element. In the present study the zinc loss during the use of conventional doses of EDTA for lead poisoning was determined.

MATERIALS AND METHODS

The calcium salt of EDTA, CaEDTA, was used for diagnostic purposes for chronic lead poisoning in three patients, and one patient who had extensive extra-skeletal calcifications received the sodium salt of EDTA, Na_2EDTA. CaEDTA (3 gm) was given in 500 ml 5% glucose in water over a 4-hour period intravenously daily on 3 consecutive days to two patients and for 2 days to one patient. The patient with extra-skeletal calcifications was in a sub-optimal nutritional state. He received 5 daily infusions of 2 gm Na_2EDTA for calcium removal. Complete daily urine collections were analyzed for zinc prior to and on the days of the EDTA infusions as well as in subdivisions, namely from 0-4 hours, i.e., during the 4-hour EDTA infusions, from 4-8 hours and 8-24 hours. Plasma zinc levels were determined before, during and after the EDTA infusions. Zinc in urine and plasma was determined by atomic absorption spectroscopy.[4]

RESULTS

Conventionally used doses of CaEDTA for the diagnosis and treatment of chronic lead poisoning as well as small doses of Na_2EDTA significantly increased urinary zinc excretion in each case. Normally, urinary zinc is less than 1 mg/day, usually 0.5 mg. The total excess zinc excretions in the 3-day infusion period ranged from 33 to 64 mg and was 73 mg in 5 days. Plasma levels of zinc decreased at the end of the 4-hour EDTA infusions in most cases and were still low normal 24 hours later. In the patient who was in a suboptimal nutritional state zinc plasma levels were very low.

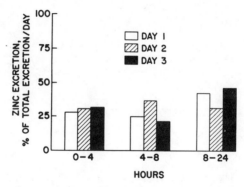

Fig. 1. Urinary Zinc Excretion During EDTA Infusion

Figure 1 shows an example of the prompt excess excretion of urinary
zinc during the 4-hour infusion of CaEDTA which was given on 3 consecutive days. Considerable and similar amounts of zinc were excreted
daily during the infusion period from 0-4 hours, accounting for about
30% of the total 24-hour urinary zinc. Following discontinuation of
the EDTA infusion similar amounts of zinc were excreted from 4-8 hours,
and in the next 16 hours, i.e., from 8-24 hours, larger amounts of zinc
were excreted.

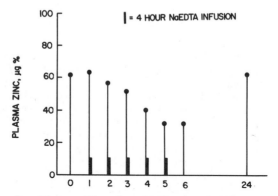

Fig. 2. Plasma Zinc Levels During NaEDTA Infusion

Figure 2 shows changes of the zinc plasma levels following infusions of
2 gm Na_2EDTA given on 5 consecutive days. The dot on top of the first
vertical line shows the zinc plasma level on the day prior to the infusions, 62 μg%, a subnormal level which was also the same prior to the
start of the first EDTA infusion, indicating good agreement of these
values. All other zinc plasma levels, determined 24 hours after the
preceding EDTA infusion decreased progressively each day and were
lowest, 32 μg%, prior to the 5th EDTA infusion and remained the same in
the next 24 hours (6th day). Serial determinations done thereafter
showed low zinc plasma levels up to Day 24, i.e., for 18 days after the
last EDTA infusion.

DISCUSSION

The stability constant of EDTA for zinc, 10^{16}, is considerably higher than that for calcium, 10^6, but lower than that for lead, 10^{18}. EDTA was highly suitable for the study of calcium homeostasis in normocalcemic[5,6] and in hypercalcemic patients.[7,8] The present studies have shown that EDTA induces an acute zinc loss which may or may not be associated with low plasma zinc levels, depending on the nutritional state. The patient, who was in a suboptimal nutritional state, had very low plasma levels of zinc, 32 μg%, following the EDTA infusions and he was clinically zinc deficient as evidenced by complete loss of appetite, which was restored with zinc supplements.

Increase in urinary zinc following EDTA infusions has been reported in extensive studies by Perry et al.[9,10] as well as by McCall.[11] The demonstration of the acutely induced zinc loss by conventional doses of EDTA for the diagnosis or treatment of lead poisoning is of clinical importance, particularly in that children suspected of lead poisoning are usually from lower socio-economic strata and may already be in poor nutritional state which would be associated with a low zinc status. This would be further compromised by the zinc loss induced by EDTA. Therefore, monitoring of the zinc status during the use of EDTA should be a routine procedure to evaluate changes in zinc metabolism and prevent manifestations of induced zinc deficiency. The excess zinc loss during EDTA testing or treatment may best be compensated by the use of supplemental zinc.

REFERENCES

1. S. P. Bessman and E. C. Layne, Jr., A rapid procedure for the determination of lead in blood or urine in the presence of organic chelating agents, J. Lab. Clin. Med. 45:159 (1955).
2. J. J. Chisolm, Jr., The use of chelating agents in the treatment of acute and chronic lead intoxication in childhood, J. Pediatr. 73:1 (1968).
3. S. Chaberek and A. E. Martell, "Organic Sequestering Agents," J. Wiley, New York (1959).
4. J. B. Willis, Determination of lead and other heavy metals in urine by atomic absorption spectroscopy, Analyt. Chem. 34:614 (1962).
5. H. Spencer, V. Vankinscott, I. Lewin, and D. Laszlo, Removal of calcium in man by ethylenediaminetetraacetic acid: A metabolic study, J. Clin. Invest. 31:1023 (1952).
6. H. Spencer, The use of chelating agents in the study of mineral metabolism in man, in: "Metal-Binding in Medicine," M. J. Steven and L. A. Johnson, eds., J. B. Lippincott, Philadelphia, p 104 (1960).
7. J. F. Holland, E. Danielson, and A. Sahagian-Edwards, Use of ethylenediamine tetraacetic acid in hypercalcemic patients, Proc. Soc. Exp. Biol. Med. 84:359 (1953).
8. H. Spencer, J. Greenberg, E. Berger, M. Perrone, and D. Laszlo, Studies of the effect of ethylenediaminetetraacetic acid in hypercalcemia, J. Lab. Clin. Med. 47:29 (1956).
9. H. M. Perry, Jr. and E. F. Perry, Normal concentrations of some trace metals in human urine: Changes produced by ethylenediaminetetraacetate. J. Clin. Invest. 38:1452 (1959).
10. H. M. Perry, Jr. and H. A. Schroeder, Lesions resembling vitamin B complex deficiency and urinary loss of zinc produced by ethylenediaminetetraacetate. Am. J. Med. 22:168 (1957).
11. J. T. McCall, K. G. McLennan, N. P. Goldstein, et al., Copper and zinc homeostasis during chelating therapy, in: "Trace Substances in Environmental Health-II," D. D. Hemphill, ed., University of Missouri (publ), Columbia, MO, p 127 (1968).

TRACE ELEMENT DEFICIENCIES AND IMMUNE RESPONSIVENESS

M. Eric Gershwin, Carl L. Keen, Mark P. Fletcher, and
Lucille S. Hurley

Departments of Internal Medicine and Nutrition
University of California
Davis, CA 95616

INTRODUCTION

In recent years, great interest has been generated regarding the
influence of nutrition upon host immunocompetence (Katz and Steihm, 1977;
Suskind, 1977). While the main focus has been on protein and energy
nutriture and their role in immune responsiveness (Malave, et al, 1980),
trace metals, many of which are critical for mammalian survival and
reproduction, have begun to be investigated in this connection
(Cunningham-Rundles, 1982). Because of the consumption of highly refined
and heavily processed food items, in which the trace element content may
be reduced significantly (Zwickl, et al, 1980) there is now concern that
marginal deficiency of trace elements may represent a public health
problem in developed nations (Klevay, et al, 1980), including marginal
trace element status in countries such as the United States (Hambidge, et
al, 1976). In this paper we will focus on the effects of deficiencies of
zinc, copper and manganese on the immune system.

ZINC

Of those trace elements essential to humans, zinc has probably been
studied more extensively than most. Appreciation for the fundamental
role of zinc in mammalian structure and function has increased since the
1930's with a focus on the role of zinc as an essential cofactor for the
activity of what is now over 100 enzymes. Zinc can function as a tightly
bound moiety, termed a metalloenzyme, as a more loosely bound entity, as
in metal-enzyme complexes or in stabilizing the tertiary structure of
nucleic acid and protein macromolecules (Vallee and Falchuk, 1981). The
effects of zinc deficiency upon any animal are most likely mediated
through factors such as decreased enzyme activity, altered membrane
structure, and aberrant DNA transcription and RNA translation.
Observations of marginal zinc status coupled with experimental findings
of altered immunocompetence and impaired response to pathogenic challenge
in zinc-deprived animals (Beach, et al, 1980; Chandra and Au, 1980;
Hansen, et al, 1982) underscore the importance of studies designed to
ascertain the effects of marginal zinc deficiency.

A significant appreciation for the essentiality of zinc for intact
immunological function has been generated. The first indication of this

interaction came from human epidemiological surveys; patients with low
levels of serum zinc had increased susceptibility to a variety of
infectious disorders (Sandstead, et al, 1976); as well as abnormal immune
parameters. In addition, patients with acrodermatitis enteropathica, an
autosomally recessive disorder of zinc metabolism, have an increased
incidence of infectious maladies (Endre, et al, 1975). This increased
susceptibility to infection is associated with profoundly altered
cellular immune responsiveness and responds to zinc administration with
nearly complete amelioration of their immunodeficiency syndrome (Oleske,
et al, 1979). More recently, protein-energy malnourished children have
been shown to have low levels of serum zinc, and the immunodeficiency
syndrome observed in such patients is partially corrected by zinc
administration (Golden, et al, 1978; Castillo-Duran, et al, 1987).
Patients undergoing long-term parenteral alimentation have impaired
immunological function (particularly cell-mediated immune responses),
associated with low serum zinc levels which is corrected in large part by
adequate zinc supplementation (Pekarek, et al, 1979; Allen, et al, 1983).
In patients with immunodeficiency diseases with low levels of serum zinc,
zinc repletion is associated with restoration of many of these immune
functions (Cunningham-Rundles, et al, 1980). In addition, the
immunodeficiency complicating Down's Syndrome, sickle-cell anemia,
obesity, infection, cirrhosis (Beisel, 1976; Tapazoglou, et al, 1985;
Yoffe, et al, 1986) and severe burns (Lennard, et al, 1974) has been
shown to be associated with low levels of plasma zinc. The full
implications of this decline in serum zinc in these syndromes remain to
be understood. Though all of these observations suggest a possible
interplay between zinc and the immune response, well-controlled studies
in appropriate animal models, involving both in vitro and in vivo systems
will be required to substantiate the nature of this interaction.

While early work on zinc and the immune function focused on the
requirements for zinc in blast transformation of lymphocytes, attention
has now shifted to the investigation of the effects of alterations in the
level of dietary zinc upon immune responsiveness. As with patients, the
initial indication of an interaction between zinc and immunological
function in experimental animals indicated that animals deficient in zinc
had an inability to effectively respond to bacterial, viral or parasitic
pathogens. Tests for immune function in a variety of zinc-deficient
animals have shown that in adult's postnatal life those immune events
mediated by T lymphocytes seemed to be most profoundly affected by zinc
deprivation. Fraker and associates have suggested that the immune defect
in zinc deprivation may be localized, at least in part, at the level of T
helper cell function (Fraker, et al, 1978). Giugliano and Millward
(1987) have reported that in the rat, zinc deficiency results in a loss
in thymus DNA and protein which in part can be explained by a reduction
in protein synthesis. While the effects of zinc deprivation seem to be
focused on the thymus and T cell-mediated functions, recent evidence
indicates that B cells may also be affected by zinc availability. In
most cases, restitution of zinc in the diet resulted in at least partial,
and at times complete restoration of immune function (Golden, et al,
1978; Zwickl, 1980).

If deprivation of zinc is imposed during fetal or early postnatal
development, the immune dysfunction that results is even more profound
than at later stages; while growth and development of zinc-deprived
offspring is markedly retarded, growth of lymphoid organs, most notably
the thymus, is more severely altered than is the growth of other organs
(Beach, et al, 1980). In addition, immune function is also altered, as
indicated by depressed mitogens responses, especially Con A and PHA

(Beach, et al, 1979), altered direct plaque-forming cell response to sheep erythrocyte immunization, and a markedly altered serum immunoglobulin profile, (no detectable serum IgM, IgG_{2a} or IgA, with elevated levels of IgG_1) (Beach, et al, 1980). Among the most remarkable observations of the studies in developing mice is the finding that even marginal deprivation of zinc, when imposed during the early postnatal period, results in a substantial reduction of immunological responsiveness, particularly T cell-mediated immunity (Beach, et al, 1980, 1982a,b, 1983). The effect of marginal zinc deficiency on ontogeny of the immune system has also been investigated using rhesus monkeys as a model. Haynes, et al (1985) reported that rhesus monkey infants fed a marginally zinc-deficient diet (4 ppm zinc) from conception through 12 months postnatal life were characterized by low responses to phytohemagglutinin, concanavalin A and pokeweed mitogen. In addition, infants in this group were characterized by low serum IgM levels. Thus, with regards to the immune system, the response of a non-human primate model to zinc deficiency during early development is very similar to that observed with rodent models. These studies serve to underscore the importance of two critical aspects in any study of zinc deprivation: (1) the timing of the deprivation with respect to the life cycle of the animal, and (2) the magnitude of the zinc deprivation. It is also important to recognize the fact that if postnatal zinc supplementation is provided to rodents prenatally deprived of zinc, there can be considerable improvement in the development of the offspring (Beach, et al, 1982b). Similarly, early zinc supplementation of pups rendered zinc deficient during the suckling period can overcome the immune defects arising as a result of the deficiency (Fraker, et al, 1984).

Looking at the efferent arm of the immune response, phagocytic cell function, the complement system, and mast cell mediator release are also affected by zinc status. Mast cells have a high local concentration of zinc perhaps because zinc binds to histamine, through a chelating action, within mast cells (Kazimierczak and Malinski, 1974). A single study concerned with the interaction of zinc and interferon production and activity failed to determine the nature of this interaction (Gainer, 1977). Much more work is needed in this area with respect to the effect of zinc upon both interferon, per se, and interferon inducers (Dion, et al, 1974).

COPPER

The importance of copper for effective response to pathogenic challenges of a wide variety has gained modest attention recently (Suttle and Jones, 1986). In addition, animals supplemented with dietary copper show significantly attentuated pathological damage due to the invading organism (Omole and Onawunmi, 1979). It has also been proposed that patients with Menkes disease (Menkes, et al, 1962), an inherited defect of copper metabolism resulting in copper deficiency, may suffer from a basic T lymphocyte defect, as reflected by increased susceptibility to T cell-mediated infections (Pedroni, et al, 1975). Studies investigating the role of copper metabolism and immune function have demonstrated effects on T and B cells, neutrophils, macrophages, as well as complement and immunoglobulin structure and function. In mice, a copper deficient diet imposed from the time of parturition throughout the early postnatal period results in a decreased antibody response to SRBC, a reduction in lymphocyte stimulation by Concanavalin A, and a reduced mixed lymphocyte reaction response (Lukasewyz, et al, 1987). Consistent with the above, Prohaska, et al (1983) have reported that copper deficiency in mice has a profound effect on thymus and spleen morphology. Lukasewyz, et al.

(1987) have suggested that one biochemical lesion underlying the effect of copper deficiency on these tissues is a reduction in Cu, Zn-superoxide dismutase activity with a resulting increase in cellular lipid peroxidation. Copper also seems to be integrally involved in the inflammatory process, at times potentiating the inflammatory reaction, and at other times inhibiting it (Milanino, et al, 1979). A severely copper deficient diet leads to a pro-inflammatory effect. Teleologically this may, in part, explain why serum copper and ceruloplasmin levels are elevated in rheumatoid arthritis patients (Grennan, et al, 1980) and why copper complexes, especially copper salicylate, are used widely in the treatment of rheumatoid arthritis and other degenerative diseases (Sorenson and Hangarter, 1977). An additional mechanism by which copper acts in altering immune responses may involve an interaction at the level of the plasma membrane, as copper has been shown to ameliorate the toxic effects of diethyldithiocarbamate upon both T cells and polymorphonuclear leukocytes, an interaction which must occur at the cell membrane (Rigas, et al, 1979).

Serum copper and ceruloplasmin levels increase in a variety of acute infectious disorders (Beisel, 1976) as opposed to depressed levels of other trace metals such as zinc and iron (Cousins, 1985). Leukocytic endogenous mediator (LEM) seems to act as a key regulatory factor, influencing host-pathogen interaction and its effect upon the metabolism of copper and other trace elements (Pekarek, et al, 1977). While many aspects of the role of copper in host defense remain to be investigated, results indicate a significant interaction.

MANGANESE

Manganese appears to play a critical role in at least three basic aspects of metabolism: (1) glycosaminoglycan synthesis (2) carbohydrate metabolism, and (3) lipid metabolism (Keen, et al, 1984). For glycosaminoglycan metabolism, manganese is essential for the activation of glycosyltransferases and thus, many of the lesions characteristic of manganese deficiency, e.g., skeletal abnormalities, congenital ataxia and impaired eggshell formation, can be attributed directly to altered glycosyltransferase activity and thus altered glycosaminoglycan levels. Manganese is required by pyruvate carboxylase and phosphoenolpyruvate carboxykinase, thus gluconeogenesis may be in part be regulated by cellular manganese concentrations. In addition, manganese is required for insulin synthesis and release. Thus, one consequence of manganese deficiency can be a diabetic-type condition. An additional effect of manganese deficiency can be an increase in cellular lipid peroxidation rates which may be due to a reduction in the activity of Mn-superoxide dismutase, and alterations in membrane lipid composition.

Very little is understood about the role of manganese in immunological function. Manganese supplemented horses, goats, hens, and rabbits produce elevated antibody titres, and maintain these elevated antibody titres over extended time periods. Manganese supplementation also results in increased levels of other non-specific resistance factors, although the precise significance of these elevated proteins is not known (Bazhora, et al, 1974; McCoy, et al, 1979). Manganese given by intramuscular injection has been shown to stimulate natural killer cell activity, potentially via the stimulation of interferon production (Smialowicz, et al, 1987). Much of this work has suffered from methodological inconsistencies, as in the case of the work done with hens, which included supplementation with cobalt and iodine simultaneously with the manganese supplementation, and also utilized a basal diet in which the levels of all trace elements were not specified. McCoy, et al (1979) reported that in rats fed marginally deficient levels of manganese, IgG agglutinins and the 19s fraction of gamma globulin were reduced, but the hemolysin titre was significantly elevated nine days after intraperitoneal injection with sheep red blood cells. Addition of three times the requirement for manganese in the diet resulted in a depression of antibody responses in rats.

In addition to the singular effects on manganese, there may also be an interaction between manganese and calcium. Calcium ions are intimately involved in many aspects of lymphocyte activation, i.e., blast transformation in response to mitogens, and manganese may interact with calcium at many of these numerous steps. Calcium and manganese are known to compete in a number of biological systems, and manganese can interfere with a number of calcium-dependent processes by competitive inhibition.

REFERENCES

Allen, J.I., Perri, R.T., McClain, C.J., Kay, N.E. (1983). J. Lab. Clin. Med. 102:577.

Amivaian, K., McKinney, J.A., and Tuchna, L. (1974). Immunol. 26:1135.

Attramadal, A. (1969). J. Peridont. Res. 4:281.

Bazhora, I.I., Shtefan, E.E., and Timoshevski, I.V. (1974). Mikrobiol. Zh. 36:771.

Beach, R.S., Gershwin, M.E., and Hurley, L.S. (1979). Develop. Comp. Immunol. 3:725.

Beach, R.S., Gershwin, M.E., Makishima, R.K., Hurley, L.S. (1980). J. Nutr. 110:805.

Beach, R.S., Gershwin, M.E., Hurley, L.S. (1982a). Science 218:469.

Beach, R.S., Gershwin, M.E., and Hurley, L.S. (1982b). J. Nutr. 112:1169.

Beach, R.S., Gershwin, M.E., Hurley, L.S. (1983). Am. J. Clin. Nutr. 38:579.

Beisel, W.R. (1976). Med. Clin. North Am. 60:831.

Castillo-Dunn, C., Haresi, G., Fisberg, M., Uavy, R. (1987). Am. J. Clin. Nutr. 45:602.

Chandra, R.K. and Au, B. (1980). Am. J. Clin. Nutr. 33:736.

Cousins, R.J. (1985). Physiol. Rev. 65:238.

Cunningham-Rundles, S., Cunningham-Rundles, C., Dupont, B., and Good, R.A. (1980). Clin. Immunol. Immunopathol. 16:115.

Cunningham-Rundles, S. (1982). Am. J. Clin. Nutr. 35:1202.

Dion, A.S., Vaidya, A.B., and Fout, G.S. (1974). Cancer Res. 34:3509.

Endre, L., Katona, Z., and Gyurkovits, K. (1975). Lancet 1:1196.

Fraker, P.J., DePasquale-Jardieu, P., Zwickle, C.M., and Luecke, R.W. (1978). Proc. Natl. Acad. Sci. U.S.A. 75:5660.

Fraker, P.J., Hilderbrant, K., Luecke, R.W. (1984). J. Nutr. 114:170.

Gainer, J.H. (1977). Am. J. Vet. Res. 38:869.

Giugliano, R., Millward, D.J. (1987). Br. J. Nutr. 57:139.

Golden, M.H.N., Golden, B.E., Harland, P.S.E.G., and Jackson, A.A. (1978). Lancet 1:1226.

Greenan, D.M., Knudson, J.M.L., Dunkley, J., MacKinnon, M.J., Myers, D.B., and Palmer, D.G. (1980). N. Zeal. Med. J. 91:47.

Hambidge, K.M., Walravens, P.A., Brown, R.M., Webster, J., White, S., Anthony, M., and Roth, M.L. (1976). Am. J. Clin. Nutr. 29:734.

Hansen, M.A., Fernandes, G., Good, R.A. (1982). Annu. Rev. Nutr. 2:151.

Haynes, D.C., Gershwin, M.E., Golub, M.S., Cheung, A.T.W., Hurley, L.S., Hendrickx, A.G. (1985). Am. J. Clin. Nutr. 42:252.

Katz, M., and Steihm, E.R. (1977). Pediatr. 59:495.

Kazimierczak, W., and Maslinksi, C. (1974). Agents Actions, 4:1.

Keen, C.L., Lonnerdal, B., Hurley, L.S. (1984). In: E. Frieden (ed.). Biochemistry of the Essential Ultratrace Elements. Plenum, New York, pp. 89-132.

Klevay, L.M., Reck, S.H., Jacob, R.A., Logan, G.M., Munoz, J.M., and Sandstead, H.H. (1980). Am. J. Clin. Nutr. 33:45.

Lennard, E.S., Bjornson, A.B., Petering, H.G., and Alexander, J.W. (1974). J. Surg. Res. 16:286.

Lukasewycz, O.A., Kolquist, K.L., Prohaska, J.R. (1987). Nutr. Res. 7:43.

Malave, I.B., and Pocino, M. (1980). Clin. Immunol. Immunopathol. 16:19.

Malave, I., Nemeth, A., and Pocino, M. (1980). Cell. Immunol. 49:235.

McCoy, J.A., Kenney, M.A., Gillham, B. (1979). Nutr. Rep. Inter. 19:165.

Menkes, J.H., Alter, M., Steigleder, G.K., Weakley, D.R., and Sung, J.H. (1962). Pediatrics 29:764.

Milanino, R., Conforte, A., Fracasso, M.E., Franco, L., Leone, R., Passarella, E., Tarter, G., and Velo, G.P. (1979). Agents Actions 9:581.

Mulhern, S.A., Vessey, A.R., Taylor, G.L., Magruder, L.E. (1985). Proc. Soc. Expt. Biol. Med. 180:453.

Oleske, J.M., Westphal, M.L., Starr, S.S., Shore, S., Gorden, D., Bogden, J., Coplen, D.B., and Nahmias, A. (1979). Am. J. Dis. Child. 133:915.

Omole, T.A., and Onawunmi, O.A. (1979). Annal Parasitol (Paris). 54:495.

Pedroni, F., Bianchi, F., Vgazio, A.G., and Burgio, G.R. (1975). Lancet 1:1303.

Pekarek, R.S., Hoagland, A.M. and Powanda, M.C. (1977). Nutr. Rep. Int. 16:267.

Pekarek, R.S., Sandstead, H.H., Jacob, R.A., and Barcome, D.F. (1979). Am. J. Clin. Nutr. 32:1466.

Prohaska, J.R., Downing, S.W., Lukasewycz, O.A. (1983). J. Nutr. 113:1583.

Rigas, D.A., Eginitis-Rigas, C., and Head, C. (1979). Biochem. Biophys. Res. Commun. 88:373.

Sandstead, H.H., Vo-Khactu, K.P., and Solomons, N.W. (1976). In: Trace Elements in Human Health and Disease, A.S. Prasad (Ed.), 1, pp. 33, Academic Press, New York.

Schrauzer, G.N. (1976). Med. Hypothesis 2:31.

Smialowicz, R.J., Rogers, R.R., Riddle, M.M., Luebke, R.W., Fogelson, L.D. (1987). J. Toxicol. Environ. Health 20:67.

Sorenson, J.R.J., and Hangarter, W. (1977). Inflammation 2:217.

Suskind, R.M. (1977). Raven Press, New York.

Suttle, N.F., Jones, D.G. (1986). Proc. Nutr. Soc. 45:317.

Tapazoglou, E., Prasad, A.S., Hill, G., Brewer, G.J., Kaplan, J. (1985). J. Lab. Clin. Med. 105:19.

Vallee, B.L., Falchuk, K.H. (1981). In: Trace Element Deficiency, pp. 185-197 (L. Fowder, G.A. Garton and C.F. Mills, eds.), London: The Royal Society.

Yoffe, B., Pollack, S., Ben-Porath, E., Zinder, O., Barzilai, D., Gershon, H. (1986). Immunol. Lett. 14:15.

Zwickl, C.M., and Fraker, P.J. (1980). Immunol. Com. 9:611.

IMMUNE RESPONSES AS PARAMETRES FOR SELENIUM

TOLERANCE DETERMINATION IN SHEEP

Knut Moksnes, Hans Jørgen Larsen and Gunnar Øvernes

Norwegian Agricultural Purchasing and Marketing Cooperative
Oslo, The Norwegian College of Veterinary Medicine, Oslo
and National Veterinary Institute, Oslo

INTRODUCTION

Field experiments were carried out to study the effect of selenium (Se) supplementation on the antibody response to tetanus toxoid, serum IgG levels and on the lymphocyte response to mitogens.

MATERIALS AND METHODS

Twenty-one 6-month old lambs of the Norwegian Dala breed, were kept indoors and divided into 7 groups, and fed a basal diet of hay an compound concentrates for 11 weeks (Moksnes and Norheim, 1983). The basal diet contained a total of 0.13 mg Se/kg (group 0). This was supplemented with 0.1, 0.5 or 1.0 mg Se/kg either as sodium selenite (group 4-6) or as seleno-DL-methionine (group 1-3) for six groups of three lambs for the whole experimental period. After a feeding period of 4 weeks, the sheep were immunized subcutaneously with 2 ml Ovivac (against Cl. perfringens type D, Cl. septicum, Cl. tetani, Cl. chauvoei) (Hoechst). A booster injection was given 3 weeks later, and after further 4 weeks the sheep were slaughtered. Blood samples were collected during the experimental period for chemical and immunological analyses.

Immunological methods. Serum and colostrum IgG concentrations were measured using a conventional single radial immunodiffusion test (Mancini et al. 1965). Specific antibodies against tetanus toxoid was assayed by a passive haemagglutination test (Avrameas et al. 1969). The antibody titre values were $\log_{10}$ transformed before statistical calculations (Nissen 1982) were carried out.

Lymphocyte transformation test. Lymphocyte responses to mitogens were assayed according to the method described for the whole blood transformation test (Larsen 1979). The final concentration of the mitogens used were 10 μg PHA/ml, 50 μg PWM/mg and 10 μg Con A/ml.

RESULTS

The mean primary antibody response values to tetanus toxoid were higher in all supplemented groups than in the control group. The same effect was seen after the booster injection. However, due to the response variation within the groups, only the secondary response in group 6 differed significantly ($p < 0.05$) from the control group.

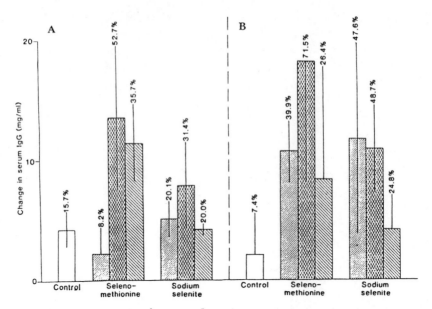

Figure 1. The effect of dietary selenium supplementation on the serum IgG concentrations in 6 month old lambs. Mean change in serum IgG concentration after 4 weeks (A) and 11 weeks (B) of feeding expressed in mg/ml (histogram) and as per cent values of the IgG concentration at the start of the feeding trial. The lambs were immunized after 4 and 7 weeks of feeding. Standard errors of mean are indicated as vertical bars.

$\boxtimes$ = 0.1 ppm Se, $\boxtimes$ = 0.5 ppm Se, $\boxtimes$ = 1.0 ppm Se

Table 1. The lymphocyte response to phytohaemagglutinin (PHA), poke weed mitogen (PWM) and concanavalin A (Con A) in groups of lambs fed different amounts of selenium for 11 weeks.
Lymphocyte response = (Mean counts per minute of stimulated culture)$^{\frac{1}{2}}$ − (Mean counts per minute of control cultures)$^{\frac{1}{2}}$. SD = standard deviation.

Groups	Dietary selenium supplementation mg/kg	Whole blood selenium µg/ml (SD)	Whole blood glutathione per-oxidase activity (SD)	Lymphocyte response to mitogens at the end of the feeding period PHA (SD)	PWM (SD)	Con A (SD)
G0	0.0	0.13 (0.04)	243 (176)	229.2 (52.7)	184.9 (26.8)	255.7 (33.8)
G1	0.1 Se-meth.	0.26 (0.09)*	586 (137)*	274.3 (44.8)	175.1 (42.6)	208.6 (78.8)
G2	0.5 Se-meth.	0.34 (0.02)***	774 (84)***	285.5 (15.5)	186.7 (16.0)	209.2 (24.7)
G3	1.0 Se-meth.	0.50 (0.02)***	864 (87)***	127.6 (13.5)*	23.4 (2.5)***	56.6 (22.7)***
G4	0.1 Selenite	0.26 (0.02)***	702 (95)**	323.9 (31.3)*	268.4 (21.6)**	290.8 (17.5)
G5	0.5 Selenite	0.33 (0.02)***	593 (58)*	315.6 (36.3)*	254.9 (16.6)**	354.6 (33.1)*
G6	1.0 Selenite	0.30 (0.02)***	604 (23)*	166.7 (47.4)	111.2 (32.2)	216.4 (34.4)

Statistical difference from the control group (G0). * $p < 0.05$, ** $p < 0.01$, *** $p < 0.005$.

CONCLUSION

The impaired mitogen response at a dietary Se level of 1.13 mg/kg and the peak response of IgG at 0.63 mg Se/kg may indicate that the maximum tolerable level of 2 mg Se/kg feed is too high.

REFERENCES

Avrameas, S., B. Tavdov and S. Chuilon. Immunochemistry, 6, 67-76 (1969).
Larsen, H. J.. Research in Veterinary Science, 27, 334-338 (1979).
Mancini, G., A. Carbonara and J. F. Heremans. Immunochemistry, 2, 235- (1965).
Moksnes, K. and G. Norheim. Acta vet. scand., 24, 45-58 (1983).
Nissen, Ø. Agricultural University of Norway. Report No. 202 (1982).

THE MOBILIZATION OF STORAGE IRON AS A

DETERMINANT OF REFEEDING INFECTION

M. J. Murray and A. B. Murray

University of Minnesota Hospital
Box 508
420 Delaware St., S.E.
Minneapolis, MN 55455

We have observed reported an increased incidence of infections during refeeding after famine. In a study of 4,382 famine victims with a mean weight loss of 29% the incidence rose from 4.9% before refeeding to 30% 2 weeks into refeeding (1). Most were infections resulting from intracellular micro-organisms of which falciparum malaria was the commonest. These infections were occurring when cellular and humoral immunity ought to be returning to normal, implying that refeeding might have a negative impact on host resistance to some infections. While there may be many reasons for this negative effect we examined only the role of mobilization of iron during refeeding on host resistance to infection.

We observed 117 undernourished Somali before and during refeeding. Before refeeding they had a mean loss of 29% body weight but no clinical or laboratory evidence of infection. Bloods were drawn before and at 2, 4 and 24 weeks of refeeding. Serum was separated and tested for ferritin, total iron binding capacity, % saturation of transferrin and C reactive protein, the latter as a marker of infection. Table 1 records our findings at 2 weeks. Ferritin was raised before refeeding and fell by 2 weeks with a significant rise in serum iron and transferrin saturation. At 4 weeks the results were similar but by 24 weeks iron measures were near normal. Infections occurred in 31 or 26% at 2 weeks, in 27 or 23% at 4 weeks and were more frequent in those who gained the most weight and had the greatest drop in serum ferritin. So far we have shown a relationship between mobilization of iron and the frequency of infection but no direct direct cause and effect.

In 5 groups of Swiss-Webster mice of about 20g each we compared the effect of 2 successive injections of 1 mg desferrioxamine or of saline on mortality after inoculation with LD_{50}'s of Listeria monocytogenes. The groups included normals, 25% undernourished, and 25% undernourished refed to 3, 6 and 9 days before inoculation with desferrioxamine and Listeria. If infection were related to mobilization of iron then desferrioxamine by chelating the iron and facilitating its excretion might reduce mortality. Mortality fell in all desferrioxamine treated groups - Table 2. On the 6th day it dropped from 86 to 36% as liver non-heme iron concentration as a measure of iron storage fell 41%. Our observations suggest that mobilization of iron from stores during refeeding may be an important determinant of infections during refeeding. The sudden availability of iron may permit rapid replication of micro-organisms before immune host-defense mechanism have returned to normal.

Table 1. Change in Iron Status ± SD and Incidence of Infections in 117 Undernourished Somali Before and After 2 Weeks of, Refeeding

	Δ%Wt	SERUM Fn**	SERUM Fe	%TF Sat***	Infections
BEFORE REFEEDING*	BASELINE	711±13	127±7.9	44±3.7	0
REFED					
UNINFECTED	+9.8	532±48	161±7.3	51±3.7	0
INFECTED	+13.1	463±61	165±6.9	58±3.9	31

* MEAN WEIGHT LOSS 29%
** SERUM FERRITIN
*** % TRANSFERRIN SATURATION

Table 2. Effect of Desferrioxamine (DFX) on Mortality of Undernourished Mice Inoculated with LD_{50}'s of Listeria monocytogenes Before and During Refeeding.

STATUS	WEIGHT G		Δ% LIVER NON-HEME IRON CONCENTRATION IN DFX	% MORTALITY	
	DFX-	DFX+		DFX-	DFX+
CONTROL	19.6	20.3	-	56.6	43.3
UNDERNOURISHED	14.2	13.9	-51	16.6	23.3
REFED. 3 DAYS	15.3	15.1	-39	76.6	33.3
6 DAYS	16.6	16.9	-41	86.6	36.6
9 DAYS	18.5	18.1	-33	63.3	30.0

REFERENCES

1. Murray MJ, and Murray AB. Mobilization of storage iron as a deter minant of refeeding infection. Clin Res 37 777A 1987.

INFLAMMATION-RELATED CHANGES IN TRACE ELEMENTS, GSH-METABOLISM,

PROSTAGLANDINS, AND SIALIC ACID IN BOVINE MASTITIS

F. Atroshi, S. Sankari, J. Työppönen, and J. Parantainen

Department of Pharmacology & Toxicology; Department of
Biochemistry, College of Veterinary Medicine, Box 6,
Helsinki 55;   Huhtamäki Oy Pharmaceuticals, Clinical
Research, Box 325, Helsinki 10, Finland

INTRODUCTION

Chronic mastitis is a severe veterinary problem due to frequent
relapses and a variable response to antibiotics. The immunological
elimination of pathogens may be impaired. Mastitis may even present as
"sterile" inflammation, without apparent involvement of bacteria (1),
which further stresses the importance of nonspecific resistance. Our
approach to the problem has been to find tissue factors that predispose
to infection and might help in the struggle against the disease.

GSH-METABOLISM, PROSTAGLANDING AND SELENIUM

We have found that milk and blood prostanoid levels were elevated
in mastitic animals, while erythrocyte glutathione peroxidase (GSH-Px)
and reduced GSH were substantially and significantly decreased suggesting that Se-related processes may have been disturbed in mastitis (2).
Lipid peroxidation and formation of oxygen free radicals, important
factors of an inflammatory reaction, are opposed by the selenoenzyme
GSH-Px. An unexpected finding was (3) that both GSH-Px and GSH were
elevated in the inflammed udder tissue (fig. 1). As GSH-enzymes have an
important role in the formation of PGs we suggested that the increase in
GSH-metabolism might cause the increase in PG-levels (fig. 2).

SILICON AND ELECTRICAL CONDUCTIVITY

We have observed a marked decline in blood and milk silicon level
associated with mastitis (4). A moderate negative correlation between Si
and electrical conductivity was also observed. Like selenium, silicon is
an essential trace element. It is considered to have protective function
in vascular intima, and has a close and unique relation to macrophages
(5) Dietary intake of silicon may be interesting in this respect.

VITAMIN E AND SIALIC ACID IN MASTITIS

We have found (6) that milk and blood contents of vitamin E may be
lower in bovine mastitis (Fig 3). Vitamin E functions as an antioxidant
as such and the actions of vitamin E and selenium are interconnected.

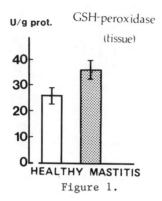

Figure 1.

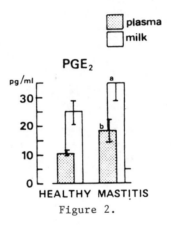

Figure 2.

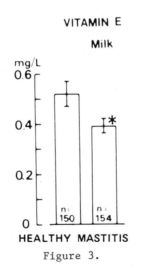

Figure 3.

Vitamin E when given together with Se may favourably influence the course of clinical mastitis (7). Likewise, in Se and vitamin E deficiency the ability of neutrophils to produce hydroxyl radical for killing of bacteria is impaired (8). Both milk and blood sialic acids were elevated in mastitic animals compared to the healthy (9). A similar pattern was also observed in mastitic udder tissue (3). In cow serum sialic acids have been found to be elevated in various inflammatory and infectious diseases (10). As constituents of mucous glycoproteins sialic acids are needed for lubrication and protection of body cavities, mucus membranes, and other surfaces. On the other hand, mucus, sputum and phlegm are typical products of infection and with their antioxidant activity they might hamper oxidative killing of pathogens.

CONCLUSIVE REMARKS

Substantial changes in parameters related to membrane charges and intracelular redox state in mastitic animal were demonstrated. Antioxidants have their tissue protective functions but they might also impair oxidative killing of pathogens. Besides pathophysiological, there are some dietary aspects as well.

REFERENCES

1. Gunther M., Lancet 1: 175-180 (1965).
2. Atroshi F. Parantainen J. Sankari. S. Österman. T., Res. Vet. Sci. 40: 361-366 (1986).
3. Atroshi F. Parantainen J. Kangasniemi R. Sankari S., J. Anim physiol. Anim. Nutr. (In press, 1987).
4. Parantainen J. Tenhunen E. Kangasniemi R. Sankari S. Atroshi F., Vet. Res. Commun. (In press, 1987).
5. Allison A. C., In: In vitro methods in cell-mediated and tumor immunity. p. 395 (B R Bloom and J R David, eds.) Academic Press, N.Y. (1976).
6. Atroshi F. Työppönen J. Sankari S. Kangasniemi R. Parantainen J., Int. I. Vit. Nutr. Res. (In press, 1987).
7. Smith K. I. Harrison I. H. Hancock D. D. Todhunter D. A. Conrad R., J. Dairy Sci. 67: 1293-1297 (1984).
8. Boyne R. Arthur J. R. J., comparat. Path. 89: 151-8 (1979).
9. Atroshi F. Parantainen J. Sankari S. Kangasniemi R. Salonen J., Vet. Med. B. 33: 620 (1986).
10. Motoi Y., J. Jap. Vet. Med. Assoc. 37, 643-649 (1984).

THE EFFECTS OF DIETARY Mo ON OVINE HOST AND INTESTINAL PARASITE

Neville Suttle, David Knox and Frank Jackson

Moredun Research Institute, Edinburgh, Scotland

The improvement of pastures by liming and reseeding raises their Mo and S concentrations, induces hypocuprosis in the lamb, and increases susceptibility to microbial but not nematode infection (1). The poor supply of Cu from the digesta to the host might also harm the parasites sharing the same nutrients. The effects of dietary Mo on the development of Trichostrongylus vitrinus (T.v.) were therefore investigated.

MATERIALS AND METHODS

Four groups of six worm-free lambs (3 male and 3 female), weighing 22.6 kg, were given either 0 or 5 mg Mo/kg DM (as Na_2MoO_4) and 0 or 2,500 T.v. larvae (3rd stage), 5 days a week for four weeks, in a 2 x 2 x 2 factorial design. The basal diet was a pelleted mixture of whole barley (0.60), whole oats (0.17), oat feed (0.10), dried skimmed milk (0.12), urea (0.01), $CaSO_4$ (0.0136), NaCl (0.010) and added Co, Zn, Fe, I and vitamin supplements: it contained 79 umol Cu/kg DM and was fed at 80 g/kg $WO.75$. After 10 days a liver biopsy sample was taken and dosing with larvae commenced. Progress of infection was indicated by faecal egg counts and albumin and globulin in plasma. After 9 weeks, all worms were retrieved from the small intestine and the liver and dressed carcase weights (DCW) recorded. Samples of small intestine were taken for histology. Cu, cytochrome oxidase (CO), superoxide dismutase (SOD), acetylcholinesterase (Ach), protease activities in worm homogenates (150 worms in 20 ml buffer) and protease and protein secretion by the worms during a 60 h incubation were measured (protease is secreted as a digestive enzyme). Means are given with the standard error of difference between them unless stated otherwise.

RESULTS

The principal effects at 9 weeks are presented. Infection increased plasma globulin (42.9 v 36.2 ± 1.92 g/l; p < 0.01) and liver size as a fraction of DCW (0.039 v 0.037 ± 0.0016; p < 0.05); reduced DCW (kg) in female (13.5 v 15.8) but not male lambs (15.3 v 15.0 ± 0.65; interaction p < 0.05). There was a three-way interaction between infection, Mo and sex affecting plasma albumin in which hypoalbuminaemia was only induced when Mo-supplemented female lambs were infected (Table 1).
Mo reduced DCW (14.2 v 15.6 ± 0.46 kg; p < 0.01) and reduced liver Cu in all but infected females (3-way-interaction; p < 0.001) but did not

affect plasma Cu (grand mean 13.3 ± 2.2 9S.D.) umol/l (Table 1); decreased the number of worms retrieved (p < 0.05). Mean (s.e.), CO and SOD activities for groups not given or given Mo were similar at 0.047 (0.0066) or 0.037 (0.0056) U and 1.968 (0.147) or 2.147 (0.340) U/homogenate, respectively; differences were not present in Ach.

Mo markedly increased protein but reduced protease activity in the worm homogenate towards Azocasein as substrate (similar results were found for Azocallagen and Elastin Orcein) and reduced the amount of soluble protein and protease activity secreted by the intact worms during culture.

Table 1. Effects of Mo, infection and sex on the lamb

	Mo	Sex	Infection M	Infection F	T. vit. M	T. vit. F	s.e.d.
Dressed	0		15.6	16.2	15.4	15.1	
Carcase (kg)	+		14.4	15.3	15.1	12.0	0.92
Plasma	0		29.4	33.3	30.0	29.4	
albumin (g/l)	+		29.9	30.8	29.5	21.2	2.2
Liver Cu	0		124	428	229	138	
umol	+		32	-91	-88	150	115.3

Table 2. Effects of Mo on some physical and chemical properties of T. vitrinus (per 150 worms*) (homogenized or cultured).

Mo	Number recovered	Soluble protein (ug) In tissue	Soluble protein (ug) Secreted	Protease (o.d. Azocasein) In tissue	Protease (o.d. Azocasein) Secreted
0	6247	172 ± 25.3	1339 ± 256	13.4 ± 4.08	11.4 ± 3.72 s.e.
	± 436				
+	4810	327 ± 73.9	786 ± 89	5.0 ± 2.17	2.4 ± 1.06 s.e.

DISCUSSION

The hypothesis that Mo would retard the development of host and parasite by inducing a state of Cu deficiency in both should be replaced with another, namely that neither effect involved Cu. We suggest that Mo directly inhibited protease activity in the parasite. Subsequent studies (D. Knox, unpublished data) have shown that addition of MoO_4^{2-} to culture fluid inhibits protease activity in T.v. This, together with the lack of association between the effects of Mo on protease and the two cupro-enzymes suggests independency from Cu. We suggest that affinity of Mo for SH groups on proteins results in inhibition of activity of the thiol-dependent worm protease, thus decreasing the amino acid supply for parasite growth. Worm metabolism was not generally impaired because Ach was not affected by Mo.

The retardation of lamb growth by Mo was surprising in that it occurred in the face of normal plasma Cu and an average liver Cu of 91 ppm. Since proteases (e.g., trypsin) are also important in mammalian digestion, the adverse effects of Mo in the lamb may also have arisen through protease inhibition.

The drastic effect of infection in Mo-treated females must be repeated in view of the small numbers. Nevertheless, it is worth noting that the female lamb is more prone to diarrhoea than the male on pastures high in Mo (1).

REFERENCE

1. J. A. Woolliams et al., Studies on lambs from lines genetically selected for low and high copper status. Anim. Prod. 43:293-317.

INFLUENCE OF HORMONES ON COPPER METALLOPROTEIN LEVELS

R. A. DiSilvestro

Department of Foods & Nutrition
Purdue University
West Lafayette, IN 47907

In the last 15 years it has become abundantly clear that hormones can influence the levels of many different proteins including those containing trace metals. However, much remains to be learned concerning hormonal regulation of metalloproteins, particularly those containing copper. For instance, the response of every copper protein in every type of cell or fluid to every hormone, acting individually or in combination, awaits complete identification. Further, the mechanisms behind even presently identified hormonal effects have not been fully clarified. Finally, the metabolic and functional consequences of hormonally induced changes require additional study. The present paper will highlight the still limited information gathered thus far concerning hormonal control of copper metalloproteins.

Hormones could conceivably influence copper holoprotein levels through at least two types of mechanisms. First, these agents could regulate actual apoprotein concentrations of a particular metalloprotein. Second, hormones could render copper more or less accessible to a particular apoprotein pool.

HORMONAL REGULATION OF COPPER APOPROTEIN LEVELS

The levels of 2 copper apoenzymes have been definately found to be influenced by hormones. Serum levels of the multifunctional protein ceruloplasmin are known to rise following administration of any of several hormones to humans or experimental animals[1]. A number of pathological and nonpathological stress states also raise ceruloplasmin concentrations through processes believed to involve hormones[1]. Many of these studies measured only ceruloplasmin enzyme activity. However, certain projects, especially those from this laboratory, have confirmed increases in ceruloplasmin immunoreactive protein levels for a number of stress states[2,3,4]. The effects of the hormone interleukin-1 on rat ceruloplasmin activity levels appear to result from increased ceruloplasmin protein synthesis[5]. Protein levels of another copper protein, rat adrenal dopamine beta hydroxylase, are also known for certain to be hormonally controlled. Protein contents are maintained by glucocorticoids which restrict enzyme degradation[6]. This enzyme comprises the terminal enzyme in norepinephrine synthesis.

Connective tissue activity concentrations of lysyl oxidase, which initiates crosslinking of collagen and elastin, are also hormonally controlled. Rat lung lysyl oxidase activity contents rise during streptozotocin induced diabetes and fall rapidly during starvation[7]. Rat skin lysyl oxidase activity contents are reduced by hypophysectomy and by glucocorticoid injection into young animals[8,9]. In all these cases only enzyme activity, not specific protein levels were measured. None the less, at least some of these observations probably involve changes in lysyl oxidase protein levels.

Hormones could control the protein levels of other copper metalloenzymes such Cu-Zn superoxide dismutase (SOD), an enzyme believed to eliminate potentially dangerous superoxide radical[10]. Thus far, no hormones have been specifically observed to alter Cu-Zn SOD apoprotein levels in vivo. However, these protein levels are capable of undergoing change. Chronic hyperoxia is known to raise rat lung Cu-Zn SOD protein contents[11]. Rat liver Cu-Zn SOD activity and protein levels were recently found in this laboratory to be higher in the adult rat than in the neonate (Table 1). A similar pattern was found for rat lung Cu-Zn SOD. Cu-Zn SOD protein levels were measured by an unpublished enzyme linked immunoadsorbant assay (ELISA). Since serum levels of thyroid hormone levels are higher in younger rats than adults[13], the possibility was tested that these hormones keep hepatic Cu-Zn SOD levels low in neonates. L-thyroxine injections to adult rats (3 daily injections sc, 2 mg/kg, sacrifice 24h after last injection) reduced rat liver Cu-Zn SOD activity contents by around 36%. However, serum concentrations of thyroid hormones in different aged young rats did not negatively correlate with liver SOD contents. Rats that were 16 days old showed higher hepatic Cu-Zn SOD activity values than 8 day olds even though serum thyroid horome levels rise between these 2 ages[13]. Possibly, thyroid hormones do exert a major impact on liver Cu-Zn SOD levels under circumstances beside simple aging.

The synthetic glucocorticoid dexamethasone, thyroid hormone T_3 and insulin can each increase Cu-Zn SOD activities by 73 to 92% in rat fetal mixed lung cells cultured in serum-free conditions[14]. However, the relationship between these actions in vitro and potential effects in vivo remains unclear. Injections of L-thyroxine (same procedure as above) or dexamethasone (2 mg/kg, ip, sacrifice 24h later) produced only 23% and 8% increases in adult rat lung Cu-Zn SOD activities, respectively. The inconsistancies between studies in cultured cells and intact rats could

Table 1. Influence of age on rat liver Cu-Zn SOD levels

| | SOD | |
	units[a]/g liver	ug/g liver
Neonate (10 days)	4,800 + 310	213 + 9
Adult (60 days)	12,115 + 406	532 + 12

[a]Units are those previously described for the pyrogallol autoxidation assay of Prohaska[12].

relate to differences in lung cell populations of adults used for the latter work versus the neonate rats used in the cell culture studies. Alternatively, in the whole animal experiments tissues other than lung may have preferentially taken up the injected hormones. On the other hand, the hormones tested here may only raise lung Cu-Zn SOD activity contents <u>in</u> <u>vivo</u> when other hormone concentrations are raised or lowered. Thus, injection of a single hormone would have only limited effectiveness. Finally, these hormones may be necessary for maintanence of normal lung Cu-Zn SOD contents but not be capable of producing increases.

RELATIONSHIP OF CERULOPLASMIN AND MT LEVELS TO COPPER METALLOENZYME ACTIVITY LEVELS

Ceruloplasmin, which is synthesized by the liver, has been proposed to transport copper to extrahepatic tissues[1]. Conceivably, increases in serum ceruloplasmin synthesis and secretion rates could decrease copper accessability to hepatic enzymes while increasing accessability to nonhepatic enzymes. To test the first hypothesis, ceruloplasmin concentrations were raised by turpentine induced experimental inflammation (0.1 ml injection, im in the leg, sacrifice 3 days later). This treatment increased ceruloplasmin activity levels by about 250% (Table 2). Increases in ceruloplasmin levels due to inflammation are believed to be mediated by the hormone interleukin-1[1] which seems to increase ceruloplasmin synthesis[5]. Rat liver Cu-Zn SOD activities fell during inflammation by 27% while Cu-Zn SOD protein values were unchanged (Table 2). In contrast, estrogen treatment (14 daily injections of estradiol benzoate, 50 ug/rat, sc, sacrifice 24h after last injection) more than doubled ceruloplasmin concentrations without affecting hepatic Cu-Zn SOD activity levels (Table 2). However, estrogen could be decreasing serum ceruloplasmin degradation rates rather than increasing hepatic ceruloplasmin synthesis. Obviously, more work is needed in this area including direct evaluations of liver copper distribution during various circumstances.

Possibly, some of the copper needed for increases in ceruloplasmin levels results from depletion and continual diversion of copper away from storage sites throughout the body. Consistant with this hypothesis, the amount of copper bound to renal metallothionein is decreased by interleukin-1 injection despite increased renal metallothionein mRNA activity concentrations[15]. The amount of zinc bound to this protein increases. Possibly, reducing agents preferentially promote copper release over that of zinc. Interestingly, ascorbic acid injections into chicks and guinea pigs increases ceruloplasmin levels as measured by its oxidase activity[16,17]. Ascorbic acid may not in itself represnt a normal physiological releaser of stored copper[17]. However, the fact that injection of a reducing agent increases ceruloplasmin levels suggests that such agents could be important to body copper redistribution.

Although liver metallothionein is induced by a several hormones, these changes as of yet have no clear role for copper redistribution. Induction of rat liver metallothionein by the glucocorticoid dexamethasone or by endotoxin, which alters hormone secretion rates, raises the amount of zinc but not copper bound to liver metallothionein[15,18].

Elevations in ceruloplasmin levels could make copper more accessible to nonhepatic enzymes. Increases in serum ceruloplasmin have been found to correlate with increased nonhepatic copper metalloenzyme activity

Table 2. Effects of inflammation and estrogen on rat serum
ceruloplasmin and liver Cu-Zn SOD activity levels

	Ceruloplasmin units[a]/dl	Cu-Zn SOD units[b]/g liver
Control	81 + 4	11,859 + 294
Inflammation	205 + 24	8,674 + 275
Estrogen	185 + 41	11,814 + 299

Rat treatments are described in the text.
[a]Units are those previously described for oxidation of
p-phenylenediamine oxidase[17].
[b]Units are those previously described for the pyrogallol
autoxidation assay[12].

concentrations under some circumstances. Aortic lysyl oxidase activities
in copper deficient chicks are restored more effectively by a single copper
injection when the rise of ceruloplasmin levels is enhanced by estrogen
injection[19]. There is indirect evidence that estrogen injections in rats
on a standard diet raises adrenal dopamine beta hydroxylase activities[20].
During chronic hyperoxia, rat serum ceruloplasmin concentrations rise
before a rise in lung Cu-Zn SOD activity contents occurs[21]. In contrast,
experimental inflammation was found to increase ceruloplasmin levels (Table
1) without altering rat lung Cu-Zn SOD activity contents. Possibly, the
influence of ceruloplasmin levels on nonhepatic tissue copper enzyme
activities depends on the degree of copper saturation of the particular
apoenzyme pool. In the inflammation study, the lung Cu-Zn SOD pool could
have already been fully saturated with copper. However, the increased SOD
protein levels occuring during hyperoxia[11] could have rendered this pool
temporarilly less than fully copper saturated. Aortic lysyl oxidase in the
copper deficient chick project could have been less than fully saturated
even after copper injection. Perhaps the adrenal dopamine beta hydroxylase
pool is less than fully copper saturated even in a copper adequate state
with no abnormal circumstances. Such is the case for atleast one copper
protein, namely serum ceruloplasmin[1].

CONCLUSION

The full range of hormonal effects on copper metalloenzyme levels and
the mechanisms behind these effects requires further attention. However,
it is certain that several copper metalloenzyme levels can be influenced by
hormones.

REFERENCES

1. R. J. Cousins, Absorption, transport, and hepatic metabolism of
   copper: special reference to metallothionein and ceruloplasmin,
   Physiol. Rev. 65:238 (1985).
2. R. A. DiSilvestro and E. A. David, An enzyme immunoassay for
   ceruloplasmin: application to serum for cancer patients, Clin.
   Chim. Acta 158:287 (1986).

3. R. A. DiSilvestro, Immunoreactive levels of ceruloplasmin and other acute phase proteins during lactation, Proc. Soc. Exp. Biol. Med. 183:251 (1986).

4. R. A. DiSilvestro, E. F. Barber, E. A. David, and R. J. Cousins, An enzyme-linked immunoadsorbent assay for rat ceruloplasmin, Biol. Trace Elem. Res., in press.

5. E. Barber, and R. J. Cousins, Induction of ceruloplasmin by interleukin-1 in copper deficient and copper sufficient rats, Fed. Proc. 45:235 (1986).

6. R. D. Ciaranello, G. F. Wooten and J. Axelrod, Regulation of dopamine-hydroxylase in rat adrenal glands, J. Biol. Chem. 250:3204 (1975).

7. A. M. Madia, S. J. Rozovski, and H. M. Kagan, Changes in lung lysyl oxidase activity in streptozotocin-diabetes and in starvation, Biochim, Biophys. Acta 585:481 (1979).

8. S. Shoshan, and S. Finkelstein, Lysyl oxidase: a pituitary hormone-dependent enzyme, Biochim. Biophys. Acta 439:358 (1976).

9. S. C. Benson, and P. A. LuValle, Inhibition of lysyl oxidase and prolyl hydroxylase activity in glucocorticoid treated rats, Biochem. Biophys. Res. Commun. 99:557 (1981).

10. J. M. McCord, and I. Fridovich, Superoxide dismutase. An enzymic function for erythrocuprein (hemocuprein), J. Biol. Chem. 244:6049 (1969).

11. J. D. Crapo, and J. M. McCord, Oxygen-induced changes in pulmonary superoxide dismutase assayed by antibody titrations, Am. J. Physiol. 231:1196 (1976).

12. J. Prohaska, Changes in tissue growth, concentrations of copper, iron, cytochrome oxidase and superoxide dismutase subsequent to dietary or genetic copper deficiency in mice. J. Nutr. 113:2148 (1983).

13. J. H. Dussaut, and F. LaBrie, Development of the hypothalamic-pituitary-thyroid axis in the nonatal rat, Endocrinology 97:1321 (1975).

14. A. K. Tanswell, M. G. Tzaki, and P. J. Byrne, Hormonal and local factors influence antioxidant enzyme activity of rat fetal lung cells in vitro, Exp. Lung Res. 11:49 (1986).

15. R. A. DiSilvestro, and R. J. Cousins, Mediation of endotoxin-induced changes in zinc metabolism in rats, Am. J. Physiol. 247:E436 (1984).

16. R. A. DiSilvestro, and E. D. Harris, A postabsorptive effect of L-ascorbic acid on copper metabolism in chicks, J. Nutr. 111:1964 (1981).

17. R. A. DiSilvestro, Effects of ascorbic acid and inflammation on ceruloplasmin activity levels in guinea pigs. Nutr. Res. 6:1009 (1986).

18. L. L. Hutchings, D. M. Scholler, J. S. Valentine, M. R. Swerdel, K. R. Etzel, and R. J. Cousins, Influence of adrenalectomy and dexamethasone on rat liver metallothionein and superoxide dismutase activity, Inorg. Chim. Acta 91:L21 (1984).

19. E. D. Harris, and R. A. DiSilvestro, Correlation of lysyl oxidase with the p-phenylenediamine oxidase activity (ceruloplasmin) in serum. Proc. Soc. Exp. Biol. Med. 166:528 (1981).

20. V. Schreiber, T. Pribyl, and J. Jahodova, Effect of dopamine-beta-hydroxylase inhibitor (disulfiram) on the response of adenohypophysis, serum ceruloplasmin and hypothalmic ascorbic acid to estradiol treatment, Endocrin. Exper. 13:131 (1979).

21. S. A. Moak, and R. A. Greenwald, Enhancement of rat serum ceruloplasmin levels by exposure to hyperoxia, Proc. Soc. Exp. Biol. Med. 177:97 (1984).

NOREPINEPHRINE AND DOPAMINE DISTRIBUTION IN COPPER-DEFICIENT MICE

Joseph R. Prohaska and Karen L. DeLuca

Department of Biochemistry
University of Minnesota, Duluth
Duluth, MN 55812

INTRODUCTION

Copper is required for development and homeostasis of the nervous system but there exists much controversy as to specific changes that occur when copper is limiting. Conflicting results have been published regarding receptor density, ligand affinities,and steady-state level changes for several neurotransmitters.[1]

Emphasis has been placed on norepinephrine (NE) and dopamine (DA), the product and substrate, respectively, of the Cu-dependent enzyme dopamine-β-monooxygenase (DBM). Perinatal Cu deficiency in rats results in lower NE levels in brain,[2-5] however, the hypothalamus does not show this deficit, apparently.[4] Dopamine levels in brain of Cu-deficient rodents have been reported to be lower,[3,4] unchanged,[5,6] or higher[5] when compared to controls. Cu-deficient rats have lower NE in adrenal gland.[7] The purpose of these studies was to compare NE and DA distribution following dietary copper deficiency in a number of organs and fluids.

METHODS AND MATERIALS

A series of seven dietary experiments (two sets) were conducted using protocol described previously.[6] Male weanling C57BL mice were divided into two groups and fed a purified diet low in copper (-Cu) (0.5 ppm) for four weeks. One group received deionized water to drink (-Cu) and the other water containing cupric sulfate, 20μg Cu/ml, (+Cu). During lactation mice were nursed by dams on their respective treatments. Mice were anesthetized with ether or halothane (when heart puncture was performed for plasma catecholamines) and killed by decapitation. Organs were frozen in liquid N_2 and kept at $-85°$ until analyzed. Liver Cu was determined by flame AAS following wet digestion in HNO_3.

Organs were homogenized with 0.05 N $HClO_4$ and NE and DA were analyzed following HPLC with electrochemical detection.[5] Plasma catecholamines were enriched on phenylboronic acid gel rather than alumina, 200μl samples were processed. Urinary catecholamines were quantified using an additional ion-exchange column (Amberlite CG-50) preceding alumina adsorption. Urinary creatinine (Cr) was determined using an alkaline-picrate method.

RESULTS AND DISCUSSION

Mice in the first set of experiments exhibited a modest degree of Cu deficiency as liver Cu was low and mild cardiac hypertrophy was evident (TABLE 1); growth was normal in the -Cu mice and mild anemia (hematocrit 42%) was detected. Evidence for impaired NE synthesis in -Cu mice was evident as spleen NE was lower and heart DA higher. In previous work,-Cu mice similar to these were found to contain lower brain NE levels whereas DA was unchanged.[6] When plasma catecholamines were analyzed an unexpected **elevation** in NE was detected in −Cu mice (TABLE 1). Urinary excretion of both NE and DA was higher in −Cu mice, most noticeably for DA (TABLE 1).

Mice in experiment 2, although reared the same as those in experiment 1, were much more Cu-deficient as evidenced by a growth effect, lower liver Cu and severe cardiac hypertrophy (TABLE 1). DBM block was more convincing in these −Cu mice as major deficits in NE levels were evident for heart, brain, and hypothalamus; 5.5 and 7 fold elevations in heart and spleen DA were measured (TABLE 1). Brain DA was not changed.

TABLE 1. Effects of dietary copper deficient in mice on catecholamine distribution.

Parameter	Exp. 1		Exp. 2	
	Cu-adequate	Cu-deficient	Cu-adequate	Cu-deficient
Body Wt. (g)	22.5 ± 1.6	22.9 ± 2.4	20.6 ± 1.8	14.1 ± 3.3 *
Heart/BW (mg/g)	5.46 ± 0.38	7.27 ± 1.12 *	5.23 ± 0.33	15.0 ± 3.75 *
Liver Cu (μg/g)	4.97 ± 0.46	2.77 ± 0.53 *	4.90 ± 0.16	0.75 ± 0.10 *
Heart NE (ng/g)	988 ± 227	867 ± 224	1023 ± 254	344 ± 61.5 *
Heart DA (ng/g)	20.7 ± 10.2	66.4 ± 52.5 †	18.9 ± 18.1	105 ± 35.3 *
Spleen NE (ng/g)	762 ± 146	546 ± 130 *	557 ± 140	445 ± 232
Spleen DA (ng/g)	38.4 ± 20.5	51.1 ± 17.9	11.7 ± 12.1	82.6 ± 27.8 *
Brain NE (ng/g)	---	---	407 ± 80.8	233 ± 73.7 *
Brain DA (ng/g)	---	---	1011 ± 45.5	1104 ± 86.6
Hypothalamus NE (ng/g)	---	---	1103 ± 238	558 ± 137 *
Plasma NE (ng/ml)	1.39 ± 0.42	3.27 ± 1.52 †	---	---
Urine NE (μg/mg Cr)	0.44 ± 0.19	0.68 ± 0.29 †	---	---
Urine DA (μg/mg Cr)	0.60 ± 0.28	2.34 ± 1.46 †	---	---

Values are means ± SD for 4-12 male 7-week-old C57BL mice, † $P < 0.05$ or * $P < 0.01$ compared to Cu-adequate by Student's t-test.

These results suggest that DBM activity is limiting in vivo in Cu deficiency despite the failure to confirm this in vitro.[5,7] Furthermore, the elevated plasma and urinary NE levels suggest another possible basis for lower organ levels, i.e. enhanced NE turnover. These data in mice also suggest that NE in the hypothalamus is not spared as it is in −Cu rat brain.[4] Lastly, these data suggest that following Cu deficiency DA is elevated in noradrenergic neurons of the sympathetic nervous system. The deficits in brain DA reported by others in −Cu rats[3,4] is likely due to loss of dopaminergic neurons during gestational development since no DA deficit occurs in brains of −Cu mice that are clearly Cu-deficient after birth.

ACKNOWLEDGMENT

Supported in part by NIH grant HD 20975

REFERENCES

1. J. R. Prohaska, Functions of Trace Elements in Brain Metabolism, _Physiol. Rev._ in press (1987).
2. J. R. Prohaska and W. W. Wells, Copper deficiency in the developing rat brain: a possible model for Menkes' steely-hair disease, _J.Neurochem._ 23: 91-98 (1974)
3. R. F. Morgan and B. L. O'Dell, Effect of copper deficiency on the concentrations of catecholamines and related enzyme activities in the rat brain, _J. Neurochem._ 28: 207-213 (1977).
4. D. J. Feller and B. L. O'Dell, Dopamine and norepinephrine in discrete areas of the copper-deficient rat brain, _J. Neurochem._ 34: 1259-1263 (1980).
5. J. R. Prohaska and T. L. Smith, Effect of dietary or genetic copper deficiency on brain catecholamines, trace metals and enzymes in mice and rats, _J. Nutr._ 112: 1706-1717 (1982).
6. J. R. Prohaska and D.A. Cox, Decreased brain ascorbate levels in Cu-deficient mice and in brindled mice, _J. Nutr._ 113: 2623-2629 (1983).
7. J. E. Hesketh, The effect of nutritional copper deprivation on the catecholamine content and dopamine-β-hydroxylase activity of rat and cattle adrenal glands, _Gen Pharmacol._ 12: 445-449 (1981).

EFFECT OF COPPER DEFICIENCY ON ENZYME SECRETION FROM ISOLATED

PANCREATIC ACINI

Michael A. Dubick[1,2] and Adhip P.N. Majumdar[3,4]

[1]VA Medical Center, Martinez, CA 94553, [2]Univ. of California
Davis, CA 95616 and [3]VA Medical Center, Allen Park, MI 48101
[4]Wayne State University, Detroit, MI 48201

INTRODUCTION

Over the past 15 years a number of reports have appeared which indicate
that copper (Cu) deficiency is accompanied by a selective and progressive
atrophy of pancreatic acinar tissue without any change in the ductal tissue[1-3]. Smith, et al[3] observed that this pancreatic atrophy was also associated with functional alterations of the organ, as indicated by a reduced
in vivo pancreatic secretory response to secretin or caerulein. To evaluate
further the changes in secretory response of the exocrine pancreas in Cu deficiency, the present study investigates the changes in enzyme secretion
from dispersed pancreatic acini in response to CCK-8 (cholecystokinin octapep-
tide), secretin and carbachol. In addition, amylase, trypsin and chymotrypsin activities (as a measure of enzyme levels) in the pancreas were also
determined.

MATERIALS AND METHODS

Groups of 12 adult female Sprague-Dawley rats, weighing 175 ± 3 g were
fed ad libitum an isocaloric diet either deficient (0.5 ppm) or sufficient
(6.2 ppm) in Cu (Zeigler, Bros., Gardners, PA) for 6 weeks, and killed by an
overdose of pentobarbital. The pancreases were removed, trimmed of fat,
weighed, and either used for isolation of acini or for determination of amylase, trypsin and chymotrypsin activities as described previously[4,5]. Freshly
isolated dispersed pancreatic acini were suspended in HEPES-Ringer buffer,
pH 7.5 and following equilibration at 37°C for 30 min, 1 ml aliquots of cell
suspension were incubated for 60 min in the absence (basal) or presence of
CCK-8 (100 pM), secretin (1 μM) or carbachol (7.5 μM). The reaction was
terminated and the supernatant and pellet (cells) were assayed for amylase
activity[4,5]. In some experiments, trypsin and chymotrypsin activities were
also determined[4,5].

RESULTS AND DISCUSSION

Although consumption of a Cu-deficient diet for 6 weeks resulted in a
significant 37% reduction in liver Cu levels (3.3 ± 0.5 μg/g vs 5.2 ± 1 μg/g in
controls), no differences in final body weight between the two groups were observed. Other gross abnormalities such as hair loss or lethargy were also

Table 1. Basal and Stimulated Amylase Release from Isolated Pancreatic Acini

	Basal	CCK-8 (100 pM)	Secretin (1 μM)	Carbachol (7.5 μM)
Control (n=4)	7.3±0.5	17.7±1.1	16.8±0.9	22.5±0.5
Cu-deficient (n=4)	7.2±0.6	6.0±0.4*	5.4±0.5*	4.4±0.6*

Data expressed as mean±SE of the percent of total amylase released. *$p<0.05$.

not observed in the Cu-deficient group[2,3], suggesting that consumption of the Cu deficient diet produced only a marginal Cu deficiency. Despite this, the functional properties of the exocrine pancreas were markedly affected by the diet. Pancreatic acini isolated from Cu-deficient rats showed a total loss of responsiveness to both cholinergic and peptide secretagogues (Table 1). Whereas secretagogue-mediated amylase secretion from isolated acini from control rats was stimulated 2- to 3-fold over basal levels, no such stimulation was observed in acini isolated from Cu-deficient rats. Release of trypsinogen and chymotrypsinogen by these secretagogues was also similarly affected (data not shown). These findings agree with in vivo observations by others[3], and indicate further that Cu-deficiency directly affects the secretory responses of the exocrine pancreas.

Table 2. Effect of Copper Deficiency on Pancreatic Enzymes

	Amylase (U/g)	Trypsin (U/g)	Chymotrypsin (U/g)
Control (n=11)	21,802±1382	98,314±3788	89,402±6749
Cu-deficient (n=12)	16,472±1282*	85,322±4737*	70,877±4625*

*Significantly different from control at $p<0.05$

Although in young rats, severe Cu deficiency has been found to cause pancreatic atrophy[2], in the present experiment neither the pancreatic weight nor its DNA or protein content were affected by diet. Nevertheless, amylase, trypsin and chymotrypsin activities decreased by 24%, 13% and 21%, respectively, in Cu-deficient rats in comparison with the controls (Table 2).

In conclusion, our current data demonstrate that consumption of a Cu-deficient diet by adult rats for 6 weeks, affects the levels of a number of pancreatic enzymes and the secretory responsiveness of the exocrine pancreas to various secretagogues, despite no apparent pancreatic atrophy.

Acknowledgements: The authors thank Dr. Carl Keen for the measurement of liver Cu concentrations. This study was supported by the Medical Research Service of the Veterans Administration and a Grant-In-Aid from the American Heart Association, California Affiliate (Alameda County Chapter).

REFERENCES

1. W.B. Muller, Virch. Arch. Abt. A Path. Anat. 350:353 (1970).
2. B.F. Fell. T.P. King and N.T. Davies, Histochem. J. 14:665 (1982).
3. P.A. Smith, J.P. Sunter and R.M. Case, Digestion 23:16 (1982).
4. A.P.N. Majumdar, G.D. Vesenka, M.A. Dubick, G.S.M. Yu, J.M. DeMorrow and M.C. Geokas, Am. J. Physiol. 250:G598 (1986).
5. G.A. Kaysen, A.P.N. Majumdar, M.A. Dubick, G.D. Vesenka, G. Mar and M. C. Geokas, Am. J. Physiol. 249:F518 (1985).

DIFFERENTIAL EFFECT OF COPPER DEFICIENCY ON PLASMA ATRIAL NATRIURETIC

PEPTIDES IN MALE AND FEMALE RATS

Sam J. Bhathena, Bruce W. Kennedy, Patricia Marsh, Meira Fields and Nadav Zamir

ARS, USDA, CNL, Beltsville Human Nutrition Research Center, Beltsville, MD 20705; NINCDS and NHLBI, National Institutes of Health, Bethesda, MD 20982

Severely copper deficient animals manifest hypertriglyceridemia, hyperuricemia, glucose intolerance, decreased hemoglobin concentration, abnormalities of the electrocardiogram and sudden death due to rupture of the heart. In addition, experimental copper deficiency in animals is also characterized by central nervous system disorders such as ataxia, motor incoordination, and brain and spinal cord degeneration. Further, feeding high levels of dietary fructose or sucrose as compared to starch has been reported to produce a more severe copper depletion syndrome in rats. Serum ceruloplasmin and superoxide dismutase activity in erythrocytes are also decreased in copper deficiency and these parameters are used as markers of copper status. Decreases in plasma enkephalins levels have also been suggested as markers of copper deficiency[1]. Recently, significant differences in symptoms of copper deficiency have been observed between male and female rats fed low copper diets[2]. Only male rats fed a copper deficient diet with fructose died of cardiac rupture. Females were protected against this syndrome, indicating sex hormones may be involved. Atrial natriuretic peptides (ANP) have been shown to play a role in cardiac function, where, their levels in plasma are higher in patients with congestive heart failure and in hypertensive subjects. It is possible that in copper deficiency, atrial peptides may be differently regulated in male and female rats. We therefore studied the effect of copper deficiency and of sex hormones on ANP from plasma and atria.

MATERIALS AND METHODS

Weanling Sprague-Dawley rats were fed copper deficient (0.6 μg Cu/g) (CuD) or copper supplemented (6.0 μg Cu/g) (CuS) diets with either 62% fructose or starch. (See table.) Males were fed the diets for 7 weeks. Some CuD fructose fed males were then fed CuS fructose diet for an additional 3 weeks. Females were fed for 11 weeks. In order to study the correlation between sex hormones and atrial ANP on cardiac function, in a second experiment, castrated and ovariectomized rats were fed CuD diets containing fructose. The rats were sacrificed by decapitation after an overnight fast, blood was collected in EDTA and Trasylol, and plasma was frozen. Atria, heart and liver were quickly removed and frozen. ANP levels were measured from plasma and atrial extracts by radioimmunoassay.

Table 1. Plasma and atrial ANP in male and female rats fed copper supplemented or copper deficient diets with either starch or fructose.*

Diet	Males Plasma ANP (pg/g)	Males Total Atrial ANP** (μg/g)	Females Plasma ANP (pg/g)	Females Total Atrial ANP** (μg/g)
Starch CuS	48.4± 5.6[a]	9.87±0.57[a]	---	---
Starch CuD	60.6± 9.8[a]	8.46±1.25[a]	---	---
Fructose CuS	40.7± 5.5[a]	9.23±0.92[a]	52.4±9.0[a]	9.29±1.04[a]
Fructose CuD	200.9±41.5[b]	5.46±1.21[b]	50.6±5.5[a]	4.36±0.56[b]
Fructose CuD to CuS	76.1±18.5[a]	7.07±0.64[ab]	---	---

* Values are means±SEM. Values with different superscripts within a column
  are significantly different by Duncan's multiple range test at $p<0.05$.
**Values from left and right atria are combined.

RESULTS

    Copper deficiency in rats fed CuD diets was ascertained by a
significant decrease in plasma, heart and liver copper levels. Plasma ANP
levels increased significantly in male rats fed CuD fructose diet but not
in those fed starch. Female rats did not show any change in plasma ANP
even after 11 weeks. When male rats fed CuD fructose diet for 7 weeks
were switched to CuS diet, plasma ANP levels decreased to values observed
in CuS rats.
    In atria of male as well as female rats fed CuD diets, ANP levels
tended to be lower than in those fed CuS. Feeding fructose as compared to
starch worsened the problem. However, when male rats fed CuD diets were
switched to CuS diets, atrial ANP increased to levels observed in CuS
rats. In copper deficient rats fed fructose, neither castration nor
ovariectomy had any effect on atrial ANP suggesting that sex hormones do
not play a major role in altering atrial ANP (data not shown).

DISCUSSION

    Dietary fructose as compared to starch increases the severity of Cu
deficiency in both male and female rats. However, Fields et al.[2]
recently showed that only male rats died of cardiac rupture when fed CuD
fructose diet and that female rats were protected. In the present study,
though copper deficiency decreased atrial ANP content in both male and
female rats, only male rats showed higher plasma ANP levels. Further,
when copper deficient male rats were fed CuS diets, copper deficiency
symptoms were reversed. Concomittantly, plasma ANP levels decreased.
Sex hormones did not appear to play an important role in modulating atrial
ANP levels in copper deficiency. It is important to note that the
castrated and ovariectomized rats were not sexually mature, as seen from
prepubertal testosterone and estrogen levels. Since, in male rats
increased plasma ANP levels preceded death due to cardiac rupture, it is
possible that increased plasma ANP levels in male rats may be directly
correlated with the severity of copper deficiency.

REFERENCES

1. S.J. Bhathena, L. Recant. Peptides and opiates in copper deficiency.
    In: Biology of Copper Complexes. Ed. J.R.J. Sorensen. Humana Press,
    1987. In Press.
2. M. Fields, C.G. Lewis, D. Scholfield, A.S. Powell, A. Rose, S. Reiser,
    J.C. Smith. Female rats are protected against fructose induced mortal-
    ity of copper deficiency. Proc. Soc. Expt. Biol. Med. 183:145 (1986).

SEXUAL DIFFERENCES IN COPPER DEFICIENCY

Meira Fields, Charles G. Lewis, Todd Beal, James C. Smith, and Sheldon Reiser

Georgetown University Medical School, Washington D.C. 20007
USDA, Beltsville Human Nutrition Research Center,
Beltsville, Maryland 20705

INTRODUCTION

Since female rats have been shown to be protected against the mortality of copper deficiency of male rats when fructose is fed, we designed a study to establish whether castration of the male rat will protect but ovariectomy of the female will exacerbate the signs of copper deficiency.

MATERIAL AND METHODS

Eighty-five male and female rats were assigned to a copper deficient diet containing 62% fructose and 0.6 μg/g copper for 8 weeks. Twelve of the males were castrated and 12 of the females were ovariectomized. Blood levels of total estrogens and testosterone were determined. In addition, hematocrit and direct copper measurements in blood and liver were measured.

RESULTS

Except for those animals that were sacrificed by us, all other male rats died. However, castration delayed mortality of males by 2 weeks. In contrast, none of the females died. Only males were anemic and exhibited hypertrophied pale hearts with gross pathology which included flabiness of the ventricles and aneurysms at the ventricular apex.

Copper deficiency was verified by the undetectable activity of ceruloplasmin and low levels of plasma copper concentrations. The copper levels in liver were lower in females as compared with males but were not affected by gonadectomy. The levels of total estrogens and testosterone were significantly reduced by castration and ovariectomy in male and female rats. However, levels of sex hormones of intact animals were also low indicationg that all animals have not yet reached puberty.

117

DISCUSSION

In agreement with our previous data[1], the present study shows that female rats are protected against the anemia, heart pathology and mortality of copper deficiency when fructose is fed. This protection is provided regardless whether females are ovariectomized or intact. In contrast, all males are susceptible to the inadequate copper intake and cannot be protected against the fructose induced mortality of copper deficiency, although castration delayed mortality of males by two weeks. Thus, the sex of the animal determines copper homeostasis. Since males had double the levels of hepatic copper concentrations than females, but females did not die of the deficiency, it is suggested that the mortality of the male rat is not solely due to the low levels of hepatic copper. Metabolites of fructose may contribute to the severity of copper deficiency of the male rat.

Table 1. Direct copper measurements in plasma and liver

	Females		Males	
	intact	ovariectomized	intact	castrated
Ceruloplasmin	ND	ND	ND	ND
Plasma copper μmol/l	1.6 $\pm$ 0.5	2.8 $\pm$ 0.3	2.4 $\pm$ 0.4	2.7 $\pm$ 0.4
Erythrocyte SOD U/ml	14 $\pm$ 4	14 $\pm$ 9	9 $\pm$ 5	10 $\pm$ 1
Hepatic copper μg/g wet wt	0.87 $\pm$ 0.17	1.02 $\pm$ 0.19	1.68 $\pm$ 0.30	1.19 $\pm$ 0.17

Each value represents the Mean $\pm$ SEM of 6 observations per group. ND-non detectable.

REFERENCES

1. M. Fields, C.G. Lewis, D. Scholfield, A.C. Powell, A.J. Rose, S. Reiser, J.C. Smith, Female rats are protected against the fructose induced mortality of copper deficiency. Proc. Soc. Expt. Biol. & Med. 183:145 (1986).

CALMODULIN CONCENTRATION IN TISSUES OF ZINC AND CALCIUM DEFICIENT RATS

Hans-Peter Roth and Manfred Kirchgessner

Institut für Ernährungsphysiologie
Technische Universität München
D-8050 Freising-Weihenstephan

INTRODUCTION

The elements zinc and calcium exhibit an antagonistic action in their metabolism: calcium stimulates whereas zinc inhibits many intracellular functions. The intracellular functions of calcium are primarily accomplished by way of calmodulin, a low molecular, calcium-binding protein with a molecular weight of 16700. Zinc is an inhibitor both of calcium-activated calmodulin and of calmodulin-stimulated enzymes.[1] The aim of the rat study presented here was to show to which extent a zinc, calcium, or combined zinc/calcium deficiency changes the calmodulin concentration in rat tissues.

MATERIALS AND METHODS

Thirty-two young, male Sprague-Dawley rats with a mean body weight of 60 g were divided into 4 groups of 8 animals each. Group I received a control diet adequate in both Zn and Ca (63 µg Zn/g and 0.88% Ca), group II a diet deficient in Ca with adequate Zn (63 µg Zn/g and 0.0043% Ca), group III a diet deficient in Zn and adequate in Ca (1.1 µg Zn/g and 0.88% Ca) and group IV a diet deficient in both Zn and Ca (1.1 µg Zn/g and 0.0043% Ca). Groups I and II were pair-fed to the groups III and IV in which the feed intake was reduced by Zn deficiency in the same way: all 4 groups received the diet in the same daily amounts. After 28 days all animals were sacrificed and samples of serum, testis, brain and skeletal muscle (quadriceps) were dissected for Zn and Ca analysis by atomic absorption spectrophotometry, and for estimation of calmodulin by radioimmunoassay.

RESULTS AND DISCUSSION

The mean body weight of group III and IV animals, which was 60 g at the beginning of the experiment, was only 82 g after 28 days, resulting from the strongly reduced feed intake accompanying Zn deficiency. The addition of Ca deficiency to Zn deficiency therefore had no effect on feed intake and weight development. In contrast, group I rats, which received the same amount of feed but adequate amounts of Zn and Ca, obtained a 35% higher mean body weight (112 g), and group II rats, which received the Ca-deficient diet, a 24% higher mean body weight (102 g) than the Zn-deficient animals of groups III and IV.

The serum Zn concentration was not affected by alimentary Ca deficiency

119

(II), but was reduced by 72% during Zn deficiency (III). Simultaneous Zn/Ca deficiency in the diet (IV) caused a 61% reduction in the serum Zn concentration; this value was significantly higher than the value for the purely Zn-deficient group (III). In contrast, the Ca concentration in serum was on the one hand not affected by Zn deficiency (III), but was, on the other hand reduced by 56% during Ca deficiency (II). Combined Zn/Ca deficiency (IV) reduced the serum Ca concentration by 40%; this value -- like that for zinc before it -- was significantly higher than the value for the purely Ca-deficient group (II).

In muscle, during Zn deficiency (III), the Zn concentration dropped by a significant 23%. Both Ca deficiency alone (II) and combined Ca/Zn deficiency (IV) had no effect on the Zn concentration in muscle. Ca deficiency (II) and Ca/Zn deficiency (IV) reduced the Ca concentration in muscle by 43% and 29%, respectively. Here, the Ca concentration of group IV animals was significantly higher than the value for group II animals. Zn deficiency alone (III) did not bring about a change in the Ca concentration of muscle. The calmodulin concentration in muscle, determined radioimmunologically, was significantly elevated to over double the control animal value during Ca, Zn, and combined Ca/Zn deficiency.

In testis, a significant 11% reduction of the Zn concentration was observed for the Zn-deficient animals (III). Although Ca deficiency (II) had no effect on the Zn concentration in testis, for simultaneous Ca/Zn deficiency (IV) the Zn concentration was significantly reduced by a further 9% in comparison to the value for the Zn-deficient animals (III). Ca deficiency (II) reduced the Ca concentration in testis by 19%, but a further reduction was not observed during combined Ca/Zn deficiency (IV). Zn deficiency alone (III) caused an improvement of the Ca concentration in the testis by 19%. During Ca deficiency (II), the calmodulin concentration in the testis, similar to that in muscle, was elevated by a significant 77%; the calmodulin concentration during Zn deficiency (III) was, however, only 28% higher. In comparison to the group I control rats, combined Ca/Zn deficiency (IV) did not have an effect on the calmodulin concentration in testis.

In the brain, Zn deficiency (III) likewise caused a significant 7% reduction in the Zn concentration. Ca deficiency (II) also reduced the Zn concentration in the brain, but a significantly reduced value could not be observed during combined Ca/Zn deficiency. Ca deficiency (II) and combined Ca/Zn deficiency (IV) reduced the Ca concentration in the brain by 24% and 29%, respectively. Zn deficiency alone raised the Ca concentration in the brain by 24%, a value similar to that observed earlier in the testis. The calmodulin concentration in the brain was, like that in the muscle and testis, elevated both during Ca and during Zn deficiency, as well as during combined Ca/Zn deficiency; here, however, the value was much smaller and the tendency toward elevation could only be confirmed in the case of Ca deficiency (II).

In conclusion, one can say that the calmodulin concentrations in muscle, testis and brain, determined by radioimmunoassay, were elevated during both Zn and Ca deficiency, although these elevated values could not be proved statistically significant in all cases. Because too little data are available and the appropriate comparison studies have not yet been done, it is not possible to interpret the meaning of these elevated calmodulin concentrations during Ca or Zn deficiency and the consequences for metabolism.

REFERENCE

1. G.J. Brewer, J.C. Aster, C.A. Knutsen, and W.C. Kruckeberg, Zinc inhibition of calmodulin: A proposed molecular mechanism of zinc action on cellular functions, Am. J. Hemat. 7:53 (1979).

EFFECTS OF CALMODULIN INHIBITORS ON THE

CELLULAR METABOLISM OF ^{45}Ca AND ^{210}Pb

J. G. Pounds and A. C. Nye

Dept. of Applied Science, Brookhaven National Laboratory

Upton, NY 11973

INTRODUCTION

Many Ca^{++}-mediated cell processes depend on the intracellular Ca^{++} receptor protein, calmodulin, to exert regulatory effects on target metabolic and physiological pathways. In many cells, Ca^{++} transport mechanisms are activated by the Ca^{++}-calmodulin complex. Several aspects of the cellular metabolism of Pb^{++} and Ca^{++} are similar. Lead is actively transported or diffuses through mitochondrial and plasma membranes via Ca^{++} transporters and gates. Cellular lead is mobilized by hormones which also mobilize cell Ca^{++} and has a kinetic distribution and behavior in cultured cells similar to Ca^{++}. Lead binds to calmodulin with greater affinity, and can substitute for Ca^{++} in calmodulin-activated processes. However, the relative concentrations of free Ca^{++} and Pb^{++} ions in cytosol and the ability of Pb^{++} to effectively alter calmodulin mediated processes in situ are not well established. The objective of this study was to characterize the regulation of Pb and Ca metabolism by calmodulin-dependent processes.

METHODS

Hepatocytes were obtained by collagenase perfusion of caudate liver lobes dissected from male Sprague-Dawley rats. Cells were plated at a density of 2×10^5 viable cells/cm^3 in Williams' E containing 10% fetal bovine serum, 2 mM glutamine, 2 mU/ml insulin, 20 µm/ml gentamicin, and 1 nM dexamethasone. Cultures were exposed to 0 or 3 µM Pb acetate for 17 hours prior to addition of calmodulin inhibitors and ^{45}Ca and ^{210}Pb. Followng overnight adaptation in culture 25 µCi/ml ^{45}Ca (1.8 mM total Ca), or 1 µCi/ml ^{210}Pb (3 µm total Pb), and 60 µM Wl2 or Wl3 were added to the cultures for a 3 hr incubation period. After the labeling period, cultures were rinsed and subjected to a 210-minute washout procedure. Data obtained from each culture were plotted as the radioactivity $R(t)$ present in the cells at washout time, t, divided by the total radioactivity in the cells at washout time 0, $R(0)$. The data from each culture were fit to the polyexponential equation, $r(t) = B_1 \exp(-\lambda_1 t) + B_2 \exp(-\lambda_2 t) + B_3 \exp(-\lambda_3 t)$. The pool sizes, rate constants, halftimes, and fluxes describing the steady state cellular metabolism of Ca and Pb, normalized to 1 mg cell protein, were derived from the coefficients and exponents as previously described (Pounds et al., 1982). The data represent the mean of four cultures from a representative experiment, expressed

---

Support: NIH ES 04040, NIH P41RR01838

as percent of the untreated control. The data were analyzed statistically by one-way ANOVA and Dunnett's multiple comparison.

RESULTS

The inactive analog W-12 had little effect on the cellular metabolism of Pb or Ca. The calmodulin inhibitor W-13 did not alter total cell Pb or Ca but did affect the subcellular kinetic distribution of both Pb and Ca.

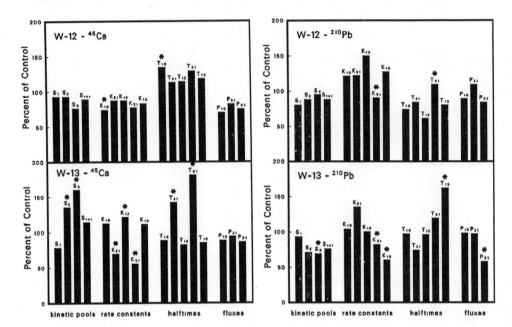

Figure 1. Effect of the calmodulin inhibitors W-12 or W-13 on the kinetic distribution and behavior of ^{45}Ca and ^{210}Pb in cultured rat hepatocytes.

DISCUSSION

Evaluation of the calmodulin-dependence of cell Ca^{++} and Pb^{++} in situ is difficult for several reasons. Altered Ca^{++} homeostasis may result from the direct inhibition of calmodulin-dependent or -independent transport processes. Changes in cell function not directly related to the transport of Ca, e.g., uncoupling of oxidative phosphorylation or altered membrane permeability also disrupt cell calcium metabolism. Thus, the effects of the calmodulin inhibitor W-13 on cell Pb^{++} metabolism may be due to its direct effects on Pb^{++} transporting Ca^{++} pumps, or indirectly as a result of changes in Ca^{++} homeostasis. Direct comparison of the effects of W-13 on the metabolism of Pb and Ca is impaired by differences in the kinetic distribution and behavior of Pb and Ca. A further complication is that the calmodulin-dependent processes are most active during periods of elevated intracellular Ca^{++}. The preliminary experiments reported here were conducted in unstimulated cells which have a low resting level of cytosolic Ca^{++}. Therefore, W-13 induced alterations in cell Ca^{++} and Pb^{++} may not reflect the changes which could occur in stimulated cells.

REFERENCES

Joel G. Pounds, Robert Wright and Ralph L. Kodell, Cellular metabolism of lead: a kinetic analysis in the isolated rat hepatocyte, Toxicol. Appl. Pharmacol. 66:88 (1982).

EFFECT OF MODERATE ENVIRONMENTAL

IODINE DEFICIENCY ON ADOLESCENT GIRLS

Rajalakshmi Krishnamachari

National Institute of Nutrition
Tarnaka, Hyderabad, India   500 007

INTRODUCTION

The effect of severe iodine deficiency in the environment on the population living in such areas is well documented. The effect of moderate
iodine deficient environment on the health status of the population is not
yet known well. The response of the thyroid gland to the onset of puberty
especially in adolescent girls is well known and this enlargement is known
as "physiological goiter." Though it tends to regress after the growth
spurt stops under normal circumstances, it increases in size in iodine deficient environments. As the adolescent girls are the future mothers, iodine
deficiency in them will affect the newborns, resulting in neonatal hypothyroidism. Hence it is important to control iodine deficiency among
adolescent girls.

The objectives of this study were two-fold:

(i) To study the effect of moderate iodine deficient environment on the
     thyroid status of adolescent and preadolescent girls.

(ii) To determine the prevalence of IDD in these populations for chalking
     out future control programmes among this population.

MATERIAL AND METHODS

Four areas with different water iodine levels were selected. School
girls aged 12-18 years from these areas were studied. The details are given
in Table 1.

All these girls were examined clinically for presence of goiter and
other nutritional deficiency signs. Anthropometric measurements were done
to determine the nutritional status. Urinary inorganic iodine (UII) excretion was done in 100 girls from each area randomly selected, fifty before
menarche, fifty after menarche.

Table 1. Water Iodine Content, Nature, Geography of the Areas and the Number of Girls Studied

Sl. No.	Area	Water I content (μg/L)	Nature	Geography	Total No.of Girls	No.of BM Girls	No.of AM Girls
1	Surat	43.5	Urban, non-tribal	Coastal plains	297	111	186
2	Bardoli	27.0	Urban mixed	Plains	195	75	120
3	Ukai	15.0	Rural, tribal	Foothill	191	90	101
4	Vyara	5.4	Rural, tribal	Hilly	249	89	160
				Total	932	365	567

BM: Before Menarche

AM: After Menarche

RESULTS

1. The following important findings were noted. The two areas with water iodine content of 15 μg/L and less (Ukai and Vyara) had significantly higher goiter prevalence rates (52.3% and 68.7%) than the two areas with water I content of 27 μg/L and above (Bardoli and Surat) (36.41% and 24.24%). The most severe forms of goiter, namely grade II and above, were noted only in girls from Vyara.

2. The relation between the onset of puberty and goiter prevalence was also clearly seen. The overall data showed that the rate was significantly higher in girls (47.98%) after menarche (AM) than in girls before menarche (BM) (38.91%). When analyzed areawise, this difference was significantly different only in Surat area.

3. Though there was no statistical significance in the difference between the tall, medium and short statured groups, there was a trend of higher prevalence among tall girls and girls with a body mass index of 0.13-0.18. There seems to be a lesser trend among shorter and fatter girls.

4. UII excretion showed a similar pattern, i.e., the lowest values seen among the girls with highest goiter prevalence and girls who had attained puberty, and tall girls. The UII excretion followed a pattern in direct proportion to the water iodine levels.

DISCUSSION

These findings among adolescent school girls help us to denominate the areas with water iodine levels of 15 μg/L or less, as deficient areas. Though the effect of puberty is seen in the different goiter rates among girls before and after menarche, the effect is marked only in the non-deficient area. This indicates that, in endemic regions, the environmental iodine deficiency plays a dominant role and so the prevalence is high even before the onset of puberty in these girls. In view of the important role played in the child-bearing by the adolescent girls, greater attention should be paid to this group and special efforts should be made to control iodine deficiency among the adolescent girls. The policy makers and health planners should be made aware of this fact. The control of I deficiency among these girls will be an important tool in the prevention of mental retardation among their offspring.

TRANSPORT, SUBCELLULAR DISTRIBUTION AND EXPORT OF Mn(II) IN RAT

LIVER AND HEPATOCYTES*

M. Brandt, C.L. Keen[1], D.E. Ash and V.L. Schramm

Department of Biochemistry
Temple University School of Medicine
Philadelphia, PA 19140

INTRODUCTION AND SUMMARY

Manganese (II) has been implicated in the function of the gluco-neogenic enzymes pyruvate carboxylase and P-enolpyruvate carboxykinase (1,2), and has been proposed to act as a regulatory factor in gluco-neogenesis (3,4). The purpose of these studies was to test the hypothesis that Mn(II) in rat liver has a regulatory function in gluconeogenesis.

Manganese (II) transport in isolated hepatocytes was characterized as a high affinity, high capacity facilitated transport system of the plasma membrane. In situ, Mn(II) taken up by the liver is excreted into the bile, however the Mn(II) of isolated hepatocytes is released only upon release of cytosolic proteins. Approximately 40% of total hepatocyte Mn(II) is associated with cytosolic components and the remainder is in compartments resistant to digitonin treatment. Hepatocytes efficiently transport Mn(II) from the incubation medium to give increased levels of both free and total cellular Mn(II). These changes are similar to those in hepatocytes from rats fed different dietary levels of Mn(II).

Rates of gluconeogenesis from lactate in hepatocytes from fasted rats were not influenced by increased intracellular free and total Mn(II) either in the presence or absence of hormonal stimulation. Hormones did not influence the rates of Mn(II) transport, the ratio of free to bound intracellular Mn(II) or the intracellular compartmentation of Mn(II). These results indicated that the Mn(II) content of isolated hepatocytes and the rates of gluconeogenesis are unrelated. Analysis of free and total Mn(II) in hepatocytes from rats maintained on diets with controlled Mn(II) content indicated that hepatocyte Mn(II) is strongly related to dietary Mn(II). Changes in dietary Mn(II) are rapidly reflected by changes in hepatocyte Mn(II).

*Supported by research grant GM36604 and training grant AM07162 from the NIH.
[1]Department of Nutrition, University of California, Davis, CA.

Mn(II) in Hepatocytes from Fed and Fasted Rats

Hepatocytes isolated by collagenase treatment of livers from rats maintained on standard laboratory chow (containing 3 µmole Mn(II)/g; 165 ppm) gave total Mn(II) content of 27 ± 3 nmol Mn(II) per mol of packed cells (5). This value was unchanged following a 36-48 hr fast of the animals. Electron paramagnetic resonance studies indicated the free Mn(II) content decreased from 0.6 ± 0.1 to 0.2 ± 0.1 nmol/ml hepatocytes in the same animals (5). These findings led to the hypothesis that free Mn(II) levels may be involved in the regulation of gluconeogenesis since Mn(II) is known to interact with both pyruvate carboxylase and P-enolpyruvate carboxykinase (1,2). During periods of gluconeogenesis, P-enolpyruvate carboxykinase is induced and more of the cellular Mn(II) was proposed to bind to the enzyme thereby lowering the free Mn(II). This hypothesis was tested by characterizing Mn(II) transport in hepatocytes and using the transport system to alter intracellular free and bound Mn(II).

Transport of Mn(II) by Liver and Hepatocytes

The early studies of Cotzias and coworkers (6) established that the liver plays a major role in manganese homeostasis, however the transport characteristics had not been quantitated. Cannulation of the bile ducts of fed rats gave a bile with 6 µM free and 35 µM total Mn(II). After fasting or feeding a manganese-deficient diet for 48 hr, the total Mn(II) content of bile is < 2 µM. Following intraperitoneal injection of 10 µmol Mn(II), the biliary free Mn(II) was 200 µM and total was 400 µM. Injection of Mn(II) into the portal vein produced bile with total Mn(II) approaching 1000 µM. Biliary Mn(II) contains both free and bound Mn(II) and reflects the presence of excess Mn(II) in the circulation.

Hepatocytes transported Mn(II) with a K_m of 1.6 µM and a V_{max} of approximately 50 nmol/min/g cells. Transport was strongly temperature dependent and undetectable at 4^0C. Specificity for Mn(II) was high with no significant inhibition by 500 µM Ca(II), Mg(II) or Fe(II). Cobalt (II) was a competitive inhibitor with an inhibition constant of 14 µM. Transport appears to be facilitated diffusion based on saturation characteristics, insensitivity to uncouplers of oxidative phosphorylation and temperature effects. Incubation mixtures containing several uM Mn(II) can be nearly depleted of Mn(II) by the transport causing hepatocyte Mn(II) to increase as much as 10 fold. Efflux of Mn(II) from hepatocytes could not be demonstrated, even in the presence of extracellular EDTA. This finding is inconsistent with a process of simple facilitated diffusion in equilibrium with the cytosol. When the plasma membrane is permeabilized to small molecules with dextran sulfate, no Mn(II) is released, indicating that free Mn(II) detected by EPR is not cytosolic. Digitonin treatment releases 40% of hepatocyte Mn(II) together with cytosolic marker enzymes. Cytosolic Mn(II) is therefore tightly bound in hepatocytes. The majority (60%) of total Mn(II) is not released by digitonin. The lack of Mn(II) efflux from hepatocytes contrasts with the results in situ and suggests that the biliary architecture is important in normal Mn(II) excretion.

Effects of Mn(II) on Gluconeogenesis

Incubation of hepatocytes with 1 uM Mn(II) for 30 min increased the intracellular Mn(II) four-fold. If alterations in intracellular Mn(II) are important in the regulation of gluconeogenesis, rates of gluconeogenesis from lactate should be altered following this treatment. The results of these experiments are summarized in Table 1.

Table 1. Effect of Mn(II) on Glucose Production in Rat Hepatocytes[a]

Addition	Glucose Production; μmol/hr/mg	
	Endogenous Mn(II)	Increased Mn(II)
None	0.18 ± 0.01 (12)[b]	0.20 ± 0.01 (5)
Glucagon, 10^{-8} M	0.27 ± 0.04 (5)	0.28 ± 0.03 (5)
Epinephrine, 10^{-6} M	0.28 ± 0.03 (5)	0.29 ± 0.03 (5)
Glucagon and		
Epinephrine	0.29 ± 0.01 (5)	0.29 ± 0.03 (5)

[a]Gluconeogenesis from 10 mM pyruvate and 1 mM lactate using hepatocytes from 18-24 hrs fasted rats. Hepatocyte Mn(II) was increased
2-4 fold by the presence of 1 or 2 μM Mn(II) in the incubation
medium.
[b]Number of determinations.

The results establish that altered intracellular Mn(II) has no
significant effect on gluconeogenesis independent of glucagon and
epinephrine.

Effects of Dietary Mn(II) on Free and Total Mn(II) in Hepatocytes

        Rats maintained on diets with controlled Mn(II) levels from 1 to
45 ppm were used to prepare hepatocytes. The free and total Mn(II)
content of the hepatocytes is shown in Table 2. Hepatocyte content
of total Mn(II) reflects the dietary Mn(II), while the free Mn(II)
was approximately 3% of the total Mn(II). Alteration of the dietary
Mn(II) results in a rapid readjustment of free and total Mn(II)
indicating that hepatocyte Mn(II) is more closely related to the
transport and excretory function of liver than a response to hormonal
state.

Table 2. Effect of Dietary Mn(II) on Mn(II) Content of Hepatocytes[a]

Dietary Mn(II) (ppm)	Free Mn(II)	Total Mn(II)
	(nmol/ml cells)	
1	0.5 ± 0.1 (3)	16 ± 2 (3)
3	0.7 ± 0.1 (4)	25 ± 2 (4)
5	0.9 ± 0.1 (4)	32 ± 1 (4)
10	1.1 ± 0.1 (4)	38 ± 5 (4)
45	1.6 (2)	46 (2)
45 to 1[b]	0.6 ± 0.1 (3)	25 ± 5 (3)

[a]Hepatocytes were prepared from adult rats maintained on diets of
the indicated Mn(II) content. Similar results were obtained with
young rats (<100g) maintained since weaning on these diets.
[b]Young rats were maintained on a diet of 45 ppm Mn(II) and switched
to a diet of 1 ppm Mn(II) 24 hr before preparation of hepatocytes.

CONCLUSIONS

        A specific, high capacity transport system for Mn(II) has been
characterized in plasma membranes of hepatocytes. Mn(II) loading of
hepatocytes has no effect on gluconeogenesis. Export of Mn(II) from

the liver to bile is efficient and produces both bound and free forms of Mn(II). Hepatocytes are blocked in the ability to secrete Mn(II) but retain the ability to absorb relatively large quantities of the metal. The free Mn(II) in hepatocytes is not cytosolic. The altered levels of free Mn(II) previously observed in rat liver hepatocytes is related to the transfer of Mn(II) from the portal circulation to the bile.

REFERENCES

1. McClure, W.R., Lardy, H.A. and Kneifel, H.P. (1971) J. Biol. Chem. 246, 3569-3578.
2. Brinkworth, R.I., Hanson, R.W., Fullin, F.A. and Schramm, V.L. (1981) J. Biol. Chem. 256, 10795-10802.
3. Williams, R.J.P. (1982) FEBS Lett. 140, 3-10.
4. Schramm, V.L. (1982) TIBS 7, 369-371.
5. Ash, D.E. and Schramm, V.L. (1982) J. Biol. Chem. 257, 9261-9264.
6. Cotzias, G.C., Manganese, in: "Mineral Metabolism," C.L. Conar and F. Bronner, eds. (1962) Academic Press, New York, 404-442.

EXTRACELLULAR TRANSPORT OF TRACE ELEMENTS

Cornelis J.A. Van den Hamer

Department of Radiochemistry
Interuniversity Reactor Institute
2629 JB Delft
The Netherlands

INTRODUCTION

The following will be largely limited to plasma (serum) and to the
trace elements (T.E.'s) copper (Cu) and zinc (Zn). It will address three
questions, viz., a. what are the "normal" T.E. carriers in plasma, b.
which are the changes in carriers in case of, e.g., disease (either due
to the disease per se or to its treatment) and c. which is the relation
between the T.E.'s in plasma and the cell membrane.

COPPER IN PLASMA

Looking at the normal distribution of Cu in plasma, three Cu-
containing compounds (not necessarily carriers) are usually recognized:
the plasma protein ceruloplasmin, accounting for most of the plasma-Cu
($\sim$95% ; about 1 μg.ml^{-1} or 15 μmol.L^{-1}; about 0.3 mg protein.ml^{-1}), a
small amount of albumin-bound Cu ($\sim$ 5%; 0.05 μg.ml^{-1} or 1 μmol.L^{-1}) and
Cu complexed with amino acids ($<$ 1%). It is unlikely that in human
plasma another component, accounting for more than a few % of the total
Cu, would be present: treatment of plasma with antibodies against
ceruloplasmin precipitates 93 - 95% of the total plasma-Cu.

Ceruloplasmin

Ceruloplasmin is a blue glycoprotein with a molecular weight of
about 130,000 D and containing some 6 atoms of Cu, both ESR visible and
ESR unvisible (for a review see Frieden, 1979). It has been isolated in
rather pure form and is quite well characterized. Notwithstanding that
much is known about its physico-chemical and enzymatic properties and
the fact that it is often almost absent from plasma of patients with
Wilson's disease and elevated in several other diseases, its physiological role is not quite clear. Over the years many functions have been
suggested. Although it cannot be excluded that ceruloplasmin has more
than one function, possessing all properties attributed to it would be
too much of a good thing. It is an oxidase, acting on artificial substrates (polyphenols and aromatic amines like p-phenylenediamine) and
natural amines like adrenaline; it is reported to catalyze the oxidation
of Fe(II) to Fe(III) prior to uptake in cells; it has been mentioned as
Cu-transport protein, particularly for the transport of Cu to extrahepatic tissues; in times of stress it is synthesized by the liver at

higher rate and is therefore called an acute phase reactant; it could be the source of the bile-Cu and thus (because the bile is a major route of Cu-excretion) would play a function in the Cu-homeostasis. In support of this last function, one can make a small computation. The biological $t_{1/2}$ of ceruloplasmin in plasma is about 5.5 days (Fig.1). Assuming that 4% of the body weight of a standard man of 70 kg is plasma, that plasma contains 0.95 µg ceruloplasmin-Cu per ml and that about 50% of the ceruloplasmin is intravascular, the total body ceruloplasmin would account for

$$0.04 \times 70,000 \times 0.95 \times 2 = 5320 \ \mu g \ Cu.$$

The $t_{1/2}$ of 5.5 d means that every day about $1/8^{th}$ of the ceruloplasmin will be degraded, viz., $1/8 \times 5320 = 650$ µg Cu. This value matches roughly the daily bile-Cu losses. Yet, if ceruloplasmin were the source of the bile-Cu, the observation that the $t_{1/2}$ of ceruloplasmin is similar in controls, patients with primary biliairy cirrhosis and with Menkes' disease, would be unexpected.

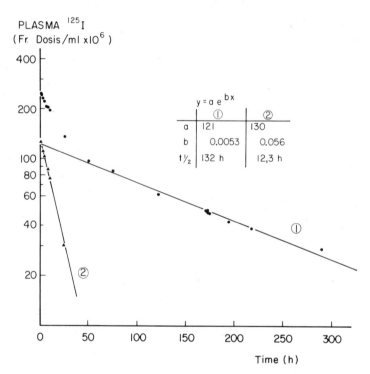

Fig. 1. The disappearance of intravenously injected ^{125}I-labeled human ceruloplasmin from the plasma of a control person as function of time post dose.

In normal plasma, but also in plasma from Wilson patients, a small amount of ceruloplasmin ($\sim$0.02 mg.ml^{-1}) is present which does not have its full complement of Cu. It reacts with antibodies against ceruloplasmin, but does not show enzyme activity.

Albumin

Albumin binds Cu very strongly: in human albumin the last three amino acids at the N-terminal end, viz., aspartic acid, alanine and histidine, contribute 4 N's surrounding the Cu in its divalent form. In contrast to ceruloplasmin, only one albumin molecule in 300 carries an atom of Cu. Like most proteins, albumin can also bind more than one atom of Cu. Those in addition of the one at the N-terminal end are, however, much less tightly bound. Moreover, those loosely bound Cu's can only be present in situations in which the plasma-Cu is higher than 30 µg Cu/ml. This highly unphysiological condition can only be seen in extreme conditions of poisoning. The high toxicity of such Cu is then explained by the low stability of the Cu-complex. Because the stability equals that of a complex of Cu with any average protein, the Cu is easily transferred from albumin to sites were it can cause damage. Such situation can occur in, e.g., patients with untreated Wilson's disease and lead to haemolysis.

Amino acids

A small amount of the plasma-Cu is bound to amino acids, particularly to histidine and cystine. This amount is in dynamic equilibrium with the albumin-bound Cu. Physiologically, the albumin-Cu and amino acid-Cu are probably indistinguishable.

ZINC IN PLASMA

A similar set of T.E.-binding compounds is known for Zn. About 1/3 of the plasma-Zn (0.3 µg.ml^{-1} or 5 µmol.L^{-1}) is present as α_2-macroglobulin, about 2/3 (0.6 µg.ml^{-1} or 10 µmol.L^{-1}) is bound to albumin and a small amount (<0.01 µg.ml^{-1} ?) is present in a low molecular weight form, predominantly bound to amino acids.

α_2-Macroglobulin

α_2-Macroglobulin is a protein with a molecular weight of about 720,000 and which contains probably 4 atoms of Zn per molecule. It has antiprotease properties. There is no evidence that the Zn is involved in this activity; the Zn probably stabilizes the protein conformation. This endogenous Zn cannot exchange with other Zn, but can be removed by dialysis against EDTA. The protein can then be reconstituted with either Zn or Mn.

Albumin

Unlike Cu, most of the plasma-Zn is bound to albumin. If this protein has a special binding site for Zn, its stability is not so extreme compared to the second and following binding sites as in case of Cu. The albumin-Zn is, like the albumin-Cu, in equilibrium with a small amount of Zn bound to amino acids, particularly to histidine and cysteine.

EXCHANGEABLE TRACE ELEMENT POOL

The transport form of most T.E.'s is the equilibrium mixture of T.E. bound to albumin and to amino acids. Seemingly, these two compounds form a distinct T.E. pool. But, apart from the albumin and the amino acids, a fairly large amount of additional T.E. may also belong to this pool. It consists of T.E.'s bound to, e.g., erythrocytes and membranes of tissue cells. The amino acids may act then as intermediates between the albumin and the aspecific binding sites on the membranes. In case of Zn a rough estimate can be made of this exchangeable or readily available pool. A patient had been given total parenteral nutrition (TPN) without added Zn. After three weeks she developed signs of Zn-deficiency: very low plasma-Zn and -alkaline phosphatase and typical skin lesions (Bos et al., 1977). Assume her normal dietary Zn intake was 10 mg.d^{-1} of which 30% was absorbed from the gut, then in 3 weeks

$$21 \times 10 \times 0.3 = 63 \text{ mg Zn}$$

would have been available to compensate for the unavoidable losses during the period in which now, because of the TPN, a state of Zn-deficiency was reached. Of course, at that stage the available Zn-pool would not be entirely emptied. On the other hand the TPN would have contained at least some Zn. An amount of 100 mg seems therefore a fair estimate of the available Zn-pool. From measurement of the specific activity of labeled Zn in plasma after an oral dose of, e.g, ^{69m}Zn, a similar value can be estimated. Of course, due to its nature, the exchangeable ("available") pool is poorly defined. Most of the body Zn, about 1.5 - 2 g, is locked away in muscle tissue and skeleton. From the latter it seems to be set free only in times of bone resorption. In the former it is present as integral part of enzyme and the like; withdrawing this Zn will impair the viability of the cell.

The exchangeable pool does not increase the amount of Zn available for transport to and from the tissues but, apart from acting as a buffer in times of Zn-influx or -depletion, it rapidly lowers the specific activity of (radio)isotopically labeled Zn, faking net uptake by the tissues.

ULTRAFILTRABLE ZINC IN DISEASE

The low molecular weight Zn-pool increases when excess amino acids, peptides or other low molecular weight materials are present in elevated concentration in the plasma. The former could be the case in, e.g., alcoholic liver cirrhosis; the latter in times of rapid tissue wasting. Because these forms of Zn are ultrafiltrable, they lead to increased urinary Zn losses. It is debated whether the extra Zn in the urine is "washed off" from the albumin or not. However, during tissue wasting, a good deal of Zn is set free from the tissues: 1 g muscle tissue will yield about 60 µg of Zn which, at that moment, will be superfluous and will also be lost in the urine (Cornelisse, 1985). Even less physiologic are losses due to some drugs. Depending on the metabolism of the drug, orally administered pharmaca may only enhance the uptake of a T.E. from the G.I.-tract or - after being absorbed - also increase the excretion via the urine in the form of low molecular weight T.E.-complexes.

UPTAKE IN CELLS

In an in vitro system, containing 0.2% human albumin but no amino acids, the Cu bound to albumin in the medium (ratio Cu:albumin = 0.5) is taken up by hepatoma cells (Van den Berg and Van den Hamer, 1984). Addition of physiologic amounts of histidine (15 µM) increased the amount of ultrafiltrable Cu, but had little effect on the uptake of Cu by the cells. Increasing the histidine ten till twentyfold, showed that the histidine-Cu can also contribute to the Cu-uptake if present in large enough amounts. The strongest uptake is seen with Cu bound to albumin in excess of the one atom of Cu bound at the N-terminal end of the albumin molecule.

CONCLUSION

The combination of albumin and amino acids acts as carrier of the T.E.'s Cu and Zn, and probably of other T.E.'s as well, to and from the tissues. The total exchangeable pool of T.E. with which the T.E. bound to the carrier is in equilibrium, should be taken into account when interpreting data of experiments with radiolabeled T.E.'s.

REFERENCES

Bos, L.P., Van Vloten, W.A., Smit, A.F.D., and Nube, M., 1977, Zinc deficiency with skin lesions as seen in acrodermatitis enteropathica, and intoxication with zinc during parenteral nutrition, Neth. J. Med., 20:263.

Cornelisse, C., 1985, Zinc Absorption and Retention in Man, with Special Emphasis on Surgical Patients, Thesis, Utrecht.

Frieden, E., 1979, Ceruloplasmin: the serum copper transport protein with oxidase activity, in: "Copper in the Environment. Part 2: Health Effects," J.O. Nriagu, ed., J. Wiley & Sons, New York.

Van den Berg, G.J., and Van den Hamer, C.J.A., 1984, Trace metal uptake in liver cells. 1. Influence of albumin in the medium on the uptake of copper by hepatoma cells, J. Inorg. Biochem., 22:73.

INCORPORATION OF COPPER INTO SUPEROXIDE DISMUTASE IN CULTURE

Edward D. Harris and Charles T. Dameron

Department of Biochemistry and Biophysics, and
The Texas Agricultural Experiment Station,
Texas A&M University, College Station, Texas 77843-2128

Transport of trace metals to cells and enzymes may be thought to occur in three phases: (1) an extracellular phase dominated by a centralized protein or binding factor that conveys the metal to the cell, (2) a membrane phase characterized by a membrane receptor or carrier protein that mediates the transfer of the metal-bound complex or the free metal through the membrane, and (3) an intracellular phase in which the metal binds to its cognate metalloenzyme giving rise to a biologically functional component. In our work, we have been interested in all three phases as they constitute a coordinated delivery system for copper.

Evidence has been provided for ceruloplasmin and/or albumin as transport agents for copper. In this study, we have tested each of these components separately in a model system, the transport of copper to superoxide dismutase (SOD) in aortic tissue.

MATERIALS AND METHODS

Chicks were made copper-deficient by established procedures (1). Generally, 10-12 days were needed to render the aortic CuZnSOD activity undetectable. Reactivation of the enzyme was accomplished by suspending the deficient aortic tissue in Waymouth's 752/1 growth medium, 24 h, 37°C, in the presence of a copper source. Prior to the incubation, the tubes were flooded with oxygen and sealed tightly. In some experiments, ^{67}Cu-labelled ceruloplasmin and albumin were used to study the direct transport of copper to the enzyme. Antibodies to purified CuZnSOD provided specific quantitation of the product. The activation of catalytic function was monitored by the pyrogallol assay (2).

RESULTS

The decrease in CuZnSOD activity in aorta is reversed rapidly by administering $CuSO_4$ i.p. to the animals. Reactivation by this method while specific for copper (3), cannot identify intermediates that take part in its transfer to the enzyme. To overcome this, we have used a defined culture medium, aortic tissue and various copper complexes as donors. Table 1 shows that this simple system was capable of restoring CuZnSOD activity to the deficient aortas, but only when a source of copper was present. That source could either be ceruloplasmin, albumin or free copper (most likely present as a complex with amino acids in the medium). Neither ceruloplasmin nor albumin appeared superior activators of the enzyme under the conditions employed.

Table 1. Factors Influencing Activation of CuZnSOD _in vitro_

ADDITIONS	SOD ACTIVITY (U/g)
None	1.1 ± 1.8
CuCl$_2$ (4°C)	3.3 ± 2.2
CuCl$_2$ (37°C)	16.3 ± 1.1
Cu-EDTA	5.0 ± 7.0
Ceruloplasmin	17.8 ± 2.3
Cu-Albumin	16.2 ± 1.4

Further comparative effects of copper donors were obtained with radioactive copper. ^{67}Cu-ceruloplasmin was prepared by an exchange reaction using 0.05% (w/v) ascorbate to assist in the displacement of the bound copper. The protein was treated with Chelex-100 and further purified by gel filtration to assure the ^{67}Cu was tightly bound to the protein. A similar procedure was used for preparing ^{67}Cu-labelled albumin. When the suspended aortic tissue was treated with ^{67}Cu-labelled ceruloplasmin for 24 hr, radioactivity was found in three soluble fractions derived from homogenates of the tissue (Fig. 1). Immunoprecipitable components responding to rabbit antichickSOD were found mainly in fractions 55 to 72. The boxes in the figure indicate the number of cpm appearing in immunoprecipitates from the peak tubes. The data show clearly that radioactivity was transferred from ceruloplasmin to CuZnSOD protein. A similar transfer of ^{67}Cu occurred when the tissues were incubated with the radiolabelled albumin (data

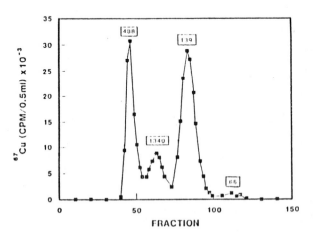

Fig. 1. Separation of ^{67}Cu-labelled Aortic Proteins. Sepadex G-75 was the medium. The proteins were labelled by reacting aortas with ^{67}Cu ceruloplasmin for 24 hr. Boxes indicate immunoprecipitable (CuZnSOD) counts in peak tubes.

not shown). The albumin immunoprecipitable cpm appeared in the second peak. No significant differences in the profiles or distribution of radioactive components was observed. In fact the assay could not determine whether ceruloplasmin, albumin or free copper was the superior donor.

DISCUSSION

The extracellular and membrane phases of copper transport bear directly on the results of this study. One may consider, for example, that copper complexes, protein or otherwise, penetrate the cell membrane via a receptor-mediated mechanism involving coated pits and vesicles or specific carrier proteins. The receptor-mediated option is consistent with the discovery of ceruloplasmin receptors in chick aorta (4), whereas the carrier option pertains to specific transport proteins that recognize and transport the free metal or low molecular weight complexes (5). Then, too, albumin is a copper-binding factor that may also take part in the transfer of the metal. This study was designed to determine which of these proteins or the free metal itself was most effective in the transport of copper to aorta CuZnSOD.

No superior transport factor emerged from the studies. While aortic CuZnSOD was observed to bind copper from ceruloplasmin, albumin and free copper, the results were equivocal regarding which was the most effective. It should be noted that ceruloplasmin levels in the activation reaction were close to the concentration of the protein in serum. The albumin-copper complex, however, was in considerable excess of what is found in serum. Thus, the system was quite artificial with regard to albumin and free copper. Nonetheless, the results show clearly that each of these components has the potential to serve in copper transport. The incorporation of copper into the enzyme apparently occurs at the apoenzyme stage and used copper from an ill-defined intracellular pool. The transport agents function to deliver copper to the pool and this mechanism does not discriminate with regard to donor. The nature of the intracellular pool is unknown.

ACKNOWLEDGEMENTS

Funding for this work was provided in part by USPHS NIH Grant DK-35920 and project H-6621 of the Texas Agricultural Experiment Station.

REFERENCES

1.   Rayton, J.K. and Harris, E.D. (1979) J. Biol. Chem. 254, 621-626.

2.   Marklund, S.L. and Marklund, G. (1974) Eur. J. Biochem. 47, 469-474.

3.   Dameron, C.T. and Harris, E.D. (Unpublished observations).

4.   Stevens, M.D., DiSilvestro, R.A. and Harris, E.D. (1984) Biochemistry 23, 261-266.

5.   Schmitt, R.C., Darwish, H.M., Cheney, J.C. and Ettinger, M.J. (1983) Am. J. Physiol. 244, G183-G191.

THE ROLE OF ALBUMIN IN COPPER UPTAKE BY HEPATOCYTES AND FIBROBLASTS

Harry J. McArdle, Sharon M. Gross, David M. Danks

The Murdoch Institute
Royal Children's Hospital
Flemington Road
Parkville, 3052, Victoria, Australia

The process of copper uptake by the liver has been extensively studied in rat liver slices, in rat hepatocytes in suspension culture and after a brief period of recovery. These studies have shown that histidine may be involved as an intermediary in the uptake process and that albumin may not be involved directly. However, little data is available on the steps involved in the process, or the mechanism that may operate in other species. Further, there are some inconsistencies in the data presented by other groups with respect to, for example, temperature dependence and metabolic specificity. Finally, in other systems, it has been clearly shown that a period of recovery is needed before hepatocytes isolated using collagenase express all the same membrane proteins as they do in vivo.

There is still considerable debate as to the mechanism of uptake by fibroblasts. Although there is data suggesting that ceruloplasmin is the copper carrier, other groups have suggested that the mechanism of uptake may be the same as in the liver, and have characterised the differences between Menkes' and normal cells using this type of system.

We are attempting to reconcile these different theses by examining different putative carriers and seeing if we can assign physiological roles to the different moieties. This abstract discusses primarily the role of albumin but also briefly considers histidine as a copper carrying amino acid.

In our hands, mouse hepatocytes would not survive in balanced salt solutions (HBSS) and required 10% FCS for optimal viability. Hence, in all these experiments, unless otherwise mentioned, the incubation medium for hepatocytes was HBSS with 10% FCS and HBSS without FCS for the fibroblasts. The results described for hepatocytes were essentially the same whether the cells are in suspension culture or on collagen-coated dishes.

Copper uptake by the hepatocytes was biphasic, with a small rapid binding component followed by uptake which was approximately linear over the next 60 min. At higher copper concentrations, the uptake deviated from linearity and tended towards a plateau. The experiments reported here relate to uptake levels after 30 min incubation.

Adding increasing amounts of unlabelled CuAlb resulted in a

decrease in the uptake of ^{64}Cu, suggesting the presence of a CuAlb receptor system. The kinetics of the system was complex, with linear transformations yielding biphasic curves. However, Hill transformations gave slopes of 1, indicating that the two sites were not interacting. Adding excess albumin without Cu also resulted in a decrease in copper uptake, suggesting that the albumin acted as a rate limiting reservoir rather than a substrate for uptake.

Adding increasing amounts of histidine gave an increase in the amount of copper taken up by the hepatocytes. This effect was not absolutely specific for histidine and other amino acids at serum concentrations could also stimulate uptake. While histidine stimulated copper uptake, copper inhibited histidine uptake, suggesting that CuHis could not be a substrate for the Na^{+}-dependent His transport system. Metabolic inhibitors had no effect on copper uptake, apart from N-ethyl maleimide, while they inhibited histidine uptake.

Thus, the model we would present for copper uptake by mouse hepatocytes is similar to that proposed by other workers for the rat. Copper is transferred from albumin to histidine and thence by a passive, carrier-mediated process to the cell. The carrier is a protein, which has a Pronase sensitive site on the exterior surface, and is stabilised in some manner by a disulphide bridge.

Copper uptake from CuAlb by fibroblasts was measured in simple HBSS. It too, was biphasic, and temperature dependent. In contrast, albumin binding was not affected by temperature, and did not increase with time. Albumin binding by the cells increased linearly with increasing concentration and the slope of the line was not altered by adding copper to the albumin. This suggested that there is no CuAlb recognition system. However, ^{64}Cu uptake from ^{64}CuAlb showed saturation, which suggests that there may be a Cu recognition system. Further, adding increasing albumin resulted in a decrease in ^{64}Cu binding and adding increasing Cu gave an increase, not a decrease in ^{64}Cu uptake from ^{64}CuAlb.

Finally, adding histidine to the incubation medium resulted in a decrease, not an increase, in the uptake of copper.

The results are very different from those found in the hepatocyte, and can be reconciled only by suggesting that the copper that is taken up by the cell is derived from the free copper pool in the incubation medium. This pool would not exist in vivo and hence these results are, we suggest, an artefact of the experimental system.

In conclusion, we have demonstrated that there are at least two different copper transport systems in mammalian systems, and that the different processes may be related to the differing ways in which the majority of the copper is treated by these cells. The mechanism of fibroblast copper uptake is not the same as the liver, and results achieved by workers using this system should be treated with extreme caution.

STRUCTURE AND FUNCTION OF TRANSCUPREIN, IN TRANSPORT OF COPPER BY

MAMMALIAN BLOOD PLASMA

Maria C. Linder, Kathryn C. Weiss, and Vu Minh Hai

Department of Chemistry and Biochemistry
California State University, Fullerton
Fullerton, CA

INTRODUCTION

Using 67Cu of high specific activity, we have shown previously in
rats that copper entering the blood from the intestine (or directly by
intraveneous injection) immediately binds to 2 components in the blood
plasma (Weiss and Linder, 1985). These are albumin, which has for some
time been known to have a single, high affinity site for Cu binding
(Lau and Sarkar, 1971), and another, new plasma protein we have named
transcuprein that has an apparent molecular weight of 270 kDa (Weiss
and Linder, 1985). Transcuprein rapidly exchanges copper with albumin,
and the specific activity of copper on both of these proteins diminishes
in parallel, and very fast, after administration of radioactive tracer.
Depletion of newly absorbed copper bound to albumin and transcuprein
with time is accompanied a deposition of copper in liver and kidney. The
kinetics of these changes are consistent with those of a precursor/pro-
duct relationship, implying that transcuprein and albumin are delivering
Cu to both of these organs after absorption from the diet. Transcuprein,
ceruloplasmin, and albumin (as well as components of low molecular weight
binding Cu) may be fractionated on columns of Sephadex G150. "Profiles"
of rat and human plasma fractionated in this manner have 10-15% of the Cu
associated with transcuprein. An approximately equal quantity is
associated with albumin, and the remainder is mainly associated with
ceruloplasmin, but about 10% is with low molecular weight components
(Wirth and Linder, 1985). Further work on the structure and function
of the new Cu transport protein is described below.

RESULTS AND DISCUSSION

Purification and Structure. Partial purification of transcuprein
has been carried out with a large variety of procedures, some of which are
much better than others. These include ammonium sulfate fractionation,
where transcuprein precipitates between 35-50% saturation; gel permeation
chromatography on Sephadex G150, where it elutes in the void volume, and
on Ultrogel AcA 34, where it elutes with an apparent molecular weight of
270 kDa; DEAE-cellulose chromatography, in phosphate, pH 7.0, where it
elutes at about 0.15M NaCl in a 0-0.5M gradient; and pseudoaffinity
chromatography (Affigel Blue), where it sticks tenaciously to Cu-chelate
affinity gels, and is difficult to remove without removing the metal from
the column. Using the best sequence of procedures so far, and tracking

the transcuprein with 67Cu (as well as by its characteristic elution in Sephadex or Ultrogel), the result is a preparation containing a major component with Rf 0.41 in disc PAGE (using a 5% separating gel and pH 8.8 Tris-glycine buffer). Three minor protein components are also present, with Rfs of 0.11, 0.24 and 0.60, respectively. The major component has subunits of Mr 80,000 after dissociation and electrophoresis in SDS. (The same results were obtained whether transcuprein was initially labeled in vivo or in vitro.) A big question is whether the major component is transcuprein, or whether we should be pursuing the minor components instead. To answer this question, we cannot use our standard nondenaturing electrophoretic techniques, as the buffers involved rapidly remove the Cu from transcuprein (and albumin). (In contrast, transcuprein Cu is very stable at pH 7-8, in phosphate buffer.)

Properties and copper binding. The specific binding of copper to partially purified transcuprein was measured in a nitrocellulose filter assay, after incubation of the protein with various concentrations of 67Cu-labeled Cu, in the presence and absence of a 100-fold molar excess of non-radioactive Cu-nitrilotriacetate (Cu-NTA), at pH 7.0, in 20mM phosphate-buffered saline. Preliminary studies indicated that binding was saturable and half maximal in the range of 10^{-9}M. However, in the system, albumin behaved quite similarly, and a Kd of 10^{-17}M has been established for this protein (at least for man) by equilibrium dialysis (Lau and Sarkar, 1971). The initial stoichiometry obtained also indicated that considerably less than 1 Cu atom per molecule was binding. This suggests that the transcuprein is more likely to be one of the minor components found in our purified preparations. The rate of Cu release from transcuprein was also studied and compared with that of albumin, in dialysis. Samples (2.0 ml) of radioactively labeled protein obtained from Sephadex G150 chromatography of ^{67}Cu-treated rat plasma (containing 1.5-2 million cpm), were placed in 500 ml buffer and dialyzed at 4° for minutes, hours and days. Aliquots (5.0 ml) were withdrawn periodically and measured for release of radioactive Cu. At the end point of dialysis, the ^{67}Cu remaining in the dialysis tubing was also measured. The results indicated that Cu was released at a very slow, linear rate, at pH 7-8, but at higher rates at lower and higher pHs. Overall, less than 3% of the Cu was released over 25 h at pH 7 and 8. The results for albumin were almost identical. The Cu off-rate was markedly accelerated by the presence of 0.25 ug Cu/ml (as Cu-NTA), for both proteins. Tris-glycine buffer, pH 8.8 (used in standard disc PAGE), also had a profound accelerating effect on the rate of Cu removal from transcuprein and albumin, consistent with repeated electrophoretic findings that ^{67}Cu does not remain with the proteins in our standard systems. It remains to be determined whether the concentration of Tris (higher than used in our pH 9-saline buffer) or the glycine is responsible. These studies confirm that the affinity of transcuprein for Cu is similar to that of albumin.

Detection of transcuprein by in vitro labeling and gel chromatography. Based on many observations that transcuprein (and albumin) can be labeled with radioactive copper in vitro, and that they rapidly exchange label (Weiss and Linder, 1985), we have begun to explore ways in which we might use in vitro labeling as a means of quantitating the amounts of transcuprein in various samples, if we know the albumin content. The approach used was first, to ascertain what amounts of Cu (labeled with ^{67}Cu or ^{64}Cu) could be added to samples in vitro that would allow detection of transcuprein in the presence of various amounts of albumin; second, to work towards using the proportion of radioactivity (versus that on albumin) as a means of assessing transcuprein content. For all of these studies, Cu(II) was added either as the chloride or as the NTA complex (in varying amounts) to 1.0 ml samples of whole plasma or serum of the rat and man, or to 1.0 ml samples

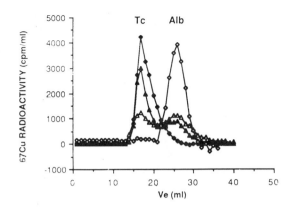

Fig. 1. Distribution of added [67]Cu radioactivity after in vitro
addition of 2-20 ng Cu to partially purified transcuprein or albumin
alone, or to various mixtures of the two. Sample applied to Sephadex
G150.

Table 1. Distribution of [67]Cu between transcuprein and albumin,
upon addition to rat plasma, in vitro

| | Percent Radioactivity on | |
Copper added(ng)[1]	Transcuprein	Albumin
1	35	65
20	31	69
200	28	72
2000	16	84
2700	2	98

[1] 0.1 ml 0.9% NaCl containing various amounts of Cu (as[67]Cu-NTA;
1:1 molar ratio) added to 1.0 ml rat plasma 30-90 min before
Sephadex G150 chromatography.

of partially purified rat transcuprein and albumin (or either alone).
As suggested by previous data (Weiss and Linder, 1985), the specific
activity of the added Cu, but more particularly the amount of Cu added,
determined what percentage bound to transcuprein versus albumin, in
whole plasma or serum (Table 1). At the lowest concentrations (in the
range of 1-2ng/ml) almost exactly one third bound to transcuprein and
two thirds to albumin, and almost none to other components that could be
separated in gel chromatography. At concentrations almost 1000-fold
higher (2 or more ug/ml), radioactivity on transcuprein was proportionately very low. (There was also a quite prominent labeling of low
molecular weight components.) When samples of partially purified transcuprein and albumin were mixed, the proportions of label varied as
expected, with almost all of the radioactivity being associated with
transcuprein at low albumin levels, and the reverse occurring with increasing albumin levels (Fig. 1). Thus the amount of radioactivity
associated with the transcuprein peak in Sephadex G150 chromatography
does indeed reflect the amount of transcuprein present in the sample.
We expect that we will be able to develop at least an approximation
assay for transcuprein (Cu binding sites) based on this approach, using
a specific quantity of added Cu (as [67]Cu), and knowing the concentration
of albumin present in the sample.

143

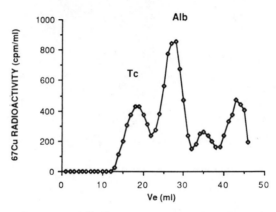

Fig. 2. Elution of copper binding components labeled with [67]Cu in portal blood plasma, 30 min after administration of radiocopper to a tied-off intestinal segment of an anesthetized rat. Plasma was applied to Sephadex G150 chromatography. The first peak elutes in the position of transcuprein, the second albumin.

Transcuprein in portal blood. Using this approach, we have also obtained evidence that transcuprein is present in portal blood plasma in similar or greater amounts to that in plasma taken from the blood of the vena cava, after passing through the liver. Fig. 2 shows the radioactive labeling of transcuprein, albumin and low molecular weight components in portal plasma, upon administration of 62ng Cu (as [67]Cu-NTA) to the lumen of a 7 cm, tied-off segment of upper small intestine in an anesthetized rat. In this example, more than one third of the radio-activity in albumin + transcuprein was with the latter, a greater proportion than is found with samples of vena caval blood labeled in vitro under optimal conditions (with very little Cu addition). (Considerable [67]Cu also appeared with the low molecular weight fraction). While the proportion on transcuprein was not as high in all samples tested so far, our results demonstrate that transcuprein is indeed present in the portal blood and that it is there in similar or in even higher amounts than in blood leaving the liver. Although this disagrees with a recent report of Gordon et al. (1987), we believe their chromatograms (especially at 30 min) do show a transcuprein peak in portal blood plasma of rats given oral [67]Cu, which they may have assumed was ceruloplasmin. In our experience, ceruloplasmin is not appreciably labeled until later times.

REFERENCES

Gordon, D., Leinart, A.S. and Cousins, R. (1987) Am. J. Physiol. 252, E1-E7
Lau, S.I. and Sarkar, B. (1971) J. Biol. Chem. 246, 5938-5
Weiss, K.C. and Linder, M.C. (1985) Am. J. Physiol. 249, E77-E88
Wirth, P.L. and Linder, M.C. (1985) J. Nat. Cancer Inst. 75, 277-284

DIFFERENCES IN CU-TRANSPORT BY HEPATOCYTES AND FIBROBLASTS

G. L. Waldrop, F. Palida, M. Hadi, P. Lonergan,and
M. Ettinger

Department of Biochemistry
State University of New York at Buffalo
Buffalo, New York 14214

INTRODUCTION

The liver rapidly and preferentially takes up Cu when administered orally or intravenously. Both kinetic and thermodynamic contributions may be involved. Kinetic factors are expected to determine the relative rates of Cu-transport by different cell-types. Thermodynamic factors may help determine the amounts of copper accumulated at steady-state. Kinetic studies with hepatocytes suggested that albumin-Cu was not directly available for Cu-uptake[1]. Histidine increased Cu-uptake from media which also contained albumin.[1] Apparently, the His_2Cu complex which forms when histidine is in excess delivers Cu to a transport protein, and Cu is transported as the free ion. Cu-transport by fibroblasts is reported here as an example of Cu-transport by an extrahepatic cell type which is relevant to Cu-metabolism and Menkes disease.

RESULTS

The Km and Vmax values from initial rate data (30S) for free-^{64}Cu(II) uptake were similar for fibroblasts and hepatocytes. K_m = 6.9 ± 0.2 and 11 ± 0.6 μM, Vmax = 2.6 ± 0.05 and 2.7 ± 0.6 nmol min^{-1} mg $protein^{-1}$, respectively. However, the Vmax for Cu-uptake by fibroblasts from His_2Cu (0.5 nmole/min/mg protein) was 5-times less than the Vmax for free Cu (2.6 nmol/min/mg protein). This effect was not observed with hepatocytes. As with hepatocytes, albumin markedly decreased the initial rate of Cu-uptake by fibroblasts by complexing extracellular Cu and decreasing the effective Cu-concentration. However, the effect of histidine on Cu-uptake in the presence of albumin differed in the two cell types. Hepatocytes and fibroblasts were incubated with equimolar Cu(II) and albumin for 1h with and without histidine. While histidine increased Cu-uptake by hepatocytes (2 fold) under these conditions, histidine decreased Cu-uptake by fibroblasts (2 fold). This dramatic difference can be attributed to the differences in Vmax for Cu-uptake from His_2Cu by hepatocytes and fibroblasts.

While kinetic differences may account for faster Cu-uptake by the liver, thermodynamic factors may contribute to the Cu-levels attained at steady-state. As a test for these contributions, hepatocytes were incubated with 5 μM Cu(II) and fibroblasts with 10 μ Cu(II) with or without 10 μM albumin for 3h. Fibroblasts accumulated only 10% as much ^{64}Cu from albumin-containing media as from albumin-free media while hepatocytes were able to take up nearly as much Cu in the presence of albumin as from albumin-free media (100%). Experiments with ^{125}I-albumin plus ^{64}Cu confirmed that both cell types took up Cu as the free ion without concomitant albumin uptake.

Since Cu-transport is passive in both directions,[2] the above results are consistent with thermodynamic equilibrium between Cu, extracellular, and intracellular Cu-ligands. It can be shown that at constant cell and albumin concentrations, the ratio of Cu retained by the cells to Cu bound to extracellular albumin should be independent of Cu-concentration if this hypothesis is valid. To test this hypothesis, Cu was varied from 2.5 to 10 μM at 10 μM albumin, and ^{64}Cu-uptake was determined after 3h incubation. The experimentally determined ratio was constant (.28 ± .04), within the experimental error.

DISCUSSION

The results were consistent with both kinetic and thermodynamic contributions to rapid, preferential Cu-uptake by the liver. Ligand displacement of Cu from His_2Cu apparently occurs more rapidly at the putative transport protein of hepatocytes than at fibroblasts. However, the similar kinetic parameters for free Cu(II) uptake suggest iso-Cu-transport proteins are present in fibroblasts and hepatocytes which are homologous. Hepatocytes and fibroblasts show equilibrium behavior with respect to Cu-uptake. The greater Cu-uptake by hepatocytes can be attributed to higher concentrations and/or higher binding constants of intracellular Cu-ligands in hepatocytes than fibroblasts.

REFERENCES

1. Darwish, H.M., Cheney, J.C., Schmitt, R.C. and Ettinger, M.J., Am. J. Physiol. 246, G72-G79, 1984.
2. Darwish, H.M., Schmitt, R.C., Cheney, J.C., Ettinger, M.J., Am. J. Phyisol. 246, G48-G55, 1984.

NEONATAL AND ADULT [64]COPPER METABOLISM IN THE PIG AND ITS RELATIONSHIP

TO COPPER METABOLISM IN WILSON'S DISEASE

Surjit K.S. Srai, Colin Bingle and Owen Epstein

Department of Medicine, Royal Free Hospital School of
Medicine, London, NW3 2PF

INTRODUCTION

The copper profiles of the neonatal mammals including man are
indistinguishable from Wilson's disease ie. high liver copper, low plasma
copper and caeruloplasmin. Soon after birth there is a switch to the adult
mode of copper metabolism, liver copper falls and both plasma copper and
caeruloplasmin concentrations increase. We have previously postulated that
Wilson's disease is caused by genetic failure to switch from the fetal to
adult mode of copper metabolism[1]. Wilson's disease patients handle radioactive copper in a characteristic manner which aids diagnosis[2]. To test
the hypothesis that WD is caused by perpetuation of the fetal mode of copper
metabolism into childhood, we have studied copper profiles and ^{64}Cu handling
in neonatal and adult pigs.

MATERIALS AND METHODS

Four adult and six neonatal large white pigs (24-48 hr) were used. At
time zero a bolus of ^{64}Cu-acetate (20 ug/kg body weight) was injected intravenously. Regular blood samples were taken and bile was collected for up
to 24 hours. At termination of the experiment, liver was removed, homogenised and centrifuged at 100,000 g for 60 min to obtain soluble supernatant.

RESULTS AND DISCUSSION

Neonatal plasma copper and caeruloplasmin concentrations (mean $\pm$ SEM;
15.00ug/100ml $\pm$ 2.41; 3.26u/100ml $\pm$ 1.01) respectively were significantly
lower than the adult (98.75ug/100ml $\pm$ 1.25; 59.60 $\pm$ 11.10). Neonatal liver
copper concentration (72.0lug/g wet wt $\pm$ 6.34) was twice the level of adult
(32.80 $\pm$ 1.29). Therefore the pig is a suitable experimental animal as
neonatal pigs have copper profiles similar to human neonates.

Following the injection of ^{64}Cu into adult pigs, plasma ^{64}Cu activity
fell to (63% of the basal activity) by 10 min and reached a nadir (14%) at
about 10 hours and then increased slowly over the next 14 hrs to 23% of the
basal level. In neonates, the fall in plasma activity was more rapid (36%)
by 10 min and 3% by 10 hrs and furthermore there was no secondary rise in

plasma activity. The secondary rise in plasma ^{64}Cu activity in adult pigs is similar to that reported in man and is due to incorporation of the isotope into newly synthesised caeruloplasmin and its secretion into blood. The secondary rise did not occur in piglets, similar to that reported in Wilson's disease patients[2].

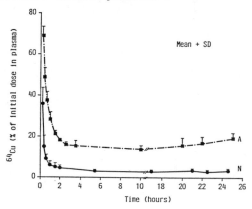
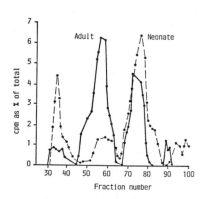

Fig. 1. Mean ^{64}Cu activity in plasma of adult and neonatal pigs.

Fig. 2. Sephadex G-75 ^{64}Cu elution profile of adult (-) and neonatal (- - -) cytosol.

Despite increased liver copper concentrations, the piglets excreted less cold copper in bile than adult. Immaturity of this export pathway is also reflected in the biliary excretion of copper isotope with seven times more ^{64}Cu excreted in adult bile compared to piglets.

After fractionation of neonatal and adult liver soluble supernatant on Sephadex G-75, ^{64}Cu activity was associated with three distinct peaks. Peak I corresponded to the void volume (MW 60,000) peak II (MW 32,000) only peak to exhibit CuZn-SOD activity, peak III corresponded to the molecular weight of metallothionein. A clear difference emerged in the association of copper with copper binding proteins. In adults, the CuZn-SOD peak contained the highest proportion of isotope (figure 2), whereas in piglets, most of the activity was associated with metallothionein and to a lesser extent, the void volume, with little activity detected in the CuZn-SOD peak. It is of interest that in Wilson's disease there is also a failure to incorporate the copper into a cytosolic copper binding protein (MW 40,000). Although this protein has not been characterised in WD its elution profile on Sephadex G-75 indicates that this is probably CuZn-SOD[2].

The striking similarities between copper profiles of the human neonate and WD and demonstration that like WD, the neonatal pig has reduced bile copper excretion, impaired caeruloplasmin secretion and impaired incorporation of ^{64}Cu into a cytosolic protein (MW +40,000) lends support to the hypothesis that Wilson's disease is caused by a regulator gene mutation causing developmental arrest in copper metabolism.

REFERENCES

1. S. K. S. Srai, A. K. Burroughs, B. Wood, and O. Epstein. The ontogeny of liver copper metabolism in the guinea pig: clues to the aetiology of Wilson's disease. Hepatology 6:427-432 (1986).
2. I. Sternlieb, C. J. A. Van den Hamer, A. G. Morell, S. Alpert, G. Gregoriadis, and I. H. Scheinberg. Lysosomal defect of hepatic copper excretion in Wilson's disease. Gastroenterology 64:99-105 (1973).

THE ROLE OF SELENIUM IN SPECIAL ENDEMIC DISEASES AND CANCER IN CHINA

Junshi Chen

Institute of Nutrition and Food Hygiene
Chinese Academy of Preventive Medicine
29 Nanwei Road, Beijing, China

INTRODUCTION

The essentiality of selenium in animals has been well documented both in experimental animals (Schwarz and Foltz, 1957) and livestocks (Muth et al., 1958) in 1950's. However, the direct evidence of an etiological relationship between selenium and human disease had not been established till the preventive effect of selenium on Keshan disease (KD), an endemic cardiomyopathy, was shown in 1975 (Keshan Disease Research Group of the Chinese Academy of Medical Sciences,1979). Recently, some evidence has emerged that another endemic disease of osteoarthropathy in nature, the Kashin-Beck disease, is also related to selenium deficiency. The reasons for China being the place of more than one endemic human disease related to selenium deficincy are that: (a) farmers (80% of the total Chinese population) live on rather monotonous dietary patterns mainly composed of plant foods, i.e. grains and vegetables; (b) the major dietary components are produced locally, and (c) the population in rural areas are non-mobile. Therefore, the low selenium status of local soil readily transfers through the diet to the local inhabitants and thus affects human health. Blood and food analysis in these endemic areas revealed that the selenium level in blood and foods are much lower than the values reported from New Zealand and Finland (Table 1). On the other hand, in consistent with studies carried out in the western countries, the interest of studying the possible effects of selenium in human cancers is growing rapidly in China and some encouraging epidemiological data have been obtained (Yu et al.,1985). This paper is going to describe some recent advances in studies on the relationship between low selenium and Kashin-Beck disease and cancer in China.

Table 1. Selenium Status of Inhabitants in Low Selenium Areas of China as Compared with New Zealand and Finland

Country	Blood (ng/ml)	Plasma (ng/ml)	Dietary Se Intake (μg/day)	References
China	18	22	7-11	Yang, 1985
New Zealand	59	48	34-70	Robinson and Thomson, 1986
Finland	69	56	20-30	Westmark, 1977

KASHIN-BECK DISEASE

Kashin-Beck disease (KBD) is usually referred as the sister disease of Keshan disease, because their geographical distribution is quite similar, Although not identical. However, the KBD involves cartilage and the KD involves myocardium, the clinical picture of these two endemic diseases are very different. According to the summary report of a World Health Organization Meeting on Kashin-Beck Disease hold in Beijing in 1985 (to be published by the International Programme of Chemical Safety/WHO), KBD is an endemic osteoarthropathy mainly involves children of 5-13 years-old and the basic pathological changes are the multiple degeneration and necrosis of articular cartilage and growth plate, which can result in permanent disabilities. The condition results in enlarged joint (especially of the fingers, toes and knees); shortened fingers, toes and extremities; and in severe cases dwarfism.About 2 million people are affected by this disease and there are more than 30 million people who live in endemic areas of the People's Republic of China are at direct risk of acquiring this disease.

Selenium deficiency is suggested to be a major cause of KBD (Yang, 1986). Studies on the selenium status in areas where KBD occurs along (without KD) are most interesting. Data shown in Table 2 and 3 indicated that the selenium status of general population in KBD endemic areas was as poor as that in KD areas and selenium contents of staple grains from KBD areas were also significantly lower than those from non-endemic areas.

Table 2.   Hair Selenium Concentration of Residents in KBD Endemic and Nearby Non-endemic Areas in Shaanxi Province [a]

Sites	Endemic	N	Hair Selenium (ppm)
Erlintu	+	19	0.080±0.011 [b]
Niujialiang	+	25	0.077±0.008
Mizi	−	25	0.195±0.011

a. Adapted from Yang, 1986.
b. Mean±SE.

Table 3.   Selenium Concentration (ppm) of Staple Grains Produced in KD, KD+KBD and Non-endemic Areas [a]

Areas	Corn	Wheat	Rice	Soyabean
KD	0.006±0.001[b] (58)[c]	0.008±0.001 (59)	0.008±0.001 (59)	0.014±0.002 (51)
KD+KBD	0.004±0.001 (143)	0.006±0.001 (72)	----	0.008±0.001 (98)
KBD	0.005±0.001 (24)	0.006±0.001 (26)	----	----
Non-endemic	0.029±0.004 (79)	0.038±0.004 (110)	0.043±0.005 (44)	0.063±0.011 (40)

a. Adapted from Yang, 1986.
b. Mean±SE.
c. Number of samples analyzed.

The therapeutic effects of selenium on KBD was first reported by Li (1979) in Gansu. Patients were treated with weekly oral doses of 0.5 to 2.0 mg of sodium selenite together with vitamin E injection for a peroid of 3 to 6 months. Out of the 224 cases, 187 (83%) showed improvement of the metaphyseal lesion in X-ray films. Later, in Shaanxi province after weekly oral administration of sodium selenite to KBD patients for 1 year, Liang (1985) reported that 81.9% of the treated group (n=166) was improved in X-ray film and none of them getting worse, while 39.6% of the control group (n=159) improved and 18.9% worsen.

However, the effects of selenium in the prevention of KBD were not consistent in the available reports. Liang (1985) used selenium fortified table salt (sodium selenite 1:60,000) for supplementation. Within 5 years, the incidence rate decreased from 65.9 to 46.8% in the selenium treated group and from 75.7% to 54.6% in the control group. No significant difference was observed betwen these two groups. On the other hand, Yin et al. (1985) increased the selenium content of wheat to 3 times by spraying sodium selenite solution on growing wheat crops. After two years, the incidence rate in children who consumed the selenium treated wheat dropped from 63 to 42% and the abnormal mataphysis rate from 59 to 32%; while the changes of both rates were not significant in the control group. In order to draw definite conclusions on the preventive effects of selenium in KBD, further intervention studies with larger sample size are needed.

SELENIUM AND CANCER

In recent years, there is a growing interest in the relationship between selenium and cancer. This is because the anticarcinogenic effect of selenium is consistently shown in both animal experiments and epidemiological studies (Combs and Combs, 1986). However, most of the ecological studies published so far have been criticised for having uncontrolled confounding factors, such as: (a) the insufficient correlaton between the selenium status of local residents and the selenium concentration of locally produced crops (Allaway, 1972), because of the diversity of food sources; (b) the ill-defined migraton habit of local population; and (c) the large number of other dissimilarities in the areas investigated, in addition to selenium. In contrast, China offers an unique setting for the study of interrelationships between selenium status and cancer mortality for the following reasons: (a) the majority of Chinese population is located in rural areas where migration is very limited; (b) they live on rather simple dietary patterns mainly comprised of locally produced foods; and (c) there is a wide variation of cancer mortality within China, thus providing an ideal model for ecological studies.

Yu et al. (1985) reported that the selenium levels of whole blood from healthy adult subjects in 24 regions were inversely correlated with age-adjusted total cancer mortality (r=-0.64, p<0.01 for males; r=-0.60, p<0.01 for females). When the data on selenium levels were compared with the age-adjusted cancer mortality rates by organ sites, statistically significant negative correlations were found, particularly for esophageal cancer (r=-0.66, p<0.01 for males; r=-0.69, p<0.01 for females) and gastric cancer (r=-0.64, p<0.01 for both sexes), which altogether comprise about 50% of total cancer deaths in China. When the blood selenium levels were classified into low, medium and high selenium areas, significantly higher mortality rates of esophagus, stomach and liver cancers were observed in areas with low blood selenium levels. In the same paper, Yu et al. (1985) showed that in Qidong county, one of the high risk areas for liver cancer, significant inverse correlations were found between the selenium contents of staple grains and liver cancer mortality rates of the 43 communes (Table 4). Maize and barley corn are the main staple foods in Qidong county and most of the dietary selenium in the local diet is derived from these two crops.

Table 4.   Grain Selenium Content in Communes with Different Age-adjusted Liver Cancer Mortality Rates in Qidong County[a]

| No. of Communes | Cancer Mortality per 100,000 | Selenium Content(ppm) | |
		Maize	Barley Corn
8	15–39	15.1±4.4[b](40)[c]	42.3±1.9(39)
17	40–49	13.5±2.2 (85)	24.7±7.8(84)
17	50–59	12.7±1.6 (81)	23.5±4.9(80)
1	61	9.7 (5)	18.1 (5)

a. Adapted from Yu et al., 1985.
b. Mean±SD.
c. Sample size.

Chen et al.(1985) carried out an ecological study in China covering 65 counties with wide ranges in cancer mortalities for 7 major cancer sites; and the relationships between a number of nutrients, including selenium, and cancer mortality were studied. Plasma selenium concentrations and glutathione peroxidase (GSHpx) activities show wide range of levels among the 65 counties (Table 5) and there is a good correlation between plasma selenium and GSHpx activity ($r=0.58$, $p<0.01$). Preliminary statistical analysis show that an inverse correlation was found between plasma selenium (but not GSHpx) and the 35-64 years-old age truncated mortality rates of esophageal and gastric cancer in both sexes. The correlation coefficients were -0.44 ($p<0.01$) and -0.28 ($p<0.02$) for esophageal cancer in males and females respectively; and -0.28 ($p<0.02$) and -0.39 ($p<0.01$) for gastric cancer. Studies on the interactions between selenium and other nutrients (e.g. lipids, vitamins, dietary fibers etc.) as well as several environmental carcinogens (e.g. N-nitroso compounds, aflatoxin etc.) are under way.

Although the results of the above Chinese ecological studies are consistent with those in the western literatures, there is still not enough evidence to prove the etiological role of selenium in cancer. For this purpose, a well controlled intervention study on esophageal cancer, which provides 50 μg of sodium selenite daily to 1700 subjects with precancerous lesions and to 15,000 general population subjects of 40-69 years-old, was initiated in 1985 in Linxian (Blot et al., 1985). Since the 40-69 years-old age truncated mortality rates of esophageal cancer in Linxian is as high as 460 per 100,000, it is expected that some preliminary results will be seen within 5 years.

Table 5.   Plasma Selenium Concentrations and GSHpx Activities in 65 Survey Counties

		Mean	Min.	Max.	SD
Plasma Se (ug/dl)	M	82	24	135	27
	F	79	20	134	27
GSHpx (unit)[a]	M	3.97	2.53	5.42	0.64
	F	3.82	2.33	5.23	0.68

a. nmol NADPH oxidised/10 μl/min.

REFERENCES

Allaway, W.H., 1972, An overview of distribution patterns of trace ele-
    ments in soils and plants, Ann. N.Y.Acad. Sci. 199:17.
Blot, J. B. and Li, J. Y., 1985, Some considerations in the design of a
    nutriiton intervention trial in Linxian, People's Republic of China,
    Natl. Cancer Inst. Monogr. 69:29.
Chen J., Peto, R., Li, J. and Campbell, T.C., 1986, Nutritional status
    and cancer mortality in China, in: "Proceedings of the XIII Inter-
    national Congress of Nutrition, 1985", T.G. Taylor and N.K. Jenkins
    eds., John Libbey, London. Paris.
Keshan Disease Research Group of the Chinese Academy of Medical Sciences,
    1979, Observations on the effect of sodium selenite in the prevention
    of Keshan disease, Chinese Med. J. 92:471.
Li, C.Z., 1979, Observations on the efficacy of the treatment of 224
    Kashin-Beck disease patients with selenite and vitamin E, Chinese
    Med. J. 59:169. (in Chinese)
Liang, S., The prophylactic and curative effects of selenium in combatting
    of Kashin-Beck disease, presented at the IPCS/WHO Meeting on Kashin-
    Beck Disease, Beijing, 1985.
Muth, O.H., Oldfield, J.E., Remment, L.F. and Schubert, J.R., 1958, Effects
    of selenium and vitamin E on white muscle disease, Science 128:1090.
Robinson, M.F. and Thomson, C.D., 1986, Selenium status of the food supply
    and residents of New Zealand, in "Proceedings of the Third Interna-
    tional Symposium on Selenium in Biology and Medicine", G.F. Combs, Jr.,
    J.E. Spallholz, O.A. Levander, and J.E. Oldfield eds., Avi Publ. Co.,
    Westport, Connecticut.
Schwarz, K. and Foltz, C.M., 1957, Selenium as an integral part of factor
    3 against dietary necrotic liver degeneration, J.Am. Chem. Soc.
    79:3292.
Westmark, T., 1977, Selenium content of tissues in Finnish infants and
    adults with various diseases, and studies on the effects of selenium
    supplementation in normal ceroid lipofuscinosis patients, Acta
    Pharmacol. Toxicol. 41:121.
Yang, G.Q., 1985, Keshan disease: an endemic selenium related deficiency
    disease, in "Trace Element Nutrition in Children, Nestle Nutrition",
    Vol.8, pp.272, R.K. Chandra ed., Vevey/Raven, New York.
Yang, G.Q., 1986, Research on selenium-related problems in human health
    in China, in: "Proceedings of the Third International Symposium on
    Selenium in Biology and Medicine", G.F. Combs, Jr., J.E. Spallholz,
    O.A. Levander and J.E. Oldfield eds., Avi Publ. Co., Westport,
    Connecticut.
Yin, P., Guo, X., Zhang, S. and Bai, C, 1985, A comparative study of
    comprehensive method and Se-fortified wheat for prevention and therapy
    of Kashin-Beck disease, presented at the IPCS/WHO Meeting on Kashin-
    Beck disease, Beijing, 1985.
Yu, S.Y., Chu, Y.J., Gong, X.C. and Hou, C., 1985, Regional variation of
    cancer mortality incidence and its relation to selenium levels in
    China, Biol. Trace Element Res. 7:21.

WILDLIFE AS INDICATORS OF ENZOOTIC SELENIUM DEFICIENCY

Duane E. Ullrey

Department of Animal Science
Michigan State University
East Lansing, Michigan 48824

The objective of this review was to assess the usefulness of
selenium analyses of tissues from wild animals in defining regional
selenium differences, particularly where commercial agriculture is not
commonly practiced.

SELENIUM IN SOILS AND PLANTS

Selenium occurs in the earth's crust with an average abundance of
about 0.09 ppm (Lakin, 1972). Although rarely found in its native state,
it is a major constituent of 40 minerals and a minor constituent of 37
others, mostly sulfides (Cooper et al., 1970). Selenium occurs in
greatest abundance in igneous rocks, but high concentrations are also
found in sedimentary rocks such as shales, sandstones, limestones and
phosphorite rocks.

Since sedimentary rocks cover more than three-quarters of the
earth's land surface, they are the principal parent materials of soils
(Lakin and Davidson, 1967). Most soils contain 0.1-2 ppm selenium
(Swaine, 1955). Seleniferous soils often contain more, but they are not
always associated with selenium toxicity in animals. Toxic seleniferous
soils are usually alkaline, contain free calcium carbonate, and occur in
regions of low rainfall. The dominant selenium form is water-soluble
selenate. Nontoxic seleniferous soils are usually acid and contain
ferric hydroxide that renders selenium insoluble and poorly available to
plants. Low-selenium soils may be formed from recent volcanic deposits,
granites, very old metamorphic rocks, or sedimentary rocks that predate
the Cretaceous peroid of selenization.

Selenium in plants is a function of selenium concentration and
availability in soils, and of plant species. If soils are low in
selenium or if soils are acid and selenium is in a relatively insoluble
form, plant uptake is minimal. Certain plant species tend to concentrate
selenium while others accumulate surprisingly low levels even when
growing on seleniferous soils (Beeson and Matrone, 1972). Thus, analyses
of soils are of limited usefulness in defining selenium-low, adequate or
toxic regions. As a consequence, a number of researchers have defined
these regional differences by analyzing domesticated forages and grains
(NRC, 1983). Regions considered low are those where approximately 80%
of all forage and grain contains <0.10 ppm selenium. Regions considered

variable have >0.10 ppm selenium in about 50% of all forage and grain.
Regions considered adequate have >0.10 ppm selenium in 80% or more of all
forage and grain. It should be mentioned that 2.0 ppm selenium in the
diet has been proposed as a maximum tolerable level for domestic animals
(NRC, 1980).

SELENIUM IN DIETS AND TISSUES OF DOMESTIC ANIMALS

The relationship between natural dietary selenium and muscle
selenium concentration was well established in swine by Ku et al. (1972).
Data gathered from 13 states in which dietary selenium levels were 27-493
ppb and swine longissimus muscle selenium levels were 118-1,893 ppb (dry
basis) revealed a highly significant (P<0.01) correlation (r=0.96)
between them. Studies with cattle and sheep (Ullrey et al., 1977) also
demonstrated a positive relationship between natural dietary selenium
concentrations and selenium concentrations in serum, stermomandibularis
muscle and liver. Others have produced similar evidence in domesticated
herbivores (NRC, 1983), including camels (Awad and Berschneider, 1980),
llamas (Espinoza et al., 1982) and water buffalo (Gazia and Wegger,
1980).

SELENIUM IN DIETS AND TISSUES OF WILD ANIMALS

It seems reasonable that tissues of wild animals would respond to
varying dietary selenium concentrations in a like manner. That this is
true in wild or semidomesticated herbivores has been established in moose
(Franzmann et al., 1976; Froeslie et al., 1984), reindeer (Froeslie et
al., 1984; Kurkela and Kaantee, 1979; Westermarck and Kurkela, 1979), red
deer (Froeslie et al., 1984), blesbok and bontebok (Turkstra et al.,
1978), and woodchucks (Fleming et al., 1977). The latter species was
trapped in a selenium-deficient area of New York state, and typical food
plants contained 0.02-0.05 ppm selenium (dry basis). Thirteen of the 24
woodchucks trapped showed varying degrees of white muscle disease.

Presumably, a dietary habit of carnivory and the high selenium
concentration in prey species consumed accounts for the relatively high
selenium levels found in tissues of alligators (Lance et al., 1983) and
mink (Norheim et al., 1984).

CERVIDS AS INDICATORS OF GEOGRAPHICAL SELENIUM STATUS

If one were to select a wildlife family that might be used to
monitor geographical selenium status of significance for terrestrial
herbivores, the family Cervidae would be a good candidate. The deer
family is widely distributed over the continents of the earth. Its
members are commonly hunted or are semidomesticated, and, as a
consequence, tissues for analysis may be easily obtained. All cervids
are herbivorous ruminants, and while dietary habits may vary somewhat,
their similarities in physiology warrant serious consideration of this
family as a related group of signal species.

Among the cervids, the white-tailed deer (Odocoileus virginianus)
may be uniquely useful. In the Western hemisphere, it is found from
southern Canada, through much of the United States, Mexico, Central
America, and into the northern regions of South America (Baker, 1984).
It is selective in its dietary habits and consumes tender, young plant
parts if it has an option. Since these plant parts are generally higher
in protein than woody, more mature plant parts, and since selenium is
largely associated with protein in plant tissues, geographical variation
in plant selenium concentration is likely to be reflected in white-tailed
deer tissues.

Table 1. Selenium in White-Tailed Deer Muscle[a]

Range of Se ppb	No. of deer	Percent of total	Mean Se ppb
0-99	9	8	73
100-199	70	64	143
200-299	27	25	224
300-399	2	2	345
400-499	1	1	490
50-490	109	100	164

[a]Dry basis.

In a study with captive white-tailed deer (Brady et al., 1978), 46 fawns consuming a diet containing 40 ppb selenium from natural sources had mean muscle selenium concentrations that were less than 250 ppb (dry basis). These fawns had received some of their dietary selenium from mother's milk while nursing does that consumed the same diet. Sixty-one percent of these fawns died prior to weaning, and 75% of the dead fawns showed evidence of white muscle disease at necropsy. These data, and those derived from the does, provided a basis for a field study in which skeletal muscle was collected from 109 deer examined by biologists of the Michigan Department of Natural Resources (DNR) during the November hunting season. The geographical location where the deer were killed was noted, and 16 counties were represented. The selenium concentrations (dry basis) found in the skeletal muscle are presented in Table 1. Only 5 deer had muscle selenium concentrations greater than 250 ppb, and 3 of these were killed in Schoolcraft County, located in the Upper Peninsula, bordering the northern limit of Lake Michigan.

Despite classification of Michigan as a low selenium state based on analyses of feed grains and forages (Kubota et al., 1967), we found, in an independent study (Ritchie et al., 1983), a region in southern Schoolcraft County that produced forages containing a high selenium concentration. Hay from 4 farms in this region contained selenium in excess of 1 ppm (dry basis), and several other nearby farms produced hay containing over 0.2 ppm. The skeletal muscle of the deer killed in this region contained 490 ppb selenium (dry basis), 5 standard deviations above the mean value for all deer.

While the preponderance of selenium values in deer muscle were low and typical of those found in captive deer and domestic livestock that are demonstrably deficient, the practical significance of the variations noted is difficult to assess in free-ranging animals. Lesions and mortality as a consequence of selenium deficiency are most likely to be seen in fawns during the first 8 weeks of postnatal life (Brady et al., 1978). However, decomposition of dead fawns in the summer would be rapid, and chances of finding an intact animal in the wild for necropsy would be remote.

Other losses may be associated with failures of conception and gestation. Michigan DNR biologists have conducted spring surveys of fetuses per doe since 1959. Twenty years of cumulative data (Friedrich, 1979) revealed 1.55 fetuses per doe, 2 years old or older (N=1,671), in a "food shortage area" of northeastern Lower Michigan as compared to 1.73 fetuses per doe (N=2,389) in the remainder of the northern half of the Lower Peninsula. The "food shortage area" was defined on the basis of small antler beam diameters in yearling bucks, and the food shortage was undoubtedly a consequence of generally inadequate intakes of energy, protein and other nutrients. A specific deficiency of selenium in free-ranging deer in this area has not been identified, but all muscle

Table 2. Selenium in Wapiti Serum[a]

Range of Se ppb	No. of wapiti	Percent of total	Mean Se ppb
0-9	4	4.3	8
10-19	18	19.4	15
20-29	18	19.4	25
30-39	44	47.3	35
40-49	6	6.5	45
50-59	1	1.1	50
60-69	2	2.2	64
6-64	93	100.2	29

[a]Serum mean alpha-tocopherol, 2.9 ug/ml (0.7-10.6).

selenium values of deer killed in this area were less than 200 ppb (dry basis), while 68% of muscle selenium values were 200 ppb or greater in other areas of the Lower Peninsula.

State-wide surveys of tissue selenium concentrations in other wild herbivores have not been conducted in Michigan. However, in 1986, blood serum was collected from 93 wapiti (Cervus elaphus) during a DNR-controlled hunt in 3 counties of the northeastern Lower Peninsula. Serum selenium concentrations varied from 6-64 ppb, and their distribution is shown in Table 2. It is apparent that many of these individuals were seriously deficient in selenium if one can extrapolate from data obtained with domestic ruminants and white-tailed deer. Since vitamin E has been shown to be protective when selenium status is low (Brady et al., 1978), deficiency lesions may have been prevented in these wapiti by their relatively high serum alpha-tocopherol concentrations (mean: 2.9 ug/ml, range: 0.7-10.6). Nevertheless, the serum selenium levels were what one would expect in this region of the State, and it appears that tissues of wapiti, like those of white-tailed deer, reflect the natural selenium concentrations of their food supply.

REFERENCES

Awad, Y.L., and Berschneider, F., 1980, Selenium content of internal organs of the camel (Camelus dromedarius, L.), Egypt. J. Vet. Sci., 14:71.

Baker, R.H., 1984, Origin, classification and distribution, pp. 1-18 in: "White-Tailed Deer Ecology and Management", Stackpole Books, Harrisburg, PA.

Beeson, K.C., and Matrone, G.M., 1972, "The Soil Factor in Nutrition. Animal and Human", Marcel Dekker, New York.

Brady, P.S., Brady, L.J., Whetter, P.A., Ullrey, D.E., and Fay, L.D., 1978, The effect of dietary selenium and vitamin E on biochemical parameters and survival of young among white-tailed deer (Odocoileus virginianus), J. Nutr., 108:1439.

Cooper, W.C., Bennett, K.G., and Croxton, F., 1970, The history, occurrence, and properties of selenium, in: "Selenium", R.A. Zingaro and W.C. Cooper, eds., Van Nostrand Rheinhold, New York.

Espinoza, J.E., McDowell, L.R., Rodriguez, J., Loosli, J.K., Conrad, J.H., and Martin, F.G., 1982, Mineral status of llamas and sheep in the Bolivian altiplano, J. Nutr., 112:2286.

Fleming, W.J., Haschek, W.M., Gutenmann, W.H., Casiik, J.W., and Lisk, D.J., 1977, Selenium and white muscle disease in woodchucks, J. Wildl. Dis., 13:265.

Franzmann, A.W., Flynn, A., and Arneson, P.D., 1976, Moose milk and hair element levels and relations, J. Wildl. Dis., 12:202.

Friedrich, P.D., 1979, Doe productivity and physical condition: 1979

spring survey results, Mich. Dept. Natl. Resour. Wildl. Div. Rep.
2843, Lansing, MI.

Froeslie, A., Norheim, G., Rambaek, J.P., and Steinnes, E., 1984, Levels
of trace elements in liver from Norwegian moose, reindeer and red
deer in relation to atmospheric deposition, Acta. Vet. Scand.,
25:333.

Gazia, N., and Wegger, I., 1980, Glutathione peroxidase and selenium in
blood from Egyptian water buffaloes, Acta. Vet. Scand., 24:137.

Ku, P.K., Ely, W.T., Groce, A.W., and Ullrey, D.E., 1972, Natural dietary
selenium, alpha-tocopherol and effect on tissue selenium, J. Anim.
Sci., 34:208.

Kubota, J., Allaway, W.H., Carter, D.C., Cary, E.E., and Lazar, V.A.,
1967, Selenium in crops in the United States in relation to selenium
responsive diseases of animals, J. Agric. Food Chem., 15:448.

Kurkela, P., and Kaantee, E., 1979, The selenium content of skeletal
(gluteal) muscle of Finnish reindeer and cattle, Zentralbl.
Veterinaermed., Reihe B 26:169.

Lakin, H.W., 1972, Selenium accumulation in soils and its absorption by
plants and animals, Geol. Soc. Am. Bull., 83:181.

Lakin, H.W., and Davidson, D.F., 1967, The relation of the geochemistry
of selenium to its occurrence in soils, p. 27 in: "Selenium in
Biomedicine: A Symposium", O.H. Muth, ed., AVI, Westport, CT.

Lance, V., Joanen, T., and McNease, L., 1983, Selenium, vitamin E and
trace elements in the plasma of wild and farm-reared alligators
during the reproductive cycle, Can. J. Zool., 61:1744.

National Research Council, 1980, "Mineral Tolerance of Domesticated
Animals," National Academy Press, Washington, DC.

National Research Council, 1983, "Selenium in Nutrition, rev.," National
Academy Press, Washington, DC.

Norheim, G., Sivertsen, T., Brevik, E.M., and Froeslie, A., 1984, Mercury
and selenium in wild mink (Mustela vision) from Norway, Nord.
Veterinaermed., 36:43.

Ritchie, H.D., Ullrey, D.E., Bergen, W.G., and Magee, W.T., 1983,
Analysis of forages fed to livestock in Michigan's upper peninsula
(AS-UP-8201), pp. 224-231 in: "Report of Beef Cattle Forage
Research (Res. Rep. 444)", Michigan State University, E. Lansing,
MI.

Swaine, D.J., 1955, The trace-element content of soils: Harpenden,
England, Common. Bur. Soil Sci. Tech. Commun., 48:91.

Turkstra, J., Devos, V., Biddlecombe, F., and Dow, R.J., 1978, The
characterization of the concentration levels of various trace
elements in liver tissue of blesbok (Damaliscus dorcas phillipsi)
and bontebok (Damaliscus dorcas dorcas) as determined by
instrumental neutron activation analysis, Z. Tierphysiolo.
Tierernaehr. Futtermittelkd., 40:149.

Ullrey, D.E., Brady, P.S., Whetter, P.A., Ku, P.K., and Magee, W.T.,
1977, Selenium supplementation of diets for sheep and beef cattle,
J. Anim. Sci., 46:559.

Ullrey, D.E., Youatt, W.G., and Whetter, P.A., 1981, Muscle and selenium
concentrations in Michigan Deer, J. Wildl. Manage., 45:534.

Westermarck, H., and Kurkela, P., 1979, Selenium content in lichen in
Lapland and south Finland and its effect on the selenium values in
reindeer, pp. 278-285 in: "Proc. Int. Reindeer/ Caribou Symp.,
2nd", Trondheim, Norway.

AN EPIDEMIOLOGICAL STUDY OF A DEFINED AREA OF N. IRELAND: SERUM Zn and Cu

Dorothy McMaster, A.E. Evans, Evelyn McCrum, M. McF. Kerr,
C.C. Patterson, and A.H.G. Love

Departments of Medicine and Community Medicine, The Queen's
University of Belfast, Northern Ireland

The MONICA Project is a WHO coordinated 10 year study of trends and
determinants of cardiovascular events. Our study area comprises Belfast,
Castlereagh, North Down and Ards Health Districts of Northern Ireland with
a total population of 499,111 and a target population aged 25-64 years of
223,575. About 65% are city dwellers, the remainder living in small country
towns and scattered farms. For the first random population screening which
took place from October 1983 to September 1984 clinics were held throughout
the day in local health centres and willing subjects were invited to attend
at a time which suited them. The overall response rate was 70%. Blood was
collected from 2,327 non fasting subjects. In addition to the assessment of
serum cholesterol, blood pressure and smoking habits, we are measuring serum
levels of Zn and Cu. The data, excluding any from women who are pregnant,
is subjected to analysis of variance and the Newman-Keuls multiple range
procedure. A p value of 5% or less is used to determine statistical
significance.

RESULTS

Interpretation of serum zinc in a non fasting population is controversial. From our other concurrent studies we now know that the ingestion
of a standardised mixed meal will cause a fall in serum zinc which is
sustained for at least 4 hours[1]. In the MONICA study subjects were asked to
state the time at which they had last had anything to eat or drink. There
was no overall change in serum zinc in the female group with post prandial
time but after 4 hours of post prandial time male serum zinc rose. It is
also probable that diurnal changes were taking place. We found statistically significant changes with the time of screening. The highest levels
of serum zinc for both males and females were found at 0900 hours. Throughout the morning and early afternoon serum zinc was gradually falling with
significant differences between successive times. Neither the time of
screening nor the time since the last intake of food or drink had any
detectable effect on serum copper.

The effect of age and sex on serum Zn and Cu is shown in Tables 1 and 2.
Serum zinc in 55-64 year old males was statistically less than the 35-44
year olds. In females, serum Zn rose steadily and significantly with age.
Below 55 years of age males had higher levels than females. Females, who
were not taking oestrogen or progesterone preparations, had higher levels of
serum Cu than age matched males. Females aged 25-34, who were taking these
preparations, had the highest levels of serum Cu at 1.62±0.05, n=52, p<0.05.

Table 1.   The effect of age and sex on serum Zn

| | MALES | | FEMALES | |
	n	Mean ± SE	n	Mean ± SE
25-34	233	0.78 ± 0.01	263	0.72 ± 0.01
35-44	276	0.80 ± 0.01	289	0.74 ± 0.01
45-54	330	0.79 ± 0.01	317	0.75 ± 0.01
55-64	312	0.77 ± 0.01	300	0.77 ± 0.01

Table 2.   The effect of age and sex on serum Cu

| | MALES | | FEMALES | |
	n	Mean ± SE	n	Mean ± SE
25-34	232	1.01 ± 0.01	193	1.16 ± 0.02
35-44	273	1.08 ± 0.01	266	1.16 ± 0.02
45-54	330	1.11 ± 0.01	305	1.23 ± 0.01
55-64	310	1.14 ± 0.01	295	1.26 ± 0.01

The prevalence of cigarette smoking was estimated from direct questioning. Just over a third of the total number screened ie 36% males and 36% females were smokers. Serum Cu in smokers was significantly higher than in non smokers. We have previously found that serum Cu was significantly increased in alcoholics when compared with controls[2]. In the present study there was a small (0.04 µg/ml) but significant increase in males who drank alcohol.

Previous studies have suggested that a poor dietary copper intake may have a profound effect on the heart[3]. Moreover, findings in animals indicate that Cu may play a role in the metabolism of lipids, cholesterol and glucose[4]. In the Belfast MONICA first survey no correlation was found between serum Cu and cholesterol or HDL-cholesterol. The ratio of Cu/Zn in serum did not correlate with cholesterol.

COMMENT

The objective of the MONICA Project is to measure trends in death rates and morbidity from coronary heart disease and to assess the extent to which they are related to long term changes in living habits. The "Big Three" risk factors are now accepted to be cholesterol, hypertension and smoking. Many other factors undoubtedly play a part. The extent to which the essential metals such as Zn, Cu and Se are involved in the patho-physiology of the human heart remains to be elucidated. Serum Zn and Cu have been measured in the Belfast population and will be reassessed on two further occasions to determine if any change is taking place.

REFERENCES

1. W. W. Dinsmore, M. E. Callender, D. McMaster, and A. H.G. Love, Am. J. Clin. Nutr., 42: 688 (1985).
2. W. W. Dinsmore, D. McMaster, M. E. Callender, K. D. Buchanan, and A. H. G. Love, Sci. Total Environ., 42: 109 (1985).
3. S. Reiser, J. C. Smith, W. Mertz, J. T. Holbrook, D. J. Scholfield, A. S. Powell, W. K. Canfield, and J. J. Canary, Amer. J. Clin. Nutr., 42: 242 (1985).
4. L. M. Klevay, Biol. Trace Element Res., 5: 245 (1983).

ZINC, IRON AND COPPER IN THE NUTRITION OF AUSTRALIAN ABORIGINAL CHILDREN

R.M. Smith, R.A. King, R.M. Spargo, D.B. Cheek and J.B. Field

CSIRO Division of Human Nutrition, Adelaide, South Australia;
Department of Health, Derby, Western Australia; Queen Victoria
Hospital, Rose Park, South Australia and CSIRO Division of
Mathematics and Statistics, Glen Osmond, South Australia

INTRODUCTION

The physical characteristics of the last Australian Aborigines to live as
nomadic hunter-gatherers were recorded by Abbie (1975). He described the
children as enjoying lively good health and noted that although stature for
age approached that of Caucasian children, weight for age was substantially
less. Today Australia's Aborigines are no longer nomadic, and most of them
live in remote, self governing communities of a few hundred inhabitants.
These settlements are sustained by welfare aid but offer neither the
survival-based disciplines of tribal life nor the full cultural constraints
of urban life and many of the children grow up with inadequate housing, poor
hygiene and frequent gastrointestinal and respiratory infections (McNeilly et
al., 1983). Several cross sectional studies of weight and stature for age of
Aboriginal children living in such settlements show that, on average, weight
for age lies close to the tenth percentile of NCHS (1977) grids, whilst
stature approximates the twenty fifth percentile (Smith et al., 1982).
Although genetic differences in growth potential have not been entirely
disproved, nutritional causes represent a strong possibility for such a
growth deficit and earlier evidence of disordered trace element metabolism
(Holt et al., 1980) led to the present studies of diet and trace element
status of these children.

DIETARY PATTERNS AND TRACE ELEMENT STATUS

Dietary intakes of 129 Aboriginal children aged from 8 to 14 years,
living in six settlements in north-western Australia were assessed over three
years. Staples were meat, flour, sugar and powdered milk with some fresh
fruit but little fresh green vegetables or dairy produce (King et al, 1985).
Compared with a group of 69 Caucasian children of similar age the Aboriginal
children's daily diet provided on average 65% of the Fe, 64% of the Cu and
85% of the Zn consumed by the Caucasians and the estimated energy intake was
only 80% that of the Caucasian children.

Mean concentrations of Cu and Zn in blood plasma and hair and of plasma
Fe are shown in the table for most of 208 Aboriginal children (110 boys, 98
girls) from four settlements in north-western Australia together with values
from 67 Caucasian children (37 boys, 30 girls) of similar mean age from a
suburban school near Adelaide, South Australia. The table also compares

Table 1. Anthropometric data and trace element status of Australian
Aboriginal children in north-western Australia compared with
a group of Caucasian children of a similar age in Adelaide.

Measurement	Aboriginal Mean ± SEM (no)	Caucasian mean ± SEM (no)	Significance of Difference
Age (years)	9.17±0.16 (208)	9.60±0.21 (67)	ns
Body Weight (Kg)	24.6 ±0.5 (208)	32.1 ±1.0 (67)	p<0.001
Stature (cm)	128.1 ±1.0 (207)	134.4 ±1.4 (67)	p<0.001
Arm Circumf. (cm)	17.5 ±0.2 (204)	20.9 ±0.3 (67)	p<0.001
Plasma Zn (mg/L)	0.73±0.01 (188)	0.82±0.01 (55)	p<0.001
Plasma Cu (mg/L)	1.64±0.02 (188)	1.03±0.02 (55)	p<0.001
Plasma Fe (mg/L)	0.63±0.02 (188)	1.05±0.04 (55)	p<0.001
Hair Zn (mg/Kg)	105.6 ±2.3 (205)	119.7 ±5.0 (64)	p<0.05
Hair Cu (mg/Kg)	33.8 ±3.7 (205)	43.5 ±3.9 (66)	ns

average weight, stature and arm circumference of the two groups. In addition
to their lower weight, height and mid-arm circumference, the Aboriginal
children showed significantly lower values for plasma Zn and Fe and for hair
Zn but hair Cu was not significantly different from the Causasian value and
plasma Cu was 60% higher (p<0.001) in the Aborigines.

Because zinc is essential both for normal growth and for the maintenance
of appetite, a primary dietary zinc deficiency was suspected and a double
blind intervention trial was conducted over 10 months with 204 children.
Although plasma Zn increased to near-normal levels in the children receiving
zinc there was no response in either height or weight to the zinc supplement
(Smith et al., 1985).

CONCLUSIONS

It is concluded that the Aboriginal children suffer a mild dietary zinc
deficiency largely associated with a low energy intake but that their low
zinc status is not primarily responsible for their depressed growth. The low
plasma iron levels are associated with relatively low intakes of dietary iron
and often with a mild anaemia, but intestinal parasites undoubtedly
contribute to the low iron status. Despite a relatively low dietary copper
intake there is no indication of a depressed copper status and the persisten-
tly high plasma copper levels may reflect a genetically determined trait combined with frequent elevations due to acute phase response on infection.

REFERENCES

Abbie, A.A. 1975 "Studies in Physical Anthropology", Australian
Institute of Aboriginal Studies, Canberra.
Holt, A.B., Spargo, R.M., Iveson, J.B., Faulkner, G.S. and Cheek, D.B.,
1980, Am. J. Clin. Nutr., 33: 119.
King, R.A., Smith, R.M. and Spargo, R.M., Proc. Nutr. Soc. Aust., 10: 173.
McNeilly, J., Cicchini, C., Oliver, D. and Gracey, M., 1983, Med. J.
Aust., 2: 547.
NCHS Growth Curves, 1977, US Department of Health Education and
Welfare, Series 11, Number 165, DHEN, Hyattsville.
Smith, R.M., King, R.A., Spargo, R.M., Cheek, D.B., Field, J.B. and Veitch,
L.G., 1985, Lancet, 1: 923.
Smith, R.M., Spargo, R.M. and Cheek, D.M., 1982, Proc. Nutr. Soc. Aust., 7: 37.

A RE-APPRAISAL OF THE RECOMMENDED ZINC REQUIREMENTS FOR

GRAZING LIVESTOCK

C.L. White

CSIRO Division of Animal Production, Private Bag, PO Wembley
WA 6014 Australia

INTRODUCTION

According to published estimates of zinc requirements (e.g. ARC 1980), breeding livestock grazing dry autumn pastures in extensive areas of south-western Australia are grossly zinc deficient. Pasture zinc concentrations in autumn of less than 10 mg/kg are not atypical, and although reproductive and growth responses to zinc supplementation have been reported (Masters and Fels, 1980), they appear to be isolated in extent and difficult to reproduce (Masters and Fels, 1985). A closer examination of the ARC (1980) recommendations of the zinc requirements for grazing animals reveals that they are almost certainly too high, and cannot be met under natural grazing conditions without some form of supplementation. The aim of the following work is to derive critical minimum values for zinc requirements of grazing livestock below which there is a high probability of reduced productivity.

ESTIMATION OF REQUIREMENTS

Total herbage intake required for each level of production and for each class of ruminant was estimated from metabolizable energy (ME) requirements using the formulae of Corbett et al. (1987). ME requirements are calculated by dividing each component of net energy by an efficiency value, k. Values for k were dependent upon the herbage quality and for a diet with dry matter digestibility (DMD) of 70%, estimates of k for maintenance, liveweight gain pregnancy and lactation were 0.7, 0.32 (ruminating), 0.13 and 0.6. Milk-fed animals had a k(gain) of 0.7. The data in table 1 are for Merino sheep (60 kg mature) and B. taurus (500 kg mature).

Net zinc requirements (mg/day) are defined as $Znn = Zne + Zng + Znw + Znc + Znl$, where Zne is zinc of enodgenous origin lost in feces and urine, and Zng, w, c, l are zinc sequestered for bodyweight gain (excluding wool), wool growth, conceptus growth and milk production. The following values derived from the literature and from experimental studies in this laboratory were used in Table 1 for sheep and cattle: Zne, 0.055 and 0.053 mg/kg body weight/day; Zng, 24 and 24 mg/kg gain in liveweight (LWG); Znw, 110 mg/kg clean wool grown; Znc, 1.2 and 6.3 mg/day in late gestation for single fetuses; Znl, 5.5 and 4 mg/kg milk from day 14 of lactation.

Gross zinc requirements are net requirements divided by the coefficient of true absorption (TA). Fractional TA values at low zinc intakes in sheep are between 0.7 and 1.0 of zinc intake (Suttle et al., 1982, C.F. Ramberg, Per. Comm.).

Table 1. Critical zinc requirements for sheep and cattle

	BW (kg)	LWG, milk yield (kg/d)	Net reqt. for Zn (mg/d)	Coeff. of true absorption (TA)	Gross dietary reqt. for Zn	
					(mg/kg DM)	(mg/MJ ME)
Growing Sheep	30	0.10	5	0.7	6 (30)[a]	0.6
Growing Cattle	300	0.50	28	0.7	6 (20)	0.5
Pregnant Ewe	60	0.07	5	0.8	5 (22)	0.5
Pregnant Cow	500	0.30	34	0.8	4 (17)	0.4
Lactating Ewe	60	1.00	10	0.8	6 (27)	0.6
Lactating Cow	500	20.00	106	0.8	8 (23)	0.8
Milk-fed Lamb	5	0.10	3	0.8	25 (41)	1.2
Milk-fed Calf	40	0.50	14	0.8	24 (28)	1.2

[a]Equivalent ARC (1980) values.

RESULTS AND DISCUSSION

Table 1 provides a summary of the factorial estimation of critical requirements for zinc by sheep and cattle of different ages and physiological state grazing herbage containing 10 MJ ME/kg. The values represent an attempt at defining the minimum amount of zinc required for the achievement of a desired level of production in a particular class of livestock. The values are in most cases lower than the amount required for the maintenance of normal blood and tissue concentrations of zinc and, in this regard, differ from other published estimates of zinc requirements. For example, the equivalent ARC (1980) estimates of gross requirement are up to five-fold higher. The main reason for this is the much lower value for TA chosen by the ARC (0.2-0.5). Other reasons for differences include the generally higher values used by the ARC for each component of the factorial equation, and the selection of values for feed intake that are not relevant in some cases to grazing animals. On poor quality pastures (e.g. 6 MJ ME/kg), ARC values are up to ten-fold higher than those estimated using the current factorial model.

The advantages of a factorial model based on ME requirements are apparent when the prediction of zinc requirements for rapidly growing sheep grazing green pasture is 10 mg Zn/kg DM, versus 2.5 mg/kg DM for the same animal attempting to maintain leveweight on poor quality pastures (6 MJ ME/kg). Most green pastures and grains contain >20 mg Zn/kg. The model helps explain why responses to zinc supplementation are generally restricted to certain classes of animals grazing dry pasture of low zinc content.

REFERENCES

Agricultural Research Council (ARC), 1980. In: The Nutrient Requirements of Ruminant Livestock. pp256-263. Comm. Agric. Bureaux, Slough, UK.
Corbett, J.L., 1987, In: Recent Advances in Animal Nutrition in Australia. D.J. Farrell, ed. University of New England Publ. Unit, Armidale, NSW. (in press)
Masters, D.G. and Fels, H.E., 1980, Biol. Trace Element Res. 2:281-290.
Masters, D.G. and Fels, H.E., 1985, Biol. Trace Element Res. 7:89-93.
Suttle, N.F., Lloyd-Davies, H. and Field, A.C. 1982, Br. J. Nutr. 47:105-112.

THE UTILIZATION OF HENS' EGGS AS A SELENIUM MONITOR

IN EPIDEMIOLOGICAL STUDIES

P. Brätter, V. E. Negretti de Brätter, U. Rösick, H. Mendez C.*

Hahn-Meitner-Institut Berlin GmbH, Berlin, FRG
Res. group: Trace Elements in Health and Nutrition

* Fundacredesa, Caracas, Venezuela

INTRODUCTION

An assessment of the regional dietary selenium (Se) intake can be made via an analysis of the selenium concentration in monitor materials which can be taken relatively easily from a large number of subjects. Hair, urine and nails are the only monitors which can be obtained without medical intervention. When hair or nails are being used the effect of contamination from external sources must be taken into consideration. In the case of urine the daily fluctuation in the elemental composition of this body fluid creates an element of uncertainty.

In order to be able to assess the physiological effect in the organism it is important to know not only the Se level but also the chemical form in which Se occurs in food. From feeding experiments it is known that the two main fractions of the hen egg reflect the ratio of the Se compounds in the diet.[1] The yolk Se fraction is more closely related to the selenite or cystine content of the diet.[2,3,4] Selenomethionine was found to be the only form of dietary Se which causes higher egg white than yolk levels.[5] The affinity of selenomethionine for egg white was found to be due to the fact that the corresponding proteins originate in the oviduct, whereas yolk proteins are synthesized in the liver.[3]

Within the framework of the epidemiological studies in seleniferous areas of Venezuela the relations between the Se content of hens' eggs and various human monitor materials were investigated in order to show how Se in egg white and yolk can be used to estimate a regional dietary Se level.

MATERIAL AND METHODS

All the eggs in the study were collected from free-range hens which were known to come from certain families being investigated in various seleniferous parts of Villa Bruzual and which had not been fed commercial foodstuffs. In addition, samples of the meals of the families were collected in order to obtain information on their Se intake. For purposes of comparison eggs were also analysed from the coastal region (Falcon, Margarita), the plains (Zaraza, El Cedral), the Andes (Tachira) and from Caracas. For the assessment of the Se status of the population only adults who had been living in the area for at least ten years were chosen as subjects. Details concerning the collection and preparation of the human monitor materials have been given elsewhere.[6] The eggs were stored deep frozen until pre-analytical treatment, as were all the other test materials.

The white and the yolk were separated completely using mechanical force after the eggs had been cooled with the help of liquid nitrogen. The separated material was freeze-dried for one week applying a temperature gradient between 0°C (the shelf) and -50°C (the condenser). Se was determined in egg, hair, toenails and the blood fractions by means of instrumental neutron activation analysis and high resolution gamma spectrometry via the decay of the long-lived radionuclide Se-75.

RESULTS AND DISCUSSION

A comparison of the white/yolk ratio of the eggs from the seleniferous regions and the results of feeding experiments (Fig. 1A) shows that the values obtained correspond to a dietary Se content between 1.0 and 4.5 ug/g. The dotted line corresponds to the Se distribution[4] in the white and the yolk as found in feeding experiments involving selenite and selenomethionine.[5] The Se levels in egg white and yolk provide evidence that nearly all the Se in the feed sources from the seleniferous regions was in the form of seleno-methionine. This pattern might be caused by plant and animal sources. The eggs from the other regions showed significantly lower Se levels (Fig. 1B). The composition of the food in these cases appears to have a lower proportion of its Se in the form of selenomethionine.

In Table 1 the results were allocated according to the estimated daily selenium intake which could be calculated from the Se content of the meals. The standard deviation of the monitor values from Caracas and Villa Bruzual may illustrate the more uniform distribution of foodstuffs in the urban area (Caracas) and the more individual differences in the locally grown food of the subjects living in the rural regions with varying soil Se content. It was found that the differences in the dietary Se intake were reflected proportionally in the blood fractions and also in hair and toenail samples, whereby significant positive correlations were also observed between the Se contents of the different compartments. The ratio of the values for region 2 to those for Caracas shows that egg white is a more sensitive indicator of the differences in Se intake than the human monitor materials.

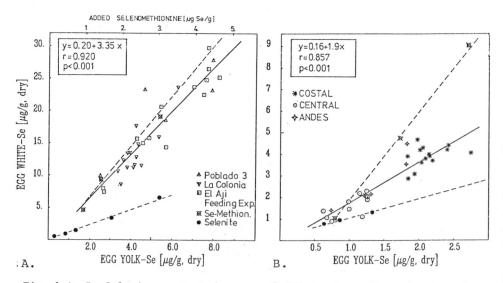

Fig. 1 A, B: Selenium content in eggs of different regions of Venezuela.

Table 1. Level of Se (ug/g dry) in monitor materials from human subjects and egg white from seleniferous regions of Villa Bruzual and from Caracas.

Location	RBC	SERUM	HAIR	TOENAILS	EGG WHITE	Se-INTAKE
Caracas n=40	1.31 (0.20)	2.17 (0.35)	0.95 (0.27)	1.47 (0.52)	(n=32) 1.63 (0.49)	220 [7]
Region 1 n=106	1.70 (0.52)	2.86 (1.01)	1.26 (0.47)	2.74 (0.84)	---	250 - 350
Region 2 n=66	3.13 (0.95)	4.86 (1.61)	1.86 (0.48)	2.68 (1.07)	(n=27) 16.12 (6.17)	500 - 600
Ratio Reg.2/Caracas	2.39	2.24	1.96	1.82	9.9	2.3 - 2.7

Region 1 (Colonia, Turen); Region 2 (Aji, Poblado 3); n = Number of subjects; ( ) standard deviation of the mean; estimated daily Se intake (ug).

CONCLUSIONS

Since the egg Se distribution is affected by the level and the chemical form of the Se in food, eggs from free-range hens from the area under investigation constitute an easily obtainable and highly suitable monitor. From the analytical point of view, compared with human monitors advantages exist in particular with regard to the number of available samples and the sample quantity. Applying enrichment procedures, selenium levels can easily be obtained which can be determined without much difficulty using the approved methods of trace element analysis.

REFERENCES

1. J. D. Latshaw and M. Osman, Distribution of selenium in egg white and yolk after feeding natural and synthetic selenium compounds, Poultry Sci. 54:1244 (1975).
2. J. D. Latshaw, Natural and selenite selenium in the hen and egg, J. Nutr. 105:32 (1975).
3. J. D. Latshaw and M. D. Biggert, Incorporation of selenium into egg proteins after feeding selenomethionine or sodium selenite, Poultry Sci. 60:1309 (1981).
4. K. Moksnes and G. Norheim, Selenium concentration in tissues and eggs of growing and laying chickens fed sodium selenite at different levels, Acta vet. scand. 23:368 (1982).
5. K. Moksnes, Selenium deposition in tissues and eggs of laying hens given surplus of selenium as selenomethionine, Acta vet. scand. 24:34 (1983).
6. P. Brätter, V. E. Negretti, U. Rösick, W. G. Jaffé, H. Mendez C. and G. Tovar E., Effects of selenium intake in man at high dietary levels of seleniferous areas of Venezuela, in: "Trace Element Analytical Chemistry in Medicine and Biology," Vol. 3, P. Brätter and P. Schramel, eds., Walter de Gruyter, Berlin-New York (1984).
7. M. C. Mondragon and W. G. Jaffé, Consumo de selenio en la ciudad de Caracas en comparacion con el de otras ciudades del mundo, Arch. latinoam. Nutr., 342 (1976).

MINERAL AND TRACE ELEMENT NUTRITION OF SHEEP

IN GANSU PROVINCE, NORTHERN CHINA

S.X. Yu*, D.G. Masters**, Q. Su*, Z.S. Wang*, Y.Q. Duang*, and
D.B. Purser**

*Beijing Institute of Animal Science, Malianwa, Haidian,
Beijing, China. **CSIRO Division of Animal Production,
Private Bag, PO Wembley, WA 6014 Australia

INTRODUCTION

Many of the 110 million sheep in China are dependent on unfertilised
natural pastures for most of their nutritional requirements. As the content
of Se and possibly other elements in these pastures may be low (Liu et al.,
1985), studies were initiated to assess the mineral status of grazing sheep.

METHODS

Huang Cheng farm is in Gansu Province, Northern China (lat 38.00°N,
long 101.80°E) and is in the Qilian mountains at an altitude of 2500 to 3500
metres above sea level. The rainfall of 360 mm/year falls mostly in the
summer and autumn and for seven months per year (winter and spring) only dry
feed is available. One hundred and forty, 2.5 year old Gansu Alpine fine
wool ewes on this farm were managed in the normal way with mating in July,
lambing in December and January and weaning in April. Sheep were grazed
throughout the year with supplementary grain, hay and silage fed in winter
and spring. Samples of blood, faeces (from 30 ewes) tissues (six ewes per
slaughter) and pasture were collected six to eight times during the year and
analysed for N, S, Ca, Na, K, Mg, Zn, Cu, Mn, Fe, Mo and Se.

RESULTS AND DISCUSSION

The concentration of Se in liver was marginal (ie 0.05-0.11 µg/g)
(Judson et al., 1987) at all times except December. The concentration of Se
in pasture was below 0.04 µg/g at all samplings except March and
supplementary feed contained only 0.021 µg/g (Table 2). Therefore intakes
of Se were probably marginal or inadequate for most of the year (Anon.,
1980). The mean concentration of Cu in liver was above 10 µg/g only in
December and mean values at all samplings were characterised by high between
sheep variability. Three to five of the six sheep slaughtered in September,
March, May and June had 6 µg Cu/g or less in the liver, therefore 50% or
more of the sheep had less than adequate Cu in the liver (ie 6-8 µg/g)
(Judson et al., 1987). Although Cu in pasture and supplements was not
exceptionally low it is possible that absorption was reduced by interaction
with molybdenum and sulphur or iron. Na in pasture was always below the
recommended level for sheep (ie 0.08-0.27%) (Anon 1980) with some

Table 1. Concentration of Se, Cu and Zn in tissues

	Sept	Dec	March	May	June
Plasma (μg/ml)					
Selenium	0.034	0.042	0.027	0.032	0.026
Copper	0.97	0.81	0.73	0.64	0.74
Zinc	0.72	0.74	0.60	0.63	1.00
Liver (μg/g wet wt.)					
Selenium	0.08	0.12	0.09	0.09	0.10
Copper	9.3	22.0	9.1	9.0	3.7

Table 2. Nutrient composition of pasture and supplements (dry basis)

	June	Sept	Dec	March	Supplements
Crude Protein (%)	12.2	9.7	4.0	5.2	11.3-13.1
Sulphur (%)	0.14	0.16	0.13	0.13	0.12-0.18
Calcium (%)	0.55	0.97	0.46	0.54	0.05-0.31
Potassium (%)	1.15	1.16	0.61	0.42	0.38-1.26
Magnesium (%)	0.14	0.22	0.12	0.13	0.10-0.14
Sodium (%)	0.011	0.006	0.015	0.025	0.004-0.013
Zinc (μg/g)	21.3	27.9	14.0	19.7	20.5-27.2
Copper (μg/g)	6.4	5.8	4.1	6.0	3.3-4.9
Molybdenum (μg/g)	1.9	2.3	1.7	2.6	0.7-1.4
Manganese (μg/g)	60.8	64.9	55.0	66.0	19.8-53.4
Iron (μg/g)	1025	789	4592	1100	112-354
Selenium (μg/g)	0.036	0.038	0.035	0.055	0.021

supplements containing 0.004% Na. The concentration of Zn in pasture was 14 μg/g in December, this is less than the variously estimated requirements of 20-51 μg/g (Anon., 1980). In conclusion the intake of Na, Se, Cu and possibly Zn may not be adequate for optimal animal production throughout the year.

ACKNOWLEDGEMENTS

    This research was supported by the Chinese Academy of Agricultural Sciences and the Australian Centre for International Agricultural Research. The assistance from the staff of Huang Cheng Farm is also acknowledged.

REFERENCES

Anonymous, 1980, "The Nutrient Requirements of Ruminant Livestock," 2nd Edition, Commonwealth Agricultural Bureaux, Slough.
Judson, G.J., Caple, I.W., Langlands, J.P. and Peter, D.W. 1987, In: "Review of Research on Pasture Production and Utilization in Southern Australia," J.L. Wheeler ed. (In Press)
Liu, C.H., Lu, Z.H., and Su, Q. 1985, Chinese J. Agric. Sci. 4:76.

THE EPISTEMOLOGY OF TRACE ELEMENT BALANCE AND INTERACTION

Berislav Momčilović

Institute for Medical Research and
Occupational Health
M. Pijade 158, 41000 Zagreb, Yugoslavia

Numerous interactions among trace elements (TE) have been observed 1-3. The aim of this paper is to review basic ideas of balance and interaction of TE from an epistemological point of view. Epistemology is that department of philosophy which critically investigates nature, grounds, limits and criteria of validity of human knowledge i.e. theory of cognition.

Each living being could be regarded as a homeostatically controlled system. Homeostasis is defined as the total of steady-states achieved by biological feedback mechanisms within the body which maintain stability against fluctuating conditions 4. Every real system has an indefinitely large number of possible inputs and outputs or responses 5. A "black box" i.e. object of experimentation (observation) is coupled with experimenter (observer) via experimental design or observational plan 5,6. Our capacity to perceive nature is limited by the very nature of our methods of observation and experimentation. Hence the limits of our cognition are set by the principle of uncertainty 7 and Gödel's incompleteness theorem 8.

BALANCE

Balance is defined as a state of being in equilibrium. To be at equilibrium each part of the body must be in a state of equilibrium in the conditions provided by the other parts. Balance tells us nothing about the pattern of connections which will be formed and which depend on the set of inputs and outputs used. If we feed a human being zinc, iron and copper according to Recommended Dietary Allowances the pattern of metal distribution would be different from tissue to tissue indicating the existance of an infinite number of particular balances as reflected by the response of chosen indicators (Table 1). The proper choice of indicator on the level of the body, organ and tissue 12, cell 13 or molecule 14 is essential in TE research, otherwise the expected reactions could not be observed 15. In studying the bioavailability of zinc from animal and plant proteins a suitable indicator was first identified (femur) and then a choice between parallel line and slope ratio assay was decided in favor of the latter 16. The balanced

Table 1. Zinc, iron and copper in human diet and tissue

	Diet(RDA)[9] mg/day	Body[10] mg/body wt	Plasma[11] µg/ml	Milk[11] µg/ml	Nails[*] mg/g dry wt
Iron	18	5000	1.2	0.5	15
Zinc	15	1300	1.2	3.5	124
Copper	2	150	1.0	0.2	6

diet is that blend or mixture of dietary ingredients which maintains physiological function of the body. Each food variety is a genetically determined final system of nutrients and of distinct bioavailability to man. Even the cell culture does not react to Cd, Zn, Mn and Ni as to a mere mixture of components [17]. Copper complexes are remarkably active pharmacological agents whereas this activity is not shared by copper ions or the parent ligands on their own [18]. Hence each variety of food could be regarded as an "alloy". Only by blending the various foods could a balanced diet be obtained which would maintain the physiological function of the body.

Dose-response nutrition may seem the more scientific but in practice it is affected by even more uncertainty than food based nutrition [19] as the estimates of minimum requirements of safe intake of an element are only valid for particular dietary conditions [11]. Even health and normality of man can be present without optimal development having taken place as balanced nutrition can occur at various levels of nutrients and sparse diets are not necessarily deficient ones, although they often are [20,21]. This may be true only if the rate of all metabolic processes in the body depend in the same way upon the intake of nutrients either in excess or deficiency [22]. In an experiment on rats fed a diet suboptimal in protein and minerals increased lead absorption was observed [23]. Under the conditions of imposed strain adaptive (metabolic) response took place and set body functions at a new level of equilibrium [24].

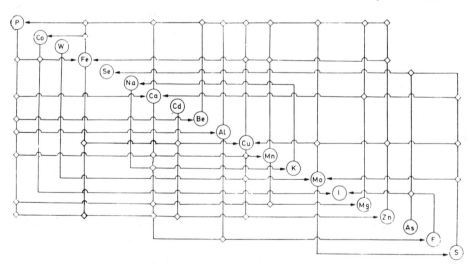

Fig. 1   Interaction chart

[*] Momčilović (unpublished)

Table 2. Zinc, iron and copper in the brain of suckling
rats fed fortified cow's milk (% dose) 28a,b,

	Milk[1]	Zn[2]	Fe[3]	Cu[4]	Zn+Fe	Zn+Cu	Fe+Cu	Zn+Fe+Cu
65_{Zn}*	1.95 a	1.80 b,c	1.81 a,b,c	1.68 c,d	1.80 b,c	1.66 d	1.86 a,b	1.56 d
59_{Fe}	1.20 a,b	1.29 a	1.13 c,d	1.20 a,b	1.13 c,d	1.07 d	1.17 b,c	1.06 d
64_{Cu}	0.58 b	0.71 a	0.71 a	0.62 a,b	0.61 a,b	0.59 b	0.58 b	0.59 b

[1] 5 µg Zn, 0.5 µg Fe and 0.05 µg Cu/ml,   [2] 12 µg Zn/ml
[3] 25 µg Fe/ml, 5 µg Cu/ml.   a - d $P < 0.05$ for the same isotope

INTERACTION

Interaction is defined as a reciprocal action. Interaction could be viewed as a state of new equilibria where each
part must be in a state of equilibrium in the conditions provided by the others 5. The interactions presented in Fig. 1
may not be up to date and they do not specify the indicator
of response, magnitude and direction of effect. Liebig's Law
of minimum was the first to state that the nutrient in the relative minimum determines the rate of growth. Later Voison
proved that nutrient in the relative maximum also determines
the yield 21. Both Laws are concerned with the extremes in the
continuum of possible states of equilibria. Several models of
dose-response curves have been proposed so far indicating the
complexity of the task 15,25,26. Dose response curves differ
from element to element, some having better homeostatic capacity than others and some with only a brief safety margin
between optimal and toxic concentrations 11. In radiation
physica dose response curves for some tumor types did not
correspond to those expected from physical theory, i.e. they
were not stohastic 27. We do not yet have evidence that all
phenomena in dose response curve of TE are stohastic as dose
rate has not been studied.
Studies with individual elements can be seriously mis-
leading unless their quantitative relationship to the other
interacting elements is known and considered 11. If any of
the kinetically significant parallel processes responsible
for the disposition of the administered TE is dose-dependent
the assumption of proportionality between external and internal dose is violated 22. In a series of experiments I
attempted to find out what the effect would be of fortifying
cow's milk with a physiological dose of zinc (15 ppm), iron
(25ppm) and copper (5ppm) added either separately or in combination. The changes in food density induced a response of
homeostatic control at the level of the gastro-intestinal
tract with a decrease in retention of isotopes in the brain
28a,b (Table 2). Also an antagonism between copper and zinc
on 65Zn retention but not 64Cu retention was observed, as
indicated by decreased deposition of zinc in the brain of
copper-fed animals. Redistribution of elements occurred and
proportionality between external and internal dose was violated. In a study on a human volunteer fed four liters of
fortified milk to supply daily 15, 18 and 2 mg of zinc, iron

Table 3. Allergogenic trace elements (electronic configuration)

	At.No.	ON=0		At.No.	ON=0
Beryllium*	4	$2s^2$	Zirconium*	40	$4d^2 5s^2$
Silica	14	$3s^2 3p^2$	Silver*	47	$4d^{10} 5s^1$
Chromium*	24	$3d^5 4s^1$	Cadmium*	48	$4d^{10} 5s^2$
Iron(w)*	26	$3d^6 4s^2$	Antimony*?	51	$5s^2 5p^3$
Nickel*	28	$3d^8 4s^2$	Iodine	53	$5s^2 5p^5$
Cobalt*	27	$3d^7 4s^2$	Tellurium	52	$5s^2 5p^4$
Zinc(w)*	30	$3d^{10} 4s^2$	Platinium	78	$5d^9 6s^1$
Copper(w)*	29	$3d^{10} 4s^1$	Gold*	79	$5d^{10} 6s^1$
Arsenic	33	$4s^2 4p^3$	Mercury*	80	$5d^{10} 6s^2$
Selenium	34	$4s^2 4p^4$	Lead*	82	$6s^2 6p^2$

* Trace elements with cation chemistry, (?) cation chemistry suspected —— essential TE, (w) weak allergen, ON oxidation No.

and copper respectively, the complex formula appeared beneficial regarding copper balance and iron status of plasma in relation to milk diet alone. The observed increase in calcium absorption tends to normalize even at a very high level of dietary intake **. Again we encountered the problem of adaptive mechanisms and proper choice of indicator of response.

The possible attributes of TE interactions have not yet been scrutinized formally and therefore the following is proposed:

Additivity	Antagonism	Synergism
$R = n_A R_A + n_B R_B$	$R < n_A R_A + n_B R_B$	$R < n_A R_A + n_B R_B$

Where R = response, n = dose, A and B are substances.

PROSPECT

The epidemiology of TE has grown 29,30 and interactions with other systems of the body are under scrutiny. Immunity 31, hormones 32, interrelationship with drugs 33 are a few examples. Allergy appears to be a particularly intriguing subject as many essential TE are proved allergens (Table 3) 34. Most of them are metals with nearly filled orbitals and many of them have cation chemistry. It is really a paradox that essential TE could become allergens and induce many adverse reactions.

Our expanding knowledge makes it clear that biology is a continuum and one cannot tamper with one area without producing shifts and changes in another 55.

REFERENCES

1. M. Kirchgessner, A.M. Reichlamyr-Lais, and F.J. Schwartz, Prog.Clin.Biol.Res. 77:189 (1981).
2. C.F. Mills, Ann.Rev.Nutr. 5:173 (1985).
3. H.H. Sandstead, J.Lab.Clin.Med. 98:457 (1981).

** This Symposium

4. I.M. Lerner, "Genetic homeostasis', Dover Publications, New York (1970).
5. W.R. Ashby, "An introduction to cybernetics", John Wiley and Sons Inc., New York (1965).
6. M. Bickis and D. Krewski, in "Toxicological risk assessment", D.B. Clayson, D. Krewski, and I. Munro, eds., CRC Press, Boca Raton (1985).
7. A. Novick and G. Cowley, The Sciences, Nov/Dec:54 (1986).
8. D.R. Hofstadter "Gödel, Escher, Bach: An internal golden braid", Penguin Books Ltd., Harmondsworth (1980).
9. A.E. Harper, Nutr. Rev. 31:393 (1973).
10. G. Feuer and A. Felix, "Molecular Biochemistry of human disease", CRC Press Inc., Boca Raton (1986).
11. E.J. Underwood, "Trace elements in human and animal nutrition", Academic Press, New York (1971).
12. S.G. Schäfer and W. Forth, Trace Elements Med., 2:158 (1985).
13. A.J. Vander, D.R. Mouw, J. Cox, and B. Johnson, Am.J. Physiol., 236:F373 (1979).
14. J.J. Dulka and T.H. Risby, Analyt. Chem. 48:640A (1976).
15. W. Mertz, Science 213:1332 (1981).
16. B. Momčilović, B. Belonje, A. Giroux, and B.G. Shah, Nutr. Rep.Intl. 12:197 (1975).
17. Y. Škreb and Vl. Simeon, Period. Biol. in print.
18. J.R.J. Sorenson, Chem. Brit., Dec.:1110 (1984).
19. G. Atherley, Am.J.Industr.Med. 8:101 (1985).
20. J.V.G.A. Durnin, Phil. Trans.R.Soc.Lond.B. 274:447 (1976).
21. K.H. Schütte, "The biology of trace elements" Crosby Lockwood and Son Ltd., London (1964).
22. E.J. O'Flaherty, in "Toxicological assessment" D.B. Clayson and D. Krewski, eds., CRC Press, Boca Raton (1985).
23. B.G. Shah, B. Momčilović and J.M. McLaughlan, Nutr.Rep.Intl. 21:1 (1980).
24. C.L. Bernard, "An introduction to the study of experimental medicine" Dover Publications Inc., New York (1957).
25. B.L. Vallee, in "New trends in bio-inorganic chemistry" R.J.P. Williams and J.R.R.F. DaSilva, eds., Academic Press, London (1978).
26. D.R. Williams, "An introduction to bioinorganic chemistry" Charles C. Thomas Publisher, Springfield (1976).
27. R.L. Ullrich, M.C. Jernigan, G.E. Cosgrove, L.C. Satterfield, N.D. Bowles and J.B. Storer, Radiat.Res. 68:115 (1976).
28. B. Momčilović, in "Trace element metabolism in man and
a/b animals (TEMA 4)", J. McC. Howell, J.M. Gawthorne, and C.L. White, eds., Austral.Acad.Sci., Canberra (1981).
29. P.C. Elwood, Clin.Endocrinol.Metabol. 14:617 (1985).
30. D. Kromhout, A.A.E. Wibowo, R.F.M. Herber, L.M. Dalderup, H. Heerdink, C. de L. Coulander, and R.L. Zielbuis, Am.J. Epidemiol. 122:378 (1985).
31. K.H. Schmidt and W. Bayer, Trace Elements Med. 4:35 (1987).
32. A.S. Prasad, Clin. Endocrinol. Metabol. 14:567 (1985).
33. A. Nagamuma, M. Satoh, and N. Imurs, Res.Comm.Chem.Pathol. Pharmacol. 46:265 (1984).
34. S. Fregert and H. Rorsman, Acta Dermatovenerol. 46:144 (1966).
35. C.A. Elvehjem and W.A. Krehl, Borden's Rev. Nutr. Res. 26:69 (1955).

SUBCELLULAR CHANGES AND METAL MOBILISATION IN THE LIVERS OF COPPER LOADED RATS

I.C. Fuentealba and S. Haywood

Department of Veterinary Pathology
University of Liverpool
P.O. Box 147 Liverpool. L69 3BX

INTRODUCTION

Rats recover from copper-induced liver damage and subsequently become tolerant to the metal; an adaptation associated with changes in the subcellular distribution of copper within the liver (Haywood, Loughran & Batt, 1985).
This study examined the ultrastructural changes within the livers of copper-loaded rats in relation to the intracellular localisation of copper and other metals, in order to more clearly define this movement.

MATERIALS & METHODS

Male rats fed a high copper diet (1500ppm) for 16 weeks were killed at intervals and their livers fixed in 4% glutaraldehyde and processed for transmission electron microscopy and electron probe x-ray microanalysis. Analysis of liver copper (Cu) and Zinc (Zn) was performed by AA spectrophotometry.

RESULTS

Liver copper concentrations rose to 2840 ± 45ug/g M+SEM dry weight at 4 weeks, falling subsequently to 2271 ± 92ug/g at 16 weeks (control 20 ± 1.0ug/g). Liver zinc content remained stable at 141.9 ± 3.0ug/g (control 129 ± 2.0ug/g).
Pericanalicular, electron-dense, membrane-bound bodies identified as lysosomes were prominent within the hepatocytes of Cu-loaded livers at 1 week. X-ray emission spectra demonstrated a markedly elevated Cu peak, also raised Fe and Zn (lysosomes - Type I). By 3 weeks many large, irregularly-shaped lysosomes were dispersed throughout periportal hepatocytes. Their emission spectra included S and P (lysosomes - Type II). At 4 weeks, lysosomes were very numerous and displayed marked diversity with respect to both electron density and emission profiles In addition to Types I & II, a Type III containing much reduced elemental residues was identified (Fig 1). Subsequently lysosomes declined in number and at 16 weeks the remaining lysosomes displayed emission spectra characteristics of Types II & III. The lysosomal integrity appeared intact at all times and degenerative changes were not observed within these organelles.
By contrast, at 3 weeks the liver cell nuclei showed chromatin

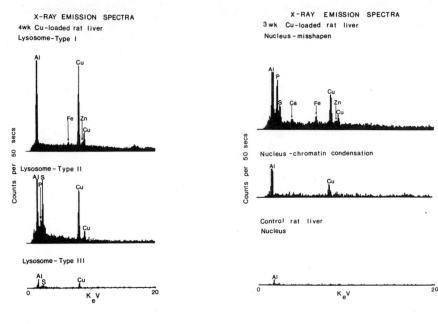

X-RAY EMISSION SPECTRA
4wk Cu-loaded rat liver
Lysosome–Type I

Lysosome–Type II

Lysosome–Type III

X-RAY EMISSION SPECTRA
3wk Cu-loaded rat liver
Nucleus–misshapen

Nucleus–chromatin condensation

Control rat liver
Nucleus

Fig. 1.   Fig. 2.

condensation and a misshapen appearance progressing to chromatolysis at
4 weeks. X-ray emission spectra identified elevated Cu with the addition
of Fe, Zn, Ca, P & S in the more disturbed nuclei (Fig. 2). At 16 weeks
the nuclei appeared normal with only very low x-ray emission activity of
residual elements. Other changes, which were reversed by 16 weeks,
included early swelling of smooth endoplasmic reticulum, swollen
mitochondria and fragmentation of rough endoplasmic reticulum. Neither
copper nor other elements were identified in these organelles.

CONCLUSIONS

1. High levels of Cu & S within the nucleus and certain lysosomes
    suggests the presence of copper-metallothionein in both these sites.
2. Elevated concentrations of Fe & Zn in lysosomes and nucleus may
    indicate displacement of these metals associated with excess uptake.
3. Copper-induced damage is probably the consequence of nuclear
    degeneration caused by the movement of Cu and other elements into
    the nucleus rather than lysosomal damage.
4. Cellular recovery is associated with a loss of copper from both
    nucleus and lysosomes.

REFERENCE

Haywood S, Loughran M, Batt R M (1985) Copper Toxicosis and Tolerance
    in the Rat III Intracellular Localisation of Cu in Liver and Kidney.
    J. Exp. & Molec. Pathol. 43 : 209-219.

ACKNOWLEDGEMENT

    We wish to acknowledge the help of Dr. J. Foster, Central Toxicology
Laboratory, ICI, Alderley Edge, UK, with the electron probe microanalysis.

THE INTERACTIONS AMONG FE, ZN AND CU AFFECTING LIVER CU AND FEMUR ZN CONCENTRATIONS IN THE RAT

Dennis T. Gordon* and Mark Ellersieck#

*Department of Food Science and Nutrition, #Agricultural Experiment Station, University of Missouri, Columbia, Missouri 65211

INTRODUCTION

Numerous studies have indicated that interactions can occur between each pair of the three transition elements, Fe, Zn and Cu (1). We have shown that a significant 3-way interaction does occur among these three elements which affected hemoglobin levels in the rat (2). The purpose of this study was to examine the effects of varying dietary concentrations of Fe, Zn and Cu on femur Zn and liver Cu concentrations in the growing rat.

MATERIALS AND METHODS

Male weanling rats (n=180) were blocked according to weight groups (n=5) in a randomized complete block design in which treatments were arranged in a 3 X 4 X 3 factorial. Rats were fed semipurified diets (AIN) ad libitum with the following level of dietary Fe: 18, 90 and 270 ug/g; Zn: 8.5, 10.5, 42.5 and 170 ug/g and; Cu: 2, 10 and 40 mg/g. After 21 days, the rats were killed and their livers and one femur were removed, dried, acid digested and analyzed for Zn and Cu by atomic absorption spectrophotometry. The linear statistical model evaluated the effects of Cu, Zn and Fe and all possible interactions of these three elements for femur Zn and liver Cu concentrations.

RESULTS AND DISCUSSION

Femur Zn decreased by 7% and liver Cu decreased by 44% when dietary Fe was increased from 18 to 270 ug/g diet, respectively (Table 1). Iron has been known to be antagonist to Zn utilization, but the adverse affect of higher dietary Fe on Cu status has only recently been reported (1, 3).

Liver Cu concentrations were reduced by 14% and 27% when dietary Zn was increased from 8.5 to 42.5 ug/g and 8.5 to 170 ug/g diet, respectively (Table 2). Increasing the level of dietary Cu from 2 to 10 ug/g diet caused femur Zn concentrations to be lowered by 11.2 ug/g, approximately 5% (Table 3).

Table 1. Effect of dietary Fe on femur Zn and liver Cu concentrations in the growing rat.

Dietary iron	n	Femur zinc	Liver copper
ug/g		----------- ug/g[1] -----------	
18	60	206.0[a]	14.1[a]
90	60	195.7[b]	10.4[b]
270	60	192.3[b]	7.9[c]

[1]Mean pooled values (dry wt) for 12 groups of rats at all dietary concentrations of Zn and Cu. a ≠ b ≠ c (P≤0.01).

Table 2. Effect of dietary Zn on femur Zn and liver Cu concentrations in the growing rat.

Dietary zinc	n	Femur zinc	Liver copper
ug/g		----------- ug/g[1] -----------	
8.5	45	170.1[d]	11.8[a]
10.5	45	185.1[c]	12.6[a]
42.5	45	211.4[b]	10.2[b]
170.0	45	225.5[a]	8.6[c]

[1]Mean pooled values (dry wt) for 9 groups of rats at all dietary concentrations of Fe and Cu. a ≠ b ≠ c (P≤0.01).

Table 3. Effect of dietary Cu on femur Zn and liver Cu concentrations in the growing rat.

Dietary copper	n	Femur zinc	Liver copper
ug/g		----------- ug/g[1] -----------	
2	60	205.9[a]	5.1[c]
10	60	194.7[b]	10.6[b]
40	60	193.5[b]	16.7[a]

[1]Mean pooled values (dry wt) for 12 groups of rats at all dietary concentrations of Zn and Fe. a ≠ b (P≤0.01).

A significant interaction (P≤0.01) was observed between Fe and Zn affecting femur Zn concentrations. Significant (P≤0.01) 2-way interactions occurred between Fe and Zn and Zn and Cu affecting liver Cu concentrations. (The authors recognize these interactions are most important and took liberty in reporting main effects on pooled data in Tables 1, 2 and 3.)

CONCLUSION

To avoid these interactions, dietary ranges which should be utilized are as follows: Fe 25-90; Zn 10.5-42.5 and Cu 5-10 ug/g diet.

REFERENCES

Gordon, D.T. (1987) Interactions among iron, zinc and copper, in: Nutrition - 1987, FASEB, Washington, D.C. (in press)

Gordon, D. T. (1985) Interactions among Fe, Zn and Cu affecting hemoglobin status in the rat. XIII International Congress of Nutrition. p. 154. (Abs.)

Bremner, I. and Price, J. (1985) Effects of dietary iron supplements on copper metabolism in rats, In: "Trace Elements in Man and Animals - TEMA 6." Mills, C. F., Bremner, I. and Chesters, J. K. ed. pp. 374-376. Commonwealth Agricultural Bureaux Royal, United Kingdom.

INTERACTIONS BETWEEN ZINC AND VITAMIN A METABOLISM IN RATS AND SWINE

Manfred Kirchgessner, Franz X. Roth and Hans-Peter Roth

Institut für Ernährungsphysiologie
Technische Universität München
D-8050 Freising-Weihenstephan

INTRODUCTION

Metabolic interactions between zinc and vitamin A have been postulated for human and animal nutrition (Smith, 1982). In particular, it has been demonstrated that zinc deficiency has a strong adverse effect on the vitamin A status. Conversely, to what degree vitamin A supplements affect the zinc supply status when inadequate zinc is provided has hardly been studied. In the work presented here, we examined such interactions between zinc and vitamin A metabolism in 4 experiments with rats and swine.

1. The effect of zinc deficiency (Exp. 1)

How severe alimentary zinc deficiency affects vitamin A metabolism, was investigated with weaned Sprague-Dawley rats (Kirchgessner et al., 1987a). After zinc depletion the serum-zinc level was 80% lower than the value for the two control groups, whereas the vitamin A and beta-carotene concentrations did not differ significantly from the pair-fed control group. In comparison to the value for ad lib.-fed control rats, however, the vitamin A and beta-carotene concentrations in the depletion animals were elevated. The zinc concentration in the livers of depletion animals was reduced only in comparison to the pair-fed control animals, whereas the vitamin A and beta-carotene concentration in the zinc-deficient animals was reduced in comparison to both control groups.

2. The effect of trace element and/or vitamin supplementation (Exp. 2)

A grain-soya based ration (36 mg Zn/kg DM) with or without trace elements and/or vitamins, respectively, was provided to 4 x 8 swine during the growth period from 55 kg to 95 kg (Kirchgessner et al., 1987b). In comparison to swine which provided the basal ration alone, the vitamin or trace element supplementation strongly increased the growth performance. Only those animals which were provided the basic diet alone exhibited strong parakeratotic skin changes which are seen normally only during extreme zinc deficiency. Those swine showed a serum zinc concentration and serum alkaline phosphatase activity that was reduced by two-thirds and 60%, respectively; the zinc binding capacity was correspondingly elevated. The vitamin A concentration in serum was increased, a result similar to that obtained in the rat experiment. Trace element supplementation brought about a decrease in the serum vitamin A concentration, whereas vitamin supplementation caused

183

an increase in the vitamin A concentration and brought about a slight improvement in the zinc supply status.

3. The effect of zinc and/or vitamin A supplementation (Exp. 3 and 4)

In experiment 3, the grain-soya basal ration provided to swine was supplemented with either trace elements and vitamins, or solely with zinc (120 mg/kg) and/or vitamin A (5510 ug/kg). The supplementation of zinc or vitamin A increased the growth rate similarly to the combined zinc/vitamin A supplementation. The basal diet drastically reduced the serum zinc concentration and alkaline phosphatase activity. The addition of vitamins or vitamin A to the diet did not only elevate the serum vitamin A concentration, but also tendenciously improved the serum zinc status.

Experiment 4 focussed upon the interaction of zinc and vitamin A in swine which received an otherwise completely supplemented basal ration. Zinc supplementation increased growth and feed intake by 60% and 30%, respectively. Vitamin A improved growth by a non-significant 10%. The absence of a zinc supplement in the diet caused strong zinc deficiency symptoms in the swine; a vitamin A supplement was not able to prevent the zinc deficiency. The zinc concentration, alkaline phosphatase activity and zinc binding capacity of serum clearly reflected the zinc supply. The serum vitamin A concentration was not significantly reduced during zinc deficiency; on the other hand, the vitamin A supplement did not bring about a changed plasma level.

CONCLUSION

With the four experiments presented here, we were not able to confirm that zinc deficiency causes a reduction in plasma vitamin A levels, which cannot be increased by high vitamin A supplements. Indeed, extreme zinc deficiency led to either increased or unchanged plasma vitamin A levels. The mobilization of vitamin A from the liver to the plasma was not disturbed during zinc deficiency.

In the experiments presented here, the fundamental interaction is the following: when sufficient dietary zinc is not provided, vitamin A supplementation tends to elevate the serum zinc concentration and alkaline phosphatase level, but cannot prevent clinical symptoms of zinc deficiency. This result could stand in relationship to the fact that during vitamin A deficiency a vitamin A supplementation brings about an elevation of zinc absorption possibly as a result of a vitamin A dependent zinc binding protein in the intestinal mucosa (Berzin and Bauman, 1987).

REFERENCES

Berzin, N. J. and Bauman, V. K., 1987, Vitamin-A-dependent zinc-binding protein and intestinal absorption of Zn in chicks, Br. J. Nutr. 57:255.
Kirchgessner, M., Plank, J. and Roth, H. P., 1987a, Zum Einfluss von alimentärem Zinkmangel auf den Vitamin-A- und α-Tocopherolstoffwechsel, Z. f. Ernährungswissenschaft, in press.
Kirchgessner, M., Roth, F. X. and Roth, H. P., 1987b, Auswirkungen einer fehlenden Spurenelement- und Vitaminergänzung zu einer getreidereichen Futtermischung auf verschiedene Leistungsparameter bei Mastschweinen, J. Vet. Med. A. 34:188.
Smith, J. C., Jr., 1982, Interrelationship of zinc and vitamin A metabolism in animal and human nutrition: A review, in: "Clinical, Biochemical and Nutritional Aspects of Trace Elements," A. S. Prasad, ed., Alan R. Liss, Inc., New York.

ZINC, COPPER AND IRON CONCENTRATIONS IN

TISSUES OF ANENCEPHALIC AND CONTROL FETUSES

Deirdre Fehily[1], Catherine Keohane[2], Frank M. Cremin[1], and David M. Jenkins[3]

Departments of Nutrition[1], Histopathology[2] and Obstetrics and Gynaecology[3], University College, Cork, Ireland

INTRODUCTION

The incidence of anencephaly and spina bifida in the Irish population is one of the highest in the world. That dietary zinc deficiency may be a factor in the aetiology of neural tube defects (NTD) in humans is supported by epidemiological studies[1], observations on pregnancy outcome in acrodermatitis enteropathica patients[2] and clinical studies which indicate lower circulating zinc concentrations in the mothers of con-genitally abnormal infants compared with control mothers[3]. However, observations on cord blood indicate elevated zinc concentrations in anencephalics[4]. The present study was undertaken to assess the actual concentrations of tissue zinc in NTD infants. Because of the well documented interactions of zinc, copper and iron metabolism, the concentrations of these elements were simultaneously analysed.

SPECIMEN COLLECTION AND ANALYSIS

Tissue samples were taken from 25 fetuses at post-mortem. These included 10 with NTD and 15 normal fetuses who had died from either abruptio placentae (6 cases), prematurity (4 cases), anoxia (3 cases), vasa previa (1 case) or pneumonia (1 case). Tissue samples were removed from liver, kidney (cortex), pancreas, muscle (psoas), bone (rib), oesophagus and nervous tissue; samples of placenta and umbilical cord were also taken. Trace element concentrations were estimated on a Pye Unicam SP9 atomic absorption spectrophotometer after wet ashing of dried tissue with a 2:1 mixture of nitric : perchloric acids. All analyses were conducted in duplicate.

RESULTS AND DISCUSSION

Results (Table 1) indicate higher concentrations of zinc in liver, kidney, pancreas and placenta from NTD fetuses, kidney copper and liver and nervous tissue iron concentrations were also elevated.

Indirect evidence suggests a link between zinc deficiency and the occur-rence of NTD in humans[1-3]; such a deficiency would be expected to result in

Table 1. Zinc, Copper and Iron Concentrations in Fetal Tissues

	Zinc (µg/g dry wt.)			Copper (µg/g dry wt.)			Iron (µg/g dry wt.)		
	N.T.D. (Mean ± ISD)	Controls (Mean ± ISD)	p value	N.T.D. (Mean ± ISD)	Controls (Mean ± ISD)	p value	N.T.D. (Mean ± ISD)	Controls (Mean ± ISD)	p value
Liver	1,119 ± 363	622 ± 162	<0.001	275 ± 91	263 ± 98	N.S.	3,860 ± 1,795	1,965 ± 744	<0.01
Kidney	168 ± 74	114 ± 32	<0.05	15.8 ± 8.2	9.3 ± 2.7	<0.05	400 ± 132	494 ± 198	N.S.
Pancreas	436 ± 189	298 ± 118	<0.05	23.5 ± 9.6	17.4 ± 6.7	N.S.	339 ± 79	381 ± 216	N.S.
Oesophagus	100 ± 33	127 ± 51	N.S.	16.1 ± 22.4	8.0 ± 4.0	N.S.	280 ± 104	364 ± 179	N.S.
Muscle	131 ± 35	154 ± 38	N.S.	3.5 ± 1.8	5.9 ± 6.7	N.S.	112 ± 51	159 ± 83	N.S.
Rib	102 ± 39	106 ± 56	N.S.	4.7 ± 2.3	6.4 ± 3.9	N.S.	172 ± 94	201 ± 136	N.S.
Nervous tissue	71 ± 36	54 ± 23	N.S.	11.7 ± 7.98	8.3 ± 3.3	N.S.	480 ± 295	166 ± 98	<0.001
Placenta	80 ± 16	64 ± 8	<0.05	8.2 ± 5.8	6.9 ± 2.4	N.S.	623 ± 344	605 ± 255	N.S.
Umbilical Cord	60 ± 20	52 ± 21	N.S.	8.8 ± 7.2	7.6 ± 4.7	N.S.	432 ± 344	231 ± 170	N.S.

lowered tissue zinc concentrations in fetuses with NTD. In this study trace element concentrations in control fetuses were similar to those reported by others[5], whereas the concentration of zinc was elevated in several tissues from NTD fetuses. These results contrast with the findings of congenital defects and reduced tissue zinc concentrations in the offspring of animals fed diets deficient in zinc[6], suggesting that the zinc-deficient animal model may not be relevant to the majority of human NTD occurrences, in Western populations at least. The observations of hyper-zincaemia in anencephaly[4] and of elevated maternal hair zinc concentrations in NTD pregnancy[7] are also not compatible with the animal model. Zinc retention in certain tissues may reflect a defect of zinc metabolism in NTD fetuses.

ACKNOWLEDGMENTS

We are grateful to Birthright for supporting this research and to Birthright and the Wellcome Trust for grants to enable the presentation of this paper.

REFERENCES

1. L. E. Sever and I. Emanual, Is there a connection between maternal zinc deficiency and congenital malformations of the central nervous system in man? Teratology 7: 117 (1973).
2. K. M. Hambidge, K. H. Neldner and P. A. Walravens, Zinc, acrodermatitis enteropathica and congenital malformations, Lancet (i): 577 (1975).
3. M. H. Soltan and D. M. Jenkins, Maternal and fetal plasma zinc concentration and fetal abnormality, Br. J. Obstet. Gynaecol. 89: 56 (1982).
4. A. W. Zimmerman, Hyperzincaemia in anencephaly: a clue to the pathogenesis of neural tube defects? Neurology 34: 433 (1984).
5. C. E. Casey and M. F. Robinson, Copper, manganese, zinc, nickel, cadmium and lead in human foetal tissues, Br. J. Nutr., 39: 639 (1978).
6. L. S. Hurley and H. Swenerton, Lack of mobilization of bone and liver zinc under teratogenic conditions of zinc deficiency in rats, J. Nutr., 101: 597 (1974).
7. K. E. Bergmann, G. Makosch and K. H. Tews, Abnormalities of hair zinc concentration in mothers of newborn infants with spina bifida, Am. J. Clin. Nutr., 33: 2145 (1980).

EFFECTS OF DIETARY BORON, ALUMINUM AND MAGNESIUM ON SERUM ALKALINE

PHOSPHATASE, CALCIUM AND PHOSPHORUS, AND PLASMA CHOLESTEROL IN

POSTMENOPAUSAL WOMEN

F.H. Nielsen, L.M. Mullen, S.K. Gallagher,
J.R. Hunt, C.D. Hunt, and L.K. Johnson

United States Department of Agriculture, Agricultural
Research Service, Grand Forks Human Nutrition Research
Center, Grand Forks, ND

INTRODUCTION

Concerns about bone disorders, especially osteoporosis, of humans has stimulated interest in calcium, phosphorus and magnesium nutrition and metabolism. Calcium has received the most attention, and intakes are now being recommended at levels difficult to achieve, up to 1500-2000 mg/day. These recommendations are made despite evidence indicating that massive intakes of calcium do not prevent bone loss (1), and that certain population groups with a low incidence of osteoporosis consume relatively low amounts of calcium (2). Thus, we decided to examine the possible effects on major mineral metabolism of some dietary substances other than cholecalciferol, calcium and fluoride; these substances were aluminum, magnesium and boron.

METHODS

Thirteen postmenopausal women were housed in a metabolic unit, and fed a diet made from conventional foods supplying about (per day) 600 mg calcium, 870 mg phosphorus, 116 mg of magnesium, and 0.25 mg boron. After 23 days of equilibration, during which the basal boron-low diet supplemented with 200 mg of magnesium/day was fed, all women participated in four dietary periods of 24 days. These periods were: 1) basal diet only, 2) +1000 mg aluminum/day, 3) +200 mg of magnesium/day, and 4) +1000 mg aluminum and 200 mg of magnesium/day. After completing this phase of the study, 12 women participated in two additional 24-day dietary periods in which the basal diet was supplemented with 3 mg of boron/day. Seven women were fed: 1) the boron basal diet only, and 2) +1000 mg aluminum/day. The other five women were fed: 1) +200 mg of magnesium/day, and 2) +200 mg of magnesium and 1000 mg aluminum/day. The study protocol was approved by the Institutional Review Board of the University of North Dakota and the Human Studies Committee of the U.S. Department of Agriculture. Calcium and magnesium were determined by using standard atomic absorption methods. Phosphorus, alkaline phosphatase activity and cholesterol were determined by using standard colorimetric procedures. Statistical treatment of data was done by using repeated measures of ANOVA.

RESULTS AND DISCUSSION

Because previous reports indicated that antacids containing aluminum affected calcium and phosphorous metabolism (3), we expected to see numerous aluminum effects in this study; however, this did not occur. The findings that high dietary aluminum depressed plasma cholesterol regardless of dietary magnesium, and elevated serum phosphorus in magnesium-low women, were the only ones of consequence (Table 1).

The magnesium treatment also did not markedly affect mineral metabolism in the first four dietary periods. In addition to the effect, through an interaction with aluminum, on serum phosphorous noted above, magnesium deprivation slightly, but significantly, elevated the alkaline phosphatase activity, and depressed the magnesium concentration in serum (Table 1). The only other magnesium finding of consequence, which is not shown, was that magnesium deprivation depressed the urinary excretion of magnesium.

Table 1. Effect of Magnesium and Aluminum on Selected Variables.

Dietary Treatment, mg/day		Serum Alkaline Phosphatase, units	Plasma Cholesterol, mg/dl	Serum P, mg/dl	Serum Mg, mg/dl	Serum Ca, mg/dl
Mg	Al					
116	0	2.63	210	3.97	2.07	9.6
116	1000	2.68	204	4.16	2.05	9.6
316	0	2.58	209	4.05	2.09	9.6
316	1000	2.54	201	3.98	2.10	9.6
Significant effects (P Value)		Mg (0.03)	Al (0.006)	Mg x Al (0.04)	Mg (0.05)	—

Perhaps the most impressive magnesium effect was that magnesium status seemed to affect the response to dietary boron. Boron supplementation induced changes in postmenopausal women consistent with the prevention of calcium loss and bone demineralization. In Table 2, the data indicate that boron supplementation elevated percent ionized calcium in serum and decreased the urinary excretion of calcium; the effect of boron seems to be more marked in the magnesium-deprived than the magnesium-adequate postmenopausal women.

Table 2. Effect of Boron and Aluminum on Selected Calcium Indices.

Dietary Treatment, mg/day		Mg-low Diet		Mg-adequate Diet	
		Serum ionized Ca, %	Urinary Ca excretion, g/24 hrs	Serum ionized Ca, %	Urinary Ca excretion, g/24 hrs
B	Al				
0.25	0	49.7	0.117	49.5	0.132
0.25	1000	49.1	0.124	49.1	0.128
3.25	0	52.1	0.065	51.0	0.104
3.25	1000	51.6	0.073	51.5	0.113
Significant effects (P Values)		B (0.002)	B (0.0004)	B (0.02)	B (0.001)

To summarize, findings were obtained indicating that dietary boron affects major mineral metabolism; magnesium status apparently affects the response to dietary boron. High dietary aluminum (1000 mg/day as aluminum hydroxide) did not markedly affect mineral metabolism.

REFERENCES

1. G.S. Gordon and C. Vaughan, Calcium and osteoporosis, J. Nutr. 116:319 (1986).
2. D.M. Hegsted, Calcium and osteoporosis, J. Nutr. 116:2316 (1986).
3. H. Spencer and L. Kramer, Antacid-induced calcium loss, Arch. Int. Med. 143:657 (1983).

TRACE ELEMENTS IN INFANCY: A SUPPLY/DEMAND PERSPECTIVE
(Eric M. Underwood Memorial Lecture 1987)

Bo Lönnerdal

Departments of Nutrition and Internal Medicine
University of California
Davis, CA 95616 USA

INTRODUCTION

The rapidly growing newborn is particularly vulnerable to imbalances in
nutrient supply. In early life, milk or infant formula will be the sole
diet for long periods of time, putting strong demands on the particular diet
to provide adequate but not excessive quantities of each nutrient. Later in
life, the variety of the diet, as well as better homeostatic regulation,
diminish this risk. While we know the protein and energy requirements of
infants reasonably well, our knowledge regarding the micronutrients is more
limited. This is particularly evident for the trace elements, which are
present in milk in comparatively low concentrations (1). For this reason,
infant formulas intended to replace human milk are supplemented with trace
elements, with the possibility of excesses to occur (2). Thus, both deficiency and toxicity of trace elements may occur during infancy (3). This
situation is further complicated by the interactions that occur among trace
elements, i.e., while a need for supplementation with one trace element is
recognized, the possibility of this element reducing the absorption of
another element is less commonly taken into account. This is illustrated by
the fact that the Committee on Nutrition of the American Academy of Pediatrics have established minima for the level of several trace elements in
formulas, while a maximum has only been recommended for iron (4). On the
other hand, it is difficult to issue strong recommendations before an incidence of trace element deficiency or toxicity is documented. This is a
difficult task since there are established clinical tools to detect only
iron deficiency. Zinc, copper, manganese and selenium status are much more
difficult to assess and blood or serum levels are not tested as a clinical
routine. Further compounding the lack of tests is the vague nature of symptoms of trace element deficiency/toxicity in their less pronounced forms.
General failure to thrive and possibly slower than normal growth are known
consequences of trace element deficiency in human infants, but are too non-
specific to trigger a suspicion of trace element deficiency. While the mentioned signs may not be of serious concern, the recent findings of marginal
trace element deficiency causing impairment in immune competence (5), in
turn leading to a higher risk for infection, should be of concern.

ISOTOPE METHODOLOGY

When considering measurements of trace element bioavailability from
infant diets, several considerations are important. The first one concerns

189

the validity of the labeling method used, if isotopes are employed. If an extrinsic label can be shown to exchange with the trace element in the diet, this provides an opportunity to study trace element absorption and retention. This can be documented by comparing a diet labeled both extrinsically and intrinsically with different isotopes. For many diets relevant to the human infant, particularly human milk, this is difficult (stable isotopes) or ethically not permissible (radioisotopes). We have explored an indirect approach to assess the validity of extrinsic labeling. We have reasoned that if an extrinsic radiolabel distributes among all binding ligands in milks and formulas that can be separated by a battery of biochemical methods in a fashion identical to that of the native element, it is reasonable to believe that isotope exchange occurs. We have shown that this occurs for zinc (6), copper (7) and manganese (8), but not for iron (9).

DIGESTION IN INFANTS

Trace elements rarely occur in free form in the diet, but rather are associated with proteins or complexed to low molecular weight ligands. While digestion of proteins in adults is assumed to be virtually complete, the situation for the newborn infant with immature digestive functions is completely different. Acid secretion is low, leading to a relatively high gastric pH which in turn minimizes the action of pepsin (10). Pancreatic enzyme output is also lower than at older age, reducing the capacity for intestinal proteolysis. This limitation is further compromised by the short transit time for ingested food in the infant's gastrointestinal tract. Thus, a second important consideration is the maturity (age) of the infant, in that it will affect release of minerals from dietary proteins. A third important consideration is that infant diets contain trace element binding proteins, such as casein and lactoferrin, which are remarkably resistant towards proteolytic degradation. Therefore, the interaction of these intact or only partially digested proteins with the mucosal uptake sites needs to be determined. Finally, the term "bioavailability" should be a measure of what is retained in the body and thus what the body can utilize for various functions. Trace element retention in infants is also likely to be different from adults as pancreatic output and biliary secretion are lower and mucosal uptake and tissue incorporation higher in infants than in adults. We have therefore chosen to use a variety of models to study bioavailability of trace elements to infants. The suckling rat is a useful tool for looking at tissue specific uptake and retention and can be used if the data obtained can be shown to correlate to data from humans. The infant Rhesus monkey has similar gastrointestinal development to that of human infants and receives a milk similar to that of human milk (11). Finally, radioisotopes can be used in adult humans, not to mimic the human infant, but at least to study absorption and retention in the same species. We have reasoned that if differences in trace element bioavailability can be demonstrated for human adults, they are likely to occur in the infant as well and perhaps to an even larger degree because of the previously mentioned considerations.

IRON NUTRITION

Iron deficiency in infants is common in many parts of the world (12). Consequences of prolonged iron deficiency are anemia, decreased immune competence, impaired energy metabolism and learning disabilities (13). The reasons for iron deficiency in this age group are several: introduction of formulas which are low in iron and/or that have a low bioavailability of iron, early introduction of solid foods and frequent infections (14). Breast-feeding, on the other hand, has been shown to result in a low incidence of anemia during the first six months of life (15), while at an older age this incidence increases. This low incidence of anemia in breast-fed infants combined with the fact that human milk has a low concentration of iron, led to the hypothesis that iron bioavailability from human milk is high compared to other infant diets. Saarinen et al. (16) used extrinsic labeling of human milk and formula and found a higher retention of iron from human milk than from formula in human infants. The validity of the extrinsic labeling method had not been investigated and it is likely that these investigators studied the absorption of iron from lactoferrin in human milk, rather than from whole human milk (9). Lactoferrin binds about one-third of all iron in human milk (17), but it has been shown that an extrinsic label virtually exclusively binds to lactoferrin (9). Nevertheless, the study indicates that iron is well utilized from human milk lactoferrin.

We have assessed the bioavailability of iron from lactoferrin in suckling pigs (18) and mice (19). These studies indicated a rapid and higher uptake of iron from lactoferrin than from an inorganic salt. We have subsequently studied the uptake of iron from lactoferrin using brush border membrane vesicles from infant Rhesus monkeys (20). Similar to human milk, Rhesus milk is high in lactoferrin concentration and lactoferrin from these two species have considerable similarity with regard to amino acid and carbohydrate side chain composition (21). Lactoferrin is a glycoprotein with a molecular weight of 80,000 daltons which can bind two ferric ions concomitant with binding of HCO_3^- or CO_3^{2-}. The binding constant for iron is very high, 10^{30}, and since lactoferrin is present in milk in a very unsaturated form ($\sim 5\%$) it has been suggested to have a bacteriostatic function (22). To exert the two suggested physiological functions in the gut of the newborn, however, the major part of its structure must be maintained. That this is the case can be demonstrated by the findings of intact human lactoferrin in the stool of breast-fed term (23) and preterm infants (24). This is most likely explained by the relative stability of lactoferrin against proteolytic degradation (25). Thus, it is likely that the structural features of lactoferrin needed for iron-binding are maintained.

A receptor for lactoferrin has been found in brush border membranes from monkey (30) as well as the rabbit (26). In the primate, the receptor appears to require terminal fucose residues on the glycan chain of lactoferrin, as fucosidase treatment as well as addition of high levels of a fucose polymer will inhibit binding (27). Monkey and human lactoferrin contain terminal fucose residues, while lactoferrins from other species, such as the cow, do not (28). Consistent with this finding, bovine lactoferrin does not deliver iron to the monkey brush border receptor, while human lactoferrin does. The receptor has a high affinity for lactoferrin, $K_m = 9 \times 10^{-6}$ M, and transferrin does not bind. In the monkey, lactoferrin receptors were found at all ages, including the fetal intestine, but the number of receptors was found to be highest during infancy. We therefore suggest that iron uptake from human milk is mediated by a brush border receptor for lactoferrin.

ZINC NUTRITION

In contrast to iron deficiency, the prevalence of zinc deficiency in infants is not known. One reason for this is the lack of common diagnostic tools for determining zinc status. While plasma zinc often is used as an index of zinc status, many other factors such as stress, infection and time since last meal will affect this parameter. Similarly, hair zinc levels are subject to many influences besides zinc status (29). Another reason may be the non-specific nature of the signs of mild zinc deficiency: slow growth, failure to thrive and impaired immune capacity.

Zinc deficiency in healthy, term infants was first described by Hambidge et al. (30), who found that male formula-fed infants were shorter and had lower plasma zinc values than breast-fed infants and that feeding the same formula with a higher level of zinc normalized plasma zinc and growth. More pronounced forms of zinc deficiency have been described in premature infants fed human milk or formula (31,32). These infants, who are likely to have a higher than normal requirement for zinc because of rapid catch-up growth and lower than normal stores, presented with severe dermatitis which was cured by zinc supplementation. These signs of zinc deficiency were similar to those of children with acrodermatitis enteropathica, an inborn error of zinc metabolism which was first described by Moynahan (33).

A difference in the bioavailability of zinc from human milk and cow's milk was suggested from the study of Hambidge et al. (30), since the formula contained a level of zinc similar to that of human milk. It has subsequently been shown in human adults that plasma zinc uptake (34) as well as whole body retention of zinc (35) is higher from human milk than from cow's milk. The reasons for this difference are not completely known, but it is likely that the high casein content of cow's milk has a negative effect on zinc absorption. We have shown in human adults that zinc absorption from casein-predominant cow's milk-based formula is lower than from whey-predominant formula (36). In contrast to cow's milk, zinc in human milk is bound to casein to a small extent and the major zinc-binding ligands are citrate, serum albumin and the milk fat globule membrane (37). In addition, zinc absorption from human casein is higher than from bovine casein in our suckling rat pup model (38). Zinc absorption is similar from human and cow's whey and it is unlikely that lactoferrin in human milk plays any role in zinc absorption as a radiolabel of zinc does not bind to lactoferrin under physiological conditions and lactoferrin isolated from human milk does not contain zinc (8). Similarly, picolinic acid, which has been proposed to be involved in zinc absorption from human milk (39), is an unlikely candidate, as this tryptophan metabolite is very low in concentration (40) and does not bind zinc in human milk (41). Citrate, on the other hand, which binds zinc in human milk, is present in high concentration in both human and cow's milk, possibly explaining the high zinc absorption from both human and bovine whey.

In our studies of zinc absorption from infant diets in humans, we found a very low absorption from soy formula (35). Since addition of phytate to cow's milk formula at a level similar to that of soy formula drastically reduced zinc absorption to the level found for soy formula (36), this highly negatively charged compound was implicated as the cause. We have subsequently studied zinc absorption from dephytinized soy formula in infant Rhesus monkeys and found a similar degree of absorption to that observed for cow's milk formula (42). Thus, removal of phytate during processing of soy protein isolates could have a substantial positive effect on zinc absorption. Low plasma zinc values in soy formula-fed infants have been shown by Craig et al. (43), suggesting impaired zinc status of infants consuming this type of formula for longer periods of time.

The infant's potential to respond to low zinc supply from the diet should also be considered. Ziegler et al. (44) have shown that percentage zinc absorption determined by stable isotopes was higher in infants fed a formula low in zinc (1 mg/L) than in those infants that were fed the same formula with a higher level of zinc (6 mg/L). The extent to which such a homeostatic regulation of zinc absorption can compensate for the lower zinc supply may, however, be limited as the total amount of zinc absorbed was lower in the group fed the low level of zinc. Another indication of limitations in homeostatic up-regulation of zinc absorption is obtained from the case reports of zinc deficiency in preterm infants fed breast milk with an unusually low concentration of zinc (45). Thus, while compensatory mechanisms may be adequate to prevent severe zinc deficiency in term infants, the increased demand for zinc in preterm infants may not be met.

COPPER NUTRITION

Copper deficiency in infants is usually only observed in preterm infants that have been fed cow's milk (46). This is most likely explained by the lower than normal copper stores in these infants combined with a low level of copper in cow's milk (1). Possibly also contributing is a low bioavailability of copper from cow's milk as compared to particularly human milk but also cow's milk formula (7). It is thus possible that processing of cow's milk formula has a beneficial effect on copper bioavailability. Absorption of copper from soy formula was found to be low, but the relatively generous level of copper in this formula (0.6 mg/L) as compared to human and cow's milk (0.1-0.2 mg/L) may prevent copper deficiency. Similar to zinc, however, the signs of mild copper deficiency, which include neutropenia, leukopenia and failure to thrive (47), are non-specific and there is no routine clinical test for copper status. Thus, the true incidence of mild copper deficiency is not known. Severe copper deficiency, however, is manifested by anemia (which does not respond to iron supplements), bone abnormalities and hair and skin discoloration and is therefore easier to recognize.

MANGANESE NUTRITION

The intake and metabolism of manganese in infants has so far received little attention. Concern was raised, however, when many formulas were found to have considerably higher levels of manganese than human milk (3,48). The reasons for these high levels are not completely clear, but may be a combination of intentional supplementation and unintentional contamination from the processing of these products. High intakes of manganese can lead to toxicity (49) and correlations of high hair levels of manganese in older children with learning disabilities have implicated manganese toxicity as a possible cause (50,51). We have shown in our suckling rat pup model that manganese is absorbed to a very high degree from all milks and formulas tested and that at a young age, a high proportion of absorbed manganese is retained by the brain (52). This is of concern since the primary target tissue for manganese toxicity is the extrapyramidal part of the brain (49). At an older age, manganese absorption is lower and the fraction of manganese retained by the brain is also low. This may be correlated to the onset of more efficient secretion of bile, the major excretory pathway for manganese. High absorption of manganese may also be caused by impaired iron status. We have shown that a human subject with low hemoglobin levels absorbed 45% of manganese from formula while subjects with adequate iron status only absorbed 5-10% from the same diet (53). This is consistent with previous findings on iron and manganese sharing the same absorptive pathway (54). Thus, it is possible that an iron-deficient infant may absorb iron to an even higher extent than normal. It is also possible that iron supplementation of formula, which can lower manganese absorption in humans (55), may be of benefit as it would reduce the amount of manganese absorbed from formulas naturally higher in manganese than human milk.

CONCLUDING REMARKS

Although our understanding of the dietary factors affecting trace element absorption has been growing considerably, much less is known about the infant's ability to respond to a situation where the body's demand for an element is not met by the supply. Can a breast-fed infant which admittedly is receiving a diet low in iron up-regulate absorptive mechanisms? Will a formula low in zinc or copper cause homeostatic regulation of zinc or copper metabolism by increasing uptake and/or decreasing endogenous losses? Is it possible that a high intake of manganese can down-regulate iron absorption? It is obvious from these questions that our understanding of adaptive mechanisms in infants is limited and that further long-term studies are needed.

REFERENCES

1. B. Lönnerdal, C. L. Keen, and L. S. Hurley, Ann. Rev. Nutr. 1:149-174 (1981).
2. B. Lönnerdal and C. L. Keen, in: "Reproductive and Developmental Toxicity of Metals," T. W. Clarkson, G. F. Nordberg and P. R. Sager, eds., Plenum Press, New York, pp. 759-776 (1983).
3. B. Lönnerdal, C. L. Keen, M. Ohtake, and T. Tamura, Am. J. Dis. Child. 137:433-437 (1983).
4. Committee on Nutrition, American Academy of Pediatrics. G. B. Forbes and C. W. Woodruff, eds., American Academy of Pediatrics, Elk Grove Village, IL, 421 pp. (1985).
5. M. E. Gershwin, R. Beach, and L. S. Hurley, "Nutrition and Immunity," Academic Press, New York, 405 pp. (1985).
6. B. Sandström, C. L. Keen, and B. Lönnerdal, Am. J. Clin. Nutr. 38:420-428 (1983).
7. B. Lönnerdal, J. G. Bell, and C. L. Keen, Am. J. Clin. Nutr. 42:836-844 (1985).
8. B. Lönnerdal, C. L. Keen, and L. S. Hurley, Am. J. Clin. Nutr. 41:550-559 (1985).
9. B. Lönnerdal, in: "Iron Nutrition in Infancy and Childhood," A. Stekel, ed., Nestle, Vevey/Raven Press, New York, pp. 95-117 (1984).
10. E. Lebenthal, P. C. Lee, and L. A. Heitlinger, J. Pediatr. 102:1-6 (1983).
11. B. Lönnerdal, C. L. Keen, C. E. Glazier, and J. Anderson, J. Pediatr. Res. 18:911-914 (1984).
12. P. R. Dallman, M. A. Siimes, and A. Stekel, Am. J. Clin. Nutr. 33:86-91 (1980).
13. D. Vyas and R. K. Chandra. in: "Iron Nutrition in Infancy and Childhood," A. Stekel, ed., Nestle, Vevey/Raven Press, New York, pp. 45-59 (1984).
14. J. D. Reeves, R. Yip, V. A. Kyley, and P. R. Dallman, J. Pediatr. 105:874-879 (1985).
15. M. A. Siimes, L. Salmenperä, and J. Perheentupa, J. Pediatr. 104:196-200 (1984).
16. U. M. Saarinen, M. A. Siimes, and P. R. Dallman, J. Pediatr. 91:36-39 (1977).
17. G.-B. Fransson and B. Lönnerdal, J. Pediatr. 96:380-384 (1980).
18. G.-B. Fransson, K. Thoren-Tolling, B. Jones, L. Hambraeus, and B. Lönnerdal, Nutr. Res. 3:373-384 (1983).
19. G.-B. Fransson, C. L. Keen, and B. Lönnerdal, J. Pediatr. Gastroenterol. Nutr. 2:693-700 (1983).
20. L. A. Davidson and B. Lönnerdal, in: "Proteins of Iron Metabolism," G. Spik, J. Montreuil, R. R. Crichton and J. Mazurier, eds., Elsevier, Amsterdam, pp. 275-278 (1985).
21. L. A. Davidson and B. Lönnerdal, Pediatr. Res. 20:197-201 (1986).
22. J. J. Bullen, H. J. Rogers, and L. Leigh, Br. Med. J. 1:69-75 (1972).
23. L. A. Davidson and B. Lönnerdal, Acta Paediatr. Scand. 76:733-740 (1987).
24. R. J. Schanler, R. M. Goldblum, C. Garza, and A. S. Goldman, Pediatr. Res. 20:711-715 (1986).
25. J. H. Brock, F. Arzabe, F. Lampreave, and A. Pineiro, Biochim. Biophys. Acta 446:214-225 (1976).
26. J. Mazurier, J. Montreuil, and G. Spik, Biochim. Biophys. Acta 821:453-460 (1985).
27. L. A. Davidson and B. Lönnerdal, Fed. Proc. 45:588 (1986).
28. G. Spik, B. Coddeville, D. Legrand, J. Mazurier, D. Leger, M. Goavec, and J. Montreuil. in: "Proteins of Iron Metabolism," G. Spik, J. Montreuil, R. R. Crichton and J. Mazurier, eds., Elsevier, Amsterdam, pp. 47-51 (1985).
29. K. M. Hambidge, Am. J. Clin. Nutr. 36:943-949 (1982).

30. K. M. Hambidge, P. A. Walravens, C. E. Casey, R. M. Brown, and C. Bender, J. Pediatr. 94:607-608 (1979).
31. P. J. Aggett, D. J. Atherton, J. More, J. Davey, H. T. Delves, and J. T. Harries, Arch. Dis. Child. 55:547-550 (1980).
32. A. W. Zimmerman, K. M. Hambidge, M. L. Lepow, R. K. Greenberg, M. L. Stover, and C. E. Casey, Pediatrics 69:176-183 (1982).
33. E. J. Moynahan, Lancet ii:399-400 (1974).
34. C. E. Casey, P. A. Walravens, and K. M. Hambidge, Pediatrics 68:394-396 (1981).
35. B. Sandström, Å. Cederblad, and B. Lönnerdal, Am. J. Dis. Child. 137:726-729 (1983).
36. B. Lönnerdal, Å. Cederblad, L. Davidsson, and B. Sandström, Am. J. Clin. Nutr. 40:1064-1070 (1984).
37. B. Lönnerdal, B. Hoffman, and L. S. Hurley, Am. J. Clin. Nutr. 36:1170-1176 (1982).
38. B. Lönnerdal, C. L. Keen, J. G. Bell, and L. S. Hurley, in: "Trace Elements in Man and Animals (TEMA) - 5," C. F. Mills, I. Bremner, and J. K. Chesters, eds., Commonwealth Agricultural Bureaux, Farnham Royal, U.K., pp. 258-261 (1985).
39. G. W. Evans and P. E. Johnson, Pediatr. Res. 14:876 (1980).
40. T. Rebello, B. Lönnerdal, and L. S. Hurley, Am. J. Clin. Nutr. 35:1-5 (1982).
41. L. S. Hurley and B. Lönnerdal, Pediatr. Res. 15:166-167 (1981).
42. B. Lönnerdal, J. G. Bell, A. G. Hendrickx, and C. L. Keen, Am. J. Clin. Nutr. 43:674 (1986).
43. W. J. Craig, L. Balbach, S. Harris, and N. Vyhmeister, J. Am. Coll. Nutr. 3:183-186 (1984).
44. E. E. Ziegler, R. Figueroa-Colón, R. E. Serfass, and S. E. Nelson, Am. J. Clin. Nutr. 45:849 (1987).
45. S. A. Atkinson and B. Lönnerdal, in: TEMA-6 (this volume).
46. A. Cordano, in: "Zinc and Copper in Clinical Medicine," K. M. Hambidge and B. C. Nichols, Jr., eds., SP Medical and Scientific Books, New York, pp. 119-126 (1978).
47. P. A. Walravens, Clin. Chem. 26:185-189 (1980).
48. D. Stastny, R. S. Vogel, and M. F. Picciano, Am. J. Clin. Nutr. 39:872-878 (1984).
49. C. L. Keen and B. Lönnerdal, in: "Manganese in Metabolism and Enzyme Function," F. C. Wedler and V. Schramm, eds., Academic Press, New York, pp. 35-49 (1986).
50. O. Pihl and M. Parkes, Science 198:204-206 (1977).
51. P. G. Collipp, S. Y. Chen, and S. Maitinsky, Ann. Nutr. Metab. 27:488-494 (1983).
52. C. L. Keen, J. G. Bell, and B. Lönnerdal, J. Nutr. 116:395-402 (1986).
53. B. Sandström, L. Davidsson, Å. Cederblad, and B. Lönnerdal, Fed. Proc. 46:570 (1987).
54. A. B. R. Thomson, D. Olatunbosun, and L. S. Valberg, J. Lab. Clin. Med. 78:643-655 (1971).
55. L. Davidsson, Å. Cederblad, B. Lönnerdal, and B. Sandström, in: TEMA-6 (this volume).

TRACE ELEMENTS IN MALNOURISHED POPULATIONS

M.H.N. Golden and B.E. Golden

Wellcome Trace Element Research Group, Tropical Metabolism
Research Unit, University of the West Indies, Kingston 7,
Jamaica

INTRODUCTION:

Malnutrition, in its various forms, is the most prevalent serious
disease in the world. It is not only encountered in areas of famine and
after natural disasters: it is also a constant feature wherever there is
poverty, pestilence or civil strife. The morbidity due to malnutrition is
grossly underestimated. In every disease process, one has to consider both
the 'soil' and the 'seed'. The usual effect of malnutrition is to alter the
'soil': in other words, there are biochemical, physiological and immunological changes, which may not, of themselves, produce overt disease but
which profoundly alter the host's response to an aetiological agent. For
example, an upper respiratory infection which is trivial in a normal child
may progress inexorably to produce death in a malnourished child; measles
kills about two million malnourished children a year. One could even
speculate that the progression to clinical AIDS, in an infected person, is
determined by his nutritional wellbeing. Trace elements have a major role
to play in these metabolic changes. The medical **profession records** morbidity on the basis of the aetiological agent, almost without reference to
the state of the 'soil.'

Overt malnutrition takes three major forms: kwashiorkor, which affects
up to 2% of the population of many poor countries; wasting, which normally
affects about 5% of a population except under exceptional circumstances,
when the local prevalence can be higher; and stunting, which affects about
30% of the population of the world. The number of individuals involved
is staggering.

The question arises as to whether trace elements are involved aetio-
logically in these conditions. Sadly, we lack data, for there have been
no epidemiologically sound studies, apart from those involving iron and
iodine. By analogy with iron and iodine, trace element deficiencies should
be extraordinarily prevalent. For much of this paper, we are forced to
extrapolate and draw inferences which may not be warranted.

Characteristically, poor populations tend to have a very restricted
number of items in their diets: indeed, one of the best indicators of
poverty and poor nutrition is a simple food frequency analysis. They often
have seasonal gluts and shortages as crops come to harvest, and almost no
animal protein intake. These populations eat foods which are grown locally:

197

many villages grow and eat their own produce almost entirely. If there is any trace element deficiency in the local soil, it is much more likely to express itself in the human population than in a developed country. The high prevalence of many trace element deficiencies in crops and domestic animals in the Third World is clearly recognised. Given the circumstances under which most Third World populations live, it would be surprising if they were trace element replete.

MINERAL AVAILABILITY FROM TRADITIONAL DIETS

Malnutrition appears to be uncommon amongst groups living in a traditional manner and maintaining their own cultural practices, whereas it is common in those displaced populations that have altered their culinary or dietary practices.

Three general strategies have evolved to increase the mineral availability in traditional diets. First, many staples are fermented by the traditional users. The Amerindians soak cassava in troughs for days; the Polynesians bury breadfruit; wheat flour is fermented with yeast and beans are ground and fermented to beancurd. Second, the seed is germinated: sprouting of beans is widely practised in the Orient. Both these culinary techniques enzymatically break down phytic acid, the major metal complexing agent of foods. The third strategy is to directly increase the mineral content by burning certain species of plant and adding the ash to the prepared food. This practice is widespread among traditional peoples of North America (Kuhnlein, 1980), South America (Levi-Strauss, 1950) and Africa (Junod, 1927). A priore, mineral element deficiencies may have initiated the evolution of similar culinary practice throughout the world. When a new food is introduced to an area, like cassava to Africa or breadfruit to the West Indies, the art of preparing it is not also transferred. Displaced foods as well as displaced people are liable to mineral deficiency, particularly when they rely on that displaced food for a very large proportion of their intake.

KWASHIORKOR

Kwashiorkor is clearly a pathological condition. It is characterised by oedema, fatty liver, skin lesions, mental changes and dyspigmentation. It is not due to protein deficiency. We have evidence that it is caused by a generalised imbalance between the production of free radicals, catalysed by iron, and the capacity of the body to safely dissipate the radicals and repair any damage done (Golden and Ramdath, 1978). Thus we find that vitamin E, zinc, ceruloplasmin, copper and glutathione are particularly depressed in children with kwashiorkor. We also find a very low level of the selenoenzyme, glutathione peroxidase. Selenium and iron may be of particular importance because we find a close association between low glutathione peroxidase and high iron (Fig. 1). Selenium status has also been found to be low in Central America (Burk et al, 1976), Thailand (Levine and Olson, 1970) and Nigeria (Smith, I.F. and Golden, M.H.N., unpublished).

The question now arises as to whether there is likely to be widespread selenium deficiency in those areas of the world where kwashiorkor is common. There are several lines of reasoning that suggest that it may be so, apart from finding low blood selenium in children with kwashiorkor from three continents.

plasma FERRITIN

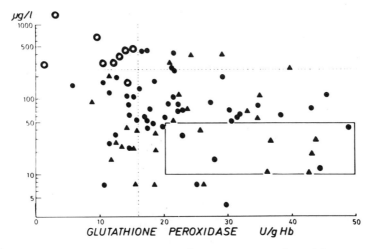

Fig. 1. Admission plasma ferritin and erythrocyte glutathione peroxidase
in children with malnutrition. Each point represents a separate child.
Those children with open circles died. The box represents normal values.

Epidemiologically, kwashiorkor occurs in: (1) regions which have a
high rainfall. (2) areas where malaria, intestinal infections and food
spoilage are common and (3) areas where the staple foods are cassav , yam,
sweet potato, plantain, sago, rice and maize. Conversely, it does not
occur in areas which are dry, where hygienic conditions pertain or where
the grains wheat, millet or sorgum are the staple foods. Of course, each
of these three factors are geographically associated with one another as
well as with kwashiorkor. There are areas where kwashiorkor occurs which
violate each of these associates. The rainfall, the diet and even the
intestinal infections may affect selenium status.

Selenite in an acid or anaerobic soil is readily reduced to selenide
and then to elemental selenium, both of which are very insoluble and unavailable to plants. In alkaline well aerated soils, selenite is oxidised
to the very soluble selenate.

Anionic selenium has an acid pK and is thus highly ionised under most
conditions. It is thus particularly easily leached out of soil by heavy
rainfalls. Selenosis tends to occur in semi arid areas. In contrast, wet
tropical environments, often with reducing soils, would be expected to
favour selenium unavailability and leaching. Two further factors will
modulate the selenium status of an area. First, selenite forms very insoluble complexes with iron and aluminium, common minerals in tropical
soils, which will further reduce their availability. Second, the selenium
content of the parent rock from which the soil is formed, varies widely.
Shield rocks and limestones have low selenium contents, whereas sedimentary
rocks, particularly carboniferous shales, that were laid down in the shallow
seas of the cretaceous period are high in selenium.

In much of the wet tropics where kwashiorkor occurs, the soil has a low ph, Eh, high leaching potential, high iron and aluminium content and low parent rock selenium; ideal conditions to promote selenium deficiency.

Most children in the tropics take feeds highly contaminated with faecal bacteria and the malnourished children have bacterial overgrowth of their stomachs and small intestines. Again, the physical chemistry of selenium may be important as with the reduced Eh of the upper intestine, selenium may be unavailable. It is noteworthy that ruminants are particularly vulnerable to selenium deficiency and unlike monogastric animals excrete most of it as elemental selenium in their faeces (Wright and Bell, 1966).

The hypothesis that kwashiorkor was due to protein deficiency was championed on the grounds that the staple food items have a low protein content, including the sulphur amino acids. It is precisely these foods which have a very low selenium content, whereas wheat has generally a high selenium content (Morris and Levander, 1966).

We postulate, on the grounds of (1) observations from malnourished children in four countries, (2) theoretical considerations on the likely availability of selenium from tropical soils and (3) the known low content of selenium in many staple foods of the tropics, that selenium nutriture is likely to be compromised throughout much of the wet tropics.

STUNTING

By far the commonest form of malnutrition is stunting in height. Recent evidence (McGregor S. - personal communication) of a relationship between the increment in height and the increment in mental development in children seems to indicate that nutritional stunting does indeed matter.

There have been several large scale supplementation studies conducted in children in developing countries (Beaton and Ghassemi, 1982). The supplement has usually been dried skim milk or various combinations of wheat and soya or local legumes. The outstanding feature of all of these studies is that the supplements made very little difference in terms of the height deficits that the children had. In all the studies, the supplement was used very inefficiently. There do not seem to have been any studies where nutrients (as opposed to food) were used, and in none of the studies was mineral nutrition considered. However, we can conclude that it is unlikely that calcium, phosphorus, protein or energy per se were the major determinants of the stunting in these studies. Of the nutrients which give rise to growth failure as the primary manifestation of their deficiency (Golden, 1987) zinc deficiency is the most likely candidate.

There have been reports of growth responses to zinc in adolescents in the Middle East, selected because they had delayed puberty (Golden and Golden, 1985). These studies, although suggestive of zinc deficiency, were very poorly designed and controlled. The results were inconsistent, and many of the data collected were not presented. Nevertheless, it appeared that there were sporadic cases in which zinc supplementation was followed by a dramatic increase in height. In the only well controlled study (Carter et al, 1969) there was no improvement with zinc supplementation.

The most carefully conducted studies of zinc supplementation have been done by Hambidge's group in the USA. They have shown unequivocally that additional zinc leads to increased rates of gain in length in male infants and stunted children (Walravens et al, 1983). The important point is whether these children are typical of Third World infants. Signs of severe zinc deficiency do occur in malnourished infants (Golden et al, 1985), and there is a suggestion that those with the worst zinc status are the most stunted.

Several conclusions are possible from these studies. First, zinc deficiency in man gives rise to limitation of longitudinal growth as it does in experimental animals. Second, males are much more susceptible to zinc deficiency than females. Third, if severe zinc deficiency is evident in children requiring admission to hospital with malnutrition, then, because of the large difference between dietary intake necessary to cause growth retardation and that necessary to give clinical disease, it is likely that zinc limited growth is quite common.

If this is true, then we can indeed generalise Hambidge's results and conclude that mild zinc deficiency affects an enormous number of children. Perhaps we should go back to adding ash to our food.

ACKNOWLEDGEMENT

Our work is fully supported by the Wellcome Trust.

REFERENCES

Beaton, G.H. and Ghassemi, H., 1982, Supplementary feeding programmes for young children in developing countries. Am. J. Clin. Nutr. 35: 864-916.
Burk, R.F., Pearson, W.N., Wood, R.P. and Viteri, F., 1967, Blood selenium levels and in-vitro red cell uptake of 75Se in kwashiorkor. Am. J. Clin. Nutr. 20: 723-733.
Carter, J.P., Grivetti, L.E., and Davis, J.T. et al., 1969, Growth and sexual development of adolescent Egyptian village boys. Am. J. Clin. Nutr. 22: 59-78.
Committee on Medical and Biological Effects of Environmental Pollutants, 1976, Selenium. National Acad. Sci. Washington.
Golden, M.H.N., 1978, The role of individual nutrient deficiencies in growth retardation of children as exemplified by zinc and protein, In: "Linear Growth Retardation in Third World Children", J.C. Waterlow and P. Goyens, ed, Raven Press, New York (in press).
Golden, M.H.N. and Golden, B.E., 1985, Problems with the recognition of human zinc-responsive conditions. In: "TEMA 5". C.F. Mills, I. Bremner and J.K. Chesters, ed, Commonwealth Agric. Bureaux, Slough, pp. 933-938.
Golden, M.H.N., Golden, B.E. and Bennett, F.I., 1985, Relationship of trace element deficiencies to malnutrition, In: "Trace Elements in Nutrition of Children." R.K. Chandra, ed, Raven Press, New York, pp. 185-207.
Golden, M.H.N. and Ramdath, D., 1987, Free radicals in the pathogenesis of kwashiorkor. Proc. Nutr. Soc. (in press).
Junod, H.A., 1927, "The Life of an African Tribe." Volume 2, McMillan, London.
Kuhnlein, H.V. 1980, The trace element content of indigenous salts compared with commercially refined substitutes. Ecol Food Nutr. 10: 113-121.
Levine, R.J. and Olson, R.E., 1970, Blood selenium in Thai children with protein calorie malnutrition. Proc. Soc. Exp. Biol. Med. 134: 1030-1034.
Morris, Y.C. and Levander, O.A., 1970, Selenium content of foods. J. Nutr. 100: 1383-1388.
Walravens, P.A., Krebs, N.F. and Hambidge, K.M., 1983, Linear growth of low income preschool children receiving a zinc supplement, Am. J. Clin. Nutr. 38: 195-201.
Wright, O.L. and Bell, M.C. 1966, Comparative metabolism of selenium and tellurium in sheep and swine, Am. J. Physiol. 211: 6-10.

THE ROLE OF ZINC IN PRENATAL AND POSTNATAL DEVELOPMENT

Carl L. Keen, Michael S. Clegg, Bo Lönnerdal
and Lucille S. Hurley

Departments of Nutrition and Internal Medicine
University of California
Davis, CA 95616 USA

INTRODUCTION

That a severe deficiency of Zn can be teratogenic for mammals was demonstrated twenty years ago by Hurley and Swenerton (1966). Rats fed diets severely deficient in Zn (< 1 ug/g) throughout pregnancy gained less weight than controls and had significantly fewer live fetuses at term than did controls. Surviving fetuses from the deficient dams were characterized by a variety of malformations including skeletal abnormalities, fused or missing digits, micrognathia, misshapen heads and numerous abnormalities of the heart, lung and urogenital system. The frequency of one or more of the above defects in litters obtained from dams fed diets severely deficient in Zn (< 6 ug/g) usually exceeds 75% (Rogers et al., 1985). In addition to the gross morphological abnormalities, biochemical and functional abnormalities can also occur as a result of maternal Zn deficiency. These include defects in pancreatic function, lung metabolism, and immune competence (Hurley, 1981; Beach et al., 1983).

Evidence that Zn deficiency could be a teratogenic agent in humans was provided by Hambidge and coworkers in 1975, who summarized the literature of pregnancy outcome in women with the autosomal genetic recessive disorder acrodermatitis enteropathica (AE), a disorder which mimics Zn deficiency. The authors noted that prior to the introduction of Zn therapy for AE, the risk of abnormal pregnancy outcome in women with this disorder was markedly higher than that for healthy "control" women. Subsequent to the introduction of Zn therapy for AE patients, Brenton et al. (1981) reported that pregnant women with AE are able to maintain normal plasma Zn levels and are characterized by normal deliveries and births.

Evidence that maternal Zn status can significantly affect pregnancy outcome in women has also been provided from a number of prospective studies. Jameson (1976) reported that in a study of 316 pregnancies, a high proportion (60%) of the women who gave birth to infants with congenital defects showed low serum Zn levels in the first trimester. In addition, women who delivered before or after normal term had low serum Zn levels in the third trimester. During the last decade, others have made similar observations of a relationship between maternal serum Zn levels and poor pregnancy outcome (Cavdar et al., 1985; Prema, 1980; Cherry et al., 1981; Soltan and Jenkins, 1982; Mukherjee et al., 1984; Buamah et al., 1984).

MECHANISMS OF TERATOGENESIS OF ZINC DEFICIENCY

It is clear that a deficiency of Zn during early development can result in significant biochemical and structural defects in the developing embryo/ fetus, However, at present, the biochemical lesions underlying these defects are, for the most part, not well understood. It is recognized that Zn deficiency can affect development even during its earliest stages. In studies with preimplantation embryos, Hurley and Shrader (1975) observed that when a Zn-deficient diet was initiated on day 0 of pregnancy, only 70% of the embryos collected from the deficient dams on day 3 of pregnancy were "normal," and by day 4 of pregnancy less than 25% of the embryos were considered morphologically normal (in contrast, over 95% of the embryos collected from control dams on day 4 of pregnancy were normal). Following implantation, rapid deleterious effects of maternal Zn deficiency can still be demonstrated with the observation of gross congenital defects occurring in the fetuses of rats fed Zn-deficient diets for only 4 days during mid-pregnancy (Hurley, 1981).

The rapid and severe effects of transitory maternal Zn deficiency on embryonic/fetal development strongly suggest the absence of mobilizable Zn pools in the dam which can be utilized during brief periods of Zn deficiency. Consistent with this idea is the observation that in rats, plasma Zn concentrations can decrease over 50% of normal values within 24 hours of the introduction of a Zn-deficient diet (Dreosti et al., 1968; Hurley et al., 1982). It is reasonable to suggest that the rapid effects of maternal Zn deficiency on embryonic/fetal development are due to the rapid reduction in the maternal plasma Zn pool, as this is the primary source of Zn for the developing embryo. That the extent of the reduction in maternal plasma Zn concentration is functionally significant is suggested by the report that uterine fluid Zn concentrations were 50% lower in dams which had been fed Zn-deficient diets for 4 days following mating compared to dams fed control diets (Gallaher and Hurley, 1980). The concentration of Zn in uterine fluid is normally about 5 times that of plasma, suggesting active transport of Zn into this fluid from the plasma pool. Thus, the observations of Gallaher and Hurley suggest that the ability of the dam to concentrate Zn in pools essential for the normal development of the embryo is compromised by the reduction in plasma Zn concentration. Additional evidence that the reduction in maternal plasma Zn is critical for the embryo is the observation that if maternal tissue catabolism is increased in the Zn-deficient dam by either withholding dietary Ca, or by reducing the food intake of the Zn-deficient dam, maternal plasma Zn levels increase and the teratogenicity of the Zn deficiency is alleviated (Masters et al., 1986).

A number of hypotheses have been advanced regarding the possible biochemical lesions which may underlie the teratogenicity of embryonic/fetal Zn deficiency. Considerable attention has been given to the idea that one of the basic defects underlying the abnormalities observed in Zn deficiency is abnormal nucleic acid metabolism. DNA synthesis is markedly lower in Zn-deficient embryos and fetuses than in controls, and this lower rate of synthesis has been linked to low activities of DNA polymerase and thymidine kinase (Eckhert and Hurley, 1977; Dreosti et al., 1985). It should be noted that while it has been suggested that the low activities of these two enzymes in Zn-deficient embryos is the result of their absolute requirement for the element, studies to date have not ruled out the possibility that the reduction in the activities of these enzymes is secondary to a more generalized metabolic block. In addition to the effects of Zn deficiency on DNA polymerase and thymidine kinase, abnormalities in chromatin structure have been reported to occur as a result of cellular Zn deficiency. Chromatin is the genetic material of eukaryotes and is composed of DNA, small amounts of mRNA, and the histone and nonhistone proteins. Falchuk et al. (1986), working with Zn-deficient _Euglena gracilis_, and Castro et al. (1986), working with Zn-deficient adult rats, have reported that Zn deficiency makes chromatin more resistant to micrococcal nuclease digestion, suggesting an increased ratio of nuclear protein to DNA and/or an alteration in the conformation of Zn-deficient chromatin. One consequence of the above changes in chromatin could be an alteration in the normal access of RNA polymerase to various genes in the Zn-deficient chromatin. Mazus et al. (1984) have also reported that the chromatin isolated from Zn-deficient Euglena gracilis is characterized by a reduction in the normal amount, and types, of histone proteins; similar findings have been reported by Castro et al. (1986) for the Zn-deficient rat. Since histones are necessary for the proper structure and function of chromatin, the Zn deficiency-induced alterations in histones may represent a significant biochemical lesion in the Zn-deficient cell. If the impairment in nucleic acid synthesis and/or expression is sufficient, this could then result in an interruption of normal cellular differentiation and in alterations in the differential rates of cellular growth necessary for normal morphogenesis. The resulting asynchrony in histogenesis and organogenesis could then result in malformations.

The reduction in cellular growth and movement observed with Zn deficiency could also be due in part to the involvement of Zn in microtubule formation. Microtubules, a major cytoskeletal component found in all eukaryotic cells, are formed through the polymerization of tubulin subunits with small amounts of other proteins and metal ions. It is known that microtubules participate in a number of cellular events including chromosome movement, cell motility, and movement of cytosolic components. These events are vital to developmental processes including morphogenesis and cellular differentiation. It has been well documented that Zn can stimulate the polymerization of purified tubulin in vitro, and Hesketh (1981) observed that a severe dietary deficiency of Zn (0.7 ug Zn/g diet) was associated with a reduction in the rate of brain tubulin assembly in adult pigs and rats. Recently we have demonstrated that brain tubulin assembly in vitro is also lower in adult lactating rats fed diets which are only marginally deficient in Zn (10 ug Zn/g) compared to rats fed control diets (50 ug Zn/g) (Oteiza et al., 1988). Furthermore, based on preliminary results obtained in our laboratory, this lower rate of brain tubulin polymerization is also observed in the suckling offspring of marginally Zn-deficient dams, and in fetuses obtained from dams fed severely Zn-deficient diets during pregnancy. In all three situations, the most pronounced effect observed for the impact of Zn deficiency on in vitro brain tubulin assembly kinetics was during the initial velocity phase. To date, we have been unable to correlate the "Zn deficiency"-induced reduction in brain tubulin polymerization to a reduction in brain tubulin concentration; thus the slower rate of polymerization in the Zn-deficient animals is probably the result of a deficiency of a cofactor essential for the polymerization process. While it is tempting to suggest that the missing cofactor in question is Zn, it must be appreciated that the intracellular pool(s) which provides Zn for tubulin polymerization in vivo has not been identified, and thus a rigorous test of the hypothesis that the reduction in tubulin polymerization is the direct result of a Zn deficit cannot yet be made. However, it is reasonable to hypothesize that the observed impairment in tubulin polymerization may explain some of the other cellular defects associated with Zn deficiency. For example, a defect in tubulin polymerization could explain the finding of chromosomal aberrations, including gaps, fragments, and terminal deletions in chromosomal spreads prepared from liver of Zn-deficient fetuses (Bell et al., 1975). Similarly, maternal Zn deficiency has been reported to affect fetal brain cell cycle kinetics with the deficiency resulting in a block in the G_0G_1 phase (Clegg et al., 1986). While this block could be the result of impairments in nucleic acid and/or protein synthesis, it could also arise as a result of defective tubulin polymerization. Finally, Zn deficiency-induced reductions in tubulin polymerization could result in an interruption of the normal exocytotic and endocytotic processes of the cell. For example,

Rogers et al. (1987) have suggested that the accumulation of iron in the Zn-deficient fetus may be secondary to an impairment in the exocytosis of transferrin-bound iron.

A third mechanism by which Zn deficiency has been suggested to affect embryonic/fetal development is through changes in the susceptibility of the embryo/fetus to peroxidative damage. It has been observed that in livers collected from fetuses of Zn-deficient rats, there are significantly elevated levels of malonaldehyde, an indicator of lipid peroxidation (Dreosti, 1987). The increase in lipid peroxidation products may reflect 1) a role of Zn in the stabilization of membranes, 2) an increase in the pool of unsaturated fatty acids available for peroxidation or 3) a reduction in the normal antioxidant potential of the cell. Regardless of the cause of the increased peroxidation products, evidence has been presented by Dreosti (1987) that the increased peroxidation products observed in the Zn-deficient embryo reflect cellular membrane damage and cellular necrosis. Premature cellular necrosis in areas in which it is destined to occur, or in areas in which it is not programmed to occur, could give rise to a number of gross developmental defects.

A fundamental question that can be asked with regard to the biochemical lesions underlying the teratogenicity of Zn deficiency is whether the effects of Zn deficiency on the embryo are direct, that is, the consequence of an embryonic Zn deficiency, or are they a result of an indirect effect of Zn deficiency on the metabolism of the mother? Evidence that some of the effects of Zn deficiency on the embryo are direct has been provided by the use of rat embryo culture systems. Using this methodology, embryos are removed from control dams during early development (typically from day 8 to day 9.5 in the rat) (New, 1966). The embryos are then placed in culture tubes containing rat serum, cultured for up to 48 hours and examined. Using this system, Meiden et al. (1986) have reported that embryos grown on serum collected from Zn-deficient dams developed abnormally, while embryos grown on serum collected from control dams demonstrated normal development. Significantly, when Zn was added to serum collected from Zn-deficient dams to a concentration equal to that in control serum, the development of embryos grown on this serum was judged to be normal. Thus, taken together, the results from the embryo culture study by Meiden et al. (1986) support the idea that the serum collected from Zn-deficient dams is teratogenic, and that the teratogenicity of the serum is in part due to its low Zn concentration. It should be pointed out that in contrast to Meiden et al. (1986), Record et al. (1985) were unable to produce the teratogenic effects of Zn deficiency in the rat embryo culture system using methodologies similar to those described by Meiden et al. (1986). However, Record et al. (1985) did demonstrate abnormal development in day 9.5 embryos collected from Zn-deficient dams which were then cultured on Zn-deficient serum for 48 hours. Thus results from both studies support the idea that embryo culture systems may prove to be valuable tools for the investigation of the sequence of biochemical defects associated with embryonic Zn deficiency.

MARGINAL ZINC DEFICIENCY

It is clear that the effects of severe Zn deficiency on early development are dramatic. However, in terms of health and reproduction in humans and domesticated animals, the effects of long term marginal deficiency of Zn may be more pertinent than those of acute severe deficiency. Beach et al. (1983) reported that one consequence of prenatal marginal Zn deficiency can be a marked influence on the ontogeny of the immune system. For example, the offspring of mice which have been fed a marginal Zn diet during the last two thirds of pregnancy were characterized by low serum IgM concentrations which persisted to at least six months of age despite their receiving Zn-adequate diets from birth on. Significantly, this effect of marginal Zn deficiency on serum IgM levels persisted through three generations, although the detrimental influence of the Zn deficiency was attenuated with each generation. Vruwink et al. (1988) have observed that, similar to the effects of prenatal Zn deficiency on the immune system, marginal prenatal Zn deficiency can have a persistent effect on the postnatal expression of the Zn-binding ligand, metallothionein. They also observed that mice which had been prenatally deprived of Zn were characterized by an amplification of metallothionein synthesis when they were challenged with a Zn load as adults. The biochemical explanation underlying the effect of prenatal marginal Zn deficiency on metallothionein metabolism is not known; however, it is reasonable to suggest that it may involve gene amplification and/or alterations in the methylation patterns around the metallothionein genes (Vruwink et al., 1988). Regardless of the biochemical lesion underlying the effect of marginal Zn deficiency on metallothionein, it is evident that marginal Zn deficiency during development can result in profound biochemical defects which can persist well into adulthood. The impact that these defects have on an animal's ability to survive and respond to environmental challenges needs to be ascertained. However, it is clear that more attention needs to be given to identifying the subtle metabolic defects which may arise due to early marginal Zn deficiency in both human and domesticated animals.

REFERENCES

Beach, R. S., Gershwin, M. E., and Hurley, L. S., 1983, Am. J. Clin. Nutr, 38:579.
Bell, L. T., Branstrator, M., Roux, C., and Hurley, L. S., 1975, Teratology 12:221.
Brenton, D. P., Jackson, M. J., and Young, A., 1981, Lancet ii:500.
Buamah, P. K., Russell, M., Bakes, M., Milford Ward, A., and Skillen, A. W., 1984, Br. J. Obstet. Gynaecol. 91:788.
Castro, C. E., Alvares, O. F., and Sevall, J. S., 1986, Nutr. Rep. Intl. 34:67.
Cavdar, A. O., Babacan, E., Asik, S., Arcasoy, A., et al., 1985, Nutr. Res. Suppl. I:331.
Cherry, F. F., Bennett, E. A., Bazzano, G. S., Johnson, L. K., Fosmire, G. J., and Barson, H. K., 1981, Am. J. Clin. Nutr. 34:2367.
Clegg, M. S., Rogers, J. M., Zucker, R. M., Hurley, L. S., and Keen, C. L., 1986, Fed. Proc. 45:1086.
Dreosti, I. E., 1987, Neurotoxicology 8:369.
Dreosti, I. E., Tao, S., and Hurley, L. S., 1968, Proc. Soc. Exp. Biol. Med. 127:169.
Dreosti, I. E., Record, I. R., and Manuel, S. J., 1985, Biol. Trace Element Res. 7:103.
Eckhert, C. D., and Hurley, L. S., 1977, J. Nutr. 107:855.
Falchuk, K. H., Gordon, R. R., Stankiewicz, A., Hilt, K. L., and Vallee, B. L., 1986, Biochemistry 25:5388.
Gallaher, D., and Hurley, L. S., 1980, J. Nutr. 110:591.
Hambidge, K. M., Nelder, K. H., and Walravens, P. A., 1975, Lancet i:577.
Hesketh, J. E., 1981, Intl. J. Biochem. 13:921.
Hurley, L. S., 1981, Physiol. Rev. 61:249.
Hurley, L. S., and Shrader, R. E., 1975, Nature (London) 254:427.
Hurley, L. S., and Swenerton, H., 1966, Proc. Soc. Exp. Biol. Med. 123:692.
Hurley, L. S., Gordon, P., Keen, C. L., and Merkhofer, L., 1982, Proc. Soc. Exp. Biol. Med. 170:48.
Masters, D. G., Keen, C. L., Lonnerdal, B., and Hurley, L. S., 1986, J. Nutr. 116:2148.
Mazus, B., Falchuk, K. H., and Vallee, B. L., 1984, Biochemistry 23:42.

Mieden, G. D., Keen, C. L., Hurley, L. S., and Klein, N. W., 1986, J. Nutr. 116:2424.

Mukherjee, M. D., Sandstead, H. H., Ratnaparki, M. V., Johnson, L. K., Milne, D. B., and Stelling, H. P., 1984, Am. J. Clin. Nutr. 40:496.

New, D. A. T., 1966, "The Culture of Vertebrate Embryos," Logos Press, London.

Oteiza, P. I., Keen, C. L., Lonnerdal, B., and Hurley, L. S., 1988, J. Nutr., in press.

Prema, K., 1980, Indian J. Med. Res. 71:554.

Record, I. R., Dreosti, I. R., and Tulsi, R. S., 1985, Aust. J. Exp. Biol. Med. Sci. 63:65.

Rogers, J. M., Keen, C. L., and Hurley, L. S., 1985, Teratology 31:89.

Rogers, J. M., Lonnerdal, B., Hurley, L. S., and Keen, C. L., 1987, J. Nutr. 117:1875.

Soltan, M. H., and Jenkins, D. M., 1982, Br. J. Obstet. Gynaecol. 89:56.

Vruwink, K., Gershwin, M. E., Hurley, L. S., and Keen, C. L., 1988, Proc. Soc. Exp. Biol. Med., in press.

TERATOGENESIS AS A FUNCTION OF MATERNAL AGE IN ZINC-DEFICIENT RATS

I.R. Record and I.E. Dreosti

CSIRO Division of Human Nutrition
Kintore Avenue, Adelaide SA 5000

Earlier studies from this, and other laboratories have shown that, in rats and mice, the anabolic or catabolic status of a pregnant animal can have profound effects on the development of the litter when the dam is fed a diet deficient in zinc. These current experiments were designed to study in greater detail the inter-relationships between maternal anabolism, embryonic development and zinc metabolism in the pregnant rat.

Materials and Methods
Female Sprague-Dawley rats allowed free access to a commercial rat diet containing 62 µg Zn/g were used throughout. After being placed overnight with males of the same strain dams were fed a soya-bean based diet containing either less than 0.5 µg Zn/g or 100 µg Zn/g from the time of detection of sperm (day 0). In the first experiment (using animals weighing 180g or 320g) dams were fed to regimes designed to elicit a maximal teratological response on day 11 or a minimum response on the same day. To compensate for differences in body-weight, the amounts of diet allowed to each animal were adjusted on a (body-weight) 0.75 basis. Dams were killed on either day 11 (to allow examination of the embryos) or day 20 (to examine) the fetuses.

In the second study, dams weighing 220g were allotted to similar dietary treatment groups, but were allowed free access to colony diet from day 12 of gestation when the dams were killed and the fetuses removed for examination.

Results
In the first study embryos from the young (Group A) zinc-deficient dams were significantly smaller in terms of crown-rump length (3.29±0.3 mm is 2.01±0.6 mm) and protein content (138±17 µg vs. 83±40 µg), as well as being developmentally retarded (somite no) and there were dramatically more malformed embryos. Group B zinc-deficient embryos did not show any signs of deformities or growth retardation. Embryos from the mature group A zinc-deficient dams had a lower protein content than their controls (166±36 µg vs. 248±55 µg), but were otherwise apparently unaffected.

Reproductive performances of rats allowed to continue their pregnancy until the 20th day of gestation was also studied. All animals examined had been pregnant at some stage, judging by the presence of either corpora lutea or resorbtion sites. The number of live fetuses was significantly reduced in both juvenile zinc-deficient groups, but not in the mature dams. The incidence of fetal malformations was negligible in the zinc-replete litters, however 89% of the fetuses from the mature zinc-deficient dams and 65% from the juvenile groups were severely malformed. Zinc deficiency resulted in a large decrease in fetal weight in both mature groups, but there was a lesser effect in the younger animals.

Correlations between various fetal and maternal parameters were also examined. Maternal serum zinc levels in the zinc-replete pair-fed groups were inversely related to both conceptual mass (Mature : $r = -0.74$; P <0.05, Juvenile : $r = -0.61$, P<0.05) and total conceptual zinc (Mature : $r = -0.81$, P<0.05, Immature : $r = -0.56$, P<0.05), although there were no similar relationships in the other groups.

In the second study it was found that, of the dams which carried their fetuses to day 20, dietary zinc deficiency during the first half of pregnancy was associated with higher maternal serum zinc levels on day 20 than similar animals fed zinc replete diets throughout gestation. Fetal and placental weights were relatively unaffected by the transient period of zinc deprivation, however when the maternal serum zinc levels of these dams were compared with fetal parameters such as total fetal weight, total placental weight, total conceptual weight and the number of live fetuses there were significant negative correlations were observed (all P<0.001).

Discussion

These studies provide further evidence to support earlier contentions that the maternal metabolic state is of great importance in both embryonic and fetal development. We have demonstrated that younger, more rapidly growing dams have a greater demand for zinc than older, more slowly growing animals, and are therefore less able to meet the demands of the litters when the availability of dietary zinc is reduced. In addition, it can be suggested that the size of the litter can influence the circulating zinc levels in the pregnant dam at the end of gestation. This could provide an explanation for the observed inverse relationship between birthweight and maternal serum zinc levels noted in several human studies. Indeed it can be proposed that elevated maternal serum zinc levels need not be indicative of a defect in the transport of zinc to the fetus.

Although extrapolation from animal studies to the human situation is fraught with danger, it can be suggested that the human female most at risk of having complications during pregnancy due to a deficit of zinc has (a) a marginal zinc intake (b) is at risk from an induced zinc deficiency and (c) is unable to catabolise her own tissues to release zinc.

ZINC SUPPLEMENTS IN INFANTS WITH FAILURE TO THRIVE: EFFECTS ON GROWTH

P.A. Walravens, K.M. Hambidge, and D.M. Koepfer

Department of Pediatrics
University of Colorado Health Sciences Center
Denver, CO   80262

Mild zinc deficiency in childhood is one of the reasons why some children grow poorly. The provision of zinc supplements at nutritional doses to preschool children with low growth centiles resulted in improved height velocity and increased dietary intakes of energy, protein and zinc in the males[1,2]. During our studies of preschool children with mild zinc deficiency it became apparent that growth centiles first started to decline during infancy, generally between 6-15 months of age. Weight for age would initially decline typically over a period of 3-6 months and would drop from a mid-percentile level to a level close to the 5th percentile. During this period of decreased weight gain, length percentiles remained at first on their original curve, but subsequently a few months later declined to lower levels. Such decreases in weight velocity out of proportion to concomitant decreases in length velocity are generally indicative of an inadequate nutritional intake but could also be an early manifestation of mild nutritional zinc deficiency. The latter hypothesis was tested in a double-blind, controlled study of zinc supplementation in infants and toddlers with declining weight percentiles. The questions to be tested included that a nutritional deficiency of zinc was partially responsible for the declining growth centiles which sometimes occur in late infancy and that zinc supplementation under controlled conditions would be accompanied by increased growth velocity, particularly in weight for age. The pertinent methods and results of this study follow.

Participants were infants who received their health care at the Westside Neighborhood Health Center in Denver and who demonstrated in the preceding months a decline in weight for age or weight for height of 20 or more centiles resulting in a weight for age below the 10th percentile. This decline in weight for age percentiles occurred in absence of detectable causes such as neglect, repeated infections or chronic illnesses. The infants were pair-matched for sex, Z-scores for weight, age to the nearest 3 months and ethnic origin. Random assignment to the test group resulted in the infant receiving 5 mg of zinc daily in cherry syrup, whereas the control infants received the syrup alone. Assignment to the supplemental or placebo group was performed by an investigator not in contact with the patients and the double-blind nature of the study was maintained throughout the 6 months of the study duration. Follow-up was scheduled at 1, 3 and 6 months.

Eighty-five children were approached regarding the study and 57 completed the six months of supplementation leaving 14 pairs of boys, 13 pairs of girls and 3 subjects for whom there was no appropriate match. The infants' ages ranged from 7 months to 27 months, with a mean age at onset of 14.9 $\pm$ 4.5 months. Anthropometric measurements were converted to standard deviation scores which are also known as weight for age (WAZ), height for age (HAZ) and weight for height Z scores (WHZ). Initial WAZ for test and control subjects was -2.00 $\pm$ 0.38 and -2.08 $\pm$ 0.39 respectively. The paired differences of changes in WAZ, HAZ, and WHZ were analyzed after 1, 3 and 6 months. The most pertinent changes related to weight gains and the mean paired differences of changes in WAZ between supplemented and control infants are summarized in the table .

TABLE 1. Paired differences in WAZ between test and control infants.(SE)

	0-1 mo	0-3 mo	0-6 mo
Males(14)	0.09 $\pm$ 0.04	0.22 $\pm$ 0.05**	0.15 $\pm$ 0.06*
Females(13)	0.09 $\pm$ 0.08	0.13 $\pm$ 0.05+	0.16 $\pm$ 0.11
Both sexes (27)	0.09 $\pm$ 0.04	0.17 $\pm$ 0.03**	0.15 $\pm$ 0.06+

*p=0.05   +p = 0.025   **p=0.005   (two-tailed)

From a clinical point of view, analysis of the results using the percentage of change in WAZ (Final WAZ-Initial WAZ/Initial WAZ x 100) provides a better perspective on the changes in weight velocity. After three months of zinc supplementation, paired analysis of the differences in percent changes in WAZ showed a mean 11.1 $\pm$ 3.0% improvement for the males (p 0.005) and a 7.8 $\pm$ 2.8% mean improvement for the girls (p 0.025).

A trend towards significant differences in changes in height for age Z-scores was present in the zinc supplemented girls at 6 months. The males however did not show differences in HAZ. Analysis of dietary records showed an inadequate daily intake of energy (800 kcal) and zinc (4.3 mg) in both test and control groups at the beginning of the study. Nutrient and caloric increases in the test group did not achieve statistical significance over the study period, but only small numbers of dietary records were provided by the participants.

This study demonstrates a beneficial effect of zinc in infants with declining growth velocity of primarily nutritional origin. Since zinc per se does not have a known pharmacologic effect on growth the observed improvements must result from the correction of a pre-existing deficiency that started in infancy. This may complicate some cases of nutritional growth deviations if the concomitant anorexia limits attempts at oral rehabilitation. In view of the difficulties in diagnosing mild zinc deficiency in young children, supplementation at nutritional doses should be considered in infants with decreasing weight velocity. (Supported by USDA grant 82-CRCR-1-1006, grant RR 69 from NIH, General Clinical Research Center, and grant 5R22-AM12432 from NIADDKD).

REFERENCES

1. PA Walravens, NF Krebs and KM Hambidge, Linear growth of low income preschool children receiving a zinc supplement, Am J Clin Nutr 38:195 (1983).
2. NF Krebs, KM Hambidge and PA Walravens, Increased food intake of young children receiving a zinc supplement, Am J Dis Child 138:270 (1984).

ZINC RETENTION IN RELATIONSHIP TO FAT-FREE BODY AND CALCIUM ACCRETION IN

EARLY AND LATE PREGNANCY

Janet C. King, Christine A. Swanson, Deborah D. Marino and
Francoise M. Costa

Department of Nutritional Sciences
University of California
Berkeley, CA 94720

INTRODUCTION

In studies of zinc retention during pregnancy, the measured rates of
retention are generally greater than the predicted needs. The predicted
daily zinc need is about 0.2 mg/day in the first half of gestation and 0.6
mg/day in the last half[1]. Zinc retentions of pregnant women tend to be 3 to
7 times the predicted need after correcting for integumental losses[2-5]. It
is unclear why measured rates of zinc retention are so much higher than predicted needs. Possibly, zinc retention was overestimated by the balance
technique, or the predicted need for zinc may be underestimated. Also, pregnant women may retain more zinc than needed if intake is in excess.

The purpose of this study was to quantitate zinc, nitrogen and calcium
retention in a metabolic study of three groups of women: 6 nonpregnant (NP),
6 early pregnant (EP, 10-20 weeks gestation) and 4 late pregnant (LP, 30-40
weeks gestation). Fat-free body (FFB), or lean tissue, gain was calculated
from the nitrogen balance data, and the zinc retention associated with the
FFB was estimated assuming a concentration of 25 µg zinc/g FFB. Calcium retention was measured to determine if there was any correlation between the
retention of these two bone minerals in nonpregnant and pregnant women.

METHODS AND MATERIALS

The women were confined to the metabolic unit for the 20-day study and
fed a constant, semipurified formula diet[6]. The diet provided 16.4 mg zinc/
day and 1.13 g calcium/day. The protein intake equaled 0.8 g protein/kg nongravid standard body weight[7] plus 30 g for pregnancy. Fecal and urinary output was collected throughout the study. The zinc and calcium content of the
urine, fecal and diet samples was determined by flame atomic absorption spectrophotometry (Perkin Elmer Model 560, Perkin Elmer Co., Mountain View, CA)
after the samples were ashed in a low temperature asher (Model LTA-604, International Plasma Corp., Hayward, CA). The total nitrogen (N) content of
the fecal, urine and diet samples was determined by the microKjeldahl method.
Balances were calculated for the last 12 days of study. The N and zinc balances were corrected for integumental losses. The gain of FFB was calculated
from the N balance data by assuming that 1 g N was equivalent to 6.25 g protein and that FFB was 20% protein in the NP group, 8% protein in the EP group
and 11% protein in the LP group. The differences among the NP, EP and LP

Table 1

Zinc, Calcium, Nitrogen and Fat-free Body Accretion

	NP	EP	LP	p
Zinc balance, mg/day	1.17 ± 0.25[1]	0.62 ± 0.49	1.71 ± 0.49	NS
Calcium balance, mg/day	30 ± 16	28 ± 54	102 ± 83	NS
Nitrogen balance, g/day	-0.04 ± 0.11	0.06 ± 0.36	1.58 ± 0.37	<0.05
FFB gain, g/day	-1.1	4.9	89.6	

[1] Mean ± SEM

groups for all variables were tested by one-way analysis of variance (ANOVA).

RESULTS

All three groups of women were in positive zinc balance, and the pregnant women retained more zinc than the predicted need (Table 1). Although the LP women tended to retain more calcium than the NP or EP women, the difference was not significant. Calcium and zinc balances were not correlated. The N retention of the LP women was significantly greater than that of the NP or EP women; the NP and EP women were essentially in N equilibrium. The N retention of the LP women was greater than the theoretical need of 0.98 g/day. Based on the N balance data, the LP women gained about 90 g FFB/day.

DISCUSSION

The retention of both zinc and N was greater than the predicted need in the LP women. Using the value of 25 μg zinc/g FFB, the predicted zinc retention of the LP women was 2.25 mg zinc/day; the measured retention was 1.7 mg/day. Thus, the gain of lean tissue could account for all of the zinc retained. The NP and EP women gained very little or no lean tissue during the study. Yet, they retained 1.2 and 0.6 mg zinc/day, respectively. Bone is the most likely tissue for deposition of this zinc. If all of the retained zinc is stored in bone, the total bone zinc content would have increased 4 and 2%, respectively, in the NP and EP women during the 20-day study. Zinc retention was not associated with calcium retention in the NP and EP women. Possibly, zinc is deposited in bone without concomitant calcium deposition when the intake of available zinc is greater than the need.

REFERENCES

1. C. A. Swanson and J. C. King, Zinc and pregnancy outcome, Am. J. Clin. Nutr. In press.
2. L. J. Taper, J. T. Oliva, and S. J. Ritchey, Zinc and copper retention during pregnancy: the adequacy of prenatal diets with and without dietary supplementation, Am. J. Clin. Nutr. 41:1184 (1985).
3. K. K. Schraer and D. H. Calloway, Zinc balance in pregnant teenagers, Nutr. Metabol. 17:205 (1974).
4. C. A. Swanson and J. C. King, Zinc utilization in pregnant and nonpregnant women fed controlled diets providing the zinc RDA, J. Nutr. 112:697 (1982).
5. C. A. Swanson, J. R. Turnlund, and J. C. King, Effect of dietary zinc sources and pregnancy on zinc utilization in adult women fed controlled diets, J. Nutr. 113:2557 (1983).
6. L. E. Nagy and J. C. King, Energy expenditure of pregnant women at rest or walking self-paced, Am. J. Clin. Nutr. 38:369 (1983).
7. National Research Council, Food and Nutrition Board, "Recommended Dietary Allowances," 9th edition, National Academy of Sciences, Washington, D.C. (1980).

ZINC (Zn) ABSORPTION IN PREMATURE INFANTS

P.L. Peirce, K.M. Hambidge, P.V. Fennessey
L. Miller, and C.H. Goss

Department of Pediatrics
University of Colorado Health Sciences Center
Denver, Colorado 80262

Traditional balance techniques have given variable data on Zn absorption by the premature infant. However, in the majority of studies undertaken prior to 36 weeks post-conception, Zn absorption has not been sufficiently positive to meet calculated requirements for growth. Use of stable isotopes can help to avert some of the potential sources of error with traditional balance studies and allows measurement of true absorption. This was a pilot study of 70 Zn absorption in premature infants, in which isotopically enriched zinc was fed and Zn stable isotope ratios in fecal samples were measured by a Fast Atom Bombardment/Mass Spectrometry (FAB/MS) technique (1).

METHODS

Five formula fed premature infants (table 1) were each studied on one occasion. An aqueous 70 Zn preparation was equilibrated with the formula for 12 hr prior to feeding in one or two consecutive feeds. The preparation provided 30-180 ug Zn or 10-25% of total Zn in the feed. The formula was thoroughly mixed and accurately weighed prior to the gavage feed(s). All feces were collected on ashless filter papers from the time of administration of the 70 Zn enriched feed until the end of the study period. Urine was kept separate by collection in plastic bags maintained under continous aspiration. Milk and fecal samples were dry ashed at 450°C, the ashed dissolved in HCl and total Zn determined by atomic absorption spectrometry. Zinc in the remaining dissolved ash was separated from other sample components by ion exchange column chromatography and m/e 70/64 Zn ratios were determined by FAB/MS. True Zn absorption was determined by subtracting the cumulative enriched isotope excreted in the feces from the enriched isotope ingested. Net Zn absorption was determined from the total Zn intake for 72 hr between markers minus total fecal Zn excretion between these markers (Fig 1).

RESULTS AND DISCUSSION

The timing of excretion of isotope and markers is shown in Fig 1.

Detectable 70 Zn enrichment of feces occured at 19 ± 10 hr and continued
until 34 ± 20 hr. Results of Zn absorption studies are summarized in
Table II. Subject 5 had a high true Zn absorption and net Zn absorption
that corresponded to estimated requirements. This observation is
compatible with the suggestion (2) that maturation of the intestinal
mucosa results in improved Zn absorption after about 36 wks
post-conceptual age. Of the other 4 subjects, who were all studied at
34-35 wks post-conception, only subject 3 had adequate true and net Zn
absorption. Subject 1 failed to absorb any of the 70 Zn. Subjects 2 and
4 had low true Zn absorption and net Zn absorption was substantially

TABLE 1

INFANT # AND SEX		WEIGHT (g) BIRTH	STUDY	WEIGHT GAIN DURING STUDY g/d	g/kg/d	POST-CONCEPTUAL AGE (wks) BIRTH	STUDY	TYPE OF FEED
1	M	1530	2230	22	9.9	30	35	A
2	M	1210	1430	24	16.8	32	35	B
3	F	1600	1750	20	11.4	31	34	A
4	F	1545	1730	26	15.0	31	34	B
5	M	1500	1720	26	15.1	34	37	A

A = humanized cow's milk 20 kcal/oz.   B = premature formula.

TABLE 2

INFANT NUMBER	TRUE ZN ABSORPTION %	ug Zn/d	ug Zn/kg/d	NET ZN ABSORPTION ug Zn/kg/d	ESTIMATED REQUIREMENT[a] ug Zn/kg/d
1	0	0	0	92	220
2	11	318	222	-148	340
3	23	615	351	369	230
4	6	274	158	-87	310
5	65	1911	1111	303	310

a. estimated requirement for net absorption based on individual weight
gains and urine losses and assuming there are no available body stores.

worse suggesting that there may have been inappropriately high fecal
losses of endogenous Zn. However no firm conclusions about endogenous
losses are possible because of the lack of a steady state and of
differences in the time course of the studies of true and net absorption.
More extensive studies will be required to determine the effects of
post-conceptual age, type of feeding and other factors on Zn absorption
in the premature infant. Meanwhile this study indicates that poor net
absorption of dietary Zn is attributable in part to poor true Zn
absorption but, in some cases, it appears that inappropriately high fecal
losses of endogenous Zn may also contribute to the low net Zn absorption.
(Acknowledgements: Supported by NIH grants, 5R22 AM12432, HDO4024 NICHD,
RR01152 and RR-69 from the General Clinical Research Center).

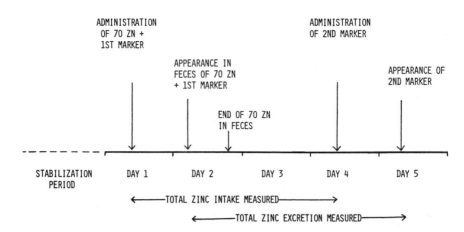

Figure 1. Timing of stable isotope studies in relation to balance studies.

REFERENCES

1. Peirce PL, et al. Anal Chem. In press, (1987).
2. Dauncey MJ, et al. Pediat Res 11:991-997, (1977).

ZINC BALANCE IN PREMATURE INFANTS FED THEIR

MOTHERS' MILK: EFFECT OF POSTNATAL AGE

Stephanie A. Atkinson, Debra Fraser,
Ruthann Stanhope and Robin Whyte

Department of Pediatrics, McMaster University
Hamilton, Ontario, Canada

INTRODUCTION

The calculated minimum zinc (Zn) requirement to cover tissue growth and obligatory losses in the growing low birthweight (LBW) infant is 350 µg dietary Zn/kg bodyweight/24 h (assuming obligatory losses = 70 µg/kg/24 h).[1] Given the reported concentration of Zn in preterm milk[2] and potential high bioavailability of Zn due to the presence of Zn binding ligand,[3] sufficient Zn should be provided by mothers' milk for LBW infants to achieve a retention of Zn which parallels intrauterine accretion. However, the few previously published studies indicate that LBW infants achieve negative[4,5] or only moderately positive[6] Zn balances in the first four weeks of life.

The purpose of the reported study was to determine Zn balance under carefully controlled conditions in LBW infants fed their mothers' non-heat treated milk and to determine if Zn absorption and retention are affected by postnatal development.

METHODS

Infants admitted to the Neonatal Intensive Care Unit, Chedoke-McMaster Hospitals, Hamilton, Ontario were selected for the study if they were fully fed with their mothers' expressed milk (MEM); they suffered no major congenital anomalies; did not require ventilatory support at the time of the study; and informed consent was provided by the parents. During balance periods aliquots of each 24-h collection of MEM were obtained for analysis and the milk for each feed pre-measured. Seventy-two hour collections of urine, stool and any regurgitations were obtained using materials previously acid-washed in nitric acid and then sterilized. Apparent retention (balance) of Zn was calculated from: INTAKE - (STOOL + URINE + REGURGITATIVE LOSSES).[7] Duplicate samples of stool were wet ashed, milk and urine were diluted with double deionized water and all samples were analyzed for Zn by atomic absorption spectrophotometry. A non-fat dried milk powder (#1549) from the National Bureau of Standards was used as analytical reference material.

RESULTS

Sixteen infants of birthweight = 1360±70 g (X±SEM) and gestational age = 30±0.5 wk were studied at two postnatal ages. Milk Zn (µg/ml) declined significantly over the first five weeks of lactation (y=4.22-.056x, r=0.54, (n=81) p<0.01), but was similar within mother over each serial three day period. Milk from one mother was determined to be "Zn deficient."[8]

Intake (I), excretion (E) in stool (S) and urine (U) and apparent retention (AR) of Zn at two postnatal ages (PNA) are summarized in the table.

Table: Zinc Balance in LBW Infants (μg/kg/24 h)

PNA,d	I	SE	UE	AR, μg/kg/d
20±1[2]	500±31[1,2]	514±56	140±21	-154±57
35±4	350±36	558±81	116±20	-324±87

[1]X±SEM, [2]p<0.01 between two time periods by ANOVA.

A small positive Zn balance (11 to 127 μg/kg/24 h) was attained in 4/16 infants at postnatal day 20 when Zn intakes were significantly higher owing to the higher milk-Zn content in early lactation. Intake but not urinary, stool or retained Zn differed between the two postnatal ages. At both postnatal ages the mean Zn balance was negative. Despite this infants grew with a mean weight gain of 17±1 g/kg/d, length gain of 0.8±0.1 cm/wk and head circumference gain of 1.0±0.1 cm/wk.

DISCUSSION

Moderate to large negative Zn balances occur in clinically stable and growing LBW infants fed MEM alone even when Zn intakes of 680 ug/kg/24 h are achieved. Most[4,5] but not all[6] previous reports had comparable results although the magnitude of the Zn deficit within and between studies is relatively large. Neither fecal nor urinary loss of Zn decreased by postnatal week five even though Zn intake had declined significantly. This suggests that regulation of Zn uptake in the intestine and reabsorption at the renal level had not improved even in the face of an accruing body Zn deficit and the occurrence of linear growth. Endogenous loss of Zn may also be elevated during early neonatal life. The physiological significance of a sustained Zn deficit on growth, metabolic function and bone mineralization during early neonatal life and the efficacy of Zn supplementation for LBW infants have yet to be assessed. Although premature infants may achieve positive Zn balance after term gestation is reached, if dietary Zn is limited as with "Zn deficient" milk, fulminant nutritional Zn deficiency can occur.[8]

REFERENCES

1. C. E. Casey and K. M. Hambidge, in: "Vitamin and Mineral Requirements in Preterm Infants," R. C. Tsang, ed., Marcel Dekker Inc., New York, pp. 153-184 (1985).
2. N. F. Butte, C. Garza, C. A. Johnson, E. O'Brian-Smith, and B. L. Nichols, Early Hum. Devel. 9:153-162 (1984).
3. B. Lonnerdal, A. G. Stanislowski, and L. S. Hurley, J. Inorg. Biochem. 12:71-78 (1980).
4. M. J. Dauncey, J. C. L. Shaw, and J. Urman, Pediatr. Res. 11:1033-1039 (1977).
5. M. Voyer, M. Davakis, I. Antener, and D. Valleur, Biol. Neonate 42:87-92 (1982).
6. R. A. Mendelson, M. H. Bryan, and G. H. Anderson, J. Pediatr. Gastroenterol. Nutr. 2:256-261 (1983).
7. M. S. Clegg, C. L. Keen, B. Lonnerdal, and L. S. Hurley, Biol. Trace Element Res. 3:107-115 (1981).
8. S. A. Atkinson and B. Lonnerdal, in: Trace Elements in Man and Animals - 6, L. S. Hurley, C. L. Keen, B. Lonnerdal and R. B. Rucker, eds., Plenum Press, New York (1988).

TRACE MINERAL AND CALCIUM INTERACTIONS

K. T. Smith and J. T. Rotruck

The Procter & Gamble Company
Cincinnati, Ohio

INTRODUCTION

The negative or competitive interaction of calcium with dietary trace mineral bioavailability is well established in the literature. However, relatively little progress has been made to identify and understand the parameters associated with such interactions or to dimension these effects with respect to their potential nutritional impact. An increased interest in calcium nutrition has recently helped to resolve this information deficiency by rekindling much needed experimental enthusiasm. We can now hope that investigators will address such issues as the long-term nutritional consequences of negative mineral-mineral interaction, adaptation, and the effect of other dietary components on final mineral balance. Our primary focus over the past several years has been an elucidation of some of these factors with respect to calcium and iron. However, additional minerals such as magnesium, zinc, copper and manganese have also been examined. The work I will describe today includes both animal model and clinical evaluations of these interrelationships.

Of course, central to all work in the field of trace element nutrition is the methodology which is used in pursuit of such investigations. Common to the field of minerals are such methods as atomic absorption spectrophotometry and stable and radioisotope tracer methodologies. This latter group, i.e. tracer techniques, coupled with a thorough understanding of their potential, strengths and limitations is critical to our ability to design, evaluate and interpret the experimental programs that we conduct. However, crucial to the valid interpretation of tracer methodology is the assumption that the tracer is truly representative of the metal in question. Of course, this raises such issues as isotope dilution and extrinsic/intrinsic tagging procedures.

I don't wish to dwell on this subject since it is not the intended focus of this paper but rather would like to point out one example where such understanding is required. We have recently worked with whole body retention techniques using isotopes such as calcium-47, zinc-65, and iron-59. First, the retention values as a function of time after dose can vary and perhaps shed some additional light on the mechanism of absorption of these minerals. For example, representative curves of post dose retention values for calcium, iron and zinc reveal the following. Zinc

and iron follow curves with a near bimodal function. The rapid phase
which occurs over the first three days is followed by the slower or second
phase which leads to a very gentle slope or point of biological
equilibrium. Such an observation is not new. Dr. William Hoekstra twenty
years ago used such a technique to illustrate the apparent absorption of
metals. Calcium on the other hand comes to rapid equilibrium after the
first twenty-four hours post dose. Thus, mucosal holding of the zinc and
iron with subsequent sloughing of intestinal enterocytes might be the
mechanistic hypothesis to explain such differences.

With respect to tagging, Figure 1 illustrates a very important point
in experimental methodology. Calcium carbonate was intrinsically or
extrinsically labeled with the radiotracer ^{45}Ca. Subsequently, this
material was stirred while small aliquots of acid (12 N HCl) were added.
The amounts of acid were calculated to be incremental amounts less than
that needed for total dissolution. Note that for the extrinsic tagged
material, radioactivity appears at a greater rate than would be predicted
based on the linear relationship of mass to isotope. Thus, specific
activity is changing as a function of added acid. These data emphasize
the importance of validating such methodology so as not to be seriously
misled. Of course, true solutions should offer no such restrictions since
the chemical kinetics would allow for rapid exchange.

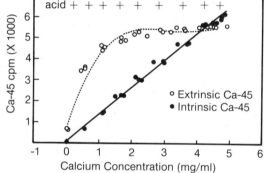

Figure 1. Intrinsic-Extrinsic Labeling of $CaCO_3$. Calcium carbon-
ate was labeled with ^{45}Ca either intrinsic or extrinsic.
Acid was added (+) to partially solubilize the carbonate.
Specific activity was measured in triplicate.

INTERACTIONS

Our particular interest in the calcium-iron interaction came as an
outgrowth of the calcium fortification technologies we were developing.
The literature contains documented evidence of such interaction as far
back as the early 1950s and such information might provide insight into
mechanisms of trace mineral bioavailability in greater detail.

One of the first experiments performed in this regard was to examine
the relationship of calcium to iron in milk. Factor(s) in human milk and
cow's milk which are attributed to the enhancement of mineral absorption
are a matter of record as are the differing calcium concentrations across

species milk. In Figure 2, the addition of calcium (as either the carbonate or phosphate salt) to human milk, significantly reduced the bioavailability of iron, as measured by whole body retention in the rat. Further addition of calcium to cow's milk lowered the iron bioavailability further, but this decrease was not statistically significant.

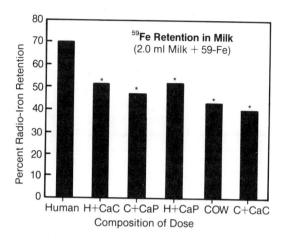

Figure 2. Iron Bioavailability from Human and Cow Milk. Radio-iron was added to milk samples, added calcium (25 mM) was from calcium carbonate (CaC) or calcium phosphate (CaP). Rats were dosed and ^{59}Fe retention measured. Mean values, * significantly different from human milk control. Values are N = 7 per group.

This observation led to additional studies in which different calcium salts were examined for their ability to interact with either iron or zinc in the whole body retention assay system. The ratios of calcium to metal in the studies was chosen to represent the human RDA weight:weight ratio of calcium to trace mineral. Thus, 60:1 approximates the current Ca:Fe or Ca:Zn ratio in the "ideal" RDA. Due to the recent and real potential for recommended increased calcium intake, a ratio of 120:1 was also examined. The data are illustrated in Figure 3. The results are expressed as percent difference in retention compared to no calcium. Clearly, the effects are variable. For zinc, the phosphate sources are more inhibitory than the carbonate sources. Nevertheless, calcium and iron as well as calcium and zinc do interact during absorption.

Recognizing that these observed effects were from a single dose of calcium when placed in conjunction with the mineral, we also carried out an eight week feeding study in weanling rats to examine dietary effects. Minerals were supplied in the diet at the NRC (National Research Council – 1978) requirements for the rat while calcium was supplied at 300% of the recommended 0.5% level using calcium carbonate in one diet and calcium phosphate in another. In addition, a group included both calcium and trace mineral feeding at elevated levels. In essence, this represents a balanced diet formulation but at higher absolute mineral amounts.

In this study, food consumption, feed efficiency and growth were similar for all groups. Iron status, as measured by hemoglobin and hematocrit showed a marked calcium-iron interaction. At both four and eight weeks, the hemoglobin and hematocrit values in the calcium supplemented groups were significantly lower. The group which received additional iron via the increased mineral mix or balanced dietary minerals had a similar serum iron profile to the control group.

The serum trace mineral profiles also showed some calcium-mineral interaction. Interestingly, the effects were not always the same for both calcium sources. This would indicate possible antagonistic anion effects as well. For example, serum magnesium was lower in the calcium carbonate group at four weeks but by eight weeks both sources were similar in their effects. Serum zinc was more affected by the additional calcium phosphate in the diet. At both four and eight weeks, the calcium phosphate group had reduced serum zinc. Once again, the group which received the additional or balanced minerals had serum mineral profiles indistinguishable from controls.

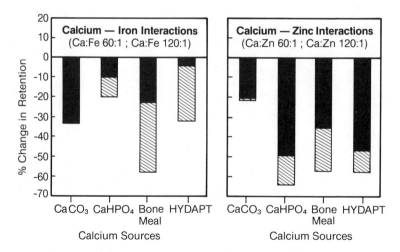

Figure 3. Calcium-Iron/Calcium-Zinc Interactions. Rats were dosed with or without calcium at weight:weight ratios of 60:1 (solid) or 120:1 (cross hatched). Results are expressed as percent change in isotope (^{59}Fe or ^{65}Zn) retention from no calcium. Values are N = 7 per group.

Additional tissues such as bone, liver, spleen, and kidney were also examined for mineral content. For the most part, these tissue analyses substantiated the view that calcium and minerals do interact under dietary situations and that this interaction can be controlled by providing a balance of one mineral to another.

Since iron in the feeding study demonstrated the most consistent and in some respects the most severe interaction coupled with the fact that iron is also consumed at marginal levels, this phenomenon was examined closer. Briefly, hemoglobin repletion trials were conducted at normal and elevated (3X) levels of calcium in the diet. The trials included three different levels of dietary iron (4, 8, 12 ppm) and the gain in hemoglobin after a three week repletion period was determined. The results were somewhat surprising. As illustrated in Figure 4, the gain in hemoglobin during high calcium feeding is greatly reduced as compared to normal calcium diets. This blunted repletion response was interesting since one would predict that the biological need for iron during anemia might overwhelm the interaction. Clearly, this was not the case. In a separate set of experiments not reported here, it was found that by providing additional dietary iron, the repletion response returned to normal.

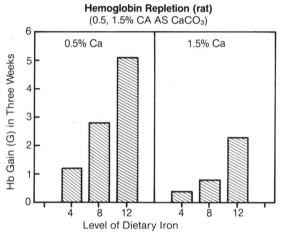

Figure 4. Hemoglobin Repletion During Elevated Calcium Intake. Rats (N = 6/group) were depleted on iron free diets. Repletion diets contained incremental levels of iron (FeSO₄) and two different calcium levels (0.5, 1.5%). Results are mean hemoglobin gained in three weeks.

MECHANISMS

At this point, we were intrigued by the potential mechanism(s) in these various interactions. For example, could we effectively separate the minerals or perhaps provide absorption enhancers to overcome these observed effects. Thus, we undertook a series of studies to examine such possibilities.

One of the first studies performed, the relationship of time of dose was examined in context of these interactions. We hypothesized that by separating the dose of interacting minerals, we should alleviate the interaction. This principle was demonstrated in an experiment which measured iron absorption by the whole body ^{59}Fe counting technique as a function of calcium dosed either simultaneously with the iron or at twenty minute intervals post calcium dose. It was determined that the interaction requires both the calcium and iron be present during the same time frame. Actually, by allowing a separation of approximately one to two hours the interaction was completely alleviated. This response was actually not surprising since logic would dictate such a separation phenomenon should exist.

Research at our laboratories has also investigated enhancers of mineral bioavailability. This work led to the identification of a new source of highly bioavailable calcium which has been used in the fortification of orange juice (Citrus Hill Plus Calcium®). Thus, intrigued by the interactions noted so far, we examined them in greater detail in this system as well. Enhancers of iron absorption were considered as a possible step toward understanding the mechanism of this interaction and the potential for providing additional relief. Naturally, ascorbic acid was one of the first enhancers considered. This well established promoter of iron absorption was tried in a series of whole body retention experiments.

Ascorbic acid could, in fact, partially alleviate the calcium-iron interaction. However, when ascorbic acid was used in conjunction with other components known or suspected to influence iron absorption, the results were far more favorable.

For example, Figure 5 illustrates the use of single and combinations of ingredients to regulate the calcium-iron interaction. Note that any one of the three ingredients tried by itself had partial alleviating effects. These materials were used at the levels that they normally would be found in an orange juice matrix, i.e. ascorbic acid (500 ppm), citric acid (0.8%), and fructose (4-5%). As we had observed in other experiments, the most effective combination was actually a synthetic orange juice (at least in relation to these three ingredients). Thus, it appears that one can overcome the calcium-iron interaction by using known affectors of iron absorption. Moreover, orange juice, due to its natural composition, is an excellent vehicle for calcium fortification since it virtually eliminates such negative interactions.

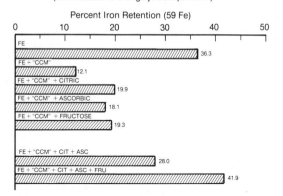

Calcium and Iron Interactions in Rats
Effects of CHPC Components on Iron Retention
(levels dosed at orange juice equivalent)

Figure 5. Calcium-Iron Interactions During Addition of Orange Juice Components. Iron retention was measured with addition of orange juice levels of ascorbic acid (500 ppm), citric acid (0.8%) or fructose (4.5%) to calcium containing (0.125%) solutions. "CCM" refers to the delivery system in Citrus Hill Plus Calcium (CHPC). Values are mean of 6 rats per group.

Since this observation, we have examined the details of calcium and iron distribution in juice by a variety of chromatographic methods. It appears that calcium and iron <u>do not</u> compete for the same ligands in the juice. In fact, iron is predominantly bound by citric acid at least under the condition used in these experiments. Further clarification of this phenomenon is currently underway.

CLINICAL STUDIES

In order to ascertain the practical significance of these findings, we have also conducted clinical studies on the interaction of calcium with iron and zinc. These studies were designed to determine the single dose effects of calcium on radio-iron and radio-zinc absorption. In one case, whole body isotope retention was used as the criterion for absorption while in the other, plasma appearance of the isotope for the first 24 hours post-dose was assessed. In both cases, the iron or zinc was given after an overnight fast under a light breakfast meal routine. The objective was to determine the likelihood of interaction under as near normal conditions as possible. Thus, 500 mg of calcium (from different sources) was consumed along with the meal and 3-5 mg of iron or zinc. The final ratio of calcium:metal was in the range of 100:1. In both studies, subjects were randomly assigned to treatment sequences and received all treatments under consideration resulting in a triple crossover design.

The results of these studies are shown in Table 1. They are expressed as the whole body isotope retention fourteen days after dose and clearly demonstrates that a calcium-iron interaction exists. In the case of calcium carbonate a near fifty percent reduction in the absorption of inorganic iron is noted. Moreover, with a second calcium source, calcium hydroxyapatite, there is an even greater apparent inhibition of iron absorption. Interestingly, when the same study was done with zinc, no interaction was noted. These data suggest that calcium does exert a significant effect on the absorption of minerals consumed at the same time under normal breakfast conditions. The lack of a zinc effect under the meal conditions in this study suggest that the meal may have exerted some ameliorating effects on zinc absorption. In fact, we have been able to show meal versus non-meal effects in zinc-calcium interactions in subsequent animal studies.

Table 1. Calcium Iron and Zinc Interactions in Humans During Calcium Supplementation. Post-menopausal females consumed a breakfast meal with radiolabeled iron (4.5 mg + ^{59}Fe) or zinc (3.0 mg + ^{65}Zn) with or without $CaCO_3$ or hydroxyapatite (500 mg). Whole body retention was measured at baseline and 2 weeks later (Dawson-Hughes et al.).

Clinical Study
Calcium Iron/Zinc Interactions

	% Whole Body Retention after 2 Weeks (mean ± SEM)	
	Iron (N = 13)	Zinc (N = 11)
Placebo	6.3 ± 2.0	18.1 ± 1.0
CaCO$_3$ (500 mg Ca)	3.3 ± 1.2*	19.8 ± 1.7
HA (500 mg Ca)-	1.6 ± 0.4*	19.3 ± 1.9

* Significantly different from placebo (P < 0.05)

In the second study, we explored the mechanistic aspects of the calcium-iron interaction a bit further. Ascorbic and citric acid containing beverages were used as the way to test for relief of interaction. Mean plasma appearance profiles for radio-iron absorption were used as the criterion for absorption. Here, the inhibition of calcium on iron absorption was noted in the reduction of peak height and

227

area under the curves as compared to the ascorbic-citric acid or placebo beverages. Once again, these data support the view that one can control interactions by use of the appropriate modifiers of mineral absorption.

CONCLUSIONS

The negative impact of calcium on the bioavailability of trace minerals represents a sampling of the complexity and diversity of the relationships of one mineral to the next. Although various pieces of this knowledge has been available in the literature for some time now, we are just beginning to scratch the surface with respect to the mechanisms and practical significance of these and similar findings.

As new information becomes available, we need to continue to integrate it into our understanding of the trace mineral absorption processes. Such information may aid in our collective quest to define mineral absorption and also should serve as a set of guidelines under which nutrition recommendations can be made. Clearly, the adverse interaction of one mineral on another with the amelioration of such effects by a correct balance would argue against single nutrient fortification without appropriate knowledge and concern about secondary effects.

In summary, calcium, by virtue of the fact that it is a macro-mineral with a particular set of chemical characteristics common to other metals, can exert an influence on the bioavailability of other essential minerals. We must be constantly aware of this and other interactions in our design, conduct and interpretation of mineral nutrition experiments. At the same time, it is not out of the realm of possiblity that some previously held theses and experimental programs might do well by their re-examination under such interaction considerations.

REFERENCES

Seligman, P., et al. Measurements of iron absorption from prenatal multivitamin - mineral supplements, Obstet. & Gynecol. 61:356, 1983.
Monsen, E. R. and Cook, J. D. Food iron absorption in human subjects. IV. The effects of calcium and phosphate salts on the absorption of non-heme iron, Am. J. Clin. Nutr. 29:1142-1148, 1976.
Barton, J. C., Conrad, M. E. and Parmely, R. T., Calcium inhibition of inorganic iron absorption in rats, Gastroenterol. 84:90, 1983.
Apte, S. V. and Venkatachalam, P. S., The influence of dietary calcium on absorption of iron, Indian J. Med. Res. 52:213, 1964.
Hoekstra, W. G. Recent observations on mineral interrelationships, Fed. Proc. 23, 1068, 1964.
Chapman, D. G. and Campbell, J. A., Effect of calcium and phosphorous on the utilization of iron by anemic rats, Br. J. Nutr. 11:127, 1957.
Grieg, W. A., The effects of additions of calcium carbonate to the diets of breeding mice, Br. J. Nutr. 6:280, 1952.
Hazell, T., Minerals in foods: dietary sources, chemical forms, interactions, bioavailability, World Review of Nutrition and Dietetics 46:1-123, 1985.
Sandstead, H. H., Trace element interactions, J. Lab. Clin. Med. 98:4, 457, 1981.
Mills, C. F., Dietary interactions involving the trace elements, Ann. Rev. Nutrition 5:173, 1985.
Dawson-Hughes, B., Seligson, F. H. and Hughes, V. A., Effects of calcium carbonate and hydroxyapatite on zinc andiron retention in postmenopausal women, Am. J. Clin. Nutr. 44:83, 1986.

DIETARY PHYTATE:ZINC AND PHYTATE X CALCIUM:ZINC RATIOS OF LACTO-OVO VEGETARIAN TRAPPIST MONKS

Barbara F. Harland, Selina A. Smith, Rex Ellis,
M. Pat Howard, and Robert D. Reynolds

School of Human Ecology, Howard University, Washington, DC
20059 and Human Nutrition Research Center, USDA, Beltsville,
MD, 20705 USA

INTRODUCTION

One of the Dietary Guidelines proposed to reap health benefits for the United States population is to increase the intake of dietary fiber. Because dietary fiber is not a discrete nutritional component, the recommendation promotes alterations in other aspects of the diet as well: a decrease in animal fat and protein and an increase in vegetable fat and protein. Implicit, though not stated, is a decrease in mineral-rich foods devoid of phytate to be replaced by foods higher in phytate. The primary concern from a trace element standpoint is that the recommended increase in whole grains, fruits and vegetables as a way of increasing total dietary fiber also increases phytate intake.

Of particular concern with recommendations of an increased dietary fiber intake are the infants, children, adolescents, pregnant and lactating women and the elderly who may not be well advised as to the proper nutrient intakes for meeting their special conditions of dietary need. A number of studies have been performed to isolate certain fiber fractions in order to test their effects on mineral nutriture. Often the fibrous components are altered by the isolation procedures themselves. As a tool for estimating the potential for a diet to create mineral deficiencies, the phytate:mineral ratios were developed. These have been employed in this study. Since national recommendations have already been made to increase dietary fiber intake, it is very important that well-controlled, long-term human studies be conducted to assess the effects of these recommendations on nutrient, particularly mineral status.

NUTRITIONAL ASSESSMENT OF TRAPPIST MONKS

Thus, after a 10-year interval, a nutritional assessment was again made of a community of 21 lacto-ovo vegetarian Trappist monks. Many of the same members of the community served as subjects for this week-long study which was closely supervised by 2 members of the Beltsville, MD, USDA Human Nutrition Research Center. Biochemical measurements were determined from blood and urinary samples. Three-day dietary intake records were obtained from all of the subjects and six of the brothers were trained to collect daily food composites for three days. The men were 48.5 ± 3.0 years of age (M+SE) their height, 175.6 ± 1.5 cm and their weight 68.9 ± 0.4 kg. Cholesterol averaged 205.3 ± 8.0 mg/dL. Hematocrits and hemoglobins were normal as were biochemical measurements of blood and urine. Complete medical histories were taken and an assessment of activity level was made.

The total dietary fiber intake was 58 ± 6 g in this vegetarian group. It is unlikely to find intakes of this high level among omnivores in the American public. The phytate intake was 927 ± 84 mg, 130% higher than the presumed American norm. With these high intakes of fiber and phytate, our concern was that the phytate:mineral ratios would also be high. Calculating from both the 3-day food records and from the analyzed food composites (there was close agreement between the two methods of analysis), we found phytate:zinc molar ratios of 9.93 and 9.81 for the 3-day food records and analyzed composites, respectively. A molar ratio of 10 or above has been shown to compromise zinc status. The phytate X calcium:zinc ratios of 0.34 and 0.29 for the 3-day food records and food composites, respectively gave us some concern. A ratio below 0.20 has been shown in animals to be acceptable.

Plasma and urinary values for zinc and calcium were within normal range (plasma zinc, 104.6 ± 3.1 ug/dL, urinary zinc, 594.0 ± 74.4 ug/24 hr; plasma calcium, 9.8 ± 0.1 mg/dL and urinary calcium 205.4 ± 15.0 mg/24 hr).

SUMMARY

According to the parameters examined, the lacto-ovo vegetarian Trappist monks appear to be in good health even without dietary supplements (used by 50% of the group). Nutritional imbalances in this population with a high mean daily intake of 3547 kcalories were not observed. The monks have developed an interest in and a knowledge of good nutrition and they experience a life-style and environment conducive to good health. Thus, despite higher than normal intakes of dietary fiber and phytate, they appear to be well-nourished.

ZINC, IRON AND COPPER INTERACTIONS IN HUMANS, RATS AND CHICKS

J.L. Greger, M.L. Storey, J.L. Stahl, M.E. Cook,
S.E. Gentry-Roberts, and J.C. Lynds

Department of Nutritional Sciences
University of Wisconsin
Madison, Wisconsin 53706

The effect of iron on zinc metabolism is of interest because iron supplements and foods highly fortified with iron are used so frequently. However, in practical situations we think that supplemental zinc is more apt to affect iron metabolism, than vice versa. We base this comment on data collected in studies conducted with humans, rats and chicks.

HUMAN STUDY

Nutritional status in regard to zinc and iron was evaluated in 40 women during both the third and trimester of pregnancy and 2 to 4 months after parturition.[1] The women's consumption of iron (14.0 mg/d) and zinc (11.8 mg/d) was fairly constant throughout the study. However, the women varied greatly in their use of iron supplements. Six refused to consume iron supplements; 12 regularly consumed multivitamins with iron; 19 regularly consumed prenatal iron supplements (>60 mg Fe/day).

Despite these large differences in iron intake, no differences were observed in plasma zinc levels of the 3 groups of women during pregnancy or during the postpartum period. These data are consistent with the observations of Yip et al.[2] but contrast with those of Solomons.[3] However, these women, like Yip's subjects, did not necessarily consume iron supplements with meals.

STUDIES WITH RATS AND CHICKS

Effect of Iron on Zinc Utilization

In three studies rats fed very high levels of iron (1408-3000 µg Fe/g diet) for three weeks had elevated levels of iron in their soft tissues but experienced no changes in zinc concentrations in livers, kidneys, tibias, spleens or thymuses.[4] These data are consistent with the observations of Bafundo et al.[5] with chicks but not with the observation of Solomons.[2] To help resolve these differences, which we thought might reflect methodological differences, we fed one meal with excess iron labelled with carrier-free Zn-65 to rats that were adjusted to a high iron intake as well as to those that were not. High levels of dietary iron whether fed for 3 weeks or in one meal had no effect on apparent absorption of Zn-65. However, the chronic consumption of excess iron, but not consumption of one meal with excess iron, significantly depressed retention of Zn-65 in tibias.

Effect of Zinc on Iron Utilization

We found in 3 separate studies that ingestion of excess zinc ($\approx$2400 µg Zn/g diet) induced anemia in weanling rats.[4] Similarly chicks fed $\approx$2000 µg Zn/g diet for 3 weeks became anemic.

In rats, ingestion of excess zinc elevated kidney, tibia, and liver levels of zinc but did not depress tissue levels of iron. In chicks, ingestion of excess zinc did not affect the level of iron in soft tissues but did depress tibia iron levels significantly in three studies.

The effect of one meal vs. chronic (3 weeks) feeding of excess zinc was compared in rats using Fe-59. Retention of Fe-59 in liver was depressed whether rats were fed excess zinc in one meal or for 3 weeks. However, ingestion of one meal with excess zinc did not depress apparent absorption of Fe-59 or retention of Fe-59 in kidneys, but chronic ingestion of excess zinc did.

Copper as a Mediator of Zinc/Iron Interactions

Ingestion of high levels of zinc depressed serum copper levels and tended to depress liver copper levels in rats and chicks. The ingestion of supplemental copper (250 µg Cu/g diet) with high levels of zinc for 3 weeks elevated hematocrits in chicks. However, the ingestion of additional copper was not sufficient to elevate plasma and liver copper levels in chicks fed excess zinc but did result in red blood cells that resisted hemolysis in vitro.

These data suggest that the gut is not the only site of zinc-iron interactions. Investigators studying Fe-Zn-Cu interactions should not rely solely on studies in which unadjusted humans or animals are given a single dose of zinc or iron because responses to a single dose do not reflect all the changes induced by chronic feeding.

ACKNOWLEDGMENT

This work was supported by the College of Agric. & Life Sci., UW-Madison project 2623, NIH grant #5T32CA09451 and the UW Graduate School.

REFERENCES

1. J.L. Greger, S.E. Gentry-Roberts, J.C. Lynds, and S.J. Voichick, Nutritional status in regard to iron and zinc during pregnancy and postpartum period, Nutr. Rep. Int. 35:(In press) (1987).

2. R. Yip, J.D. Reeves, B. Lonnerdal, C.L. Keen, and P.R. Dollman, Does iron supplementation compromise zinc nutrition in healthy infants?, Am. J. Clin. Nutr. 42:683 (1985).

3. N.W. Solomons, Competitive interaction of iron and zinc in the diet: consequences for human nutrition, J. Nutr. 116:927 (1986).

4. M.L. Storey, and J.L. Greger, Iron, zinc and copper interactions: chronic versus acute responses, J. Nutr. 117:(In press) (1987).

5. K.W. Bafundo, D.H. Baker, and P.R. Fitzgerald, The iron-zinc interrelationship in the chick as influenced by Eimeria acerbulina infection, J. Nutr. 114:1306 (1984).

INHIBITION OF IRON ABSORPTION IN MAN BY MANGANESE AND ZINC

Leif Hallberg, Lena Rossander, Mats Brune, Brittmarie
Sandström and Bo Lönnerdal

Depts. of Internal Medicine and Clinical Nutrition,
University of Gothenburg, S-413 45, Gothenburg, Sweden, and
Dept. of Nutrition, University of California, Davis, CA 95616

INTRODUCTION

Studies on the interaction of Mn and Zn with Fe absorption are important not only from a nutritional point of view but also to get insight into
the mechanisms regulating Fe absorption. Interactions among these trace
elements have been observed by other investigators but our knowledge is
fairly limited in man, especially at physiological dose levels. A main
reason is the difficulty involved in measuring absorption and retention of
trace elements in humans. Fortunately, the absorption of Fe can be
measured with great accuracy, as two suitable radioiron isotopes are
available (^{55}Fe and ^{59}Fe).

METHODS

Healthy human volunteers were given $FeSO_4$ in a fasting state on 4 consecutive mornings, either alone or with Mn (as $MnCl_2$) or Zn (as $ZnSO_4$).
The two Fe radioisotopes were used in an alternating design and 14 days
later a blood sample was drawn. ^{59}Fe absorption was measured in a whole
body counter and ^{55}Fe with a liquid scintillation counter (1).

RESULTS

Mn-Fe Interaction

Addition of Mn (7.5 or 15 mg) reduced the absorption of Fe both from
$FeSO_4$ and from non-heme iron in a meal (Table 1). The _fraction_ of Fe
absorbed from a 3 mg dose and a 0.01 mg dose was the same when the latter
dose was given with 2.99 mg Mn. Normally the fraction absorbed would be
much higher from the lower Fe dose.

Zn-Fe Interaction

Addition of Zn (15 or 45 mg) interfered with the absorption of Fe
given as $FeSO_4$. Of special interest is the comparison of the fraction of
Fe absorbed from 3 mg and 0.01 mg when adding 2.99 mg Zn to the latter
dose. Contrary to the corresponding study with Mn, the fractional

233

Table 1. Effect of Mn on Fe-absorption

Study	Number of subjects	Iron absorption %		Absorption Ratio without/with
		without Mn	with Mn	
3 mg Fe ± 7.5 mg Mn	10	20.9 ± 3.43	16.3 ± 3.02	0.72 ± 0.07*
3 mg Fe ± 15 mg Mn	9	30.5 ± 7.40	18.1 ± 4.09	0.58 ± 0.08*
3 mg Fe** ± 15 mg Mn	10	14.0 ± 2.87	8.24± 1.8	0.56 ± 0.04*
18 mg Fe/3 mg Fe + 15 mg Mn	10	18.0 ± 3.12	15.3 ± 2.53	0.84 ± 0.07
3 mg Fe/0.01 mg Fe + 2.99 mg Mn	10	27.5 ± 3.98	26.0 ± 2.69	0.94 ± 0.11

** 3 mg Fe as <u>nonheme</u> iron in a hamburger, * $p < 0.001$.

Table 2. Effect of Zn on Fe-absorption

Study	Number of subjects	Iron absorption %		Absorption Ratio without/with
		without Zn	with Zn	
3 mg Fe ± 15 mg Zn	10	22.9 ± 3.94	12.6 ± 4.01	0.38 ± 0.07*
3 mg Fe** ± 15 mg Zn	10	11.0 ± 2.51	11.5 ± 2.27	1.06 ± 0.08
3 mg Fe ± 45 mg Zn	10	24.9 ± 4.08	11.1 ± 2.52	0.30 ± 0.07*
3 mg Fe/0.01 mg Fe + 2.99 mg Zn	10	18.7 ± 4.23	40.7 ± 6.04	0.44 ± 0.05*

** 3 mg Fe as <u>nonheme</u> iron in a hamburger, * $p < 0.001$.

absorption of Fe from the lower dose was high, indicating no effect of Zn. This is similar to what has been observed for the effect of Fe on Zn absorption (2,3). An important difference in the effects of Mn and Zn was that the addition of Mn reduced the absorption of nonheme Fe from a hamburger meal, whereas Zn had no such effect.

CONCLUSIONS

Mn and Fe are likely to compete on equal terms for uptake into the mucosal cell and thus share the same mechanism. At physiological levels, Zn does not seem to interfere in this uptake step and probably has another uptake mechanism. The interaction of higher doses of Zn on Fe absorption in the absence of organic compounds may therefore be related to some competition for the pathways of Fe absorption. When organic ligands are present such as when a meal is ingested, Zn is utilizing its specific pathway and is not interfering with Fe absorption.

REFERENCES

1. L. Hallberg, Bioavailability of dietary iron in man, <u>Ann. Rev. Nutr.</u> 1:123 (1981).
2. L. S. Valberg, P. R. Flanagan, and M. J. Chamberlain, Effects of iron, tin and copper on zinc absorption in humans, <u>Am. J. Clin. Nutr.</u> 40:536-541 (1984).
3. B. Sandström, L. Davidsson, A. Cederblad, and B. Lönnerdal, Oral iron, dietary ligands and zinc absorption, <u>J. Nutr.</u> 115:411-414 (1985).

LIPID PARAMETERS IN WILSON'S DISEASE PATIENTS ON ZINC THERAPY

G.M. Hill, G.J. Brewer, M.M. McGinnis and W.D. Block

University of Missouri, Columbia, MO 65211 and
University of Michigan, Ann Arbor, MI 48109, USA

INTRODUCTION

While the relationship of zinc and copper nutriture to cardiovascular health was hypothesized by Klevay some time ago (1973), only a small amount of evidence has been available in humans to test this theory. Wilson's disease (WD) patients abnormally handle copper as a result of a genetic defect and therefore accumulate this trace element. Brewer et al (1983) reported that copper balance in this special population could be controlled with adequate zinc therapy. Therefore, WD patients on zinc therapy present a special population in which to study Klevay's theory.

STUDY POPULATION AND RESULTS

Twelve hour fasted blood samples were obtained from WD patients who had a clinically confirmed diagnoisis of WD and no known congenital cardiovascular disease (CVD). Of the two males and seven females studied before and 6 months after zinc therapy, three were newly-diagnosed and the others had been "decoppered" with penicillamine. Of these, two males and three females were studied again after 1 year of zinc therapy. In a second experiment, seven males and thirteen females were studied only after receiving zinc (>75 mg/day) for 1 year or more.

HDL-chol was significantly less (P$<$0.05) after 6 mo (46.7 vs 39.9 mg/dl), but it was not lower after 1 year of Zn therapy (Table 1). Fractionation of HDL-cholesterol indicates the change was in the HDL_2 fraction. Total triglycerides were significantly higher after 6 months of Zn therapy but were not statistically different after 1 year of Zn therapy. The VLDL fractions reflected this change in a similar pattern.

The measured parameters of the patients taking Zn for over 1 year did not differ significantly by sex (Table 2) nor did they differ from the population values usually accepted in clinical laboratories. For example, WD patients had an approximate mean concentration of about 150 mg total chol/dl compared to the accepted maximum of 240 mg/dl. While a maximum of 170 mg LDL chol/dl is an accepted norm, the WD mean did not exceed 85 mg/dl. The accepted minimum HDL-chol concentration is usually 40 mg/dl

Table 1. Lipid Parameters of Wilson's Disease Patients Before and After Zinc Therapy (>75 mg/day)

Fraction	Cholesterol Parameters			Triglyceride Parameters		
	Before Zn	After 6 mo. Zn	After 1 yr. Zn	Before Zn	After 6 mo Zn	After 1 yr
Total	148.1 ±20.7	143.4 ±28.4	155.6 ±31.9	91.9*±27.2	105.8*±21.0	109.0±49.9
Chylomicron	5.2*± 2.1	7.2 ± 2.8	7.8*± 2.3	8.1 ± 3.5	11.8*± 4.6	11.2± 4.9
VLDL	15.6 ± 5.6	18.1 ± 6.3	26.4 ± 5.7	44.1*±19.1	54.0 ±13.6	58.0±43.0
LDL	80.3*±20.8	78.1 ±28.6	79.4 ±29.2	25.4 ± 7.2	26.6 ± 5.1	24.6± 3.0
HDL	46.7*± 7.1	39.9*±10.0	42.0 ±13.2	14.0 ± 6.8	13.3 ± 3.7	15.4± 8.3
HDL_2	12.8 ± 3.4	8.8 ± 5.1	14.0 ±11.0			
HDL_3	33.9 ± 4.1	31.1 ± 5.8	30.0 ± 3.2			

* Significantly different (P<0.05)

Table 2. Lipid Parameters in Wilson's Disease Patients after >1 year Zinc Therapy (>75 mg/day)

Fraction	Cholesterol Parameters		Triglyceride Parameters	
	Males	Females	Male	Females
Total	151.6±18.6	152.5±26.4	91.9± 8.6	86.4±19.7
Chylomicron	4.3± 1.0	7.0± 2.8	8.6± 4.9	8.8± 1.9
VLDL	15.9± 9.1	17.5± 7.4	48.1±39.6	40.5±16.6
LDL	84.3±15.2	81.0±22.9	21.6± 4.6	24.1± 4.5
HDL	48.1±12.6	48.4±11.7	13.6± 3.7	12.7± 5.5
HDL_2	16.0± 1.7	15.4±11.5		
HDL_3	31.0±10.9	29.5± 9.5		

which was exceeded by WD patients (48 mg/dl). Desirable mean total/HDL chol ratios (3.24-3.93) which predict risk of CVD, are low for this population.

CONCLUSION

While it appears that a short term (<6 mo) effect of zinc therapy on WD patients may be temporarily reduced HDL-chol, the mean concentrations do not drop below population norms. Long term (>1 year) Zn therapy does not appear to increase the risk of CVD due to changes in lipid parameters.

REFERENCES

Klevay, L.M., 1973, Hypercholesterolemia in rats produced by an increase in the ratio of zinc to copper ingested, Am J Clin Nutr, 26:1060.

Brewer, G.J., Hill, G.M., Prasad, A.S., Cossack, Z.T. and Rabbani, P. 1983. Oral Zinc Therapy for Wilson's Disease, Ann Int Med 99:314.

EFFECTS OF MODEST AMOUNTS OF WHEAT BRAN AND

DIETARY PROTEIN ON MINERAL METABOLISM OF HUMANS

H.H. Sandstead, F.R. Dintzis, J.R. Mahalko,
L.K. Johnson, and T.P. Bogyo

UT Medical Branch, Galveston, Texas
USDA-NRRC, Peoria, Illinois
USDA-HNRC, Grand Forks, North Dakota

Adverse effects of whole meal wheat bread on retention of calcium and iron by men were reported by McCance and Widdowson in the early 1940's.[1,2] Since then, their findings have been confirmed by others. One of these later investigators was Reinhold who showed that common whole meal wheat bread of Iranian villagers and Mexican tortilla impaired retention of various essential minerals.[3-6] His findings provided substantial support for the hypothesis that habitual consumption of large amounts of such breads is an important factor in the occurrence of zinc deficiency among the poor of Third World countries. Because the poor of Third World countries subsist in large part on cereals, the relevance of these findings for persons who consume western type diets which include only modest amounts of whole-grain cereal products is unclear.

With the above question in mind, we fed 45 male volunteers mixed diets of conventional foods low in dietary fiber under carefully controlled metabolic ward conditions and measured the effect of the addition of 26 g daily of various sources of dietary fiber, including wheat brans, on their retention of various minerals. This report describes findings in 20 men who were fed several varieties of wheat bran. The results of this project will be presented in more detail elsewhere.

Experimental intervals were for 26-30 days. The last 12 days of each experimental interval, duplicate diets and all excreta were collected for chemical analysis and subsequent calculation of balance. Because level of protein intake had been reported to influence retention of calcium,[7] protein was fed at 8% and 15% of energy. Because the diets were designed to meet the energy requirements of the volunteers, and nutrients were provided proportionately, the mineral intakes of the volunteers differed according to their energy needs. The addition of wheat bran to the diets significantly increased the intake of all minerals except calcium ($P<0.05$).

Examination of the raw data showed that wheat bran reduced the mean retention of calcium from about 20% to 7% when protein provided 8% of energy ($P<0.055$), while retention of iron was reduced from about 17% to 6% ($P<0.18$) and phosphorus retention was reduced from about 8% to 0% ($P<0.13$). Zinc retention was reduced by wheat bran from about 15% to 4% of intake when protein provided 15% of dietary energy ($P<0.10$). Copper and magnesium retention were not adversely affected. In no instance was a negative balance statistically different from zero.

The data were then evaluated by ANCOVA, a model that assumes linear relationships, using mineral content of the diet as the independent covariant to remove the confounding effects of differences in mineral intakes. When this was done, relationships which had been "marginally significant" became "highly significant" (Table 1), and consistent with previous reports of the effects of large amounts of whole meal bread on mineral retention of humans.

Table 1.   Effects of Protein and Wheat Bran
on Mineral Retention

Mineral	Protein n=36		Bran n=36		8% Pro + Bran n=18		15% Pro + Bran n=18	
	%	p<	%	p<	%	p<	%	p<
Ca	-111	0.01	-50	0.01	-34	0.001	-980	NS
P	-148	0.001	-138	0.001	-69	0.001	-367	0.001
Fe	+26	NS	-30	NS	-65	0.01	+15	NS
Cu	-159	0.05	-174	0.01	-108	0.05	-319	0.01
Zn	-125	0.01	-90	0.01	-32	NS	-504	0.001

These findings suggest that the addition of relatively small amounts of wheat bran to western diets may reduce the retention of some essential minerals and that the level of dietary protein influences this phenomenon. Because these studies were relatively short term, the findings must be interpreted with caution. It is unknown if persons "adapt" to higher intakes of bran and protein so as to maintain mineral homeostasis. Prudent recommendations might include consumption of amounts of protein that approximate the RDA and avoiding substantial increases in the intake of wheat bran until potential adverse affects of habitual high intakes and adaptive responses have been clarified.

REFERENCES

1. R.A. McCance and E.M. Widdowson, The digestability and absorption of calories, protein, fat and calcium in whole meal wheat breads, Br. J. Nutr. 2:26 (1948).
2. E.M. Widdowson and R.A. McCance, Iron exchange of adults on white and brown bread diets, Lancet 1:588 (1942).
3. J.G. Reinhold, H. Hedayati and A. Lahimgarzadeh, Zinc, calcium, phosphorus, and nitrogen balances of Iranian villagers following change from phytate-rich to phytate-poor diets. Ecol. Food Nutr. 2:157 (1973).
4. J.G. Reinhold, B. Faradji, and P. Abadi, Decreased absorption of calcium, magnesium, zinc and phosphorus by humans due to increased fiber and phosphorus consumption as wheat bread, J. Nutr. 106:493 (1976).
5. F. Ismail-Beigi, B. Faraji and J.G. Reinhold, Binding of zinc and iron to wheat bread, wheat bran and their components. Am. J. Clin. Nutr. 30:1721 (1977).
6. J.G. Reinhold, L.J. Salvadar-Garcia and P. Garzon, Binding of iron by fiber of wheat and maize. Am. J. Clin. Nutr. 34:1384 (1981).
7. H.M. Linkswiler, C.L. Joyce and C.R. Anand, Calcium retention of young adult males as affected by level of protein and of calcium intake, Ann. NY Acad. Sci. 36:333 (1974).

FIRST DESCRIPTION OF A VARIANT OF *E.coli* LACKING SUPEROXIDE

DISMUTASE ACTIVITY YET ABLE TO GROW EFFICIENTLY ON MINIMAL,

OXYGENATED MEDIUM

James A. Fee, Eric C. Niederhoffer, and Cleo Naranjo

Isotope and Structural Chemistry Group, INC-4, C-345
Los Alamos National Laboratory
Los Alamos, New Mexico 87545 U. S. A.

INTRODUCTION

   E. coli contains two loci coding for proteins which catalyze superoxide dismutation

$$2 \ O_2^- \ + \ 2 \ H^+ \ -> \ H_2O_2 \ + \ O_2 \qquad\qquad [1].$$

sodA lies near 87 min. on the chromosome and codes for a Mn-requiring enzyme; the gene has been cloned[1] and sequenced[2]. sodB lies near 37 min on the chromosome and codes for the Fe-requiring enzyme; this gene has also been cloned[3,4] and sequenced[5]. The three dimensional structure of the Fe-protein is known at ~3 Å resolution[6], and recent work suggests that the Fe- and Mn-proteins probably have identical three dimensional structures[5]. Considerable work has been done on the mechanism of their superoxide dismutase activity[7].

   Under normal laboratory growth conditions, the two proteins are synthesized in approximately equal amounts, but their biosynthesis depends, in part, on the concentration of the individual metal ions in the culture medium[8]. Thus, increasing the iron concentration causes more Fe-protein to be synthesized while increasing the Mn^{2+} concentration causes more of the Mn-protein to be synthesized. The expression of the Mn-protein usually depends on the presence of air and is enhanced by a higher oxygen concentration or by treatment with redox active agents such as methylviologen(*cf* Ref. 9 and references therein). Some workers have suggested that this is due to a O_2^- mediated induction of the protein(*cf* Ref. 3). However, this cannot be true as Moody and Hassan[10] have shown that the Mn-protein is readily synthesized under anaerobic conditions when the amount of Fe in the culture medium is selectively diminished. By contrast, the amount of Fe-protein synthesized generally decreases with increased oxygenation of the culture medium[3]. A satisfactory interpretation of these observations has not been formulated.

It is widely assumed that superoxide dismutases protect organisms from oxygen toxicity. Fridovich[11] has maintained that superoxide dismutases are necessary for aerobic life because they reduce the concentration of toxic O_2^- by promoting Reaction [1]. Fee[12] has objected to this interpretation, pointing out that O_2^- is an extremely unreactive species, and has argued that the true biological function of these proteins is not known. The tools of molecular biology, as applied to *E. coli* provide a logical approach to the solution of this controversy.

In 1986, Carlioz and Touati[13] described a double mutant of *E. coli* in which the sodA gene was inactivated by insertion of a Mu transposon, MudIIPR13, carrying chloramphenicol resistance and sodB was inactivated by a Mu transposon, MudIIPR3, carrying kanamycin resistance. This strain (QC774) is completely devoid of superoxide dismutase activity. It can thus be used to examine the biological function of the sodAB gene products.

The properties of strain QC774 are listed as follows:

- ability to grow normally on oxygenated, rich medium (*cf* Ref. 14)

- high sensitivity to pro-oxidants (H_2O_2, and methylviologen)

- high, spontaneous rate of mutation (*xth* and oxygen dependent)

- unable to grow aerobically in minimal glucose/salts medium

We describe here a variant strain of sodAB (QC774-LA) which is able to grow efficiently on oxygenated, minimal glucose/salts media. In March of 1985, we received from Dr. Touati strains GC4468 (parent, deleted for the lactose operon) and QC774 (sodAB). The latter was shown to carry resistance to both chloramphenicol and kanamycin and to be deficient in Fe- and Mn-dismutases Fig. 1). The cytosol of these cells is completely deficient in superoxide dismutase activity. In contrast to the observations of Carlioz and Touati[13], however, this double mutant grew remarkably well on glucose minimal media under normal conditions of aeration (Fig. 2A). Receipt of a new culture of QC774 from the Touati laboratory showed the prescribed properties and suggested that we had, by good fortune, received a variant strain in the first shipment.

We have examined the effects of diverse perturbations on the new strain, its parent, and the new strain into which we introduced the plasmid pHS1-4 which carries the sodB gene. Some of the phenotypes of QC774-LA are:

- able to grow on oxygenated, minimal glucose/salts medium (Fig. 2B)

- high sensitivity to methylviologen

- relatively low sensitivity to H_2O_2 and able to develop resistance by pre-treatment with 50 μM H_2O_2

- growth completely inhibited by low concentrations of diethylenetriaminepentaacetic acid (DETAPAC)

- high sensitivity to Co^{2+}

- does not spontaneously transform to Co^{2+} resistance (*cf* Ref. 15 and references therein)

- unable to grow on glycerol

The latter four phenotypes are distinct from the parent strain (GC4468), and they are not transformable with pHS1-4. We therefore presume that this strain (QC774-LA) has suffered an additional, spontaneous mutation which complements the absence of the sodAB gene products. We have recently shown that it is possible to isolate strains similar to QC774-LA after mutagenesis with nitrosoguanidine, and we are now attempting to identify the modified loci.

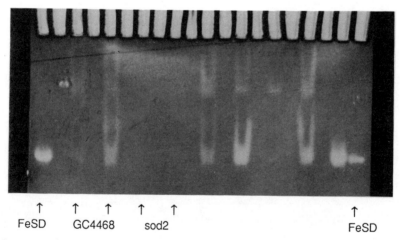

Fig. 1.   Stain for superoxide dismutase activity in electrophoretograms of *E. coli* strains: GC4468 and QC774-LA (sod2). The staining was carried out according to the method of Beauchamp and Fridovich.

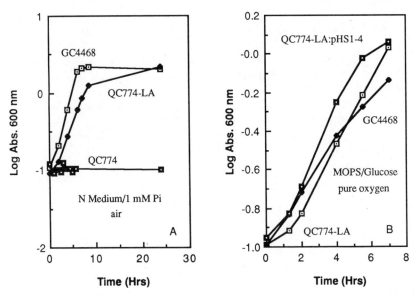

Fig. 2. Growth curves for GC4468, QC774-LA, and QC774-LA:pHS1-4 on MOPS minimal medium with normal aeration(A) and sparging with pure oxygen(B).

CONCLUSIONS

A strain of *E. coli* has been obtained which is completely lacking in superoxide dismutase activity and is able to grow efficiently on glucose minimal media being sparged with pure oxygen. While its genotype is unknown, it appears to have arisen from mutation(s) which complements the absence of sodAB gene products without introducing superoxide dismutase activity.

ACKNOWLEDGEMENTS

This work was supported by U. S. P. H. S. grant GM35189 and the U.S.D.O.E./O.H.E.R. Stable Isotopes Program. We thank Drs. D. Touati and T. Kogoma for valuable discussion and assistance.

REFERENCES

1. D.Touati, Isolation of the Mn-superoxide dismutase gene (sodA) from *E. coli* K12, J.Bacteriol.155:1078-1085 (1983).
2. Y. Takeda and H. Avila, Structure and expression of the *E. coli* Mn-superoxide dismutase gene, Nucleic. Acids Res. 14: 4577-4589 (1986).
3. C. Nettleton, C. Bull, and J. A. Fee, Isolation of the *E. coli* iron superoxide dismutase gene, Proc. Nat'l. Acad. Sci. 81:4970-4973 (1984).

4. H. Sakemoto and D. Touati, Cloning of the iron superoxide dismutase gene (sodB) in *E. coli* K12, J. Bacteriol. 159;418-420 (1984).

5. A. Carlioz, M. L. Ludwig, W. C. Stallings, J. A. Fee, H. M. Steinman, and D. Touati, Iron superoxide dismutase: nucleotide sequence of the gene from *E. Coli* and correlations with crystal structure, Submitted for publication.

6. W. C. Stallings, T. B. Powers, K. A. Pattridge, J. A. Fee, and M. L. Ludwig, Iron superoxide dismutase from *E. coli* at 3.1 Å, Proc. Nat'l. Acad. Sci. 80:3884-3888 (1983).

7. C. Bull and J. A. Fee, Steady-state kinetic studies of superoxide dismutases, J. Am. Chem. Soc. 107;3295-3304 (1985).

8. S. Y. R. Pugh and I. Fridovich, Inductions of superoxide dismutases in *E. coli* by manganese and iron, J. Bacteriol. 160:137-140 (1984).

9. S. Y. R. Pugh and I. Fridovich, Induction of superoxide dismutases in *E. coli* by metal chelators, J. Bacteriol. 162:196-202 (1985).

10. C. S. Moody and H. M. Hassan, Anaerobic biosynthesis of Mn-containing superoxide dismutase in *E. coli*, J. Biol. Chem. 259:12821-12825 (1984).

11. I. Fridovich, The biology of superoxide and of superoxide dismutases - in brief,in "Oxygen and Oxy-Radicals in Chemistry and Biology," M. A. J. Rodgers and E. L. Powers, Academic Press, New York, page 197ff (1981)

12. J. A. Fee, Is superoxide important in oxygen poisoning?, Trends in Biochemical Sciences 7;84-86 (1982).

13. A. Carlioz and D. Touati, Isolation of superoxide dismutase mutants in *E. coli*, EMBO J. 5:623-630 (1986).

14. E. Chan and B. Weiss, Endonuclease IV of *E. coli* is induced by paraquat, Proc. Nat'l. Acad. Sci. 84:3189-3193 (1987).

15. S. P. Hmiel, M. D. Snavely, C. G. Miller, and M. E. Maguire, Magnesium transport in *S. typhimurium*, J. Bacteriol. 168:1444-1450 (1986)

EFFECT OF SELENIUM (Se) on PEROXIDATION AND GLUTATHIONE (GSH) LEVELS IN MURINE MAMMARY CELLS

H. W. Lane, L. G. Wolfe, and D. Medina

Depts. of Nutrition and Foods and Pathology and Parasitology
Auburn University, AL and Dept. of Cell Biology
Baylor College of Medicine, Houston, TX

Dietary Se (2.0 ppm) inhibited chemical carcinogen induced murine mammary tumorigenesis, however, the mechanism of Se's effect is unknown (1). Studies on the mechanism of inhibition of mammary tumorigenesis have included measurements of lipid peroxidation and GSH levels. The purpose of this study was to determine the effect of Se on lipid peroxidation, cell numbers (in vitro), and GSH levels in mammary epithelial cells.

Peroxidation. The effect of 2.0 ppm Se on two indices of peroxidation, thiobariturate reactants (TBA), malonyldialdehyde; and conjugated dienes levels in mammary gland microsomal-mitochondrial membranes was determined. At seven weeks of age BALB/c virgin female mice were fed the AIN-76 diet containing no added Se (0.03 ppm Se) or 2.0 ppm Se as sodium selenite. At 8 weeks of age, all mice received a pituitary isograft to increase mammary epithelial cell numbers. At eleven weeks of age, the mice were sacrificed and microsomal-mitochondrial membranes were prepared from the glands by centrifugation (2). TBA reactants were 0.145±0.03 and 0.275±0.025 nmoles/g of tissue for 0.03 and 2.0 ppm Se, respectively (an 89% increase). The level of conjugated dienes was determined by measuring the absorbance before and after in vivo peroxidation between wavelengths of 220 and 250. These data show that there was a greater difference between in vivo and in vitro absorbance with the 2 ppm Se treatment than with the 0.03 ppm Se treatment (Fig. 1). At the 225λ, there was a 33% difference in absorbance between the two dietary treatments. This study suggests that under these conditions selenium did not promote a decrease in peroxidation but in contrast promoted an increase in the level of peroxidation in these membranes. Selenite's pro-oxidant property may relate to its inhibition of mammary tumorigenesis.

GSH. Se is known to affect the level of GSH in various types of cells. Two cell lines were studied: a murine mammary cell line (non tumor), FA-5; and a canine mammary tumor cell line, CMT-12. The FA-5 cells were exposed from 0 to 10^{-6} M Se as sodium selenite added to the media for 48 hours. The CMT-12 cells were exposed to 10^{-8} to 10^{-5} M Se as sodium selenite added to the media for 48, 72, or 96 hours. After incubation, the cells were harvested and cell numbers determined. The cells were rinsed to remove excess extracellular Se and the centrifuged cell pellet was lyophilized. GSH and protein levels were determined on the recon-

stituted sample. The total GSH was determined by incubating the sample
with 5,5'-dithiobis-2-nitrobenzoic acid, NADPH, buffer, and glutathione
reductase. A linear increase in absorbance at 412 nm was determined.
Various concentrations of GSH were used for determination of a standard
curve.

Cell counts for the FA-5 cell line decreased with increasing Se
concentration, demonstrating Se toxicity with these cells. Concurrently, GSH concentration decreased from 1.3 nmoles GSH/g protein
without Se to 0.55 nmoles GSH/g protein at 10^{-6} M Se. In contrast, with
the CMT cells, cell counts increased with increasing Se concentration
from 2.0×10^6 cells/flask at 10^{-8} M Se to 3.0×10^6 cells/flask at 10^{-6} and
10^{-5} M Se after a 48 hours incubation. After 96 hours of incubation,
cell counts had increased from 2×10^6 to 5.0×10^6 cells/flask with 10^{-8}
and 10^{-5} M Se, respectively. GSH concentration also increased from 1.8
nmoles GSH/g of protein with no added Se to 4.5 and 5.3 nmoles GSH/g of
protein for 10^{-8} and 10^{-5} M Se, respectively, after a 48 hour incubation. There was no effect of Se on GSH levels when GSH concentration
was expressed per cell counts.

It has been postulated that GSH may have a role in Se's inhibition
of mammary tumorigenesis. However, Se increased cell numbers with the
CMT cell line, but was toxic with the FA-5 cell line while GSH levels
varied with cell numbers. Thus, it is
unlikely that the effect of Se on GSH
levels provides an avenue of research
on the mechanism of tumor inhibition.

In summary, Se affected peroxidation, cell numbers and GSH levels. It
is unclear if these effects are
related to Se's role in inhibition of
mammary tumorigenesis.

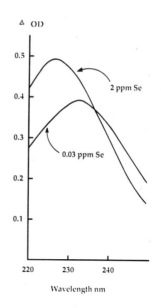

Fig. 1. Difference Spectrum for
conjugated diene measurement of
mammary cell microsomal-mitochondrial
membranes of mice consuming 0.03 or
2.0 ppm Se.

ACKNOWLEDGEMENTS

Supported in part by American Institute for Cancer Research,
PHS #CA-11944, and the Alabama Agriculture Experiment Station

REFERENCES

1. D. Medina, H.W. Lane, F. Shepherd. Effect of dietary selenium
   levels on 7, 12-dimethylbenzanthracene-induced mouse mammary tumorigenesis, Carcinogenesis 4:1159 (1983).
2. H.W. Lane, J.S. Butel, C. Howard, F. Shepherd, R. Halligan, and D.
   Medina. The role of high levels of dietary fat in 7, 12-dimethyl-
   benzanthracene-induced mouse mammary tumorigenesis: Lack of an
   effect on lipid peroxidation. Carcinogenesis 6:403 (1985).

THE RELATIONSHIP BETWEEN MANGANESE INTAKE AND THE ACTIVITY OF

MANGANESE SUPEROXIDE DISMUTASE IN TISSUES OF SHEEP

David G. Masters* and David I. Paynter**

*CSIRO Division of Animal Production, Private Bag, PO Wembley,
WA 6014 Australia, ** Department of Agriculture and Rural
Affairs, Regional Veterinary Laboratory, PO Box 388, Benalla,
Vic 3672 Australia

INTRODUCTION

The relationship between manganese superoxide dismutase (MnSOD)
activity and manganese intake appears to be both tissue and species
specific. In manganese deficient mice and chickens the activity of MnSOD in
the liver is reduced (De Rosa et al., 1980). In manganese deficient rats
the activity of MnSOD decreases in the heart and kidney (Paynter, 1980a) but
may only change in the liver after prolonged deficiency throughout gestation
and the early post-natal period (Zidenberg-Cherr et al., 1983). Post-natal
changes in the activity of MnSOD in sheep have been related to an increased
susceptibility to peroxidation in some tissues (Paynter and Caple, 1984).
The aim of this study was to determine the effects of both low and high
intakes of manganese on the activity of MnSOD in tissues of sheep.

METHODS

In experiment 1, two groups of six, 12-month old merino wethers were
fed a lupinseed diet containing either 8.7 or 30 μgMn/g for six weeks. In
experiment 2, four groups of 10-month old merino rams (six or seven rams per
group) were fed a 20% lupinseed and 80% hay diet with either 13, 20, 30 or
45 μgMn/g for 16 weeks. In experiment 3 25, 6-month old sheep were fed
lucerne chaff supplemented with $MnCl_2$ to contain 70, 300, 600, 1200 or
2400 μgMn/g (five sheep per group). All sheep were slaughtered at the end
of the experiments. Manganese concentration in the diets and tissues was
determined by flame or flameless atomic absorption spectroscopy. Tissue
MnSOD activities were determined (Paynter and Caple, 1984) after storage at
-40°C. Groups in experiment 1 were compared using Students 't'-test. In
experiment 2 and 3 analysis of variance was used after the treatment sums of
squares were partitioned into linear and quadratic components or after $\log_{10}$
transformation of the results. Regression analysis was on individual values
rather than treatment means.

RESULTS AND DISCUSSION

The heart showed the greatest relative change in the activity of MnSOD
in response to changing the concentration of manganese in the diet at both
high and low intakes on manganese. Low dietary manganese (experiment 2)

Table 1. Effect of dietary manganese on tissue Mn concentrations (μg/g wet wt) and MnSOD activities (Units/g wet wt[a]). (Mean ± SEM)

Expt	Dietary Mn (μg/g)	Heart		Liver	
		Mn	MnSOD	Mn	MnSOD
1	8.7	0.31±0.02[b]	226±17	2.3±0.1[b]	326±37
	30.0	0.38±0.02[b]	240±17	2.9±0.1[b]	302±21
2	13	0.26±0.02	209±14	2.8±0.2	414±57
	20	0.31±0.01	239±16	3.0±0.1	475±33
	30	0.29±0.02	231±16	2.9±0.3	367±78
	45	0.34±0.02[c]	263± 9[c]	3.2±0.2	388±37
3	70	0.39±0.03	326±28	2.6±0.1	474±48
	300	0.44±0.02	383±21	3.2±0.2	589±37
	600	0.55±0.01	439±13	4.0±0.2	627±39
	1200	0.62±0.02[d]	448±21[d]	5.4±0.3	608±66[d]
	2400	0.67±0.03[d]	483±16[d]	7.7±0.5[c]	631±30[d]

[a] One unit = activity of 1 μg purified bovine CuSOD
[b] Significant difference between groups (P<0.05)
[c] Significant linear relationship (P<0.05) or [d] significant log/linear relationship (P<0.05) with Mn in the diet

caused no change in the activity of MnSOD in liver, kidney, lung, testes or skeletal muscle, whereas there was a linear relationship between manganese in the diet and both MnSOD activity (r^2 = 0.19, P<0.02) and concentration of manganese (r^2 = 0.18, P = 0.02) in the heart. High intakes of manganese did not influence the activity on MnSOD in lung or muscle however there was a log/linear relationship between manganese in the diet and both the activity of MnSOD (r^2 = 0.64, P<0.001) and manganese concentration (r^2 = 0.74, P<0.001) in the heart. While there was also a linear relationship between manganese in the diet and the concentration of manganese in the liver (r^2 = 0.84, P<0.001), there was a log/linear relationship between manganese in the diet and MnSOD activity in the liver (r^2 = 0.23, P<0.02). Therefore, the increase in MnSOD activity in the liver was small in comparison to the increase in manganese concentration.

The changes in MnSOD activity in the heart associated with changes in the manganese intake indicates this tissue may be more susceptible to damage caused by peroxidation during manganese deficiency. Previous research with rats has shown that selenium and vitamin E dependent tissue peroxidation is increased by deficiencies of copper and manganese (Paynter, 1980b). In heart tissue of sheep, copper superoxide dismutase is relatively low and the selenium containing enzyme glutathione peroxidase may be reduced to very low activities by selenium deficiency (Paynter and Caple, 1980). The role of dietary manganese, through changes in activity of MnSOD may be an additional factor involved in the etiology of nutritional muscular dystrophy of the sheep.

REFERENCES

De Rosa, G., Keen, C.L., Leach, R.M. and Hurley, L.S. 1980, J. Nutr. 110:795-804.
Paynter, D.I. 1980a, J. Nutr. 110:437-447.
Paynter, D.I. 1980b, Biol Trace Element Res. 2:121-135.
Paynter, D.I. and Caple, I.W. 1984, J. Nutr. 114:1909-1916.
Zidenberg-Cherr, S., Keen, C.L., Lonnerdal, B. and Hurley, L.S. 1983, J. Nutr. 113:2498-2504.

EFFECTS OF SELENIUM INTAKE ON THE ACTIVITY OF SUPEROXIDE DISMUTASE AND GLUTATHIONE PEROXIDASE AND ON MALONDIALDEHYDE EXCRETION DURING THE DEVELOPMENT OF RAT MAMMARY CARCINOGENESIS

Mary R. L'Abbé, E.R. Chavez, P.W.F. Fischer and K.D. Trick

Nutrition Research Division, Health and Welfare Canada, Ottawa, Ont. K1A OL2 and Dept. of Animal Science, Macdonald College, McGill Univ., Ste Anne de Bellevue, P.Q. H9X 1C0

INTRODUCTION

Human prospective studies showed that serum selenium levels were significantly lower in individuals who subsequently developed cancer compared to matched controls (1-4). The purpose of this study was to investigate the protective enzymes: selenium-dependent glutathione peroxidase (Se-GSHPx) and superoxide dismutase (SOD) in plasma and erythrocytes at biweekly intervals before and during the development of DMBA-induced tumors in rats. In addition, urinary malondialdehyde (MDA) was determined.

METHODS

Animals and Diets

110 weanling female Sprague-Dawley rats (Charles River, Canada Inc.) were block randomized according to body weight into 4 groups: 20 controls and 3 treatment groups of 30. Each group of rats was fed an AIN-76A (5) casein-based diet, modified to contain 20% fat (3:1, lard: stripped corn oil), and either 0.1 (Control), 0.035, 0.1 or 3 ppm Se. After 5 weeks, mammary tumors were induced in groups 2 to 4, by intragastric administration of 3 mg of 7,12-dimethylbenz(a)anthracene (Sigma Chemical Co.), dissolved in corn oil. Control rats (group 1) were dosed with corn oil only. Rats were palpated weekly for tumors.

Table 1. SOD Activity in Erythrocytes and Plasma Before Tumor Development

| | RBC CuZnSOD (U/mL cells x100) | | | Plasma SOD (U/mL) | | |
	Control[1]	NT[2]	WT[3]	Control	NT	WT
Week 5	218	230	234	451	407	474
Week 9	247	237	247	572[a]	497[b]	369[c]
Week 11	240[a]	249[a]	228[b]	556[a]	344[b]	365[b]
Week 13	255[a]	238[b]	226[b]	485[a]	346[b]	300[c]

[1]Control rats received corn oil; NT and WT rats received 3 mg DMBA; [2]NT rats remained free of tumors at wk 21; [3]WT rats with tumors at wk 21. Values with different superscripts are significantly different (p<.05).

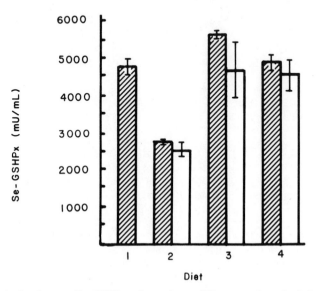

Fig. 1. Week 9 plasma Se–GSHPx. Open bars WT rats; hatched bars NT rats.

<u>Biochemical Determinations</u>

Animals were fasted overnight and 75 µL of blood was collected from the tail vein in a heparinized microhematocrit tube, centrifuged and the plasma and packed cells used for the assays. Blood samples were removed every 2 weeks from weeks 5 to 21. SOD was determined by the xanthine-xanthine oxidase-cytochrome C coupled assay (6), and Se-GSHPx by a modification of the method of Paglia and Valentine (7), using an automated discrete sample analyzer (ABA-200, Abbott Laboratories Ltd.). Urine samples were collected during weeks 10 and 20, and analyzed for lipid peroxides using the TBA method (8); creatinine was also determined.

RESULTS AND DISCUSSION

Both erythrocyte Cu,ZnSOD and plasma SOD activities decreased with DMBA treatment (Table 1). A further reduction was seen in those rats which later developed tumors (WT) compared to DMBA-treated animals remaining free of tumors for 21 weeks (NT). Plasma Se-GSHPx activity was significantly ($p<.05$) lower in the WT group compared to the NT group. This reduction was observed at all dietary Se levels before the appearance of tumors (Figure 1). These data suggest that the Finnish (4,5) and US (3) epidemiological data may not be inconsistent with each other, as the reduction seen in Se-GSHPx activity in the WT animals was observed at widely differing selenium intakes.

At week 10, MDA excretion was not affected by diet nor by tumor status. The reduction seen in blood GSHPx and SOD activities suggests that those animals that eventually developed tumors had a reduced antioxidant status. This reduction, however, was not reflected in an increase in lipid peroxide excretion.

REFERENCES

1. W.C. Willett, B.F. Polk, J.S. Morris, M.J. Stampfer, S. Pressel, B. Rosner, J.O. Taylor, K. Schneider, and C. Hames, Lancet 2:130 (1983).
2. J.T. Salonen, G. Alfthan, J.H. Huttenen and P. Puska, Am. J. Epidem. 120:342 (1984).
3. J.T. Salonen, R. Salonen, R. Lappetelainen, P.H. Maenpaa, G. Alfthan and P. Puska, Br. Med. J. 290:417 (1985).
4. F.J. Kok, A.M. deBruijn, A. Hofman, R. Vermeeren and H.A. Valkenburg, Am. J. Epidem. 125:12 (1987).
5. American Institute of Nutrition, J. Nutr. 107:1340 (1977).
6. M.R. L'Abbé and P.W.F. Fischer, Clin. Biochem. 19:175 (1986).
7. D.E. Paglia and W.N. Valentine, J. Lab. Clin. Med. 70:158 (1967).
8. H. Ohkawa, N. Ohishi and K. Yagi, Anal. Biochem. 95:351 (1979).

TRACE ELEMENTS, FREE RADICALS AND LIPOPEROXIDATION IN RATS

Ivor E. Dreosti and Eric J. Partick

Division of Human Nutrition
CSIRO (Australia)
Adelaide 5000

INTRODUCTION

Previous studies from this laboratory have demonstrated increased levels of malondialdehyde (MDA) in livers from adult and fetal rats rendered zinc deficient or receiving alcohol during pregnancy - with evidence of potentiated lipoperoxidation when the treatments occurred concurrently (1). Also relevant was the finding (2) that hepatic superoxide dismutase increased in both tissues following gestational alcoholism, which suggested a measure of alcohol metabolism in fetal rat livers, and pointed to the involvement of superoxide-related lipoperoxidation in the raised MDA levels reported earlier. Substantial evidence exists for the generation of superoxide during the metabolism of alcohol via the microsomal ethanol oxidising system, and a protective role for zinc has been proposed which relates to a diminution of the iron-catalysed Haber-Weiss conversion of superoxide into highly reactive hydroxyl radicals.

The relationship of other antioxidant defence mechanisms to the zinc/alcohol interaction has received little attention, but is nevertheless of interest because of the large part played by several micronutrients in these systems. The present paper describes studies performed with pregnant, and weanling male rats in which attention was paid to the effects of zinc deficiency and selenium supplementation on intracellular levels of reduced (GSH) and oxidized glutathione (GSSG), and on the activity of the selenoenzyme glutathione peroxidase (GSH-Px).

METHODS

Pregnant dams received diets containing either 100 ppm zinc (+Zn) or <0.5 ppm (-Zn) from day 0 to day 20 of gestation. Some animals were also given 20% alcohol in their drinking water. Weanling male rats were fed the same diets, supplemented in some cases with 0.5 ppm of sodium selenate, for a period of 3 weeks (+Se).

RESULTS

Data obtained from pregnant dams and from 20-day-old fetuses

253

confirmed the trend to increased levels of MDA in liver tissues taken
from zinc-deficient and alcohol-treated animals, with greatest
peroxidation evident when the treatments occurred concurrently. Levels
of GSH were sharply lower in zinc-deficient tissues, but were not
affected by gestational alcoholism.

With weanling male animals (Table 1) levels of MDA were again
higher in testes from zinc deficient animals, with a non-significant
trend towards less peroxidation when animals were supplemented with
selenium. Ratios of GSH:GSSG reflected increased oxidation of GSH in
the zinc-deficient animals, which was not affected by supplementary
selenium. Activities of GSH-Px levels were significantly raised
($p < 0.01$) in the livers of all animals receiving selenium.

Table 1. The Effect of Zinc Deficiency and Selenium Supplementation on
MDA Levels, GSH:GSSG ratios and the Activity of GSH-Px in
Young Male Rats

Treatment	MDA (μmole/mg protein)		GSH:GSSG		GSH-Px (μmole NADPH/min/ 100 mg protein)	
	Liver	Testes	Liver	Testes	Liver	Testes
+Zn	1.37±0.15	0.75±0.05	29.4	27.2	29.5±1.6	4.98±0.28
		$p < 0.05$			$p < 0.05$	
-Zn	1.50±0.02	0.95±0.09	20.6	24.3	35.4±1.4	4.48±0.16
+Zn, +Se	1.48±0.06	0.57±0.10	30.8	31.5	43.9±1.6	5.15±0.18
		$P < 0.05$				
-Zn, +Se	1.43±0.20	0.85±0.10	19.4	20.1	43.1±1.9	5.06±0.21

DISCUSSION

Zinc deficiency has again been linked with increased
lipoperoxidation and with accelerated oxidation of GSH to GSSG.
Supplementary selenium does not appear to be protective in this regard
nor does it seem to influence the GSH:GSSG ratio despite an increase in
the activity of GSH-Px.

REFERENCES

1. I.E. Dreosti, I.R. Record and S.J. Manuel, 1985, Zinc and the
    embryo: A review, Biol. Trace Element Res., 7: 103.

2. I.E. Dreosti, S.J. Manuel and I.R. Record, 1982, Superoxide
    dismutase, manganese and the effect of ethanol in adult and fetal
    rats. Br. J. Nutr. 48: 205.

CONTRASTING EFFECTS OF SELENIUM AND VITAMIN E DEFICIENCY

ON THE ANTI-MALARIAL ACTION OF QINGHAOSU IN MICE

O.A. Levander, A.L. Ager, V.C. Morris, and R. May

USDA Human Nutrition Research Center, Beltsville, MD 20705;
Center for Tropical Parasitic Diseases, University of Miami,
Miami, FL 33177

The herb Artemisia annua (also known as sweet wormwood) has been used for many centuries in Chinese traditional medicine as a treatment for malaria and fever (1). In 1971, Chinese chemists isolated from the leafy portions of the plant the substance responsible for its reputed medicinal action. This substance, called qinghaosu (QHS) is an endoperoxide of a sesquiterpenoid lactone. It occurred to us, therefore, that the therapeutic efficacy of QHS might be influenced by the antioxidant status of the malarial host. For that reason, we decided to investigate the effect of dietary selenium and vitamin E deficiency on the suppressive and curative efficacy of QHS in a murine model of malaria.

METHODS

Animals and diets. In experiment 1, weanling mice were fed the basal Torula yeast diet supplemented with 50 mg all-rac-alpha tocopheryl acetate/kg plus either 0 or 0.5 ppm Se as sodium selenite for six weeks. In experiment 2, weanling mice were fed the basal Torula yeast diet supplemented with 0.1 ppm Se as sodium selenite plus either 0 or 100 mg all-rac-alpha-tocopheryl acetate/kg for nine weeks. The basal Torula yeast diet contained (%): torula yeast, 30; AIN-76 salt mix (no selenium), 3.5; AIN-76 vitamin mix (no vitamin E), 1; DL-methionine, 0.3; corn oil, 5; choline dihydrogen citrate, 0.225; and sucrose to 100. In experiment 1, ordinary corn oil was used, whereas in experiment 2 tocopherol-stripped corn oil was used.

Parasites and QHS tests. After the mice were fed their respective diets for the stated period, they were all inoculated by intraperitoneal injection with 2.5×10^6 erythrocytes parasitized with Plasmodium yoelii. The mice were then divided into groups of 6 to 8 and given 0, 4, 16 or 64 mg QHS/kg bid on days 3, 4 and 5 after the inoculation. The QHS was given orally by gavage as an aqueous suspension in 0.5% hydroxyethylcellulose-0.1% Tween-80. The chemosuppressive activity of QHS was estimated by reading blood films for percent parasitemia on day 6 after inoculation, while curative activity was assessed by noting survival after 60 days.

RESULTS AND DISCUSSION

   Dietary Se deficiency had no potentiating effect on the anti-malarial
activity of QHS as determined either in the chemosuppressive or curative
assay of the mouse model despite the fact that plasma Se levels and
hepatic GSH-Px activities in the deficient group were less than 1% that
in the supplemented group (Table, experiment 1). On the other hand,
vitamin E deficiency (verified by depressed plasma tocopherol levels) did
potentiate the anti-malarial potency of QHS as judged by lower
parasitemias in the groups receiving 4 or 16 mg/kg of the drug or by
improved survival in the groups receiving 16 or 64 mg/kg of the drug
(Table, experiment 2). These results demonstrate that the antioxidant
status of the host as altered by nutritional manipulation can influence
the therapeutic efficacy of QHS against murine malaria. Our data are in
accord with the work of others who have shown that the malarial parasite
exerts a pro-oxidant stress on the host (2).

Effect of Dietary Se or Vitamin E Deficiency on
the Suppressive and Curative Anti-malarial Activities of QHS

Dose of QHS (mg/kg)	Experiment 1 Dietary Se (ppm)		Experiment 2 Dietary Vitamin E (IU/Kg)	
	0	0.5	0	100
Chemosuppressive assay (percent parasitemia):				
0	50 + 6	36 + 5	27 + .5	26 + 2
4	54 + 2	53 + 5	11 + 2	26 + 3
16	9 + 3	8 + 3	3 + 1	13 + 3
64	0 + 0	0 + 0	0 + 0	0 + 0
Curative assay (60 - day survival):				
0	0/8	2/8	1/6	0/8
4	0/7	0/6	0/7	0/7
16	2/8	0/7	7/8	0/8
64	3/8	4/7	7/8	3/8
plasma Se (ng/ml)	4 + 1	532 + 70	464 + 41	552 + 28
hepatic GSH-Px (mu/mg protein)	10 + 0	1450 + 60	1247 + 58	1188 + 103
plasma tocopherol (ug/100 ml)	not determined	not determined	< 50	446 + 40

REFERENCES

1. Klayman, D.L. (1985) Qinghaosu (Artemisinin): an antimalarial drug
    from China. Science 228: 1049-1055.

2. Clark, I.A., Hunt, N.H., and Cowden, W.B. (1986) Oxygen-derived free
    radicals in the pathogenesis of parasitic disease. Adv. Parasitology
    25: 1-44.

THE EFFECT OF SELENIUM AND VITAMIN E DEFICIENCIES ON THE SPIN TRAPPING OF FREE RADICALS IN HOMOGENATES OF RAT HEART

John R. Arthur, *Donald B. McPhail and *Bernard A. Goodman

Rowett Research Institute, and *Macaulay Land Use Research
Bucksburn,   Institute,
Aberdeen, AB2 9SB, U.K.   Aberdeen, AB9 2QJ, U.K.

Tissue damage mediated by the formation of chemically reactive free radicals, often derived from oxygen, has been implicated as the cause of many diseases in man and animals. However, in many instances the evidence for the involvement of free radicals in disease processes is indirect. This evidence includes formation of breakdown products from free radical mediated destruction of essential cell constituents and the protective effect of free radical scavengers and antioxidant enzymes. Electron spin resonance (ESR) spectroscopy has been used for the characterisation of free radicals in both chemical and biological systems, either directly in the case of stable radicals or after reaction of unstable radicals with a "spin trap" to form a more long lived radical which can be detected. In this report we describe the application of both direct ESR spectroscopy and ESR spectroscopy after spin trapping to investigate free radicals in heart from Se- and vitamin E-deficient rats.

Experiments were carried out with male Hooded Lister rats of the Rowett Institute strain. For 6 weeks from weaning the rats consumed control or Se/vitamin E deficient diets (Abdel-Rahim et al., 1986) containing respectively 0.1 mg Se and 200 mg α-tocopherol/kg or <0.01 mg Se and/or <1 mg α-tocopherol/kg. Rats were anaesthetised with ether and tissues removed and immediately frozen in liquid nitrogen before use in the ESR studies. Se-containing glutathione peroxidase (Se-GSPHx) activities and vitamin E concentrations were determined as described by Duthie and Arthur, 1987.

Figure 1 shows an ESR spectrum recorded at 77K of a heart from a control rat. The major component at $g = 2.004$ is from a free radical, with additional features from high spin Fe(III) being seen at $g \simeq 6$ (haem) and at $g \simeq 4.3$ and from low spin Fe(III) near $g = 2$. Similar spectra were observed when the rats had consumed Se-deficient and/or vitamin E-deficient diets for 6 weeks (not shown). Furthermore, an ip injection of $FeCl_3$, to provide an oxidant stress (Arthur and Morrice 1985), did not increase the intensity of any of the components in ESR spectra from hearts.

When heart homogenates were incubated with 45 mM α-(4-pyridyl-1-oxide)-N-tert-buty nitrone (4-POBN) as spin trap in the presence of 6.8 μM $FeSO_4$, 2.3 mM ADP and 0.22 mM NADPH (final concentrations in phosphate buffer) to initiate peroxidation, ESR signals characteristic of a free radical/4-POBN adduct developed over 120 minutes. Spectra were triplets of peaks due to interaction of the unpaired electron with a ^{14}N with the triplets further

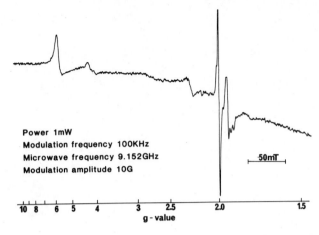

Power 1mW
Modulation frequency 100KHz
Microwave frequency 9.152GHz
Modulation amplitude 10G

Fig. 1. ESR spectrum at 77K of whole heart from a control rat.

split into doublets by interaction with a ^{1}H. Hyperfine coupling constants were A(^{14}N) = 1.56 mT and A(^{1}H) = 0.253 mT. In heart homogenate from control and Se-deficient rats the 4-POBN/radical adduct ESR signals after 90 min were similar in height but these were significantly greater in vitamin E-deficient hearts and greater still in heart from rats low in both Se and vitamin E (Table 1).

Table 1. Se-GSHPx Activities, Vitamin E Concentrations and ESR
         Signal Heights in Rat Heart Homogenates

Variable	Group			
	+Se+vitE	+Se-vitE	-Se+vitE	-Se-vitE
Se -GSHPx	372[a]	282[a]	14[b]	16[b]
(mU/mg protein)	±84	±26	±2	±1
Vitamin E	151[a]	25[b]	151[a]	28[b]
(ng/mg protein)	±13	±3	±2	±7
ESR signal height	1.48[a]	5.90[b]	1.80[a]	10.32[c]
(arbitrary units)	±0.17	±1.02	±0.17	±0.85

Results are means ± SEMs of at least 5 animals/group. Within rows groups without a common superscript differ significantly, $p < 0.01$.

Se and vitamin E deficiencies in rats did not affect the stable ESR free radical signal in heart and other tissues. Formation of free radicals which react with the spin trap 4-POBN was however increased by low vitamin E concentrations and further exacerbated by Se deficiency (Table 1). This is consistent with the hypothesis that tissue damage in combined Se and vitamin E deficiency is initiated by free radicals. Absolute proof of this will await the demonstration of increased free radical formation in vivo prior to the onset of disease in deficient animals.

REFERENCES

Abdel-Rahim, A.G., Arthur, J.R. and Mills, C.F., 1986, Effects of dietary copper, cadmium, iron, molybdenum and manganese on selenium utilisation by the rat, J. Nutr., 116:403.
Arthur, J.R. and Morrice, P.C., 1985, Effects of Se and vitamin E deficiencies on the response of rats to iron injection. in "Trace Elements in Man and Animals - TEMA5", C.F. Mills, I. Bremner and J.K. Chesters, eds. C.A.B. Slough, UK. pp 109.
Duthie, G.G. and Arthur, J.R., 1987, Antioxidant status and plasma pyruvate kinase activity of halothane-reacting pigs, Am. J. Vet. Res., 48:309.

TRACE ELEMENTS IN CALCIFIED TISSUES AND MATRIX BIOLOGY

Robert B. Rucker

Department of Nutrition
University of California
Davis, CA 95616

The roles for trace minerals in soft and hard tissue matrix development
and maintenance have been investigated only in a limited and selected
fashion. For example, it is known that Cu deprivation can influence markedly
the structural integrity of collagen and elastin matrices by causing reduced
crosslinking (1). It is also known that Zn deprivation may cause changes in
the composition of basal laminae or influence the functional activity and
distribution of cell surface proteins (2-5). Further, Mn deprivation can
influence proteoglycan or structural glycoprotein deposition (1). These
examples are important in that it is now clear that most of the abnormal
morphogenetic effects of matrix molecules are associated with alteration of
their physical properties or at cell surfaces (6).

For example, when one considers phenomenon associated with Cu
deficiency, one is struck not only by the disorganization of collagen and
elastin fibers, but by other changes that occur in cellular distribution,
e.g., such as thickening of the endothelium of major vessel walls (7), or the
altered differentiation and distribution of pulmonary cells important to the
transition from a saccular to alveolar lung (8,9). Indeed, even pursuing the
epiphenomena associated with trace element deficiencies can lead to interesting end points, cf. the discussion which follows on Cu, elastin cross-
linking and susceptibility to inappropriate proteolysis. The case for the
importance of minerals to matrix development and differentiation is also
easily made when cartilage development is considered as a model, e.g., the
chondrodystrophic changes that occur in manganese deficiency versus the
chondrodysplasia that occurs in other trace mineral deficiencies (10).
Likewise, Zn can regulate matrix membrane phenomenon. Zinc is an inhibitor
of membrane-bound enzymes. A number of growth factors also undergo self-
activation when zinc is removed, e.g., nerve growth factor (2).

Some of these relationships are depicted in the following diagrams. In
Figure 1 various possibilities are outlined that are related specifically to
what happens when copper is limiting in important stages of vascular matrix
development. Figure 2 depicts possibilities related to abnormal cell
differentiation following zinc deprivation or the abnormal distribution of
proteoglycans. The copper example is taken from work by D. Tinker and
coworkers, who have shown that when collagen and elastin are inappropriately
crosslinked they become substrates, because of lysyl enrichment, for a number
of proteinases that normally do not act on these proteins (Figure 1). This
can lead to abnormal remodeling, disruption of fibers, and subsequently

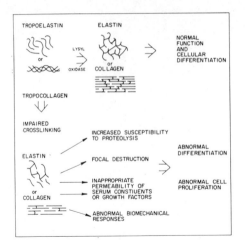

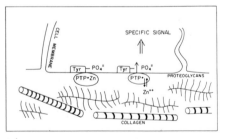

Figure 1   Figure 2

increased permeability through abnormally altered lamilae. Cells who are normally protected from serum-derived growth factors may come in contact with such factors when collagen and elastin fibers are disrupted. The expression of many protein-derived products from cells is also altered with changes in cell shape due to abnormal biomechanical responses.

The zinc example is taken from work by Donahoe, Huston, Chandler, Trelstad and co-workers (3-5), who have speculated that zinc may be important in the regulation of cells by influencing phosphotyrosine phosphatase (PTP) activity, a membrane bound phosphatase which also contacts proteoglycans located in the basement membrane. The dissociation of zinc from PTP by binding to proteoglycan is suggested to activate the enzyme producing a signal for subsequent messages. Clearly many of the recent advances in cell biology and our expanding understanding of cell-matrix interactions should allow us to proceed in better defining specific and functional roles that trace minerals play when they in turn interact with the molecules that comprise both calcifying and non-calcifying extracellular matrices.

REFERENCES

1. D. Tinker and R. Rucker, Phys. Revs. 65:607-657 (1985).
2. M. Young and M. J. Koroly, Biochem. 19:5316-5321 (1980).
3. G. Budzik, S. Powell, S. Kamagata and P. Donahoe, Cell 34:307-314 (1983).
4. J. A. Chandler, F. Sinowatz, B. G. Timms and C. G. Pierrepoint, Cell Tissue Res. 185:89-103 (1977).
5. P. K. Donahoe, G. P. Budzik, R. L. Trelsted, B. R. Schwartz, M. E. Fallat and J. M. Huston, in: "Role of the Extracellular matrix in Development," pp. 573-595, R. Trelstad, ed., Alan Liss, New York.
6. E. D. Hay, in: "Role of the Extracellular Matrix in Development," pp. 1-31, Robert Trelstad, ed., Alan R. Liss, New York.
7. H. A. Hunsaker, M. Michio and K. Allen, Atherosclerosis 51:1-19 (1984).
8. M. Dubick, C. Keen and R. Rucker, Exp. Lung Res. 8:227-241 (1985).
9. B. L. O'Dell, K. H. Kilburn, W. N. McKenzie and R. J. Thurston, Am. J. Pathol. 91:413-432 (1978).
10. R. M. Leach and C. V. Gay, J. Nutrition 117:784-790 (1987).

THE EXTRACELLULAR MATRIX

Marcel E. Nimni

Laboratory of Connective Tissue Biochemistry
University of Southern California School of Medicine and
Orthopaedic Hospital of Los Angeles
2400 S. Flower Street
Los Angeles, California 90007

INTRODUCTION

Collagen is the single most abundant animal protein in mammals accounting for about 30% of all proteins. The collagen molecules, after being secreted by the cells, assemble into characteristic fibers responsible for the functional integrity of tissues such as bone, cartilage, skin, and tendon. They contribute a structural framework to other tissues such as blood vessels and most organs. Crosslinks between adjacent molecules are a prerequisite for the collagen fibers to withstand the physical stresses to which they are exposed. Significant progress has been made towards understanding the functional groups on the molecules that are involved in the formation of such crosslinks, their nature, and location. Collagen does not exist in a vacuum, but it is surrounded by a large number of macromolecules which all together form the extracellular matrix. The fundamental properties of these macromolecules, such as the proteoglycans, elastin, laminin, fibronectin and others will be briefly described. For a more detailed discussion and specific references the readers are referenced to comprehensive reviews from which this information was derived.[1,2,3]

The Collagen Molecule, Its Structure and Biosynthesis

In order for the organism to develop an extracellular network of collagen fibers, the cells involved in the biosynthetic process must first synthesize a precursor known as procollagen. This molecule is later enzymatically trimmed of its nonhelical ends, giving rise to a collagen molecule that spontaneously assembles into fibers in the extracellular space. The sequence of events associated with the biosynthesis of collagen is summarized in Figure 1.

Crosslinking

Crosslinking renders the collagen fibers stable, and provides them with an adequate degree of tensile strength and visco-elasticity to perform their structural role. The degree of crosslinking, the number and density of the

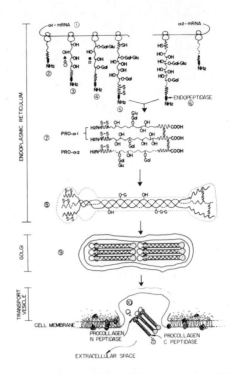

Figure 1. (1) Synthesis of specific mRNAs for the different procollagen
chains. (2) Translation of the message on polysomes of the rough E.R.
(3) Hydroxylation of specific proline residues by 3-proline hydroxylase (Δ)
and 4-proline hydroxylase (▲) and of lysine by lysyl hydroxylase (o).
(4) Glycosylation of hydroxylysine by galactosyltransferase (■) and addition
of glucose by a glucosyltransferase (□). (5) Removal of the N-terminal
signal peptide. (6) Release of completed α-chains from ribosomes. (7) Recognition of 3 α-chains through the C-terminal prepeptide and formation of
disulfide crosslinks. (8) Folding of the molecule and formation of triple-
helix. (9) Intracellular translocation of the procollagen molecules and
packaging into vesicles. (10) Fusion of vesicles with the cell membrane and
extrusion of the molecule accompanied by the removal of the C-terminal nonhelical extensions and part of the N-terminal nonhelical extensions by
specific peptidases.

fibers in a particular tissue, as well as their orientation and diameter,
combine to provide this function. Crosslinking begins with the conversion
to peptide-bound aldehydes of specific lysine and hydroxylysine residues in
collagen. It involves the oxidative deamination of the ε-carbon of lysine
or hydroxylysine to yield the corresponding semialdehydes (allysine or hy-
droxyallisine) and is mediated by lysyloxidase, a Cu^{++} requiring enzyme.
Its activity is inhibited by β-aminopropionitrile, chelating agents such as
EDTA and D-penicillamine, and isonicotinic acid hydrazide and other carbonyl
reagents. Lysyloxidase exhibits particular affinity for the lysines and
hydroxylysines present in the nonhelical extensions of collagen, but can, at
a slower pace, also alter residues located in the helical region of the
molecule.

The events that lead to the formation of crosslinks are summarized in Figure 2.

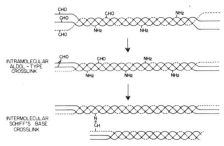

Figure 2. Formation of intramolecular and intermolecular crosslinks in type I collagen. Intramolecular crosslinks occur in the nonhelical regions and involve an aldol condensation reaction between lysine or hydroxylysine derived aldehydes within a single molecule. Intermolecular crosslinks on the other hand involve aldehydes and ε-amino groups of lysine present in different molecules.

Different Types of Collagen

Over fifteen years have passed since we first realized that all collagen fibers within a particular organism are not made up of identical molecules. Since 1970 a great deal of experimental work has been devoted to understanding these various collagen types, their molecular structure, biosynthesis, cells of origin, distribution, and turnover. The different collagen types have been identified using Roman numerals, which were assigned as they became characterized.

Table I summarizes the chain composition and distribution of these collagens.

Type	Disulfide Crosslinked	Tissue Distribution	Chain Composition	Molec. Wt. of each chain
I	-	Skin, bone. blood vessels and most organs	$\alpha1(I)_2 \alpha2(I)$	100,000
I-trimer	-	cell culture, tumors	$\alpha1(I)_3$	100 000
II	-	cartilage	$\{\alpha1(II)\}_3$	100.000
III	+	Blood vessels, granulation tissue, skin	$\{\alpha1(II)\}_3$	100.000
IV	+	Basement membrane	$\{\alpha1(IV)\}_3$	175.000 - 185,000
V	-	Most tissues except hyaline cartilage, usually pericellular.	$\{\alpha1(V)\}_2 \alpha2(V)$	Unknown. globular and helical domains?
VI	+	Uterus, placenta and skin		Short triple helical domains 40,000-70.000
VII	+	Epithelial basement membrane		Seems larger than any known collagen
VIII	-	Aortic endothelial cells corneal endo cells	$\{\alpha_1(VIII)\}_3$	180,000
IX	+	Cartilage	$\alpha1(IX)\alpha2(IX)\alpha3(IX)$	69-85,000
X	-	Hypertrophic Cartilage	$\{\alpha_1(X)\}_3$	59,000

***1α ,2α and 3α may be the type V equivalent for hyaline cartilage

Noncollagenous Proteins Found in the Connective Tissues

Basement membranes contain, in addition to significant amounts of type IV collagen, noncollagenous glycoproteins that presumably account for their positive periodate Schiff reaction.

Laminin

Laminin comprises almost 50% of the matrix proteins of the EHS tumor, most of which can be extracted in neutral buffers of moderate ionic strength. It has also been isolated from cultured endodermal and teratocarcinoma cells, in some instances in native form after collagenase treatment. Immunohistochemical localization of laminin indicates that it is abundant in all basement membranes. The amino acid composition of laminin distinguishes it from fibronectin. It contains approximately 12%-15% carbohydrate. While the native protein migrates in electrophoresis as a narrow band, reduction of the disulfide bonds produces two broad, faster migrating bands with Mr = 200-220,000 and 400-440,000. Ultracentrifugal analyses indicate that laminin has a molecular weight in the range of 800,000 - 1,000,000 daltons both in neutral buffer and under dissociating conditions (6 M guanidine). Laminin interacts with heparin, heparan sulfate, and type IV collagen, and among other things links endothelial cells to basement membranes.

Fibronectin

Fibronectin is a major biosynthetic product of cultured fibroblasts. It is similar to the cold insoluble globulin of human plasma described earlier. It is also produced by a variety of cells, including endothelial and smooth muscle cells and some epithelial cells. Immunohistology studies have shown that fibronectin is produced early in development and is associated with most embryonic basement membranes, but is not always detectable in fully developed basement membranes. Fibroblasts adhere to collagen substrates through attachment to fibronectin. Fibronectin has been visualized by electron microscopy of shadowed specimens as long, thin flexible strands that have recently been shown to bind to actin and that have many features in common with actin-binding protein and laminin.

Collagen Degradation

The changing patterns of the connective tissue matrix during growth, development, and repair following injury, require a delicate balance between synthesis and degradation of collagen and proteoglycans. Under normal circumstances this balance is maintained, while in many diseased states it is altered, leading to an excessive deposition of collagen or to a loss of functional tissue. The first animal enzyme capable of degrading collagen at neutral pH was isolated from the culture fluid of tadpole tissue. This was shown to cleave the native molecule into two fragments in a highly specific fashion at a temperature below that of denaturation of the substrate. These fragments were characterized by electron microscopy and reflect the cleavage of a native collagen molecule at a specific site closer to the C-terminal end of the molecule, yielding segments of 25% and 75% the length of the native collagen molecule; the larger fragment was termed TC-A and the smaller fragment TC-B.

Collagenolytic enzymes have been obtained following cell and organ culture from a wide range of tissues from animal species in which collagen is present. In general, these enzymes have a number of fundamental properties in common: they all have a neutral pH optima; they are not stored within the cell but, rather, are secreted either in an inactive form or bound to inhibitors. Figure 3 summarizes schematically their mode of action. They appear to be zinc metalloenzymes requiring calcium, and are not inhibited by agents that block serine or sulphydryltype proteinases. They are inhibited by chelating agents such as EDTA, 1,10-o-phenanthroline and cysteine, which may inactivate zinc and perhaps other metals required for enzymatic activity. The zinc in the latent enzyme can be replaced by other divalent cations such as Co, Mn, Mg, and Cu.

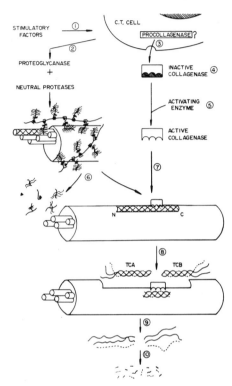

Figure 3. Sequence of events which can lead to the degradation of collagen fibers by the enzyme collagenase. (1) A variety of factors have been described which stimulate connective tissue cells to synthesize collagenase, glycosydases and neutral proteases. (2) The proteoglycan degrading enzymes remove the mucopolysaccharides which surround collagen fibers and expose it to collagenase. (3) Inactive collagenase is secreted. (4) The enzyme is usually found in the extracellular space bound to an inhibitor. (5) An activating enzyme removes the inhibitor. (6) Glycosidases complete the degradation of the proteoglycans. (7) The active collagenase binds to fibrillar collagen. (8) Collagenase splits the first collagen molecule into two fragments (TC-A and TC-B) which denature and begin to unfold at body temperature. The enzyme now moves on to an adjacent molecule. (9) The denatured collagen fragments are now susceptible to other proteases. (10) Nonspecific neutral proteases degrade the collagen polypeptides.

Proteoglycans

In order to understand the physical properties of connective tissues it is important to have an understanding of the salient features of the proteoglycan molecules. These comprise a well organized network of proteins and associated carbohydrates (glycosaminoglycans) (Fig. 4). There are essentially two types of glycosaminoglycans, those with weak negative charges, such as hyaluronic acid, and those with strong negative charges such as the chondroitin sulfates, heparins, and dermatan sulfate, the latter comprising the largest of proteoglycan species. Their distribution and physicochemical characteristics, which contribute to distinct functions, are also unique. Hyaluronic acid, with weak negative charges associated with the carboxylic acid residues present in glucuronic acid, has a tendency to form hydrated gels. It can therefore contribute significantly to the viscoelastic fluidity of synovial fluid and to the turgency of the skin of an infant. On the other hand, the negatively charged polysaccharides that contain sulfonic acid residues are able to develop strong ionic bonds with the positively charged amino acids on the surface of the collagen fibers, particularly lysine, hydroxylysine, and arginine. Such tissues are more compact, resilient, less hydrated, and exhibit the viscoelastic behavior typified by haline articular cartilage.

COLLAGEN STRUCTURE AND METABOLISM

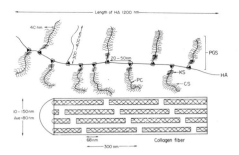

Figure 4. Collagen fibers do not exist in a vacuum. They are usually closely associated with the proteoglycans of the ground substance. This diagram depicts a collagen molecule in cartilage, adjacent to a proteoglycan aggregate containing hyaluronic acid (HA), proteoglycan subunits (PGS) and link proteins (A) which help to stabilize the structure. The PGS consists of a protein core (PC) from which the negatively charged glycosaminoglycan chains of chondroitin sulfate (CS) and keratan sulfate (KS) radiate.

CONCLUSION

Connective tissues, although heterogeneous in nature, have in common structural-functional properties that contribute to the shape of tissues and organs. They provide an adequate environment for cells to proliferate, attach and interact with each other. They retain in their midst a variety of macromolecules, and harbor smaller ions which are used to regulate intracellular and extracellular activities and contribute to specialized processes such as mineralization. The composition of bone tends to reflect the ionic milieu in which it was formed. Not much is known about the storage, transport and turnover of trace elements in this extracellular matrix, and I believe this will become an area of fruitful investigation.

REFERENCES

1. M. E. Nimni and R. D. Harkness, Molecular structure and functions of collagen, in: "Collagen: Biochemistry, Biotechnology and Molecular Biology," M. E. Nimni, ed., CRC Press, Boca Raton, FL (1987).
2. M. E. Nimni, Collagen: structure, function, and metabolism in normal and fibrotic tissues, Semin. Arthritis Rheum. 13:1 (1983).
3. M. E. Nimni, D. Cheung, B. Strates, M. Kodama, and K. Sheikh, Chemically modified collagen: A natural biomaterial for tissue replacement, J. Biomed. Mat. Res. 21: (1987).

THE ROLE OF TRACE ELEMENTS IN THE DEVELOPMENT OF CARTILAGE MATRIX

Roland M. Leach, Jr.

Department of Poultry Science, The Penn State University
205 Henning Building
University Park, PA   16802

INTRODUCTION

Several trace elements are essential for normal skeletal development in young animals and man. These include copper, manganese, silicon, and zinc. Deficiencies of these nutrients produce specific pathological lesions in the epiphyseal growth plate, a tissue which plays a key role in endochondral bone formation. The morphology and biochemistry of this tissue has been recently reviewed (Leach and Gay, 1987). Briefly, the tissue is comprised of chondrocytes which can be divided into zones based upon stage of maturation: resting, proliferative, prehypertrophic and hypertrophic. These zones represent specific stages of maturation or differentiation of the chondrocyte with each type having specific biochemical characteristics. This is especially true for the hypertrophic chondrocyte which is removed by metaphysial blood vessels and replaced by trabecular bone. The chondrocytes of the epiphyseal growth plate are responsible for synthesizing the extracellular matrix typical of the cartilage, the major constituents being proteoglycans and collagen. This tissue is sensitive to the endocrine system as well as nutrition, with the growth hormone dependent IGF-I being a key factor in cartilage growth and development. The purpose of this paper is to discuss the influence of trace elements on cartilage metabolism with emphasis upon extracellular matrix formation and metabolism.

Copper

A number of investigators have observed lameness to be associated with copper deficiency. Although not totally responsible for the lameness, there are specific changes observed in the growth plate and adjacent metaphysis. In mammals, the main effect is upon the metaphysis where spicules of calcified cartilage persist along with reduced trabecular bone formation, probably due to the virtual cessation of osteoblastic activity (Baxter et al., 1953; Follis et al., 1955). In young chicks, the epiphyseal cartilage is thickened due to an increase in immature uncalcified chondrocytes. Vascular invasion of the cartilage is greatly reduced (Carlton and Henderson, 1964).

At the present time, little is known about the biochemical role of copper in cartilage metabolism. However, we can gain some insight by examining the extensive literature dealing with tibial dyschondroplasia.

This is a condition that occurs spontaneously in many rapidly growing avian species. This cartilage abnormality can also be induced by a toxin from Fusarium roseum (Walser et al., 1982) and the pesticide, Thiuram (Vargas et al., 1983). The pathological changes observed in the chondrocytes found in this lesion are indistinguishable from those observed with copper deficiency. However, the tissue is much firmer than copper deficient cartilage which tends to disassociate this lesion from the well-known role of copper in lysyl oxidase, an enzyme essential for collagen and elastin cross-linking.

Extensive biochemical and ultrastructural studies on the lesion associated with tibial dyschondroplasia have been conducted (Hargest et al., 1985; Freedman et al., 1985; Hargest et al., 1985). The chondrocytes fail to undergo the biochemical and morphological changes associated with hypertrophy. Eventually the cells exhibit the classical symptoms of coagulative cell necrosis. Like copper deficiency, the cells exhibit reduced oxidative activity (Lilburn and Leach, 1980).

The lesion associated with copper deficiency has received little attention, thus we can only speculate on the role of copper in the metabolism of this tissue. One possible hypothesis relates to copper as an angiogenic factor. This has recently been reviewed by Folkman and Klagsbrun (1987). Since poor vascularization of the cartilage is characteristic of the lesion, such a hypothesis has promise.

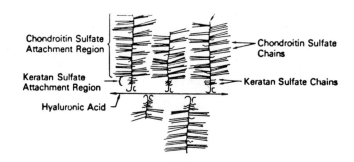

Fig. 1. A model for cartilage proteoglycan
aggregate structure taken from Hascall (1977).

Manganese

Unlike copper, the role of manganese in cartilage formation is much more clearly defined. The effect of manganese on cartilage development can be related to its role in proteoglycan biosynthesis. Proteoglycans are complex macromolecules which are major constituents of the cartilage matrix (Hascall, 1977). The basic structure of this molecule is depicted in Fig. 1. It has three major constituents: proteoglycan monomer, link protein, and hyaluronic acid. These occur in tissues as an aggregate and contribute significantly to the chemical and physical properties of the cartilage. Manganese is required for the formation of proteoglycans via its role in glycosyl transferases. These enzymes are involved in the addition of sugars to the core protein as well as elongation of the glycosaminoglycan side chains. The role of manganese in the activity of these enzymes has recently been reviewed (Leach, 1986).

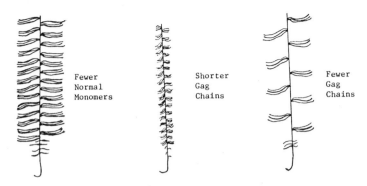

Fig. 2.   Potential effects of Mn deficiency
upon proteoglycan structure

In order to gain a better understanding of mechanisms of proteoglycan biosynthesis, we have been examining the characteristics of the proteoglycans in cartilage obtained from manganese deficient animals. It
was of interest to determine if incomplete molecules were synthesized
under these conditions (see Fig. 2). This does not appear to be the
case. Although reduced in quantity, the proteoglycans synthesized by manganese deficient tissues share most of the properties with those isolated
from normal tissue. This conclusion is based upon labeling of the glycosaminoglycan side chains as well as the properties of the isolated proteoglycans when subjected to ultracentrifugation and gel filtration.

Another laboratory has examined the possibility that the effect of
manganese was mediated via an effect upon insulin-like growth factors
(somatomedins). These growth hormone dependent peptides play a key role
in cartilage metabolism. Originally, they were known as the "sulfation
factor," a substance found in serum which would stimulate cartilage explants to incorporate radioactive sulfate into proteoglycans. Studies
with manganese deficient animals led to the conclusion that the effect of
manganese on cartilage metabolism could not be explained by a reduction
in IGF activity (Bolze et al., 1985).

Silicon

Silicon has been implicated in skeletal development based upon reports that silicon deficiency results in alterations in the skull and
long bones of experimental animals (Carlisle, 1972; Schwarz, 1972). A
role for silicon in cartilage matrix metabolism is suggested based upon
the observation that there are decreased amounts of articular cartilage
as well as changes in the epiphyseal growth plate. In this tissue, there
was a decrease in the width of the proliferating zone and a decrease in
staining intensity of the matrix in the hypertrophic zone. Chemical analysis revealed a decrease in proteoglycan content of this tissue
(Carlisle, 1976, 1980).

At this time, a specific mechanism for the role of silicon proteoglycan has not been elucidated. One possibility would a direct function of
this element in proteoglycan biosynthesis. A second would be an involvement in proteoglycan structure. This was suggested by Schwarz (1973) who
found isolated proteoglycans to be rich in silicon. It was proposed that
silicon, in the form of silanolate, was acting by forming bridges which

were important in the structural organization of glycosaminoglycans. More research is needed in order to clearly define the role of silicon in cartilage matrix formation.

Zinc

Zinc deficiency has profound effects upon skeletal development (O'Dell et al., 1958; Young et al., 1958). Externally, the bones are shortened and thickened, with enlargement of the joints. Reduced cell division was associated with the reduction in width of the epiphyseal growth plate. An excess of lightly staining matrix was also observed in the proliferative zone, while the cells in the zone of hypertrophy were smaller and less active. The effects on the proliferative zone were most pronounced where the cells were remote from blood vessels. Proximity to blood vessels also appeared to be a factor in the changes found in the zone of hypertrophy (Westmoreland and Hoekstra, 1969). Histamine and indomethacin did not alter these lesions although these substances had been found to alleviate the lameness associated with zinc deficiency (Nielsen et al., 1968).

At this time, no specific role for zinc in cartilage matrix formation has been defined. There are decreases in cartilage alkaline phosphatase activity associated with zinc deficiency (Westmoreland, 1971). This is a zinc containing enzyme which is widespread in connective and other tissues and fluids. The need for zinc in cell division may also account for some of the changes observed in the proliferative zone.

Similar to manganese, the possibility that zinc is mediating its effect via growth hormone and/or IGF has also been explored. Although IGF is greatly reduced under conditions of zinc deficiency, this appears to be due to a reduction in food intake rather than a direct effect upon IGF metabolism (Oner et al., 1984).

SUMMARY

Copper, manganese, silicon and zinc are required for normal skeletal development, specifically for normal endochondral bone formation. The role of manganese is most clearly defined. The element is needed for the synthesis of proteoglycans, a major constituent of cartilage extracellular matrix. Manganese is needed for the activity of glycosyl transferases, enzymes involved in the synthesis of these macromolecules. Silicon also appears to play a role in proteoglycan metabolism. It may be involved in this synthesis or molecular structure of these cartilage matrix components. Although they are required for normal endochondral bone formation, the specific roles of copper and zinc are poorly defined at this time.

REFERENCES

Baxter, J. H., Van Wyk, J. J., and Follis, Jr., R. H., 1953, A bone disorder associated with copper deficiency. II. Histological and chemical studies on the bones, Bull. Johns Hopkins Hosp., 93:25.
Bolze, M. S., Reeves, R. D., Lindbeck, F. E., Kemp, S. F., and Elders, J. J., 1985, Influence of manganese on growth, somatomedin and glycosaminoglycan metabolism, J. Nutr., 115:352.
Carlisle, E. M., 1972, Silicon: An essential element for the chick, Science, 178:619.
Carlisle, E. M., 1976, In vivo requirement for silicon in articular cartilage and connective tissue formation in the chick, J. Nutr., 106:478.

Carlisle, E. M., 1980, Biochemical and morphological changes associated with long bone abnormalities in silicon deficiency, J. Nutr., 110:1046.

Carlton, W. W, and Henderson, W., 1964, Skeletal lesions in experimental copper-deficiency in chickens, Avian Dis., 8:48.

Folkman, J., and Klagsbrun, M., 1987, Angiogenic factors, Science, 235:442.

Follis, Jr., R. H., Bush, J. A., Cartwright, G. E., and Wintrobe, M. M., 1955, Studies on copper metabolism. 18. Skeletal changes associated with copper deficiency in swine, Bull. Johns Hopkins Hosp., 97:405.

Freedman, B. D., Gay, C. V., and Leach, R. M., 1985, Avian tibial dyschondroplasia. II. Biochemical changes, Am. J. Pathol., 119:191.

Hargest, T. E., Gay, C. V., and Leach, R. M., 1985. Avian tibial dyschondroplasia. III. Electron probe analysis, Am. J. Pathol., 119:199.

Hargest, T. E., Leach, R. M., and Gay, C. V., 1985. Avian tibial dyschondroplasia. I. Ultrastructure, Am. J. Pathol., 119:175.

Hascall, V. C., 1977, Interaction of cartilage proteoglycans with hyaluronic acid, J. Supramolecular Structure, 7:101.

Leach, R. M., 1986, Chapter 6, Mn(II) and glycosyltransferases essential for skeletal development, in: "Manganese in Metabolism and Enzyme Function," V. L. Schramm and F. C. Wedler, ed., Academic Press, Inc., New York.

Leach, R. M., and Gay, C. V., 1987, Role of epiphyseal cartilage in endochondral bone formation, J. Nutr., 117:In press.

Lilburn, M. S., and Leach, R. M., 1980, Metabolism of abnormal cartilage cells associated with tibial dyschondroplasia, Poultry Sci., 59:1892.

Nielsen, F. H., Sunde, M. L., and Hoekstra, W. G., 1968, Alleviation of the leg abnormality in zinc-deficient chicks by histamine and by various anti-arthritic agents, J. Nutr., 94:527.

O'Dell, B. L., Newberne, P. M., and Savage, J. E., 1958, Significance of dietary zinc for the growing chicken, J. Nutr., 65:503.

Oner, G., Bhaumick, B., and Bala, R. M., 1984, Effect of zinc deficiency on serum somatomedin levels and skeletal growth in young rats, Endocrinology, 114:1860.

Schwarz, K., 1973, A bound form of silicon in glycosaminoglycans and polyuronides, Proc. Nat. Acad. Sci. USA, 70:1608.

Schwarz, K., and Milne, D. B., 1972, Growth-promoting effects of silicon in rats, Nature, 239:333.

Vargas, M. I., Lamas, J. M., and Alvarenga, V., 1983, Tibial dyschondroplasia in growing chickens experimentally intoxicated with tetramethylthiuram disulfide, Poultry Sci., 62:1195.

Walser, M. M., Allen, N. K., Mirocha, C. J., Hanlon, G. F., and Newman, J. A., 1982, Fusarium-induced osteochondrosis (tibial dyschondroplasia) in chickens, Vet. Pathol., 19:544.

Westmoreland, N., 1971, Connective tissue alterations in zinc deficiency, Fed. Proc., 30:1001.

Westmoreland, N., and Hoekstra, W. G., 1969, Pathological defects in the epiphyseal cartilage of zinc-deficient chicks, J. Nutr., 98:76.

Young, R. J., Edwards, Jr., H. M., and Gillis, M. B., 1958, Studies on zinc in poultry nutrition. 2. Zinc requirement and deficiency symptoms of chicks, Poultry Sci., 37:1100.

Authorized for publication on _____ 1987 as Paper No. _____ in the Journal Series of the Pennsylvania Agricultural Experiment Station.

ENHANCED MATERNAL TRANSFER OF FLUORIDE IN THE MAGNESIUM-DEFICIENT RAT

Florian L. Cerklewski and James W. Ridlington

Oregon State University
Department of Foods and Nutrition
Corvallis, OR 97331

INTRODUCTION

Although placental and mammary tissue have been shown to be formidable barriers to fluoride,[1] there is some evidence to suggest that even small amounts of transferred fluoride have significance for health of developing offspring.[2] In the present study our intent was to determine if the maternal barrier to fluoride could be influenced by nutritional status. We chose magnesium as a factor based upon our previous findings of a strong relationship between magnesium and fluoride in weanling rats.[3]

METHODS

Virgin female Sprague Dawley albino rats, initial age 79 days, were fed a purified diet containing 500 ppm magnesium and distilled-deionized water for two weeks. Pregnant rats, obtained by overnight breeding following this adjustment period, were fed the same purified diet containing either 200 or 1000 ppm (control) magnesium as the carbonate and 2 or 10 ppm fluoride as sodium fluoride throughout gestation and the first 15 days of lactation. We terminated the study at this time because it coincides with a period at which offspring begin to consume solid food.

Fluoride content of maternal urine and ashed femur was directly determined with a fluoride ion-selective electrode.[4] Fluoride was determined with the same instrumentation on unashed diet, maternal milk and feces, and offspring ashed femur and molar teeth after isolation of fluoride by perchloric acid diffusion. Tissue magnesium and calcium were determined by atomic absorption spectrophotometry. All analyses were evaluated in a 2x2 factorial design with 5 replicates per treatment

RESULTS

Food intake and body weight gain were unaffected by differences in dietary treatments. True maternal magnesium deficiency occurred in both low magnesium groups evidenced by obvious hyperemia of ears during early lactation. Magnesium deficiency was also suggested by significantly reduced magnesium concentration of milk, maternal plasma and femur, and offspring serum as well as maternal hypercalcemia. Accumulation of calcium in maternal

Table 1. Influence of Maternal Magnesium Deficiency
on Fluoride Bioavailability[a]

Measures	Dietary treatments, ppm				FLSD (0.05)
	2 F		10 F		
	200 Mg	1000 Mg	200 Mg	1000 Mg	
Milk F ug/ml	0.18 ±0.06	0.11 ±0.03	0.20 ±0.03	0.08 ±0.02	0.05
F absorption %	73 ± 3	66 ± 4	76 ± 1	69 ± 4	4.5
Dam Femur F ug/g ash	320 ± 27	272 ± 12	531 ± 9	478 ± 51	40
Pup Femur F ug/g ash	6.08 ±1.66	5.41 ±1.11	13.81 ±1.46	10.56 ±2.36	2.3
Pup 1st Molar ug F/g ash	3.01 ±1.25	1.79 ±0.52	7.03 ±1.25	5.23 ±1.22	1.5

[a]Means ± SD (n=5)

kidney and depression of offspring femur magnesium in these same groups was
blunted by the higher dietary fluoride level (P<0.05). As shown in Table
1, maternal magnesium deficiency significantly enhanced fluoride absorption
which was reflected in the fluoride concentration of milk, maternal femur,
offspring femur, and offspring first molar teeth.

CONCLUSION

The results of this study confirm that the maternal barrier to fluoride
is incomplete. Furthermore, we have demonstrated for the first time that
maternal transfer of fluoride to developing offspring bone and molar teeth
can be significantly influenced by maternal nutritional status at least with
respect to magnesium. (Supported by NIH (NIDR) DE 05628).

REFERENCES

1. E.J. Underwood, Fluorine, in: "Trace Elements in Human and Animal
Nutrition," 4th edition, p. 347, Academic Press, NY.
2. F.B. Glenn, W.D. Glenn III, and R.C. Duncan, Fluoride tablet supplementation during pregnancy for caries immunity; a study of the
offspring produced, Am. J. Obstet. Gynecol. 143: 560(1982).
3. F.L. Cerklewski, Influence of dietary magnesium on fluoride bioavailability in the rat, J. Nutr. 117:496(1987).
4. F.L. Cerklewski and J. W. Ridlington, Influence of zinc and iron on
dietary fluoride utilization in the rat, J. Nutr. 115:1162(1985).

DIETARY BORON AFFECTS BONE CALCIFICATION IN MAGNESIUM AND CHOLECALCIFEROL

DEFICIENT CHICKS

Curtiss D. Hunt and Forrest H. Nielsen

Department of Anatomy, University of North Dakota and
USDA, ARS, Human Nutrition Research Center
Grand Forks, ND 58202

INTRODUCTION

The etiologies of several bone diseases can be traced to either
deficient or excessive nutriture and/or improper metabolism of specific
inorganic substances. The marked effects on bone morphology of several
dietary inorganic substances, including calcium, magnesium, and copper,
were described in the early classical literature. Several dietary organic
compounds, including the essential fatty acids and cholecalciferol, also
affect bone morphology. However, the effects of various inorganic-, or
organic- inorganic interactions on bone morphology are not well known.
Further study of the effects of those interactions may be helpful in
determining the etiology of various bone disorders, including
osteoporosis. Thus, in this study, we examined the effects of interactions
among three natural, ubiquitous, dietary inorganic substances, ie., boron,
magnesium and molybdenum, on bone morphology in the
cholelcalciferol-deficient chick.

METHODS

Day-old cockerel chicks were weighed individually upon arrival and
housed in all plastic environmental chambers. The chicks were assigned to
groups of 19 each in a fully-crossed, three factor, two by two by two
experiment with a completely randomized factorial arrangement of
treatments. The basal diet contained 125 IU cholecalciferol (inadequate),
0.420 µg Mo, 0.465 µg B/kg and 1% calcium (Ca:P=1:0.56). The treatments
were the supplementation of the basal diet (mg/kg) with boron (as boric
acid) at 0 or 3; magnesium (as magnesium acetate) at 300 (inadequate) or
500 (normally adequate); and molybdenum (as ammonium molybdate) at 0 or
20. Chicks were provided 24 hours of light daily by using fluorescent
lighting filtered through acrylic plastic and 1/4" plate glass. The
chicks were fed their respective diets for 28 days, weighed, and
decapitated subsequent to cardiac exsanguination. Elemental analyses were
obtained by inductively coupled argon plasma spectroscopy following a wet
ash procedure. The right proximal tibiae were fixed by immersion in mixed
aldehydes, postfixed in 2% OsO_4, dehydrated in a graded series of absolute
ethyl alcohol and embedded in plastic. One µm thick sections were cut and
stained with toluidine-blue. The distance between the proximal end of the
marrow sprouts and the most proximally located calcification front (MS-CF)
was determined using an ocular micrometer at 16X and transformed to the

RESULTS AND DISCUSSION

None of the chicks exhibited optimal growth because of inadequate dietary cholecalciferol. Dietary boron enhanced growth in chicks fed inadequate levels of magnesium and depressed growth in chicks fed adequate levels of magnesium (see table). Femur concentrations of boron, magnesium, and molybdenum were a reflection of the dietary status of those elements but, in the case of boron and molybdenum, were also dependent upon interactions between the dietary treatments. For instance, dietary boron supplementation depressed femur molybdenum more in molybdenum-supplemented than molybdenum-deprived chicks.

Effect in Cholecalciferol Deficient Chicks of Boron, Molybdenum, and Magnesium on Selected Bone Indices

Dietary Treatment, mg/kg			Body Weight, 28 days	Femur, µg/g			MS-CF Length
Mg	Mo	B	g	Mg	Mo	B	µm
300	0	0	418	1906	0.525	0.610	-53.0
300	0	3	426	1942	0.601	0.396	134.8
300	20	0	431	2050	3.862	0.364	-123.5
300	20	3	481	2325	2.006	0.445	260.2
500	0	0	552	2913	0.745	0.251	131.0
500	0	3·	491	2706	0.508	0.863	64.7
500	20	0	559	2726	5.667	0.527	871.7
500	20	3	503	2750	3.426	1.037	252.1

Analysis of Variance - P Value

B			NS	NS	0.030	0.022	NS
Mg			0.0002	0.0001	NS	0.045	0.034
B x Mg			0.05	NS	NS	0.004	0.015
Mo			NS	NS	0.0001	NS	NS
B x Mo			NS	NS	0.044	NS	NS

In normal chick bone histology, a discrete, but minimal MS-CF length is maintained and quantifies the initiation of cartilage calcification. In this study, the calcification process was affected by the stress induced by the dietary treatments that altered indices of cholecalciferol deficiency. Thus, calcification apparently was delayed in chicks whose growth was stimulated by dietary magnesium. Dietary boron may have enhanced growth at the expense of cartilage calcification in the magnesium-inadequate chicks and slowed growth to the benefit of calcification in the magnesium-adequate chick. In summary, the findings suggest a relationship between boron and cholecalciferol. Boron possibly participates in the hydroxylation of, or extends the half life of cholecalciferol, through its known affinity for hydroxyl groups (1), and thus may be important in bone diseases such as osteoporosis.

REFERENCES

1. S.L. Johnson, and K.W. Smith, The interaction of borate and sulfite with pyridine nucleotides, Biochem. 15:553 (1976).

THE ROLE OF COPPER AND CROSS-LINKING IN ELASTIN ACCUMULATION

D. Tinker, N. Romero and R. Rucker

Dept. of Nutrition, University of California, Davis, CA 95616

INTRODUCTION

Trace elements have been shown to alter the expression of several proteins (1). Our research has focused on the changes in elastin metabolism that occur during, and immediately after, copper deficiency. A copper-dependent enzyme, lysyl oxidase, links elastin and copper metabolism. Dysfunction of this enzyme disrupts elastin cross-link formation, resulting in accumulation of the soluble elastin precursor and decreased amounts of the insoluble product. In previous investigations, we confirmed that there are fewer total elastin products (soluble and insoluble forms) in copper-deficient aortas than in copper-sufficient aortas (2). The diminished elastin content of copper-deficient arteries resulted from plasma protease activity (3). Elastin synthesis rates were unaffected by copper status (2).

Starcher et al. (4) observed that copper-deficient aortas replenish elastin to near normal values within 16 days, but the details of this process were not investigated. Given the size of the deficit (40%), the rapidity of the recovery (16 days) and the nature of the lesion (proteolytic), details of elastin metabolism during recovery should provide valuable information concerning regulation of elastin metabolism. Questions that a copper repletion study may resolve include: Is elastin expression specifically induced during recovery, even though copper deficiency failed to influence synthesis?; and What influence do the proteolytic activities exert on elastin recovery? This communication presents our findings.

METHODS

Day-old white leghorn chicks were fed either copper-deficient or -sufficient diets for 14 days (1). Then a portion of the copper-deficient chicks were transferred and recovery initiated by feeding a copper-supplemented diet. Aortas were taken from 3 chicks in each of the 3 groups at intervals over the next 3 weeks. Insoluble elastin was measured by a hot alkali method (5). Soluble elastin was determined by competitive ELISA. Western Blots of arterial extracts were also evaluated (2,3).

RESULTS AND DISCUSSION

At the beginning of repletion, the deficient chicks were 80% (102 vs 127 g) of the control weight. Three weeks into the recovery phase, deficient chicks were only 60% (201 vs 334 g) of control weight. Repleted chicks grew like the deficient chicks for the first 4 days. However, at +14 days, though smaller, repleted chick weights were not statistically different from the controls.

Soluble elastin concentrations changed more rapidly than growth. Within 2 days of repletion, soluble elastin levels in recovering aortas returned to normal. This represents an 18-fold decrease in soluble elastin (360 vs 20 ug). Western Blots revealed that the soluble elastin patterns from the repleted arterial extracts looked like control patterns. The elastin fragments characteristic of copper-deficient extracts were essentially absent. This suggests that insoluble elastin in repleted arteries loses susceptibility to plasma protease during a very early phase of recovery. Exposing samples of insoluble elastin from each of the 3 diet groups to plasma confirmed that the day +4 repleted elastin exhibited the same protease resistance as the control sample while the deficient elastin was solubilized by the plasma proteases.

Changes in insoluble elastin accumulation lagged behind changes in the soluble elastin. At day +4, elastin content of the repleted aortas was identical to copper-deficient aortas, averaging 57% of controls (6 vs 10.5 mg). By day +7, the difference between the control and repleted aortas was no longer statistically significant. This indicates that most of the recovery occurs in a very short period. Comparison of the change in repleted elastin content with arterial weight showed a strong correlation (r=0.97) that was highly significant (p<0.01). Therefore, increased accumulation of arterial elastin appears to coincide with a general growth response by the recovering artery. Consequently, the increase in elastin content does not reflect an exclusive induction of elastin expression.

To summarize the changes in elastin metabolism that occur during recovery from copper deficiency: 1) Soluble elastin levels normalize and evidence of proteolytic damage ceases within 2 days of repletion; 2) Arterial weight, and consequently elastin content, begins to increase after 4 days of repletion and approaches normal weight by 7 days of repletion; and 3) Body weight of repleted birds required more than 7 days to deviate from deficient birds but was not significantly different from the controls after 14 days of repletion.

REFERENCES

1. R. Rucker and D. Tinker, The role of nutrition in gene expression: a fertile field for the application of molecular biology, J. Nutr. 116:177-189 (1986).
2. D. Tinker, J. Geller, N. Romero, C. E. Cross, and R. B. Rucker, Tropoelastin production and tropoelastin messenger RNA activity: relationship to copper and elastin cross-linking in chick aorta, Biochem. J. 237:17-23 (1986).
3. N. Romero, D. Tinker, D. Hyde, and R. B. Rucker, Role of plasma and serum proteases in the degradation of elastin, Arch. Biochem. Biophys. 244:161-168 (1986).
4. B. Starcher, C. H. Hill, and G. Matrone, Importance of dietary copper in the formation of aortic elastin, J. Nutr. 82:318-322 (1964).
5. B. L. O'Dell, K. H. Kilburn, W. N. McKenzie, and R. J. Thurston, The lung of the copper deficient rat: a model for developmental pulmonary emphysema, Am. J. Pathol. 90:413-432 (1978).

A SILICON AND ALUMINUM INTERACTION IN THE RAT

Edith M. Carlisle and Matthew J. Curran

School of Public Health
UCLA
Los Angeles, CA 90024

INTRODUCTION

Since beginning the work on establishing silicon's essentiality and function in higher animals,[1] all tissues analyzed for Si have been analyzed simultaneously for Al and a number of other elements which resulted in the establishment of a relationship between Si and Al.[2]

Because of the interesting observations in our laboratory with respect to Si and Al and the reports implicating Si and Al in human conditions, such as Alzheimer's disease,[3-5] a preliminary study was undertaken to further examine this Si Al interaction by investigating the long term effect of dietary Si and Al supplementation on the levels of these elements in the brain and other body tissues of rats.

MATERIALS AND METHODS

Two ages of Fischer female 344 rats were fed one of four diets, Si-low and Si-supplemented, with and without the addition of Al over approximately a two-year period. At the time of sacrifice rats were 23 months old in experiment[1] and 28 months old in experiment[2]. Twelve regional brain areas and more than 20 other body tissues were collected and analyzed for Si and Al content.

RESULTS

Experiment 1

Regional brain Si concentrations varied widely from a high of 9.7 ppm in the hippocampus to a low of 2.9 ppm in the spinal cord. This variation was independent of whether the diet was low in Si or supplemented with Si. In contrast to Si, brain Al concentrations appeared to be rather constant over different regions.

Dietary Al supplementation had no effect on brain Al content, however, it had a significant effect on the Si content of certain regions of the brain. In more than 6 of the 12 regions sampled, addition of Al to the control low-Si or Si-supplemented diet significantly ($p < 0.05$)

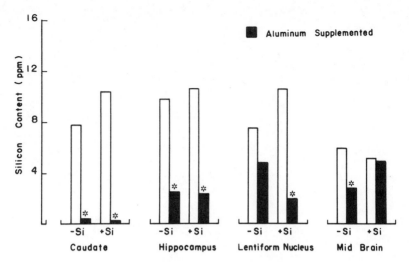

Fig. 1. The effect of Al supplementation on Si content in 4 areas of the brain of rats (n=5) fed a low-Si and Si-supplemented diet. Mean values expressed in ppm/dry wt. *Significant differences at $p < 0.05$.

decreased brain Si content. Four of these regions represented in Fig. 1 show that the greatest decrease in Si content occurs in the caudate and hippocampus, a 21-fold and 2.7-fold decrease occurring respectively in the low-Si, Al-supplemented group compared to the low-Si basal group.

Experiment 2

Upon termination of experiment 2, rats were 28 months old compared to 23 months in experiment 1. In contrast to experiment 1, where Al supplementation produced no significant increase in brain Al content, in experiment 2, Al supplementation of the low Si basal diet produced significant increases in brain Al content in most of the regions examined. No increase was seen in Si-supplemented groups of the same age. For example, Al concentrations in the hippocampus and posterior cortex were 100% and 132% greater, respectively, in animals maintained on a low-Si, Al-supplemented diet as compared to a high Si, Al-supplemented diet. By contrast, Al-supplementation did not increase brain Al content in any of the regions in 23-month-old rats, whether or not the diet was supplemented with Si. In 18 tissues other than brain a preferential accumulation of Al was also shown in rats fed the low-Si diets.

REFERENCES

1. E.M. Carlisle, Silicon as an essential trace element in human nutrition, in: Silicon Biochemistry, CIBA Foundation Symposium 121. John Wiley and Sons, New York (1986).
2. E.M. Carlisle, Unpublished data.
3. S. Duckett, P. Galle, Electron microscope microprobe studies of aluminum in the brains of cases of Alzheimer's disease and aging patients, J. Neuropath Exp Neurol 39:350 (1980).
4. T. Nikaido, J. Austin, L. Trueb, R. Rinehart, Studies in aging of the brain. II. Microchemica. Analyses of the nervous system in Alzheimer patients, Arch Neurol 27:549 (1972).
5. D.P. Perl and A.R. Brody, Alzheimer's disease: X-ray spectrophotometric evidence on aluminum accumulation in neurofibrillary tangle-hearing neurons, Science 208:297 (1980).

COORDINATE REGULATION OF ZINC METABOLISM AND METALLOTHIONEIN GENE

EXPRESSION BY cAMP, INTERLEUKIN-1 AND DIETARY COPPER AND ZINC

Robert J. Cousins, Michael A. Dunn, Teresa L.
Blalock, and Annette S. Leinart

Food Science and Human Nutrition Department,
University of Florida,
Gainesville, Florida 32611 U.S.A.

Our laboratory has placed considerable effort toward understanding regulation of mammalian zinc and copper metabolism at the cellular
and molecular level. It is clear from more recent experiments that
these metabolic controls are complex, involve a variety of hormones
and exhibit tissue specificity. The literature has been reviewed
recently and only most relevant past references will be cited here
(1,2).

HORMONAL REGULATION

Glucocorticoid hormones were the first agents definitively shown
to influence zinc metabolism at the cellular level (3,4). With
hepatocytes in primary monolayer culture, glucocorticoids increase
zinc uptake by a process that represents enhanced exchange with
increased intracellular ligands (5). Metallothionein synthesis
appears to account for the increased pool of zinc (copper or cadmium)-
binding sites (4). The general regulatory scheme is similar to that
associated with stimulation of gluconeogenesis. In addition to
glucocorticoids, glucagon stimulates zinc uptake and metallothionein
synthesis by hepatocytes. Both effects are inhibited by insulin.
Epinephrine and cAMP (dibutyryl derivative; Bt_2cAMP) also increase
cellular metallothionein levels, and zinc uptake/exchange (6).

Experiments at the cellular level, while capable of demonstrating
primary effects of hormones separated from confounding factors, they
do not present the complete integrated view found in the intact
animal. Nevertheless, it has been possible to demonstrate in rats
that glucagon, epinephrine and Bt_2cAMP are independently able to mimic
effects observed in various forms of stress. Specifically, depression
of the serum zinc concentrations (7). When the glucocorticoid,
dexamethasone was also given, the depression was not accentuated. At
10 hr into the induction cycle, liver metallothionein levels had increased significantly. Suppression of metallothionein synthesis by
actinomycin D treatment prior to hormones prevents induced
hypozincemia. Adrenalectomy prevented these responses to individual
hormones unless glucocorticoid hormones were also given. The number

of metallothionein mRNA (MTmRNA) molecules per liver cell increased
from a basal level of 235 to over 911 in rats treated with glucagon.
This level of gene expression resulted in a four-fold increase in
hepatic metallothionein. Virtually no copper was bound to liver
metallothionein induced by these methods.

Hybridization of MTmRNA in the hormonal regulation experiments
was to an synthetic oligonucleotide of the sequence:
5' GCAGGAGCAGTTGGGGTCCAT 3'. This 21-mer contains the 21-base
sequence of the 5' end of the genes for both metallothionein I and II
(8). Northern blot analysis demonstrates that hybridization is to a
single RNA species of ca. 500 bases which corresponds to the size of
MTmRNA (8). The successful use of the 21-mer as a DNA probe for
MTmRNA led to experiments to measure transcription rates for the gene
(9). Livers from rats treated with Bt_2cAMP (7) were removed 3 hr
after treatment and homogenized in 0.3 M sucrose buffer and centifuged
at 800xg. The crude nuclei were detergent-treated and isolated nuclei
were passed through a discontinuous gradient of 1.65 and 2.0 M
sucrose. The nuclei were added to a transcription mixture containing
^{32}P-UTP. RNA synthesis was terminated after 60 min and the RNA was
extracted with phenol/chloroform. The ^{32}P-labeled RNA was added to
the 21-mer DNA fragment bound to nitrocellulose for hybridization.
There was about a two-fold increase in ^{32}P-MTmRNA from the Bt_2cAMP-
treated rats. These data suggest that cAMP mediates metallothionein
gene transcription. This is supported by actinomycin D-inhibition
experiments as well (7). Regulation of metallothionein gene
expression by glucagon is probably explained on that basis also.

The mechanism that accounts for transient depression of serum
zinc levels after some hormonal treatments has not been delineated.
In order to investigate the sequence of events involved, kinetic
experiments have been undertaken. Metallothionein expression was
stimulated in rats by Bt_2cAMP. After 8 hr ^{65}Zn was administered
intravenously and they were killed at various times up to 24 hr later.
Kinetics of uptake by some organs was affected by Bt_2cAMP
administration and clearance of ^{65}Zn from the plasma compartment was
greater. Within 60 min after ^{65}Zn, 50% of the dose was recovered in
the liver in Bt_2cAMP- treated rats compared to 25% in controls.
Chromatographic separation of the liver cytosol showed that the
increased hepatic uptake was associated with metallothionein. Bt_2cAMP
administration resulted in less ^{65}Zn uptake in thymus, skin and bone,
but considerably more uptake by bone marrow. It appears from these
data that metallothionein induction accounts for a redistribution of
metabolically active zinc and a change in zinc flux from the plasma
compartment (9). Experiments to define the teleological basis for
these metabolic shifts are currently in progress.

REGULATION BY INTERLEUKIN-1

Zinc and copper metabolism have always been closely linked to
host defense processes (reviewed in 1). The effect of copper has been
relegated to ceruloplasmin in recognition of its antioxidant,
Cu-donating and/or iron oxidizing properties. In contrast, the
effect(s) of zinc have been described only vaguely in terms of an
effect on zinc-requiring metalloenzymes or on membranes.
Interleukin-1 has been shown to possess a multiplicity of effects on
systems related to host defense (10). Factors called leukocytic
endogenous mediators (now called interleukin-1) have been used for
nearly two decades to influence copper, iron and zinc metabolism (1).

Now that highly purified interleukin-1 is available, produced by
recombinant DNA technology, it is possible to study the regulation of
zinc metabolism by this lymphokine. We have examined many parameters
of zinc and copper metabolism in rats using human interleukin-1α (11).
Following administration, serum zinc levels were transiently
depressed, but to an extent greater than observed with hormones.
Ceruloplasmin levels increased over 1.5 fold. Within 48 hrs of the
dose both parameters have returned within the normal range. Liver
metallothionein levels increased about 9 fold by 36 hr after
interleukin-1. These changes correspond to increases in MTmRNA.
Small changes were observed in kidney and intestine. Considering the
many functions of metallothionein including scavenging of free
radicals (6), the significance of induction of the protein by a
lymphokine that is produced when active oxygen radicals are generated
as a cellular response to injury cannot be overlooked.

Interleukin-1 increases expression of both metallothionein I and
II genes. In order to measure the amount of zinc bound to each
isoform, metallothionein-containing fractions prepared by standard gel
filtration were fractionated by HPLC using an ion exchange column.
Isoforms were individually eluted with a linear Cl gradient. The
distribution is 27% and 73%, for isoforms I and II, respectively. It
is of interest that maximal MTmRNA levels following interleukin-1
stimulation occurred 6 hr after administration of a single dose. This
chronology agrees with peak levels found after acute treatment with
hormones. Two lines of evidence suggest that interleukin-1 actually
activates transcription: 1) interleukin-1 has been shown to initiate
metallothionein gene expression in cultured liver and kidney cells,
but with a longer than usual induction period (12) and 2) the
stimulation of expression is greater than can be observed with
hormonal stimulation.

DIETARY REGULATION

Only limited data are available on metallothionein gene regulation by dietary means. In early experiments, depletion and
repletion of the dietary zinc supply was demonstrated to produce major
changes in the amount of zinc bound to the protein (13). No
metallothionein-bound metal was detected in tissues of the depleted
rats suggesting that the gene is responsive to zinc and adequate
dietary amounts are needed to start transcription. Refeeding zinc-
containing diets to rats of depleted zinc status produced a rapid
increase in metallothionein synthesis in both liver and intestine
(14,15). MTmRNA levels based on cell-free translation were increased
to maximal levels 6-8 hr after zinc was refed. A reciprocal
relationship was demonstrated between metallothionein synthesis and
processing of ^{65}Zn added to the diet for refeeding (repletion).

The close relationship between the dietary zinc level and
metallothionein gene expression was examined in dot blot and Northern
blot hybridization experiments using the 21-mer DNA probe described
above. In order to characterize sensitivity of the native
metallothionein promoters to both copper and zinc, a 3x3 factorial
design was used where diets containing 1,6 or 36 mg Cu/kg and 5,30 or
180 mg Zn/kg were fed for two weeks (16). Total RNA was obtained from
a number of tissues and used for hybridizations. The lowest dietary
levels of copper and zinc depressed the respective serum
concentrations. The kidney represented the organ that responded to
the dietary treatments to the greatest extent. Furthermore, dietary
zinc had a far greater effect on kidney MTmRNA than did dietary

copper. At 6 mg Cu/kg the number of MTmRNA molecules per kidney cell averaged 6,16 and 24 at zinc intakes of 5,30 and 180 mg/kg, respectively. These data demonstrate the sensitivity of the metallothionein promoter system to rather subtle differences in the dietary copper and zinc supply.

Sensitivity of metallothionein gene expression to the dietary zinc supply has implications in terms of maximizing synthesis of the protein (i.e. as a beneficial, acute phase protein), as a regulatory agent via the promoter for transgenic animals and activation of fusion genes carrying the promoter for therapeutic purposes. Strategy to elucidate the mechanism of gene regulation by dietary zinc was developed as a result of that potential. We reasoned that for zinc to extent transcriptional regulation of this gene, some trans-acting mediating factor is responsible for binding intranuclear Zn^{2+} as a sensor of dietary zinc status. A generalized mechanism for dietary regulation of metallothionein gene expression via a nuclear regulatory protein (NRP) is shown in Figure. 1. The overall hypothesis is as described by our laboratory at TEMA-3 (17).

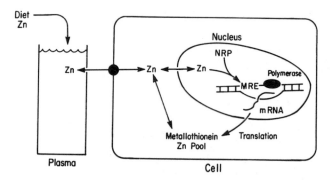

Figure 1. Hypothesis showing how dietary zinc can interact with a trans-acting nuclear regulatory protein (NRP) which in turn recognizes a DNA sequence that provides metal regulation and initiates transcription of the metallothionein gene.

The first steps in characterizing a nuclear regulatory protein for zinc were to separate nuclear proteins by polyacrylamide gel electrophoresis, transfer them to nitrocellulose by Western blotting and examine ^{65}Zn-binding properties in vitro. The property of zinc-binding is one of the criteria for a zinc responsive NRP. For these experiments nuclei were prepared as described above. A number of ^{65}Zn-binding bands were identified upon autoradiography but three were particularly intense. A second criterion was that the NRP would bind to a DNA sequence (metal regulatory element, MRE) that provides metal regulation (18). A 12-base oligonucleotide was constructed for one of these sequences, end labeled with ^{32}P-ATP and incubated with nuclear proteins bound to nitrocellulose. One major ^{32}P-MRE and ^{65}Zn-binding band was detected. It has a molecular weight of ca. 40,000 Da. Purification and characterization of this Zn-binding NRP and its interaction with dietary zinc in selected tissues is now in progress.

ACKNOWLEDGEMENTS

Research in the senior author's laboratory is supported by National Institutes of Health Grants DK 31127 and DK 31651 and the Institute of Food and Agricultural Sciences.

REFERENCES

1. Cousins, R.J. (1985). Absorption, transport, and hepatic metabolism of copper and zinc: special reference to metallothionein and ceruloplasmin. Physiol. Rev. 65:238-309.
2. Dunn, M.A., Blalock, T.L., and Cousins, R.J. (1987). Metallothionein: Minireview. Proc. Soc. Exp. Biol. Med. 185:107-119.
3. Cox, R.P. (1969). Hormonal induction of increased zinc uptake in mammalian cell cultures: requirement for RNA and protein synthesis. Science, 165:196-199.
4. Failla, M.L., and Cousins, R.J. (1978). Zinc accumulation and metabolism in primary cultures of rat liver cells: regulation by glucocorticoids. Biochem. Biophys. Acta, 543:293-304.
5. Pattison, S.E., and Cousins, R.J. (1986). Kinetics of zinc uptake and exchange by primary cultures of rat hepatocytes. Am. J. Physiol. 250:E677-E685.
6. Cousins, R.J., and Coppen, D.E. (1987). Regulation of liver zinc metabolism and metallothionein by cAMP, glucagon and glucocorticoids and suppression of free radicals by zinc. In: Metallothionein II, (Ed. Kagi, J.H.R., and Kojima, Y.) Birkhauser Verlag, Basel (in press).
7. Cousins, R.J., Dunn, M.A., Leinart, A.S., Yedinek, K.C. and DiSilvestro, R.A. (1986). Coordinate regulation of zinc metabolism and metallothionein gene expression in rats. Amer. J. Physiol. 251:E688-E694.
8. Hamer, D.H. (1986). Metallothionein. Ann. Rev. Biochem. 55:913-951.
9. Dunn, M.A., and Cousins, R.J. (1987). Regulation of metallothionein gene transcriptional and kinetics of zinc metabolism by dibutyryl cAMP. Fed. Proc. 46:1643 abs.
10. Dinarello, C.A. (1984). Interleukin-1. Rev. Infect. Dis. 6:51-95.
11. Cousins, et. al., unpublished data.
12. Karin, M., Imbra, R.J., Heguy, A., and Wong, G. (1985). Interleukin-1 regulates human metallothionein gene expression. Mol. Cell. Biol. 5:2866-2869.
13. Richards, M.P., and Cousins, R.J. (1976). Zinc-binding protein: relationship to short term changes in zinc metabolism. Proc. Soc. Exp. Biol. Med. 153:51-56.
14. McCormick, C.C., Menard, P.M., and Cousins, R.J. (1981). Induction of hepatic metallothionein by feeding zinc to rats of depleted zinc status. Am. J. Physiol. 240:E414-E421.
15. Menard, P., McCormick, C.C., and Cousins, R.J. (1981). Regulation of intestinal metallothionein biosynthesis in rats by dietary zinc. J. Nutr. 111:1353-1361.
16. Blalock, T.L., Dunn, M.A., and Cousins, R.J. (1987). Sensitivity of native metallothionein promoters to dietary copper and zinc. Fed. Proc. 46:3313 abs.
17. Cousins, R.J. (1978). Synthesis and degradation of zinc-thionein. In: Trace Element Metabolism in Animals - III. Ed. M. Kirchgessner pp. 57-63. Arbeitskreis fur Tierernahrungsforschung, Weihenstephan, W. Germany.
18. Sequin, C., and Hamer, D.H. (1987). Regulation in vitro of metallothionein gene binding factors. Science, 235:1383-1387.

METALLOTHIONEIN AND CERULOPLASMIN GENES

David M. Danks and Julian F.B. Mercer

Murdoch Institute
Royal Children's Hospital
Melbourne, Australia. 3052

For a long time, metallothioneins (MT) were regarded by most protein chemists as peculiar little proteins whose physical properties made them particularly unattractive to study. Those of us who are interested in the trace elements saw them as of obvious importance, but had difficulty analysing them. Conservation through evolution marked them as of obvious importance, but there was no agreement about the reason for this importance.

Then it all changed. Over a period of just two or three years, Richard Palmiter made the mouse metallothionein gene promoter one of the best understood among mammalian genes[1]. It became the favorite of all those who wish to produce transgenic mice and of many who are interested in gene expression in cultured cells. The laboratories of Michael Karin[2] and Dean Hamer[3] have also contributed to this knowledge.

Although the metallothionein genes have been studied very intensively, most of this effort has been directed towards understanding the regulatory DNA sequences for their own sake. Less interest has been taken in the role of metallothionein in relation to trace elements.

Our paper describing the isolation of a cDNA for rat metallothionein[4] was submitted just as Richard Palmiter's paper describing the isolation of the mouse gene was published. Our interest in the genes has been directed principally to their function in relation to trace elements using normal rats and mice, mottled mouse mutants, Bedlington terriers with copper toxicosis and cells from patients with Menkes' disease. Most recently, we have studied the metallothionein genes in sheep.

Naturally, we have spent long hours debating the functional roles of metallothioneins. We cannot claim to have reached unanimity of opinion even amongst the four senior people in our own research group - DMD, JFBM, Jim Camakaris, and Harry McArdle. One of us (DMD) believes that metallothionein plays a key role in the distribution and availability of zinc for those myriad vital functions which require this element. The range of factors which can induce the production of metallothionein, including zinc itself, and the level of affinity of the protein for zinc seems just appropriate for it to play this key role. On the other hand, it seems likely that metallothionein serves as a detoxifying back-up system for copper. The displacement of zinc from the zinc metallothionein present in most cells provides one option which is available immediately to mop up excess copper, and the induction of metallothionein by copper, although weaker than that by zinc, would be adequate to back up the system by additional production of the protein. The very high affinity for copper makes for re-utilization of this metal that has been bound to metallothionein more difficult to envisage. It is good to read that Dean Hamer shares this view of metallothionein function[3].

Ceruloplasmin (CP) genes have been cloned much more recently by Dr. Bowman's group in Houston[5] and in our own laboratory[6]. There is much less to be said about this gene at this stage. The function of this protein remains debatable.

METALLOTHIONEIN GENES IN THE SHEEP

We chose to study sheep metallothioneins because it seemed possible that an abnormality of their structure or function might be responsible for the peculiar copper phenotype of normal sheep. The excretion of copper in the bile is much less efficient in sheep than in other mammals and incorporation of copper into ceruloplasmin also proceeds more slowly. First studied in Australia because of the occurrence of copper deficiency in sheep grazing on pastures with low copper content or on pastures with high molybdenum content, the interest in recent years has swung to copper toxicity in sheep. This is because the remedies introduced to prevent copper deficiency - i.e. copper dosing or copper injection - have proved quite hazardous through the production of copper toxicity in sheep. The syndrome of chronic copper toxicosis in sheep has many features in common with Wilson disease in humans[7], and indeed the copper phenotype just described is very like that seen in humans with Wilson disease.

These were our reasons for studying the sheep metallothioneins. Our results are discussed today partly because of some interesting, albeit unexpected, findings, relevant to copper toxicosis, but also because the sheep metallothionein genes have turned out to be very interesting in their structure and arrangement.

It is not clear why genes which are so highly conserved in their essential structure should differ so greatly between species in the extent of gene duplication. Mice have only two metallothionein genes (MT-I and MT-II). Only these two genes have been clearly identified in the rat and the hamster, but the analysis has not been as exhaustive as in the mouse. In man, four functional genes and four pseudogenes have been identified, but these do not explain all of the sequences observed to hybridize to MT gene probes on Southern gels of the human genomic DNA[2,3]. Greg Peterson, a Ph.D. student in our group has identified four functional genes and one pseudogene in the sheep. These explain only four of the eight bands seen on a Southern gel to hybridize with a probe from the third exon of the MT-Ia gene used under low stringency hybridization conditions.

He has sequenced 67 kilobases (kb) of DNA which contains the cluster of metallothionein genes. Four of these five genes have the classical metallothionein gene structure with three exons, but the size of the second intron, in particular, varies quite strikingly. The fifth gene comprises just the first exon and no further exons were found in 17 kb of DNA sequenced in the 3' direction. It is interpreted to be a pseudogene. The amino acid sequence encoded by the first exon of the three active genes and of the pseudogene includes no negatively charged residue at position 10 or 11, and these genes are therefore classified as MT-I genes. The fourth active gene has a glutamic acid residue at position 11 designating it an MT-II gene.

When we examined the amino acid composition of the four MT peptide peaks that were separated by Whanger,[8] we found that a mixture of the three MT-I gene products could explain the amino acid composition observed in peak 2, and the MT-II gene product would explain the amino acid composition of peak 3. Peaks 1 and 4 are left unexplained, because their amino acid composition does not fit with any of the genes we have isolated. Perhaps these minor peaks contain the products of the metallothionein genes we have not yet isolated.

The arrangement of the genes in this cluster, with the pseudogene at the 3' end, could indicate that a larger series of metallothionein genes was broken by a major chromosome rearrangement which moved the second and third exons of the pseudogene, plus the other genes located 3' to it, to another location.

By examination of exon sequences, similarity of the coding sequences, and of the introns between the different members of this gene cluster and certain other features of the promoter regions, one can propose with reasonable confidence that the series of duplications which must have occurred to create this gene cluster began with duplication of a primordial gene to give the MT-II gene and a primordial MT-I gene. Next, the Ic must have differentiated from Ib and then Ib from Ia and finally Ia gave rise to the gene which was later disrupted in the way just mentioned to become a pseudo gene.

Analysis of the 5'- regions of mouse and human metallothionein genes has identified sequences which serve a variety of regulatory functions. In addition to the TATA box common to all transcribed genes, it has been possible to identify a sequence which is involved in determining the basal rate of expression of the gene; several repeats of a sequence which is responsive to metals; a sequence which is responsive to glucocorticoids and has a strong consensus with glucocorticoid responsive sequences in the 5'- regions of other genes which are inducible by this hormone; and another sequence which is responsive to α-interferon. Finally, a conserved sequence has been identified in the 3'- untranslated regions of all metallothionein genes. It may play a role in glucocorticoid stabilization of messenger. Analysis of the effect of glucocorticoids on metallothionein has shown that this hormone influences both transcription and translation.

Analysis of these sequences reveals some interesting correlations with the levels of expression in cultured fibroblasts of the four active metallothionein genes which we have studied in the sheep. Several examples will be given.

The MT-II gene shows the highest level of basal expression and the relevant 5'- sequence shows the best agreement with the consensus sequence derived from mouse and human genes. Only MT-II is responsive to glucocorticoid, and its glucocorticoid response element has the exact sequence of the consensus whereas those attached to the MT-I genes differ from the consensus at several bases.

No striking differences in metal responsiveness were observed between the genes, and this accords with a very similar arrangement of metal responsive sequences beside each gene. The levels of MT-Ib are very low under all circumstances, and this can probably be explained by a single base substitution in a critical base in the TATA box. The TATA sequences are completely lacking from the pseudogene giving yet another reason for its functional failure.

It seems worth discussing these findings in the sheep for their own intrinsic interest and as an illustration of the sophistication that is now available in analysing such genes. However, none of these findings has explained the peculiar copper phenotype of the sheep.

There is just one peculiarity in the actual coding sequences of the sheep metallothionein genes, but we do not think that is the explanation of the phenomenon. Each of the genes has a sequence which codes for a proline residue at position 27. However, this proline is also found in bovine metallothionein, and cattle do not seem to share the peculiarities of copper handling that have been observed in the sheep.

We do have some other findings, presented in detail elsewhere at this meeting,[9] which we regard as of considerable interest in trying to explain the copper phenotype of sheep. In brief, we have observed that sheep have much higher levels of zinc and of metallothionein messenger RNA in the liver than rats. This excess of zinc and metallothionein messenger is seen at all ages, but is striking during fetal life, when the levels are 20 times those observed in rats. At all ages, there is a very strong positive correlation between the levels of zinc and of MT mRNA and no significant correlation between levels of copper and MT mRNA. The lack of correlation with copper content is true even in sheep suffering from copper toxicosis. We are setting up experiments to test the possibility that the unusual relationship between zinc and metallothionein in the sheep might be responsible for the tendency of sheep to retain copper in the liver, perhaps by displacing zinc from an excessively large pool of zinc metallothionein.

CERULOPLASMIN GENES

The isolation of cDNAs for human and rat ceruloplasmin[5,6] has been a much more recent event than the isolation of metallothionein genes. The cDNA sequences isolated in both laboratories code for the entire sequence of the ceruloplasmin protein, and the Southern gels show just one gene locus, localised, by in situ hybridization, at 3q25[5]. The gene appears to have arisen by triplication of an ancestral gene.

The acute phase responsiveness of ceruloplasmin synthesis has been confirmed at the messenger RNA level in rats[10] in which it proves to be the most briskly responding of all the acute phase proteins.

Analysis of the tissue distribution of the messenger RNA revealed some interesting findings - highest levels of all were observed in the choroid plexus[10]. Other tissues with high levels included yolk sac, placenta, and testis.

We have been interested to determine whether ceruloplasmin production is induced by copper or whether copper is merely necessary for the production of ceruloplasmin message. At first, we felt our results were confirming the latter situation. At this moment, we are not able to interpret our results with certainty. It is proving difficult to disentangle the effects of starvation, stress, and copper supply. We look forward to many interesting studies in rats, mice, sheep, and possibly in toxic milk mice, which we hope may help us in determining the function of ceruloplasmin.

ACKNOWLEDGMENTS

The results presented have been drawn from the work of several members of our research group (Dr. Greg Peterson, Ms. Jenny Smith, and Mr. Andrew Grimes) and of collaborators in the Veterinary School at Murdoch University in Perth (Professor John Howell, Mr. Paul Gill) and in the Department of Biochemistry, University of Melbourne (Professor Gerhardt Schreiber, Ms. Angela Aldred). The work was supported by a Program Grant from the National Health and Medical Research Council and by the Scobie and Claire Mackinnon Trust.

REFERENCES

1. G. W.Stuart, P. F. Searle and R.D. Palmiter, Identification of multiple metal regulatory elements in mouse metallothionein-I promoter by assaying synthetic sequences, Nature 317:828 (1985).
2. M. Karin, Metallothioneins: Proteins in search of function, Cell 41:9 (1985).
3. D. H. Hamer, Metallothionein, Ann Rev Biochem 55:913 (1986).
4. J. F. B. Mercer and P. Hudson, Cloning of metallothionein cDNA from neonatal rat liver, Bioscience Reports 2:761 (1982).
5. F. Yang, S. L. Naylor, J. B. Lum, S. Cutshaw, J. L. McCombs, K. H. Naberhaus, J. R. McGill, G. S. Adrian, C. M. Moore, D. R. Barnett and B. H. Bowman, Characterization, mapping, and expression of the human ceruloplasmin gene, Proc Natl Acad Sci USA 83:3257 (1986).
6. J. F. B. Mercer and A. Grimes, Isolation of a human ceruloplasmin cDNA clone that includes the N-terminal leader sequence, FEBS Letters 203:185 (1986).
7. J. McC. Howell, S. R. Gooneratne and J.M. Gawthorne, Copper poisoning in sheep: Wilson's disease, Comp Path Bull 16:3 (1984).
8. P. D. Whanger, S. H. Oh and J.T. Deagen, Ovine and bovine metallothioneins - purification, number of species, zinc content and amino acid composition, J Nutr 111:1207 (1981).
9. J. F. B. Mercer, J. Smith, A. Grimes, J. McC. Howell, P. Gill and D.M. Danks, Zinc, copper and metallothionein mRNA in sheep liver during development, published in these proceedings.
10. A. R. Aldred, A. Grimes, G. Schreiber and J.F. B. Mercer, Rat ceruloplasmin: Molecular cloning and gene expression in liver, choroid plexus, yolk sac, placenta, and testis, J Biol Chem 262:2875 (1987).

METALLOTHIONEIN M RNA LEVELS IN NORMAL AND MOTTLED MOUSE

MUTANTS DURING DEVELOPMENT

T. Stevenson, J. Camarakis, *J Mercer and *D.M. Danks

Department of Genetics, University of Melbourne, Parkville
Victoria, 3052, Australia
*Murdoch Institute, Royal Children's Hospital, Parkville
Victoria, 3052, Australia

Mottled mouse mutants have an X-linked recessive disorder of copper metabolism with features very similar to Menkes' disease in humans (Cama karis et al, 1979). The pathology in these disorders is due to gross Cu deficiency, as a result of mal-absorption of Cu. However Cu accumulates in gut and kidney bound to metallothionein (MT). The purpose of the current studies was to determine whether the basic defect in mottled mouse mutants is partially constitutive synthesis of MT mRNA.

MATERIALS AND METHODS

Mottled mouse mutants used were "brindled" (severely affected, die at 13-16 days) and "blotchy" (mildly affected, die at 150 days). Tissue Cu was estimated by atomic absorption spectrophotometry. Total RNA was extracted using guanidine-HCl and separated by ultracentrifugation using a 5.7 M CsCl cushion. RNA was analysed on agarose gels under denaturing conditions and subsequently transferred to nitrocellulose filters (Northern blots). Quantitation was by dot-blot hybridisation to ^{32}P-labelled (nick-translated) cDNA probes (MT 1 cDNA from Dr. Richard Palmiter, and tyrosine-aminotransferase (TAT) cDNA from Dr. Gunther Schutz).

RESULTS AND DISCUSSION

Cu and MT mRNA were determined during the first two weeks of neonatal development in brindled mutants and normal male siblings. Tissues investigated were liver and brain which are Cu deficient, and kidney which shows elevated Cu levels in mottled mouse mutants (Camakaris et al, 1979). A regulatory mutation would result in elevated levels of MT mRNA, irrespective of Cu levels in the same tissue.

Kidney

Although kidney Cu and Cu-MT concentrations are elevated in mottled mouse mutants, MT mRNA levels were normal. It is possible that the high MT levels are sustained by a longer half-life MT (Prins and Van den Hamer, 1981).

Brain

In normal mice there was an increase in brain Cu concentration during the first two weeks, whilst in brindled mutants there was a slight decline. The MT mRNA levels in both normal and brindled mice were low throughout the same period, with the notable exception of 13 day old brindled brain which exhibited significantly elevated levels. The absence of a correlation between Cu and MT mRNA levels suggests that Cu is not the major regulator of brain MT synthesis.

Liver

Cu concentrations in the liver of neonatal mice are elevated compared to adult mice. No consistent changes were observed in the two week neonatal period. Cu levels in brindled mutants were 5%-10% of normal. MT mRNA showed a peak at 6 days in normal mice. Two highly significant differences were observed in brindled mutants: (i) absence of a peak at 6 days, (ii) a large elevation at 13 days. Blotchy mutants followed the normal developmental pattern. Nutritional Cu-deficient mice, which have the same degree of liver Cu deficiency as brindled and which also die at about 2 weeks, resembled normal in having the peak at 6 days but resembled brindled mutants in having elevated levels of MT mRNA at 13 days in both liver and brain (Wake, 1987). The peak of MT mRNA at 6 days does not appear to be related to rises in either Cu or Zn concentrations in the liver. However it is possible that an increased Cu concentration in the "MT-inducing pool" is responsible for the peak at 6 days and brindled mutants may not show an increase in this pool.

The rise in MT mRNA at 13 days in the brain and liver of brindled mutants and nutritionally Cu deficient mice is likely to be stress induced as at two weeks of age these mice display severe symptoms of Cu deficiency (e.g. staggered gait, hypothermia) and are close to death. This explanation was supported by the finding that the brindled mutants which showed elevated levels of MT mRNA in liver and brain also showed elevated levels of TAT mRNA. TAT mRNA is induced by stress/glucocorticords in the liver of normal mice.

In summary, our data does not support the hypothesis that mottled mouse mutants have a basic defect in the regulation of MT mRNA synthesis. Abnormal variations observed in MT mRNA levels at particular stages of development can be attributed to secondary effects of the mutation(s).

REFERENCES

Camakaris, J., Mann, J.R., and Danks, D.M., 1979, Copper metabolism in mottled mouse mutants: copper concentrations in tissues during development. Biochem. J., 180:597.

Prins, J.W., and Van den Hamer, C.J.A., 1981, Degradation of 35S-labelled metallothionein in the liver and kidney of brindled mice: model for Menkes' disease. Life Sci., 28:2953.

Wake, S.A., 1987, The regulation of metallothionein gene expression. Ph.D. thesis, University of Melbourne.

AGE DEPENDENCE OF METALLOTHIONEIN

INDUCIBILITY IN MOUSE LIVER

David J. Thomas, Sheldon Morris, and P.C. Huang

Department of Pediatrics, University of Nebraska Medical
Center, Omaha, NE, and the Department of Biochemistry, School
of Hygiene and Public Health, The Johns Hopkins University,
Baltimore, MD.

There is ontogenic variation in the organismic response to Cd exposure.
In the rat, neonates are more resistant than adults to the hepatotoxic
(Goering and Klaassen, 1984) and nephrotoxic (Braunlich and Otto, 1984)
effects of this metal. It is likely that resistance of neonatal rats to Cd-
induced liver injury is due to sequestration of Cd by the large endogenous
pool of metallothionein (MT), a transition metal binding protein, in this
organ. In the work reported here, ontogenic variation in the lethality of
Cd in C57Bl/6J mice has been examined in terms of developmental changes in Cd
distribution and as related to changes in the inducibility of MTmRNA in the
liver.

Over the age interval of 7 to 56 days old, the sensitivity of mice to
the lethal effects of Cd were found to decrease. The LD_{50} for parenterally
administered Cd in 7 day old mice was estimated as 1.65 mg/kg; in 56 day
old mice, it was estimated as 4.08 mg/kg. Developmental changes in MT
concentrations in liver did not correlate with variation in sensitivity
to the lethal effects of Cd treatment. The age dependence of MTmRNA
inducibility in liver was examined in mice given 2 mg of Cd/Kg ip. Mice
were killed 1 to 8 hours after treatment and the levels of MTmRNA and alpha
tubulin (AT) mRNA in liver were quantitated by cytodot hybridization assays.
Because Cd treatment did not alter ATmRNA levels, data on MTmRNA were
normalized to the amount of ATmRNA. In livers of 56 day old mice, relative
MTmRNA levels depended upon the Cd dosage level and were highest at 4 hours
post treatment. Relative levels of MTmRNA were also highest in the livers
of 14 & 28 day old mice at 4 hours post treatment; however, peak levels in
these age groups were lower than those found in 56 day old mice. In 7
day old mice, Cd treatment produced little change in the relative level of
MTmRNA in liver. Age dependent differences in the induction of MTmRNA by
Cd were not related to differences in the kinetics of Cd accumulation in
the liver. Similar patterns of Cd accumulation in liver were found in 7,
28 and 56 day old mice. The approximate LD_{50}'s for Cd in mice of various
ages were correlated with the magnitude of the CD-dependent inducibility of
MTmRNA.

Taken together, these results suggest that, in the neonatal mouse,
the presence of large endogenous stores of MT in the liver are not adequate
to protect against Cd-induced lethality. The ontogenic variation in
response to Cd exposure may relate to differences in the inducibility of
MTmRNA by this metal.

REFERENCES

Braunlich, H., and Otto, R., 1984, Age-related differences in the nephro-
    toxicity of cadmium, Exp. Pathol., 26:235.
Goering, P.L., and Klaassen, C.D., 1984, Resistance to cadmium-induced
    hepatotoxicity in immature rats, Toxicol. Appl. Pharmacol., 74:321.

CYTOSOLIC CU-BINDING COMPONENTS AND THE BRINDLED MOUSE DEFECT

F. Palida, G. Waldrop, P. Lonergan and M. Ettinger

Department of Biochemistry
State University of New York at Buffalo
Buffalo, New York 14214

INTRODUCTION

The components that Cu binds to when it first enters cells are likely to have important functions in Cu-metabolism. Menkes disease is an inborn error of Cu-metabolism which apparently involves low activity of an intracellular factor. Thus, studies with the brindled mouse model of Menkes disease may help elucidate the function of an intracellular Cu-binding component. Cu-uptake and the intracellular distribution of Cu were studied with hepatocytes and fibroblasts from normal and brindled mice to identify Cu-binding components and their functions.

RESULTS

Subcellular distribution experiments showed that steady-state levels of Cu were attained rapidly. This suggested that a cytosolic component(s) delivers Cu to the rest of the cell. Mouse hepatocytes were incubated with $10 \mu M$ ^{64}Cu (II) for 5 or 90 min, and the cell-cytosols were applied to Sephadex G150 columns (1.5 x 100 cm). Four protein fractions accounted for all of the cytosolic ^{64}Cu. The lowest molecular weight fraction contained metallothionein(s) (MT). Unlike with the rat, very little Cu was detected in this fraction from mouse hepatocytes. A 38 kd ^{64}Cu-binding fraction did not have SOD-activity. A major ^{64}Cu fraction (88 kd) eluted after the void volume from these columns. This fraction would elute in the void volume of G50 or G75 columns. When cells were preloaded with ^{35}S-cystine, the ^{64}Cu-fractions were found not to co-elute with the major ^{35}S-protein fractions. Zn bound to a fraction which eluted between the 88 and 38 kd Cu-binding fractions. The 88 kd and MT-fractions were also detected in the cytosol from normal fibroblasts or whole kidney. However, no 38 kd ^{64}Cu-containing fraction was detected with these cytosols. Cu is elevated in these two cell types in Menkes disease and low in the liver.
The extra Cu-binding component detected in hepatocytes

may contribute to the different response of the liver than the kidney or fibroblasts to the disease.

Fibroblasts from the brindled mouse are known to accumulate excess Cu when incubated for several hours with ^{64}Cu[1]. To determine if this was due to the elevated MT-levels in these cells, MT was induced to comparable levels in normal fibroblasts by incubating with 400μm Cu(II) for 48h. Initial rates of Cu-transport were normal for both brindled and Cu-treated fibroblasts. Both cell types also accumulated normal amounts of Cu when Cu-uptake was high from serum-free media. Interestingly, the Cu-treated fibroblasts accumulated normal amounts of ^{64}Cu under the conditions that brindled fibroblasts accumulated excess ^{64}Cu from media containing 10% serum. Thus, excess Cu-accumulation seems to be associated with the primary defect rather than with elevated MT.

SDS-gel electrophoresis on the G150 fractions revealed one component (55 kd, apparent) from normal hepatocytes which was barely detectable by visual inspection or by laser-densitometry with samples from the brindled mice. A second component (42 kd) may also be absent. Preliminary results with kidney samples showed the same results. Whether these electrophoretic differences represent primary or secondary aspects of the defect remains unknown.

DISCUSSION

The results are consistent with one function of the defective component in the brindled mouse being to determine how much Cu is accumulated by fibroblasts. Irrespective of whether the defective component is the 55 or 42 kd species, the results suggest that extracellular Cu is in reversible equilibrium with the component involved. Since lysyl oxidase activity is low in Menkes fibroblast cultures[2], the defective component may also deliver Cu to apo-Cu-proteins. The results suggest that when this component has low activity, Cu cannot be used, and MT-synthesis is induced by the excess Cu which accumulates. According to this hypothesis, Cu-treated fibroblasts behave normally because they still contain the active component. When Cu cannot be used by brindled hepatocytes, that Cu may be lost from the cell from the hepatocyte-specific, 38 kd fraction which also rapidly equilibrates with extracellular Cu. While these hypotheses are consistent with the results, the components involved remain to be purified and identified.

REFERENCES

1. Beratis, N.G., Price, P., LaBadie, G. and Hirschhorn, K., Pediat. Res. 12, 699-702 (1978).
2. Royce, P.M., Camakaris, J. and Danks, D., Biochem. J. 192, 579-586 (1980).

5S RIBOSOMAL RNA SYNTHESIS IN ZINC-DEFICIENT RATS

John K. Chesters and Linda Petrie

Rowett Research Institute
Bucksburn, Aberdeen AB2 9SB, U.K.

Within a few days of being offered a Zn-deficient diet, young rats show an abrupt decline in growth rate and much effort has been directed to determining which function of Zn is most important in this context. Having reviewed the accumulated evidence, Chesters (1978) suggested that the critical requirement for Zn was associated with the alterations to the genetic expression of cells which occur during the induction of enzymes and the differentiation of tissues. However, low availability of Zn in EDTA-treated lymphocyte cultures resulted in delayed maturation of 32S ribosomal RNA precursor into 28S rRNA and reduced survival of 28S compared to 18S rRNA (Chesters, 1975). This sensitivity of post-transcriptional events to lack of Zn was hard to reconcile with the above hypothesis.

Recently, however, Hanas et al. (1983) reported that a protein, TFIIIA, which is essential for the transcription of the 5S rRNA gene by RNA polymerase III contains nine Zn ions per molecule. These appear to stabilize "finger-like" loops within the protein through which it interacts with the DNA (Miller et al., 1985). Lack of Zn can impair the function of TFIIIA in vitro and differentially inhibit synthesis of 5S rRNA (Wingender et al., 1984). This could lead to incomplete formation of the large ribosomal subunits which contain both 5S and 28S rRNA and thus result in increased susceptibility of the latter to attack by ribonuclease. The reduced stability of 28S rRNA in Zn-deficient lymphocytes would thus be linked to a Zn-related failure of transcription of the 5S rRNA gene. Furthermore, "finger-like" structures probably involving Zn have now been recognized in a number of other proteins, all of which are concerned with the regulation of genes (Berg, 1986). The synthesis of 5S rRNA has therefore been investigated in Zn-deficient rats in order to assess the adequacy of TFIIIA function.

Both 5S rRNA and tRNA are synthesized in the same cells by the same enzyme and require the same cofactors apart from TFIIIA. The latter is active only in the synthesis of 5S rRNA. Variations in the adequacy of TFIIIA will therefore affect incorporation of radioactivity into 5S rRNA but not into tRNA whereas changes in any of the other components would influence both nucleic acids. The ratio of incorporation ^{32}P-phosphate into 5S rRNA should therefore be reduced relative to tRNA if inadequate Zn supply limits TFIIIA function but should be comparatively insensitive to changes in precursor specific activity and polymerase activity.

Table 1.   Relative Rates of Synthesis of 5S rRNA and tRNA in
Liver and Spleen from Zn-deficient and Control rats

| | Ratio of ^{32}P in 5S rRNA to ^{32}P in tRNA | | | |
| | Liver | | Spleen | |
Group	Mean	SE	Mean	SE
+ Zn diet for 4d	.032	.004	.045	.005
- Zn diet for 4d	.032	.003	.043	.005
- Zn diet for 19d after high food intake	.034	.002	.069	.003
+ Zn restricted-fed after low food intake	.050	.003	.051	.004

The depression in growth of Zn-deficient rats is accompanied by a marked reduction in food intake and a large increase in its day-to-day variability. These are sufficient to inhibit the growth of pair-fed control rats to a comparable extent and this necessitates a careful choice of controls to resolve specific effects of Zn deficiency. The present experiments used rats after only 4d on the Zn-deficient diet before growth and food intake were affected, Zn-deficient rats after high intake of deficient diet when availability of Zn is minimal and pair-fed rats at low intake when food effects are likely to be maximal.

The rats were offered semisynthetic diets containing either 0.5 or 40 mg Zn/kg. After the periods indicated in Table 1, groups of 8 control or Zn-deficient rats were injected with 2 mCi ^{32}P-phosphate and killed 24h later. Total RNA was extracted from both liver and spleen by a guanidinium/- phenol/chloroform method and the low molecular weight RNAs were separated by differential solubilization in 3M sodium acetate. 5S rRNA was then separated from tRNA by acrylamide gel electrophoresis. The gels were scanned at 265 nM to locate the positions of the nucleic acids and they were then minced in 2 mm segments and counted. The ratio of the ^{32}P in 5S rRNA to that in tRNA was calculated and the results are shown in Table 1. There was no suggestion of impaired TFIIIA function in the Zn-deficient rats.

In view of the likelihood that addition of EDTA to lymphocyte cultures produced a lower availability of Zn than that induced in rats by dietary deficiency, it is still possible that TFIIIA function was inadequate in the lymphocyte cultures and this is being investigated.

REFERENCES

Berg, J. M., 1986, More metal binding fingers, Nature, Lond., 319:264.
Chesters, J. K., 1975, Comparison of the effects of zinc deprivation and actinomycin D on ribonucleic acid synthesis by stimulated lymphocytes, Biochem J., 150:211.
Chesters, J. K., 1978, Biochemical functions of zinc in animals, World Rev. Nutr. Diet., 32:135.
Hanas, J. S., Hazuda, D. J., Bogenhagen, D. F., Wu, F. Y-H. and Wu, C-W., 1983, Xenopus transcription factor A requires Zn for binding to the 5S RNA gene, J. Biol. Chem., 258:14120.
Miller, J., McLachlan, A. D. and Klug, A., 1985, Repetitive Zn-binding domains in the protein transcription factor IIIA from Xenopus oocytes, EMBO J., 4:1609.
Wingender, E., Dilloo, D., Seifert, K. H., 1984, Zinc ions are differentially required for the transcription of 5S RNA and tRNA in a HeLa cell extract, Nucleic Acid Res., 12:8971.

REGULATION OF THE GENES INVOLVED IN THE UTILIZATION OF MOLYBDENUM

Nancy K. Amy and Jeffrey B. Miller

Department of Nutritional Sciences
University of California
Berkeley, CA 94720

Mo is an essential trace element for virtually all organisms since it is required for the activities of nitrate reductase, formate dehydrogenase, nitrogenase, sulfite oxidase, xanthine dehydrogenase, aldehyde oxidase, and a variety of other enzymes. All molybdoenzymes, with the exception of nitrogenase, have molybdenum incorporated into a pterin-containing Mo-cofactor (1) and the structure of this cofactor has been conserved throughout evolution from bacterial, plant and animal sources. Therefore, an understanding of Molybdenum-cofactor metabolism in one organism will have far-reaching implications for the general understanding of molybdenum metabolism.

A number of genes involved in molybdenum metabolism have been identified in E. coli, and mutations in these genes result in the loss of the activities of all molybdoenzymes. The proposed functions of the genes (designated chl) catalyzing the steps in Mo metabolism are: the chlD gene is involved in the transport or accumulation of molybdenum in a functional form; the chlA and chlE genes appear to function in the synthesis of the pteridine portion of Mo-cofactor; the chlG gene functions for the incorporation of Mo into the cofactor; and the chlB gene product functions for the assembly of the Mo-cofactor with the apo protein of the molybdoenzymes (2,3).

The goal of our present experiments is to understand the factors regulating these genes involved in Mo-cofactor metabolism. We constructed strains carrying operon fusions of the lac genes to the promoters of the various chl genes using the specialized Mu dI phage developed by Casadaban and Cohen (3). The activity of β-galactosidase in these fusion strains reflects transcription from the chl promoters. We measured β-galactosidase activity and Mo-cofactor and nitrate reductase activities under a variety of growth conditions.

The chlD-lac fusion was highly expressed when the cells were grown with less than 10 nM molybdate (4). Increasing concentrations of molybdate caused loss of activity, with less than 10% of activity remaining at 500 nM molybdate. When tungstate replaced molybdate, it had an identical effect on chlD expression. Expression of chlD-lac was increased in cells grown with nitrate. Strains with chlD-lac plus an additional mutation were constructed to test whether the regulation of the chlD-lac required the concerted action of gene products involved with Mo-cofactor or nitrate reductase synthesis. We found that the regulation of Mo incorporation was intimately co-regulated with the regulation of the synthesis of the molybdoenzyme nitrate reductase.

Table 1

Gene	β-galactosidase	Nitrate Reductase	Mo-cofactor
chlA			
JBM 226	constitutive	−	−
JBM 259	↑ with nitrate	−	−
JBM 230	constitutive	+	+
JBM 237	↑ with nitrate	+	+
chlB			
JBM 15	constitutive	−	−
chlE			
JBM 249	constitutive	−	−
JBM 12	↑ with nitrate	−	+
JBM 240	↑ with nitrate	+	+
chlG			
JBM 254	constitutive	+	+

We constructed fusions of chlA and chlE (which are involved in the synthesis of the cofactor) with the lac genes and found that these genes were expressed under all growth conditions tested. We found 4 classes of chlA-lac fusion strains and 3 classes of chlE-lac fusions (Table 1) which differ in β-galactosidase, Mo-cofactor, and nitrate reductase activity. In one class of chlE-lac strains, the lac expression was increased 2-fold when the cells were grown anaerobically with nitrate. Expression of this gene is mediated by the narL gene product which is necessary for the induction of nitrate reductase. The occurrence of several phenotypes of activity of these fusions may indicate that these loci code for more than one protein.

REFERENCES

1. J. L. Johnson, B. E. Hainline, and K.V. Rajagopalan, Characterization of the molybdenum cofactor of sulfite oxidase, xanthine oxidase, and nitrate reductase, J. Biol. Chem. 255:1783-1786 (1980).
2. N. K. Amy, Identification of the molybdenum cofactor in chlorate-resistant mutants of Escherichia coli, J. Bacteriol. 148:274-282 (1981).
3. J. B. Miller and N. K. Amy, Molybdenum cofactor in chlorate-resistant and nitrate reductase-deficient insertion mutants of Escherichia coli, J. Bacteriol. 155:793-801 (1983).
4. M. J. Casadaban and S. N. Cohen, Lactose genes fused to exogenous promoters in one step using a Mu-lac bacteriophage: in vitro probe for transcriptional control sequences, Proc. Natl. Acad. Sci. USA 76: 4530-4533 (1979).
5. J. M. Miller, D. J. Scott and N. K. Amy, Molybdenum-sensitive transcriptional regulation of the chlD locus in Escherichia coli, J. Bacteriol. 169:1853-1860 (1987).

MECHANISMS AND NUTRITIONAL IMPORTANCE OF TRACE ELEMENT INTERACTIONS

Ian Bremner

Rowett Research Institute
Bucksburn
Aberdeen AB2 9SB, U.K.

INTRODUCTION

The existence of complex interactions between trace elements has been recognized for many years. Imbalances in the dietary supply of any one element can have both nutritional and toxicological consequences with regard to the metabolism of other metals. They can be responsible for the development of clinical signs of trace element deficiencies or can modify susceptibility to metal toxicities. Hill and Matrone (1970) were first to rationalise the occurrence of such interactions and suggested that elements which have similar chemical and physical properties will often interact biologically. However in the experiments that were initially set up to test this hypothesis, diets that contained unrealistically high amounts of the antagonist metal and were severely deficient in the agonist, or affected, metal were often fed. It is therefore difficult to assess from such studies which interactions have any nutritional or physiological relevance.

This information can only be obtained from experiments in which the concentration of the antagonist metal bears a reasonable relationship to that encountered under normal physiological or environmental conditions. The level of the agonist metal should also be in a realistic range, and it is often advantageous to set it at a level which, under normal circumstances, would be just adequate to meet the animal's dietary requirements for the metal. Subtle effects of the antagonist on the metabolism of the other element are then more readily detected. Other important aspects of experimental design which merit consideration in studying trace metal interactions are listed in Table 1. By careful attention to such matters a better appreciation of the biological significance of interactions is now being obtained.

CADMIUM-ZINC-COPPER INTERACTIONS

The interactions between cadmium, zinc and copper have been shown to be of environmental and clinical significance for both humans and grazing animals. They are highly complex, involve both 2- and 3-way interactions, and are also influenced by iron status. However two main features are of relevance in this paper. First, the absorption and distribution of cadmium can be modified by changes in copper and zinc status. Second, increased cadmium intakes can result in disturbances in copper and zinc metabolism.

Table 1. Design of Experiments to Assess Significance of
          Trace Element Interactions

Use realistic levels of antagonist and agonist metals
Consider possibility of 3- or 4-way interaction
Allow for adaptation mechanisms to operate
Consider species differences in response
Consider likelihood that different responses may occur
    when isolated tisues or organs are studied
Consider possibility that other dietary components may
    influence response
Biochemical investigations needed to elucidate mechanism
    of interaction

Slight decreases in the dietary concentrations of copper and zinc below requirement levels can increase the accumulation of cadmium in the liver and kidneys of experimental animals (Fox et al., 1984). Since the onset of cadmium-induced renal dysfunction in man is closely related to the levels of cadmium in the kidneys, even marginal copper and zinc deficiencies could have long-term consequences for the development of renal lesions in individuals occupationally exposed to cadmium. Conversely cadmium accumulation should be restricted by ensuring adequate copper and zinc status. These effects are generally thought to be related to changes in the intestinal absorption of cadmium, although there is in fact only limited experimental evidence to substantiate this. The distribution of cadmium between liver and kidneys can also be disturbed by these trace element deficiences (Fox et al., 1984).

The second aspect of the cadmium-zinc-copper interactions concerns the effects of cadmium on the metabolism of these elements. This may not be a major problem at the level of exposure generally suffered by humans, but can be important in farm animals. Copper deficiency symptoms have been reported in animals grazing in the neighbourhood of industrial complexes and experimental studies have confirmed that dietary cadmium intakes of only 1.5 mg/kg are sufficient to reduce copper status in lambs (Bremner and Campbell, 1980). One interesting aspect of these studies is the finding that any associated increase in zinc intake will tend to exacerbate the copper deficiency, because of independent zinc-copper interactions. However increased zinc intake will still tend to reduce renal accumulation of cadmium and will therefore be beneficial in terms of delaying the development of renal damage. Although it has been assumed that cadmium limits zinc accumulation at the intestinal level, this was not substantiated in studies by Hamilton et al. (1978). Cadmium-induced inhibition of zinc uptake and transfer by the intestinal mucosa could only be demonstrated in iron-deficient mice; no inhibition occurred in iron-replete animals. This provides further illustration that interactions are not always explicable in simple 2-way terms.

Effects of zinc on copper metabolism have been recognised for many years and are of some clinical relevance. Thus, treatment of sickle cell disease patients with zinc can result in development of hypocupraemia. Further examples are provided by the successful treatment of Wilson's disease patients and of the control of copper poisoning in sheep by zinc supplementation. These beneficial effects of zinc result from the reduction in the concentration and cytoxicity of copper in the liver. Some changes in the intracellular distribution of copper may also occur, with an increase in the proportion bound to metallothionein, but the primary effect of zinc is probably on the intestinal absorption of copper. This is one of the few examples where progress has been made in elucidating at the molecular level the probable mechanism of interaction. Contrary to the early assumption the zinc caused displacement of copper from some transport protein, the effect of zinc is actually to stimulate the intestinal synthesis of metallothionein (Bremner and Campbell, 1980). Copper then binds preferentially to the mucosal metallothionein and this is associated with a reduced rate of transfer of the copper across the serosal membrane into the plasma. Good correlations were found between the degree of inhibition of copper absorption and the amount of metallothionein synthesised in the mucosa. A similar mechanism is probably involved when cadmium inhibits copper absorption.

INTERACTIONS BETWEEN IRON AND OTHER TRACE METALS

One of the most common nutritional disorders is iron deficiency. Homeostatic control of iron metabolism results in increased efficiency of iron absorption in iron deficient animals. This is accompanied by increased absorption of a wide range of other metals, including zinc, cadmium, manganese, lead and cobalt: it is possible that these metals share a common absorption pathway with iron, although the precise nature of this has yet to be established. Such interactions between iron and these other elements have been readily demonstrated in studies using isolated intestinal segments and by intestinal perfusion experiments (Hamilton et al., 1978). That they are also of relevance to man was convincingly demonstrated in a study of radiocadmium absorption in humans of varying iron status (Flanagan et al., 1978). Cadmium absorption was greatest in individuals with low serum ferritin levels and therefore low body iron stores. It is significant that these individuals were not anaemic, indicating that even marginal iron deficiency can result in increased uptake of other metals.

Recent studies on lead-iron interactions in rats (Morrison & Quarterman, 1987) show that iron and lead absorption are initially increased in rats given a diet that is marginally iron deficient. Some form of adaptation occurs thereafter and lead absorption reduces to the level found in iron-adequate rats, even though the enhancement in iron absorption continues. There was no simple correlation between effects on lead absorption and degree of anaemia and no enhancement in lead absorption was found in hypoxic rats, although iron absorption was stimulated. Such results may help provide an explanation for the lack of agreement in the literature as to the effects of iron deficiency on lead uptake by humans.

It has been shown in intestinal perfusion experiments that the absorption of zinc is also increased in iron-deficient animals. Conversely, excessive iron supplementation can reduce zinc availability in infants receiving formula feeds and in adult humans given simple solutions of zinc salts. A modest increase in iron:zinc ratio in these solutions to 2.5 or more resulted in decreased apparent zinc absorption (Valberg et al., 1984). However in other studies with adults given complete diets with much higher iron:zinc ratios, no effect on zinc availability could be detected using whole body counting techniques (Sandstrom et al., 1985). It may be that ligands released on digestion of the food prevented iron from interacting with zinc. Regardless of the explanation, these results highlight the problems in extrapolating to the whole animal level results from experiments in which only a small part of the absorption process is examined.

COPPER-IRON-MOLYBDENUM INTERACTIONS

Copper is another metal that interacts with iron in a variety of ways. However unlike the examples cited above, the intestinal absorption of copper is not increased in iron-deficient animals, whether measured using intestinal loops or at the whole animal level. Nevertheless, liver copper levels are increased in iron-deficient rats and in their offspring (Sherman and Tissue, 1981), indicating that other aspects of copper metabolism are affected by iron. This has been confirmed in studies on the effects of iron supplements on copper metabolism, since tissue copper concentrations are decreased and the incorporation of copper into intestinal cytochrome oxidase is inhibited in iron-treated rats (Bremner & Price, 1985). Although some of these effects are evident within a few days, they can be relatively short-lived, particularly at high dietary iron intakes, which implies again that forms of adaptation may occur in certain interactions.

The effects of iron on copper metabolism are of particular importance in ruminant nutrition. Slight increases in dietary iron content from 100 to 250 mg/kg are sufficient to reduce significantly liver and plasma copper concentrations (Bremner et al., 1987). Dietary iron concentrations of over 500 mg/kg cause the same changes in tissue copper levels and in activities or copper-dependent enzymes as does a supplement of 5 mg molybdenum/kg (Humphries et al., 1983). Such iron levels are typical of those found in many feedstuffs, particularly if there is any contamination with soil. In calves given diets with supplements of 500 mg iron/kg, pancreatic damage was detected and the ability of neutrophils to kill ingested organisms was impaired. However no other clinical effects were found and growth rates, food intake, bone conformation and reproductive performance were all normal, despite the fact that the animals were regarded as severely copper-deficient on the basis of their tissue copper concentrations.

Molybdenum-treated animals on the same experiments invariably exhibited all the classical signs of copper deficiency, even though they were not ostensibly more copper-deficient than the iron-treated animals. It is significant that when diets were interchanged, so that animals initially fed iron were given molybdenum, and vice-versa, there were rapid reversals in fertility.Introduction of molybdenum-containing diets caused infertility, whereas removal of molybdenum restored fertility to normal levels (Phillippo et al., 1987). Since these diet changes did not cause obvious changes in copper status, it seems that the clinical lesions observed in the molybdenum-treated animals are a reflection of direct effects of that metal on biological processes and are not necessarily due to an induced copper deficiency (although this may still contribute to the observed changes). Ever since the beneficial effects of copper in treating or preventing molybdenum-induced lesions were discovered, it has been assumed that the copper supplements act by overcoming a conditioned deficiency in copper. However, our recent experiments suggest that the important effect of copper may be to prevent the uptake of molybdenum and so prevent molybdenosis. Inhibition of molybdenum retention has been noted in sheep given copper supplements and tetrathiomolybdate inhibits copper absorption in rats (see Bremner and Mills, 1986).

Considerable advances have been made in recent years in understanding the complex nature of the copper-molybdenum interaction. It is one of the few interactions where the probable mechanism has been elucidated at a molecular level. There seems little doubt that thiomolybdates are responsible for the development of copper deficiency in molybdenum-supplemented ruminant animals (Bremner and Mills, 1986). Convincing evidence for the formation of thiomolybdates in the gastrointestinal tract of sheep has now been obtained (Price et al., 1987) and the effects of thiomolybdates on the performance of rats and on their copper metabolism closely resemble changes seen in ruminants. It is significant, however, that the effects of tetrathiomolybdate on the haemopoeitic and skeletal systems in rats are much more severe than those induced by simple copper deficiency; this reinforces the view that molybdenum can have effects independent of copper deficiency.

The influence of molybdenum on copper metabolism in ruminants has always been regarded as a classic example of a metal-metal interaction that is of real biological significance. It is therefore ironic that what was attributed to a molybdenum-induced copper deficiency may simply have been a form of molybdenosis. The fact that copper can overcome this condition is not in question but the important reaction now appears to be one of copper on molybdenum metabolism rather than the reverse.

REFERENCES

Bremner, I. and Campbell, J. K., 1980, The influence of dietary copper intake on the toxicity of cadmium, Ann. N. Y. Acad. Sci., 355:319-332.

Bremner, I., Humphries, W. R., Phillippo, M., Walker, M. and Morrice, P. C., 1987, Iron-induced copper deficiency in calves: dose-response relationships and interactions with molybdenum and suphur, Anim. Prod., 45: (in press)

Bremner, I. and Mills, C. F., 1986, The copper-molybdenum interaction in ruminants: the involvement of thiomolybdates, in "Orphan Diseases and Orphan Drugs", I. H. Scheinberg and J. M. Walshe, eds., pp68-75, Manchester University Press, Manchester.

Bremner, I. and Price, J., 1985, The effects of dietary iron supplements on copper metabolism in rats, In "Trace Elements in Man and Animals - 5", C. F. Mills, I. Bremner and J. K. Chesters, eds., pp374-376, CAB Press, Slough.

Flanagan, P. R., McLellan, J. S., Haist, J., Cherian, M. G., Chamberlain, M. J. and Valberg, L. S., 1978, Increased dietary cadmium absorption in mice and human subjects with iron deficiency, Gastroenterology, 74:841-846.

Fox, M. R. S., Tao, S-H., Stone, C. L. and Fry, B. E., 1984, Effects of zinc, iron and copper deficiencies on cadmium L-tissues of Japanese quail, Envir. Hlth. Persp., 54:57-65.

Hamilton, D. L., Bellamy, J. E. C., Valberg, J. D. and Valberg, L. S., 1978, Zinc, cadmium and iron interaction during intestinal absorption in iron-deficient mice, Can. J. Physiol. Pharmac., 56:384-388.

Hill, C. H. and Matrone, G., 1970, Chemical parameters in the study of in vivo and in vitro interactions of transition elements, Fed. Proc., 29:1474-1481.

Humphries, W. R., Phillippo, M., Young, B. W. and Bremner, I., 1983, The influence of dietary iron and molybdenum on copper metabolism in calves, Br. J. Nutr., 49:77-86.

Morrison, J. N. and Quarterman, J., 1987, The relationship between iron status and lead absorption in rats, Biol. Trace Element Res. (in press).

Phillippo, M., Humphries, W. R., Atkinson, T., Henderson, G. D. and Garthwaite, P. H., 1987, The effect of dietary molybdenum and iron on copper status, puberty, fertility and oestrous cycle in cattle, J. Agr. Sci., (in press)

Price, J., Will, A. M., Paschaleris, G. and Chesters, J. K., 1987, Identification of thiomolybdates in digesta and plasma from sheep after administration of ^{99}Mo-labelled compounds in the rumen, Br. J. Nutr., 58: (in press)

Sandstorm, B., Davidsson, L., Cederblad, A. and Lonnerdal, B., 1985, Oral iron, dietary ligands and zinc absorption, J. Nutr., 115:411-414.

Sherman, A. R. and Tissue, N. T., 1981, Tissue iron, copper and zinc levels in offspring of iron-sufficient and iron-deficient rats, J. Nutr., 111:266-275.

Valberg, L. S., Flanagan, P. R. and Chamberlain, M. J., 1984, Effects of iron, tin and copper on zinc absorption in humans, Am. J. Clin. Nutr., 40:536-541.

EFFECTS OF DIETARY MOLYBDENUM AND IRON

ON COPPER METABOLISM IN CALVES

W.R. Humphries, M.J. Walker, P.C. Morrice and I. Bremner

Rowett Research Institute
Bucksburn,
Aberdeen, AB2 9SB, U.K.

INTRODUCTION

The availability of Cu to ruminant animals is greatly influenced by the Mo, Fe and S content of the diet. The importance of Fe as a Cu antagonist has only recently become apparent and it has been found that dietary supplements of only 250 mg Fe/kg can reduce tissue Cu reserves in calves (Humphries et al., 1983, 1985). In one experiment with Hereford-Friesian calves dietary supplements of 800 mg Fe/kg or 5 mg Mo/kg induced similar changes in liver and blood Cu concentration and in activities of Cu-dependent enzymes (Humphries et al., 1983). When calves were given both supplements, the changes in Cu status and in clinical condition were no more severe than in calves given Mo alone. The lack of any additive action of these antagonists implies either that they affect Cu metabolism by some common mechanism or that they were individually having the maximum possible effect on Cu availability. To distinguish between these possibilities, a further experiment has been carried out in which calves were given lower amounts of dietary Mo and of Fe.

MATERIALS AND METHODS

Twenty Hereford-Friesian heifers were reared until 10-12 weeks old on a low-Cu milk-substitute ration, whereupon they were given a diet based on barley and barley straw, containing 4 mg Cu, 100 mg Fe, 0.1 mg Mo and 2.8 g S/kg. The calves were allocated on the basis of liver Cu content in a randomized block design into 4 groups and given for 41 weeks the basal diet alone (Group Con) or supplemented with 150 mg Fe/kg as saccharated ferrous carbonate (Group Fe), 2 mg Mo/kg as sodium molybdate (Group Mo), or both supplements (Group FeMo). Liver biopsies and blood samples were collected at regular intervals for Cu analysis and assay of Cu-dependent enzymes using methods described previously (Humphries et al., 1983). Statistical analysis of results was carried out by analysis of variance.

RESULTS AND DISCUSSION

Significant effects of the treatments on growth rate were detected after 23 weeks, with weight gains in Group Fe significantly greater than control values, and gains in the Mo-treated calves reduced by about 15% (Table 1). The two groups of Mo-treated calves showed changes in hair texture

Table 1. Effects of Dietary Fe and Mo on Growth and Cu Status of Calves after 30 Weeks

Group	Con	Fe	Mo	FeMo	SEDM	Significance
Weight gain (kg)[a]	284	318	262	259	20	*
Liver Cu (mg/kg DM)[b]	75	22	11	9		***
Liver Cytox (U/g protein)	67	60	25	17	15	**
Plasma Cu (mg/l)	0.70	0.57	0.05	0.04	0.07	***
Blood SOD (mg/l)	173	138	80	82	12	***
End/Shaft ratio[a]	1.55	1.55	1.58	1.70	0.04	*

[a] Weight gain and end/shaft ratio in metacarpals measured at 41 weeks.
[b] Statistical analysis carried out after log transformation.

typical of Cu deficiency but only the FeMo calves showed significant changes in bone conformation, as indicated by changes in end/shaft ratios in the metacarpals.

Liver Cu concentrations were significantly reduced in all treatment groups after 9 weeks and declined at an exponential rate thereafter (Table 1). The fractional rates of loss of liver Cu in Groups Fe, Mo and FeMo were 0.025 ± 0.002, 0.043 ± 0.004 and 0.049 ± 0.002, equivalent to half-lives of liver Cu of 12, 7 and 6.1 weeks respectively. These differences between Groups Mo and FeMo were not significant but Cu concentrations in the former group were nevertheless consistently greater than in the latter and at certain time points the differences were statistically significant. Only in Groups Mo and FeMo were cytochrome oxidase levels significantly reduced.

Plasma Cu concentrations were also greatly reduced in the 2 groups of Mo-treated calves. Significant decreases were evident after 9 weeks in Group FeMo and after 17 weeks in Group Mo. Levels in Group Fe remained in a normal range throughout. Erythrocyte SOD activities were reduced by 50% in Groups Mo and FeMo and by 20% in Group Fe after 30 weeks.

These results show that moderate increases in dietary Fe intake of only 150 mg/kg are sufficient to decrease liver Cu retention in calves and emphasise the potential importance of this antagonist in inhibiting Cu utilisation under practical conditions. Supplements of only 2 mg Mo/kg were sufficient to reduce growth rates by 15% and reduce liver and plasma Cu to levels indicative of severe Cu deficiency. Although liver and plasma Cu concentrations in Groups Mo and FeMo were not always different, the initial rates of decline in these parameters were greater in Group FeMo, indicating the additive action of these antagonists is indeed possible under appropriate conditions.

REFERENCES

Humphries, W.R., Phillippo, M., Young, B.W., and Bremner, I., 1983, The influence of dietary iron and molybdenum on copper metabolism in calves, Br. J. Nutr., 49: 77-86.
Humphries, W.R., Bremner, I., and Phillippo, M., 1985, The influence of iron on copper metabolism in calves, In: "Trace Elements in Man and Animals - TEMA5", C.F. Mills, I. Bremner, and J.K. Chesters, eds., pp 176-180, CAB; Slough.

ACCUMULATION OF Cu AND Mo BY THE FOETUS(ES) AND CONCEPTUS OF SINGLE AND

AND TWIN BEARING EWES

N.D. Grace, J. Lee, and P.L. Martinson

Biotechnology Division, D.S.I.R.

Palmerston North, New Zealand,

The importance of dietary Cu, Mo and S interrelationships in the nutrition of the ruminant have been well documented[1]. However there is little information on the rate of accumulation and distribution of Cu and Mo in the conceptus of the ewe.

MATERIALS AND METHODS

Animals. Romney ewes, aged 5 years and weighing on average 56 kg, were mated in early March to Romney rams.

Experimental design. 56 ewes of known date of conception and pregnancy status were divided into 7 groups of 8 ewes (4 single and 4 twin bearing). Another group of 8 ewes were slaughtered just prior to mating. The pregnant groups were then slaughtered at gestational ages of 62, 81, 100, 115, 125, 135 and 143 days. All ewes were grazed on a ryegrass/white clover pasture.

Slaughter and sampling procedure. The ewes were shorn, weighed and slaughtered to remove the conceptus. Each conceptus was dissected and the placental fluids, uterine tissue including the placental membranes and the foetus(es) were removed, weighed and subsampled for Cu and Mo determinations. The liver was also removed from the single foetus.

Analytical. Tissue samples, after thawing, were wet ashed in a HNO_3-$HCLO_4$ mixture. Mo was extracted from the acid digest as a Mo-dithiol complex in chloroform. The chloroform was then evaporated off, the residue digested in HNO_3, 2M HCl added and the Mo determined by inductively coupled plasma emission spectrometry (ICP). Copper and S were determined on other samples by ICP, after wet ashing and the addition of 2M HCl.

Statistics. Information from the single and twin bearing ewes was analysed separately. Data was transformed using natural logarithms and a Gompertz equation was used to relate the net mineral content of the conceptus (Y) to gestational age (X). The general equation used was $\ln Y = A-Be^{-ct}$ where A, B and C are constants determined by a least-squares fit to the data. The rate of mineral accumulation was determined from the slope of the curve relating mineral content to gestational age.

RESULTS

Mean pasture Cu, Mo and S contents per kg DM were 5.0 mg, 0.14 mg and 3.95g respectively. Mean amounts of Cu and Mo associated with the foetal liver, foetus(es), placental fluids, uterus and conceptus are given in Table 1.

Table 1. Mean Cu and Mo contents of the foetal liver, foetus(es),
placental fluids, uterus and conceptus of single and twin bearing ewes

	Copper (mg)			Molybdenum (µg)		
Gestational age (d)	62	115	143	62	115	143
Single foetus						
foetal liver	0.13	2.8	5.9	1.34	8.9	15.4
foetus	0.17	6.5	14.4	1.87	20.7	40.3
placental fluid	0.009	0.06	0.09	0.53	1.4	7.0
uterus	0.67	0.8	1.2	28.28	37.8	57.7
conceptus	0.85	7.4	15.7	30.68	59.9	105.0
Twin foetuses						
foetuses	0.29	9.3	21.8	2.98	36.1	58.9
placental fluid	0.012	0.15	0.16	1.15	3.8	14.8
uterus	0.83	1.2	1.8	37.31	23.5	64.9
conceptus	1.13	10.6	23.7	41.44	63.4	138.6

Foetal liver is the major storage organ for both Cu and Mo as about 70%
of these elements were associated with the liver at day 62, and 40% at day
143 of gestation. As the pregnancy progressed increased amounts of Cu and
Mo are taken up by the developing foetus(es). Just prior to parturition
about 90% of the Cu and 38% of the Mo found in the conceptus was associated
with the foetus(es). The daily rates of accumulation of Cu and Mo into the
conceptus of single and twin bearing ewes are shown in Table 2.

Table 2. The daily rate of accumulation of Cu and Mo into the conceptus of
single and twin bearing ewes

Gestational age	Copper (mg/d)		Molybdenum (µg/d)	
(days)	Single	Twin	Single	Twin
60	0.03	0.05	0.13	0.24
100	0.14	0.23	0.51	0.72
120	0.29	0.40	1.19	1.43
143	0.65	0.70	3.95	4.16

For most of gestation the rate of uptake of Cu and Mo into the conceptus
was greater in the twin bearing ewe compared to the single bearing ewe.
However near parturition the rates of uptake of both Cu and Mo were similar
regardless of the pregnancy status of the ewes.

DISCUSSION
Data from this study was used to determine the rates of Cu and Mo accumulation and distribution in the conceptus of ewes of an adequate Cu status on
low Mo intakes. The Cu and Mo metabolism of the conceptus, however can be
influenced by the Mo intake of the ewe. For example swayback, a nervous
disorder caused by Cu deficiency, has been observed in lambs born to ewes
given very high intakes of Mo (50 mg/d)[2]. The high Mo intake reduces the
amount of Cu and increases the amount of Mo absorbed by the ewes, thus
making less Cu and more Mo (thiomolybdates) available to the conceptus and
foetus. The extent of the reduced Cu uptake by the foetus together with a
possible interference of the Cu at various metabolic sites by thiomolybdates,
needs further investigation in the ewe.

REFERENCES
[1] N.F. Suttle, Recent studies of the copper-molybdenum antagonism.
Proc. Nutr. Soc. 33: 299 (1974).
[2] B.F. Fell, C.F. Mills, and R. Boyne. Cytochrome oxidase deficiency
in the motor neurones of copper-deficient lambs: a histochemical study.
Res. Vet. Sci. 6: 170 (1965).

INFLUENCE OF BREED AND DIETARY Cu, Mo AND S LEVELS ON BILIARY Cu EXCRETION IN CATTLE

S.R. Gooneratne, D.A. Christensen, J.V. Bailey*,
and H.W. Symonds**

Department of Animal and Poultry Science, *Western College of
Veterinary Medicine, University of Saskatchewan, Saskatoon,
Sask. S7N 0W0, Canada. **Department of Animal Physiology and
Nutrition, University of Leeds, Leeds LS2 9JT, U.K.

INTRODUCTION

Bile is considered to be the major route of copper (Cu) excretion.
But studies on dietary components which interact with Cu to produce Cu
deficiency have rarely focussed attention on this route of Cu excretion.
Thiomolybdates (TM) formed in the rumen are primarily responsible for the
induction of Cu deficiency in ruminants fed high levels of molybdenum (Mo)
and sulfur (S). We (Gooneratne et al, 1985) have shown that intravenous (iv)
administration of TM increases biliary Cu excretion and lowers liver Cu
level in sheep. But bile Cu excretion in cattle, following TM administration
or fed excess Mo and S to induce synthesis of TM in the rumen, is not known.
Breed related Cu deficiency in sheep is well recognized. In Saskatchewan
Cu deficiency may be observed more frequently in Simmental than in other
breeds. In rats biliary Cu excretion is strain dependent (Nederbragt and
Lagerwerf 1986). The present investigation was carried out to determine
biliary Cu excretion in 2 breeds of cattle Simmental and Angus, in response
to changes in Cu (low (L) = 5 or high (H) = 40 mg/kg), and/or Mo (L = 1 or
H = 10 mg/kg) and/or S (L = 0.2% or H = 0.5%) in the diet. Bile Cu
excretion was also monitored following iv administration of TM.

MATERIALS AND METHODS

Eight, 10 month old heifers (4 Simmental, 4 Angus) had their duodenum
modified surgically so that bile could be collected and returned to the
duodenum (Symonds et al. 1982). Six weeks later, 4 animals (2/breed) were
transferred to a HCu diet for 2 months. The remaining animals continued to
receive a LCu diet during this period. All animals were next transferred
to diets, LCu – HS, HCu – HS, HCu – HMo, and HCu – HMo – HS in that order
for periods of 1, 2, 2, and 2 months respectively. Twelve, 30 minute bile
samples per day were taken on 2 consecutive days of each month. Plasma Cu
(1:4 & TCA) levels were monitored on a monthly basis. All animals were
next allowed to recover for 1 month on a LCu – LMo – LS diet, prior to
injection with TM at 1 mg TM/Kg body weight. Bile samplings were similar
to as described above but more frequent.

Table 1. Total Cu excreted in bile (mg/6h) (Mean + SD)

DIET DURATION (Months)	LCu	HCu	LCu,HS	HCu,HS	HCu,HMo	HCu HS,HMo	TM (IV)
	2	2	1	2	2	2	---
SIMMENTAL	0.17	0.28	0.67	0.40	0.34	1.4	2.9
	+0.009	+0.02	+0.42	+0.06	+0.09	+0.08	+0.02
ANGUS	0.08	0.14	0.17	0.15	0.11	0.72	1.8
	+0.02	+0.03	+0.04	+0.07	+0.007	+0.52	+0.41

RESULTS

The level of Cu excreted in bile on day 1 of month 2 sampling (except for LCu - HS diet) is given in Table 2. In all treatments, the Simmental cattle excreted more Cu via bile than Angus. During all periods the bile flow and Cu excretion via bile was lower on day 2 of sampling. Increasing the dietary Cu concentration or a combination of Cu, Mo and S resulted in a significant increase in bile Cu excretion in both breeds of cattle. Increasing S or Mo in the diet also increased bile Cu excretion but this was observed only in the first month of sampling during the respective dietary regimes. High levels of Cu-Mo-S initially increased plasma Cu levels with 1:4 Cu > TCA soluble Cu, indicating the presence of an unavailable form of TCA insoluble fraction of Cu during this time. The changes in bile and plasma Cu levels following iv TM was similar to that seen when excess Mo and S was fed in the diet, but the changes were more rapid and marked.

CONCLUSION

i) Excess Cu, or a combination of Mo and S in the diet enhances biliary Cu excretion in cattle. ii) Intravenous administration of TM results in a similar, but a more rapid and a marked increase in bile Cu excretion. iii) Bile Cu excretion was higher in Simmental than in Angus cattle.

REFERENCES

Gooneratne, S.R., Christensen, D., Chaplin, R., and Trent, A. (1985) Copper: Biliary excretion in copper-supplemented and thiomolybdate treated sheep. In Proc. of 5th Int. Symp. on trace elements in man and animals. p. 342-345 C.F. Mills, I. Bremner, and J.K. Chesters, eds., Commonwealth Agricultural Bureau, UK.

Nederbragt, H., and Lagerwerf, A.J. (1986). Strain related patterns of biliary excretion and hepatic distribution of copper in the rat. Hepatology 6: 601-607.

Symonds, H.W., Mather, D.L. and Hall, E.D. (1982). Surgical procedure for modifying the duodenum in cattle to measure bile flow and the clinical variation in biliary manganese, iron, copper and zinc excretion. Res. Vet. Sci. 32: 6-11.

INTERACTIONS BETWEEN PROTEINS, THIOMOLYBDATES AND COPPER

J. D. Allen and J. M. Gawthorne

School of Veterinary Studies
Murdoch University
Murdoch, Western Australia, 6150

INTRODUCTION

The concentration of copper (Cu) is elevated in the kidneys of sheep[1] fed diets containing high levels of molybdenum (Mo) and sulfur (S) or given supplements of tetrathiomolybdate (TTM). A similar effect occurs in liver and kidney when TTM is administered to rats.[2] In both species these elevated Cu concentrations may occur together with signs of Cu deficiency such as decreased activity of Cu-dependent enzymes. The accumulated Cu elutes in the high molecular weight range (>100,000 MW) on Sephadex G-75, and is usually accompanied by Mo. It apparently accumulates in forms that do not readily equilibrate with the pools of metabolically available Cu.

The mechanism of this Cu accumulation is unknown. One possibility is that Cu-metallothionein may polymerise or cross-link with proteins under the influence of Mo-containing compounds and thereby chromatograph in the high MW range. Another possibility is that Mo-containing compounds may form complexes with certain proteins, and these complexes may then chelate Cu.

In the present study extracts of liver and kidney from rats and sheep were exposed to TTM in-vitro in order to investigate the effects on the chromatographic profile of Cu. In one experiment ^{35}S-labelled metallothionein (MT) was added to extracts in order to monitor changes in the distribution of MT and in another, bovine serum albumin was used as a model protein to determine whether TTM associates with sulfhydryl groups on proteins.

MATERIALS AND METHODS

Samples of liver and kidney were homogenised in 10mM tris-acetate buffer pH 7.4, containing 1% merceptoethanol to extract the soluble Cu. In one experiment ^{35}S-labelled MT was added to the buffer. Extracts were obtained by centrifuging the homogenates at 105,000 x g for 1h at 4°.

TTM was added to extracts to give a final concentration of 80ug Mo/ml; the preparations were allowed to stand at room temperature for 1h; were centrifuged at 105,000 x g for 30min; and 4 ml of the supernatant was chromatographed on Sephadex G-75.

The possibility that TTM binds to proteins via accessible sulfhydryl groups was investigated by using bovine serum albumin as a model protein. Sulfhydryl groups were masked by reaction with iodoacetamide (IAA), samples of masked and unmasked protein were exposed to TTM, dialysed, and protein-bound TTM was determined from a scan of the electronic spectrum between 300nm and 600nm.

RESULTS AND DISCUSSION

The chromatographic profile of tissue extracts on Sephadex G-75 exhibited three Cu-containing peaks designated Cu-1, Cu-2 and Cu-3 in decreasing order of MW.

When TTM was added to extracts of liver or kidney there was a shift in the chromatographic profile of Cu away from MT, towards proteins of high MW (peak Cu-1). This was accompanied by a re-distribution of Zn in the opposite direction. The proportion of Cu in peak Cu-1 was increased from 4% to 63% in rat liver, from 26% to 75% in sheep liver, and from 26% to 88% in rat kidney. There was no detectable Cu in the elution position of MT after TTM treatment. Significantly, when ^{35}S-labelled MT was incorporated into extracts before addition of TTM, all radioactivity remained within the MT peak and did not move with the Cu. If Cu and TTM were combined in tris-acetate buffer at concentrations that occurred in the above preparations and a sample was then chromatographed, no Cu appeared in the position of peak Cu-1, thereby ruling out any significant involvement of Cu-TTM polymers in the changes. Taken together, the results suggest that Cu is stripped from MT by stronger chelators that are formed by the association of TTM with proteins in peak Cu-1. This conclusion was confirmed by the fact that addition of TTM-treated Cu-1 proteins to extracts caused a similar change in the Cu profile as addition of TTM itself.

Reaction of bovine serum albumin with IAA masked 93% of the sulfhydryl groups accessible to DTNB but this did not decrease the capacity of the albumin to bind TTM as evidenced by the similar optical density of the masked and unmasked protein at 470nm. Thus it seems unlikely that sulfhydryl groups are the site of attachment of TTM to proteins.

It is evident that complexes between proteins and TTM will form spontaneously and that at least some of the complexes bind Cu very strongly, with a greater affinity than MT. The "thiomolybdate hypothesis" for explaining the interaction of Cu, Mo and S in ruminants needs to be modified. We postulate that the formation of protein-TTM complexes distorts the equilibrium between the physiological Cu chelators of transport processes and apoenzymes. Cu is diverted away from metabolically useful forms in the gut and tissues, and into abnormal protein-TTM-Cu complexes that accumulate (kidney), or are excreted (gut).

REFERENCES

1.   J. D. Allen and J. M. Gawthorne, Involvement of Organic Molybdenum compounds in the interaction between Copper, Molybdenum and Sulfur, J. Inorg. Biochem., 27:95 (1986).
2.   C. F. Mills, T. T. El-Gallad and I. Bremner, Effects of molybdate, sulphide and tetrathiomolybdate on copper metabolism in rats, J. Inorg. Biochem., 14:189 (1981).

ANTAGONISTIC EFFECTS OF A HIGH SULPHUR, MOLYBDENUM AND CADMIUM CONTENT OF DIETS ON COPPER METABOLISM AND DEFICIENCY SYMPTOMS IN CATTLE AND PIGS

M. Anke, T. Masaoka, A. Hennig,and W. Arnhold

Karl-Marx-Universität Leipzig
Wissenschaftsbereich Tierernährungschemie
6900 Jena, Dornburger Straße 24, GDR

INTRODUCTION

The energy production from S-rich coal leads to a considerabel S accumulation in the feedstuffs of ruminants which can reach 15 g/kg ration dry matter. Furthermore, Mo and Cd are emitted in certain areas. All 3 elements have antagonistic influences on the Cu metabolism of ruminants. The effect of a supplementation with 10 g S/kg ration dry matter, 10 g S and 10 mg Mo or 10 g S and 3 mg Cd/kg on live weight gain, skeleton and the macro and trace element status was investigated in growing fattening bulls and pigs.

Growth

The S, S and Mo or S and Cd supplementation reduced the growth of bulls significantly within 56 days. After 196 days, the bulls exposed to S and S and Mo gained 15 % less weight per day and those with S and Cd exposure 19 % less than control animals (table 1). On the average, the bulls exposed to S and Cd grew worse than all other groups after the 84th experimental day (p > 0.05).

Skeleton damage

During the experiments, mainly the bulls with S and Cd supplementation had difficulties to get up. The animals exposed to S or S and Mo reacted similarly, but less intensively. Measurements of the fore tarsal joint showed that the bulls fed on S and Cd had significantly enlarged joints (table 1).
The other bulls exposed to S also had enlarged fore tarsal joints though, on an average, they weighed 15 - 19 % less. Related to their body weight, they should have been expected to have smaller fore tarsal joints. In control bulls, there was a positive relation between the body weight and the size of the fore tarsal joint (r 0.65).
This was also true for bulls with S and Mo supplementation (r 0.70).

In animals supplemented with S or S and Cd, there was no significant relation between the size of fore tarsal joints and body weight. There was even the tendency that the heavy animals of this group had smaller fore tarsal joints.

Table 1. The influence of S, Mo and Cd supplementation on growth, skeleton development and the Cu, Mo and Cd status

		control bulls	+S	+S+Mo	+S+Cd	$SD_{0.05}$
growth	$\bar{x}$	906	766	774	738	
(g/day)	s	148	107	92	109	159
	%	100	85	85	81	-
size of the fore	$\bar{x}$	33.4	35.2	35.1	35.6	
tarsal joints	s	1.6	1.6	0.93	0.86	1.9
(cm)	%	100	105	105	107	-
Cu mg/kg	liver	61	36	22	46	38
dry matter	cerebrum	8.5	7.7	7.4	7.3	-
Mo μg/kg	blood serum	20	22	55	16	29
dry matter or μg/l	hair	177	181	595	216	139
Cd μg/kg	kidneys	648	870	758	5018	1621
dry matter	liver	289	310	280	998	272

Trace element status

The expected great influence of the Cu antagonists S, Mo and Cd on the Cu status of fattening bulls could not be fully demonstrated. Though the animals exposed to Cu antagonists stored significantly less Cu in the liver than control animals, the Cu depletion did not reach such an extent which might have been expected due to the live weight gain reduced by 15 to 19 %. The reduced Cu content of the cerebrum also points to a disturbance of Cu metabolism, but it does not fully explain the fast and extensive reduction of live weight gain. Apart from 10 g S/kg ration dry matter, the supplementation of 10 mg Mo/kg led to a significantly increased Mo content in blood serum and hair.

The same is true for the Cd status. The kidneys of bulls with Cd and S supplementation stored about 7 times more Cd than control animals. The liver of these animals contained the threefold Cd concentration. Thus, the Cd accumulation in both organs stayed within limits.

The supplementation of 10 g S/kg ration dry matter led to a significantly reduced growth. Fattening pigs also reacted on the supplementation of 20 g S/kg dry matter with greatly reduced growth, pareses of the hind legs and exitus. In pigs, the supplementation of 10 g S/kg dry matter did not influence growth. Hence it follows that they react less sensitively on S exposure than ruminants.

The great influence of S exposure on growth cannot only be explained with the secondary S, Mo and Cd-induced Cu deficiency, since there was no complete depletion of Cu supplies in the liver $<$15 mg/kg dry matter and no Cu content below 6 mg Cu/kg in the cerebrum.

HELIOTROPE ALKALOIDS AND COPPER

John McC Howell, Harjit Patel, and Peter Dorling

School of Veterinary Studies
Murdoch University
Western Australia 6150

INTRODUCTION

The ingestion of toxic pyrrolizidine alkaloids by sheep may result in the development of a clinical syndrome in which distinctive damage to the liver plays an important part. The same can be said of the ingestion of copper (Cu). However the detail of the syndromes and the nature of the changes in the liver are different. In Australia the ingestion by sheep of pasture containing hepatotoxic pyrrolizidine alkaloids has resulted in outbreaks of disease known as "The Yellows" or "Haemolytic Jaundice" due to the importance of jaundice as a clinical sign (Bull et al. 1956). In these sheep the concentration of Cu in the liver was excessive and it was presumed that the damage caused by the hepatotoxic alkaloids was responsible for this. The hepatotoxic alkaloids and the Cu in the pasture were thought to be acting synergistically.

White et al. (1984) investigated the effects of feeding sheep pyrrolizidine alkaloid containing tansy ragworth (Senecio jacobaea) and supplemental Cu. All animals fed tansy ragwort with or without supplemental Cu died. They did not develop haemolysis, the lesions reported were those of pyrrolizidine alkaloid intoxication and there were no significant differences in concentration of Cu in the liver of the various treatment groups. These interesting findings merited further experimental investigation.

RESULTS AND CONCLUSIONS

The results reported here are of an investigation of possible synergism following ingestion by sheep of the pyrrolizidine alkaloid containing plant Heliotropium europaeum and Cu. Merino sheep were divided in to a control group of 4 and three groups of 5 each. All sheep were fed a cubed diet based on lucerne and oats. The cubes fed to 2 of the groups also contained dried Heliotrope to give an alkaloid content of 0.13%. The cubes without Heliotrope contained 6, 52 and 268 ppm (dry wt.) of Cu, zinc and iron respectively, those with Heliotrope contained 12, 68 and 372 ppm (dry wt.) of those elements. On 5 days of the week one group fed control cubes and one fed cubes containing Heliotrope were also given copper sulphate orally in a gelatin capsule at a dose rate of 20 mg Cu SO_4 $5H_2O$/ kg body wt./day.

The concentration of Cu in the liver and kidney of the four groups is shown in Table 1.

Table 1.   Concentration of copper in liver and kidney μg Cu/g dry matter.

	Control	Copper only		Heliotrope only	Copper and Heliotrope
Liver	824.42 ±68.91	1394.32 ±182.15	1286.01* ±189.06	941.47 ±102.93	2783.31 ±173.72
Kidney	20.24 ±1.92	131.59 ±78.96	58.76* ±39.36	18.61 ±1.21	321.49 ±28.28

Mean ± SE

*Results for 4 animals not in haemolysis.

Animals in the control group were normal throughout the experiment. Feeding Heliotrope alone did not significantly increase the amount of Cu in the tissues, it did not produce clinical signs but there was histological evidence of alkaloid induced liver damage. The administration of Cu alone led to a significant increase in Cu concentration in the tissues with $P<0.05$ for Cu in the liver of the whole group. All animals showed histological evidence of liver damage compatible with Cu loading (Howell and Gooneratne 1987). In the group which received Heliotrope plus Cu there was a significant increase of Cu in the tissues when compared to the sheep which received Cu alone with $P<0.001$ for liver. All animals in the Heliotrope plus Cu group showed clinical signs of toxicity. One was found dead with mild jaundice and the other 4 were killed. One was ill but was not jaundiced, one had shown marked jaundice for 3 days, and the remaining 2 were killed shortly after haemolysis developed. Histological examination showed marked liver damage.

These results indicate that when Heliotrope is fed alone there may not be a marked elevation of tissue Cu concentration, but when Heliotrope is fed together with excess Cu the concentration of Cu in the tissues is markedly elevated, liver damage is severe and clinical illness is produced.

ACKNOWLEDGEMENTS

This work was supported by the Australian Wool Corporation.

REFERENCES

Bull, L.B., Dick, A.T., Keast, J.C. and Edgar, G., 1956, An experimental investigation of the hepatotoxic and other effects on sheep of consumption of Heliotropium europaeum L : Heliotrope poisoning of sheep, Aust. J. Agric. Res., 7:281.
Howell, J. McC., and Gooneratne, S.R., 1987, The pathology of copper toxicity in animals, in: Copper in Animals and Man, J. McC. Howell and J.M. Gawthorne, eds. C.R.C. Press, Boca Raton.
White, R.D., Swick, R.A., and Cheeke, P.R., 1984, Effects of dietary copper and molybdenum on tansy ragwort (Senecio jacobaea) toxicity in sheep, Am. J. vet. Res., 45:159.

INTERRELATIONSHIPS BETWEEN MOLYBDENUM AND COPPER IN FEMALE RATS

Jau-Jiin Chen, Meiling T. Yang and Shiang P. Yang

Department of Food and Nutrition
Texas Tech University
Lubbock, TX   79409, USA

INTRODUCTION

The antagonism between Mo and Cu in ruminants has been extensively studied for many years. The relationship between these two trace minerals in nonruminants has not been well defined. Ljutakova et al. reported that injection of 5 mg Cu/kg body weight (bw) into rats increased hepatic Cu content and superoxide dismutase (SOD) activity 48 h after the injection[1]. The objective of this study was to examine the possibility of Mo interference on Cu utilization by simultaneous injections of moderate levels of Mo with the Cu injection in female rats.

METHODS

Three-week-old weanling female Sprague-Dawley rats were fed ad libitum AIN-76A containing 0.026 ppm Mo and 6.3 ppm Cu, and deionized drinking water until they reached an average body weight of 200 g. They were then randomly divided into 5 groups of 10 animals each. Group 1 was subcutaneously injected with saline as the control while groups 2 to 5 were simultaneously injected with a saline solution containing 5 mg Cu from cupric sulfate/kg bw, and a saline solution containing 0, 5, 10 or 50 mg Mo from sodium molybdate/kg bw, respectively. The animals were sacrificed 48 h after the injections. Hepatic xanthine dehydrogenase/oxidase (XDH)[2], sulfite oxidase (SOX)[3], and SOD[4], as well as plasma ceruloplasmin (CP)[5] activities were determined. The Mo and Cu contents in tissues were determined by polarograph[6] and flame atomic absorption, respectively.

RESULTS AND DISCUSSION

As shown in Table 1, XDH activity was significantly increased by the Cu injection. The simultaneous injections of Mo further elevated the enzyme activity, but there was no significant difference among the different levels of Mo. The Cu injection had no significant effect on SOX activity. The injection of 5 mg Mo/kg bw raised SOX activity to a significantly higher level than the control, but further increase of Mo lowered the SOX activity to the level of those receiving no Mo. XDH thus appeared to be more sensitive to both Cu and Mo treatments than SOX. Although 5 mg Cu/kg bw raised SOD activity, the increase was not significant. The significant elevation of SOD

Table 1. Effect of copper and molybdenum treatments on the activities of hepatic xanthine dehydrogenase/oxidase (XDH), sulfite oxidase (SOX), superoxide dismutase (SOD), and plasma ceruloplasmin (CP) of female rats [1,2]

Group No.	XDH[3]	SOX[4]	SOD[5]	CP[6]
1	$0.591 \pm .073^a$	$7.52 \pm .67^a$	$5.38 \pm .55^a$	$0.117 \pm .026^a$
2	$0.718 \pm .057^b$	$7.85 \pm .52^a$	$5.73 \pm .88^{a,b}$	$0.203 \pm .027^{b,c}$
3	$0.893 \pm .108^c$	$8.84 \pm .73^b$	$5.66 \pm .83^a$	$0.229 \pm .052^{b,c}$
4	$0.890 \pm .079^c$	$7.98 \pm .78^a$	$5.97 \pm .46^{a,b}$	$0.241 \pm .090^c$
5	$0.838 \pm .058^c$	$8.14 \pm .53^a$	$6.32 \pm .27^b$	$0.189 \pm .033^b$

[1] Values are mean $\pm$ SD of 10 animals.
[2] Means with different superscripts are significantly different ($P < .05$).
[3] μmoles uric acid formed/min/g liver.
[4] μmoles ferricyanide reduced/min/g liver.
[5] units SOD/mg liver.
[6] μmoles o-dianisidine dihydrochloride oxidized/min/ml.

activity over the control was observed only at 50 mg Mo/kg bw. Injection of Cu significantly increased plasma CP activities of all groups. Mo exhibited no antagonistic effect on Cu in SOD and CP. Mo contents in liver, spleen and kidney was not affected by the Cu injection, but increased with the Mo injection level (data not shown). Cu in liver or kidney was significantly elevated by the Cu injection, but was not affected by the simultaneous injection of Mo at all the levels. Blood and spleen Cu contents remained unaffected by Cu and Mo injections. The Cu injection lowered Mo in liver, kidney, spleen and blood although not significant; the simultaneous injection of Mo more than compensated for this effect.

In summary, under the present experimental conditions, Mo had no effect on Cu utilization, whereas Cu improved Mo utilization. (Supported by PHS grant No. 1 RO1 CA 39418 awarded by the National Cancer Institute, DHHS; AMAX, Inc.; and Texas Tech University Graduate School.)

REFERENCES

1. S. G. Ljutakova, E. M. Russanov, and S. I. Liochev, Copper increases superoxide dismutase activity in rat liver, Arch. Biochem. Biophys. 235:636 (1984).
2. K. H. Schosinsky, H. P. Lehmann, and M. F. Beeler, Measurement of ceruloplasmin from its oxidase activity in serum by use of o-dianisidine dihydrochloride, Clin. Chem. 20:1556 (1974).
3. F. Stirpe and E. D. Corte, The regulation of rat liver xanthine oxidase, J. Biol. Chem. 244:3855 (1969).
4. H. J. Cohen and I. Fridovich, Hepatic sulfite oxidase, J. Biol. Chem. 246:359 (1971).
5. H. P. Misra and I. Fridovich, Superoxide dismutase: a photochemical augmentation assay, Arch. Biochem. Biophys. 181:308 (1977).
6. C. C. Deng, N. S. Wang, and C. H. Chen, Study of catalytic polarography. III. Catalytic current of molybdenum-chlorate in mandelic acid or benzilic acid medium, Acta. Sci. Natl. Univ. Fudan 11:197 (1966).

EFFECTS OF TRANSSULFURATION DEFECTS ON SELENIUM STATUS

M.A. Beilstein,[a] W.A. Gahl[b] and P.D. Whanger[a]

[a]Dept. of Agric. Chemistry, Oregon State Univ., Corvallis,
OR 97331 and [b]Section on Human Biochemical Genetics, Human
Genetics Branch, Nat. Inst. of Child Health and Human De-
velopment, N.I.H., Bethesda, MD 20892

INTRODUCTION

The identification of selenomethionine (SeMET) as a major form of selenium (Se) in several plant protein foods[1] and the retention of SeMET, as
such, in animal tissues[2] suggests that SeMET may be a major form of dietary Se. All nutritionally available Se must be metabolized to selenocysteine (SeCYS), the form of Se in glutathione peroxidase (GPx),[3] but the
mechanism of SeCYS formation and introduction into GPx is unknown. The
transsulfuration (TS) pathway is a possible mechanism for SeCYS formation
from SeMET. SeMET and its metabolites are substrates for all steps of the
TS pathway in vitro,[4] but the in vivo importance of this pathway to SeMET
utilization is not known. If the TS pathway were important in the utilization of SeMET, impairment of TS activity due to metabolic defects might
be expected to result in lowered activity of GPx and accumulation of SeMET.

EXPERIMENTAL

Se status parameters were examined in blood samples from patients with
deficiencies of three TS pathway enzymes. Cystathione β-synthase (CBS)
and methionine adenosyltransferase (MAT) deficiency result in impaired TS
activity and physiological accumulation of methionine. Deficiency of
homocysteine remethylation (HRM) activity forces methionine sulfur through
the TS pathway and thus depletes physiological methionine.

Previously published methods were used to determine GPx activities[5]
and Se concentrations[6] of erythrocyte (RBCs) and plasma and concentrations
of RBC Se as GPx and hemoglobin (Hb)[1] as presented in Table 1. Plasma
GPx activities were high compared to controls in CBS deficient patients.
Se levels were low in RBCs and plasma of CBS and MAT deficient patients
and high in the HRM deficient patient and heterozygote. Concentrations
of RBC Se as GPx were similar to controls in all patients. Se as Hb was
low in RBCs of both the CBS and MAT deficient patients and high in the
HRM deficient patient and heterozygote.

DISCUSSION

Se in Hb probably occurs as SeMet substituted for methionine during
protein synthesis.[1,2] The alteration in Hb Se observed in these
patients is consistent with competitive inhibition of SeMET protein in-

TABLE 1. BLOOD SELENIUM STATUS PARAMETERS

	CONTROLS(n)	CBS-DEF(n)	MAT-DEF (n=1)	HRM-DEF (n=1)	HRM-DEF HETEROZYGOTE (n=1)
GPx^a					
PLASMA	4.28 ± 0.93(8)	6.45 ± 1.83*(4)	4.25	4.53	6.45†
RBC	44.2 ± 6.5(7)	48.5 ± 4.2 (6)	59.1†	36.2	47.9
Se^b					
PLASMA	2.04 ± 0.28(8)	1.53 ± 0.15*(6)	1.49	3.62†	2.65†
RBC	.882 ± .092(7)	.637 ± .087*(6)	.622	1.672†	1.140†
RBC Se^c					
as GPx	.199 ± .068(5)	.199 ± .055 (6)	.242	.203	.199
as Hb	.545 ± .060(5)	.361 ± .056*(6)	.305†	1.370†	.732†

^{a} - nmole NADPH oxidized/min/mg protein (plasma) or Hb (RBC).
^{b} - ng Se per mg protein (plasma) or Hb (RBC).
^{c} - ng Se in particular fraction per mg Hb in whole RBCs.
* - Different from controls, P < .05.
† - Difference from control average > 2 S.D.

corporation by altered physiological levels of methionine. The similarly altered plasma Se levels could arise by the same mechanism.

The hypothesis that deficiency of TS activity might impair utilization of SeMET for GPx synthesis is not supported by these results. The elevated plasma GPx activity of the CBS deficient patients suggests exactly the opposite, i.e., impaired TS activity results in more efficient utilization of Se. It is likely that a decrease in SeMET protein incorporation due to competition with elevated methionine levels results in greater catabolism to inorganic forms which are available for incorporation as SeCYS in GPx.

REFERENCES

1. M.A. Beilstein and P.D. Whanger, Deposition of dietary organic and inorganic selenium in rat erythrocyte proteins, J. Nutr. 116:1701 (1986).
2. M.A. Beilstein and P.D. Whanger, Chemical forms of selenium in rat tissues after administration of selenite or selenomethionine, J. Nutr. 116:1711 (1986).
3. R.J. Kraus, S.J. Foster and H.E. Ganther, Identification of selenocysteine in glutathione peroxidase by mass spectroscopy, Biochemistry 22:5853 (1983).
4. N. Esaki, T. Nakamura, H. Tanaka, T. Suzuki, Y. Morino and K. Soda, Enzymatic synthesis of selenocysteine in rat liver, Biochemistry 10:4492 (1981).
5. D.E. Paglia and W.N. Valentine, Studies on the quantitative and qualitative characterization of erythrocyte glutathione peroxidase, J. Lab. Clin. Med. 70:158 (1967).
6. M.W. Brown and J.H. Watkinson, An automated fluorimetric method for the determination of nanogram quantities of selenium, Analyt. Chim. Acta 89:29 (1977).

ANTIOXIDANT SUPPLEMENTATION FOR ELDERLY LIVING AT A NURSING HOME

A DOUBLE-BLIND RANDOMIZED ONE YEAR CLINICAL TRIAL

T. Westermarck* , M. Tolonen** , M Halme** , S. Sarna** ,
M. Keinonen***, and U.-R. Nordberg*

*Helsinki Central Inst. Mentally Retarded, Kirkkonummi;
Dept. Publ. Hlth.*Dept. Psychol., Univ. of Helsinki, Finland

INTRODUCTION

Therapy with selenium (Se) and other antioxidants may be warranted in [1,2] various degenerative diseases associated with increased lipid peroxidation, which also may contribute to degenerative changes of aging. Tolonen et al.[3] recently reported a reduction of several geriatric symptoms and an improvement of general condition among elderly after one year supplementation with selenium and vitamin E. The aim of the present study was to develop further the antioxidant supplementation for elderly.

MATERIAL AND METHODS

Seventy elderly living at an old people´s home were matched into verum and placebo groups. The mean age was 80.5 yrs (66-95). The daily supplementation consisted of ß- carotene (9mg), vitamin B6 (2mg), vitamin C (90mg), vitamin E (460mg), zinc (15mg), selenium (100 µg as organic Se and 800 µg as inorganic Se) i.e., Bioselen+Zink (R) capsules 1x1, Ido-E (R) pills 2x2, sodium selenate caps. 1x1. Blood selenium (B-Se) was measured using the hydride generation AAS; plasma α-tocopherol and retinol were simultaneously analyzed by HPLC. Erythrocyte glutathione peroxidase (GSH-PX) activity was measured according to a modified method of Beutler et al.[4] Serum lipid peroxide level was analyzed using thiobarbituric acid (TBA) reaction by a modified method of Yagi.[5] Psychological tests were used to study the mental functions: Wechsler memory scale (personal and actual information, time orientation), WAIS number series, clock recognition and clock drawing, logical learning (immediate and delayed feed-back), Luria 10 words learning test and other cognitive tests.

RESULTS AND DISCUSSION

Mean B-Se (µg/l) increased from an initially low value 84 (range 59-128) up to 245 (186-364) in the verum group. In the placebo group the corresponding values were 85 (57-131) and 162 (116-202), respectively. The mean E-GSH-PX activity (U/g Hb) increased from 17 (12-25) to 27 (18-40) in the verum group, and from 20 (10-32) to 28 (17-37) in the placebo group. The increase of Se and GSH-PX in the placebo group are attributed to Se enrichment of fertilizers in Finland (see Tolonen et al. in this volume).

The mean plasma α-tocopherol (μmol/l) increased from 14 (7-18) to 50 (6-67) in the verum group, in which two persons had not taken the vitamin pills; and from 15 (6-27) to 20 (11-37) in the placebo group. The mean plasma retinol content (1.7 μmol/l) did not change during the study. The lipid peroxide levels (expressed in terms of malondialdehyde μmol/l of serum) were initially high: $\bar{x}$= 2.7+0.7 (n=20) in the verum group, and $\bar{x}$=2.7+0.7 (n=25) in the placebo group. An ad hoc group of young healthy adults (n=31) had x= 2.3+0.6. The lipid peroxide level of the placebo group did not change, despite the elevated Se and α-tocopherol status whereas the lipid peroxide level of the treated group decreased significantly to $\bar{x}$=2.2+0.6 (n=20), i.e. down to the level of healthy young adults. Psychological tests indicated that the placebo group deteriorated or remained unchanged, whereas the verum group improved in many parameters. Several elderly elderly persons were not able to perform all the different test. Significant improvement was obtained for the verum group (n=10) in the clock drawing test for which the scores were $\bar{x}$= 6.5+6.2 in the beginning and x= 7.5+2.4 after supplementation. The corresponding values of the placebo group (n=12) were $\bar{x}$= 7.0+2.0 and 6.9+2.9, respectively. Clinical improvements were also reported by the nurses; no toxic side effects were observed.

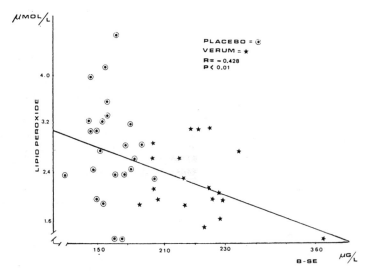

Fig. 1. Progression line of the serum lipid peroxide and whole blood selenium levels at the end of the study.

REFERENCES

1. J. M. C. Gutteridge, T. Westermarck, and B.Halliwell, Oxygen Radical Damage in Biological Systems, in: "Free Radicals, Aging, and Degenerative Diseases", J. E. Johnson, Jr, R. Walford, D. Harman, J. Miquel eds. , Alan R. Liss, New York (1985).
2. P. Santavuori, T. Westermarck, J.Rapola, M. Lappi, R. Moren, U. Vuonnala, Antioxidant treatment in Spielmeyer-Sjögren´s disease, Acta Neurol Scand. 71:136 (1985).
3. M. Tolonen, M. Halme, and S. Sarna, Vitamin E and Selenium Supplemen tation in Geriatric Patients, Biol. Tr. Elem. Res. 7:161 (1985)
4. E. Beutler, K. D. Blume, J. C. Kaplan, G. W. Löhr, B. Ramot, and W. N. Valentine, Recommended methods for red cell enzyme analysis, Br. J. Haematol. 35, 331 (1977).
5. K. Yagi, Assay for serum lipid peroxide level and its clinical significance in: "Lipid Peroxides in Biology and Medicine". Academic Press New York (1982).

EVIDENCE OF SELENIUM DEFICIENCY IN MICE FED A SOY-BASED DIET

Karin Ploetz and Edith Wallace

Biology Department
William Paterson College of New Jersey
Wayne, NJ 07470

The selenium (Se) status of humans and animals receiving total nutrition from single source diets has been investigated by measuring Se levels in blood, liver, nail or hair samples or by determining Se-dependent glutathione peroxidase in tissue, plasma, erythrocytes or platelets. Our objective was to determine if a soy-based formula (SB) contains adequate levels of Se using mouse spermatogenesis as a bioassay.

One group of Swiss Webster mice was fed SB alone (ProSobee, Mead Johnson, Evansville, IN) for three generations using the protocol of Wallace et al. (1983a). The SB fed to the control mice was supplemented with 0.1 ppm Se as sodium selenite. All animals were sacrificed at two months of age, weighed, and epididymal sperm were examined by phase microscopy. The testes were weighed and prepared along with the epididymides for light and transmission electron microscopy. Body and testes weights were compared by analysis of variance followed by Duncan's New Multiple Range Test.

The SB, with or without added Se, provided adequate nutrition for the mice as shown in Table I. Epididymal sperm from the second generation SB-alone mice were less motile and had more deformed midpieces than did sperm from the Se-supplemented animals. These defects in sperm motility and morphology were more common in third generation mice receiving unsupplemented SB. The testes of both second and third generation mice fed the SB alone weighed less (Table I) and contained seminiferous tubules smaller in diameter and with fewer maturing spermatids (Fig. 1A) than did the testes of mice fed the Se-supplemented SB (Fig. 1B). Transmission electron micrographs of midpieces from unsupplemented mice show mitochondria that are irregularly shaped and in disarray (Fig. 2A), in comparison with mitochondria from the controls (Fig. 2B).

While small seminiferous tubules with fewer maturing spermatids and reduced motility of morphologically aberrant epididymal sperm are not specific to Se deficiency they are consistent observations (Wu et al. 1979; Wallace et al. 1983a, 1987). The abnormalities of shape and arrangement of sperm mitochondria are characteristic of Se deficiency in mice and are not found in Se-supplemented mice (Wallace et al. 1983b). As this pattern was found in mice fed SB alone and not in their controls given added Se, we find that SB did not provide adequate levels of available Se to maintain normal mouse spermatogenesis. Further studies should be made of the Se status of infants receiving SB as their only source of nutrients for extended periods.

Table I.

	MEAN BODY WEIGHTS (GM)		MEAN TESTIS WEIGHTS (MG/PAIR)	
	Soy formula +0.1 ppm SE	Soy formula unsupplemented	Soy formula +0.1 ppm Se	Soy formula unsupplemented
First Generation	35.2 ± 0.6 (12)	34.3 ± 1.0 (9)	235.9 ± 9.0[a] (12)	229.6 ± 11.9[a] (9)
Second Generation	34.0 ± 0.9 (6)	34.4 ± 0.7 (7)	287.5 ± 12.7[b] (6)	221.1 ± 8.4[a] (7)
Third Generation	37.3 ± 1.2 (3)	34.8 ± 1.1 (11)	314.7 ± 14.8[b] (3)	110.1 ± 4.1[c] (11)

Mean body weights (±SEM) of the different groups were not significantly different. Mean testis weights with no common superscripts are significantly different (P<0.5). Number of animals per group is in ().

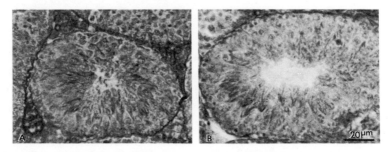

Fig. 1. Light micrographs of paraffin embedded iron hematoxylin-stained third generation mouse seminiferous tubules at stage I (Oakberg 1959). A: SB. B: SB+Se.

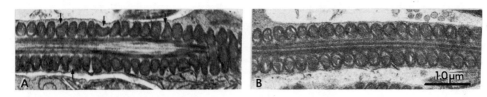

Fig. 2. Transmission electron micrographs of midpieces of epididymal sperm from third generation SB (A) or SB+Se (B) mice. Several irregularly shaped mitochondria are indicated by arrows.

REFERENCES

Oakberg, E., 1959, Am.J.Anat., 99:399.
Wallace, E., Calvin, H.I., and Cooper, G.W., 1983a, GameteRes., 4:377.
Wallace, E., Cooper, G.W., and Calvin, H.I., 1983b, GameteRes., 4:389.
Wallace, E., Calvin, H.I., Ploetz, K., and Cooper, G.W., 1987, in: "Selenium in Biology and Medicine," Coombs, G.F., Spallholz, J.E., Levander, O.A., Oldfield, J.E.eds., AVI, New York.
Wu, A.S.H., Oldfield, J.E., Schull, L.R., and Cheeke, P.R., 1979, Bio. Reprod., 20:793.

DETERMINATION OF SELENIUM AND MULTIPLE STABLE ISOTOPE ENRICHMENT OF

SELENIUM IN BIOLOGICAL MATERIALS BY INDUCTIVELY COUPLED PLASMA MASS

SPECTROMETRY

W.T. Buckley[1], D.V. Godfrey[1], K.M. Koenig[2] and J.A. Shelford[2]

[1]Agriculture Canada, Research Station, Agassiz, B.C. Canada, VOM 1AO and [2]Dept. of Animal Science, University of British Columbia, Vancouver, B.C. Canada, V6T 2A2

INTRODUCTION

Stable isotopes have many potential applications in the study of nutrition and metabolism of trace elements. Se stable isotope enrichment has been determined by neutron activation analysis (Janghorbani et al. 1981) and by gas chromatography/mass spectrometry (Reamer and Veillon 1983). A very low detection limit for Se when analyzed by inductively coupled plasma mass spectrometry (ICPMS) with sample introduction by hydride generation was recently reported (Powell et al. 1986). The object-ive of the present study was to determine the applicability of ICPMS to Se stable isotope analysis.

MATERIALS AND METHODS

The ICPMS system used was an ELAN 250 (Perkin-Elmer SCIEX, Norwalk, CT) with a continuous-flow hydride generator (Thompson and Walsh 1983). The phase separator of the hydride generator was replaced with a glass Y made of 6.2 mm (OD) tubing. The sample flow (1.0 ml/min) was mixed with the borohydride flow (0.8 ml/min; 0.26 M NaBH$_4$ in 0.1 M NaOH) and was introduced by means of a teflon tube (1.5 mm OD) into the stem of the Y. The liquid with a small percentage of the generated gas was pumped off the bottom of the Y and Ar gas (1.4-1.6 L/min) passing through the top of the Y carried hydrides to the ICP. The new phase separator was required because optimum injector Ar flow for ICPMS was found to be at higher pressure than could be achieved with the Thompson and Walsh (1983) phase separator.

Samples containing 8-600 ng Se in up to 2 g dry matter were pre-digested with 15-25 ml of 3% Mg (wt/vol) in concentrated nitric acid using 200-ml Berzelius beakers with watchglass covers. A final temperature of 115 C in the mixture was required to prevent subsequent loss of Se. The samples were taken to dryness and ashed at 500 C for 4 h. Five milliliters of water and then 15 ml concentrated HCl were added to dissolve the ash.

Stable isotope enrichment was expressed as 100 X the mass of tracer divided by the mass of tracee; i.e. tracer/tracee mass percentage (TTMP). Factors affecting the limitations of the analytical method were invest-igated.

Table 1. Detection limits of Se stable isotope enrichment determined by ICPMS[†]

Sample	TTMP$_{76}$	TTMP$_{77}$	TTMP$_{82}$
Selenite (400 ng Se)	0.19	0.15	0.24
Blood serum (172 ng Se)	0.61	0.34	0.42
Grass hay (16 ng Se)	2.2	0.60	0.80

[†]Triplicate determinations were done each day for 5 days for each sample type. Se in serum and orchard grass was 86 ng/ml and 8 ng/g. Detection limit is the TTMP above which the probability of obtaining a measurement of natural abundance Se is <0.05 for a single analysis of a sample.

RESULTS AND DISCUSSION

Signal intensity and signal noise were optimum with introduction of the sample/borohydride flow at the bottom of the stem of the phase separator. This position appeared to minimize the incorporation of a sample solution aerosol in the injector Ar flow. Se-80, the most abundant isotope of natural Se, could not be determined because of interference by a molecular ion of Ar. Se-76, Se-77, Se-78 and Se-82 ions could be determined. Se-74 was not investigated.

A final sample concentration of 9 M HCl prevented interference from up to 10 microg Cu/ml, which would exceed the Cu concentration of most biological samples. Vijan and Leung (1980) suggested the use of 7.5 M HCl to control chemical interference in Se analysis by hydride generation.

Accuracy of Se stable isotope enrichment was determined by regressing predicted (pred, independent variable) versus observed (obs) isotope enrichments. Preliminary results yielded the following regression equations for Se-76 and Se-77, respectively: obsTTMP = 1.03 X predTTMP - 0.133; obsTTMP = 0.96 X predTTMP - 0.002.

Enrichment of Se-77 was consistently determined with lower detection limit than the other two isotopes, and would appear to be the first isotope to chose for tracer kinetic procedures (Table 1). No statistically significant cross contamination, or "memory", for isotope enrichment was found. Stable isotope dilution analysis for Se using Se-76 as the internal standard provided quantitative analysis in good agreement with published values for reference materials from National Bureau of Standards, Washington, D.C. (citrus leaves 1572, bovine serum 8419, and bovine liver 1577) and International Atomic Energy Agency, Vienna (milk powder A-11).

Determination of Se stable isotope enrichment by ICPMS with sample introduction by hydride generation is a rapid, precise and accurate procedure suitable both for stable isotope dilution analysis of selenium and for tracer kinetic studies.

REFERENCES

Janghorbani, M., Ting, R.T.G., Young, V.R., 1981, Am. J. Clin. Nutr., 34:2816-2830.
Powell, M.J., Boomer, D.W., McVicars, R.J., 1986, Anal. Chem., 58:2867-2869.
Reamer, D.C., Veillon, C., 1983, J. Nutr., 113:786-792.
Thompson, M., Walsh, J.N., 1983, A Handbook of Inductively Coupled Plasma Spectrometry, pp. 149-170, Blackie & Son, Glasgow.
Vijan, P.N., Leung, D., 1980, Analy. Cnim. Acta, 120:141-146.

GLUTATHIONE PEROXIDASE, VITAMIN E AND POLYUNSATURATED FATTY ACID CONTENT

OF TISSUES OF PIGS WITH DIETETIC MICROANGIOPATHY

Desmond Rice and Seamus Kennedy

Veterinary Research Laboratories
Stormont
Northern Ireland, BT4 3SD

INTRODUCTION

Selenium and vitamin E deficiency and/or polyunsaturated fatty acid (PUFA) excess in pigs is associated with various diseases including; dietetic microangiopathy, hepatosis dietetica, myopathy, steatitis, still births and weak piglets. In Northern Ireland dietetic microangiopathy is the most frequent manifestation. This disease still occurs, albeit at a reduced incidence, in spite of at least 4 fold increases in dietary vitamin E and selenium in recent years.

Affected pigs die suddenly with myocardial haemorrhage and occlusion of myocardial capillaries with dense PAS-positive microthrombi. Since microthrombosis is not a unique response to selenium and vitamin E deficiency, this study was undertaken to determine whether current cases of dietetic microangiopathy are associated with tissue deficiencies of GSH-Px and/or vitamin E and/or PUFA excess. Preliminary results are shown below.

MATERIALS AND METHODS

A study was made of fattening pigs with a history of sudden death. Tissues were collected from 27 pigs with histologically confirmed dietetic microangiopathy and 27 control pigs which had died suddenly due to other causes. Feed being consumed by affected pigs was collected where possible.

Liver, kidney, heart were analysed for alpha tocopherol, GSH-Px and long chain fatty acids up to $C22:6;(n-3)$. Feed concentrations of vitamin E, selenium, oil and long chain fatty acids were also measured.

RESULTS AND DISCUSSION

All pigs were in good body condition. Pigs demonstrating accumulations of serofibrinous fluid in the pericardial sac, pleural and peritoneal cavities, fibrin clots, pulmonary oedema and congested livers, together with myocardial haemorrhage and multiple microthrombi of intramural cardiac arterioles and capillaries, were classified as dietetic micro-angiopathy positive.

Liver and heart concentrations of alpha tocopherol were lower in pigs with dietetic microangiopathy compared with controls (Table 1). This was most significant in the case of heart tissue. The correlation between heart and liver alpha tocopherol was higher in control (R^2 = 0.66) than in affected pigs (R^2 = 0.28) affected pigs. Vitamin E appears therefore to have an important pathogenetic role in the aetiology of the disease.

Table 1   Alpha tocopherol, selenium and GSH-Px concentrations of tissues of pigs with dietetic microangiopathy compared with controls

		Alpha tocopherol (μg/g)		Selenium (μg/g)	GSH-Px (IU/mg protein)
		Liver	Heart	Heart	Heart
Group*					
	Mean	2.02	2.63	0.25	216.10
A	SD	0.84	1.34	0.06	161.96
	N	25	23	22	20
	Mean	3.12	4.93	0.25	167.67
B	SD	1.78	2.29	0.03	76.76
	N	26	24	24	24
	t-test	****	*****	NS	NS

* A = Positive dietetic microangiopathy; B = Controls.

Selenium concentrations of heart and liver were similar in both groups of pigs. GSH-Px concentrations of heart tissue did not differ between groups although the concentration in kidneys was lower in pigs with microangiopathy. These do not appear therefore to be of primary aetiological importance.

The dietary intakes of alpha tocopherol, selenium and oil were not different between groups.

The tissue deficiency of vitamin E could be due to a genetic predisposition of individual pigs to maintain sub-optimum concentrations of alpha tocopherol in sub-cellular membranes in spite of high dietary intakes. This could be due to mal-absorption or inadequate transport mechanisms. Alternatively there may be an increased rate of lipid peroxidation in sub-cellular membranes increasing the anti-oxidant requirement. The study has not determined whether the microthrombosis is due to a vitamin E deficiency induced Prostacyclin I2/Thromboxane A2 imbalance as described by Gilbert et al., (1983) or to peroxidative damage to vascular endothelium leading to thrombosis.

Coronary vascular endothelium appears to be the target organ affected by deficiency in this disease. It would be of interest to examine the antioxidant status of tissue of other species including man where microthrombosis occurs.

ACKNOWLEDEGMENTS

We acknowledge financial assistance from Roche, Basel.

REFERENCES

Gilbert,V.A., Zebrowski, E.J., and Chan, A.C., 1983, Differential effects
    of megavitamin E on prostacyclin and thromboxane synthesis in
    streptozotoxin-induced diabetic rats.   Horm. metabol. Res, 15: 320.

IDENTIFICATION OF SELENIUM CONTAINING PROTEINS FROM SOYBEANS

April C. Mason, Shridhar K. Sathe,
Rosemary Rodibaugh, and Connie M. Weaver

Department of Foods and Nutrition
Purdue University
West Lafayette, IN 47907 USA

Abstract

Soybeans provide a high quality, economical source of vegetable protein. The selenium content of soybeans is variable, dependent on selenium levels in the soil where crops are produced. The majority of selenium in soybeans is contained in protein. This study identifies specific soybean proteins where selenium is associated. Hydroponically grown soybean plants were intrinsically labeled with sodium (^{75}Se) selenite added to the nutrient solution. Harvested mature soybeans were ground into a meal and cold acetone defatted. Samples of 100 mg meal were extracted with 1.0 ml 50 mM TRIS-HCl (pH 8.5) containing 0.1% β-ME at room temperature for 2.0 hrs with constant shaking, centrifuged in an Eppendorf centrifuge for 5 minutes, and the radioactivity of the supernatant determined. The supernatant was then subjected to preparative sodium dodecyl sulfate polyacrylamide gel electrophoresis (3 mm thick, 8-25% linear acrylamide gradient), briefly stained and destained, the gel cut into appropriate sections, and the radioactivity of each section determined. The major amount of recovered radioactivity was accounted for by the major storage proteins 7S and 11S (60%). The 11S proteins contained more radioactivity than the 7S proteins. A significant amount of radioactivity (26%) was also associated with polypeptides with molecular weights $\leq$ 30,000 daltons excluding the 11S protein. These data show that the major storage proteins of soybeans are a good source of selenium and have a potential as a source of selenium for human food purposes.

Materials and Methods

Hydroponic growth of soybean plants Soybeans (Glycine max L. Merr, 'Century') were germinated in pearlite and, after three weeks of growth, transferred to 2 liter pots containing aerated Hoagland-Arnon (1950) nutrient solution. Plants were dosed for five weeks with 10 μCi/pot/week of sodium ^{75}Se selenite (sp. act. 6.9mCi/mg Se). Plants were grown to maturity, harvested and maintained at room temperature.

Polyacrylamide gel electrophoresis of soybean protein Cold acetone defatted soy meal (100 mg) was extracted at room temperature for two hours with constant shaking with 1.0 ml 50 mM Tris–HCl pH 8.5, 0.1% β–mercaptoethanol. Extracts were centrifuged and the supernatant was added to 0.7 ml sodium dedecyl sulfate sample buffer containing β–mercaptoethanol and boiled for 5 minutes. Polyacrylamide gel electrophoresis was conducted at 8 mA constant current for 18 hours on a 3 mm thick 8–25% linear acrylamide gradient (Fling and Gregerson, 1986). After electrophoresis, gels were stained and destained briefly and cut into 13 slices from the running gel/stacking gel junction to the dye front marker. Each gel piece was analyzed for ^{75}Se content. The radioactivity of each gel piece was expressed as the percent of total recovered radioactivity.

Results and Discussion

Polyacrylamide gel electrophoresis of soybean proteins   Figure 1 shows the distribution of radioactivity in soybean proteins analyzed by gel electrophoresis. In the figure, major soybean proteins are labeled along a representative stained gel. Over 60% of the recovered radioactivity was accounted for by the 7S and 11S storage proteins. The 11S protein contained more ^{75}Se than the 7S protein. The 11S acidic subunits contained more radioactivity than the basic subunits. A significant amount of radioactivity (approximately 26%) was recovered in the area of the gel containing proteins $\leq$ 30,000 daltons. The trypsin inhibitor proteins would migrate in this region of the gel. These proteins contain very high levels of cysteine, but account for only a small percentage of the total soybean protein content.

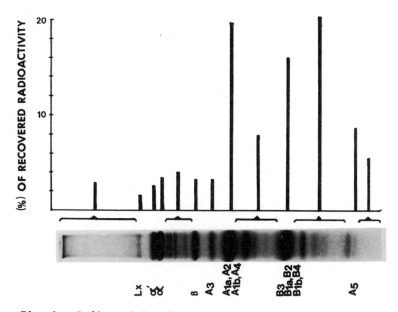

Fig. 1.   Radioactivity distribution in soybean proteins.

References

1. Hoagland, D.R., Arnon, D.I. California Agricultural Experiment Station Circular 347, 1950.

2. Fling, S.P., Gregerson, D.C. Analytical Biochemistry 155:83–88, 1986.

INFLUENCE OF DIETARY METHIONINE ON SELENOMETHIONINE METABOLISM IN RATS

Judy A. Butler and Philip D. Whanger

Department of Agricultural Chemistry
Oregon State University
Corvallis, OR 97331

INTRODUCTION

The sulfur analogs of selenium (Se) compounds have the greatest influence on Se metabolism. For example, methionine (MET) has a greater influence upon selenomethionine (SEM) than upon selenocysteine metabolism (1,2). Sulfur supplementation of Se-deficient animals can lessen the effects of Se deficiency (3), possibly by diverting Se compounds to glutathione peroxidase (GPX) precursors. MET supplementation has been shown to increase erythrocyte (RBC) GPX activity in humans (4) and plasma, liver and heart GPX activity in rats (5). The magnitude of difference in tissue Se deposition between selenite and SEM is dependent upon the level of these Se forms fed (6). The purpose of this study was to investigate the influence of different levels of MET on SEM metabolism as measured by Se deposition in tissues, GPX activity, Se excretion and percentage of Se associated with GPX tissues.

EXPERIMENTAL

Sixteen male weanling Sprague-Dawley rats were divided into 4 groups of 4 rats and fed diets for 8 weeks with 2.0 ppm Se as DL-SEM as the Se source and additions of either 0, 0.3, 0.6 or 1.2% DL-MET. The composition of the basal diet (in percent) was: torula yeast, 30; cerelose, 51.5; solka floc, 9.0; corn oil, 5.0; AIN-76 vitamins, 1.0; and AIN-76 minerals without Se, 3.5 (7). Three days before the rats were killed, urine and feces were collected and analyzed for Se. Se content and GPX activity were determined in plasma, RBCs, liver, kidney, muscle, testes, brain and lung. Se content only was determined in heart, spleen and whole blood. Se content for all tissues except heart and spleen was determined in both tissue supernatants and in whole tissues. Se content was determined by a semi-automated fluorimetric method (8) and the GPX by the coupled enzyme procedure with hydrogen peroxide as the substrate (9). Protein concentrations in the samples were determined with the Folin phenol reagent (10).

RESULTS

Rats fed the basal diet gained significantly ($P < 0.01$) less weight than did those fed the other diets. The sulfur amino acid content of this diet was calculated to be about 70% of that required by the growing rat (0.41% vs 0.60%). Dietary MET resulted in a decrease in Se deposition in all tissues examined. The greatest reduction of total tissue Se occurred in muscle (40%), kidneys (60%) and brain (47%) when 0.3% MET was compared to the basal diet. Additions of MET beyond 0.3% resulted in further Se depletion only in the whole blood, muscle, heart, spleen, and brain. In general, the same pattern was found for the tissue supernatant Se content but further Se depletion occurred with MET levels above 0.3% only in the muscle and kidneys. Dietary MET resulted in increased GPX activity in the testes, liver and lungs (1.27, 1.33 and 1.25-fold respectively) when 1.2% MET was compared to the basal diet, but other tissues were not significantly affected. MET at the 1.2% level resulted in a greater ($P <$ 0.05) excretion of Se in urine as compared to the basal diet but no differences were observed in fecal excretion. When the percentage of Se associated with GPX in the blood components and tissue cytosols was calculated, a positive correlation between this ratio and the MET level of the diet was found in most tissues, suggesting that dietary MET affects the amount of Se associated with GPX. The correlation coefficients between dietary MET and percent Se associated with GPX was greatest for kidney, lung and RBCs (0.90, 0.85 and 0.83 respectively). The sum of the Se content in all 8 tissues and whole blood were calculated for each dietary group. The total weight of tissues examined was 50% of the whole body weights and included tissues with the greatest expected Se stores. Thus, Se accumulation would be representative of that in the whole animal. Addition of 0.3% MET to the diet resulted in a marked (2.1-fold) decrease in calculated total body Se content and further additions of 0.6 and 1.2% resulted in a linear decrease of 1.2 and 1.4-fold respectively in body Se. In conclusion, MET was shown to decrease the deposition of Se in tissues when SEM was the Se source. However, MET resulted in a greater percentage of Se associated with GPX in various tissues, perhaps due to an increased diversion of SEM from protein incorporation to production of the GPX precursor, selenocysteine. This is supported by unpublished work from our lab showing a higher content of selenocysteine in tissues of ^{75}Se-SEM injected rats fed additional MET than in tissues of rats fed a diet with no added MET.

ACKNOWLEDGEMENTS

Supported by National Institutes of Health Public Health Service Research Grant NS 07413.

REFERENCES

1. Ganther, H.E., 1965, World Rev. Nutr. Diet, 5:338.
2. Beilstein, M.A. and Whanger, P.D., 1987, J. Inorgan. Biochem., 29:137.
3. Ullrey, D.E., Combs, G.F., Conrad, H.R., Hoekstra, W.G., Jenkins, K.J.W., Levander, O.A., and Whanger, P.D., 1983, Selenium in Nutrition, National Academy Press, Washington, D.C.
4. Luo, X., Wei, H., Yang, C., Xing, J., Liu, X., Qiao, C., Feng, Y., Lui, J., Lui, Y., Wu, Q., Liu, X., Guo, J., Stoecker, B.J., Spallholz, J.E., and Yang, S.P., 1987, in: Proceedings of the Third International Symposium Selenium in Biology and Medicine, AVI Book, Van Nostrand Reinhold Co., New York, N.Y.
5. Sunde, R.A., Gutzke, G.E., and Hoekstra, W.G., 1981, J. Nutr. 111:76.
6. Whanger, P.D. and Butler, J.A., 1987, Biol. Trace Elem. Res.,submitted.
7. AIN Ad Hoc Committee, 1977, J. Nutr., 107:1340.
8. Brown, M.W. and Watkinson, J.H., 1977, Anal. Chim. Acta, 89:29.
9. Paglia, D.E. and Valentine, W.N., 1967, J. Lab. Clin. Med., 70:158.
10. Lowry, O.H., Rosebrough, N.J., Farr, A.L., and Randall, R.J., 1951, J. Biol. Chem. 193:265.

PURIFICATION AND PROPERTIES OF THE HUMAN PLASMA SELENOENZYME, GLUTATHIONE PEROXIDASE

J.T. Deagen, D.J. Broderick, and P.D. Whanger

Department of Agricultural Chemistry
Oregon State University
Corvallis, OR 97331

INTRODUCTION

About 80% of the selenium (Se) present in ovine erythrocytes is associated with glutathione peroxidase (GPx).[1] In contrast, less than 4% of
the Se present in rat plasma was calculated to be present as GPx.[2] Consistent with this, evidence has been obtained in our laboratory that the
majority of the Se present in human plasma is not associated with GPx.[1]
Other workers have shown that the antigenicity of human erythrocyte GPx
differs from this selenoenzyme in plasma.[3] Therefore, the human plasma GPx
was purified to determine how its properties may differ from the erythrocyte enzyme.

EXPERIMENTAL

About 500 ml of blood was collected from the anticubital vein of 10
volunteers after a 12 hr fast. After centrifugation to obtain the plasma,
it was diluted to 20 mg protein per ml with 0.1 M phosphate buffer (pH
7.2). This was made 33% saturation with ammonium sulfate, centrifuged,
and the precipitate dissolved in buffer. After dialysis, this was chromatographed on Sephadex G-150. The appropriate fractions were pooled, concentrated and chromatographed on DEAE Sephacel. The fractions were again
combined, concentrated and applied to a chromatofocusing column. Finally,
the fractions containing GPx were chromatographed on Sephadex G-75. Slab
gel electrophoresis indicated that the preparation had been purified to
homogeneity.

RESULTS AND DISCUSSION

A summary of the purification of human plasma GPx is shown in Table 1.
Five steps were used to obtain the enzyme in pure form. It was purified
about 5,400 fold from fresh plasma with a 32% recovery of activity. Determination of Se on purified enzyme revealed a content of 3.8 g atoms per
mole GPx which is very similar to the human erythrocyte enzyme.[4] Slab
gel electrophoresis using SDS with standard proteins revealed a molecular
weight of about 23,000 for the subunits, indicating a molecular weight of
about 92,000 for the native enzyme. Amino acid analyses of the purified
GPx indicated aspartate, glutamate, proline, glycine, alanine and leucine
as the predominant amino acids, with cysteine, methionine, tryptophan and
histidine as the minor ones.[5]

Two peaks of Se are obtained when human plasma is subjected to gel
filtration using Sephadex G-150, but GPx activity does not cochromatograph

Table 1. Purification of human plasma GPx.

Step	Total Activity Units	Specific Activity (nm/min/mg)	Purification Fold	%Activity Yield
Diluted plasma	112,767	5	0	100
Ammonium sulfate	86,009	25	5	76
Sephadex G-150	80,336	143	27	71
DEAE Sephacel	58,610	9,073	1,731	52
Chromato-focusing	49,355	16,928	3,230	44
Sephadex G-75	36,366	28,106	5,364	32

with either of these peaks.[1] About 60% of the plasma Se is associated with the first peak and about 35% with the second peak. The remaining 5% of the Se is associated with GPx. The chemical form of the Se in the first peak has been shown to be selenocysteine, which is the same as that in GPx. The first Se peak is suspected to be similar to the P-protein,[6] but the nature of the second one is unknown.

In conclusion, the low amount of Se associated with GPx in comparison to the total plasma Se content is not due to a lower Se content in comparison to GPxs isolated from other sources, but instead due to low levels of GPx itself in plasma. The present results indicate that plasma GPx contains about 4 g atoms Se per mole enzyme, which is similar to that in GPxs from other sources.

REFERENCES

1. M.A. Beilstein, and P.D. Whanger. Distribution of selenium and glutathione peroxidase in blood fractions from humans, rhesus and squirrel monkeys, rats and sheep. J. Nutr. 113:2138 (1983).
2. J.T. Deagen, J.A. Butler, M.A. Beilstein and P.D. Whanger. Effects of dietary selenite, selenocystine and selenomethionine on selenocysteine lyase and glutathione peroxidase activities and on selenium levels in rat tissues. J. Nutr. 117:91 (1987).
3. K. Takahashi and H.J. Cohen. Selenium-dependent glutathione peroxidase protein and activity: Immunological investigations on cellular and plasma enzymes. Blood. 68:640 (1986).
4. Y.C. Awasthi, E. Beutler and S.K. Srivastava. Purification and properties of human erythrocyte glutathione peroxidase. J. Biol. Chem. 250:5144 (1975).
5. D.J. Broderick, J.T. Deagen and P.D. Whanger. Properties of glutathione peroxidase isolated from human plasma. J. Inorgan. Biochem. (in press), 1987.
6. R.F. Burk and P.E. Gregory. Some characteristics of ^{75}Se-P, a selenoprotein found in rat liver and plasma, and comparison of it with selenoglutathione peroxidase. Arch. Biochem. Biophys. 213:73 (1982).

EFFECT OF A SHIPMENT OF HIGH-SELENIUM WHEAT ON SELENIUM STATUS OF OTAGO

(N.Z.) RESIDENTS

M.F. Robinson and C.D. Thomson

Department of Nutrition
University of Otago
Dunedin, New Zealand

INTRODUCTION

Otago is a unique low selenium (Se) area of New Zealand where residents have low blood Se levels, unlike the North Island where Se status is affected by the periodic importation of Australian wheat (1). Because of poor wheat harvests in 1984, a shipment of Australian wheat was imported into Dunedin, and was used in wheat products such as bread and flour from September 1984 till March 1985.

METHODS

Bread and other wheat-based products were sampled from supermarkets in Dunedin, Otago during December 1984 to February 1985, and also during the 1970s, 1985 and 1986. Dietary intakes of Se were calculated using values for Se concentrations in bread and other foods sampled in Dunedin during the 1980s and food intake data obtained from the NZ National Diet Survey 1977 (J.A. Birkbeck).

Se concentrations were determined in blood samples taken from the following groups of Otago people: blood donors (1972, n=29; 1977, n=85; 1986, n=40); Milton residents (1978, n=230; 1985, n=20); Se supplementation study, control (1984-85, n=27); Exercise study (1985, n=27). Glutathione peroxidase (GSHPx) activities were assayed in the later studies.

Table 1. Se concentrations in wheat products sampled in Dunedin 1973-1986

Food	1970s	Dec 1984 – Feb 1985	Apr 1985	May 1985	1986
Whole wheat	15*	24	8		
White flour	14	44	10 - 12	10 ± 3	
Wholemeal flour			10	13 - 17	
Wheatgerm	25	91	26 - 39		
White bread	9 ± 2	29 ± 1	9 - 12	9 ± 2	10 - 12
Wholemeal flour	11 ± 3	44 ± 8	27 ± 7	14 - 20	20 ± 4

ng/g fresh weight ; mean ± sd.
*range given if n < 3

341

RESULTS

Se concentrations of bread and wheat products were raised two to four-fold during December 1984 to February 1985 and had fallen to the previously low levels by May 1985 (Table 1). The higher-Se bread contributed an extra 2-4 µg Se to the daily dietary intake, approximately 10% of the normal in-takes of about 23 µg Se and 32 µg Se for female and male Otago residents.

Blood Se concentrations of Otago blood donors were similar at all three times of sampling. However Se concentrations and GSHPx activities were raised in Milton residents and in the Exercise study subjects in 1985 for several months after the end of the Australian wheat blend. At the end of 1985, the GSHPx activity of the Exercise study subjects had fallen from 21 ± 5 to 19 ± 5 units/g Hb.

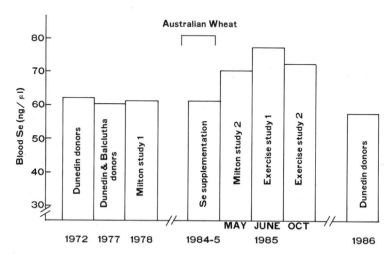

Figure 1. Blood Se concentrations of Otago residents 1972-1986

DISCUSSION

The increase in Se intake due to 6-month supplementation with high-Se Australian wheat influenced Se status as measured by blood Se levels and GSHPx activities. The calculated increase of only 10% of the normal intake was probably an underestimate as no account was taken of the use of flour in home-baking. The imported wheat was not used in the commercial manufacture of biscuits.

With the lifting of controls on wheat imports into New Zealand future importation is likely. This may have a lasting influence on the Se status of Otago residents as it has done in the North Island of New Zealand and in Finland (1,2). Therefore we shall continue to monitor Se status of Otago residents.

ACKNOWLEDGEMENTS

This work was supported by the Medical Research Council of New Zealand.

REFERENCES

1. J.H. Watkinson, Changes in blood selenium in New Zealand adults with time and importation of Australian wheat, Am. J. Clin. Nutr., 34:936 (1981).
2. P. Koivistoinen, and J.K. Huttunen, Selenium in food and nutrition in Finland, an overview on research and action, Ann. Clin. Res., 18:13 (1986).

TRANSFER OF SELENIUM FROM DIFFERENT DIETARY SOURCES INTO PLASMA

AND MILK OF DAIRY COW

Pentti Aspila and Liisa Syrjälä-Qvist

Department of Animal Husbandry
University of Helsinki
00710 Helsinki, Finland

INTRODUCTION

The results of several studies point to differences between the metabolism of inorganic and organic selenium sources (Jenkins and Hidiroglou, 1971; Fuss and Godwin, 1975). Seleno amino acids are more easily utilized and excreted into milk as compared to inorganic seleno compounds. Selenium in plant material is mostly in the form of seleno amino acids (Olson et al., 1970). The purpose of this experiment was to investigate the possibility of increasing the selenium content of cow milk.

MATERIALS AND METHODS

The experiment was carried out on 51 dairy cows according to a 2x2 factorial model. The factors were sodium selenite and grass silage sprayed with sodium selenite a week before cutting. The groups were thus: 0: no selenium supplementation; 1: sodium selenite; 2: selenited grass silage; and 3: both sodium selenite and selenited grass silage. Sodium selenite was given mixed in concentrates and selenited grass silage as only roughage. The experiment lasted 77 weeks and consisted of six experimental periods each lasting 8 to 17 weeks. Periods I and VI were standardization periods and there was no selenium supplementation during these periods. Feed consumption and milk yield were recorded daily. Blood samples from Vena jugularis were taken at the end of each period. Selenium content in feeds, milk, plasma and whole blood and GSH-Px activity in whole blood and plasma were analysed.

RESULTS AND DISCUSSION

Plasma selenium content reached its maximum concentration, 0.1 mg/1, at the dietary selenium level of 0.45 mg/kg DM. Selenited silage increased plasma selenium content more rapidly than did selenite, but there were no significant differences in the final plasma selenium concentration. Plasma selenium concentration also decreased less rapidly with decreasing dietary selenium content when selenited silage was the selenium source than when sodium selenite was the source.

Selenited silage increased milk selenium concentration more compared to sodium selenite (P < 0.001, table 1). Plasma selenium transfer into milk (Y, percent of plasma selenium in daily milk) was enhanced for selenited silage when dietary (P < 0.05) or plasma (P < 0.001) selenium levels increased but was decreased for selenite (P < 0.001) with increasing plasma selenium concentration. The transfer equation from plasma (X, mg Se/1) to milk was for selenite 0.284 - 1.717 X and for selenited silage 0.151 + 1.011 X. The transfer equations (Y, mg Se/1 milk) from diet (X, mg Se/kg DM) to milk were 0.007 + 0.0160 X for selenite and 0.011 + 0.0265 X for selenited silage. With higher dietary selenium levels (X, mg/kg DM) the transfer coefficients (Y, percent of daily selenium intake in milk) decreased (P < 0.001) both for selenite (Y = 0.132 - 0.155 X) and for selenited silage (Y = 0.157 - 0.109 X).

Selenited silage increased whole blood selenium concentration more than sodium selenite supplementation (P < 0.05). During the period IV when selenium intake was highest, differences between selenium supplementation groups were not significant (P > 0.05) indicating that the whole blood selenium plateau was reached above a dietary selenium level of 1 mg/kg DM. The correlation between milk selenium and protein content was 0.226 (P < 0.01). There were no differences between these two sources on the GSH-Px activity in either plasma or whole blood.

Table 1. Selenium content in the diet (mg/kg DM) and milk (mg/1)

Period	Group 0 Diet	Group 0 Milk	Group 1 Diet	Group 1 Milk	Group 2 Diet	Group 2 Milk	Group 3 Diet	Group 3 Milk
I	0.20	0.006	0.31	0.008	0.21	0.007	0.21	0.007
II	0.07	0.008	0.17	0.011	0.20	0.021	0.29	0.025
III	0.08	0.008	0.42	0.016	0.45	0.029	0.78	0.034
IV	0.06	0.008	0.68	0.020	1.20	0.040	1.81	0.051
V	0.03	0.008	0.11	0.011	0.09	0.023	0.17	0.027
VI	0.04	0.005	0.04	0.005	0.04	0.005	0.04	0.006

The results prove that there are differences in metabolism between inorganic and organic selenium sources. Selenium is mostly excreted in milk bound to proteins. The major precursors in milk protein synthesis are amino acids (Mather and Keenan, 1983). Following absorption in the small intestine selenium is metabolized mostly by erythrocytes and liver cells. The capacity of these tissues to metabolize selenium is limited, however. Thus rising plasma selenium concentration results in higher free seleno amino acid level when seleno amino acids are the source of dietary selenium. Seleno amino acids can be easily taken up by the mammary gland (Fuss and Godwin, 1975) resulting in a higher percentage of plasma selenium excreted into the milk. Selenite on the other hand needs to be metabolized before its excretion into milk however the capacity of tissues to metabolize selenite into compounds plasma selenium is excreted into the milk.

REFERENCES

Fuss, C.N. and Godwin, K.O., 1975, A comparison of the uptake of 75-Se-selenite, 75-Se-selenomethionine and 35-S-methionine by tissues of ewes and lambs. Austr. J. Biol. Sci. 28:239-249.
Jenkins, K.J. and Hidiroglou, M., 1971, Transmission of selenium as selenite and selenomethionine from ewe to lamb via milk using selenium-75. Can. J. Anim. Sci. 51:389-403.
Mather, I.H. and Keenan, T.W., 1983, Function of endomembranes and the cell surface in the secretion of organic milk constituents. In: Biochemistry of Lactation (ed. Mepham, T.B.). Elsevier, Amsterdam.
Olson, O.E., Novacek, E.J., Whitehead, E.I. and Palmer, I.S., 1970, Investigations on selenium in wheat. Phytochemistry 9:1181-1188.

ERYTHROCYTE AND PLASMA GLUTATHIONE PEROXIDASE: TWO DISTINCT SELENIUM

CONTAINING PROTEINS

Harvey Cohen, Nelly Avissar, Kazuhiko Takahashi, and Peter Allen

Departments of Pediatrics and Microbiology and The Cancer Center
University of Rochester Medical Center
Rochester, New York 14642

Selenium deficiency can develop in patients receiving intravenous hyperalimentation. This depletion can result in a decrease in both cellular and plasma glutathione peroxidase. Previous investigations have revealed that with repletion of selenium, there is a rapid increase in the recovery of plasma glutathione peroxidase and a more slow recovery in red blood cell glutathione peroxidase. This suggests that the plasma enzyme is not present as a result of hemolysis of red blood cells.

We were able to investigate the early and late time course for repletion of glutathione peroxidase in both plasma and red blood cells in patients who had become selenium deficient. Five patients that had been receiving home hyperalimentation treatment, as a result of various disorders, were given selenium in the form of selenious acid (400 mcg/day) when the plasma or red blood cell glutathione peroxidase was found to be < 15% of normal. We measured plasma glutathione peroxidase activity prior to, and every six hours after, the institution of selenium supplementation for one day and then daily for 5 days. There were statistically significant increases in plasma glutathione peroxidase within 6 hours after instituting selenium replacement. Recovery of red blood cell glutathione peroxidase was delayed for 2-3 weeks prior to demonstrable increase in activity and was not complete for up to 3-4 months. Thus it appeared that both enzyme activities were unrelated to each other.

We purified both red blood cell and plasma glutathione peroxidase to homogeneity using multiple chromatographic techniques. Each protein was found to have a molecular weight of approximately 100,000 on Sephadex G200. The mobility of the enzymes in polyacrylamide gel electrophoresis however, was different. On sodium dodecyl sulfate gel electrophoresis, the plasma enzyme was found to move more slowly. The apparent subunit molecular weight for the red blood cell enzyme was found to be 22,000 and for the plasma enzyme 23,000. The specific activity for each enzyme was also different. The pure plasma enzyme had a specific activity of approximately 25 units/mg. Heating the red blood cell enzyme to 60° for 10 minutes resulted in approximately 90% loss of enzyme activity, whereas the same type of treatment of the plasma enzyme had no effect on activity. Thus, there appeared to be both structural and functional differences between the red blood cell and plasma enzymes.

Purified proteins were injected into rabbits in the presence of Freund's complete adjuvant, and after repeated injections, the IgG fraction from the rabbit serum was found to precipitate, but not inhibit, the respective enzyme activities. Antibodies against the red blood cell enzyme also precipitated the neutrophil, platelet, and liver selenium-dependent glutathione peroxidase activity. Of note, is that the antibodies against red blood cell enzyme did not precipitate nor inhibit plasma glutathione peroxidase activity. Antibodies against plasma glutathione peroxidase protein did not inhibit nor precipitate the red blood cell enzyme. We were able to demonstrate that the antibodies against plasma glutathione peroxidase precipitated >90% of the plasma enzyme activity, indicating that plasma glutathione peroxidase activity is due to the protein we purified.

Red blood cell hemolysates from selenium deficient patients were unable to prevent the precipitation of red blood cell glutathione peroxidase by anti-red blood cell glutathione peroxidase antibodies. Similarly, plasma samples from selenium deficient patients were unable to inhibit the precipitation of plasma glutathione peroxidase. This data indicate that in the absence of selenium, neither the red cell nor plasma glutathione peroxidase protein is present. The mechanism by which the absence of selenium results in a decrease in the proteins of glutathione peroxidase from at least two sources is currently under investigation.

Because many plasma proteins exist as glycoproteins, we sought to determine whether the plasma glutathione peroxidase was a glycoprotein using ^{125}I-Con A. We found that the plasma, but not the red blood cell enzyme, binds to Con A and is therefore a glycoprotein. Treatment of the plasma enzyme with glycopeptidase resulted in a marked diminution of Con A binding, but no change in apparent subunit molecular weight.

The data presented herein are consistent with two structurally and functionally distinct selenium-dependent glutathione peroxidase proteins. One protein appears to be present within cells and has a high specific activity. The other protein appears to be present only in plasma and has a low specific activity. The roles that plasma glutathione peroxidase plays in maintaining glutathione homeostasis or protection against oxidative stresses need to be determined. The source for the plasma enzyme is also currently being intensively investigated, since knowing its source may give us important information concerning its function.

GSH-Px ACTIVITY IN THE BLOOD COMPARED WITH SELENIUM CONCENTRATION IN TISSUES

AFTER SUPPLEMENTATION OF INORGANIC OR ORGANIC SELENIUM

E. Johansson[*], S-O Jacobsson[**], and U. Lindh[*]

* Gustaf Werner Institute, Box 535, S-751 21 Uppsala
** Dep. of cattle and sheep diseases, Swedish University of
Agriculture, S-750 07 Uppsala, Sweden

The availability of selenium for sheep is dependent on its chemical form Jacobsson (1966). Compounds differ in their nutritional value and are stored in different forms in tissues and organs. This is important for animals in which a balanced intake of selenium is desired. This report gives the selenium content in tissues and the GSH-Px activity of the blood in lambs given a low supplementary dose of either sodium selenite or selenium yeast.

MATERIAL AND METHODS

Two groups (each containing 3 rams) of 10 Swedish, Landrace lambs aged 5 months were studied. Each lamb was fed for 3.5 months on a diet of 1.2 - 1.5 kg hay and 0.3 kg of a commercial mixture of vitamins and minerals. To the diet of each lamb, 0.100 mg Se/day was added either as sodium selenite for Group I and as selenium yeast for Group II. Each lamb recieved 150 ug Se/day.

Analytical methods

Samples of liver, pancreas, thyroid and the extensor carpi radialis muscle after collection were frozen at -80°C awaiting selenium analysis and after wet digestion the analysis was done by atomic absorption spectrophotometry. The blood GSH-Px was estimated by the method of Paglia & Valentine (1967) using cumenehydroperoxide as a substrate.

RESULTS AND DISCUSSION

Selenium uptake from sodium selenite and selenium yeast as indicated by B-GSH-Px activity.
Fig. 1 Total B-GSH-Px activity (selenium dependent enzyme and S-transferases) at different sampling times.

The increased activity was about the same in both groups and consistent with the findings in the liver by Moksnes and Norheim (1982): they also found plateauing which we did not. The liver contains GSH-Px and S-transferases but whole blood is thought to have less S-transferases, hence selenium may be metabolised differently in liver than in peripheral blood cells because of these differences. Selenium yeast was used in our study but Moksnes used dl-selenomethionine: it is not known how liver cells deal with selenium from d-selenomethionine though selenium yeast is not fully characterised it is thought to consist of about 50% l-selenomethionine, the residue being selenocystin and unknown selenium compounds. The life of erythrocytes in lambs of 6 months is 64 days. Sufficient time for two exchanges of erythrocytes occurred during the supplementary selenium period. It is not possible to show differences of B-GSH-Px when feeding low doses of sodium selenite and selenium yeast.

After 3.5 months of supplementary selenium, the liver in Group II is higher than found by Moksnes et al. who used d,1-selenomethionine which may be metabolised differently from selenium yeast. In both groups selenium in the pancreas and thyroid is higher than in the muscle and may indicate unknown functions of selenium in these organs.

Distribution of selenium in liver, pancreas, thyroid and muscle and the relation to the chemical form

The increased selenium in these tissues after supplementary Se was not significant (Wilcoxon rank-sum test). Thyroid specimens were obtained from 7 lambs in Group I, 9 in Group II and in all 10 of both groups for liver, pancreas and muscle.

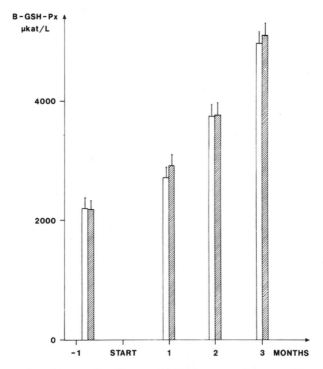

BLOOD GSH-Px ACTIVITY OF SHEEP SUPPLEMENTED SODIUM SELENITE OR SELENIUM YEAST

Figure 1 Group I sodium selenite, unfilled bar and group II shaded bar selenium yeast and the standard error (SE).

B-GSH-Px activity	group I	group II
Initial	2182 ukat/L	2199 ukat/L
3.5 months later	4966 "	5110 "

The increase in body weight did not vary much in either group.

Mean value is given with standard error. Sodium selenite and selenium yeast ended in about the same value. Low doses of these compounds may be used without discernable effects on the weight. In ruminants the chemical form of selenium intermediates. The nutritional value may differ because the different chemical forms in tissues are not known.

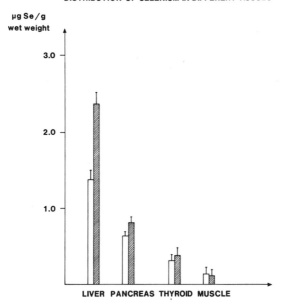

DISTRIBUTION OF SELENIUM IN DIFFERENT TISSUES

Fig 2 Each bar represents mean, given with bars of standard error (SE). Group I unfilled bars, group II shaded bars.

ACKNOWLEDGEMENT

The supply and help of selenium yeast from Pharmacia AB and Cell life is gratefully acknowledged.

REFERENCES

Jacobsson S-O., Metabolism of selenium, in sheep and mice studied with ^{75}Se-sodium selenite, ^{75}Se-selenomethionine and ^{75}Se-selenocystine, Thesis, Royal Veterinary College, Stockholm, 1966.
Moksnes K., Selenium supplementation in animal production in Norway, Thesis, National Veterinary Institute, Oslo, 1986.
Paglia D.E., and Valentine W.N., 1967, Studies on the quantitative and qualitative characterization of erythrocyte glutathione peroxidase, J.Lab.clin.Med. 70:991.

SELENIUM AND GLUTATHIONE PEROXIDASE STATUS IN TERM INFANTS FED HUMAN MILK,

WHEY-PREDOMINANT, OR SOY-BASED INFANT FORMULA.

RE Litov, VS Sickles, GM Chan, IR Hargett, and A Cordano

Mead Johnson Nutritional Group, Evansville, IN, University of
Utah Medical Center, Salt Lake City, UT, and Welborn Baptist
Hospital, Evansville, IN.

Selenium (Se) is an essential mineral for humans. Although a number of
selenoproteins have been isolated, only the metabolic function of glutathione peroxidase (GSH-Px) is understood.

Commercially available infant formulas in the USA contain Se provided
as an intrinsic part of the ingredients, primarily as protein, and are not
supplemented with additional Se. Since infants may depend on infant formulas
for a substantial part of their daily nutrient intake or as the sole item of
diet for long periods of time, it is important to determine if infant formulas with intrinsic Se and with intrinsic Se plus added Se can maintain adequate Se status in formula-fed infants relative to human milk-fed infants.

MATERIALS AND METHODS

Two studies were conducted in healthy term infants whose mothers had
elected to breast-feed or formula-feed. Institutional Review Board approval
and parental informed consent were obtained. The first study had a human
milk group and two formula-fed groups randomized to receive one of two whey-
predominant formulas. The whey-predominant formula contained 13 µg Se/L as
intrinsic Se and the whey-predominant with added sodium selenite contained a
total of 34 µg Se/L (16 µg Se/L of added Se plus 18 µg Se/L as intrinsic Se
determined by difference). The second study had a human milk group and two
formula-fed groups randomized to receive one of two soy-based formulas. The
soy-based formula contained 15 µg Se/L as intrinsic Se and the soy-based
with added sodium selenite formula contained a total of 19 µg Se/L (12 µg
Se/L of added Se plus 7 µg Se/L as intrinsic Se determined by difference).
Seven to sixteen infants were enrolled in each of the feeding groups. Blood
was drawn within one week of birth and at intervals up to 180 days of age.
Plasma, RBC, human milk, and infant formula Se were analyzed by electrothermal atomic absorption spectroscopy or by the modified procedure of Olson, et
al (1). Plasma and RBC GSH-Px activity were measured by a modified method of
Paglia and Valentine (2) using t-butyl hydrogen peroxide as the substrate.

RESULTS AND DISCUSSION

In the first study, mean levels of plasma and RBC Se ranged from 85 to
90 and 240 to 300 ng Se/ml, respectively. GSH-Px activity in plasma and RBC

ranged from 0.34 to 0.67 and 7.6 to 8.7 nmoles NADPH/min/mg protein, respectively. The means of the four blood Se indices were statistically equivalent among all three groups at birth and at 2 months of age and within each group at birth and at 2 months of age ($p \leq 0.05$).

Mean blood Se from various studies reported in the literature for USA subjects range from 78 to 157 ng Se/ml plasma and 73 to 200 ng Se/ml RBC (3). In the first study, the adequacy of Se status in infants at birth and 2 months in all three treatment groups is demonstrated by mean levels of plasma Se (85 to 90 ng Se/ml) within these published ranges and mean levels of RBC Se (240 to 300 ng Se/ml) which are higher than these published ranges. Both whey-predominant infant formulas and the human milk feedings maintained the same blood Se and GSH-Rx activity levels in the infants from birth to 2 months of age.

In the second study, mean levels of plasma and RBC Se ranged from 93 to 97 and 339 to 370 ng Se/ml, respectively and were statistically equivalent among all three groups at birth ($p \leq 0.05$). The rate of change from birth to 180 days of age was +0.031 to -0.040 and +0.019 to -0.088 µg Se/day for plasma and RBC Se, respectively. The rate of change of GSH=Px activity ranged from -0.004 to -0.009 and +0.046 to -0.0007 nmoles NADPH/min/mg protein/day for plasma and RBC, respectively. The rate of change for the four Se status indices was not significantly different among the three groups, except for the rate of change of RBC Se from birth to 180 days which was significantly higher ($p \leq 0.05$) in the human milk group (+0.019 µg Se/day) than in the soy-based with added sodium selenite group (-0.088 µg Se/day).

The concentration (mean $\pm$ S.D.) of human milk Se in the first study was 23 $\pm$ 4 µg Se/L (N=12) and in the second study was 25 $\pm$ 5 µg Se/L (N=10). These values are in the upper end of the 7 to 33 µg Se/L range (mean 18 µg Se/L) of human milk Se reported in a survey of 241 subjects from 17 states across the USA (4).

In conclusion, infants fed a whey-predominant infant formula containing at least 13 µg Se/L, --intrinsic Se alone or in combination with added sodium selenite, maintained their Se status from birth to two months of age and had Se status equivalent to infants fed human milk. Infants fed a soy-based infant formula containing at least 15 µg Se/L, --intrinsic Se alone or in combination with added sodium selenite, from birth to 180 days of age had Se status comparable to infants fed human milk. The Se concentration of human milk from two geographical locations, Evansville, IN and Salt Lake City, UT, were within the range of a geographical survey of human milk Se (4).

REFERENCES

1. O.E. Olson, I.S. Palmer, and E.E. Cary. Modification of the official fluorometric method for selenium in plants. J. Am. Oil Chem. Soc. 58:117 (1975).

2. D.E. Paglia and W.N. Valentine. Studies on the quantitative and qualitative characterization of erythrocyte glutathione peroxidase. J. Lab. Clin. Med. 70:158 (1967).

3. G.F. Combs, Jr. and S.B. Combs. In: The role of Selenium in Nutrition. Academic Press, Inc. p. 331 (1986).

4. T.R. Shearer and D.M. Hadjimarkos. Geographic distribution of selenium in human milk. Arch. Environ. Health 30:230 (1975).

INFLUENCE OF CYSTEINE ON MUCOSAL UPTAKE OF 75-Se-SELENITE

BY SHEEP JEJUNUM

Siegfried Wolffram, Regula Würmli, and Erwin Scharrer

Institute of Veterinary Physiology
University of Zürich, Winterthurerstr. 260
CH-8057 Zürich, Switzerland

INTRODUCTION

The inorganic selenium (Se) salt selenite is used for Se-supplementation of feedstuffs in livestock production. However, knowledge about the intestinal absorption of Se from selenite is very limited. Unlike selenate which is actively transported across the intestinal brush border membrane by the sulphate carriers[1], selenite seems to be absorbed by simple diffusion[2]. The goal of the present study was to investigate the influence of the amino acid L-cysteine (Cys) on selenite uptake by sheep jejunal mucosa since selenite is spontaneously reduced in the presence of SH-groups, resulting in the formation of Se-trisulfides[3].

MATERIALS AND METHODS

The midjejunum of 5 sheep (mean body weight: 24.5 kg) was surgically removed. Pieces of cleaned jejunum were mounted on plastic stoppers so that only the mucosal surface was exposed to the incubation medium. The in vitro method for measurement of short-term (3 min) mucosal substrate uptake was described in detail elsewhere[4]. The incubation media consisted of Krebs-Henseleit bicarbonate buffer containing 0.01 mmol/l 75-Se-labeled selenite with or without further additions. The results are presented as means with the standard error of mean ($\bar{x} \pm$ SEM).

RESULTS

In the presence of 1.0 mmol/l Cys uptake of Se from selenite was significantly stimulated compared with the uptake under control conditions (Tab.1). This stimulatory effect of Cys was Na^+-dependent (Tab.1). Lowering the pH of the incubation medium from 7.4 to 5.0 abolished the stimulatory effect of Cys on the uptake of Se from selenite whereas in the absence of Cys uptake of Se was not affected (109.9 ± 9.9 and 109.9 ± 17.7 pmol/cm^2·3min at pH 7.4 and 5.0, respectively). Furthermore, the amino acids L-alanine and L-lysine but not L-glutamic acid significantly reduced Cys-stimulated uptake of Se from selenite (uptake, % of control: 64.3 ± 5.6 with alanine, 62.0 ± 3.6 with lysine and 98.2 ± 7.3 with glutamic acid).

Table 1. Na$^+$-Dependence of L-Cysteine-stimulated Uptake
of Selenium from selenite (0.01 mmol/l)

Incubation condition	Se-Uptake, pmol/cm^2·3min[a]	
	with Na$^+$	without Na$^+$
controls	134.5 ± 10.1	108.2 ± 7.5
+ 1.0 mmol/l L-cysteine	824.3 ± 53.7[b]	77.1 ± 7.4

[a]Values are means ± SEM from 8-20 preparations.
[b]Significant different (p < 0.05) from the other values.

Some preparations were preincubated for 3 min in a Se-free medium containing 1.0 or 10.0 mmol/l Cys. The subsequent incubation in Cys-free medium containing 0.01 mmol/l selenite resulted in a clear stimulation of Se-uptake by the preparations preincubated with 10.0 but not by those preincubated with 1.0 mmol/l Cys (pmol/cm^2·3min: 101.2 ± 12.8 for controls, 138.5 ± 10.3 for 1.0 mmol/l Cys, 882.9 ± 66.8 for 10.0 mmol/l Cys).

DISCUSSION

The results presented clearly show that uptake of Se from selenite is strongly stimulated by Cys at pH 7.4. This stimulation is Na$^+$-dependent and does not occur at pH 5.0. Furthermore the amino acids L-alanine and L-lysine significantly reduced uptake of Se from selenite in the presence of Cys.

These findings can be explained by the extracellular formation of the Se-trisulfide Se-dicysteine from selenite and Cys. Se-dicysteine may be transported across the intestinal brush border membrane by the Na$^+$-dependent transport system for cationic amino acids which is also shared by various neutral amino acids[5]. Furthermore, intracellular formation of Se-dicysteine may contribute to the stimulatory effect of L-cysteine on the uptake of Se from selenite by maintaining the concentration gradient for diffusive uptake of selenite.

REFERENCES

1. F. Ardüser, S. Wolffram, and E. Scharrer, Active absorption of selenate by rat ileum, J. Nutr. 115:1203 (1985).
2. S. Wolffram, B. Grenacher, and E. Scharrer, Transport of selenate and sulphate across the intestinal brush border membrane of pig jejunum by two common mechanisms, Quart. J. Exp. Physiol. in press (1987).
3. H.E. Ganther, Selenotrisulfides. Formation by the reaction of thiols with selenious acid, Biochem. 7:2898 (1968).
4. E. Scharrer, W. Raab, W. Tiemeyer, and B. Amann, Active absorption of hypoxanthine by lamb jejunum in vitro, Pflügers Arch. 391:41 (1981).
5. S. Wolffram, H. Giering, and E. Scharrer, Na$^+$-gradient dependence of basic amino acid transport into rat intestinal brush border membrane vesicles, Comp. Biochem. Physiol. 78A:475 (1984).

DISTRIBUTION OF SELENIUM AND GLUTATHIONE PEROXIDASE IN PLASMA FROM HEALTHY SUBJECTS AND PATIENTS WITH RHEUMATOID ARTHRITIS

Mats Borglund[1], Anita Åkesson[2] and Björn Åkesson[1]

Departments of Clinical Chemistry[1] and
Rheumatology[2], University Hospital, University of
Lund, Lund, Sweden

INTRODUCTION

A low plasma selenium level has been demonstrated among patients with rheumatoid arthritis (1) or with abnormal plasma protein patterns (2), though neither the mechanism of this anomaly nor its possible functional significance is known. A low plasma selenium is also a risk indicator for cancer death (3). In tissues, selenium is covalently bound mainly in selenocysteine moieties in proteins, and in human tissue glutathione peroxidase is the only selenoprotein hitherto identified. It is important to identify other seleno-compounds, and this communication presents some data on the distribution of selenium and glutathione peroxidase in plasma from patients with rheumatoid arthritis and healthy controls.

METHODS

Plasma was obtained from healthy women or women with active rheumatoid arthritis (45-66 years old). It was subjected to gel filtration on Sephadex G-150 at 4°, and absorbance at 280 nm, selenium and glutathione peroxidase activity were determined in the eluted fractions.

RESULTS AND DISCUSSION

Gel chromatography of healthy plasma resulted in one predominant selenium peak, which eluted close to the second of the three major protein peaks, containing mainly IgG. The apparent molecular weight for the selenium peak was 174 ± 11 kDa in plasma from healthy subjects. Negligible amounts of the recovered selenium were found in components with molecular weights of under 30 kDa. The main peak of glutathione peroxidase activity had an apparent molecular weight of 99 ± 2 kDa, and it eluted between the IgG peak and the albumin peak, but there was also an incipient second peak among the high molecular weight proteins. No consistent correspondence could be seen between selenium content and enzyme activity, indicating that only a small proportion of selenium in plasma was bound to glutathione peroxidase (probably less than 10%). The elution

Table 1. Distribution of selenium in plasma of healthy subjects and patients with rheumatoid arthritis. Data are expressed as ng selenium in four parts of the eluate. Means (SD) from 5-7 subjects are given. The significance of intergroup differences was calculated with the Mann-Whitney U-test.

Part of eluate	Rheumatoid arthritis	Controls	Significance of difference
I, 280 kDa	20.8(4.3)	21.1(6.0)	n.s.
II, 280-125 kDa	83.9(23.9)	108.8(15.8)	$p < 0.02$
III, 125- 80 kDa	52.0(13.5)	48.6(5.8)	n.s.
IV, 80- 40 kDa	39.9(7.7)	40.6(7.2)	n.s.

pattern of selenium was also compared to that of IgG, IgA, ceruloplasmin and transferrin. The main selenium peak eluted between IgG and ceruloplasmin. No significant correspondence was evident between selenium content and any of the proteins analysed.

Plasma selenium concentrations were lower in patients with rheumatoid arthritis than in controls, and the distribution of selenium among plasma proteins differed from that in healthy subjects (Table 1). The chromatograms were divided in four parts, and the proportion of selenium migrating in the IgG region (part II) was significantly lower ($p < 0.01$) in patients with rheumatoid arthritis than in controls, whereas the proportion in the glutathione peroxidase region (part III) was significantly higher in the patients ($p < 0.01$). Hence, for patients with rheumatoid arthritis also the total amount of selenium in the IgG region was significantly lower ($p < 0.02$) for patients with rheumatoid arthritis (Table 1), but not in the other three parts of the chromatogram. The lack of correlation between the levels of selenium and IgG, indicated that most of the selenium in the major selenium peak was not bound to IgG.

(The study was supported by the Swedish Medical Research Council (project no. 3968) and the Swedish Council for Planning and Coordination of Research.)

REFERENCES

1. U. Johansson, S. Portinsson, A. Åkesson, H. Svantesson, P. A. Öckerman, and B. Åkesson, Nutritional status in girls with juvenile chronic arthritis, Hum Nutr Clin Nutr 40C:57 (1986).
2. B. Åkesson, Plasma selenium in patients with abnormal plasma protein patterns, Trace Elem Med (in press).
3. G. Fex, B. Pettersson, and B. Åkesson, Low plasma selenium as a risk factor for cancer death in middle-aged men, Nutr Cancer (in press).

PLASMA SELENIUM AND GLUTATHIONE PEROXIDASE IN RELATION TO CANCER, ANGINA PECTORIS AND SHORT-TERM MORTALITY IN 68 YEAR OLD MEN

Björn Åkesson[1] and Bertil Steen[2]

Department of Clinical Chemistry[1], University Hospital, Lund, and Department of Community Health Sciences[2], Värnhem Hospital, Malmö, Sweden

The relation of selenium status to the prevalence of a variety of diseases is under debate. Some case-control studies have indicated that subjects, who develop cancer (1) or cardiovascular diseases, have lower prediagnostic levels of plasma selenium. In addition, studies in animals on selenium-depleted diets indicate that such diets can give rise to increased occurrence of some forms of experimentally induced tumors.

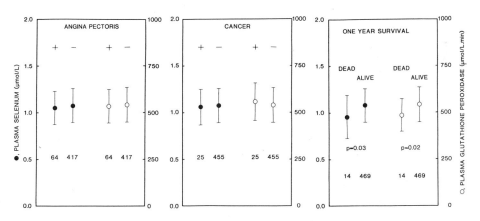

Fig. 1. Plasma selenium and glutathione peroxidase activity in 68 year-old men with angina pectoris or cancer, and in men dying during the first year after screening.

In some epidemiological studies in this field, the populations studied have been heterogenous with respect to age, sex and place of living of the probands. In the population study 'Men born in 1914' going on in Malmö, a representative sample of the population has been chosen, as described elsewhere (2,3,4). This report summarizes data on the relation of selenium status to the prevalence of angina pectoris and cancer, and to short-term mortality.

In 483 68 year-old men, plasma selenium was 1.08 ± 0.18 umol/L and plasma glutathione peroxidase activity 540 ± 90 umol/L·min (mean$\pm$SD). Plasma selenium was correlated to glutathione peroxidase (r=0.49, p < 0.001), but not significantly to plasma albumin levels (r=0.18). In subjects, who reported angina pectoris or cancer in a questionnaire, plasma selenium and glutathione peroxidase were not significantly different from the values for subjects without these diseases (Fig. 1). Plasma selenium and glutathione peroxidase in subjects, who died within one year after screening, was 89% and 88%, respectively, of the values among survivors (Fig. 1). For subjects dying during the second year after screening, plasma selenium and glutathione peroxidase were not significantly different from the values among survivors. Since the relation of low plasma selenium and glutathione peroxidase activity to mortality was observed only for the first year, this was probably due to a deranged nutritional status or abnormal plasma protein pattern (5) prior to death, rather than a long-term negative influence of low selenium status on the relative risk of mortality.

REFERENCES

1. G. Fex, B. Pettersson, and B. Åkesson, Low plasma selenium as a risk factor for cancer death in middle-aged men, Nutr Cancer (in press).
2. L. Janzon, B. S. Hanson, S. O. Isacsson, S. E. Lindell, and B. Steen, Factors influencing participation in health surveys. Results from the prospective population study 'Men born in 1914' in Malmö, Sweden, J Epid Comm Health 40:174 (1986).
3. B. Åkesson, and B. Steen, Plasma selenium and glutathione peroxidase in relation to cancer, angina pectoris and short-term mortality in 68 year-old men, Contemp Gerontol (in press).
4. B. Åkesson, B. Bengtsson, and B. Steen, Are lens opacities related to plasma selenium and glutathione peroxidase in man? Exp Eye Res (in press).
5. B. Åkesson, Plasma selenium in patients with abnormal plasma protein patterns, Trace Elem Med (in press).

BIOLOGICAL FUNCTIONS OF SILICON, SELENIUM, AND GLUTATHIONE PEROXIDASE (GSH-Px) EXPLAINED IN TERMS OF SEMICONDUCTION

J. Parantainen, S. Sankari, and F. Atroshi

Huhtamäki Oy Pharmaceuticals, Clin. Research. P.O.Box
325, 00101 Helsinki; Depts Biochem., Pharmacol., Toxicol.,
Coll. Vet. Med. P.O.Box 6, Helsinki, Finland

INTRODUCTION

Semiconduction makes the basis for control, amplification, and transformation of electric currents in electronics. In this paper it is suggested that functions of some common trace elements in a living system might also be understood in terms of semiconduction. Semiconduction was introduced in biology by Albert Szent-Györgyi in the forties but the idea was generally rejected. Since that time the knowledge has increased fundamentally. Proteins do conduct electricity at least in small distances and certain biopolymers and polycyclic aromatic compounds reveal semiconductor properties.

PRINCIPLE OF SEMICONDUCTION

A given insulating material can be made conductive by braking the electronic structure with an element having different valence number. When "doped" in the structure, trace amounts of these "valence impurities" cause dramatic changes in conductivity by generating extra electrons (-) or electron holes (+). In electronics the amounts of these "impurities" are held constant. In biology the concentrations of trace elements vary allowing a possible way to control the currents.

BIOLOGY OF SILICON

Silicon in an essential element but its biological roles are not known. Semiconduction-like functions may be suggested for silicon-containing material in some cases. When introduced into lungs, silicate polymerizes forming needle-like aggregates (with P, Fe, Al) having a marked catalytic surface and a strong negative charge. Oxygen in reduced to superoxide and singlet radicals (1). In diatoms (plankton) silicon deficiency leads to an unexplained drop in photosynthesis, photophosphorylation, and use of phosphate energy (2). Silicon is also needed in bone formation, accumulating as "hot spots" in the metabolically active osteoid cells (3). We have observed (4) a decline in blood and milk silicon during bovine mastitis (figure 1).

SILICON AND SELENIUM IN LIPID PEROXIDATION

Selenium, another semiconductor material of electronics, is a trace element needed in the action of glutathione peroxidase (GSH-Px). This selonoenzyme controls the formation oxygen free radicals during lipid peroxidation. In essence this means one-electron step in the reduction of the radical to water. As silicon-containing structures (like phospho-silicates in the lung) may perform the two early stages in oxygen reduction and selenium the later, the actions of silicon and selenium could be complementary (figure 2). In bovine mastitis the erythrocyte GSH-Px and GSH declined while their content in the inflammatory tissue increased (5). We suggested that the changes may be related to tissue defence against oxygen free radicals, and an increase in the production of E-type prostaglandins in the inflammatory tissue.

POSSIBLE WIDER APPLICATIONS

Besides silicon and selenium the list of potential "doping impurities" to be added in organic structures seems almost indefinite: Au, Al, Ag, As, B, Be, Ca, Cu, Fe, Hg, K, Li, Mg, Mn, Na, Ni, P, Pb, S, Sb, Sn, Te, and Zn. Semiconduction may have some connections to pharmacology and toxicology: electrolytes that regulate conductivity by definition, toxicity of heavy metals, silicate, asbestos etc., therapeutic action of lithium, gold, zinc etc., as well as "the bioelectric theory of cancer" presented by Szent-Györgyi (6) are possible examples.

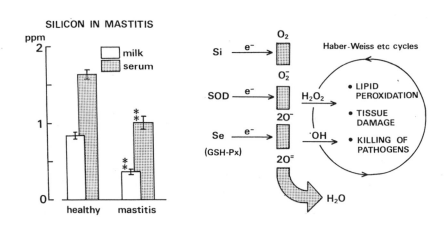

REFERENCES

1. Weiss A., In: Biochemistry of silicon and related problems. Eds G Bendz and I Lindqvist (Nobel symposium 40) pp 297-307. Plenum, New York (1978).
2. Coombs J., Spanis C., Volcani B. E., Plant Physiol 42:1607-11 (1967).
3. Carlisle E.M., Fed Proc 33:1758-66 (1974).
4. Parantainen J, Tenhunen E., Kangasniemi R., Sankari S.,Vet.Res.Commun. (in press, 1987).
5. Atroshi F., Parantainen J., Kangasniemi R., Sankari S., J.Anim. Physiol. Anim. Nutr. (In press, 1987).
6. Holden C., Science 203:522-4 (1979).

INFLUENCE OF GLUTATHIONE DEPLETION ON HEPATIC COPPER UPTAKE AND

BILIARY COPPER EXCRETION IN THE RAT

H. Nederbragt

Department of Pathology, Faculty of Veterinary Science
State University, Utrecht, The Netherlands

Although it was demonstrated[1] that glutathione (GSH) depletion reduced biliary copper (Cu) excretion in rats, the relation between GSH and Cu excretion has not been fully investigated. The purpose of this study was to contribute to the understanding of the process of biliary Cu excretion and the role of GSH in it.

Male inbred WAG/Rij rats were used because they show a very effective excretion of Cu after intravenous Cu administration[2].Their bile duct was canulated under nembutal-anaesthesia. Thirty minutes before injection of Cu (10 or 30 µg/100 g body weight) they were pretreated with diethylmaleate (DEM; 0.25 ml/100 g bw) or phorone (25 mg/100 g bw), dissolved in corn oil and administered intraperitoneally. Control rats received corn oil only. Bile samples of 30 min. were collected during 4h after Cu injection. In another experiment DEM, phorone and Cu were administered similarly to conscious rats which were killed after 0, 1 and 2h after Cu administration for determination of Cu in plasma and liver.

DEM decreased liver GSH from a control value of 6.0 µmol/g to 1.6 µmol/g (i.e. 27%) whereas phorone reduced liver GSH from 6.4 to 2.6 µmol/g (53%).

The pattern of biliary Cu excretion in control rats was similar to that described before[2] and could be resolved into a slow and a rapid component (SCuE and RCuE respectively) when a dose of 30 µg/100 g bw had been given; after a dose of 10 µg only the SCuE was observed.

The effects of DEM and phorone on SCuE and RCuE can be summarized as follows:
1. DEM had no effect on the excretion of Cu after a dose of 10 µg/100 g bw.
2. Phorone reduced the excretion of Cu after a 10 µg Cu dose to 50%.
3. Both DEM and phorone abolished the RCuE, i.e. the peak of Cu excretion emerging directly after injection of 30 µg/100 g bw, this peak being superimposed on SCuE.
4. After this 30 µg Cu dose DEM retarded the rise in Cu excretion to levels of normal SCuE with 2 hours.
5. After this same 30 µg Cu dose phorone inhibited the rise in Cu excretion; under this condition SCuE was reduced to 50% of normal SCuE levels during the total duration of the experiment.

Thus, although DEM depleted GSH levels more than phorone, phorone substantially reduced the SCuE, which was hardly affected by DEM. Both substances abolished RCuE.

Since reduced Cu uptake by the liver may explain the decrease in biliary Cu excretion, two experiments were performed to investigate this possi-

Table 1. Effect of DEM and phorone pretreatment on liver and plasma Cu concentrations after a iv Cu dose of 30 µg/100 g bw.

time (h)	liver Cu,µg/g control	DEM	plasma Cu,µg/ml control	DEM	liver Cu,µg/g control	phorone	plasma Cu,µg/ml control	phorone
0	4.56[a] ±0.14	4.57 ±0.21	1.20 ±0.03	1.20 ±0.07	5.16 ±0.13	5.15 ±0.10	1.11 ±0.05	1.11 ±0.04
1	10.15 ±0.30	8.67[+] ±0.23	1.19 ±0.01	1.98[++] ±0.10	10.34 ±0.05	9.26[+] ±0.37	1.13 ±0.02	1.99[++] ±0.14
2	10.29 ±0.43	11.02 ±0.49	1.19 ±0.01	1.41[++] ±0.03	10.48 ±0.20	10.04 ±0.08	1.07 ±0.02	1.63[++] ±0.04

a. values ± SEM. +: p<0.05. ++: p<0.01

bility. In the first experiment Cu in liver and plasma was determined after a dose of 30 µg/100 g bw in DEM- and phorone-pretreated rats. The results are given in Table 1 and show that both DEM and phorone retarded the uptake of Cu by the liver and the release of Cu from the plasma. In the second experiment rats were again canulated and DEM and phorone were administered but now after the administration of 30 µg/100 g bw. An interval of 30 min. was chosen since most of the injected Cu has been taken up by the liver in that time period. Again SCuE was affected by both treatments. DEM reduced Cu excretion temporarily after which it returned to the control SCuE level; phorone reduced Cu excretion to less than 50% of the control level during the total canulation period. This indicates that DEM and phorone affect Cu excretion in the bile, also when Cu has been taken up by the liver.

It seems unlikely that the effects of DEM and phorone on biliary Cu excretion should be explained by depletion of GSH: phorone, which had a limited effect on liver GSH concentration was much more effective in reducing Cu excretion.

The inhibition of Cu uptake by the liver by the GSH-depleting agents is an interesting finding that warrants further investigation.

References

1. J.Alexander and J.Aaseth, Biliary excretion of copper and zinc in the rat as influenced by diethylmaleate, selenite and diethyldithiocarbamate, Biochem.Pharmacol.29:2129 (1980).
2. H.Nederbragt and A.J.Lagerwerf, Strain-related patterns of biliary excretion and hepatic distribution of copper in the rat, Hepatology 6:601 (1986).

EFFECT OF ZINC DEFICIENCY ON LIPID AND PROTEIN PROFILES OF THE RAT

ERYTHROCYTE MEMBRANE

Gary L. Johanning, Daniel S. Miller, and Boyd L. O'Dell

Department of Biochemistry, 322 Chemistry Building
University of Missouri, Columbia, Missouri 65211

INTRODUCTION AND METHODS

Numerous studies indicate that zinc plays a role in preserving the structural integrity of cell membranes[1]. The osmotic fragility of rat erythrocytes is increased by dietary zinc deficiency[2], and the purpose of this study was to determine the effect of zinc deficiency on the protein and phospholipid (PL) fatty acid (FA) compositions of the rat erythrocyte membrane as they relate to fragility.

Male rats (90–120 g) were caged individually and fed for four weeks a low-zinc (<1 ppm Zn) or a zinc-adequate (100 ppm) control diet[2] supplied both ad libitum (AL) and pair fed (PF). A fourth group was fed the low-zinc diet AL for four weeks followed by a three day repletion period. Whole blood was collected and one aliquot was tested for osmotic fragility[2], while another was used to prepare plasma membranes[3]. PL were resolved from the lipid fraction of membranes by thin-layer chromatography, and the FA composition of the PL classes, as well as the cholesterol content of the neutral lipid fraction, was determined by gas-liquid chromatography. Erythrocyte membranes were also subjected to SDS-polyacrylamide slab gel electrophoresis (16.5 % gels).

RESULTS AND DISCUSSION

Plasma zinc, rate of gain and erythrocyte fragility were altered by zinc deficiency (Table 1), but upon zinc repletion for three days, these indicators of zinc status returned to normal. SDS-polyacrylamide gel electrophoresis revealed no alterations in the major protein components due to dietary zinc deficiency, but the fatty acid composition of the phosphatidylcholine (PC) and phosphatidylethanolamine (PE) fractions was changed by zinc deficiency. In the PC fraction (Table 2), zinc-deficient rat erythrocyte membranes had higher proportions of 18:0 and 18:2 ω6 and lower proportions of 16:0 and 18:1 ω9 than PF controls. The only alteration due to zinc deficiency observed in the PE fraction was a decrease in the 22:4 ω6 level from 6.56 to 5.70 % ($P<0.05$). The higher proportion of 18:2 ω 6 and lower proportion of 20:4 ω6 in the PF compared to the AL group indicates that food restriction may inhibit the desaturation of linoleic acid. The alterations in membrane FA composition that accompanied zinc deficiency were not reversed by the three-day zinc repletion, whereas the increased red cell fragility was returned to AL control values.

Table 1.   Zinc Status, Red Cell Fragility and Membrane Lipids*

	Low Zn	Low Zn–Repleted	PF Adequate Zn	AL Adequate Zn
Weight gain (g/day)	0.29 ± 0.05^a	$6.80\pm0.26^{b,+}$	0.73 ± 0.14^a	5.04 ± 0.19^c
Plasma Zn (ug/ml)	0.26 ± 0.03^a	1.29 ± 0.03^b	1.13 ± 0.04^c	$1.21\pm0.04^{b,c}$
Hemolysis (%) in 0.38 % NaCl	83.5 ± 3.3^a	56.9 ± 5.4^b	74.7 ± 3.8^a	56.3 ± 4.1^b
Total PL (umol/ mg protein)	0.77 ± 0.04	-------	0.84 ± 0.05	0.83 ± 0.03
Cholesterol (umol/mg)	1.21 ± 0.07	1.30 ± 0.08	1.25 ± 0.06	1.14 ± 0.10
Cholesterol/PL	1.62 ± 0.13^a	-------	$1.51\pm0.11^{a,b}$	1.29 ± 0.08^b

*Values are means ± SEM for eight or more animals. Values within each row with the same letter are not significantly different (P<0.05).
+Rate of gain during three day repletion period only.

Table 2.   Fatty Acid Composition of PC from Erythrocyte Membranes*

Fatty Acid	Percent Fatty Acid Composition			
	Low Zn	Low Zn–Repleted	PF Adequate Zn	AL Adequate Zn
16:0	38.08 ± 1.34^a	$39.12\pm0.56^{a,c}$	$41.07\pm0.53^{b,c}$	41.80 ± 0.67^b
18:0	20.56 ± 0.55^a	$19.50\pm0.31^{a,c}$	$19.24\pm0.33^{b,c}$	18.38 ± 0.18^b
18:1	9.25 ± 0.16^a	9.49 ± 0.14^a	10.10 ± 0.22^b	10.04 ± 0.16^b
18:2	18.66 ± 0.24^a	18.44 ± 0.30^a	17.60 ± 0.22^b	15.64 ± 0.17^c
20:4	$13.19\pm0.54^{a,b}$	$13.54\pm0.19^{a,b}$	12.24 ± 0.35^a	14.37 ± 0.50^b
22:6	1.30 ± 0.21^a	0.77 ± 0.06^b	0.81 ± 0.06^b	0.86 ± 0.13^b

*Values are means ± SEM for five or more animals.

The zinc repletion results suggest that the altered FA composition of the membrane is not responsible for the increased fragility of erythrocytes in zinc-deficient rats.

REFERENCES

1. W. J. Bettger and B. L. O'Dell, Life Sciences 28:1425 (1981).
2. B. L. O'Dell, J. D. Browning, and P. G. Reeves, in: "Trace Elements in Man and Animals," C. F. Mills, I. Bremner, and J. K. Chesters, eds., pp 79–83, Commonwealth Agricultural Bureaux, Slough S12 3BN, UK (1985).
3. T. L. Steck, R. S. Weinstein, J. H. Straus, and D. F. H. Wallach, Science 168:255 (1970).

EFFECTS OF LEAD DEFICIENCY ON LIPID METABOLISM

Anna M. Reichlmayr-Lais and Manfred Kirchgessner

Institut für Ernährungsphysiologie
Technische Universität München
D-8050 Freising-Weihenstephan

INTRODUCTION

In earlier model experiments with rats symptoms of deficiency could be induced by extremely low alimentary lead supply. Offspring of lead depleted female rats showed reduced growth (Reichlmayr-Lais and Kirchgessner, 1981a) and microcytic hypochrome anemia (Reichlmayr-Lais and Kirchgessner, 1981b) connected with biochemical changes (Reichlmayr-Lais and Kirchgessner, 1981c; Kirchgessner and Reichlmayr-Lais, 1982). Hereby, especially lipid metabolism was affected. Therefore it was the aim of newer experiments to study criteria of lipid metabolism in dams as well as in their offspring.

MATERIALS AND METHODS

Experiment 1:
Twenty female rats (lifeweight 30 ± 1 g) were fed either with depletion diet (Pb content 20 ± 5 ppb) or with control diet (supplemented with 800 ppb Pb[#]) during breeding, gravidity and lactation. After a 21 day long lactation dams and offspring were used for determinations. Details of these experiments are described in Reichlmayr-Lais and Kirchgessner, 1986a,b.

Experiment 2:
It was designed two-factorial. Not only lead content of diet (depletion 20 ± 5 ppb Pb[#], control 800 ppb Pb[#]) varied but also the lipid component of diet which was either 1% thistle oil or 1% thistle oil, 3% beef suet and 8% coconut fat. After breeding, gravidity and lactation dams and their offspring were killed. At the 13th day of lactation 5 ml milk was collected after a two hour separation of mother and litter.

RESULTS AND DISCUSSION

Experiment 1: The results are demonstrated in table 1. These results show that triglyceride metabolism, and thus the energy balance is particularly affected by lead depletion. The reduced total body lipid concentration is compatible with the observed reduced live mass of these depleted animals; therefore it is possible that growth depression during lead depletion results from the changes in lipid metabolism, possibly as a result of reduced triglycerides in the milk of lead depleted mothers.

Table 1. Changes in Lipid Metabolism Resulting from Pb Deficiency

Generation	P_0	f_1	P_1	f_2
Activity of lipase	↓	↓	↑	↓
Activity of amylase	—	↑	↑	↑
Triglycerides in serum	↓	↓	↓↓	↑
Triglycerides in liver	*	↓	↑	—
Phospholipids in liver	*	↑	—	—
Cholesterol in liver	*	↓	—	—
Whole lipids in carcass	—	*	—	↓

↑ increased (depletion > control); ↓ decreased (depletion < control); — no difference between depletion and control; * no samples.

Experiment 2: The triglyceride content in milk was only influenced by lead supply. In the milk of depleted mothers the triglyceride content was 15.3 ± 2.8% if fat content of diet was 1% and 14.4 ± 2.0% if fat content of diet was 12%. The milk fat content of control mothers was 20.4 ± 4.2% and therefore about 20% higher than in depletion groups. This result verifies our hypothesis drawn from Experiment 1. Besides, glucose and lactose content of milk was determined. Glucose content was neither influenced by lead supply nor by fat component of the diet. Lactose content was influenced by both factors. In the milk of depleted animals lactose content was lower than in control milk. As further parameters the activity of lipase and α-amylase in pancreas as well as the free fatty acid content in serum of dams was determined. The lipase activity in pancreas was increased in depleted mothers and that only in the case of low fat supply. The activity of α-amylase was increased in depleted mothers, slightly if fat content of diet was low and more intensely if fat content was high. The free fatty acids in serum were influenced only by lead supply. Depleted animals showed higher values than control dams.

CONCLUSION

Lead deficiency induces disturbances in lipid metabolism of dams causing an energetical deficit in offspring connected with growth retardation. To find the causes of these disturbances is the aim of future experiments.

REFERENCES

Kirchgessner, M. & Reichlmayr-Lais, A.M., 1982, Konzentrationen verschiedener Stoffwechselmetabolite im experimentellen Bleimangel, Ann.Nutr.Metab., 26:50.
Reichlmayr-Lais, A.M & Kirchgessner, M., 1981a, Zur Essentialität von Blei für das tierische Wachstum, Z.Tierphysiol.Tierernährg.u.Futtermittelkde., 46:1.
Reichlmayr-Lais, A.M & Kirchgessner, M., 1981b, Hämatologische Veränderungen bei alimentärem Bleimangel, Ann.Nutr.Metab., 25:281.
Reichlmayr-Lais, A.M. & Kirchgessner, M., 1981c, Eisen-, Kupfer- und Zinkgehalte in Neugeborenen sowie in Leber und Milz wachsender Ratten bei alimentärem Blei-Mangel, Z.Tierphysiol.Tierernährg.u.Futtermittelkde., 46:8.
Reichlmayr-Lais, A.M. & Kirchgessner, M., 1986a, Aktivität der Enzyme Lipase, α-Amylase und Carboxypeptidase im Pankreas von Ratten bei Pb-Mangel, J.Anim.Physiol.a.Anim.Nutr., 56:123.
Reichlmayr-Lais, A.M. & Kirchgessner, M., 1986b, Effects of lead deficiency on lipid metabolism, Z.Ernährungswiss., 25:165.

AN ASSOCIATION OF ZINC WITH OMEGA-6-FATTY ACID AND cAMP

SYNTHESIS IN MELANOMA CELLS

Noel S. Skeef and John R. Duncan

Department of Biochemistry
Rhodes University
Grahamstown, South Africa

INTRODUCTION

There is substantial evidence for a regulatory role of zinc in the control of growth of transformed cells. Of possible therapeutic interest is the reduction in tumour cell proliferation by a dietary excess of zinc[1,2]. The biochemical mechanism of action of supplemental zinc in bringing about reduced cell growth remains unclear. Recent suggestions have implicated control of intracellular cAMP synthesis as a possible mechanism[3,4]. It is suggested that zinc may regulate the activity of enzymes such as delta-6-desaturase, which control omega-6-fatty acid synthesis. The intracellular levels of these fatty acids in turn influence the synthesis of certain prostaglandins such as PGE_1 which is known to be a potent stimulator of cAMP accumulation in cells. Earlier studies in our laboratory (unpublished) have demonstrated that addition of zinc to in vitro cultured tumour cells resulted in elevated levels of cAMP which was correlated with reduced cell proliferation. In this study the effect of zinc on omega-6-fatty acid synthesis and delta-6-desaturase activity was determined in in vitro cultured melanoma cells.

METHODS

Melanoma cells (BL6) were cultured (1.5 million cells/ml) in medium containing 0-10ug/ml of added zinc. ^{14}C-linoleic acid was added to the cultures as a substrate for omega-6-fatty acid synthesis and to cell-free extracts for determination of delta-6-desaturase activity. Fatty acids were determined by a modified argentation TLC technique.

RESULTS and DISCUSSION

Addition of zinc to the culture medium resulted in reduced cell growth at all levels of zinc addition (Table 1) and at concentrations of 5 and 10ug/ml actually resulted in cell death (Note: Non-supplemented medium contained 0.63ug/ml of zinc). The effect of zinc on the synthesis of omega-6-fatty acids from absorbed ^{14}C-linoleic acid (LA) is also shown in Table 1. Addition of 1ug zinc/ml resulted in a significant reduction in gamma linolenic acid (GLA) synthesis but had little effect on dihomo-gamma linolenic acid (DGLA) and arachidonic acid (AA) production. Higher concentrations of added zinc significantly increased the synthesis of all three of these fatty acids. This stimulation of fatty acid synthesis by high concentrations of zinc may, through an influence on prostaglandin metabolism, account for the increased cellular cAMP concentrations found in an earlier study when zinc was added to tumour cell cultures.

Table 1. Effect of added zinc on in vitro proliferation of BL6 melanoma cells and omega-6-fatty acid synthesis in the cells.

Zinc Added (ug/ml)	Cell Count ($\times 10^{-6}$)	Fatty Acid Synthesis GLA (cpm ^{14}C-LA incorp/ug prot)	DGLA+AA
0	3.10 ±0.30	100 ±18	165 ±20
1	2.70 ±0.23	150 ±10*	170 ±18
3	1.95 ±0.21*	173 ±12*	207 ±31*
5	1.05 ±0.16*	217 ±19*	253 ±22*
10	0.15 ±0.09*	-	-

*$p < 0.05$ when compared to control (0 added zinc)

A possible site of action for zinc in fatty acid synthesis has been suggested to be associated with the initial enzyme in this pathway, delta-6-desaturase. Results from this study support this since the addition of zinc to the culture medium significantly increased the activity of this enzyme in the melanoma cells (Table 2).

Table 2. Effect of added zinc on delta-6-desaturase activity in BL6 melanoma cells.

Zinc Added (ug/ml)	Delta-6-desaturase activity (cpm ^{14}C-LA utiliz/ug prot/hr)
0	520 ±32
1	612 ±39*
3	850 ±49*
5	1057 ±95*

*$p < 0.05$ when compared to contol (0 added zinc)

These data suggest that the findings in this and other studies of an inhibition of tumour cell growth as a result of zinc supplementation may be mediated through an initial effect on delta-6-desaturase activity with consequent effects on omega-6-fatty acid and cAMP metabolism.

REFERENCES

1. J.R.Duncan and I.E.Dreosti, Zinc intake, neoplastic DNA synthesis, and chemical carcinogenesis in rats, J. Natl. Cancer Inst. 55:195 (1975).
2. J.Borovansky, P.A.Riley, E.Vrankova and E.Necas, The effect of zinc on mouse melanoma growth in vitro and in vivo, Neoplasma 32:401 (1985).
3. J.R.Sheppard, Difference in the cAMP levels in normal and transformed cells, Nature New Biology 236:14 (1972).
4. S.C.Cunnane, Y.S.Huang, D.F.Horrobin and J.Davignan, Role of zinc in linoleic acid and prostaglandin synthesis, Prog. LipidRes.20:157(1981)

EFFECT OF NON-STEROIDAL ANTI-INFLAMMATORY DRUGS AND

PROSTAGLANDIN E_2 ON ZINC ABSORPTION IN HUMANS

W.S. Watson[1], D.A. Pitkeathly[1], G. Sinclair[1], M.I.F. Bethel[1], T.D.B. Lyon[2], and I. Pattie[3]

Departments of Clinical Physics, Medicine and Dietetics, Southern General Hospital[1], Department of Biochemistry, Royal Infirmary[2] and Department of Biochemistry, Gartnavel General Hospital[3], Glasgow, Scotland, U.K.

INTRODUCTION

It has been shown in the rat that prostaglandin E_2 (PGE_2) can stimulate zinc absorption while indomethacin, a potent inhibitor of PG synthesis, reduces absorption (Song and Adham, 1979). High doses of PGE_2 and indomethacin were used in this study; other workers using lower doses of PGE_2 and indomethacin have been unable to confirm these findings (eg Meydani et al, 1983).

Studies in humans, however, suggest that therapeutic doses of non-steroidal anti-inflammatory drugs (NSAIDs), such as indomethacin and aspirin, do have an effect on zinc metabolism. For example, patients with rheumatoid arthritis (RA) treated with NSAIDs have lower plasma zinc levels than similar patients on other drugs, while healthy subjects taking NSAIDs have reduced serum zinc and increased urine zinc.

In order to help clarify the situation, we have measured zinc absorption with and without PGE_2 supplementation in subjects on NSAID treatment.

MATERIALS AND METHODS

Zinc absorption was measured twice in 15 patients (6M, 9F) on long term NSAID treatment for RA. A standard meal of 100g mince and 60g mashed potatoes (zinc content = 106 umol), extrinsically labelled with the radiotracer zinc-65, was taken 2.5 to 3 hours after a light breakfast of tea and toast. The subjects then fasted for a further 2 hours. Whole-body counting techniques were used to measure tracer retention, and hence absorption, 14 days after ingestion of the labelled meal with or without simultaneous ingestion of 0.5 mg PGE_2 (dinoprostone, Upjohn). The patients were given a total of 45 kiloBecquerels zinc-65.

In addition to routine haematological and biochemical screening, plasma zinc, plasma copper and 24 hour urine zinc were also measured. In 9/15 subjects, dietary zinc was estimated from seven day diet histories taken approximately six months after the absorption measurements.

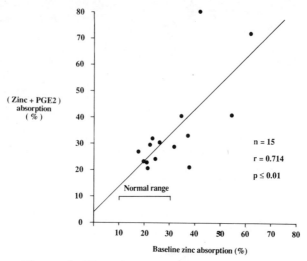

Figure 1. Zinc absorption in patients with
NSAID-treated rheumatoid arthritis.

RESULTS AND CONCLUSIONS

The zinc absorption results are shown in Figure 1. There was a
significant positive correlation between baseline absorption and absorption
plus PGE_2 (p < 0.01), but PGE_2 had no significant effect on absorption (mean
baseline absorption = 31.5 $\pm$ 13.2% (mean $\pm$ 1 standard deviation); mean
absorption plus PGE_2 = 35.2 $\pm$ 17.8%). Zinc absorption plus or minus PGE_2 was
found to be significantly higher (p < 0.01) than normal as previously
determined in 20 healthy controls (mean absorption = 20.3 $\pm$ 5.1%).

Only 2/15 of the RA subjects had abnormally low plasma zinc levels
(normal range 12-18 umol/l), although the group mean plasma zinc was slightly
but significantly below normal at 13.3 + 2.6 umol/l (p < 0.02). Urine zinc
was low in 3/15 subjects, while no subject excreted abnormally high levels
(group mean 24 hour urine zinc = 8.2 + 4.8 umol/24 h).

There was no correlation between baseline zinc absorption and any other
parameter apart from absorption plus PGE_2.

The higher than normal absorption observed in the NSAID-treated RA
subjects may be artefactual as the standard meals used for the RA group and
the control group were prepared from different batches of mince and potatoes
and, in addition, the control group was given a glass of "orangeade" with the
meal, while the RA group was given only water. We are currently
investigating this problem but feel that, irrespective of the results of
further study, it is unlikely that NSAID therapy causes malabsorption of zinc
in NSAID-treated subjects with RA.

REFERENCES

Meydani SN, Meydani M and Dupont J, 1983, Effects of prostaglandin modifiers
    and zinc deficiency on possibly related functions in rats, J Nutr,
    113:494.
Song MK and Adham NF, 1979, Evidence for an important role of prostaglandins
    E_2 and F_2 in the regulation of zinc transport in the rat, J Nutr,
    109:2152.

ZINC DECREASES LIPID AND LONG CHAIN FATTY ACID RELEASE

FROM THE RAT MESENTERIC VASCULAR BED PERFUSED IN VITRO

Stephen C. Cunnane, Bassam A. Nassar*, Kelly R. McAdoo* and
David F. Horrobin *

Department of Nutritional Sciences, University of Toronto
Toronto, Canada M5S 1A8 and *Efamol Research Institute
Kentville, Nova Scotia, Canada B4N 4H8

INTRODUCTION

The rat mesenteric vascular bed when perfused in vitro has been shown
to release prostaglandins. The vasconstrictor effects of nerepinephrine
in this preparation are sensitive to the presence of exogenous prostaglandins
as well as to the presence of 1nM-1uM zinc (1). The effects of zinc appear
to be related in part to modulating release of endogenous prostaglandins or
their precursor fatty acids (1). In view of the known inhibitory effect of
zinc on triglyceride lipase (2) and phospholipase A_2 (3), we decided to
investigate the lipid composition and metabolism of long chain fatty acids
in the rat mesenteric vascular bed when perfused in vitro with
physiological concentrations of zinc.

METHODS

Male Wistar rats (250g) were fed a semi-synthetic diet containing a
fat source of 8% evening primrose oil plus 2% marine fish oil (to maximize
prostaglandin fatty acid precursors, 20:3n-6, 20:4n-6 and 20:5n-3). After 6
weeks, the rats were anesthetized and the mesenteric vascular bed removed
according to the method previously described in detail (1). The
preparations were mounted in an organ bath and perfused in vitro at 3.5
ml/min with Krebs-Henseleit buffer at $37^{\circ}C$, pH 7.4. The total perfusion
period was 150 min; zinc was added after the first 30 min at 3, 6, or 9nM
zinc (different preparations were used for each zinc dose). The perfusion
effluent was sampled at 30 min intervals.

Unperfused and perfused preparations were analysed for phospholipid
and triglyceride fatty acid composition and the perfusion effluent was
assayed for protein, total phospholipid, triglyceride, free fatty acids and
fatty acid profile using standard methods (4).

RESULTS

Table 1 shows that the fatty acid composition of the phospholipids of
the mesenteric vascular bed changes significantly after 150 min perfusion
in vitro and that 6 nM zinc largely prevents this effect. The fatty acid

371

composition of the mesenteric triglycerides was not significantly affected by perfusion with of without zinc.

Table 2 shows the effect of 6 nM zinc on the composition of the perfusion effluent near the beginning of the perfusion period (30 min) and near the end of perfusion (120 min). Protein in the effluent was significantly lower after 120 min. perfusion but was unaffected by zinc. Total phospholipids in the effluent were unaffected by time but were significantly lower in preparations perfuse with 6 nM zinc (p<0.01). Triglycerides and free fatty acids in the perfusion effluent were unaffected by duration of perfusion but free fatty acid release was lower in zinc-perfused preparations (120 compared to 30 min perfusion, p<0.05).

The composition of long chain fatty acids (18:2n-6, 20:4n-6 and 22:6n-3) in the total lipids extracted from the perfusion effluent were affected by both zinc and time such that 6nM zinc approximately doubled the rate of decay from a total of approximately 250 ug/100 ml (ZINC (-) at 30 min) to approximately 60 ug/100ml (ZINC (+) at 120 min).

DISCUSSION

Zinc has been shown in this study to prevent the changes in phospholipid fatty acid composition of the rat mesenteric vascular bed which would normally be induced by 150 min perfusion with an oxygenated iso-osmotic buffer. This was demonstrated in both the normal fatty acid composition of mesenteric phospholipids in preparations perfused with zinc (Table 1) and in the lower amounts of free fatty acids, intact phospholipids and long chain fatty acids in the effluent of preparations perfused with zinc (Tables 2 and 3). This would suggest an important endothelial membrane stabilizing role for zinc in this vascular preparation as has been indicated for other membrane models elsewhere (5).

TABLE 1

Long chain fatty acid composition (mg%) of the mesenteric vascular bed phospholipids before (UNPERFUSED) and after perfusion in vitro without added zinc - ZINC(-) or with 6 nM added zinc - ZINC(+).

	UNPERFUSED	ZINC(-)	ZINC(+)
16:0	20.1+1.7#	13.5+2.3 *	19.7+2.0
18:0	6.5+1.7	2.0+0.7 *	6.0+1.4
18:1n-9	13.8+1.3	23.1+4.0 *	16.0+0.8
18:2n-6	35.7+7.3	32.9+7.3	37.6+3.5
20:3n-6	0.9+0.1	0.4+0.1 *	0.9+0.2
20:4n-6	8.6+2.1	2.2+0.7 *	4.7+1.0
22:6n-3	2.3+0.3	3.9+0.9 *	2.4+0.2

mean+SD, n=8/group   *p<0.01 versus UNPERFUSED or ZINC(+)

TABLE 2

Composition of the perfusion effluent (ug/ml) from the mesenteric vascular bed perfused without added zinc - ZINC(-) or with 6 nM zinc - ZINC(+).

		ZINC(-)	ZINC(+)
Protein	30 min:	38.7+3.2 #	31.1+10.2
	120 min:	7.8+3.1 +	7.7+4.5 +
Phospholipid	30 min:	194+19	143+39
	120 min:	187+14	124+4 *
Triglyceride	30 min:	9.5+5.0	7.4+3.4
	120 min:	5.1+2.9	4.1+2.9
Free Fatty Acid	30 min:	1.8+0.5	1.9+0.8
	120 min:	1.3+0.4	1.6+0.3

mean $\pm$ SD, n=8/group;   + p$<$0.01 v. 30 min.;   *p$<$0.01 v. ZINC(-)

ACKNOWLEDGEMENTS

SCC was the recipient of an Industrial Research Fellowship from the National Science and Engineering Research Council of Canada. Excellent technical assistance was provided by D.K. Jenkins.

REFERENCES

1. M.S. Manku, D.F. Horrobin, M. Karmazyn, and S.C. Cunnane, Prolactic and zinc effects on vascular reativity: Possible relationship to dihomo-gamma-linolenic acid and to prostaglandin synthesis, Endocrinology, 104: 774 (1979).
2. T. Rebello, D.J. Atherton, and C. Holden, The effect of oral zinc administration on sebum free fatty acids in acne vulgaris, Acta Dermatol. Venereol. 66:305 (1986).
3. M.A. Wells, Spectral pertubations of Crotalus adamanteus phospholipase A$_2$ induced by divalent cation binding, Biochemistry 12:1080 (1973).
4. S.C. Cunnane, K.R. McAdoo, and J.R. Prohaska, Fatty acid and lipid composition of organs from copper dificient mice, J. Nutr., 116: 1248 (1986).
5. W.J. Bettger, and B.L. O'Dell, A critical physiological role of zinc in the structure and function of biomembranes, Life Sci. 28: 1425 (1981).

RAT HEART PERFORMANCE AND LIPID METABOLISM IS AFFECTED BY COPPER INTAKE

Stephen C. Cunnane, K.R. McAdoo* and M. Karmazyn**

Department of Nutritional Sciences, University of Toronto,
Toronto, Canada M5S 1A8, *Efamol Research Institute, Kentville,
Nova Scotia, Canada B4N 4H8 and **Department of Pharmacology,
Dalhousie University, Nova Scotia, Canada.

INTRODUCTION

Copper is an essential trace metal with an important role in the
cardiovascular system; its function is related to the integrity of both
cardiac rhythm and structure. Experimental copper deficiency is associated
with cardiac rhythm disturbances and a greater risk of myocardial infarction
in both humans (1,2) and animals (3), and with aneurysms and cardiac and
aortic rupture in animals (3). Cardiac rhythm instability induced by
copper deficiency or prostaglandins can be reversed by copper supplementation (3,4).

In humans with either acute myocardial infarction (5) or made
experimentally copper deficient (1), similar changes in serum phospholipid
fatty acid profile are observed; levels of arachidonic acid (20:4n-6) and
docosahexaenoic acid (22:6n-3) being increased significantly in both
conditions. Since copper depletion in animals is known to alter tissue
fatty acid and lipid composition (6,7) as well as energy metablolism (8,9),
and fatty acids are the preferred energy substrate in the heart, it was
hypothesized that experimental copper deficiency in the rat may affect
heart performance in part by interfering with cardiac lipid metabolism.
The rat heart perfused in vitro was the model used and an ischemic episode
was included in the perfusion protocol to determine the possible role of
copper in the ability of the ischemic myocardium to regain normal performance during reperfusion.

METHODS

Male Sprague-Dawley rats (60g) were housed in groups in stainless
steel wire-bottomed cages and fed a semi-synthetic diet formulated
according to American Institute of Nutrition recommendations (10). Copper
content of the diet was either 1.3 (Cu-) or 243 (Cu+) mg/kg copper. Both
food and distilled water were available ad libitum for 10 weeks. Hearts
were perfused with Krebs-Henseleit buffer according the Langendorff technique (11). The perfusion period was 90 minutes; 0 - 30 minutes at 10 ml/
minute (normoxia), 30 - 60 minutes at 1ml/minute (ischemia) and 60 - 90
minutes at 10 ml/minute (reperfusion). Coronary perfusion pressure, force
of contraction and heart rate were monitored constantly. Unperfused hearts
(BEFORE, n=4) and perfused hearts (AFTER, n=6) were analysed for

triglyceride, cholesterol and phospholipid composition and fatty acid profiles in triglycerides and phospholipids by standard methods (7).

RESULTS

Plasma copper in Cu- rats was 41% of that in Cu+ hearts. Post-perfusion, hearts were 18% heavier in the Cu- group (p<0.01). Coronary perfusion pressure was 15-30% higher in the normoxic period and 70% higher during reperfusion in the Cu+ compared to Cu- group (p<0.01). During reperfusion the Cu- hearts developed 50-60% less tension than the Cu+ hearts (p<0.01). Spontaneous heart rate did not differ in either group at any stage in the perfusion protocol.

TABLE 1

Lipid   composition (mg/g) in hearts of rats fed for 10 weeks on diets con-taining copper at 243 mg/kg (Cu+) or 1.3 mg/kg (Cu-).

		Cu+	Cu-
Triglyceride	Before: (1)	2.1+0.4	5.5+2.4*
	After: (2)#	0.3+0.2	0.6+0.3*
Cholesterol	Before:	1.2+0.1	1.2+0.1
	After: #	1.0+0.1	0.9+0.04
Phospholipid	Before:	17.6+0.8	17.8+0.4
	After: #	14.0+1.8	14.8+0.6
Cholesterol/Phospholipid (x 10)			
	Before:	0.70+0.07	0.66+0.04
	After:	0.69+0.09	0.59+0.03 *

(1)   Unperfused hearts (n=4/group)
(2)   Hearts perfused for 90 min, including 30 min ischemia (n=6/group)
p<0.01 (After v. Before, ANOVA)
*   p<0.01 (Cu- v. Cu+, ANOVA)

Cardiac total triglyceride was decreased 85% in all hearts after perfusion; 50% less triglyceride remained in the Cu+ compared to Cu- hearts (p<0.01). Both total phospholipid and cholesterol were 20% decreased in all hearts after perfusion but cholesterol/phospholipid remained higher in Cu+ hearts (Table 1).

After perfusion, stearic acid (18:0) and 20:4n-6 were peferentially retained incardiac triglyceride with greater retention of both fatty acids in Cu+ hearts)p<0.01). Only minor differences in cardiac phospholipid fatty acid profiles were noted before or after perfusion in Cu- compared to Cu+ hearts, notably a significant decrease in 20:3n-6/20:4n-6 in Cu- hearts.

DISCUSSION

The results confirm previous observations of impaired heart function
in vitro of Cu- rats but indicate that, in contrast to previous reports
(3,8), only marginal Cu depletion is required to effect significant changes
in coronary perfusion pressure and force of contraction. The lipid changes
in these hearts suggested significantly reduced homeostatic control of
membrane lipid composition (phospholipid and cholesterol) as well as energy
stores (triglyceride) in Cu- hearts, both of which would be expected to
contribute to impaired viability of Cu- hearts.

REFERENCES

1. L.M. Klevay, L. Inman, L.K. Johnson, M. Lawler, J.R. Mahalko, D.E.
    Milne, H.C. Lukaski, W. Bolonchuk, H.H. Sandstead, Increased
    cholesterol in a young man during experimental copper depletion,
    Metabolism 33:1112 (1984).
2. S. Reiser, J.C. Smith Jr., W. Mertz, J.T. Holbrook, D.J. Scholfield,
    A.S. Powell, W.K. Canfield, J.J. Canary, Indices of copper status
    in humans consuming a typical American diet containing either
    fructose or starch, Amer. J. Clin. Nutr., 42:242 (1985).
3. K.E. Vienstenz and L.M. Klevay, A randomized trial of copper therapy
    in rats with electrocardiographic abnormalities due to copper
    deficiency, Amer. J. Clin. Nutr., 35:258 (1982).
4. A. Swift, M. Karmazyn, D.F. Horrobin, M.S. Manku, R.A. Karmali, R.
    Morgan, A.I. Ally, Low prostaglandin concentrations cause cardiac
    rhythm disturbances. Effect reversed by low levels of copper or
    chloroquine, Prostaglandins, 15:651 (1978).
5. G. Skuladottir, T. Hardarson, N. Sigfusson, G. Oddsson, and S.
    Gudbjarnason, Arachindonic acid levels in serum phospholipids of
    patients with angina pectoris or fatal myocardial infarction, Acta
    Med. Scand., 218: 55 (1985).
6. S.C. Cunnane, D.F. Horrobin, and M.S. Manku, Contrasting effects of
    low or high copper intake on rat tissue lipid essential fatty acid
    composition, Ann. Nutr. Metab.. 29:103 (1985).
7. S.C. Cunnane, K.R. McAdoo, and J.R. Prohaska, Lipid and fatty acid
    composition of organs from copper deficient mice, J. Nutr., 116:
    1248 (1986).
8. J.R. Prohasha and L.J. Heller, Mechanical properties of the copper
    deficient rat heart, J. Nutr., 112:2142 (1982).
9. C.H. Gallagher and V.F. Reeve, Copper deficiency in the rat. Effect
    on adenine nucleotide metabolism, Aust. J. Biol. Med. Sci., 54:
    593 (1971).
10. American Institute of Nutrition, report of the AIN Ad Hoc Committee on
    standards for nutritional studies, J. Nutr., 107:1340 (1977).
11. M. Karmazyn, D.F. Horrobin, M.S. Manku, S.C. Cunnane, R.A. Karmali,
    A.I. Ally, R.O. Morgan, K.C. Nicolaou, and W.E. Barnette, Effects
    of prostacyclin on perfusion pressure, electrical activity, rate
    and force of contraction in isolated rat and rabbit hearts, Life
    Sci., 22: 2079 (1978).

TISSUE FATTY ACID RESPONSES TO SELENIUM AND VITAMIN E DEFICIENCY AND

ORAL C18:3,n-3 IN CATTLE

Desmond Rice, Seamus Kennedy, and John Blanchflower

Veterinary Research Laboratories
Stormont
Northern Ireland, BT4 3SD

INTRODUCTION

We have previously shown in TEMA 5 (Rice et al., 1985) that myopathy in cattle can result from a vitamin E and selenium deficient diet alone. The rate of onset of myopathy and the severity of the pathological sequelae can be increased by adding protected linolenic acid (C18:3,n-3) to the ration in the form of protected linseed oil. This causes an increased concentration of C18:3,n-3 in plasma and tissues and a simultaneous increase in plasma CK. We have used this experimental model for spontaneous and induced myopathy to define the fatty acid changes in tissues, particularly in phospholipids resulting from these dietary manipulations in order to help explain the pathogenetic mechanisms.

MATERIALS AND METHODS

Three groups of calves were used which received different dietary treatments, all based on our experimental vitamin E and selenium deficient diet based on sodium hydroxide treated barley (Rice and McMurray, 1986): Group A - The vitamin E and selenium deficient barley diet with added vitamin E and selenium. Group B - The vitamin E and selenium deficient diet alone. Group C - The vitamin E and selenium deficient barley diet + protected C18:3,n-3 for 6 days. The dose of protected fat was sufficient to simulate linolenic intake from grass.

Animals were slaughtered after 113-190 days on their diets. Liver, heart and muscle (triceps femoris) were analysed for their fatty acid content by capillary gas chromatography. Additionally, the various lipid components of muscle of calves in groups B and C were separated by thin layer chromatography and the fatty acid profiles determined.

RESULTS AND DISCUSSION

The most significant effects were: feeding of a vitamin E and selenium deficient diet increased the concentration of arachidonic acid in both skeletal (Table 1) and cardiac muscle; oral protected linolenic acid (C18:3,n-3) increased the concentration of this fatty acid in the total lipids of skeletal (Table 1) and cardiac muscle, there was not however any increase in the chain elongated desaturated n-3 derivatives of this fatty acid (C20:4,n-3 - C22:6,n-3); feeding C18:3,n-3 resulted in increases in this fatty acid in the phospholipids (0.68 to 2.8%) and in the triglycerides and non-esterified fatty acids of skeletal muscle. Again there were no increases in the ratios of its chain elongated derivatives, nor were there any compensatory decreases in the proportions of n-6 fatty acids in muscle.

Table 1 The fatty acid profile (%) of lipid extracted from the triceps femoris muscle of cattle on three different diets

Group		18:0	18:1	n-6 Series 18:2	20:4	22:5	n-3 Series 18:3	20:4	22:5	22:6
A(6)**	Mean	15.30	26.26	21.48	7.46	0.13	0.59	0.79	1.42	0.22
	SD	1.46	8.21	6.06	3.64	0.14	0.21	0.47	0.70	0.29
B(4)	Mean	17.01	16.95	24.88	12.25	0.20	0.63	1.47	2.32	0.47
	SD	3.71	2.39	0.64	1.66	0.27	0.47	0.14	0.32	0.36
C(8)	Mean	15.63	19.82	21.45	9.54	0.05	4.27	1.09	1.88	0.12
	SD	1.51	4.00	4.42	2.80	0.09	1.97	0.55	0.72	0.17
Signif.*	AVB	NS	2	NS	2	NS	NS	3	2	NS
	BVC	NS	NS	1	1	NS	5	1	NS	NS

** Number in parenthesis indicates the number of animals/treatment.
 * Significance: 1, p<0.05; 2, p<0.025; 3, p<0.001; 4, p<0.005; 5, p<0.001.

Our findings of increased concentrations of arachidonic acid in muscle, as a result of vitamin E and selenium deficiency, agree with results of Chan et al., (1979) in rabbits. This may well explain the occurrence of spontaneous myopathy in cattle maintained indoors on a vitamin E and selenium deficient diet; ie, the increasing concentration of this highly peroxidisable substrate may overcome the protective effects of GSH-Px and alpha tocopherol in sub-cellular membranes.

Our previous work (Rice et al., 1986) showed that C18:3,n-3 is present in tissues of calves within 6 days of feeding protected linseed oil - which is the time at which myopathy occurs under field conditions. This work demonstrates that C18:3,n-3 appears to be myopathic per se. It does not cause mobilisation of n-6 fatty acids from other sites to muscle. It is not chain elongated and desaturated over this 6 day time period - therefore its longer chain family members do not appear to be the peroxidisable substrate for initiation of myopathy.

REFERENCES

Chan, A.C., Allen, C.E. and Hegarty, P.V.J., 1979, Am. J. Clin. Nutri., 32: 1456.
Rice, D.A. and McMurray, C.H., 1986, Vet. Rec., 118: 173.
Rice, D.A., McMurray, C.H. and Kennedy, S. 1985, "Trace elements in man and animals", C.F. Mills., J. Bremner and J.K. Chesters, eds., CAB Press, Slough.

PANCREATIC SUPEROXIDE DISMUTASE ACTIVITY

IN THE COPPER-DEPLETED RAT

A. Mylroie, A. Boseman and J. Kyle

Department of Physical Sciences
Chicago State University
Chicago, IL 60628

Since Cu deficiency in rats is accompanied by progressive atrophy of pancreatic tissue (2,6-8), this study was designed to investigate the role of the Cu-dependent enzymes superoxide dismutase (CuSOD) and cytochrome oxidase and the Mn-dependent enzyme superoxide dismutase (MnSOD) in this process. Superoxide dismutase catalyzes the dismutation of the superoxide radical anion (O_2^-) to O_2 and H_2O_2 and is considered to have an important role in protecting cells against direct and indirect oxidative damage (4,5). The cytotoxicity of O_2^- has been ascribed to the secondary production of highly reactive intermediates, such as hydroxy radicals (4).

Groups of male Sprague Dawley rats were fed either Cu sufficient (6 ppm Cu) or Cu deficient (0.5 ppm Cu) purified AIN diet (1). At the end of each experimental period (3,5,7,9,11 weeks), four rats from each group were sacrificed. Ceruloplasmin, hemoglobin values, packed cell volume, and Cu levels in blood and tissues were determined to assess Cu status. Pancreatic homogenates were assayed for CuSOD, MnSOD, cytochrome oxidase and amylase activity (7).

In the first experiment, the degree of Cu deficiency was "moderate". Growth rate of rats fed Cu deficient diet was only slightly decreased; by week 13 the final body weights were 10% lower relative to controls. Hemoglobin values and packed cell volume decreased 23% and 16% respectively by week 3 and remained at these depressed levels. Serum Cu levels decreased to about 0.04 ug Cu/mL blood by week 3 and did not change significantly with increased length of exposure. Pancreatic weights decreased 50% by week 5 and 76% by week 7 in Cu-deficient animals. Protein content of the pancreas (mg protein/g pancreas) decreased 45% by week 7 and 80% by week 11 as a result of Cu deficiency. CuSOD activity decreased 50% by week 9 and was undetectable by week 11. In contrast, MnSOD activity increased initially by 30%, then decreased and was negligible by week 9.

Cu deficiency was less pronounced in the second experiment. Hemoglobin and hematocrit values were only slightly reduced (8% and 5%, respectively; p<.02) in animals fed Cu deficient diet relative to the control group. Serum Cu levels of Cu depleted rats were higher (0.1 - 0.2 ug Cu/ml) than in the first experiment. Pancreatic amylase activity decreased only 28% by week 9 as a result of Cu deficiency. There was no significant difference in pancreatic weights, protein content or pancreatic CuSOD activity between rats fed Cu deficient and control diets. However, cytochrome oxidase activity decreased (50%) and MnSOD increased (50%) by weeks 7-9 as a result of Cu deficiency. The increase in MnSOD appears to be associated not with a decrease in CuSOD as expected, but rather with a decrease in cytochrome oxidase. A decrease in cytochrome oxidase activity could result in a more reduced state of the components of the electron transport chain with a resultant increase in the production of O_2^- (3) and possible resultant induction of MnSOD.

Further studies are required to determine whether the increase in MnSOD activity is the result of an increase in superoxide anion (O_2^-) production and whether O_2^- contributes to pancreatic atrophy in the absence of a compensatory increase in MnSOD activity. (Supported by NIH grant RR-08043.)

REFERENCES

1. American Institute of Nutrition. 1977. Ad hoc committee on standards for nutritional studies. J. Nutr. 107: 1340-1348.
2. Fell, B.F., King, T.P. and N.T. Davies. 1982. Pancreatic atrophy in copper-deficient rats: histochemical and ultrastructural evidence of a selective effect on acinar cells. Histochem. J. 14: 665-680.
3. Forman, H.J. and A. Boveris. 1982. Superoxide Radical and Hydrogen Peroxide in Mitochondria. In: Free Radicals in Biology V. Academic Press, New York pp 65-90.
4. Fridovich, I. 1981. Role and Toxicity of Superoxide in Cellular Systems. In: Oxygen and Oxy-Radicals in Chemistry and Biology. Academic Press, New York.
5. Halliwell, B. 1978. Biochemical mechanisms accounting for the toxic action of oxygen on living organisms: The Key Role of Superoxide Dismutase. In: Cell Biol. Int. Reports 2: 113-128.
6. Muller, H.B. 1970. Der einfluss kupferarmer kost an das pancreas. Virchows Arch. Abt. A Path. Anat 350: 353-367.
7. Mylroie, A.A., Tucker, C., Umbles, C. and J. Kyle. 1987. In: Trace Substances in Environmental Health, Hemphill, D.D. ed., University of Missouri, Missouri. In press.
8. Smith, P.A., Sunter, J.P. and R.M. Case. 1982. Progressive atrophy of pancreatic acinar tissue in rats fed a copper-deficient diet supplemented with D-Penicillamine or triethylene tetramine: morphological and physiological studies. Digestion 23: 16-30.

CONTROLLED EXERCISE EFFECTS ON CHROMIUM, COPPER AND ZINC IN THE URINE OF TRAINED AND UNTRAINED RUNNERS CONSUMING A CONSTANT DIET

Richard A. Anderson, Marilyn M. Polansky and Noella A. Bryden

Beltsville Human Nutrition Research Center, ARS, USDA
Beltsville, MD 20705 USA

INTRODUCTION

Aerobic exercise is a popular pastime often leading to improved physical fitness, weight control and an overall general feeling of improved health. However, strenuous acute aerobic exercise, especially for poorly trained individuals, may be a form of stress leading to acute changes in blood parameters and an overall loss of the trace elements chromium and zinc as well as altered copper metabolism. The present study was designed to determine the effects of acute controlled exercise on trace element metabolism of trained runners and sedentary controls.

SUBJECTS AND METHODS

Eight trained and five sedentary control subjects participated in the study. Training state of the subjects was determined by measuring the maximal oxygen consumption VO_2 max) on a treadmill with progressive increases in running according to the Bruce Protocol (American Heart Association, 1972). A subject's VO_2 max (ml/kg/min) was defined as the highest value noted before voluntary termination of the exercise test. Subjects with VO_2 max values in the good or high range, based upon their age, were assigned to the trained group and subjects in the average or below comprised the sedentary control group.

Sample collection, trace metal analyses and statistical treatment of the data were similar to those reported previously (Anderson et al., 1984).

RESULTS

To minimize changes in trace element losses, subjects were fed a constant nutritious well-balanced diet containing 8.8 ± 1 mcg Cr, 0.48 ± 0.07 mg Cu and 5.8 ± 0.2 mg Zn per 1,000 kilocalories. Caloric intake was based upon the size of the subjects and their exercise practices. Basal urinary Cr losses of the trained subjects were significantly lower than those of the untrained subjects (Table 1). Urinary Cu losses were also significantly lower in the trained than in the untrained subjects. However, since urinary Cu losses represent only a small precentage of the total Cu losses, the overall significance of decreased urinary Cu losses of the trained subjects compared to the sedentary control subjects is difficult to evaluate. In contrast, urinary Cr losses represent the major route of excretion of absorbed Cr, therefore decreases in Cr losses of approximately 50 percent represent a significant physiological change. Basal urinary Zn losses of the trained and untrained subjects were similar (Table 1).

Table 1   Basal Urinary Losses of Cr, Cu and Zn of Trained and Untrained Subjects

Element (mcg/da)	Trained (8)*	Untrained (5)*
Cr	0.09 ± 0.01^a	0.21 ± 0.03^b
Cu	9.5 ± 0.06^a	12.2 ± 0.09^b
Zn	828 ± 85	754 ± 71

*Number in parenthesis denotes number of subjects
[a,b]Values in same row with different superscripts are significantly different at $p < 0.05$
All values are Mean $\pm$ SEM

Controlled exercise to exhaustion at 90% VO$_2$ max with 30 second exercise and 30 second rest periods resulted in a significant increase in daily urinary Cr excretion from 0.09 ± 0.01 to 0.12 ± 0.02 mcg for the trained subjects but no change in urinary Cr losses of the untrained subjects (Table 2). Increased Cr losses of trained subjects following a strenuous 6-mile run have been reported (Anderson et al., 1984). In previous experiments, we have shown that Cr losses of sedentary subjects also increase in response to acute exercise. The urinary losses of Cu and Zn tended to be higher on the day of controlled exercise compared to the sedentary days but increases were not significant (Table 2).

Table 2. Exercise Effects on Daily Urinary Losses of Cr, Cu and Zn

	Chromium (mcg)	Copper (mcg)	Zinc (mcg)
TRAINED (8)			
Basal	0.09 ± 0.01^a	9.5 ± 0.06	828 ± 85
Exercise*	0.12 ± 0.02^b	11.4 ± 1.3	911 ± 173
UNTRAINED (15)			
Basal	0.21 ± 0.02^b	12.2 ± 0.9	759 ± 71
Exercise*	0.21 ± 0.06^b	14.6 ± 1.8	687 ± 81

*Exercise at 90% VO$_2$ max to exhaustion (30" exercise 30" rest periods)
[a,b]Values in same column with different superscripts are significantly different at $p < 0.05$.

In summary, basal urinary Cr and Cu losses of trained runners were significantly greater than those of untrained subjects. Basal urinary zinc losses were not significantly different. Controlled exercise led to significant increases in daily urinary Cr losses for trained subjects but not for untrained subjects. Increases in urinary Cu and Zn losses due to controlled acute exercise were not statistically significant.

REFERENCE

Anderson, R.A., Polansky, M.M. and Bryden, N.A., Strenuous running:acute effects on chromium, copper, zinc and selected variables in urine and serum of male runners. Biol. Trace Element Res. 6:327 (1984).

THE ROLE OF ZINC ON THE TROPHIC GROWTH FACTORS,

NERVE GROWTH FACTOR AND GUSTIN

R. I. Henkin, J. S. Law, and N. R. Nelson

Taste and Smell Clinic
Washington, DC   20016

Growth factors can be divided into two major groups, mitotic and
trophic. Mitotic growth factors are those which activate cell division and
include platelet derived growth factor, epithelial growth factor and fibroblast growth factor. Trophic growth factors, on the other hand, do not enhance cell mitosis but affect cellular differentiation. These latter growth
factors are fewer in number than mitotic growth factors. The major trophic
growth factors are nerve growth factor (NGF), gustin, and to some extent,
calmodulin. It is the purpose of this short summary to review some of the
functions of these growth factors, to detail the role of zinc and other
metals in these factors and attempt to specify some aspects of how these
factors regulate cellular growth and differentiation.

NGF, gustin, and calmodulin are all secreted by the major salivary
glands, in humans, the parotid gland. Each growth factor is synthesized in
the salivary gland and each has a metal cofactor. NGF (7.5S) has zinc as
its cofactor as does gustin. NGF has one mole of zinc per mole of 7.5S protein. Gustin (37KD) also has one mole of zinc per mole protein[1] but another
mole of zinc is less tightly bound, bridged to its carrying protein, lumi-
carmine, a 34KD phosphoprotein also synthesized and secreted by the human
parotid gland.[2] For NGF and gustin, the salivary glands are the major
source of the proteins, the metals inserted into the proteins posttrans-
lationally. For calmodulin (17KD), although it is also secreted by the
human parotid gland,[3] its major site of synthesis is the brain and testis.
The major metal in calmodulin is calcium and there are four calcium molecules per mole protein.

The role of the endogenous metals in each of these growth factors differs. In NGF, the β subunit (2.5S) does not have a metal cofactor and can
perform its characteristic growth function without an associated metal. On
the other hand, both gustin and calmodulin require a metal for their cellular activity. For gustin, chelation of zinc by 1,10 phenanthroline inhibits
it enzymatic function of activating cAMP phosphodiesterase (PDE).[4] Replacement of zinc restores activity to normal. Chelation of zinc in gustin has
recently been shown to inhibit the ability of this protein to activate cAMP
PDE whereas zinc replacement restores activity to normal. For calmodulin,
treatment with 1,10 phenanthroline produced no effect on enzymatic activation of cAMP PDE although chelation with EGTA abolished this activity;
chelation with EGTA had no effect on gustin activity. These results define
the Ca^{++} dependence of calmodulin and the Zn^{++} dependence of gustin.

There are several interesting physiological similarities between gustin, NGF and calmodulin (Table 1). Both NGF and gustin are sensitive to heat whereas each protein is resistant to acid inactivation; gustin retains its protein structure but zinc is removed at low pH, producing an apoprotein. Both NGF and calmodulin activate several enzymes, including myosin light chain kinase, phosphorylase kinase and Ca dependent phosphorylation but gustin activates none of these enzymes.

TABLE 1

GROWTH FACTOR	GUSTIN	NGF	CALMODULIN
PHYSIOLOGICAL CHARACTERISTCS			
SOURCE			
SALIVARY GLAND	YES	YES	YES
MAJOR GENERATING TISSUE	PAROTID GLAND	SUBMAXILLARY GLAND (MOUSE)	BRAIN, TESTIS
BLOOD	YES	YES	YES (PTS)
MAJOR REGULATORY ACTIVITY	TASTE	SYMPATHETIC NEURITE SPROUTING	Ca++ DEPENDENT PROCESSES
REGULATORY SITE			
EXTRACELLULAR	YES	YES	YES
SURFACE MEMBRANE	YES	YES	YES
INTRACELLULAR	YES	YES	YES
SELF REGULATION	YES	YES	YES
HORMONE DEPENDENT ACTIVITY	YES	YES	?
THYROID	YES	YES	?
TESTOSTERONE	PROBABLE	YES	?
PATHOLOGY ASSOCIATION			
DECREASED IN HYPOGEUSIA	YES	?	YES
DECREASED IN HYPOSMIA	YES	?	YES
DECREASED IN Zn DEFICIENCY	YES	?	YES
AUTOIMMUNE FUNCTION			
ANTI NGF ANTIBODY INHIBITION	?	YES	?
IL1, IL2 ACTIVITY	?	NO	?
OTHER EFFECTS			
MICROTUBULE DEPOLARIZATION	?	PROBABLE(?)	?
ACTIN INDUCTION	?	YES	?
PC 12 CELL SPROUTING	NO	YES	?
LAMININ INHIBITION OF SPROUTING, CELL PRESERVATION	?	YES	?
STIMULATES TYROSINE KINASE	?	NO	?
BENZODIAZEPINE ACTION	NO EFFECT (cAMP PDE)	INCREASES (c fos mRNA)	INHIBIT (cAMP PDE)

Each growth factor has an effect extracellularly on the surface membrane of its target cell and intracellularly as well. Each factor appears to regulate its own activity. Both NGF and gustin are hormone dependent, both reactive to the presence of thyroid hormone and testosterone. Each is present in blood and in saliva. Both gustin and calmodulin are decreased in patients with loss of taste (hypogeusia) or smell (hyposmia) and in zinc deficiency although the exact mechanisms for these changes are unclear.

Drug inhibitors which influence enzymatic function of these growth factors can be used to differentiate the molecular characteristics of these factors. Benzodiazepenes inhibit the action of calmodulin on cAMP PDE activation, increase the activity of NGF on cfos mRNA but have no effect on cAMP PDE activation by gustin. Triflurophenazine inhibits the action of calmodulin on cAMP PDE activation but has no effect on gustin activation of cAMP PDE. These differentiating effects allow some dissection of the molecular characteristics of each of these growth factors.

Metals play their role in each of these factors in a characteristic manner. For NGF zinc appears to play a structural role in maintaining the integrity of the protein trimer. For gustin, zinc plays a specific role in cAMP PDE activation as does calcium for calmodulin. One prominent effect of calcium in calmodulin activity relates to changes in fluorescence in the apoprotein albeit zinc has a similar effect. These effects promote conformational changes in calmodulin which induce the active form of the protein and initiate enzyme and perhaps growth factor activity.

These results lead to the speculation of the role of metals in these growth factors. Specifically zinc function growth factors, particularly zinc in gustin, promotes the active form of the factor and by this activation the trophic effects of this substance (i.e., the growth and development of taste buds) are initiated.

REFERENCES

1. R. I. Henkin, R. E. Lippoldt, J. Bilstad, and H. Edelhoch, 1975, A zinc protein isolated from human parotid saliva, Proc. Natl. Acad. Sci. USA, 72:488.
2. A. R. Shatzman and R. I. Henkin, 1983, Proline, glycine, glutamic acid rich pink-violet staining proteins in parotid saliva are phosphoproteins, Biochem. Med., 29:182.
3. J. S. Law and R. I. Henkin, 1986, Low parotid saliva calmodulin in patients with taste and smell dysfunction, Biochem. Med. Met. Biol., 36:118.
4. J. S. Law, N. Nelson, K. Watanabe, and R. I. Henkin, 1987, Human salivary gustin is a potent activator of calmodulin-dependent brain phosphodiesterase, Proc. Natl. Acad. Sci. USA, 84:1674.

THE EFFECTS OF HIGH AND LOW CHROMIUM YEAST SUPPLEMENTATION ON GLUCOSE

METABOLISM OF NONINSULIN DEPENDENT DIABETIC PATIENTS

Cheng Nan-Zheng[1], Jiang Gui-Rong[1], Xu Xing-Yiou[1], Zhang Mei-Fang[1], Hu Xiao-Lin[1], Zhao Zhong-Liang[1], Zhang Guo-Wen[1], Wang Jin-Fu[1], Yin Chang-Rong[1], Wang Xiou-Fang[2], Lian Xi-Zhen[3], and Bei Zhen-Zhu[4]

[1]Institute of Aviation Medicine, [2]Yongding Rd. Hospital, [3]721st Hospital, [4]304th Hospital
Beijing, China

INTRODUCTION

Chromium in the trivalent state is an essential element required for normal carbohydrate and lipid metabolism; in the form of hexavalent compounds it is a carcinogen. Chromium deficient animals display impaired glucose metabolism. Chromium deficiency in humans is also characterized by impaired glucose tolerance. Insufficient dietary Cr leads to signs and symptoms similar to those associated with diabetes and cardiovascular diseases. It has been reported that chromium supplementation can improve glucose metabolism. Most studies of chromium deficiency have been in Western countries, but it is not known whether Chinese have chromium deficiency. The purpose of this study was to investigate whether chromium supplementation would improve glucose tolerance of Chinese diabetic patients.

MATERIALS AND METHODS

Sixty-three diabetic patients (31 men and 32 women) ranging in age from 22-69 years participated in the study. All were type II diabetics without ketosis. Potential subjects were excluded if they had diabetes secondary to other endocrine problems, liver and kidney disorders, evidence of gastrointestinal malabsorption or chronically ingested yeast. The patients were divided into three groups randomly: Group A (23: 10 men; 13 women) was administered high-Cr yeast 1 g/day (100 µg Cr/day) which was from Nutrition 21 in America; Group B (20: 11 men; 9 women) was administered placebo; Group C (20: 10 men; 10 women) was administered natural brewer's yeast 10 g/day (10 µg Cr/day) which was from China. The brewer's yeast was dried by vacuum from 60° to 80° C and ground. The yeast powder was mixed and compressed into biscuits without sugar. There were no imposed restrictions on eating, drinking or living patterns except (1) the dose of medicine could not be changed; (2) every morning a compressed biscuit (62.5 g) must be eaten; (3) subjects were requested to fast from 10 p.m. in the evening before testing. All the patients completed the two month study and all the blood samples were drawn. After the fasting blood sample was drawn (time zero), a steamed bread meal was ingested, and additional blood samples were taken at 60, 120, and 180 min. The steamed bread test meal contained approximately the same amount of carbohydrate as a 75 g glucose load.

Blood glucose was measured by the glucose-oxidase method. The area of glucose was calculated as:

$$\frac{X_0 + 2X_{60} + 2X_{120} + X_{180}}{2}$$

where subscripts denote glucose value at 0, 60, 120, 180 min. Results are expressed as the mean $\pm$ SEM. The paired T, D and F tests were used for statistical analysis.

RESULTS

Before the experiment and after one month, the blood glucose values were not significantly different between each group. After 2 months, the fasting blood glucose and glucose area in Group A were decreased significantly ($P < 0.01$). Neither Group B nor Group C changed significantly (Table 1). After 2 months the glucose tolerance curve in Group A was significantly improved while glucose tolerance of Groups B and C hadn't changed significantly. There were no differences due to age or sex.

CONCLUSIONS

Noninsulin-dependent diabetic patients were given supplements of 100 μg Cr/day in the form of high-Cr yeast. After 2-mo the fasting blood glucose concentrations were decreased significantly and glucose tolerance was improved. Results suggested that some diabetic patients may be short of Cr. So Cr supplementation can be used as a treatment for these kind of diabetic patients.

Table 1. The blood glucose concentrations of diabetic patients before and after 2 months supplementation.

Group*	n	Fasting blood glucose			Blood glucose area		
		Before	After	P	Before	After	P
A	23	191 $\pm$ 78	167 $\pm$ 72	<0.01	938 $\pm$ 351	809 $\pm$ 272	<0.01
B	20	183 $\pm$ 66	172 $\pm$ 64	>0.05	839 $\pm$ 330	857 $\pm$ 309	>0.05
C	20	185 $\pm$ 56	185 $\pm$ 65	>0.05	877 $\pm$ 217	880 $\pm$ 241	>0.05

*Group A = High-Cr Yeast
Group B = Placebo
Group C = Natural Brewer's Yeast

STRESS INDUCED CHANGES IN INDICES OF ZINC, COPPER,

AND IRON STATUS IN U.S. NAVY SEAL TRAINEES

A. Singh, B.A. Day, J.E. DeBolt, and P.A. Deuster

Department of Military Medicine
Uniformed Services University of the Health Sciences
Bethesda, MD 20814

INTRODUCTION

Decreased serum zinc (Zn) and iron (Fe) concentrations and increased serum copper (Cu) concentrations are some of the acute phase responses to infection, inflammation, and physical trauma (1). Lower serum Zn but higher Cu concentrations have also been observed in highly trained endurance athletes as compared to untrained men (2,3). Whether similar patterns are observed in response to physical training programs or when physically active, healthy individuals are exposed to chronic stress has not been previously studied.

METHODS

The study was conducted at the Basic Underwater Demolition/ SEAL Special Warfare Training Center, Coronado, CA. Blood samples and 24 hour urine collections were obtained from 38 men (mean age 22.4±0.3 years) prior to training (BS), after 5 weeks of training but before Hellweek (Pre-HW), and after Hellweek (Post-HW). Hellweek is a 5 day period when trainees undergo several rotations of diverse activities, such as field exercises, obstacle course runs, and swimming, while being sleep deprived. Twenty-four hour diet records were obtained at BS and Pre-HW and dietary intake during Hellweek was determined by direct observation.

Plasma, red blood cell (RBC), and urine zinc and plasma copper were analyzed by flame atomic absorption spectrophotometry. Urine creatinine (cr) was measured by centrifichem. Plasma ferritin was measured by radioimmunoassay. Plasma ceruloplasmin, and the Zn binding proteins, albumin, and alpha 2-macroglobulin were measured by radial immunodiffusion. Dietary data (food and supplements) were analyzed using the Intake Nutritional Analysis System. The data were statistically analyzed by ANOVA with repeated measures.

RESULTS

Zn intakes were similar at BS, Pre-HW and during Hellweek (Table 1). Plasma Zn concentrations remained unchanged after 5 weeks of training but decreased significantly Post-HW (Table 2). Albumin concentrations showed a pattern similar to plasma Zn (BS=4.5±0.1; Pre-HW=4.5±0.1; Post-HW= 4.1 ± 0.1 g/dl; p<0.05). Although alpha 2-macroglobulin increased with training there was a significant decrease Post-HW (BS=295.5±10.2; Pre-HW = 304.8 ± 10.5; Post-HW=273.1±10.3 mg/dl; p<0.05). RBC Zn decreased significantly with training but was not affected by Hellweek (Table 2). Urinary Zn excretion increased with training and increased even further Post-HW.

Table 1. Mean daily (± SD) Zn, Cu and Fe intakes

	BS	Pre-HW	During HW
Zn (mg)	25.2 ± 16.3	22.2 ± 13.1	23.6 ± 6.8
Cu (mg)	3.2 ± 1.7	2.9 ± 1.7	3.0 ± 1.0
Fe (mg)	27.1 ± 15.5	25.4 ± 11.1	35.4 ± 7.7

Table 2. Biochemical indices of Zn, Cu, and Fe status (Mean ± SEM)

	BS	Pre-HW	Post-HW
Zn:			
Plasma Zn (μg/dl)	81.6 ± 2.5^a	83.7 ± 2.4^a	54.6 ± 1.6^b
RBC Zn (μg/g Hb)	46.1 ± 1.9^a	38.2 ± 1.8^b	37.5 ± 2.1^b
Urine Zn (μg/g cr)	291.0 ± 22.3^a	418.5 ± 52.6^{ab}	497.8 ± 93.9^b
Cu:			
Plasma Cu (μg/dl)	99.7 ± 3.3^a	116.2 ± 7.0^b	111.0 ± 2.4^b
Ceruloplasmin (mg/dl)	26.8 ± 0.6^a	28.3 ± 0.7^b	30.6 ± 0.7^c
Fe:			
Ferritin (ng/ml)	57.8 ± 5.8^a	46.1 ± 4.3^b	73.9 ± 7.7^c

($p < 0.05$; Means with different letters are significantly different)

Despite similar Cu intakes at BS and Pre-HW, plasma Cu increased significantly with training (Tables 1 and 2). Ceruloplasmin concentrations increased with training and also Post-HW (Table 2). Fe intakes, like Cu and Zn intakes, were similar at BS and Pre-HW but plasma ferritin dropped with training. Ferritin rose significantly Post-HW.

DISCUSSION

In the present study the effects of physical training on indices of Zn, Cu, and Fe status were studied in men. After 5 weeks of training, urinary Zn excretion increased, probably due to increased muscle turnover (4). Although lower plasma Zn concentrations have been reported in physically trained men as compared to untrained men (2), plasma Zn remained unchanged in the present study after a physical conditioning program. This finding may be related to the decrease in RBC Zn in that Zn could have been redistributed from RBC into plasma. The significant increase in plasma Cu and ceruloplasmin with training may reflect Cu's role in cytochrome c oxidase activity and Fe transport (3). Despite apparently adequate Fe intakes, ferritin decreased significantly indicating that training reduced Fe stores. This decrease may reflect increased Fe utilization and/or loss.

The changes observed in blood minerals and proteins Post-HW as compared to Pre-HW were characteristic of the acute phase response (1). Plasma albumin and Zn decreased while ceruloplasmin and ferritin concentrations increased significantly. The drop in plasma Zn may be explained either by sequestration of Zn by the liver induced by glucocorticoids and/ or interleukin 1, or by tissue depletion (4). It is unlikely that the observed decrease reflected inadequate intake in that Zn intake during Hellweek was well above the Recommended Dietary Allowances. However, urinary Zn excretion was elevated Post-HW. Thus, the possibility of tissue depletion due to accelerated Zn loss is a likely explanation. Future studies should address the mechanisms responsible for the observed changes induced by chronic training and stress.

REFERENCES

1. Pepys MB, Baltz ML. Acute phase proteins with special reference to C-reactive protein and related proteins (pentaxins) and serum amyloid proteins. Advances in Immunology 1983; 34:141-212.

2. Dressendorfer RH, Sockolov R. Hypozincemia in runners. Phys Sportsmed 1980; 8(4): 97-100.

3. Lukaski HC, Bolonchuk WW, Klevay LM, Milne DB, Sandstead HH. Maximal oxygen consumption as related to magnesium, copper, and zinc nutriture. Am J Clin Nutr 1983; 37: 405-415.

4. Cousins R.J. Toward a molecular understanding of Zinc metabolism. Clin Physiol Biochem 1986; 4:20-30.

ZINC-DEPENDENT FAILURE OF THYMIC HORMONE

IN HUMAN PATHOLOGIES

Nicola Fabris, Eugenio Mocchegiani, and Rita Palloni

Immunology Center, I.N.R.C.A. Res. Dept.
Ancona, Italy

It is generally accepted that the thymus produce hormonal factors, such as the facteur thymique serique (FTS), more recently called thymulin in its zinc-bound form , which induce proliferation and differentiation of T-cells, responsible for cell-mediated immunity[1]. The plasma level of thymulin is high during ontogeny, both in animals and man, and progressively declines thereafter to reach nearly undetectable levels by the 5th decade in man[1] and by the 12-15th month in mice[2].

The causes for such an age-dependent decline of plasma thymic hormone level are still undefined; both an intrinsic failure of the thymus[2] or an age-related alteration in microenvironmental factors relevant for the endocrine function of the thymus[2] have been taken into consideration. Among these microenvironmental factors, a peculiar role of zinc ion bioavailability has been recently hypothesized, since it has been demonstrated that at least one of the best thymic hormones, i.e. the facteur thymique serique, is biologically active only when bound to zinc ions[1,3]. With these premises we have measured the plasma level of thymulin in different human pathological conditions, characterized by more or less pronounced deficiency of zinc. Thymulin has been measured by the rosette-inhibition assay[2] or by the modified assay which takes into account the role played by zinc[4]: the first method measured only biologically active thymulin (Zn-FTS); the second method measures total thymic hormone concentration, both active and inactive (Zn-FTS +FTS).

The serum level of zinc has been measured by atomic absorption spectrometry.

The human conditions investigated have been various congenital diseases, such as trysomy 21, Duchenne muscular distrophy, acrodermatitis enteropathica and cystic fibrosis. The table reports the data on plasma thymulin activity (total and active thymulin) and on serum zinc level.

Table 1

Pathology	Zinc serum level (ug/dl)	Total thymulin (log$_2$ of the titre)	Active thymulin (log$_2$)
Normal	113 ± 2.4	5.8 ± 0.5	5.3 ± 0.4
Down	86 ± 3.9	4.8 ± 0.4	1.0 ± 0.2
Cystic fib.	71 ± 3.0	4.7 ± 0.5	1.7 ± 0.3
Duchenne	67 ± 9.4	3.5 ± 0.5	1.2 ± 0.2
Acroderm. e.	58 ± 1.3	5.4 ± 0.4	1.4 ± 0.3

From these data it can be deduced that in all pathological conditions tested the total thymic hormone concentration is within normal range when compared with the values observed in age-matched normal children, whereas the active fraction is strongly reduced. It appears also that the defect in zinc activation of thymic hormone is not strictly correlated with the deepness of serum zinc deficiency. If the data on zinc serum level from each subject were plotted against the corresponding values of active thymulin, the lack of correlation was even more evident.

These findings on one hand demonstrate that zinc deficiency is present in all the congenital diseases studied, though the reduction is deeper in acrodermatitis enteropathica, and that such a deficiency causes a loss of activity of thymic hormone, in spite of a nearly normal production by the thymus. On the other hand, they pinpoint that the plasma zinc values may not reflect the real bioavailability of the element and that the determination of the ratio total/active thymulin may represent a better measure of zinc bioavailability for thymulin activation and more in general for the detection of marginal zinc deficiencies.

This work was supported by Health Ministery-INRCA targeted project on "Nutrition in elderly". We thank Mr. M. Marcellini and Mrs. N. Gasparini for his technical assistance.

REFERENCES

1. M. Dardenne, J. M. Pleau, B. Nabama, P. Lefancier, M. Denien, J. Choay, and J. F. Bach, Contribution of zinc and other metals to the biological activity of the serum thymic factor, Proc. Natl. Acad. Sci. USA 79:5370 (1982).
2. M. A. Bach, and G. Beaurain, Respective influence of extrinsic and intrinsic factors on the age-related decrease of thymic secretion, J. Immunol. 122:2505 (1979).
3. N. Fabris, E. Mocchegiani, L. Amadio, M. Zannotti, F. Licastro, and C. Franceschi, Thymic hormone deficiency in normal ageing and Down's syndrome: is there a primary failure of the thymus?, Lancet 1:983 (1984).
4. N. Fabris, E. Mocchegiani, S. Mariotti, F. Pacini, and A. Pinchera, Thyroid function modulates thymic endocrine activity. J. Clin. Endocrinol. Metab. 62:474 (1986).

EFFECT OF COBALT DEFICIENCY ON THE IMMUNE FUNCTION OF RUMINANTS

Allan MacPherson, George Fisher and Jessie E. Paterson

The West of Scotland Agricultural College
Auchincruive
Ayr, Scotland

Cobalt deficiency has been shown to lead to impaired immune function in ewes[1] and calves[2,3]. This has caused enhanced susceptibility to infection in sheep[4] and reduced viability in newborn lambs[1,5]. This study was designed to monitor the effects and consequences of cobalt depletion and subsequent repletion on the immune function of ruminants. Two experiments with ewes and one with calves were set up as follows.

In Experiment 1, sixty Scottish Blackface x Swaledale ewes were assigned to three treatment groups: 1. Co-deficient intake throughout pregnancy "NS"; 2. Initially Co-sufficient intake but deficient from mid-pregnancy "HS", and 3. Co-sufficient intake throughout pregnancy "FS". A Co-deficient diet of Timothy hay, flaked maize and prairie meal at 0.06, 0.026 and 0.03 mg Co/kg DM respectively was fed from tupping. Treated animals received a weekly oral dose of 0.7 mg Co/head. Ewes were bled and serum vitamin B_{12} concentrations measured. Eight sheep from each group were monitored for neutrophil function (NF). Lambing details were recorded. Experiment 2 was identical to Experiment 1 with the exception that the HS group was initially Co-deficient but then repleted from mid-pregnancy. In Experiment 3, six Friesian calves were maintained on a Co-deficient diet as described above until their serum vitamin B_{12} concentrations were <100 ng/1. They were then supplemented with Co and/or vitamin B_{12} until the end of the experiment. NF and vitamin B_{12} were monitored every two weeks.

Results from the three experiments are presented in Tables 1-2. In Expt. 1 the mean % kill by neutrophils in the NS and HS groups fell significantly below that of the FS group some four weeks after serum vitamin B_{12} concentrations reached the lower normal limit of 400 ng/1. Nine neonatal lamb mortalities occurred in the NS group compared to only one in each of the other two groups. In Expt. 2 both Co-depleted groups exhibited rapid decline in % kill after vitamin B_{12} values had dropped to 400 ng/1, falling from values in excess of 50% to around 25% in 80 days. After Co repletion, NF in the HS group returned to normal within 30 days; faster than the improvement in vitamin B_{12} concentration. Lambing has not yet been completed but to date there have been 4, 1 and 0 deaths in the NS, HS and FS groups respectively. In Experiment 3 the calves had initial NF values around 30% kill. These declined during Co depletion to range from 7-12% and paralleled a decline in serum vitamin B_{12} from 200 ng/1 to around 60 ng/1. On repletion NF values increased to 20% within 8 weeks and to almost 30% in some of the calves in a further 4 weeks. Although there was a concurrent response in vitamin B_{12}

concentration in all calves in some it was insufficient to restore values to the putative normal range of >200 ng/1. During depletion the calves continued to grow in size but not in condition and their coat lacked lustre and they were decidedly nervous when handled. Repletion reversed all these features.

Table 1. Ewe Serum Vitamin B_{12} (ng/1) and NFT

Weeks on trial	Treatment					
	NS		HS		FS	
	Vit B_{12}	% Kill	Vit B_{12}	% Kill	Vit B_{12}	% Kill
Expt. 1 4	369	51.0	594	50	281	46.0
10	353	28.8	843	36.3	867	36.8
17	275	30.8	419	34.5	554	43.8
21	258	24.0	237	27.0	791	44.6
27	149	21.8	199	20.5	651	46.5
Expt. 2 8	389	50.8	363	49.3	466	53.3
14	176	29.3	192	34.3	428	51.8
17	159	23.5	193	28.0	411	50.8
25	109	25.0	261	35.3	403	56.0
31	193	27.5	408	49.5	613	52.0

Table 2. Calf Serum Vitamin B_{12} and NFT

Weeks Pre and Post-Repletion	Serum Vitamin B_{12} (ng/1)	Neutrophil Function Test (% Kill)
Start	220	34
-30	122	21
-24	108	16
-4	57	10
0	65	11
+4	136	15
+6	169	18
+8	195	20

The results presented indicate that Co deficiency adversely affected immune function of both sheep and cattle with particularly severe consequences for lamb viability. Cobalt repletion restored neutrophil function in both species.

REFERENCES

1. G. Fisher and A. MacPherson, Co deficiency in the pregnant ewe and lamb viability, in "Proc. VI Int. Conf. Production Disease in Farm Animals" Belfast (1986).
2. C.L. Wright, A. MacPherson and C.N. Taylor, The effects of Co deficiency in calves, in "Proc. XII World Congress on Disease of Cattle", Amsterdam (1982).
3. A. MacPherson, D. Gray, G.B.B. Mitchell and C.N. Taylor, Ostertagia infection and neutrophil function in cobalt deficient and sufficient cattle, Brit. Vet. J. (1987) In Press.
4. A. MacPherson, F.E. Moon and R.C. Voss, Biochemical aspects of Co deficiency in sheep, Brit. Vet. J. 132, 294 (1976).
5. W.R.H. Duncan, E.R. Morrison and G.A. Garton, Effects of Co deficiency on pregnant ewes and their lambs, Brit. J. Nutr. 46, 337 (1981).

THE EFFECT OF EXCESS ZINC ON cAMP SYNTHESIS AS A REGULATORY FACTOR IN TUMOUR CELL PROLIFERATION

Noel S Skeef and John R Duncan

Department of Biochemistry
Rhodes University
Grahamstown, South Africa

INTRODUCTION

Dietary zinc excess has been shown to result in reduced tumour growth in certain types of cancers[1]. Recent studies indicating a function for zinc in essential fatty acid metabolism have led to proposals of a mechanism of cell growth regulation involving control of cAMP synthesis[2]. In this mechanism zinc is suggested to regulate the activity of delta-6-desaturase, a rate controlling enzyme in the synthesis of the omega-6-fatty acids. These fatty acids serve as the precursors for the synthesis of prostaglandins (PG's) such as PGE_1 which is an important stimulator of cAMP accumulation in cells. cAMP in turn appears to have a regulatory role on cell proliferation[3]. The purpose of this study was to investigate the effect of zinc supplementation on cell proliferation in both normal and malignant-tumour cells and to relate this effect to a role in one aspect of the above relationship, an involvement in cAMP synthesis.

METHODS

Animals and diets. Rats were fed diets containing 50 ug zinc/g (control) or 500 ug zinc/g (zinc- excess). Four weeks after injection with hepatoma(350) cells in the ip cavity the animals were sacrificed and the tumours removed, weighed and cAMP concentrations determined. As controls, some animals were not injected with hepatoma cells. Partial hepatectomies were performed on these animals. The fraction of liver removed at that time was saved and is referred to as normal resting liver. After 24 hrs the regenerating liver was removed.

Cell culture. Type 350 hepatoma cells and non-malignant monkey kidney (MDBK) cells were cultured (0.5×10^6 cells/ml) in medium supplemented with 0-5 ug/ml of zinc. (The initial zinc concentration of the growth medium was 0.63 ug/ml). After a period of 4-5 days the cells were trypsinised, counted and cAMP concentration determined.

RESULTS and DISCUSSION

Dietary zinc excess resulted in reduced _in vivo_ tumour growth which was inversely correlated with a significant increase in cAMP concentration (Table 1). In the control liver tissue (normal resting and regenerating) the effect of dietary zinc was non-significant.

Table 1. Effect of dietary zinc on in vivo tumour growth and cAMP concentration in hepatoma (350) and normal liver cells.

Group	Tumour Wt (g)	cAMP conc (pM/mg prot)		
		Hepatoma	Resting Liver	Regenerating Liver
Control	4.13+0.94	10.05+0.94	3.30+0.17*	3.64+0.21*
Zinc excess	3.71+0.56	12.41+1.15*	3.63+0.17*	4.25+0.23*

*p<0.05 when compared to control

In hepatoma cells grown in vitro a significant reduction in tumour cell proliferation was observed when zinc was added to the medium (Table 2). As in the case of the in vivo studies, reduced cell growth was inversely correlated with increased cAMP concentration in these cells. Addition of zinc at a level of 3 ug/ml resulted in a slight stimulation of cell proliferation and cAMP concentration in the non-malignant MDBK cells but there was no statistically significant effect of zinc addition in either case.

Table 2. Effect of zinc on growth and cAMP concentration in hepatoma and MDBK cells grown in vitro.

Cell Type	Zinc Added (ug/ml)	Cell Number ($\times 10^{-6}$)	cAMP Conc (pM/mg prot)
Hepatoma	0	4.45+0.50	6.75+1.2
	1	2.45+0.50*	10.00+0.7*
	3	2.40+0.45*	13.05+0.8*
	5	1.25+0.25*	24.50+5.0*
MDBK	0	0.45+0.15	10.00+2.0
	1	0.85+0.40	10.30+2.2
	3	1.05+0.35	17.50+4.5
	5	0.95+0.35	8.30+1.3

*p<0.05 when compared to 0 added zinc

The results of this study support the proposal that there is a close relationship between zinc, cAMP and cell proliferation in rapidly dividing cells such as tumour cells. The fact that a similar relationship was not found in either non-transformed cells in vivo or in vitro suggests that the association between zinc and cAMP synthesis may be cell specific. The mechanism of zinc effect on cAMP synthesis is uncertain but may be the result of an involvement with essential fatty acid and protaglandin metabolism as outlined in the introduction.

REFERENCES

1. J.R.Duncan and I.E.Dreosti, Zinc intake, neoplastic DNA synthesis, and chemical carcinogenesis in rats, J. Natl. Cancer Inst. 55:195 (1975).
2. D.F.Horrobin, A biochemical basis alcohol and alcohol induced damage includingfetal alcohol syndrome and cirrhosis, Med.Hypoth.6:929(1980)
3. M.M.Burger, B.M.Bombick, B.M.Breckenridge and J.R.Sheppard, Growth control and cAMP in the cell cycle, Nature New Biol. 239:162(1972).

ZINC AND THE BACTERIOSTATIC QUALITY OF HUMAN MILK

N.F. Krebs, A.M. Novacky, R.T. Ellison III, and K.M. Hambidge

University of Colorado Health Sciences Center

Department of Pediatrics, Denver, CO 80262

INTRODUCTION

Zinc has been proposed to be an important antibacterial factor in human amniotic and prostatic fluids. The objectives of this study were to investigate the possible role of zinc as an antibacterial factor in early milk in which the zinc concentration is elevated compared with that of more mature milk, and to determine if any such observed effects are concentration related within a physiologic range for zinc in human milk.

METHODS

This was an in vitro study in which the bacteriostatic effects of early and mature human milk were compared. Inorganic zinc was added to mature milk samples to levels comparable to those in early milk.

With informed consent, manually expressed milk samples were obtained from healthy mothers, 10 within 4 days of delivery and 10 at $\geq$ 4 months of lactation. The mean ($\pm$ S.D.) stage of lactation for the 2 groups were 63 $\pm$ 22 h and 10 $\pm$ 6 months. Initial processing of samples included centrifugation for cell and lipid removal and ultrafiltration for sterilization. Aliquots of processed milk were analyzed for zinc concentration by atomic absorption spectrophotometry (1). Mean zinc concentrations for the early and mature milks, respectively, were 8.07 $\pm$ 3.46 and 0.87 $\pm$ 0.55 µg/ml.

Bacteriostatic activity against Escherichia coli 01-11 strain was determined by incubation in 70% milk and trypticase soy agar at 37°C, with plate counts at 0 and 5 h. Zinc sulfate was added to mature milk samples to provide concentrations of 5, 10, and 15 µg zinc/ml of milk. Control cultures consisted of identically processed heated mature milk incubated with organisms as described for test samples. A bacteriostatic index (BI) was computed by calculating the difference between the log growth for the heated milk control and log growth for each milk sample. Mean BI's of the early and mature milk samples were compared to zero and to each other by paired comparison and 2-sample t-tests, respectively. Pearson correlations were computed between the zinc concentration and BI of the milk samples.

RESULTS AND DISCUSSION

The early milk samples demonstrated significant bacteriostatic activity, with a mean BI of 1.95 ± 0.33 (p < 0.001). The mean BI of the mature milk samples without added zinc was 1.40 ± 0.40, which was also significantly greater than zero but was significantly less than the BI of the early milk. Addition of zinc to the mature milk samples did not result in significantly enhanced bacteriostasis. The highest mean BI of 1.54 ± 0.42 for the milk plus 15 µg/ml zinc was still significantly less than that of the early milk (p < 0.001). Correlations between the zinc concentrations and the BI for both early and mature milk samples were not significant.

The zinc concentration in human prostatic fluid is very high (> 150 µg/ml) (2), while that of amniotic fluid is much lower (< 0.5 µg/ml) (3). In both of these fluids, zinc has been proposed to have an antibacterial function. The mean zinc concentration of the early milk samples in this study was intermediate to those of prostatic and amniotic fluids, and was approximately tenfold greater than the concentration of the mature milk samples. Estimates of the newborn infant's nutritional zinc requirement suggest that only about 30% of the zinc present in early milk would need to be absorbed to meet nutritional needs (4). This raised the possibility of the high zinc concentration in early milk having some additional non-nutritional function. Although the greater antibacterial activity of human colostrum compared to mature milk has been attributed to differences in IgA and lactoferrin concentrations, the marked decline in zinc concentration over the early days of lactation was hypothesized to also contribute to the differences in antibacterial activity.

The results of this study demonstrated the anticipated greater antibacterial effect of early human milk, as well as significant bacteriostatic activity of mature milk. In contrast to findings with amniotic fluid (3), the addition of inorganic zinc to mature milk did not significantly enhance bacteriostasis. These results do not preclude the possibility of zinc interacting with other known antibacterial systems in human milk. Recent work with lactoferrin suggests that its antibacterial activity is greater in the presence of zinc compared to that of lactoferrin alone (5). Additional research will be required to elucidate the possible interaction of zinc and lactoferrin over the course of lactation which may partially account for observed longitudinal changes in bacteriostatic activity of human milk. (Supported by NIADDKD, 5R22 AM12432, RR69 NIH, General Clinical Research Center).

REFERENCES

1. N.F. Krebs, K.M. Hambidge, M.A. Jacobs, et al. The effects of a dietary zinc supplement during lactation on longitudinal changes in maternal zinc status and milk zinc concentrations. Am J Clin Nutr 41:560 (1985)
2. W.R. Fair and R.F. Parrish. "The Prostatic Cell: Structure and Function," Alan R. Liss, New York (1981).
3. P. Schlievert, W. Johnson. R.P. Galask. Bacterial growth inhibition by amniotic fluid. V. Phosphate-to-zinc ratio as a predictor of bacterial growth-inhibitory activity. Am J Obstet Gynecol 125:899 (1976).
4. N.F. Krebs and K.M. Hambidge. Zinc requirements and zinc intakes of breast-fed infants. Am J Clin Nutr 43:288 (1986).
5. J. Stuart, S. Norrell, and J.P. Harrington. Kinetic effect of human lactoferrin on the growth of Escherichia coli O111. Int J Biochem 16:1043 (1984).

TRACE ELEMENT AND MACRO ELECTROLYTE BEHAVIOUR

DURING INFLAMMATORY DISEASES IN CATTLE AND SHEEP

G.M. Murphy[1], T.D. St. George[2], V. Guerrini[1], R.G. Collins[1],
A.C. Broadmeadow[1], M.F. Uren[2] and D.L. Doolan[2]

1. Queensland Department of Primary Industries, Animal
    Research Institute, Yeerongpilly, Q 4105 Australia

2. CSIRO Division of Tropical Animal Science, Long Pocket
    Laboratories, Indooroopilly, Q 4068 Australia

Ephemeral fever is a disease of cattle caused by bovine ephemeral
fever (BEF) virus, a rhabdovirus. It occurs in a wide band of tropical
and sub-tropical Asia, Africa and Australia and is spread by insect
vectors (St. George et al., 1984). In susceptible cattle, BEF causes high
morbidity and variable mortality. Effects are worst in prime, fat cattle
and high producing cows. St. George et al. (1986) have argued that the
viraemia should be considered as an inflammatory/toxic response as shown
by the marked neutrophilia and elevated fibrinogen levels. We (Murphy
et al., 1986) have provided physiological support for this hypothesis by
measuring changes in circulating levels of Fe, Zn and Cu during the
viraemia. As well, we have confirmed an earlier report (St. George et al.,
1984) that in affected animals (a) uncompensated hypocalcaemia (plasma
Ca < 2.0 mM L^{-1}) is commonplace; and (b) the cardinal signs (tachycardia,
tachypnea, ruminal stasis and sternal recumbency) are consistent with the
gross disturbance of calcium homeostasis. More importantly, phenyl-
butazone treatment of BEF affected cattle has shown that the trace
element changes and hypocalcaemia are independent of fever per se (Murphy
et al., 1986). Overall, this physiological expression of BEF is consistent
with an Interleukin-1 initiated sequence.

Blowfly strike in sheep (myiasis) costs the Australian Wool Industry
in excess of A$150 million each year. Myiasis occurs when larvae of
Lucilia cuprina (primary Australian sheep blowfly) infest the fleece.
Struck sheep rapidly become pyrexic and anorexic, heart and respiration
rates increase markedly as does the level of circulating neutrophils
(Broadmeadow et al., 1984). Again, well conditioned animals are often
worst affected, and even moderate strike (ca. 4000 larval burden) can
reduce exercise tolerance. Sternal recumbency is common in severely
affected sheep. Broadmeadow et al. (1984) concluded that the changes
during fly strike were due to severe toxic challenge. Our results
(Fig. 1: plasma Fe, Cu, Zn) support this conclusion. As well, we found
elevated fibrinogen (> 8g L^{-1}) and again, as in BEF viraemia, disturbed
calcium homeostasis (Fig. 1) consistent with the observed cardinal signs.

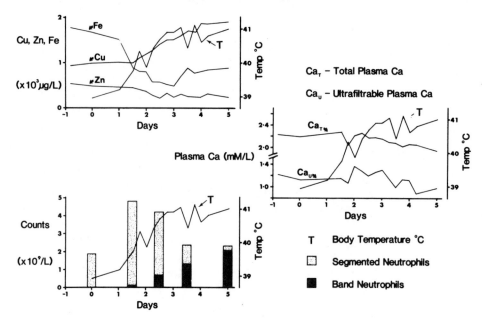

Day 0 - Start of Flystrike

Fig. 1. Physiological response during myiasis.

We are continuing to investigate the mechanism(s) which enable(s) these two pathogens, one viral, the other parasitic, to initiate the same cascade of physiological dysfunction and similar, though not identical, clinical disease. Ephemeral fever is largely self-limiting hence its synonym, 3 day sickness, whereas myiasis continues while fly larvae remain in significant numbers.

REFERENCES

Broadmeadow, M., Gibson, J. A., Dimmock, C. K., Thomas, R. J., and O'Sullivan, B. M., 1984, The pathogenesis of flystrike in sheep, Wool Technol. Sheep Breed., 32:28.

Murphy, G. M., St. George, T. D., Uren, M. F., and Collins, R. G., 1986, The biochemistry of ephemeral fever in cattle, Arbobirus Research in Australia, Proc. 4th Symp., 307.

St. George, T. D., Cybinski, D. H., Murphy, G. M., and Dimmock, C. K., 1984, Serological and biochemical factors in bovine ephemeral fever, Aust. J. Biol. Sci., 37:341.

St. George, T. D., Uren, M. F., and Zakrzewski, H., 1986, Pathogenesis and treatment of bovine ephemeral fever, Arbovirus Research in Australia, Proc. 4th Symp., 303.

GENETIC AND BIOCHEMICAL ANALYSIS OF COPPER TRANSPORT

IN *ESCHERICHIA COLI*

*D. Rouch, J. Camakaris , G. Adcock and B.T.O. Lee

Department of Genetics, University of Melbourne, Parkville
Victoria, 3052, Australia
*Department of Microbiology, Monash University, Clayton
Victoria, 3168, Australia

Copper is an essential trace element yet it is also potentially toxic.
Complex cellular Cu management systems should exist to ensure that supply of
Cu to Cu-dependent enzymes occurs without associated toxicity. We predict
that these Cu management systems should include: (i) uptake, (ii) intracellular transport and storage, (iii) repair and nullification of Cu damage,
(iv) efflux, (v) regulation of cellular Cu levels. Our aim is to use a
genetic approach in the microorganism *Escherichia coli* to investigate these
systems. The findings in *E. coli* will provide leads for research in
mammalian systems.

RESULTS AND DISCUSSION

Isolation and Characterisation of Chromosomal Cu Transport Mutants

Cu sensitive and Cu sensitive/Cu dependent mutants were isolated following mutagenesis using N'-methyl-N'-nitro-N-nitrosoguanidine. In order to
permit isolation of mutants in essential genes, conditional lethal mutants
were isolated on the basis of Cu tolerance at the permissive temperature
(30°C) and Cu sensitivity/Cu dependence at the non-permissive temperature
(42°C). The mutants were characterised for: (i) accumulation of ^{64}Cu over
30 minutes, (ii) initial rates of uptake over a 10 second period which
yielded apparent Km and apparent Vmax values, (iii) efflux. The properties
of three of the mutants will be described. *GME111* showed normal uptake
kinetics but an increased Cu accumulation which suggests that Cu efflux is
reduced in this strain leading to Cu sensitivity. This was confirmed in
efflux studies. *GME115* has an increased apparent Vmax for Cu uptake and
therefore an increased uptake capacity. Cu sensitivity is consistent with
this as intracellular Cu control systems would presumably become saturated.
GME135 was isolated as a temperature sensitive Cu-sensitive mutant. It is
also Cu-dependent at the non-permissive temperature. However uptake, efflux
and accumulation are apparently normal at both temperatures. These
properties are consistent with an alteration in a storage function. Defective storage function means that Cu would bind to a number of intracellular
materials causing damage and therefore Cu sensitivity. Cu dependence would
result from decreased efficiency of a putative transport pathway for
donation of Cu to essential Cu-dependent functions.

Plasmid-determined Cu Resistance

Tetaz and Luke (1983) have shown a plasmid-determined Cu resistance among isolates of *E. coli* from piggery effluent where animals were fed a diet supplemented with copper sulphate. The plasmid, pRJ1004, has a molecular weight of 78.0×10^6. Presence of the plasmid extends the resistance of *E. coli* K12 from 8 mM CuSO4 to 22 mM CuSO4 in complex medium. The resistance is inducible and is proportional to the inducing dose of Cu (Rouch et al, 1985). The basis of the resistance is reduced accumulation of Cu. Two determinants contribute to the resistance. The major CuR determinant, *pCo*, has been cloned and contains at least four genes.

The aim of studies with pRJ1004 is to determine the molecular basis of the resistance, its regulation, and the nature of interaction between plasmid genes and chromosomal genes involved in Cu resistance.

An inducible 25,000 molecular weight protein was detected in mini-cells programmed by *pCo* (method of Jackson and Summers, 1982).

Cell-free extracts of uninduced and Cu-induced parental and plasmid containing strains were prepared by ultrasonication and ^{64}Cu-binding proteins analysed by gel filtration on Sephadex G-75. It was found that Cu induction of the plasmid containing strain resulted in production of two Cu-binding proteins, one with a molecular weight of 11,000 and the other with a molecular weight of 25,000. As the latter protein was also observed in mini-cells it is presumably the product of the *pCo* C gene. Cu induction of the parental strain resulted in production of only the 11,000 molecular weight protein suggesting that it is chromosomally encoded. No evidence was obtained for the existence of a metallothionein-like protein.

In summary: (i) copper resistance in *E. coli* can be mediated by both plasmid and chromosomally encoded gene products. The resistance is inducible thus allowing Cu homeostasis of normal Cu levels, (ii) the existence of a carefully regulated Cu transport pathway is postulated. Cu-sensitive and Cu sensitive/Cu dependent chromosomal mutants were isolated to dissect such a pathway. Mutant classes included uptake mutants, efflux mutants, and storage function mutants. Mutant properties are as predicted by the Cu transport pathway. Current studies are aimed at using these mutants in strategies to clone the genes for the affected functions and to determine the nature of the gene products and their regulation.

REFERENCES

Jackson, W.J., and Summers, A.O., 1982, Biochemical characterisation of HgCl2-inducible polypeptides encoded by the mer operon of plasmid R100. J. Bacteriol., 151:962.

Rouch, D., Camakaris, J., Lee, B.T.O., and Luke, R.K.J., 1985, Inducible plasmid-mediated copper resistance in *E. coli*. J. Gen. Microbiol., 131:939.

Tetaz, T.J., and Luke, R.K.J., 1983, Plasmid controlled resistance to copper in *E. coli*. J. Bacteriol., 154:1263.

EFFECT OF PRIOR LUMINAL WASHING ON IRON ABSORPTION FROM DUODENUM AND JEJUNUM IN MICE: DO LUMINAL FACTORS MEDIATE THE ABSORPTION OF IRON?

Peter R. Flanagan, James Haist and Leslie S. Valberg

Department of Medicine
The University of Western Ontario
London, Ontario, Canada, N6A 5A5.

INTRODUCTION

The mechanism of iron absorption remains unclear. Several steps are probably involved in the passage of inorganic iron from the intestinal lumen to the circulation: 1) solubilization of iron and its uptake into the absorptive cell; 2) intracellular transport; and 3) transfer of the iron across the basal membrane and into the blood stream. The presence of transferrin in intestinal absorptive cells, apparently sensitive to body iron stores, has aroused interest (Savin and Cook, 1980; Johnson et al., 1983). Also, Huebers et al., (1983) proposed that luminal transferrin, secreted either from intestinal mucosal cells or from bile may act as a shuttle protein for iron absorption. If luminal factors, e.g. transferrin, are required for iron absorption then their removal should inhibit the process. In this paper we examined the effect of prior luminal washing on subsequent iron absorption in mice.

METHODS

Groups of 8 adult female Swiss mice were fed an iron-deficient diet (Flanagan et al., 1983) for 3 weeks to enhance their capacity to absorb iron. The duodenum (5cm proximal to the Ligament of Treitz) or the jejunum (5cm distal to the Ligament) was cannulated (Flanagan et al., 1980). The segments were prewashed by perfusing either isotonic saline or this solution containing 0.3mM Hepes-NaOH, pH7 or 0.3mM phosphate, pH7 for 10 min at 0.5ml/min. Subsequently, 0.2ml of 3.6μM ^{59}FeCl$_3$ in isotonic saline at pH2 was introduced into the segments which were then tied off for 10min. In a second experiment the segment was perfused with an isotonic saline containing either 50μM ^{59}Fe(III)-citrate or 2mg/ml diferric [^{59}Fe] mouse transferrin (Cappel Laboratories). Subsequently, iron uptake (^{59}Fe in the washed segment plus that in the carcass) and iron transfer (^{59}Fe in the carcass only) were measured (Flanagan et al., 1980).

RESULTS

Prewashing the duodenum or jejunum of mice with saline-phosphate or saline-Hepes buffers did not significantly affect the uptake of iron from a ^{59}FeCl$_3$ placed in either segment (Table 1). Prewashing with saline-phosphate significantly reduced iron transfer in duodenum (Table 1). In the second experiment, prewashing with isotonic saline did not affect the

Table 1.   Effect of Prewashing on Uptake and Transfer of $^{59}FeCl_3$.[a]

Prewash	Duodenum		Jejunum	
	Uptake	Transfer	Uptake	Transfer
None (Control)	0.64±0.03	0.47±0.02	0.38±0.07	0.06±0.02
Phosphate	0.58±0.01	0.37±0.02[b]	0.49±0.03	0.03±0.0
Hepes	0.64±0.02	0.44±0.02	0.42±0.03	0.04±0.01

[a] Data are nmol±SE   [b] indicates $p<0.05$ by ANOVA

absorption of iron from Fe(III)-citrate or from diferric transferrin perfused through an open loop of duodenum (results not shown). In the latter case the absorptive advantage of chelated iron over transferrin-bound iron, demonstrated by ratio of iron transfer from Fe(III)-citrate/diferric transferrin was 19 and 13 in mice with and without prewashing, respectively.

DISCUSSION

These experiments indicate that luminal factors in general, and transferrin in particular, are not necessary for inorganic iron absorption in mice. The single significant lowering of iron transfer following prewashing with phosphate (Table 1) is unlikely to be due to specific removal of an essential luminal factor because uptake was not similarly affected.

The kinetic advantage of iron absorption from simple chelates over transferrin has been confirmed by us in other experiments (Flanagan et al., 1987) and by others (Simpson et al., 1986). Although transferrin is unlikely to be a carrier of iron for the entire pathway of iron from the intestinal lumen to portal blood, it may be involved in a portion of the route, e.g. the intracellular step. We have demonstrated transferrin in the terminal web and apical cytoplasm of human intestinal absorptive cells (Banerjee et al., 1986). Transferrin receptors, apparently sensitive to body iron stores, were also present in basal and lateral membranes. The role of these proteins, if any, in the mechanism of iron absorption awaits clarification. (Supported by MRC of Canada)

REFERENCES

Banerjee, D., Flanagan, P. R., Cluett, J., and Valberg, L. S., 1986,
    Transferrin receptors in the human gastrointestinal tract. Relationship
    to body iron stores.   Gastroenterology, 91:861-9.
Flanagan, P. R., Haist, J., and Valberg, L. S., 1980, Comparative effects
    of iron deficiency induced by bleeding and a low-iron diet on the
    intestinal absorptive interactions of iron, cobalt, manganese, zinc,
    lead and cadmium. J. Nutr., 110:1754-63.
Flanagan, P. R., Haist, J., and Valberg, L.S., 1987, The mechanism of
    intestinal iron absorption in the mouse: Does luminal transferrin play
    a role? Fed. Proc., 46:1161.
Huebers, H. A., Huebers, E., Csiba, E., Rummel, W., and Finch, C. A.,
    1983, The significance of transferrin for intestinal iron absorption.
    Blood, 61:283-90.
Johnson, G., Jacobs, P., and Purves, L. R., 1983, Iron binding proteins of
    iron-absorbing rat intestinal mucosa. J. Clin. Invest., 71:1467-76.
Savin, M. A., and Cook, J. D., 1980, Mucosal iron transport by rat
    intestine. Blood, 56:1029-35.
Simpson, R.J., Osterloh, K. R. S., Raja, K. B., Snape, S. D., and Peters,
    T. J., 1986, Studies on the role of transferrin and endocytosis on the
    uptake of Fe^{3+} from Fe-nitrilotriacetate by mouse duodenum. Biochim.
    Biophys. Acta, 884:166-71.

PATHOBIOCHEMICAL ASPECTS TO THE MECHANISM OF ZINC ABSORPTION

Jürgen D.Kruse-Jarres, Eva-Maria Hecht and Wolfgang Hecht

Insitute of Clinical Chemistry, Katharinenhospital, D-7000
Stuttgart, and Department of Surgery, University of Freiburg
D-7800 Freiburg, Fed. Rep. of Germany

INTRODUCTION

The chemical compound of nutritional zinc has an intense influence on
the absorption rate. ZnS or mixed oxides with Fe or Mn are excreted in an unchanged shape, while ZnO, $ZnCl_2$, $ZnCO_3$, and $ZnSO_4$ are absorbed very well. Or-
ganically bound zinc being part of the normal nutrition and zinc complexes
(e.g. zinc bound to amino acids) are reported to be absorbed better than inorganic compounds [1,2], that are absorbed less than 10% in average. Thus high
content of proteins in the meals have a positive influence on the absorption
rate.

In order to avoid the influence of a most heterogeneous complex binding
of zinc to different sorts of food, we first applied the various zinc compounds directly to the duodenum via an external fistula and than removed the
proximal part of the intestinum.

MATERIAL AND METHODS

The studies have been done in 14 young pigs (2-4 months of age) with an
average weight of 33 kg. The animals' standard food was 55% grain, 10% wheat
bran, 8% soya, 7% rye, 4% maize, and the rest miscellaneous including the
following minerals and trace elements: zinc (120 mg/kg), iron (72 mg/kg), manganese (72 mg/kg), copper (29 mg/kg), and cobalt (0,6 mg/kg). Water was at an
unlimited disposal. The pigs did only get water on the days of experiments.

Zinc was applied using an experimental duodenal fistula (and in 2 cases
a gastric fistula): $ZnCl_2$ (111 mg/10 ml equivalent to 53,2 mg Zn^{++}), $ZnSO_4$
(235 mg/10 ml equivalent to 53,4 mg Zn^{++}), and Zn-aspartate $C_8H_{12}O_8N_2Zn$ (270
mg/10 ml equivalent to 53,5 mg Zn^{++}). Increasing amounts of the three different solutions were applied on different days: $ZnCl_2$ 111-222 mg, $ZnSO_4$ 235-
470 mg, and Zn-aspartate 270-540 mg.

After these experiments the pigs were jejunectomized by removing the
proximal half of the small intestine (approx.length of resection 3,5 m). The
experiments of zinc loading have been repeated not earlier than a fortnight
after the jejunectomy. The decisive criterion for the repetition of the zinc
loading experiments was the recovery of original zinc values after passing
through a common post-surgical hypozincaemia.

Zinc concentrations were determined in serum from venous blood. The collection of blood was carried out by using a jugular vein catheter. The zinc determinations were carried out by flameless atomic absorption spectrometry.

RESULTS

While no gastric absorption of any of the tested solutions has been observed, the analysis of variance for the duodenal absorption show the following results.

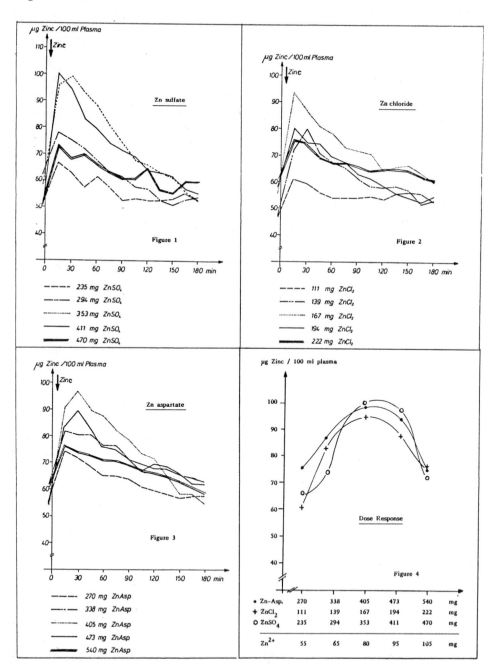

Figure 1

Figure 2

Figure 3

Figure 4

1. The direct application of zinc without linkage to food causes a duodenal absorption (p $\leq$ 0,01).
2. No differences between the three Zn-compounds are observed (p $>$ 0,005).
3. Using changing dosages significant differences in the absorption behaviour are obvious (p $\leq$ 0,01). Average dosage (80 mg Zn^{++}) show optimal results of absorption with regard to all three compounds (Figure 1-4).
4. Significant individual differences with regard to zinc concentrations in venous blood are remarkable (p $\leq$ 0,001).
5. The main absorption site is the proximal intestinum, as can be demonstrated after jejunectomy (p $\leq$ 0,01).

DISCUSSION

Opposite to our findings in rabbits (3) and other authors' results (4) no significant differences have been found between the three compounds $ZnCl_2$, $ZnSO_4$ and Zn-aspartate in pigs. Under the described conditions the absorption of zinc does not seem to be dependent on the compound of zinc applied to the organism by a duodenal fistula. First of all this is due to the fact, that the conditions in rabbits are hardly to be compared with those in pigs with respect to the absorption conditions. The latter are more similar to the conditions in human beings and therefore more relevant to the human zinc metabolism. The degree of zinc absorption is distinclty less under conditions of poor plant food than under an uptake of a mixed nutrition. The reason for this different availability of zinc for the absorption seems to be the high portion of plants' phytane, a hexaphosphate ester of inositol. The food of the pigs in our experiments was rich of phytane as well; but we stopped feeding one day before the experiment and applied the zinc compound directly to the site of absorption without mixing it to the food. Phytane or other complexes cannot be involved in the kind of preparation being more or less decisive for the quantity of the absorbed zinc. Thus these experiments have a superiority in value with respect to the function of the intestinal wall. They do not give answer to the question, if – under physiological conditions of oral intake – a special complex binding by a pancreatic ligand can differentiate between the various zinc compounds and its linkage to the food.

The experiments in the jejunectomized pigs clearly prove the essential site of the proximal intestine for the absorption of zinc. Next to nothing of the zinc will be absorbed by the gastric or the lower intestinal wall. These results have – as already be discussed above – to be seen in the context of a non-physiological application, that means without help from a hypothetic tracer, that probably could support the binding to the intestinal wall and its passing through.

CONCLUSION

No differences between the three tested Zn-compounds can be observed. Using changing dosages significant differences in the absorption behaviour are obvious. An average dosage of 80 mg Zn^{++} shows optimal results of absorption. The same experiments in the pigs jejunectomized afterwords clearly prove the essential site of the proximal intestine for the absorption of zinc. No absorption has been observed in the gastric or the lower intestine wall.

REFERENCES

(1) Kirchgeßner,M. et al. in "Spurenelemente" (Ed.I.Staib), Schattauer Stuttgart, 12-61 (1982); (2) Kirchgeßner,M. and Weigand,E. in "Spurenelemente" (Ed.E.Gladtke et al.), Thieme Stuttgart, 30-46 (1985); (3) Herr,T.N. Dissertation University of Freiburg, 1-58 (1977); (4) Schwarz,F.J., Kirchgeßner,M. in "Zinkstoffwechsel" (Ed.J.D.Kruse-Jarres), TM Bad Oyenhausen, 129-137 (1979)

THE EFFECT OF SAPONINS ON MINERAL AVAILABILITY

S. Southon, A.J.A. Wright, I.T. Johnson, J.M. Gee,
K. Price, and S.J. Fairweather-Tait

AFRC Institute of Food Research
Colney Lane
Norwich NR4 7UA, U.K.

INTRODUCTION

It has been suggested that increased consumption of saponins might
be beneficial to certain sectors of the population because of their
ability to lower serum cholesterol.[1] Some saponins, however, are able to
form insoluble complexes with Fe and Zn in vitro[2] and so may reduce the
amount of mineral available for absorption and metabolism in vivo. The
present investigation was undertaken to determine the effect of dietary
saponins on mineral utilisation in the rat. In order to evaluate the
effects under conditions of suboptimal mineral intake, the study included
feeding diets which were low in Fe and Zn, as well as control mineral-
replete diet.

METHODS

Two experiments were performed. The saponin used in Experiment 1
was a crude commercially-available extract from Gypsophila sp. (60.4%
pure w/w: Sigma, Dorset, U.K.). For Experiment 2, this crude extract was
further purified by a reversed-phase flash chromatographic technique.[3]

Experiment 1.

Male, Wistar rats (70g) were divided into 3 groups and given diets
containing 26mg Fe and 55mg Zn (Basal), 26mg Fe and 5mg Zn (Low Zn), or
13mg Fe and 55mg Zn (Low Fe). Groups were further divided so that half
the rats received diets with 20g/kg crude Gypsophila saponins. After 3
wks Fe and Zn status was assessed by Hb, PCV, liver Fe and femur Zn.

Experiment 2.

Male, Wistar rats (80g) were given basal semi-synthetic diet ad lib
for 7d, meal-fed for 2d, fasted overnight and given 3g cooked
starch:sucrose (1:1) paste containing 120µg Fe or 139µg Zn labelled with
18.5kBq ^{59}Fe or 37kBq ^{65}Zn, and varying amounts of crude and purified
Gypsophila saponins. Estimates of absorption were obtained by whole-body
counting techniques.

RESULTS

Expt. 1 Fe and Zn status of rats fed diets containing 20g/kg crude
         Gypsophila saponins

Dietary treatment	Body wt gain (g)		Total food intake (g)		Hb (g/100ml)		PCV (%)		Liver Fe (µg/g dry wt)		Femur Zn (µg/g dry wt)	
	mean	se	mean	se	mean	se	mean	se	mean	se	mean	se
Basal	158	4	399	14	13.8	0.3	42	1	169	6	192	6
Basal + saponin	137*	3	343*	5	12.9*	0.2	40	1	161	5	201	5
Low Fe	142	3	364	9	10.5	0.3	32	1	118	4	196	5
Low Fe + saponin	138	3	347	7	9.3*	0.3	29*	1	105*	2	203	3
Low Zn	133	4	356	8	13.5	0.2	41	1	188	6	61	3
Low Zn + saponin	131	5	339	9	13.5	0.2	40	1	162*	7	68	3

*Mean values significantly different ($P<0.05$) from those for rats given a
similar diet without saponin.

Expt. 2 Effect of Gypsophila saponins on Fe and Zn absorption at
         various saponin:mineral molar ratios

Saponin:Fe (Zn) molar ratio	^{59}Fe absorption (%)		
	n	mean	se
0	10	62	2
8+	8	51*	4
8‡	7	49*	2
16+	5	52*	4

Saponin:Fe (Zn) molar ratio	^{65}Zn absorption (%)		
	n	mean	se
0	8	70	1	10	45	2
0.5+	10	67	2	10	44	2
0.5‡	10	69	2	10	44	4
1+	10	64*	2	10	51	2
2+	10	62*	1	9	45	3
4+	10	58***	3	9	42	4

*Mean values within a column significantly different from control group
 (sap:Fe molar ratio 0)   *$P<0.05$, ***$P<0.001$.
+Purified saponins   ‡Crude saponins

CONCLUSIONS

Rats consuming Gypsophila saponin had reduced Fe status compared to their controls, shown by reductions in Hb concentration (Basal Group), liver Fe (Low Zn Group) and Hb, PCV and liver Fe (Low Fe Group). Increasing amounts of saponin given in a single meal progressively reduced the amount of Fe available for absorption in the rat. The effect appeared to plateau at a saponin:Fe molar ratio of 4. Values for crude and purified saponin extracts, given at the same saponin:Fe molar ratio were similar, indicating that the non-saponin fraction of the crude preparation was not responsible for the effect on Fe status. Zn absorption and status were not affected by consumption of the saponin.

In terms of human dietary intakes the lower saponin:Fe molar ratios used in the second experiment would not be excessive but, as yet, the relative importance of the molar ratio, compared to absolute amounts, is not known.

Further work is in progress to ascertain whether saponins found in the human diet have a similar effect.

REFERENCES

1. D. G. Oakenfull, Dietary fibre, saponins and plasma cholesterol, Food Technol. Aust. 33:432 (1981).
2. L. G. West, J. L. Greger, A. White and B. J. Nonnamaker, In vitro studies on saponin-mineral complexation, J. Food Sci. 43:1342 (1978).
3. K. R. Price, C. Curl and G. R. Fenwick, Flash chromatography - A simple technique of potential value to the food chemist, Food Chem. in press.

RATE OF PASSAGE OF CHROMIUM III THRU

THE SMALL INTESTINE

D. Oberleas and J.C. Smith Jr.

Food and Nutrition Section, ENRHM, Texas Tech Univ.,
Lubbock, TX 79409-4170 and Vit. Min. Nutr. Lab., USDA,
Beltsville, MD 20705, USA

INTRODUCTION

Chromium (Cr)(III), the biologically active form of Cr, is poorly
absorbed from the G.I. tract. Absorption has been reported to be in the
range of 0.4-3% of the dietary intake regardless of dose or chromium
status (1-4). Because of poor absorption, Cr has been used as a marker
for the passage of food and nutrients thru the G.I. tract. Both chromic
oxide and Cr(III) chloride (Cl) have been utilized as dietary markers.
The low level of absorption was tolerable as a marker however, an appropriate marker should transcend the G.I. tract at approximately the same
rate as the diet. There are few reports that have studied the rate of
passage of Cr(III) Cl added to the diet (5).

METHODS

The current experiment was designed to utilize Cr(III)Cl as a
dietary marker to estimate food passage in relation to pancreatically
secreted zinc. Rats were fed four similar diets for 4 weeks, differing
only between egg albumin or soybean protein and either with or without
52 mg/kg of zinc supplement. The animals were fasted for 12 hours then
fed 2 g of their respective diets containing (51)Cr(III)Cl. Radioactivity was calculated to be sufficient to provide 10,000 counts/g of diet
(Mean actual counts were 7826/g diet). Rats were allowed 1 hour to
consume the diet then were sacrificed either at 1 hour (baseline), 2, or
4 hours after the initiation of feeding. The small intestine was
removed, and segmented into 4 sections approximating the duodenum, 2
segments of jejunum, and ileum. Cr radioactivity was measured in each
segment by gamma counter.

RESULTS

At baseline (about 1 hour), 62% of the counts were recovered in the
small intestine. Of the recovered counts, only 2% were in the duodenum
and 68% (42% of original counts) were in the ileum. In one study
(unpublished) gavaged Cr was found in the feces within 1 hour. At 1
hour, all of the food was still in the stomach. At two hours after

feeding, 58% of the counts were still in the small intestine with 72% in the ileum. At 4 hours, much of the food was still in the stomach but only 32% of the counts were in the G.I. tract; 0.1% in the duodenum and 92% of the recoverable counts (29% of the counts added to the diet) were in the ileum.

Table 1. Net Counts of Chromium (III) in Various Segments of the Small Intestine

Time / Segment	Duodenum	Upper Jejunum	Lower Jejunum	Ileum
Baseline	171	506	2231	6097
Two Hours	150	268	2088	6290
Four Hours	12	109	335	4915

SUMMARY

Results indicated that Cr(III) transcends the G.I. tract much faster than organic food components making it an inappropriate dietary marker. The rapid rate of passage may also decrease the probability of absorption. In searching for a mechanism, it appears that this passage is related to the flux of water through the intestine.

REFERENCES

1.   E.J. Underwood, Chromium, in: "Trace Elements in Human and Animal Nutrition, 4th Ed.", Academic Press, New York (1977).
2.   L.L. Hopkins Jr. and K Schwarz, Chromium (III) binding to serum proteins, specifically siderophilin, Biochem. Biophys. Acta 90:484 (1964).
3.   W. Mertz, E.E. Roginski and R.C. Reba, Biological activity and fate of trace quantities of intravenous chromium (III) in the rat, Am. J. Physiol. 209:489 (1965).
4.   J.S. Borel and R.A. Anderson, Chromium, in: "Biochemistry of the Essential Ultratrace Elements", E. Frieden, ed., Plenum Press, New York (1984)
5.   D. Oberleas, Y.C. Li, and B.J. Stoecker, Intestinal transit of (51)Cr in the rat, Fed. Proc. 46:904 (Abstract 3424) (1987).

CD-BINDING PROTEINS ANCHORING TO MUCOSAL BRUSH BORDER MEMBRANE FROM RAT SMALL INTESTINAL TRACT

N. Sugawara and C. Sugawara

Department of Public Health
Sapporo Medical College
S-1, W-17, Central Ward, Sapporo, 060, Japan

Introduction

Cd absorption is influenced by several endogenous or exogenous factors such as aging, sex, food components and dietary pollutants. To resolve the mechanism of Cd absorption from the gastrointesinal tract, metallothionein (MT) is one of the most interesting endogenous substances. It seems reasonable to assume that the protein is involved in the prevention of the transport of Cd from the lumen to the body (1,2). However, the reason why Cd or Cd-MT is only absorbed slightly (3) is unclear. To get a clue understanding this point, we estimated the Cd binding to mucosal brush border membranes (BBMs) isolated from control and Cd-exposed rats.

Materials and Methods

Wistar male rats (5 weeks old) were maintained for 10 days before allocation, with food and water ad libitum. After 10 days, they were divided into two groups, control- and Cd-groups. Cd group was given deionized water containing 100 ppm Cd (CdCl2), acidified with acetic acid for 10 days. Control-group was given only deionized water. At 10 day, they were killed 12 hr after starvation. Small intestines (about 30 cm from the pylorus) were excised from the each group. To get mucosal brush border membrane (BBM), the method of Victery et al (4) was prefered here.

To estimate an affinity of Cd to BBM in vitro, cadmium chloride or Cd-MT (II) was added in the BBM suspension from the control- or Cd-group. The solution was incubated with or without Cd compounds for 20 min at 25C. After the incubation, the reaction mixture was washed by several centrifugations (35,000 g for 20 min). The final precipitate was sonicated with a sonicator, or solubilized by deoxycholate to measure Cd and protein concentrations, and to estimate Cd distribution by a Sephadex G-75 column. Total content of Cd added to the incubation solution was 5 ug in each case.

Results and Discussion

When the control BBM was incubated with CdCl2, 72% of added

419

Cd was yielded at the BBM fraction. Furthermore, Cd anchoring to the BBM was recovered only in the HMW region on Sephadex G-75 column. In the LMW region, namely MT region, Cd was not found. When Cd-MT was added, a recovery of the Cd was only 6.1%. The Cd anchoring to BBM was not MT-form. The Cd anchoring to the BBM was yielded only in the HMW region on the column, even by solubilization with deoxycholate. The Cd released from the Cd-MT may directly associate to the HMW substances. These data suggest that intact BBM isolated from the control rats doses not possess anchoring sites for MT.

Selenke and Foulkes (5) reported previously that Cd-MT (hepatic 109Cd-MT) binds to isolated renal proximal tubular BBM, in vitro. Furthermore, they mentioned the existence of two classes of binding sites with different affinities for Cd-MT. Even though some techniques and materials, such as binding assay of MT, source of BBM and purity of MT, were different from each other, their results were markedly different from our data.

Cd was detected in the BBM isolated from the Cd group. The Cd existed in the two regions, HMW- and MT-region, on the Sephadex column. Even though the induced Cd-MT existed richly in the supernatant obtained from the first centrifugation contained, MT was not a large component for Cd in the BBM fraction. The MT peak could not be detected in the supernatant obtained by the final centrifugation. Accordingly, the BBM MT peak was not a artifact contaminant from the supernatant fraction. When CdCl2 was added to the BBM suspension obtained from the Cd group, 42% of the added Cd was recovered in the BBM. All of the Cd was found in the HMW region on this column. On the other hand, when Cd-MT(II) was incubated with the BBM, Cd was not yielded as MT in the BBM. The result suggests that the BBM's sites for binding to MT was not stimulated by the exposure of Cd. When the mucosa was exposed to Cd, Cd-MT is induced in the mucosal cells. In its cytoplasma, Cd existed in the form of binding to HMW-proteins and MT, respectively. A small part of these MTs may be anchored to the BBM to transport into the body and then BBM may be saturated rapidly with a small amount of MT. Intact MT is certainly transported into the body at a very low level (3). Our data picked up the reason why Cd is transported at a very low level.

References

1. E.C.Foulkes, and D.M.McMullen, Endogenous metallothionein as determinant of intestinal cadmium absorption : A reevaluation. Toxicol. 38: 285-291(1986).
2. N.Sugawara, and C.Sugawara, Role of mucosal metallothionein preinduced by oral Cd or Zn on the intestinal absorption of a subsequent Cd dose. Bull. Environ. Contam. Toxicol. 38: 295-299 (1987).
3. M.G.Cherian, R.A.Goyer, and L.S.Valber, Gastrointestinal absorption and organ distribution of oral cadmium chloride and cadmium-metallothionein in mice. J.Toxicol. Environ. Health 4: 861-868 (1978).
4. W.Victery, C.R.Miller, and B.A.Fowler, Lead accumulation by rat renal brush border membrane vesicles. J. Pharmacol. Exp. Therap. 231: 589-596 (1984).
5. W.Selenke, and E.C.Foulkes, The binding of cadmium metallothionein to isolated renal brush border membranes. Proc. Soc. Exp. Biol. Med. 167: 40-44 (1981).

ZINC ACCUMULATION BY HEPATOCYTES ISOLATED FROM

MALE RATS OF DIFFERENT ZINC NUTRITIONAL STATUS

William A. House, Ross M. Welch and Darrell Van Campen

USDA, ARS, U.S. Plant, Soil and Nutrition Laboratory
Tower Road
Ithaca, N.Y.

INTRODUCTION

Some characteristics of Zn uptake and effects of various multivalent cations, metabolic inhibitors, hormones, and sulfhydryl-blocking agents on Zn accumulation by rat hepatocytes in monolayer culture[1] or in suspensions[2] have been reported. Hepatocytes used in these studies were isolated from rats fed Zn-adequate (ZA) diets. The study reported here was conducted to evaluate the effect of rat Zn status on Zn uptake by isolated hepatocytes.

METHODS

Male rats (ca. 200 g each) served as liver donors in three experiments. In experiment 1, all rats (nine/group) had free access to deionized water and either a Zn-deficient (ZD) (1 ppm Zn) or a ZA (40 ppm Zn) diet for 7 d. The rats were anesthetized and hepatocytes were isolated.[1] Suspensions of hepatocytes were incubated[2] in medium containing $^{65}ZnCl_2$ (360 nCi/3.6 ml); final Zn concentrations were either 25, 100, 250 or 500 uM. Aliquots (200 ul) of cells were removed from incubation vessels at 15, 30, 45 and 60 min after adding ^{65}Zn. Samples were filtered, washed three times with suspension buffer, and assayed for ^{65}Zn and protein content. Zn uptake was expressed as nmol/mg protein. In experiment 2, 10 rats were paired by weight. One rat in each pair was fed the ZD diet for 10 d, and the other rat was fed the ZA diet but food intake was limited to the amount eaten by the ZD rat. Hepatocytes were then isolated, incubated in medium containing 250 or 500 uM Zn, and Zn uptake was determined as described above. Experiment 3 was conducted to determine effects of ZD followed by Zn repletion on Zn uptake by isolated hepatocytes. Rats (n=16) were paired by weight and fed for 10 d as described in experiment 2. ZD rats were then repleted with Zn for 0, 1, 2 or 4 days. Hepatocytes were isolated from two pairs of rats on each day, and Zn uptake by cells incubated 1 h in 100 uM Zn was determined.

RESULTS AND DISCUSSION

In experiment 1, hepatocytes from ZD rats accumulated less Zn (P 0.05) than cells from ZA rats (Table 1). Food consumption by ZD rats was cyclic and lower than that of ZA rats fed ad libitum. Therefore, differences in Zn accumulation by ZD and ZA hepatocytes possibly resulted from nutritional

Table 1. In Vitro Uptake of Zn by Hepatocytes Isolated from
either Zn-adequate (ZA) or Zn-deficient (ZD) Rats[a]

Rat Zn status	Zn in medium	Minutes after adding radiozinc			
		15	30	45	60
	μM	nmoles Zn/mg protein			
ZA	25	1.4 (0.1)	1.6 (0.1)	1.8 (0.1)	1.9 (0.1)
	100	5.8 (0.6)	7.1 (0.6)	7.6 (0.6)	8.4 (0.6)
	250	14.5 (1.4)	19.0 (1.8)	22.8 (1.2)	25.0 (1.0)
	500	30.5 (2.8)	38.0 (2.4)	44.4 (2.3)	48.4 (2.4)
ZD	25	1.3 (0.1)	1.4 (0.1)	1.5 (0.1)	1.6 (0.1)
	100	5.1 (0.3)	5.8 (0.4)	6.3 (0.4)	6.6 (0.4)
	250	13.6 (0.6)	14.9 (0.7)	15.7 (0.7)	17.1 (1.1)
	500	26.8 (1.2)	28.9 (1.3)	31.1 (1.3)	33.4 (1.2)

[a]Mean (±SEM) of preparations from nine rats in each group.

factors other than Zn per se. Subsequently, food intake by ZA rats was
limited to the amount eaten by ZD rats. When incubated 1 h in 500 uM Zn,
Zn uptake by cells from pair-fed ZD and ZA rats averaged 23 and 46 nmol/mg
protein, respectively. Experiment 3 established that the effect of Zn deficiency on Zn uptake was reversible. Zn uptake by cells from ZD rats that
were repleted with Zn for 0, 1, 2 or 4 d averaged about 68, 80, 91 and 99%
of respective amounts accumulated (8 nmol/mg protein) by cells from ZA rats.

Antithetic to expectations, cells from ZD rats accumulated less Zn than
cells from ZA rats. ZD hepatocytes may have accumulated less Zn because of
reduced intracellular metallothionein content[3] or changes in cell membranes.
Zn contributes to the integrity of biomembranes,[4] possibly by protecting
sulfhydryls.[5] Notably, sulfhydryl-blocking agents inhibited Zn uptake by
hepatocytes.[1] When extracellular Zn declines from feeding a ZD diet, Zn may
dissociate from membrane sulfhydryl groups, and oxidation of these groups
may result in loss of Zn binding sites with concomitant changes in membrane
fluidity and stability.[5] Also, since Zn affects the fragility of hepatic
lysosomes,[4] the integrity of cells from ZD and from ZA rats may have been
affected differently by the cellular isolation procedures we used. However,
no effect of Zn on hepatocyte integrity was evident from ability of cells to
exclude trypan blue dye. In all experiments, apparent cellular viability
averaged 90 and 91% in preparations from ZD and ZA rats, respectively.

Our results do not directly provide evidence that Zn is necessary to
maintain the functional integrity of hepatocyte membranes. Our results do
indicate that Zn status should be considered when isolated hepatocytes are
used as models to study ion transport processes and cellular metabolism.

REFERENCES

1. M. L. Failla and R. J. Cousins, Zinc uptake by isolated rat liver
    parenchymal cells. Biochim. Biophys. Acta 538:435 (1978).
2. N. H. Stacy and C. D. Klaassen, Zinc uptake by isolated rat
    hepatocytes. Biochim Biophys. Acta 640:693 (1981).
3. R. J. Cousins, Regulatory aspects of zinc metabolism in liver and
    intestine. Nutr. Rev. 37:97 (1979).
4. J. C. Ludwig and M. Chvapil, Reversible stabilization of liver
    lysosomes by zinc ions. J. Nutr. 110:945 (1980).
5. B. L. O'Dell, Metabolic functions of zinc - a new look, in "Trace
    Element Metabolism in Man and Animals, 4" J. Howell, J. Gawthrone
    and C. White, eds., Australian Academy of Science, Canberra (1981).

IDENTIFICATION OF FIVE SITES OF REGULATION OF HUMAN

ZN METABOLISM

M. E. Wastney, R. L. Aamodt and R. I. Henkin

Georgetown University Medical Center,
Washington D.C. 20007 and NIH, Bethesda, MD 20892

INTRODUCTION

Zn metabolism in humans has been investigated through mathematical modeling of kinetic data (Foster et al., 1979; Babcock et al. 1982; Wastney et al., 1986). Through this approach Zn absorption, distribution, secretion and excretion have been measured in normal volunteers and in clinical disorders. The present studies involved tracer studies in a normal and a perturbed state to investigate sites of Zn metabolism in normal volunteers.

METHODS

^{65}Zn was administered to normal volunteers (N=32) orally, while in the fasting state. Activity was measured in plasma, red blood cells (RBC), urine and feces and over whole body, liver and thigh for 270 d while subjects consumed their regular diet (~ 10 mg Zn/d, basal state). Studies continued for a further 270 days while subjects added 100 mg Zn/d to their diets (Zn loading).

Data were analysed using SAAM/CONSAM (Berman and Weiss, 1978; Berman et al., 1983) and a model for Zn metabolism (see Wastney et al., 1986).

DEFINITION OF SITES OF REGULATION OF ZN METABOLISM

Tracer was lost more rapidly from the body during Zn loading compared to the basal state. To determine which parameters changed during Zn loading to cause the loss of tracer, basal and Zn loading data of each individual were fitted simultaneously. Changes were introduced in the basal parameters until the model solution predicted the data obtained during Zn loading. It was necessary to introduce changes in five parameters to fit the Zn loading data.

First, during Zn loading plasma Zn increased by only two-fold while Zn intake increased by 10-fold. It was necessary to reduce the amount of Zn entering plasma by decreasing the fraction of Zn absorbed from the gut.

Second, ^{65}Zn excreted in urine increased within days of Zn loading. To fit these data it was necessary to increase the fraction of Zn lost in urine.

The appearance of tracer in urine continued to increase for several weeks after the start of Zn loading. These data were not fitted by the increase in fraction excreted and it was necessary to change a third parameter to fit these data. The data could only be fitted by releasing more tracer into plasma, from a slowly turning over compartment. The only compartment whose kinetics were slow enough to be consistent with the rate of release was muscle. The third change on Zn loading was increased release of Zn from muscle.

Increasing the turnover of muscle fitted the second part of the urine curve during Zn loading but did not account for the end of the curve when tracer was still being released into urine. These data could only be fitted by reducing loss of Zn into gut. The fourth change was reduced secretion of Zn into gut.

While changes in four parameters fitted data from most tissues on Zn loading these changes were not sufficient to fit data from RBC. It was necessary to reduce the uptake of Zn by the RBC to fit these data. The fifth site of regulation was uptake of Zn by RBC.

Effect of Regulation at Five Sites on Zn Metabolism

The effect of regulation at these five sites appeared to be maintenance of long-term Zn homeostasis. When Zn intake increased by 10-fold and plasma Zn increased by 2-fold the model predicted that Zn mass in other tissues would not change significantly.

SUMMARY AND CONCLUSION

When Zn intake increased by 10-fold changes occured at five sites representing five sites of Zn regulation. Two sites were previously described, absorption and urine excretion, (Babcock et. al., 1982). The three new sites were release of Zn by muscle, secretion of Zn into gut and exchange of Zn with RBC. Regulation at these five sites appears to maintain Zn homeostasis in humans.

REFERENCES

Babcock, A.K., Henkin, R. I., Aamodt, R. L., Foster, D. M., and Berman, M., 1982, Effects of oral zinc loading on zinc metabolism in humans II. In vivo kinetics, Metab., 31:335.
Berman, M., and Weiss, M.F., 1978, SAAM Manual. Washington, DC: U.S. Printing Office, [DHEW Publication No. (NIH)78-180].
Berman, M., Beltz, W. F., Greif, P.C., Chabay, R. C., and Boston, R. C., 1983, CONSAM User's Guide. Washington, DC: US Govt. Printing Office, 1983-421-132:3279.
Foster, D. M., Aamodt, R. L., Henkin, R. I., and Berman, M., 1979, Zinc metabolism in humans: a kinetic model, Am. J. Physiol., 237(Regulatory Integrative Comp. Physiol. 6):R340.
Wastney, M. E., Aamodt, R. L., Rumble, W. F., and Henkin, R. I., 1986, Kinetic analysis of zinc metabolism and its regulation in normal humans, Am. J. Physiol. 251(Regulatory Integrative Comp. Physiol. 20):R398.

SATURABLE ZINC UPTAKE BY SYNCYTIOTROPHOBLAST MICROVILLOUS PLASMA

MEMBRANE VESICLES FROM HUMAN PLACENTA

G. Quinn[1], A. Flynn[1], and B. Lonnerdal[2]

Departments of Nutrition, [1]University College, Cork,
Ireland, and [2]University of California, Davis, U.S.A.

INTRODUCTION

An adequate supply of zinc is essential for normal fetal development.
Zinc from maternal plasma is transported across placental membranes, but
the mechanism is poorly understood. The syncytiotrophoblast microvillous
plasma membrane is the effective primary interface between mother and
fetus and vesicles prepared from this membrane constitute a relatively
simple experimental system for investigating the uptake of zinc from
maternal plasma into the syncytiotrophoblast without interference from the
subsequent stages of transport through the cell to the fetal plasma (Flynn
et al., 1986). The aim of this study was to investigate the kinetics of
zinc uptake by these vesicles.

MATERIALS AND METHODS

Syncytiotrophoblast microvillous plasma membrane vecicles (SMPMV)
were prepared from frozen normal full-term human placenta (70–100g) by
cold saline extraction and differential centrifugation (Smith et al.,
1974). Zinc uptake was determined by suspending vesicles at a protein
concentration of 80 µg/ml with ^{65}Zn in 10 mM HEPES buffer, pH 7.4, containing 138 mM NaCl, 5.2 mM KCl, 0.9 mM $MgCl_2$ and 1.0 mM $CaCl_2$. ^{65}Zn
uptake was stopped by diluting vesicles in 50 ml ice-cold 0.5 mM $ZnNO_3$
in 10 mM HEPES buffer, pH 7.4, containing 0.15M NaCl and vesicles were
then collected by filtration on Whatman GF/C filters followed by washing
with five 10 ml washes of ice-cold 10 mM HEPES buffer, pH 7.4, containing
0.15M NaCl. The filters were then counted in a well gamma counter.

RESULTS AND DISCUSSION

The uptake of zinc by SMPMV at 2°C was linear with time between 0.5
and 5 min and binding of zinc to membrane surfaces and filters was determined by extrapolation of this time progress curve back to zero time.
After allowing for this background, the uptake of zinc at 37°C was linear
up to ~1 min and the initial uptake rate was determined from values
obtained at 0.5 min.   Zinc uptake was strongly dependent on temperature
and intial uptake rate at 37°C was 15–30 times greater than at 2°C.

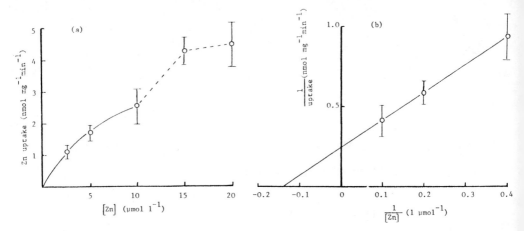

Figure 1. The effect of Zn concentration on the initial rate of Zn uptake
by SMPMV. Vertical bars represent ± 1 SD from the mean of 5
individual determinations on the same SMPMV preparation.

The initial rate of zinc uptake into vesicles at $37^{\circ}C$ increased with
increasing zinc concentration (Fig. 1a) and uptake appeared saturable at
zinc concentrations of 2.5 - 10 µM. At concentrations of 15 and 20 µM
there appeared to be a break in the uptake curve and it is possible that
damage to vesicle membranes may have occurred at these high zinc concentrations. Kinetic constants were calculated from data obtained at zinc
concentrations in the range 2.5 - 10 µM (Fig. 1b). The apparent K_m was
7.0 ± 0.6 µM and V_{max} was 4.1 ± 0.2 nmol/mg protein/min.

These results indicate that zinc is taken up rapidly into SMPMV by a
saturable mechanism and suggest that the syncytiotrophoblast microvillous
plasma membrane of human placenta contains a carrier-mediated system for
zinc transport.

ACKNOWLEDGMENT

This work was supported by a grant from the Medical Research Council
of Ireland.

REFERENCES

Flynn, A., Glazier, C. and Lonnerdal, B., 1986, Zinc uptake by syncytiotrophoblast microvillous plasma membrane vesicles from human
placenta, Am. J. Clin. Nutr., 43:675.
Smith, N.C., Brush, M.G. and Luckett, S., 1974, Preparation of human
villous surface membrane, Nature, 252:302.

NICKEL ABSORPTION AND ELIMINATION IN HUMAN VOLUNTEERS

F. William Sunderman, Jr.[1], Sidney M. Hopfer[1], Thomas Swift[1], Linda Ziebka[1], Allan H. Marcus[2], Bernard M. Most[3], and John Creason[2]

[1]University of Connecticut School of Medicine, Farmington, CT 06032; [2]U.S. Environmental Protection Agency, Research Triangle Park, NC 27711; [3]Northrup Services, Inc., Research Triangle Park, NC 27709, USA

INTRODUCTION

This study was performed to confirm and extend the observations of Solomons et al. (1982) concerning the inhibitory effect of food on intestinal absorption of nickel in human volunteers. The protocol included (a) stringent precautions against nickel contaminination, (b) quantitation of nickel elimination in feces, and (c) analyses of nickel in body fluids and excreta by sensitive and specific techniques of electrothermal atomic absorption spectrometry.

METHODS

This study was approved by the Human Experimentation Committee of the University of Connecticut. The subjects (6 men, 4 women, age 22 to 55 y) met the following criteria: (a) no serious illness during the past 5 years, (b) no present illness, based on medical history and physical examination, (c) no recent use of drugs, medications, or tobacco, (d) no nickel sensitivity, based on negative history of contact dermatitis and negative dermal patch test, (e) negative pregnancy test in women, and (f) comsumption of a typical North American diet, based on an itemized dietary log for one week. In Experiment #1, subjects fasted 12 h prior to drinking water (1.5 mL/kg) containing $NiSO_4$. In Experiment #2, subjects ingested a standard American breakfast (Solomons et al., 1982), containing the same dose of $NiSO_4$ (added to two scrambled eggs prior to cooking). $NiSO_4$ dosages were: (a) 50 µg Ni/kg in one man, (b) 18 µg Ni/kg in 4 subjects (2 men, 2 women), and (c) 12 µg Ni/kg in 4 subjects (2 men, 2 women). Six weeks elapsed between the experiments. One man dropped out of the study after Experiment #1 and was replaced in Experiment #2 by a man of equal age, diet, and habitus.

Blood samples were collected by venepuncture, using polyethylene iv cannulae and polypropylene syringes, at 24 and 1 h pre-treatment and 1, 3, 7, 10, 24, 48, and 72 h post-treatment. Urine specimens were collected in polypropylene jars during 15 intervals, including four 3-h samples after the Ni dose, plus twelve 12-h samples (during 48 h pre-treatment and from 12 to 96 h post-treatment). Feces was collected in polyethylene bags during 48 h pre-treatment and 96 h post-treatment. Nickel was analyzed by Zeeman-EAAS with a Perkin-Elmer 5000-Z spectrometer (Sunderman et al., 1984, 1986a).

RESULTS

During 4 control days, Ni levels (mean ± SD and range) were:- serum Ni 0.32 ± 0.17 µg/L (0.1-0.6); urine Ni 2.4 ± 1.1 µg/day (0.9-4.6); fecal Ni 158 ± 75 µg/day (69-289). After subjects drank Ni in water, peak levels of serum Ni (8-56 µg/L) averaged 33 (range 12-62) times the corresponding values when Ni was ingested in food. When Ni was ingested in water, peak levels of serum nickel occurred 3.3 ± 1.3 h post-treatment, versus 7.7 ± 6.1 h when Ni was ingested in food (P < 0.05). After subjects drank Ni in water, peak urine Ni levels (39-1044 µg/L) averaged 22 (range = 5 to 63) times the corresponding values when Ni was ingested in food. Recovery of Ni in urine during 4 days post-treatment averaged 24 ± 16% of the dose in water (range = 8 to 50%), versus 1.7 ± 1.2% (range = 0.6 to 4.5%) of the dose in food (P < 0.01). Compensatory changes in fecal elimination of Ni were noted, so that the total Ni recovery in urine plus feces during 96 h post-treatment averaged 100 ± 8% in Experiment #1 and 104 ± 21% in Experiment #2.

CONCLUSIONS AND DISCUSSION

This study shows that < 5% of Ni added to scrambled eggs was absorbed from the gut and excreted in urine, compared to 8 to 50% of Ni ingested in water after an overnight fast. These findings corroborate the report of Solomons et al (1982) that dietary constituents markedly reduce the bioavailability of orally administered Ni^{2+}. The results of the present study are being fitted to a multicompartment model of Ni kinetics, to derive parameters for Ni^{2+} absorption, distribution, and elimination in humans. Studies are underway to determine whether diminished absorption of nickel added to food reflects (a) complexation of Ni^{2+} by dietary constituents, (b) presence of dietary factors that inhibit Ni^{2+} absorption, and/or (c) preferential intestinal uptake and absorption of other elements that compete with Ni^{2+} for mucosal trapping and intestinal transport (Foulkes and McMullen, 1986).

REFERENCES

Foulkes, E.C., and McMullen, D.M., 1986, Toxicology, 38: 35.
Solomons, N.W., Viteri, F., Shuler, T.R., and Nielsen, F.H., 1982, J. Nutr., 112: 39.
Sunderman, F.W., Jr., Crisostomo, M.C., Reid, M.C., Hopfer, S.M., and Nomoto, S., 1984, Ann. Clin. Lab. Sci., 14: 232.
Sunderman, F.W., Jr., Hopfer, S.M., Crisostomo, M.C., and Stoeppler, M., 1986a, Ann. Clin. Lab. Sci., 16: 219.

Influence of Feed Components in Semisynthetic Diets on Cadmium Retention in Chicks.

W.A. Rambeck[*], R. Hettich[*], D. Berg[**] and W.E. Kollmer[**]

[*]Institut für Physiologie, Physiologische Chemie und Ernährungsphysiologie der Ludwig-Maximilians-Universität München, Germany

[**]Gesellschaft für Strahlen- und Umweltforschung mbH, Abt. Nuklearbiologie, München, Germany

Introduction

The finding that certain substances in the diet as well as the nutritional status can increase Cd toxicity in man suggests that interactions between Cd and food components play a major role. We recently studied the influence of calcium (Winkler et al. 1984), zinc (Bundscherer et al. 1985) and other components on cadmium absorption and retention in growing chicks. Furthermore we investigated the effect of different binding forms, especially Cd-phytate, on the bioavailability of this heavy metal (Jackl et al. 1985, Rambeck et al. 1987). The present studies were undertaken to study the effects of varying contents of fiber, protein, fat and phytate on renal Cd-retention in growing chicks.

Material and Methods

One week old broiler chicks were fed either a corn-soybean diet or a semisynthetic diet containing 3 ppm of Cd. After three to five weeks the animals were killed and renal Cd-concentration was determined by atomic absorption spectroscopy.

Results and Discussion

Feeding a corn-soybean diet containing 3 ppm $CdCl_2$ to chicken for five weeks results in a Cd content of 1100 ppb. The bioavailability of Cd_3-

phytate and Cd_6-phytate does not differ very much from $CdCl_2$ since similar renal Cd-values (1250 ppb, 1400 ppb) are obtained. However, when the chickens are given 3 ppm $CdCl_2$, Cd_3-phytate or Cd_6-phytate in a semisynthetic diet the renal Cd-retention is about 4000 ppb.

The main difference between the two types of feed is the low content of dietary fiber in the semisynthetic diet. In order to find out if this induces the three-fold higher Cd-retention, the fiber content was increased in the next experiment. By adding soybean hulls, the dietary fiber content was brought to 5 % in the semisynthetic diet and to 9 % in the corn-soy bean diet. Though addition of soybean hulls to the semi-synthetic diet reduces renal Cd-retention, it is still three times as high as with the corresponding corn-soybean diet. Raising the fiber content from 5 to 9 % has no effect on Cd-retention.

Increasing the protein content (30 instead of 20 % crude protein) reduces Cd-concentration slightly in the semisynthetic diet, while a higher fat content (10 instead of 3 % soybean oil) increases Cd-reten-tion by one third. Effects like these are known from the literature (Schenkel 1986).

Another difference between the two types of feed is the lack of phytate in the semisynthetic diet. Since it is well known that phytate is able to reduce the bioavailability of a number of essential dietary minerals, and since this effect is greatest for Zn, it might well be that the chemically related Cd might also be insoluble or otherwise unavailable for intestinal absorption. Most surprisingly the addition of phytate to the semisynthetic diet did not reduce Cd-retention and in corn-soybean diets, it even increased Cd-concentration in the kidney by 50 %. An explanation might be that phytate binds dietary calcium. A low calcium content however increases Cd-bioavailability.

References

Bundscherer, B., Rambeck, W.A., Kollmer, W.E. & Zucker, H. (1985, Z. Ernährungswiss. 24, 73-78

Jackl, G.A., Rambeck, W.A. & Kollmer, W.E. (1985), Biol. Trace. Res. 7, 69-64

Rambeck, W.A., Meiringer, G., Kollmer, W.E. & Zucker, H. (1987) in preparation

Schenkel, H. (1986), Schriftenreihe des Bundesministeriums für Er-nährung, Landwirtschaft und Forsten, Angewandte Wissenschaft, 335, 137-158

Winkler, C., Rambeck, W.A., Kollmer, W.E. & Zucker, H. (1984), Z. Tierphysiol. Tierernährg. u. Futtermittelkde. 51, 250-256

THE EFFECT OF DIETARY FIBER AND DIFFERENT CONCENTRATIONS OF ZINC ON THE INTESTINAL ABSORPTION OF CADMIUM IN RATS.

Anncatherine Moberg, Göran Hallmans,
Rolf Sjöström and Kenneth Wing

Biophysics Laboratory and the Departments
of Nutritional Research, Oral Roentgenology
and Pathology, University of Umeå
S-901 87 Umeå, Sweden

INTRODUCTION

The following questions were addressed in this study:
1) Do binding factors such as dietary fiber and phytic acid present in bran and whole wheat reduce the absorption and accumulation of Cd?
2) Does the concentration of Zn in the diet affect the uptake of Cd?
3) Can the accumulation of Cd be estimated by the absorption of Cd-109 from a single meal of the diet?

MATERIAL AND METHODS

Five groups of six rats each were fed deionized water and one of five composite diets (50% crisp bread, 50% basic diet) with various concentrations of cadmium, zinc, dietary fiber and phytic acid (Table 1). After 3 weeks on these diets and a 12 hour fast the rats were given Cd-109 in 5g of their respective diets. Five hours later the radionuclide-labelled diets were replaced with the unlabelled diets which were continued until the rats were killed 3 weeks later. The Cd and Cd-109 concentrations in the test meals, liver and kidneys and the Cd-109 excreted in faeces were measured and the amount of Cd retained from 1g of test meal was calculated.

Table 1. Composition of the diets

Diets	Cadmium μg/kg	Zinc mg/kg	Dietary fiber g/kg	Phytic acid mmol/kg
Endosperm	16	24	23	−
Whole wheat	35	29	59	42
Bran	29	42	82	76
Endosperm+Cd	35	23	22	−
Endosperm+Zn	12	43	19	−

RESULTS AND CONCLUSIONS

The relative accumulation of cadmium (% Cd-109) was
reduced in rats fed whole wheat and bran diets high in fiber
and phytic acid (Table 2). The amount of Cd in the liver and
kidneys which derived from 1 g of the test meal (calculated
from Cd-109 retention) and the total Cd contents after 6
weeks on the diets were increased in the group given
endosperm wheat bread with Cd added to the whole wheat
level (Table 2). The Cd accumulation in the whole wheat
group was between those in the endosperm and endosperm+Cd
groups.

The Zn concentrations in the diets at these levels had
no influence on the relative Cd-109 or the total Cd
accumulation (Table 2).

There were strong correlations between the amount of Cd
accumulated during 6 weeks on the diets and the amount of Cd
derived from 1 g of the test meals both in the liver
($r=0.811$; $p<0.001$) and in the kidneys ($r=0.716$; $p<0.001$).
This indicates that both methods are reliable in measurements
of the relative accumulation of small amounts of Cd from
different diets.

Table 2.   The relative accumulation of Cd-109, the amount of
Cd derived from 1 g of the test meal and the
accumulated amount of Cd during 6 weeks on the
diets in the liver and kidneys.
The results are given as means (S.E.)

Groups	Endo	Whole wheat	Bran	Endo+Cd	Endo+Zn
		LIVER			
Relative accumulation (% Cd-109)	0.121 (0.005)	0.083 (0.008)	0.076 (0.005)	0.132 (0.007)	0.141 (0.009)
Amount of Cd from 1 g test meal (ng Cd)	0.019 (0.001)	0.029 (0.003)	0.022 (0.001)	0.046 (0.002)	0.016 (0.001)
Accumulated amount of Cd (ng Cd)	16.75 (2.28)	25.50 (1.30)	21.25 (2.31)	33.66 (1.66)	14.90 (0.61)
		KIDNEYS			
Relative accumulation (% Cd-109)	0.181 (0.009)	0.118 (0.011)	0.107 (0.007)	0.191 (0.008)	0.209 (0.011)
Amount of Cd from 1 g test meal (ng Cd)	0.028 (0.001)	0.042 (0.004)	0.031 (0.002)	0.067 (0.003)	0.024 (0.001)
Accumulated amount of Cd (ng Cd)	5.91 (0.82)	7.69 (0.63)	6.46 (1.10)	13.03 (1.12)	5.90 (0.67)

TRANSFERRIN RECEPTORS AND IRON UPTAKE OF RAT

MAMMARY GLAND MEMBRANES DURING LACTATION

M. Sigman and B. Lönnerdal

Department of Nutrition, University of California
Davis, CA 95616

INTRODUCTION

Human infants require iron (Fe) for proper growth and development. Although of high bioavailability, breast milk contains a low concentration of Fe, causing some pediatricians to question whether human milk contains enough Fe for the rapidly developing infant. It has been suggested that Fe supplementation of the mother may be a means of increasing milk Fe concentrations. However, most human studies have failed to demonstrate a correlation between maternal Fe intake or impaired iron status and breast milk Fe concentration (1).

Milk Fe concentration declines throughout the course of lactation, with the highest concentration in colostrum and early lactation and a lower, relatively constant concentration in mature milk. These observations suggest a tightly regulated mechanism controlling Fe influx from maternal plasma into breast milk. As the Fe concentration of rat milk exhibits a pattern similar to that of human milk, the rat was chosen as an animal model. Using rat mammary cell plasma membranes, we previously have demonstrated a transferrin (Tf) mediated binding mechanism of Fe (2).

METHODS AND MATERIALS

Rat mammary cell plasma membranes were prepared using a modification of Maeda et al. (3). 5' nucleotidase was used as a marker enzyme for membrane purification. Binding studies were performed with ^{59}Fe-labeled rat Tf using vacuum filtration. Isolated membranes were incubated with the labeled Tf at 37°C. Aliquots were removed at 5, 15, 30 and 60 minutes and passed through .45 um filters. Bound ^{59}Fe Tf was measured on a gamma counter. Nonspecific membrane binding was determined using a 20-fold excess of unlabeled Tf.

RESULTS AND DISCUSSION

We identified that Fe binding onto rat mammary plasma membrane appears to be via a Tf mediated mechanism similar to that found in other cells, with no ^{59}Fe binding when presented as ^{59}FeCl$_3$, ^{59}Fe-labeled bovine serum albumin or in the presence of a 20-fold excess of cold Tf (Table 1). These results suggest a specific binding mechanism for Fe entry into the mammary cell.

Table 1. Competitive Binding in Day 14 Rat Mammary Plasma Membrane	
	pmoles ^{59}Fe bound
^{59}Fe Transferrin	20
^{59}Fe Transferrin + 20-fold excess cold	0
^{59}Fe Bovine Serum Albumin	0
^{59}FeCl$_3$	0

Table 2. ^{59}Fe Transferrin Binding Kinetics During Course of Lactation in the Rat	
Stage of Lactation	nM ^{59}Fe Tf bound (Tf conc. 18 uM)
Day 1	115
Day 7	198
Day 14	201
Day 21	125

The dissociation constant for d14 of lactation ($Kd = 5.4 \times 10^{-6}$ M) appears to be higher than those values reported for many other cell types, indicating that mammary tissue may have a lower affinity for Tf.

One potential mechanism for higher milk Fe content in early lactation would be an increase in receptor binding capacity. Our data do not show higher capacity at d1 although values are higher for tissue from d7 and d14 than for d21 (Table 2). It should be emphasized that isolation of purified plasma membranes from rat mammary tissue is confounded by the presence of high amounts of adipose and connective tissue. Since these problems are most pronounced at the initiation of lactation, we do not have confidence in the value obtained for d1. Further investigations using improved membrane separation techniques as well as isolated cells should provide a clarification of the phenomena observed.

Other possible explanations for high milk Fe in early lactation include higher cell number, higher turnover of membrane receptors, and larger gap junctions between cells allowing non-receptor mediated Fe to enter the milk.

CONCLUSION

1) We have demonstrated specific binding of Fe mediated via Tf in rat mammary cell plasma membrane.
2) Receptor affinity ($K_d = 5.4 \times 10^{-6}$ M) appears lower for this tissue than that determined from other cell types.
3) Our data suggests variations in this binding capacity for various stages of lactation.

REFERENCES

1. B. Lönnerdal, Effect of maternal iron status on iron in human milk, in: "Human Lactation 2: Maternal and Environmental Factors," M. Hamosh and A. S. Goldman, eds., Plenum Press, New York (1986).
2. M. Sigman and B. Lönnerdal, Identification of a transferrin-receptor mediating iron uptake into rat mammary tissue plasma membranes, Fed. Proc. 46:438 (1987).
3. T. Maeda, K. Balakrishnan and S. Q. Mehdi, A simple and rapid method for the preparation of plasma membranes, BBA 731:115 (1983).

PREDICTION OF DISORDER IN Cu-DEFICIENT LAMBS FROM DIFFERENT GENOTYPES

N.F. Suttle, D.G. Jones, Moredun Research Institute, Edinburgh
J.A. Woolliams and C. Woolliams, AFRC Institute for Animal
Physiology and Genetics Research, Edinburgh Research Station
N.F. Suttle*, D.G. Jones*, J.A. Woolliams**, And C. Woolliams**

*Moredun Research Institute, Edinburgh
**AFRC Institute for Animal Physiology and Genetics Research,
Edinburgh Research Station

Growth retardation and susceptibility to infection are important recent additions to the recognised clinical consequences of Cu deficiency in lambs (1,2): their recognition increases the importance of accurate diagnosis and prognosis of clinical Cu deficiency (hypocuprosis).

MATERIALS AND METHODS

Most of the data and the analytical methods used to obtain them are given elsewhere (1,2). Briefly, the animals were from two lines genetically selected for low (L) and for high (H) concentrations of Cu in plasma, within an interbred Scottish Blackface (B) x Welsh Mountain (W) population, and from contemporary groups of unselected B and W lambs. Half of the lambs of both lines and both pure breeds (410 in all) were given Cu supplements generally from 6 weeks of age. Blood samples were taken at 6-weekly intervals and analysed for plasma (P) Cu, haemoglobin (Hb) and activity of superoxide dismutase (SOD) (described P_6, Hb_{12}, SOD_{24} etc re. time of sampling).

The linear relationships between biochemical criteria of Cu status at 6 (for prediction) or 24 weeks (for diagnosis) and performance traits were investigated. Regressions of mortality from 6 to 24 weeks were made using GLIM. The linear models had binomial errors with logit link functions. Regression of gain in weight (GLW) or Hb (GHb) weeks from Cu supplementation on P and SOD were made using standard analysis of variance techniques, before and after log transformation, on L and H lambs alone. Analyses were made (i) on individuals within selection line by set subclasses for each year and (ii) between breed type by set by year subclasses, giving 8 df: only the latter are presented.

RESULTS

Mortality (logit of probability of death) between 6-24 weeks was predicted by:

$$\text{Logit P} = -0.48 \pm 0.148\ P_6 - 0.70\ (P < 0.05)$$
$$\text{and} = -0.00815 \pm 0.00299\ SOD_6 - 0.438\ (P < 0.05)$$

Table 1

Index	Age (wks)	Trait	Slope	Intercept	r
Plasma Cu (log umol/l)	6	GHb	-3.377	2.31	0.96
		GLW	-1.506	2.19	0.62
	24	GHb	-2.749	2.43	0.66
		GLW	-2.398	3.07	0.75
SOD (log U/ml blood)	6	GHb	-3.863	13.02	0.67
		GLW	-1.58_{ln}	9.89	0.89
	24	GHb	-0.667	2.56	0.28
		GLW	-3.129	10.50	0.76

The two indices showed similar precision and a logarithmic reduction in mortality (as a fraction) from 0.30 to 0.01 as P_6 increased from 0 to 8 umol/l and as SOD increased from 50 to 500 U/ml whole blood. The relationship for GHb (g/dl) and GLW (kg) after 24 weeks is given in the table.

The predictive relationships between P_6 and Hb and between SOD_6 and GLW were particularly strong. The diagnostic relationships for GLW showed no advantage of one index over another and for GHb, SOD_{24} was a poor index. Again the best relationship was logarithmic with GHb and GLW increasing rapidly with SOD_{24} < 1.2 U/ml Hb and P_{24} < 6 umol/l. The confidence limits for prediction of mortality (95% risk of value greater than the probable baseline of 0.05) were P_6 < 3.3 umol/l and SOD_6 < 230 U/ml. The corresponding limits for probability of a weight advantage from Cu supplementation were 5.0 umol/l and 310 U/ml, respectively, suggesting that mortality was marginally the more sensitive index of disorder.

DISCUSSION

SOD did not fulfil its promise as an alternative to P in the diagnosis of hypocuprosis (as GLW or GHb) in lambs but it was a good predictor of mortality. This may have arisen because most deaths occurred in weeks 6-12 when SOD was a good predictor of Cu status. SOD_{24} was more highly correlated with SOD_6 than was P_{24} with P_6 (r 0.92 v 0.70). This probably reflected the slowness with which the erythrocyte-based index can change. Plasma Cu fluctuated markedly during the study and may have given a better reflection of the supply of Cu for growth during the later weeks.

The sensitivity of survival to Cu deficiency means that greater emphasis must be placed on prediction and the use of preventative measures. While the existing norm for plasma Cu (9 umol/l) provided a generous margin of safety on this particular farm, it should be retained, for general use given the inevitable variation from farm to farm and the natural constraints on frequency of assessment.

The relationships seemed to hold for the parental breed which was most sensitive to Cu deficiency, the Scottish Blackface, which like L showed high mortality and growth retardation. The general lack of disorder in the other parent breed, Welsh Mountain, grazing the same pasture but showing consistently higher P and SOD values indicates the superiority of animal over herbage based indices of hypocuprosis in sheep.

REFERENCES

1. C. Woolliams, N.F. Suttle, J.A. Woolliams, D.G. Jones and G. Wiener, Studies on lambs from lines genetically selected for low and high copper status. 1. Differences in mortality, Anim. Prod. 43:293 (1986).

2. J.A. Woolliams, C. Woolliams, N.F. Suttle, D.G. Jones and G. Wiener, Studies on lambs from lines genetically selected for low and high copper status. 2. Incidence of hypocuprosis on improved hill pasture, Anim. Prod. 43:303 (1986).

METAL ANALYSIS OF HUMAN LIVER AS AN AID TO DIAGNOSIS

Dorothy McMaster, M.E. Callender, and A.H.E. Love

Department of Medicine, The Queen's University of Belfast
Northern Ireland

A simple procedure for the direct estimation of Fe, Zn and Cu in fresh human liver obtained by percutaneous sampling has been developed. After drying, the sample of 3-5 mg is dissolved in nitric acid to give approximately 2 mg/ml. Fe and Zn are measured by flame atomic absorption and Cu in a graphite furnace.

The control group of 9 was made up of 2 post mortem, 2 Gilbert's disease and 5 psoriasis patients. Haemachromatosis patients provided 13 samples, alcoholic liver disease patients 14 samples and primary biliary cirrhosis patients 4 samples. The means are given in Table 1 and the individual values in Figures 1-3.

In haemachromatosis the store of liver iron is greatly increased. Zn is lower in patients suffering from alcoholic liver disease than in controls $p < 0.0001$, haemachromatosis $p\ 0.001$ and primary biliary cirrhosis $p\ 0.02$. Cu is also lower in the alcoholic patients than in the controls $p\ 0.006$, but increased in primary biliary cirrhosis.

Needle biopsy is often carried out in patients undergoing routine investigation for liver disease. Direct estimation of Fe, Zn and Cu in such specimens is feasible and may become a useful aid to diagnosis.

Table 1.   Mean values ± standard error per g dry weight of liver (n)

	Fe mg	Zn µg	Cu µg
Controls	0.70 ± 0.11 (8)	307 ± 43 (8)	29.7 ± 4.4 (8)
Haem	12.85 ± 1.51 (13)	220 ± 20 (13)	28.3 ± 3.6 (11)
Alcohol	1.08 ± 0.19 (14)	113 ± 17 (12)	15.6 ± 2.4 (14)
PBC	0.50 ± 0.14 (4)	198 ± 23 (4)	394 ± 230 (4)

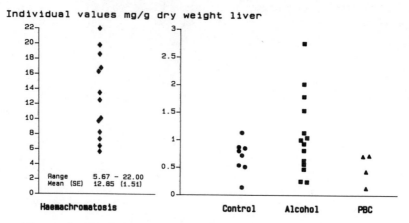

Figure 1. Fe levels in fresh human liver needle biopsies

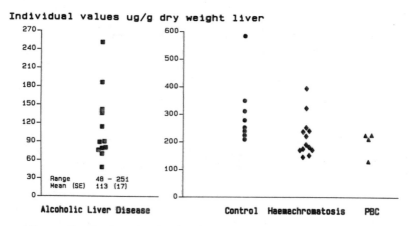

Figure 2. Zn levels in fresh human liver needle biopsies

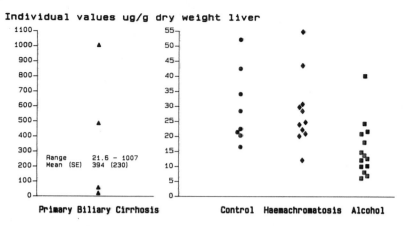

Individual values ug/g dry weight liver

Figure 3. Cu levels in fresh human liver needle biopsies
PBC - Primary biliary cirrhosis

ACKNOWLEDGEMENTS

We thank Mr. D. J. O. Gill for the illustrations.

THE USE OF MULTIVARIATE ANALYSIS TO IDENTIFY LIMITING TRACE

ELEMENTS IN SHEEP

D. B. Purser*, D. G. Masters*, P. L. Payne* and R. Maller**

*Division of Animal Production, CSIRO, **Division of
Mathematics and Statistics, CSIRO;   Private Bag, PO
Wembley, WA, 6014, Australia

INTRODUCTION

Deficiencies of trace elements in the diets of grazing animals
result in reduced production of wool and meat or reproduction (Judson et
al. 1987). A variety of measures and techniques have been proposed for
diagnosis and identification of deficient elements with varying degrees
of precision and success. However, the diagnosis of subclinical
deficiencies, in which no overt symptoms are apparent but production is
depressed, is not possible. Until the diagnosis of limiting trace
elements can be readily achieved, undetected and costly losses in
production will occur and, in practice the extent of the problem will be
largely unrecognised. The aim of this work was to examine the usefulness
of a modified form of multivariate analysis to discriminate between
groups of sheep having known deficiencies in their diet, with a view to
subsequently using the analysis for diagnostic purposes. The advantage
of such an approach lies in using interactions between elements as part
of the diagnosis; interactions between elements are therefore useful for
diagnostic purposes rather than a major obstacle to interpretation and
diagnosis.

METHODS

Groups of six sheep received diets low in only S, Cu, Zn, or Cu and
S, Zn and S, Cu and Zn, or Cu, S and Zn and one group received a control
diet containing all elements. Sulphur was provided as an addition to the
diet at 0.15% of the dry matter. Manipulation of the various
combinations of the trace elements was achieved by providing the sheep
with Controlled Release Devices (Laby, 1980) with the appropriate
combination of the elements in them. The quantities provided as
supplements per day, (estimated to provide 50 per cent of the daily
requirement) where appropriate were Cu(3.0 mg/d) and Zn(20.0 mg/d).
After six weeks of treatment (depletion phase) all remaining sheep were
fed the control diet for four weeks (repletion phase). Three sheep from
each of the treatment groups were slaughtered after both the depletion
and repletion phases.

Mineral concentrations in faeces and a number of tissues, including blood and wool, were determined after both the depletion and repletion phase. Values from the depletion phase, the repletion phase and the ratio of the depletion and repletion phases were used in a principal components analysis (PCA) of the data, initially more than 300 observations per sheep were used. This technique results in eigenvector scores describing the relative position of respective treatment groups of sheep. Only the first four eigenvectors produced have been used in the evaluation of the data. Further evaluation of the variables used in the PCA has allowed separation of treatment groups using variables derived only from blood, wool and faeces, i.e. sheep do not need to be slaughtered to allow a test to be made.

RESULTS AND DISCUSSION

In Figure 1, the relative position of the treatment groups, and for the three sheep per group continued through the repletion phase, as described by vectors 1 and 3 is shown. The treatment groups were further separated by vectors 2 and 4.

This work has shown that a certain technique of multivariate analysis can be used to discriminate between treatment groups, where elements are marginally limiting and no overt clinical symptoms are apparent; changes in dry matter intake were the only indications of a limitation. It is proposed to develop the test further (other diets and physiological states) in order to use the relative position of the treatments as described by vector scores, for the diagnosis of limiting trace elements.

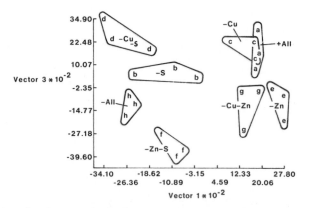

Fig. 1. Separation of treatment groups by eigenvectors 1 and 3. The location of the individual sheep in each of the treatment groups is shown. Only results of analysis of wool, faeces and blood were used.

REFERENCES

Judson, G.J., Caple, I.W., Langlands, J.P., and Peter, D.W., 1987, Mineral nutrition of grazing ruminants in southern Australia in: "Review of Research in Pasture Production and Utilization in Southern Australia ", J.L. Wheeler, ed. (In Press.)

Laby, R.H., 1980, Modern Technological Developments, Proceedings Aust. Soc. Anim. Prod. 13:6.

444

CONCENTRATION AND DISTRIBUTION OF ZINC,

COPPER AND IRON IN "ZINC DEFICIENT" HUMAN MILK

Stephanie A. Atkinson[1] and Bo Lönnerdal[2]

[1]Department of Pediatrics, McMaster University,
Hamilton, Ontario, Canada and [2]Department of Nutrition,
University of California, Davis, CA 95616 USA

INTRODUCTION

Acrodermatitis, diarrhea, hyperirritability and delayed growth in association with low plasma zinc levels have been documented in case reports of
premature infants exclusively fed human milk (1-3). In all cases the etiology of the nutritional zinc deficiency was attributed to levels of zinc in
the infants' mothers' milk that were below the usual range of values reported for milk at similar lactational stages. However, there has only been
speculation as to whether this is caused by defective mammary gland zinc
secretion or whether it simply represents the lower end of a normal spectrum
(3). In the present report, we describe the quantitation of citrate, a
major zinc binding ligand in human milk (4), and the distribution of zinc,
copper and iron in the casein, whey, fat and whole milk fractions of milk
which we determined to be "zinc deficient."

METHODS

Milk samples were obtained by SAA from mothers whose infants were
patients at Chedoke-McMaster Hospitals, Hamilton, Ontario. One infant (milk
A), born at 31 weeks gestation and 1470 g birthweight, was investigated for
abnormal zinc absorption because of a sibling history of the congenital disorder of acrodermatitis enteropathica. The second infant (milk B), born at
33 weeks gestation and 1760 g birthweight, presented with nutritional zinc
deficiency associated with 4 1/2 months of exclusive breast feeding. A
third infant (milk C), born at 28 weeks gestation and 1430 g birthweight,
was part of a nutritional study and the "zinc deficient" milk was identified
retrospectively. A 72 h metabolic balance was completed in infants receiving milk A and C.

Milk samples were obtained by complete emptying of the breast using
acid-washed breast pumps. An aliquot of milk was separated and analyzed for
zinc, copper and iron in whole milk, milk fat, pellet (casein) and whey (5).
Citrate was quantitated by the citrate lyase method (4).

RESULTS

For milk A at lactation day 25 zinc concentration in ug/ml (% of total
Zn) for whole milk, fat, whey and pellet (casein) was 1.17, 0.21 (18%), 0.92

(78%) and 0.08 (7%), respectively. Predicted normal values were 2.82 ug/ml, 14%, 54% and 15%, respectively. Values for milk B at lactation day 135 for whole milk, fat, whey and pellet were 0.32, 0.12 (38%), 0.14 (47%) and 0.04 (12%), respectively. The predicted value for whole milk zinc was 1.05 ug/ml. For milk C at lactation day 16 whole milk zinc was 1.14 ug/ml, whereas 3.32 ug/ml was predicted. Milk was termed "zinc deficient" if the zinc concentration was >2 S.D. below the predicted (lactational stage-adjusted) value as calculated from linear regression analysis for "preterm" milk zinc (y=4.22-0.56x, r=-0.54, p<0.01). Milk A and C met this criterion. Milk B was collected later in lactation than our reference population so we simply report it to be one-third the zinc value for term milk (6). Iron and copper concentrations and distribution were normal. Citrate concentration was 2.01 mM for milk B compared to 2.06 mM for a control milk sample.

Zinc balance in the infants fed milk A and milk C were negative (-372 and -392 ug/kg/24 h) at 26 and 16 days postnatally. This represents the low "normal" range observed for other human milk fed premature infants at similar postnatal ages (7).

DISCUSSION

The reason for individual mothers to have such distinctly low levels of milk zinc and yet normal iron and copper concentrations and normal distribution of trace elements remains an issue of great speculation but limited scientific explanation. Previous investigators have attempted to provide maternal zinc supplements which served to elevate maternal serum but not milk zinc (3). Zinc is probably transported into milk via an active process, possibly together with a low molecular weight ligand such as citrate (5), since in early lactation zinc is nearly ten times higher in milk relative to plasma (6). However, a defective mammary transport mechanism for zinc into milk is likely not the cause of low milk zinc content in one mother since citrate in her milk was normal. This information refutes previous statements that mammary secretion of zinc in such cases must be abnormal (3). It appears more likely that mechanisms which accrue zinc from the circulation (plasma) into the mammary gland may be abnormal. A systemic low uptake of zinc into the gland but normal synthetic and secretory mechanisms would explain a low concentration but normal distribution of zinc in the milk. The reasons for poor mammary zinc uptake need further investigation.

REFERENCES

1. P. J. Aggett, D. J. Atherton, J. More, J. Davey, H. T. Delves, and J. T. Harries, Arch. Dis. Child. 55:547-550 (1980).
2. I. Blom, S. Jameson, F. Krook, B. Larson-Stymno, and L. Wranne, Br. J. Dermatol. 104:459-464 (1980).
3. A. W. Zimmerman, K. M. Hambidge, M. L. Lepow, R. K. Greenberg, M. L. Stover, and C. E. Casey, Pediatrics 69:176-183 (1982).
4. B. Lönnerdal, A. G. Stanislowski, and L. S. Hurley, J. Inorg. Biochem. 12:71-78 (1980).
5. G. B. Fransson and B. Lönnerdal, Nutr. Res. 3:845-853 (1983).
6. L. A. Vaughan, C. W. Weber, and S. R. Kimberling, Am. J. Clin. Nutr. 32:2301-2306 (1979).
7. S. A. Atkinson, D. Fraser, R. Stanhope, and R. K. Whyte, in: "Trace Element Metabolism in Man and Animals -- 6," L. S. Hurley, C. L. Keen, B. Lönnerdal and R. B. Rucker, eds., Plenum Press (this volume).

EFFECTS OF HEAT TREATMENT OF DEFATTED SOY FLOUR

FED TO YOUNG JAPANESE QUAIL

S.-H. Tao,[1] M. R. S. Fox,[1] B. E. Fry, Jr.,[1] and W. H. Stroup[2]

Division of Nutrition[1] and Division of Food Chemistry and Technology[2]
Food and Drug Administration, Washington, DC,[1] and Cincinnati, OH[2]

INTRODUCTION

Soy protein products, with versatile functional properties and good nutritional quality, are increasingly used in human foods. It is important to develop animal bioassays to assess the adequacy of processing to assure the safety and nutritive quality of these products. We observed that feeding a commercial fully toasted, defatted soy flour to young Japanese quail as the sole protein source resulted in poor growth, even with supplemental zinc (Fox et al., 1983). We investigated the effects of graded heat treatment of soy flour on growth, development, and mineral utilization in quail.

METHODS

Day-old Japanese quail (Coturnix coturnix japonica) of both sexes, 10 per group, were fed for 7 d nutritionally adequate purified diets containing 100 ppm Zn and 31.5% protein supplied by casein-gelatin (CG) or one of the soy flours (Table 1). Body weight and food intake were measured. Tissue minerals were measured by atomic absorption spectrophotometry. Results were evaluated by analysis of variance and Duncan's multiple range test and are reported as means ± SEM. Values with the same superscript letter in a column are not significantly different ($P<0.05$).

RESULTS

None of the quail fed commercial minimally toasted soy flour (CMTF) survived to d 7 (Table 1). Heating CMTF for 20 min improved the survival rate to 60%, but growth was still below that of CG controls. Heating for 35 min increased the survival rate to 90% and provided normal growth equal to that of the CG controls. Food efficiency data showed that quail utilized soy proteins less efficiently than CG, especially when soy protein was heated for only 20 min. An increase in heating time markedly decreased trypsin inhibitor (TI) activity in soy flour. Fifty-minute heating reduced the TI activity to the level of that in commercial fully toasted soy flour (CFTF). However, quail fed CMTF heated for 50 or 70 min grew better than those fed CFTF, indicating that one or more factors other than TI are also involved. Relative pancreas weights were increased by CMTF heated for

Table 1. Growth, Survival, and Food Efficiency in Relation to Heated Soy Flour and Its Trypsin Inhibitor Activity

Dietary Protein[a]	Trypsin Inhibitor	Heat Treatment[b]	Survival	Body Weight	Food Efficiency[c]
	g/kg diet	min	%	g	g gain/g food
CG	--	0	100	20.2 ± 0.60^{BC}	0.717 ± 0.0284^{A}
CMTF	22.9	0	0	--	--
CMTF	17.6	20	60	13.9 ± 1.30^{D}	0.387 ± 0.0927^{C}
CMTF	5.6	35	90	20.4 ± 1.30^{ABC}	0.537 ± 0.0553^{B}
CMTF	3.9	50	100	22.9 ± 0.53^{A}	0.598 ± 0.0175^{AB}
CMTF	2.0	70	100	22.5 ± 0.74^{AB}	0.562 ± 0.0290^{B}
CFTF	3.9	0	90	18.6 ± 0.99^{C}	0.536 ± 0.0601^{B}

[a]CG: casein-gelatin (4:1, w/w); CMTF, CFTF: commercial minimally and fully toasted soy flours with 6% moisture (Cargill Inc. Protein Products, Cedar Rapids, IA).
[b]Steamed at 100°C in an autoclave (come-up time about 10 min), followed by grinding and freeze-drying.
[c]Measured during d 3 to 7.

Table 2. Effects of Defatted Soy Flour with Graded Heat Treatment on Pancreas Size and Tissue Minerals

| Dietary Protein[a] | Pancreas | | Tibia[b] | | |
	Weight	Zn	Zn	Fe	Ash
	mg/100 g body weight	µg/g	µg/g	µg/g	%
CG (0)	432 ± 21.1^{CD}	123 ± 6.2^{A}	311 ± 16.7^{A}	94 ± 7.2^{C}	45.7 ± 0.45^{A}
CMTF (20)	661 ± 42.9^{A}	47 ± 6.8^{D}	224 ± 4.8^{D}	132 ± 8.2^{B}	42.7 ± 0.68^{B}
CMTF (35)	572 ± 23.5^{C}	65 ± 7.7^{C}	240 ± 4.9^{CD}	145 ± 6.5^{AB}	43.6 ± 0.57^{B}
CMTF (50)	446 ± 11.5^{D}	72 ± 2.9^{BC}	257 ± 4.9^{BC}	142 ± 5.9^{AB}	43.2 ± 0.76^{B}
CMTF (70)	381 ± 10.4^{D}	86 ± 4.6^{B}	270 ± 7.6^{B}	159 ± 7.6^{A}	42.5 ± 0.37^{B}
CFTF (0)	457 ± 16.0^{C}	78 ± 3.5^{BC}	263 ± 6.9^{BC}	153 ± 10.1^{AB}	43.8 ± 0.47^{B}

[a]Time (min) of heat treatment at 100°C is given in parentheses.
[b]Values are on a fat-free dry weight basis.

either 20 or 35 min over those of CG controls, but were normal with quail fed CFTF, or CMTF heated for 50 or 70 min (Table 2). All soy flour-fed quail had significantly lower pancreas and tibia Zn, less tibia ash, and higher tibia Fe than did the CG-fed controls. Among the soy flour-fed quail, both pancreas and tibia Zn were negatively correlated with the relative pancreas weight (r = -0.6683 and -0.5934, respectively; $P<0.001$) over the entire range of heat treatment.

CONCLUSIONS

Young quail are very sensitive to inadequately heated soy flour. Moist heat (100°C, 50 or 70 min) improved the nutritive quality of CMTF to support good growth and increased zinc utilization. Pancreas and tibia Zn and relative pancreas weight appear to be very good criteria for evaluating the adequacy of heat processing of soy flour.

REFERENCE

Fox, M. R. S., Tao, S.-H., Fry, B. E., Jr., Johnson, M. L., Stone, C. L., and Hamilton, R. P., 1983, Effects of soy on development and mineral utilization in quail, Fed. Proc., 42:391.

COMPARISON OF INDICES OF COPPER STATUS IN MEN AND WOMEN

FED DIETS MARGINAL IN COPPER

David B. Milne, Leslie M. Klevay, and Janet R. Hunt

United States Department of Agriculture, Agricultural
Research Service, Grand Forks Human Nutrition Research
Center, Grand Forks, ND

Although the effects of Cu deficiency are well documented in
several animal species (1), there are relatively few data on the effects
of controlled marginal intakes of Cu on different indices related to Cu
nutriture in adult humans (2, 3).

To study this effect, eight men were maintained on a metabolic unit
and fed a diet, supplemented as outlined in Table 1, of conventional
foods believed to be marginal in Cu (0.89 ± 0.09 mg/d).

Table 1. Effect of Dietary Cu on Indices of Cu Status in Men[a]

	Control	Depletion	Repletion
Diet Cu, mg/d	1.41 ± 0.15	0.89 ± 0.09	5.05 ± 0.72
Cu Balance, mg/d	0.015 ± 0.068	0.025 ± 0.081	0.51 ± 0.30
Plasma Cu, mg/dl	79.1 ± 8.1[b]	74.8 ± 12.0 (64.8 ± 11.5)[c]	77.8 ± 12.7
CP (ENZ), mg/dl	40.9 ± 6.3 (7[d])	33.9 ± 6.0 (7[d])	33.2 ± 5.5 (7[d])
CP (RID), mg/dl	23.8 ± 5.8	20.6 ± 3.3	21.9 ± 2.5
RBC SOD, U/g Hgb	3406 ± 478	2872 ± 1007	3686 ± 565

[a]8 men studied in 4 separate experiments. Polycose added to the diet
to meet energy needs of 6 of the men. Durations: 3-4 wks control; 16
wks depletion; 3-4 wks repletion.   [b]Mean ± SD at end of each diet
period.   [c]Lowest value during depletion.   [d]Number of observations.

Copper balance, exclusive of surface losses, was slightly positive during
all dietary intakes of Cu. However, if measured surface losses of 0.12
and 0.15 mg/d for control and depletion periods were included, the men
were in negative Cu balance. Overall, when compared with the ends of
control or repletion periods there was a slight tendency toward low
plasma Cu, ceruloplasmin (Cp), and red cell superoxide dismutase (RBC SOD)
at the end of the period of low Cu intake. Lower plasma Cu was often seen
before the end of depletion rather than at the end of depletion; this
reflects a cyclical pattern of plasma Cu adjustment with time on the low

451

Cu diet presumably the consequence of homeostatic adjustments. One of the men who did not receive polycose for energy needs, showed definitive Cu depletion signs of lower plasma Cu, 55 µg/dl at the end of depletion vs 77 µg/dl during control and 70 µg/dl at the end of repletion, and low RBC SOD, 705 U/g Hgb (depletion) vs 3619 U/g Hgb (repletion). Two other men exhibited similar but not significant trends that were within the normal range.

In a separate experiment, eight women were maintained in a metabolic unit and fed a diet lower in Cu (0.67 ± 0.05 mg/d) and for shorter dietary periods than those described for the men (Table 2).

Table 2. Effect of Dietary Cu on Indices of Cu Status of Women[a]

	Control	Depletion	Repletion
Diet Cu, mg/d	1.47	0.67 ± 0.05	2.65 ± 0.07
Cu Balance, mg/d	-	-0.04 ± 0.09	0.02 ± 0.13
Plasma Cu, mg/dl	98 ± 24[b]	89 ± 19	95 ± 19
CP (ENZ), mg/dl	47.8 ± 8.9	38.5 ± 8.2	48.4 ± 9.8
CP (RID), mg/dl	26.7 ± 5.3	26.6 ± 3.8	27.1 ± 5.3
CP (ENZ/RID)	1.81 ± 0.32	1.44 ± 0.40	1.80 ± 0.27
RBC SOD, U/g Hgb	4826 ± 823	4732 ± 940	4107 ± 438
Platelet CCO, U/10^6	-	5.31 ± 0.78	6.76 ± 1.31
MNC CCO, U/10^6	-	0.63 ± 0.19	0.89 ± 0.08

[a]8 women. Energy intake regulated by porportional diet adjustment 14 d control, 42 d depletion, 37 d repletion.
[b]Mean ± SD at end of each diet period.

Measured indices of Cu nutriture were higher for the women than they were for the men. If surface losses of Cu were included, the women were in negative Cu balance. Plasma Cu, monoamine oxidase, and RBC SOD apparently were not affected by Cu intake. Enzymatically measured Cp at the end of the low Cu intake period was 20% lower than control or repletion ($p < 0.05$). No differences were seen in immunoreactive Cp. Consequently, the specific activity of Cp (Cp ENZ/Cp RID) was reflective of Cu intake. Cytochrome c oxidase (CCO) activity in platelets and mononucleated white cells (MNC) was significantly higher at the end of repletion than at the end of the low Cu period. These findings indicate that the specific activity of Cp and the activity of CCO in platelets and MNC are more sensitive indicators of Cu status than plasma Cu or RBC SOD, and that dietary intakes of 0.8 mg Cu or less/day are insufficient for healthy adults to maintain normal Cu status.

REFERENCES

1. E.J. Underwood, "Trace Elements in Human and Animal Nutrition", 4th ed., Academic Press, New York (1977) pp. 56-108.
2. L.M. Klevay, L. Inman, L.K. Johnson, M. Lawler, J. R. Mahalko, D.B. Milne, H. Lukaski, W. Bolonchuk and H.H. Sandstead, Increased cholesterol in plasma in a young man during experimental copper depletion, Metabolism 33:1112 (1984).
3. L.M. Klevay, W.K. Canfield, S.K. Gallagher, L.K. Henriksen, H.C. Lukaski, W. Bolonchuk, L.K. Johnson, D.B. Milne and H.H. Sandstead, Decreased glucose tolerance in two men during experimental copper depletion, Nutr. Rep. Intl. 33:371 (1986).

BEER INCREASES THE LONGEVITY OF RATS FED A DIET DEFICIENT IN COPPER

Leslie M. Klevay

USDA, ARS, Human Nutrition Research Center
Grand Forks, ND 58202

The origin of ischemic heart disease, the leading cause of death in the industrialized world, remains obscure. It has been suggested that copper deficiency or abnormal metabolism of copper is of prime importance in the etiology and pathophysiology of this disease (1-3). As moderate intakes of alcoholic beverages, especially beer, often are associated with decreased risk of death from ischemic heart disease or coronary artery occlusion (4-6), it was decided to test the hypothesis that consumption of beer could have a favorable effect on rats fed a diet deficient in copper.

METHODS

Male, weanling rats (Sprague-Dawley, Indianapolis, IN) were fed a diet based on 62% sucrose, 20% egg white and 10% corn oil (7). As it is deficient in both copper and zinc, finely ground zinc acetate was added to increase dietary zinc by 13 mg/kg. In each of three experiments, two groups of 15 rats were matched by weight (51g overall mean); half the rats were given demineralized water to drink and half were given beer. Beer (Budweiser, Anheuser Busch, St. Louis, MO) was purchased in quarts which were opened and allowed to stand at room temperature overnight to minimize foaming. Animals were housed under standard conditions (8). Cholesterol in plasma was measured by fluorescence (9); copper, zinc and iron in organs and diet were measured by atomic absorption after destruction of organic matter with nitric and sulfuric acids and hydrogen peroxide (10). After one death in experiment three, organs were obtained under pentobarbital anesthesia. Means were compared by "t" test (11).

RESULTS AND DISCUSSION

There was no difference in growth in any experiment at the time of first death (28-33 days); after this, mortality made further comparison useless. Median longevity (Table 1A) of rats drinking beer was more than 4 times that of rats drinking water. At this writing 80% of the rats drinking beer have not died; beer also increased this measure of longevity.

Consumption of beer was associated with lower plasma cholesterol (Table 1B) in two of the three experiments. Hematocrits for all groups of rats in experiments one and two were less than 33%; beer was without effect. Copper deficiency was verified by the presence of ventricular aneurysms, anemia and low liver copper.

Table 1A. Longevity, days[a]

	Exp. 1 Beer	Exp. 1 Water	Exp. 2 Beer	Exp. 2 Water
median	204	62	>242	42
80% dead	>375	103	>242	74

[a]After weaning and travel.

Table 1B. Cholesterol in plasma, mg/dl, after 4-5 weeks

Exp.	Beer	Water	p
1	102(4.0)	101(3.0)	n.s.
2	98(4.1)	143(14.4)	<0.02
3	85(6.1)	107(3.3)	<0.005

[a]Mean (S.E.)

Ventricular aneurysms were numerous in experiments one and two; beer seemed without effect. Cardiac enlargement in copper deficiency has been found many times (e.g. 12,13); however, hearts of rats that drank beer were approximately 20% lighter (Table 2).

Copper in liver (Table 2) was increased approximately 3-fold in rats that drank beer. Even the higher value, 3.9 µg/g, is substantially lower than the 10 to 13 found when the present diet is supplemented with copper (14-16). In contrast, liver zinc was increased only 9% (p < 0.02) and liver iron was unaffected (p > 0.1). Copper and iron in hearts of rats that drank beer were increased approximately 20% (p=0.02, 0.0008, respectively). Dietary copper ranged from 0.74 to 0.85 µg/g during these experiments. Copper in beer was 25 ng/ml.

Table 2. Heart Weight and Organ Analyses in Experiment 3

	Beer	Water	p
Heart, wet, g	1.12 (0.04)	1.41 (0.07)	<0.003
Liver Cu, dry, µg/g	3.91 (0.49)	1.34 (0.17)	0.0001
Liver Zn, dry, µg/g	73.0 (1.7)	67.0 (1.6)	<0.02

The hypothesis was tested successfully as rats that drank beer lived longer. Approximately a dozen cholesterotropic and cuprotropic chemicals have been found to reciprocally alter the metabolism of cholesterol and copper (17-19). Although beer is a complex mixture, it shares properties of 5 of these that decrease cholesterol in plasma and enhance copper metabolism. The active ingredient in beer is unknown; 4% ethanol and 5 µg chromium/ml were without beneficial effect in my similar experiments.

2. L.M. Klevay, The role of copper, zinc, and other chemical elements in ischemic heart disease, In: Rennert OM, Chan W-Y, Metabolism of Trace Metals in Man, Vol I. Boca Raton, FL: CRC Press, 1984:129-57.
3. L. M. Klevay, Ischemic heart disease. A major obstacle to becoming old, Clin Geriatric Med, in press.
6. T.B. Turner, V.L. Bennett, H. Hernandez, The beneficial side of moderate alcohol use, Johns Hopkins Med J 1981; 148:53-63.
7. L.M. Klevay, Hypercholesterolemia in rats produced by an increase in the ratio of zinc to copper ingested, Am J Clin Nutr 1973;26:1060-8.
19. L.M. Klevay, Cholesterotropic and cuprotropic chemicals, In: Mills, C.F., Bremner, I., Chesters, J.K., ed., Proceedings of the Fifth International Symposium on Trace Element Metabolism in Man and Animals - TEMA 5, United Kingdom: CAB, 1984:180-3.
A complete reference list can be obtained from the author.

THE INFLUENCE OF ZINC DEFICIENCY ON THE STORAGE OF ZINC IN BONE

D. Berg and W.E. Kollmer

Abt. für Nuklearbiologie
Gesellschaft für Strahlen- und Umweltforschung
D 8042 Neuherberg, Germany

INTRODUCTION

Bone which contains about 27% of total body zinc[1] has been considered
as a storage organ for this element which may supply Zn to the soft tissues
in periods of nutritional deficiency. We have shown previously that radioactive Zn which had been deposited in "deep" bone was not mobilized in Zn
deficiency and furthermore that Zn deficiency does not lead to an enhanced
resorption of bone mineral.[2] On the other hand it was found that the level
of Zn in bone decreases in Zn deficiency.[2] This study was designed to
examine the influence of Zn deficiency on the turnover of radioactive Zn in
bone.

MATERIAL AND METHOD

Male rats of a Wistar derived strain weighing 330 g were fed a commercial Zn deficient diet containing 2 µg Zn per g. The animals were distributed to 5 experimental groups. Distilled water was supplied ad libitum for
drinking. In 3 groups (K1,K2,B2) this water was supplemented with 50 µg Zn
per ml, 2 groups (C1,C2) remaining Zn deficient. 14 days after the start of
the experiment all animals received a tracer dose of 370 kBq Zn 65 and 370
kBq Sr 85 intravenously as a chloride. A series of whole body measurements
covering the whole experimental period was started immediately after the
injection of the tracer. One day after this Zn supplement the water in one
group (B2) was exchanged for distilled water. At the same time 4 animals
were killed in the control (K1) and in the Zn deficient group (C1). All
others were killed 14 days later (K2,B2,C2). Blood plasma, tibia, femur,
parietal bone, liver, kidneys, spleen, heart, muscle and fur were collected
and their radioactivity determined by γ-spectroscopy. The radioactivities
of the tracer are reported as a percentage of the dose. Zn was analysed by
flame atomic absorption in the tissues and by ICP in plasma. Statistical
significance was calculated by students t-test.

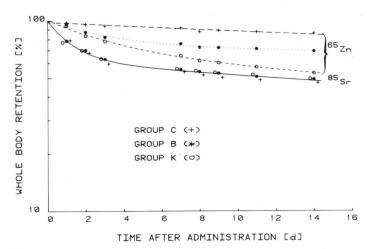

FIG. 1. WHOLE BODY RETENTION IN RATS AFTER INTRAVENOUS
INJECTION. CONTROLS (K), Zn DEFICIENCY STARTED
1d AFTER (B) OR 14d BEFORE (C) TRACER INJECTION

TABLE 1. Zn 65 CONCENTRATION IN PLASMA, LIVER AND SKELETAL MUSCLE
1d AND 14d POST ADMINISTRATION ($\bar{x} \pm s_{\bar{x}}$)

	GROUP	PLASMA	LIVER	SKELETAL MUSCLE	TIBIA
		\% DOSE PER G FRESHWEIGHT			
1d P.A.	K1	0.063±0.004	0.809±0.011	0.134±0.009	0.439±0.040
	C1	0.041±0.002*)	1.125±0.018+)	0.157±0.005*)	0.431±0.009
14d P.A.	K2	---	0.144±0.002	0.074±0.007	0.412±0.021
	B2	---	0.273±0.005+)	0.122±0.005+)	0.472±0.040
	C2	---	0.379±0.005+)	0.152±0.008+)	0.365±0.012

SIGNIFICANCE COMPARED TO CONTROL (K1 OR K2)

*) $2p < 0.05$

+) $2p < 0.001$

RESULTS AND CONCLUSIONS

As in our earlier experiment the level of Sr 85 in the whole body (Fig. 1) and in bone as well as the freshweight of bone were not influenced by Zn deficiency. In contrast to this finding in adult animals, Hurley et al.[3] have in young animals reported a decrease of the accretion rate of Ca. The whole body retention of Zn 65 (Fig. 1) was different in all experimental groups and demonstrates that there was an increasing restriction of Zn 65 excretion during the period of Zn deficiency. Furthermore the distribution of Zn 65 within the body of Zn deficient animals was not the same as in the controls (Table 1). Whereas the tracer in the soft tissues of the deficient animals just as in the whole body was elevated compared with the controls, it was significantly decreased in plasma and equal to or lower (not significant) compared with the controls in the bone of the animals on the longer lasting Zn deficiency (group C2). At the same time the level of stable Zn in plasma was only 0.9 ± 0.1 µg/ml in the deficient rats compared with 1.6 ± 0.1 in the controls. These findings indicate that in Zn deficiency the physiological control mechanisms give preference to the Zn supply of the soft tissues while that of bone becomes throttled. In the animals where Zn deficiency was initiated only one day after the administration of the tracer (B2) its level in bone after 14 days was rather elevated (not significant) than decreased relative to the controls in spite of a less conspicuous restriction of Zn 65 excretion compared with group C2. However, this increase above the controls in bone was not significant and furthermore it was much lower than the simultaneous increase in the soft tissues. This confirms the above conclusions drawn from the data of the other experimental groups. There was no clear evidence for an enhanced mobilization of the element from bone in the deficient animals. Nevertheless the lower balance in bone relative to the controls combined with the inhibited lower excretion contributes to maintaining the level of the element in the soft tissues of the deficient animals at the cost of its level in bone.

REFERENCES

1. I.G.F. Gilbert, D.M. Taylor, The behaviour of zinc and radiozinc in the rat, Biochim. Biophys. 21:545 (1956).
2. D. Berg, W.E. Kollmer, Mobilization of zinc and strontium from bone in Zn deficient rats, in: "Trace Elements-Analytical Chemistry in Med. & Biol." P. Brätter, P. Schramel, Walter de Gruyter, Berlin-New York (1987).
3. L.S. Hurley, J. Gowan, G. Milhaud, Calcium metabolism in manganese-deficient and zinc-deficient rats, Proc. Soc. Exp. Biol. Med. (N.Y.) 130:856 (1969).

EVALUATION OF SOME FACTORS THAT MAY AFFECT PLASMA OR SERUM
ZINC CONCENTRATIONS

J.L. English, K.M. Hambidge, and M. Jacobs Goodall

Department of Pediatrics
University of Colorado Health Sciences Center
Denver, Colorado 80262

INTRODUCTION

The objective of this study was to evaluate some selected factors which may influence zinc concentrations during or subsequent to sample collection. The factors assessed were 1) tourniquet pressure, 2) quantity of heparin used, 3) length of time between collection and centrifugation, 4) centrifuge speeds and 5) reaming of serum clots.

METHODS

Blood samples were collected from a peripheral vein of healthy volunteers. For tourniquet pressure study, a catheter was inserted in the subject and samples were taken at seven minute intervals, alternating pressures. This was repeated in triplicate using the same catheter. For the remaining studies, a sufficient volume of blood for any one comparison was drawn at one time. Samples were immediately transferred to covered polypropylene tubes for subsequent separation and storage. Fourteen USP units of sodium heparin/ml blood was routinely used as anticoagulant. All materials and reagents were found to be free of detectable zinc before use.

Plasma and serum zinc analyses were performed using flame atomic absorbtion spectrophotometry. Student's and paired t-tests were used in determining significant differences.

RESULTS AND DISCUSSION

Results are summarized in table below. With the exception of time intervals between collection and separation, no significant differences were observed for any of the comparisons in the table. Plasma and serum zinc concentrations were significantly ($p < 0.01$) higher when centrifugation was delayed for 120 min compared with centrifugation at 0 min and 10 min respectively. Serum values were approximately 3 ug/dl higher than plasma samples (collected for same subjects at same time) whether centrifugation was immediate or delayed.

TABLE 1

FACTORS INVESTIGATED	Zinc (ug/dl)	
	(mean + SEM)	
TOURNIQUET PRESSURE (n=3)		
No Cuff	80.55 + 0.69	
"Loose" Penrose Tubing	81.44 + 0.53	
"Tight" Penrose Tubing	81.75 + 0.50	
Sphygmomanometer Cuff (> diastolic BP x 3 min)	83.00 + 0.99	
AMOUNT OF HEPARIN USED AS ANTICOAGULANT (n=10)		
0.75 USP units/ml blood	80.4 + 1.86	
14 USP units/ml blood	78.9 + 1.29	
TIME BETWEEN COLLECTION AND CENTRIFUGATION (n=9)		
	Plasma	Serum
>10 minutes	86.46 + 1.55	88.71 + 1.94
120 minutes	91.91 + 2.02	94.63 + 2.30
CENTRIFUGATION SPEED (n=9)		
200 x G (platelets: 163 x 10^6/ml + 45)	79.51 + 3.68	
18,000 x G (platelets: 7x 10^6/ml +1)	77.33 + 3.97	
REAMING OF SERUM SAMPLES		
Reamed (n=10)	88.00 + 0.83	
Non-reamed (n=15)	89.18 + 0.73	

Initial and 120 min time samples for both serum and plasma were also assayed for copper, magnesium, sodium, potassium and chloride, calcium and albumin. No significant increase over time between collection and centrifugation was observed for these parameters.

There was reason for concern that each of the factors examined in this study could contribute to erroneous data for plasma and/or serum zinc concentrations. Hence, in general, these results are reassuring. Contrary to a previous report (1) tourniquet pressure did not have a significant effect. Heparin concentrations more than an order of magnitude greater than those recommended recently (2) were not associated with a detectable decrease in plasma zinc concentrations. Thus, the proposed osmotic effects of heparin were not detectable over this range of heparin concentrations. Centrifuge speeds that were too slow to remove a substantial percentage of platelets were not associated with any measureable increase in plasma zinc concentrations. Reaming serum samples prior to and after centrifugation did not cause any noticeable increase in zinc concentrations as was previously proposed (3).

One notable exception to these negative results was the steady and significant increase in both plasma and serum zinc concentrations with increasing time between sample collection and centrifugation. This observation was unique for zinc; no significant increase under the same experimental conditions was observed for any other mineral or trace element including those that, like zinc, have a substantially higher intracellular than extracellular concentration, e.g. potassium and magnesium. Nor were there any significant changes in serum albumin.

It is concluded that time between collection and centrifugation should be standardized and that it is preferable to restrict this interval as much as possible for both plasma and serum. (Supported by NIADDKD, 5R22 AM12432, and RR69 from NIH, General Clinical Research Center).

REFERENCES

1. S Kiilerich, MS Christensen, J Naestoft, et al.,   Clin Chim Acta 105:231-239 (1980).
2. JC Smith Jr, S Lewis, J Holbrook, et al., Clin Chem, (1987) in press.
3. JC Smith, JT Holbrook, DE Danford, J Amer Col Nutr 4:627-638 (1985).

CERULOPLASMIN AND GLUTATHIONE PEROXIDASE

AS COPPER AND SELENIUM INDICATORS IN MINK

Jouko T. Työppönen

College of Veterinary Medicine, Department of Biochemistry
P.O.Box 6, SF-00551 Helsinki, Finland, and Finnish Fur
Breeders Association, P.O.Box 5, SF-01601 Vantaa, Finland

INTRODUCTION

Indirect analysis of a given trace element through its enzyme activity
has some advantages over direct analysis: (i) smaller amount of sample is
needed, (ii) simpler analytical equipment is required and (iii) the enzyme
assay determines the biologically active form of the element. Ceruloplasmin
(Cp) has been successfully used as copper indicator in cattle and sheep
(Blakley and Hamilton, 1985). Serum glutathione peroxidase (GSH-Px) has been
shown to correlate with blood and dietary selenium in pigs (Hakkarainen et
al., 1978). In the present study, the usefulness of plasma Cp and GSH-Px as
Cu and Se indicators, respectively, in mink was studied.

MATERIALS AND METHODS

Feeding and tissue sampling of healthy and anemic minks used in the
study are described elsewhere in this book (Työppönen and Lindberg, 1987).

Ceruloplasmin determination in plasma was based on its o-dianisidine
oxidase activity (Schosinsky et al., 1974). Glutathione peroxidase was
analyzed in plasma samples by the method of Lawrence and Burk (1978) where
Se-dependent (Se-GSH-Px) and Se-independent (Non-Se-GSH-Px) activities are
distinguished by using 0.25 mM H_2O_2 or 1.5 mM cumene hydroperoxide as
substrates for the assay. The copper content in plasma, liver and spleen was
analyzed by plasma emission spectroscopy. Selenium content in plasma and
liver was determined as described by Lindberg (1968).

RESULTS AND DISCUSSION

There was a close correlation between Cu content and Cp activity in the
plasma of the minks studied (Fig. 1.; y = 48.3x - 9.7; r=0.920; P<0.001). In
anemic mink the plasma Cu content was higher (+41 % ; P<0.01) than in the
control mink (0.74 ± 17 mg/1). Plasma Cp activity which was 28 ± 11 U/1 in
control mink increased with increasing plasma Cu content (+39 % ; n.s.).

Plasma or whole blood Se correlated highly with plasma Se-GSH-Px
(Fig. 1.). The correlation in these cases can described as y = 5.2x + 33.4;
r=0.627; P<0.01 and y = 7.1x - 1300; r=0.709; P<0.001, respectively.

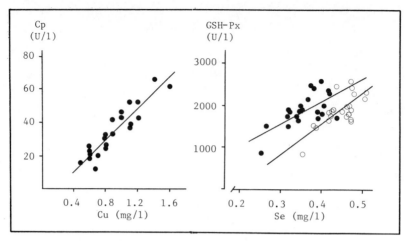

Fig. 1. Correlation between Cu content and Cp activity in mink (left). Correlation between plasma Se (●) or whole blood Se (○) and plasma GSH-Px activity in mink (right).

Iron deficiency seemed to have no significant influence on plasma Se, whole blood Se or plasma Se-GSH-Px. The mean values of these parameters were 0.36 ± 0.05 mg/1, 0.45 ± 0.04 mg/1 and 2130 ± 330 U/1, respectively. The plasma samples exhibited exclusively Se-GSH-Px activity.

Plasma Cu did not correlate with hepatic or splenic content of this trace element. The Cu concentrations in these organs are given elsewhere (Työppönen and Lindberg, 1987). Plasma Se content did not correlate with liver Se content (0.71 ± 0.16 μg/g). Therefore, the use of Cp activity as Cu indicator is limited to plasma as it was also in cattle and sheep (Blakley and Hamilton, 1985). Plasma GSH-Px activity can be used as plasma or whole blood Se indicators in mink as previously demonstrated in pigs (Hakkarainen et al., 1978).

REFERENCES

Blakley, B.R. and Hamilton, D.L., 1985, Ceruloplasmin as an indicator of copper status in cattle and sheep, Can. J. Comp. Med., 49:405.
Hakkarainen, J., Lindberg, P., Bengtsson, G. and Jönsson, L., 1978, Serum glutathione peroxidase activity and blood selenium in pigs, Acta vet. scand., 19:269.
Lawrence, R.A. and Burk, R.F., 1978, Species, tissue and subcellular distribution of non Se-dependent glutathione peroxidase activity, J. Nutr., 108:211.
Lindberg, P., 1968, Selenium determination in plant and animal material, and in water, Acta vet. scand., suppl. 23.
Schosinsky, K.H., Lehmann, P.H. and Beeler, M.F., 1974, Measurement of ceruloplasmin from its oxidase activity in serum by use of o-dianisidine dihydrchloride, Clin. Chem., 20:1556.
Työppönen, J.T. and Lindberg, P.O., 1987, Hepatic and splenic content of iron, copper, zinc, and manganese in anemic mink, This book.

COPPER AND IRON STATUS OF CALVES

IN RELATION TO PLASMA PROTEIN POLYMORPHISM

Inger Wegger

Department of Veterinary Physiology and Biochemistry
Royal Veterinary and Agricultural University
Copenhagen, Denmark

INTRODUCTION

Interactions between various trace elements regarding their absorption and metabolism in animals have been known for several decades. Another fact is that the metabolic function of inorganic micronutrients is closely related to their interplay with enzymes and other functional proteins. Furthermore, several studies have shown that trace element status of domestic animals to a certain extent is genetically controlled (Wegger, 1985). The results presented here lend further support to these statements. In addition our studies on calves show a rather close correlation between an animals copper and iron status and its phenotype regarding the polymorphic plasma proteins ceruloplasmin (Cp) and amylase (Am).

ANIMALS AND METHODS

The investigations were carried out on a progeny testing station in order to eliminate the influence of environmental factors. A total of 499 bull calves belonging to 59 half-sib groups each with 6-10 calves were used. Three breeds were represented: Black and White Danish Milk Race (226), Red Danish Milk Race (116) and crosses between RDM and American Brown Cattle (147). In order to minimize the influence of age variations blood samples were taken when the calves were 60-75 days old. The following quantitative parameters were determined as described earlier (cf. Wegger, 1985): plasma Cu content and Cp activity; plasma Fe, total iron binding capacity (TIBC) and per cent transferrin iron saturation (Tf sat. %). The qualitative parameters plasma Cp and Am phenotypes were determined by means of starch gel electrophoresis (Andreasen and Larsen, 1976).

RESULTS

Table 1 summarizes the quantitative results. The overall range of the various parameters is very wide. However, as shown by analysis of varians, a considerable amount of this variation in apparently normal calves can be ascribed to genetic influence since the variation between half-sib groups (S ext.) is significantly higher than the variation within groups of half-sib calves (S int.).

463

Table 1. Copper and iron status in bull calves (n = 499) aged 2-2½ month, genetic variation of parameters

Parameter	Range	Mean ± s.e.	Analysis of varians $S_{ext.} > S_{int.}$, P <
Plasma Cu, µg %	61 – 169	96 ± 0.9	0.01 – 0.001
Plasma Cp, U/1	22 – 111	51 ± 0.8	0.05
Plasma Fe, µg %	13 – 291	93 ± 2.2	0.001
TIBC, µg %	213 – 618	392 ± 3.4	0.001
Tf sat. %	4.3 – 57.4	23.3 ± 0.5	0.001

Table 2. Plasma Cu and Cp (mean ± s.e.) in relation to Cp phenotype, number of calves shown in brackets. Figures in a line not followed by the same letter are significantly different, P < 0.001

Parameter	Ceruloplasmin phenotype		
	AA (282)	AC (135)	CC (76)
Plasma Cu, µg %	103 ± 1.2 a	90 ± 1.6 b	82 ± 1.6 c
Plasma Cp, U/1	57 ± 1.0 a	46 ± 1.3 b	38 ± 1.3 c

Table 3. Parameters of copper and iron status (mean ± s.e.) in relation to Am phenotype, number of calves shown in brackets. Figures in a line not followed by the same letter are significantly different, P < 0.05 – 0.001

Parameter	Amylase phenotype		
	1-1 (254)	2-1 (161)	2-2 (84)
Plasma Cu, µg %	102 ± 1.3 a	92 ± 1.5 b	86 ± 1.7 c
Plasma Cp, U/1	55 ± 1.1 a	49 ± 1.3 b	43 ± 1.4 c
Plasma Fe, µg %	87 ± 3.0 a	93 ± 4.2 a,b	110 ± 5.3 c
Tf sat. %	22.1 ± 0.6 a	24.0 ± 0.9 a,b	25.4 ± 1,0 b,c

The copper containing plasma protein ceruloplasmin shows electrophoretically detectable polymorphism in cattle. Three phenotypes, namely Cp AA, Cp AC and Cp CC, occur in Danish cattle breeds. As shown in Table 2 calves with different phenotypes have different copper status as measured by their plasma Cu concentration and Cp activity. Calves with Cp type AA have the highest and those with Cp type CC the lowest plasma Cu content and Cp activity while the heterozygous type AC calves are intermediary regarding copper status.

The copper status of calves is also related to their phenotype of another plasma enzyme, namely amylase, as shown in Table 3. Again the two groups of homozygous calves have the highest (Am 1-1) and lowest (Am 2-2) plasma Cu content and Cp activity and heterozygotes (Am 2-1) show intermediary values.

Cp and Am phenotypes are apparently not inherited independently of each other. The highest frequency of Cp AA is found in calves with Am type

1-1 while Cp AC and CC predominate in animals with the Am types 2-1 and 2-2. Consequently calves with the combination Cp AA/Am 1-1 have the highest copper status and those with the combined type Cp CC/Am 2-2 the lowest status regarding this trace element.

As shown in Table 3 the iron status of the calves, measured by their plasma Fe and Tf sat. %, also varies with Am type. Calves with Am 1-1 have the lowest and those with Am 2-2 the highest iron status and again the heterozygous animals show intermediate values. The statistical significance of these differences is not as high as in the case of the relations between Cu status and protein type.

DISCUSSION

Investigations on sheep (Wiener et al., 1978) and cattle (Flagstad, 1976) have shown that genetic factors can influence the intestinal absorption of copper and zinc respectively and thereby interfere with the health and manifestation of life of the animals. In other studies on man, swine, calves and sheep it was found that the activity of some trace element dependent enzymes is genetically controlled (cf. Wegger, 1985). The present investigation shows that the great "normal" variation in copper and iron status of calves even under strictly standardized rearing conditions may be due to genetic influence. This assumption is also supported by the finding of significant differences in trace element status of calves with different phenotypes regarding the two plasma proteins Cp and Am. Whether this phenomenon should be considered as a biochemical-genetic "curiosity" or it has some functional significance for the development and health of the calves is not clear. In this context it can be mentioned that Reetz and Feder (1974) found a positive correlation between ceruloplasmin activity, plasma iron concentration and slaughter weight in pigs.

ACKNOWLEDGEMENTS

The investigations were supported by a grant from the Danish Agricultural and Veterinary Research Council.

REFERENCES

Andreasen, B. and Larsen, B., 1976, Ceruloplasmin polymorphism and parentage test in Hereford cattle, Acta vet. scand., 17:264.
Flagstad, T., 1976, Lethal trait A 46 in cattle. Intestinal zinc absorption. Nord. Vet.-Med., 28:160.
Reetz, I. and Feder, H., 1974, Ceruloplasmin, Kupfer- und Eisengehalt im Blutplasma beim Schwein und ihre Beziehungen zu Leistungseigenschaften, Proc. 1st World Congr. Genetics appl. Livestock Prod., Madrid, 3:1159.
Wegger, I., 1985, Variations in trace element status of calves and their relation to trace element dependent enzymes and ascorbic acid, Ann. Rep. Sterility Res. Inst., 28:125.
Wiener, G., Suttle, N.F., Field, A.C., Herbert, J.G. and Woolliams, J.A., 1978, Breed differences in copper metabolism in sheep, J. Agric. Sci., 91:433.

FURTHER DATA ON THE BIOLOGICAL ESSENTIALITY OF NICKEL

M. Anke, B. Groppel, Ute Krause, and M. Langer

Karl-Marx-Universität Leipzig
Wissenschaftsbereich Tierernährungschemie
6900 Jena, Dornburger Straße 24, GDR

INTRODUCTION

Ni belongs both to the essential and toxic trace elements. In the following, the results of Ni deficiency experiments with male and female goats which have been repeated 15 times since 1971 are summarized.

Furthermore, the findings of Ni exposure experiments with 125 to 1000 mg Ni/kg ration dry matter are presented.

Effects of Ni deficiency

An Ni intake of < 100 μg/kg semi-synthetic ration led to a significantly lower feed-intake in growing, pregnant and lactating young goats as well as in adult ones. This reduced feed-intake caused by Ni deficiency already resulted in a worse growth of kids (table 1).

Table 1. The influence of nickel deficiency on growth, reproduction and life expectancy during the experiments (89 control and 96 nickel deficiency goats)

parameter		+Ni	−Ni	p
growth	1st day of life, kg	3.1	2.8	< 0.05
	91st day of life, kg	19.1	16.3	< 0.001
	168th day after suckling, g/d	99	88	< 0.05
reproduction performance	success of first insemination, %	70	55	< 0.05
	conception rate, %	83	71	< 0.05
	services per gravidity	1.4	1.9	< 0.05
	abortion rate, %	1	9	< 0.001
life expectancy	dead kids, %	9	38	< 0.001
	dead mothers, %	24	43	< 0.01

After the suckling period, Ni deficiency goats grew significantly more slowly than control animals.

Ni deficiency had a significantly negative influence on the reproduction performance of female goats. Intrauterinely Ni-depleted male goats had smaller testicles, a reduced semen production and Libido sexualis.

Ni deficiency also led to a significantly reduced milk, milk fat and milk protein production of Ni deficiency goats.

The most striking result of the experiments was that the mortality of Ni deficiency goats was higher than that of control animals. As a rule, Ni deficiency goats had shaggy hair and suffered from cutaneous eruptions. There were individual cases of dwarfism. Due to the disturbed Ca metabolism induced by Ni deficiency, the latter leads to Zn deficiency which manifests itself in these symptoms and damaged testicles. Ni deficiency goats absorbed less ^{65}Zn than control animals and also transferred fewer amounts of this element into their milk. Their Fe metabolism was disturbed as well. Thus, Ni deficiency resulted in secondary disturbances of the Ca, Zn and Fe metabolism.

Effects of Ni exposure

The effects of a Ni supplementation of 125 to 1000 mg/kg ration dry matter were systematically investigated. When Ni intakes up to 250 mg/kg ration dry matter were administered to hens, chicken and fattening bulls, feed consumption was not significantly influenced. Pigs reacted more sensitively. The egg production of hens and the growth of chicken, fattening bulls and pigs remained uninfluenced up to 250 mg Ni/kg ration.

The Ni supplementation reduced the Zn and Mg content of the skeleton and the egg significantly (table 2). Thus, Ni offers have an antagonistic influence on the Zn status. The Zn content is reduced to such an extent that the hatching capacity of chicken is impaired.

Ni supplementation also led to a significantly reduced Mg content of ribs (table 3) and eggs.

Table 2. The influence of different nickel intakes and of time on the zinc content of eggs (mg/kg dry matter)

day of laying	0	125	250	500	1000	Fp offer
7th	58	55	58	56	48	
14th	58	54	54	45	44	
21st	53	54	49	37	35	
28th	63	62	49	39	38	< 0.001
35th	60	54	51	40	31	
42nd	60	50	54	44	39	
Fp time		< 0.001				-

Table 3. The influence of rising nickel offers on the magnesium content of ribs (mg/kg dry matter)

		0	125	250	500	1000	$SD_{0.05}$	%
pigs	$\bar{x}$	3002	3096	2942	3199	2101	709	70
	s	230	536	453	220	588		
chicken	$\bar{x}$	3166	2644	2404	2349	2337	509	74
	s	333	142	255	307	445		
hens	$\bar{x}$	2706	2782	2325	2420	2069	469	76
	s	253	375	138	295	266		

ALTERED GLUCURONYL TRANSFERASE ACTIVITY IN MAGNESIUM DEPLETED RATS

R.C. Brown and W.R. Bidlack

Department of Pharmacology and Nutrition
USC School of Medicine
Los Angeles, CA 90033

INTRODUCTION

Magnesium intake for the population often appears to be suboptimal. However, the occurrence of magnesium deficiency in man is most frequently
associated with a complicating disease state, such as alcoholism, malabsorption syndromes, burns, and acute or chronic renal disease. Drug therapy may
also exacerbate magnesium deficiency, and conversely magnesium depletion may
affect the efficacy of drug therapy.

In the rat, magnesium deficiency has produced a significant reduction in
both cytochrome P-450 and in vitro microsomal drug metabolism (Becking and
Morrison, 1970). Recently, we used a shorter depletion period (10 days) and
reported that the decrease in drug metabolism occurred without changes in the
enzyme components of the mixed function oxidase (Brown and Bidlack, 1987).
However, when the magnesium depletion was extended to 18 days, both cytochrome P-450 and drug metabolism were diminished. At this time, glucuronyl
transferase (GT) activity was also reduced by 50% in the deficient animals.

Using hepatocytes isolated from livers of magnesium depleted animals, drug
metabolism and conjugation were examined further. p-Nitroanisole metabolism
was decreased as expected. The metabolite, p-nitrophenol, appeared to be
conjugated by GT only at higher drug concentrations, suggesting an alteration
in the activity of GT or its regulation by magnesium.

An evaluation of the GT activity and its regulation was examined in the
microsomal membranes isolated from control and magnesium deficient animals.
The results of these experiments are provided in this report.

METHODS

Male Sprague-Dawley rats weighing 106 +5 grams were maintained on a synthetic rat chow diet either adequate (48 mg/100 g) or deficient (3 mg/100 g)
in Mg and allowed to drink distilled water ad libitum. The control animals
were pair fed to the dietary intakes of the experimental animals.

Livers were perfused and homogenized (10% w/v) in Tris (3 mM)- sucrose
(0.25 M) buffer, pH 7.4. The homogenate was centrifuged at 28,000 x g for
20 minutes, and the resulting supernatant was recovered and centrifuged at
105,000 x g for 30 minutes to obtain the microsomal membrane fraction. The

pellet was carefully rinsed and resuspended in Tris (50 mM)–KCl (150 mM) buffer, pH 7.4, containing 0.1 mM EDTA. The pellet was again recovered after centrifugation at 105,000 x g for 30 minutes. The membrane was resuspended in the Tris-KCl buffer without EDTA.

Glucuronyl transferase activity was determined in a reaction system containing UDPGA (2.5 mM), p-nitrophenol (0.0-0.4 mM), and 0.5 mg of microsmal protein in the Tris-KCl buffer, with or without 5 mM magnesium chloride. The reaction was carried out at 37°C for 15 or 30 minutes. Maximal GT activity was determined following addition of Triton X-100 (0.1% final). GT activity was reported as n moles of p-nitrophenol conjugated/ 15 or 30 minutes/0.5 mg of microsomal protein.

RESULTS AND CONCLUSION

The microsomal membrane was washed with EDTA to assure removal of any residual cations. The addition of magnesium to the reaction system produced the expected cation activation, ca. 60-100% increase, in the control GT enzyme. However, the magnesium activation was much less in the GT obtained from the depleted animals (Table 1). The GT activities from both the control and depleted animals were increased 100% in the presence of the detergent (Table 1). However, the total GT activity in the magnesium depleted animals was 55% of the total activity determined for the control animals. The addition of magnesium to the reaction system again enhanced the control GT activity ca. 60%, while the depleted GT enzyme was enhanced only 30%.

Thus, magnesium deficiency affects the total GT activity at a point in depletion not yet reflected in the drug metabolism enzymes. In addition, the cation regulation site would appear to be altered, or a specific isozyme form of GT has been lost. These alternatives are currently under further investigation.

Table 1. Effect of Magnesium Depletion on Microsomal Glucuronyl Transferase Activity

	Glucuronyl Transferase Activity[1] (n moles conjugated/30 minutes)			

Non-Activated

pNP (n moles)	Control	Depleted (0 mM Mg)	Control	Depleted (5 mM Mg)
10	9.6 ± 0.4	6.8 ± 2.8**	9.6 ± 1.8	6.4 ± 1.8**
50	11.4 ± 2.0	8.4 ± 0.7**	14.2 ± 1.7	11.8 ± 2.5
100	13.1 ± 3.2	10.4 ± 2.8	21.5 ± 3.9	15.5 ± 2.3**
200	21.3 ± 3.1	20.5 ± 2.7	42.8 ±23.8	27.7 ± 4.6
400	48.6 ± 18.3	29.9 ± 2.6**	67.1 ± 3.4	31.8 ± 8.5**

Triton X-100 Activated　　　　　　(n moles conjugated/15 minutes)

10	9.2 ± 0.2	7.9 ± 0.6**	9.8 ± 0.2	7.5 ± 0.2**
50	15.7 ± 3.0	10.3 ± 1.2**	25.1 ± 4.8	13.4 ± 2.9**
100	25.8 ± 3.8	15.7 ± 1.7**	44.3 ± 8.9	21.4 ± 2.7**
200	41.8 ± 4.2	29.4 ± 4.0**	82.0 ±19.2	38.4 ± 3.5**
400	84.3 ± 30.9	54.7 ±19.2	125.8 ±19.2	59.6 ±10.5**

N= 4; **significantly different from control, $p < 0.01$
1. EDTA washed microsomes incubated with 0 or 5 mM Mg

THE PREDICTION OF IMPAIRED GROWTH DUE TO Cu, Co AND Se DEFICIENCIES IN

LAMBS ON IMPROVED HILL PASTURES IN SCOTLAND

Neville Suttle[1], Cliff Wright[2], Alan MacPherson[2], Ron Harkess[2]
Gordon Halliday[3], Keith Miller[4], Pat Phillips[5], Colin Evans[1]
and Desmond Rice[6]

Scottish Agricultural Research Institutes[1], East[2], North[3] and
East[4] of Scotland Agricultural Colleges, and Unit of
Statistics[5], c/o Department of Agriculture and Fisheries for
Scotland, Chesser House, Edinburgh, and Veterinary Research
Laboratories, Stormont, Belfast[6]

INTRODUCTION

Cu supplements can increase growth and survival amongst lambs grazing
hill pastures improved by liming and reseeding, but how often such conditions occur in practice is unknown. Cu response trials were therefore conducted while monitoring soil, herbage and lamb blood samples for criteria
which might predict Cu status and growth responsiveness. Since the same
questions apply to Co and Se deficiencies, these elements were also monitored and growth responses to Co and Se were sought in successive years.

MATERIALS AND METHODS

Twenty improved hill sites were selected for above average risk of
hypocuprosis (chiefly herbage Mo > 2.5 mg/kg DM). In May, 1985 (year 1), 25
pairs of lambs, matched for sex and bodyweight, were all give Se (25 mg as
barium selenate) and vitamin B_{12} (1 mg) by injection and one member received
1.3 g CuO 'needles' orally. A further group of 10 received no supplements.
In year 2, the B_{12} and Cu treatments were interchanged. Herbage was sampled
and lambs were bled and weighed at dosing (May), late June and late August.
Blood samples were analysed for Cu, B_{12} and methylmalonic acid (MMA) in
plasma, superoxide dismutase (ESOD) and glutathione peroxidase (GSHPx) at
37° in haemolysed erythrocytes. Soil samples taken in August were analysed,
like the herbage, for a large number of macro and trace elements, generally
using an ICP method.

RESULTS

Mean herbage Mo was high in May (3.3) and higher in August (4.9 mg/kg
DM) but herbage Cu and Se were unremarkable in 1985 (means 6.8 mg and 2.9 g/
kg DM, respectively). Growth responses in Cu-treated lambs were small (+4 ±
1.8 g/d; p < 0.06) and significant on only one farm. Hypocupraemia was
rare: on only four farms were mean plasma values subnormal (< 9 umol/l) and
they generally increased in untreated lambs as the season progressed to a
mean of 13.0 (s.e.) umol/l while SOD decreased from 11.9 to 9.0 U/mg Hb:
treatment increased August values to 16.1 umol/l and 10.6 U/mg Hb.

Table 1. Relationships between Year 1 Variables in Soils (S), Herbage (H) and Animals (see text for units).

Dependent Variable	Month	Regression Equation	Month	Regression sums of squares (% total)
ESOD	May	3.82+1.583HCu - 0.583HMo	May	38.7
Plasma Cu $\log 10$	May	0.69+0.53HCu - 0.013HMo	May	34.9
LWG	Aug	155.6 - 0.0041 ESOD	May	39.2
Plasma B_{12}	Aug	1171 - 44.7 MMA	Aug	48.2
	June	530 + 424 Co_S	Aug	49.0
GSHPx	Aug	36 + 0.56 GSHPx	May	47.0

Some significant relationships are shown in Table 1. There was wide variation (157-279 g/d) in growth rate between farms and a between-farm relationship between LWG (g/d) and ESOD (U/mg Hb) which accounted for 39% of the variation initially but less as the season progressed. There were no relationships between LWG and plasma Cu but both ESOD and plasma Cu were positively related to herbage Cu and negatively related to herbage Mo, particularly in May.

Mean GSHPx and plasma B_{12} in the lambs varied greatly from farm to farm, declined from May to August but generally remained above 100 u/g Hb and 300 ng/l respectively in 1985. Nevertheless the mean MMA was high in June and August, with 8 farms having values > 5 umol/l, and related to plasma B_{12}. The predominant determinant of plasma B_{12} and blood GSHPx in August was the level in May: herbage Se and Co were unimportant but extractable soil Co did correlate with plasma B_{12}.

In year 2 the growth response to Co was small (3.7 ± 13.0 g/d) but correlated with the plasma B_{12} in May.

DISCUSSION

A feature of the trials has been the importance of initial Cu, B_{12} and Se status of the lamb as a determinant of status at the end of grazing. In the case of Cu, as measured by ESOD, there was a correlation between ESOD in May and LWG between farms, yet no growth response to Cu. It seems unlikely that the treatment was insufficient or that growth responses are attainable in normocupraemic animals. We suggest that initial Cu status influenced growth, via the placental and mammary transfer of Cu. Since Cu is absorbed from milk with 40x the efficiency of that in high Mo pasture (1), milk probably represents the predominant source of available Cu from May to August. B_{12}, like Cu, is well absorbed from milk (1). Thus the prediction of Cu, Co and Se responsiveness from herbage composition during the summer is compromised by the maternal influence. Provided that the ewe receives home-grown feed during winter, soil indices may yet have predictive value. It is noteworthy that extractable Mo and Co in soils were correlated with the Cu and B_{12} status, respectively, of the lambs. The values recorded for GSHPx were generally high and growth responses to Se in year 3 (in progress) seem unlikely.

REFERENCE

1. N. F. Suttle, 1983, The nutritional basis of trace element disorders in ruminants, Br. Soc. Anim. Prod. Oc. Publ. No. 7, pp. 19-25.

ZN DEFICIENCY IN THE RAT: RELATIONSHIP BETWEEN GROWTH AND METAL CONTENT

AND DISTRIBUTION.

D. H. Petering and B. Fowler

Department of Chemistry, University of Wisconsin-Milwaukee
Milwaukee, WI 53201 and
The National Institute of Environmental Health Sciences
Research Triangle Park, NC 27709

INTRODUCTION

Zinc$_1$deficiency causes many physiological changes in mammalian organisms[1]. The inhibition of growth is both a rapid and conspicuous effect. However, the underlying relationships between site of metal loss and its biological effects are not well understood. It is appreciated that cells and tissues need not lose large amounts of Zn to display signs of Zn deficiency. The present study was undertaken to examine carefully the distribution of Zn and Cu in the rat after it is placed on a Zn-deficient diet.

RESULTS AND DISCUSSION

The procedures for maintaining Sprague Dawley CD male rats on semipurified diets are described elsewhere[2]. The control, Zn-normal (Zn+) diet contained 20 ppm Zn, whereas the deficient diet (Zn-) had <1 ppm Zn. Both contained 15 ppm Cu. Young male rats weighing about 380 g in each experiment were placed on the special diets and distilled water at time 0. They were kept on these regimes for 1-5 weeks. Weights and feed consumption were monitored during the experiments. At their conclusion, parameters related to the growth and metal status of the two groups were measured.

Animals transferred to the Zn- diet at time 0 stopped growing immediately within the 1-2 day uncertainty of the measurement. Their weights did not change for 3 weeks. A small decrease was noted after 5 weeks. Decrease in feeding did not occur until week two. Organ weights were normalized to animal weight. The specific organ weights of kidney, heart, and lung of Zn- animals did not change relative to Zn+ controls. Only liver showed a decline during the five week period. Among the four organs there was also no change in Zn content during this time (liver was measured after 2 weeks). However, in kidney alone there was a 50% decrease in Cu concentration.

To assess the effects of Zn deficiency on organ distribution of Zn and Cu, liver and kidney were fractionated. Liver homogenate was separated in membrane, mitochondrial, lysosomal, microsomal, and cytosolic compartments by differential centrifugation in conjunction with careful washing of the fractions. Except for lysosomes, which were not examined, the Zn and Cu concentrations in each compartment of the tissue were not altered during two weeks of Zn deficiency. Furthermore, when liver cytosol was chromatographed over Sephadex G-75, the distributions of Zn and Cu were the same in the two sets of animals.

Kidney was also separated into subcellular fractions--membrane, mitochondria, and cytosol. As previously described, cytosol has uninduced Zn,Cu-metallothionein (Zn,Cu-Mt)[2]. The other fractions also contain a substantial amount of a Zn,Cu binding protein, presumably also metallothionein. After 3 and 5 weeks of Zn deficiency, the major sites which lose Zn are Mt and the particulate, Mt-like species. The same is true for Cu. Because the Mt band constitutes the major site for Cu binding in contrast to Zn, the loss of metal from this pool is readily observed at the whole organ level.

It is clear that despite the continual turnover of organ proteins in the cycle of synthesis and degradation, Zn content and distribution is generally resistent to change in the face of a deficit in nutrient Zn. Only plasma and kidney Zn,Cu-Mt respond by losing Zn. Both of these pools are depleted of labile Zn within 24-48 hours of the imposition of the Zn deficient diet[2]. Hence, at this level of analysis, therefore, there few candidates for the critical site of Zn depletion, which leads to a coordinated, rapid cessation of growth of the organism. Plasma Zn, in particular, drops with kinetics which are fast enough to permit it to function as the pool which determines Zn dependent growth status[2].

ACKNOWLEDGEMENTS

Supported by NIH grant ES-05223.

REFERENCES

1. C. J. McClain Jr., E. J. Kasarskis, Jr., and J. S. Allen, Functional consequences of zinc deficiency. Prog. Food Nutr. Sci., 9:185 (1985).
2. D. H. Petering, J. Loftsgaarden, J. Schneider, and B. Fowler, Metabolism of cadmium, zinc and copper in the rat kidney: the role of metallothionein and other binding sites. Environ. Health Persp. 54:73 (1984).

ACTIVATION OF RAT ERYTHROCYTE SUPEROXIDE DISMUTASE BY COPPER

R. A. DiSilvestro

Department of Foods & Nutrition
Purdue University
West Lafayette, IN 47907, U.S.A.

Red blood cell superoxide dismutase (SOD) activity levels have been used to assess human copper status[1,2]. Initially, there was concern that these values may not accurately reflect the progress of copper repletion of people previously in poor copper status. This concern arose because red cell proteins do not undergo the usual turnover and synthesis processes[3]. A preexisting copper-free SOD apoprotein will not necessarily bind copper newly introduced into the body. Such is the case for rat serum ceruloplasmin which will only incorporate copper before secretion by the liver[4]. For these reasons, administration of copper to people in poor copper status might not be expected to quickly raise erythrocyte Cu-Zn SOD activity contents to copper adequate values. Possibly, a new red cell population would have to be generated. However, in practice, copper repletion restores human red cell SOD activities faster than could be accounted by normal production of new red cells with normal SOD activities[1,2]. The present study used rats to elucidate the mechanisms by which copper repletion increases erythrocyte SOD activities.

A two week feeding of adequate amounts of copper to copper deficient rats elevated erythrocyte SOD activity contents to levels occurring in rats continuously fed adequate copper. SOD activity was assessed using the pyrogallol autoxidation assay as modified by Prohaska[5]. Several lines of evidence strongly suggested that the copper induced increase in SOD activity values resulted in part from activation of a preexisting SOD apoprotein pool.

First, red cells from copper deficient rats were shown to contain normal amounts of SOD protein by employing an unpublished enzyme linked immunoadsorbent assay. In addition, feeding copper to deficient rats raised erythrocyte SOD activities without raising SOD protein concentrations. This result implied that new red cells in the copper repleted rats were not synthesizing Cu-Zn SOD protein at higher than usual rates.
Further studies demonstrated that copper repletion increased SOD activity contents of both young and old red blood cells. If no copper activation of preexisting erythrocyte SOD apoprotein had occurred, then increased activities would only have been seen in the young cells. Finally, copper was shown capable of increasing SOD activity levels in vitro in red blood

cells from copper deficient rats. Cells were incubated with shaking at 37°C in Hepes buffered saline plus glucose with or without 0.2 ppm copper ion for 2, 4, 8 or 16h. Cu-Zn SOD activity levels were elevated in the cells exposed to copper at all time points. The increase at 16h was about 2.5 fold. The rat serum protein ceruloplasmin, even at copper levels of 1 ppm, was very ineffective at raising SOD activities compared to inorganic copper.

SUMMARY

Erythrocyte Cu-Zn SOD activities in copper deficient rats were elevated to copper adequate levels by feeding adequate copper levels for only two weeks. Several lines of evidence strongly suggested that the increase resulted in part from activation of SOD apoenzyme.

REFERENCES

1. S. Reiser, J. C. Smith, W. Mertz, J. T. Holbrook, D. J. Scholfield, A. S. Powell, W. K. Canfield, and J. J. Canary, Indices of copper status in humans consuming a typical American diet containing either fructose or starch, Am. J. Clin. Nutr. 42:242 (1985).
2. R. Uauy, C. Castillo-Duran, M. Fisberg, N. Fernandez, and A. Valenzuela, Red cell superoxide dismutase activity as an index of human copper nutrition, J. Nutr. 115:1650 (1985).
3. E. Beutler, "Red Cell Metabolism, a Manual of Biochemical Methods," 2nd Ed., Cruse and Stratton, New York (1979).
4. I. Sternlieb, A. G. Morell, W. D. Tucker, M. W. Greene, and I. H. Scheinberg, The incorporation of copper into ceruloplasmin in vivo: studies with copper^{64} and copper^{67}, J. Clin. Invest. 40:1834 (1961).
5. J. Prohaska, Changes in tissue growth, concentrations of copper, iron, cytochrome oxidase, and superoxide dismutase subsequent to dietary or genetic copper deficiency in mice. J. Nutr. 113:2148 (1983).

KINETICS OF CHROMIUM IN PATIENTS ON CONTINUOUS AMBULATORY PERITONEAL DIALYSIS

B. Wallaeys[1], R. Cornelis[1*], N. Lameire[2], and M. Van Landschoot[2]

[1] Laboratory for Analytical Chemistry, Rijksuniversiteit Gent
Proeftuinstraat 86, B-9000 Gent, Belgium

[2] Renal Division, Department of Internal Medicine, University
Hospital, De Pintelaan 185, B-9000 Gent, Belgium

INTRODUCTION

In a previous study[1] we investigated trace element concentrations in
serum, packed blood cells and in dialysate of patients on continuous ambulatory peritoneal dialysis (CAPD). In the serum of the patients extremely
high Cr levels were found (20 to 50 times the normal serum value of 0.16
ng/ml) pointing to a substantial absorption of Cr from the dialysate. The
Cr content of the fresh dialysate amounted to 1 ng/ml. Half of it was absorbed by the patient. We found that due to the presence of lactate the
chemical species of Cr present in the dialysate is solely the trivalent
form.

The present study reports on the kinetics involved during the transport of Cr(III) from the dialysate to the body. Applying a modified functional model for Cr(III) metabolism the impact of long-term CAPD treatment
on the Cr levels in various body compartments could be estimated.

MATERIAL AND METHODS

The kinetics of Cr were followed in dialysate, plasma and urine of
three patients on CAPD.

Two liters of fresh dialysate (Dianeal[R] 137, Travenol, Deerfield, Il.
USA) was labelled with $3.7 \cdot 10^5$ Becquerel of $^{51}CrCl_3$ (Amersham) and a sample of 10 ml was taken. Subsequently the dialysate bag was connected with
the patient's Thenchhoff catheter with three way valve. The ^{51}Cr labelled
dialysate was instilled in the peritoneal cavity. The time at which half
of the solution was instilled was taken as the starting point. The actual
sampling of 10 ml dialysate was performed through the three way valve after withdrawal of 20 ml of fluid. This was done immediately after instillation and repeated every 30 minutes throughout the dwell period (6 hours).

---

* R. Cornelis is a Senior Research Associate of the National Fund for
Scientific Research.

Blood samples were taken before the instillation of the dialysate and at point of time 5, 15, 30 min, 1, 2, 4 and 6 hours after the start of the session. The blood was heparinized and the plasma was separated from the packed cells by centrifugation.

In the follow-up period of 12 days, blood of the patients was sampled at 0.5, 1, 2, 4, 6, 8 and 10 days after the start of the ^{51}Cr peritoneal dialysis (PD) session. The drained dialysate of the four exchanges following the acute ^{51}Cr PD session were sampled as well. During the follow-up period 24 hours urine was collected.

The γ-radioactivity of the ^{51}Cr isotope (E_γ = 320 keV) was counted on a 5" x 5" well type NaI(Tl) crystal.

RESULTS AND DISCUSSION

The experimental data were used to modify a functional model for Cr(III) metabolism developed by Lim et al. [2]. Such a modeling allowed to describe the disappearance of ^{51}Cr(III) from the dialysate, after correction of the experimental data for an estimated average ultrafiltration course.

The subsequent uptake, distribution and elimination pattern of ^{51}Cr in the plasma of the CAPD patient during the 12 day follow-up period were simulated and matched very well the experimental data. Modeled urinary excretion appeared to be slightly underestimated at the beginning of the follow-up period. This may be possibly due to strongly individually dependent urinary excretion rates and to the different pathologies.

It appears most interesting to use the compartmental model to estimate the impact of long-term CAPD not only on the Cr content of plasma but also on that of the different body compartments. The calculated effect of 10 years CAPD treatment on the levels of Cr in plasma and in fast, medium and slowly exchanging compartments were spectacular.

The values in plasma increase very quickly after the start of CAPD treatment. A similar observation was made in a previous study [1] : the serum Cr concentration of a uremic patient on conservative treatment was 0.28 ng/ml and increased to 4.06 ng/ml after 2 months on CAPD. At that moment an accumulation up to 17 μg Cr in the total plasma compartment could be derived from the model, giving rise to a Cr concentration of 4 to 5 ng/ml plasma. The effect of long-term CAPD can be anticipated in the other compartments as well. In the slowly exchanging one (mainly characterized in liver and spleen) an increase of Cr levels with a factor up to 100 might be expected, which may give rise to toxic effects.

REFERENCES

1. B. Wallaeys, R. Cornelis, L. Mees, N. Lameire, Trace elements in serum, packed cells, and dialysate of CAPD patients, Kidney Int., 30:599 (1986).

2. T.H. Lim, T. Sargent III, N. Kusubov, Kinetics of trace element chromium(III) in the human body, Am. J. Physiol., 244:R445 (1983).

THE INFLUENCE OF BODY MASS INDEX AND HYPOGLYCAEMIC DRUGS ON
THE CIRCULATING AND EXCRETED LEVELS OF CHROMIUM FOLLOWING AN
ORAL GLUCOSE CHALLENGE, IN SUBJECTS WITH AND WITHOUT DIABETES
MELLITUS

K.E. Earle, A.G. Archer, J.E. Baillie, and A.N. Howard

University of Cambridge, Department of Medicine
Addenbrooke's Hospital, Hills Road,
Cambridge CB2 2QQ, England.

The possibility of an altered chromium metabolism in subjects with diabetes has been a topic of discussion for many years. The aims of this study were (a) measure the plasma chromium profile and excretion following a glucose challenge in controls, non-insulin dependent (NIDD) and insulin dependent diabetics (IDD), (b) observe the effect of obesity, (c) assess the effect of hypoglycaemic drugs on chromium. Subjects included 34 controls, 22 NIDD's and 23 IDD's. Groups subdivided by BMI; BMI<30 as lean, BMI>30 as obese. Food and drugs were withdrawn overnight and fasting urine was collected prior to the test. 75g glucose were given and blood collected every 30 mins. for 3 hrs. Urine was collected for 4 hrs. post glucose. Chromium was analysed by EAAS. Lean controls had significantly lower plasma chromium and insulin than obese controls at all time points except zero (1hr; 0.66 ± 0.35 vs 1.16 ± 0.69ng.ml^{-1}, $p<0.020$). No significant differences were seen between lean and obese NIDD's and IDD's. Lean controls had lower chromium than lean NIDD's (fasting; 0.68 ± 0.30 vs 1.08 ± 0.44ng.ml^{-1}, $p<0.01$) but were not different from lean IDD's. No differences were seen in the chromium response of the obese groups. After subdivision of NIDD's (i) medication, (ii) diet only; those taking drugs had higher chromium than lean controls (0.68 ± 0.03 vs 1.16 ± 0.26ng.ml^{-1}, $p<0.05$), whereas lean NIDD's diet only, were not different from lean controls. Mean chromium conc. of oral drugs was 22.4ng/tablet; the soluble insulins had $0.012 \pm .003$ng.U^{-1}. No significant differences in chromium excretion either between or within groups. Results suggest chromium metabolism is influenced by BMI in controls; this was not seen in diabetics. Higher chromium was recorded in diabetics on hypoglycaemic drugs.

SERUM ALUMINUM LEVELS IN DIALYSIS PATIENTS

York Schmitt and Jürgen D. Kruse-Jarres

Institute of Clinical Chemistry
Katharinenhspital
Stuttgart, FRG

Introduction

Despite the ubiquitousness of aluminum in the environment, serum and blood aluminum levels in healthy non-occupationally exposed individuals are low. This is achieved by gastrointestinal barriers to aluminum absorption and by renal elimination of the metal. In comparison, the patients with chronic renal failure have higher aluminum levels as an effect of oral therapy with aluminum-containing phosphate binders, which result in a higher risk for aluminum-related disorders like osteodystrophy, anemia and encephalopathy by disturbing the normal metabolism (1).

Material and Methods

Serum and plasma aluminum levels were determined in dialysis patients, in patients after kidney transplantation, and in healthy laboratory staff using electrothermal atomic absorption spectrometry (Zeeman 3030 with graphite furnace HGA 600 and autosampler AS 60; Perkin-Elmer Corp.). The pyrolytically coated platform was used in connection with the max-power heating program and Argon gas flow stop for atomization in order to improve atomization and sensitivity. Long ramp times for drying and ashing were chosen. Calibration was performed of human-like material and checked by use of standard addition of aqueous aluminum (Sigma Corp.). The serum and plasma was measured after twofold dilution with purified water. Low background was well compensated by Zeeman-correction. The measurement of samples were done within 4 hours to minimize interferences with sample cups during storage and reduce high contamination risk (2).

Results

For 66 blood samples from hemodialysed patients with a mean hemodialysis period of 4 years the aluminum content in serum and plasma ranged from 12.0 to 152.5 ug/l with an average of 55.5 ug/l. 95 % of the patients had aluminum levels below 120 ug/l. There was a positive correlation with dialysis duration ($y = 24.8 + 0.66 * x$) and dosage of aluminum-containing phosphate binders given to the patients. In patients after kidney transplantation serum and plasma aluminum values decreased significantly within a few months. Healthy laboratory people had values about 5 ug/l.

Discussion

In acid milieu of stomach soluble aluminum chloride is derived from hardly soluble aluminum compounds and can be resorbed. In lower intestine (pH 6 to 8) hardly soluble aluminum phosphate is formed and excreted via intestine. Blood and tissue aluminum increases also because of lowered elimination by hemodialysis, which is caused by the neutral pH of the dialysate (3).

The agreement with the results of McCarthy (4) is good. The positive correlation with the dialysis duration has been shown in spite of most of the hemodialysed patients having had therapy with desferrioxamine to eliminate aluminum from the tissues into the intravascular space and into the dialysate.

There was no difference between serum and plasma aluminum levels, so that both materials could be used for monitoring patients with renal chronic failure. Monitoring the aluminum levels is important in order to prevent the aluminum-related disorders. Blood monitoring of aluminum level in dialysis patients should be done every two or three months and is important for further diagnostic and therapeutic procedures, but limited because low aluminum levels do not exclude the presence of aluminum-associated diseases.

REFERENCES

1. Vyver van de, F.L. et al: Serum, blood, bone and liver aluminum levels in chronic renal failure. Trace elements in medicine 3, 52-61 (1986)

2. Cornelis, R. and P.Schutyser: Analytical problems related to Al-determination in body fluids, water and dialysate. Contr. Nephrol., Vol 38, 1-11 (1984)

3. Wronski, R. et al: Die Bedeutung des Aluminium in der Inneren Medizin. In: Fortschritte in der atomspektrometrischen Spurenanalytik. Editor: R. Welz. Verlag Chemie, Weinheim (FRG). Band 1, 285-295 (1984)

4. McCarthy, J.T. et al: Interpretation of serum aluminum values in dialysis patients. AJCP 86, 629-636 (1986)

DIETARY INTAKES AND SERUM LEVELS OF SELECTED TRACE ELEMENTS IN INSTITUTION-
ALIZED ELDERLY

M. Allegrini °*, E.Lanzola*, A.Tagliabue*, G.Turconi* and
C.Tinelli*

*Ist. Scienze Sanitarie Applicate°Centro Analisi Attivazione
via Bassi, 21 - Pavia, Italy

Pb and Cd accumulate in the body with age and their effects in the elderly have been poorly studied. Studies have indicated that Zn, Cu and Se
provide protection against Cd and Pb toxicity[1] and that the elderly receive less than the dietary recommendation for these elements[2]. Our purpose in
the present study was to estimate the dietary intakes of Pb,Cd,Zn,Cu and Se
in institutionalized elderly subjects; to determine their serum levels and to
examine the interrelationships among them and other parameters.

MATERIALS AND METHODS

Twenty healthy subjects (14 females and 6 males; mean age = 80.8 Yr,
range 68-92 Yr) selected from 180 institutionalized elderly persons took part
in the study. Dietary intakes of Pb, Cd, Zn, Cu and Se were calculated from
food consumption data and by analyzing the dried foods according to previously described procedures.[3] Pb, Cd and Cu were determined by atomic absorption spectrophotometry (AAS) with the stabilized temperature platform furnace
system(STPF). Zn was measured with flame (AAS) and Se via hydride evolution-
AAS technique and by radiochemical neutron activation analysis. Mean dietary
intakes of energy, protein and fat were calculated using Italian food composition tables.[4] Fasting blood samples were drawn from the subjects. Serum Zn
and Cu were analyzed by flame-AAS and STPF procedures, respectively.

RESULTS AND DISCUSSION

Table 1 shows the 7-day average daily intake of Pb,Cd,Zn,Cu,Se and ma-
cronutrients, and the mean serum concentration of Zn and Cu for the 20 elderly subjects. The mean Pb and Cd intakes of 89 and 30 ug/day, respectively,
were below the Provisional Tolerable Daily Intakes (PTDI) proposed for adults
by FAO/WHO[5] (Pb=430 ug/d; Cd=70 ug/d).A significant correlation was observed
between the intake of Cd and that of Zn (r=0.61; p 0.001). The major food

Table 1.Average trace element and macronutrient dietary intakes and
       Zn and Cu serum concentratiions

Element		Intake		%PTDI		Serum	
		Mean	Range	Mean	Range	Mean	Range
Pb	ug/d	89+49	25-308	20	6-72		
Cd	" "	30+16	6- 85	43	8-122		
				%RDA	or ESADDI		
Zn	mg/d	8.2+2.4	4.1- 22	54	28-147	0.80+0.1	0.57-1.04
Cu	" "	1.5+0.6	0.4-3.3	60	16-132	1.00+0.13	0.74-1.29
Se	ug/d	57+36	12-181				
Energy(Kcal)		1887+509					
Protein(g/d)		61+20					
Fat(g/d)		71+23					
Carb.(g/d)		268+88					

sources for Pb were cereals (36%), wine (21%) and meats (16%), and cereals
(84%) for Cd. The average Zn intake was 8.2 mg/d, 54% of the US recommended
dietary allowance (RDA)[b] of 15 mg/d. The mean Cu intake of 1.5 mg/d was 60%
of the US Estimated Safe and Adequate Daily Dietary Intake (ESADDI) of 2-3
mg. The mean Se intake of 57 ug/d fell within the ESADDI(50-200 ug/d) with
most of the subjects below the lower limit (50 ug).The main food sources were
meats (31%), cereals (31%), milk and cheese (27%) for Zn; cereals (35%),
wine (20%) and vegetables (27%) for Cu; meats (52%) and cereals (31%) for Se.
Mean serum Zn was lower than the suggested reference mean of 1.14 ug/ml[7] and
reflected the low level of intake. Significant correlations were found between the intakes of Zn and Cu (r=0.61; p 0.01), between Zn and energy
(r=0.75; p 0.001), Zn and   protein (r=0.71; p 0.001), Cu and energy (r=0.88
p 0.001), Se and protein (r=0.46; p 0.05). No significant correlations related the serum Zn and Cu values or serum and intake levels of Zn and Cu.

AKNOWLEDGEMENTS

We are grateful to S. Comizzoli, S. Gardinale, P.A. Cattaneo and R. Bazzano for their tachnical assistance.

REFERENCES

1. FL Cerklenwiski,R.M.Forbes. Influence of Dietary Zn on Lead Toxicity in
   the Rat J. Nutr.106,689 (1976)
2. J.L.Greger. Dietary intake and Nutritional Status in regard to Zn institutionalized aged. J.Gerol.,32,549 (1977)
3. M. ALlegrini,M.Gallorini, E.Lanzola,E.Orvini. Toxic and Essential Element
   Intakes of Formula Fed infants Int. Conf. Heavy metals environment Athens
   2,76 (1985)
4. E. Carnevali and F.C.Niuccio. Tavole di Composizione, Ist.Naz.Nutr.(1987)
5. WHO, Tech. Rep. Ser. n.505, 16th Rep. of the FAO/WHO (1972)
6. Recommended Dietary Allowances 9th Ed.Nat.Acad.of Science, Wash. (1980)
7. G.Peterson, E.Lee, D.A. Christiansen and D.Robertson. Zinc levels of Hos-
   pitalized elderly. J.Am.DIet.Ass., 186 (1985).

MINERAL STATUS OF ADULTS GIVEN A FIBER SUPPLEMENT

Susan J. Fairweather-Tait, Zoe Piper, and Susan Southon

AFRC Institute of Food Research
Colney Lane
Norwich NR4 7UA, U.K.

INTRODUCTION

Certain sources of dietary fiber are thought to have an adverse effect on mineral absorption. Results from an earlier study in rats [1] suggested that the mixed fiber supplement ("fiber-filler"), advocated in the F-plan diet [2], caused a slight reduction in iron status.

MATERIALS AND METHODS

Forty volunteers were recruited for the study, and each subject was given 12g fiber daily (Bran flakes (Kellogg's) 14g, bran 14g, All-bran (Kellogg's) 14g, almonds 7g, prunes 7g, apricots 7g, sultanas 14g), half with breakfast and half with their evening meal, for 12 weeks.

Fasting blood samples were taken at the start and at regular intervals throughout the study, and various haematological parameters measured.

RESULTS

Analytical results for blood samples taken before the fiber supplement are shown in Table 2.

There was a small rise in Hb concentration during the study in the male and female omnivores ($P < 0.05$). The iron-depleted individuals (initial plasma ferritin $< 20\mu g$/litre) showed a small increase in iron stores ($P < 0.05$): mean initial ferritin 13, final ferritin 17μg/litre.

Table 1. Composition of fiber supplement

	per kg	per daily amount
Dietary fiber (g)	160.1	12.3
Fe (mg)	155.1	11.9
Zn (mg)	33.3	2.6
Ca (mg)	840.0	64.7
Phytate (g)	6.0	0.5
Fructose (g)	95.2	7.3
Ascorbic acid (mg)	350	27
Citric acid (mg)	572	44

Table 2.   Subject data and analysis of initial blood samples
           Mean (± SEM)

	Male omnivores	Male vegetarians	Female omnivores	Female vegetarians
No	14	5	14	7
Age (yrs)	35.1 (3.2)	33.4 (3.4)	31.8 (2.9)	25.1 (2.7)
PCV (%)	46.3 (0.5)	45.6 (1.0)	42.2 (0.5)	40.6 (1.4)
Hb (g/100ml)	15.0 (0.2)	14.8 (0.4)	13.2 (0.2)	12.8 (0.6)
Plasma Fe (μg/ml)	1.18*(0.06)	1.62*(0.17)	1.08(0.12)	1.08(0.18)
Transferrin sat'n (%)	49.6 (2.1)	57.2 (8.9)	49.2 (4.2)	42.7 (6.7)
Ferritin (μg/litre)	52.8 (10.4)	52.2(13.0)	18.0 (1.9)	12.3 (2.1)
Plasma Zn (μg/ml)	0.96 (0.06)	1.00(0.07)	0.85(0.03)	0.90(0.04)
Alkaline phosphatase (U/litre)	33.7 (2.1)	31.1 (1.5)	20.1 (1.7)	25.5 (3.3)

*Significant difference between omnivores and vegetarians of the same sex
 (P<0.05).

   Plasma alkaline phosphatase remained steady, but plasma zinc showed
significant fluctuations during the fiber supplementation period,
suggesting the maintenance of homeostasis via adaptive mechanisms.   No
other changes were observed.

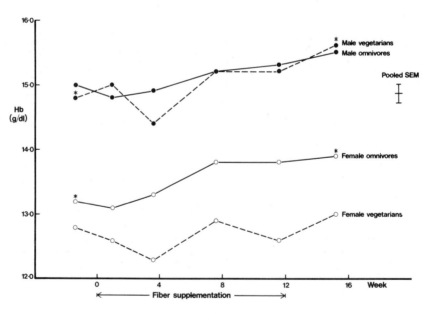

Fig. 1.   Changes in Hb concentration

*_____
 Final Hb conc. significantly different from initial value (P<0.05)

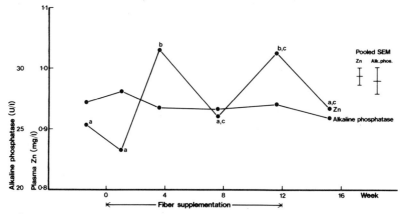

a,b,c Values with different superscripts are significantly different
(P<0.05)

Fig. 2.   Changes in plasma Zn and alkaline phosphatase (means for all
subjects)

CONCLUSIONS

Iron status was not impaired as a result of consuming the fiber
supplement as part of the normal diet. The additional iron (12mg/day)
provided by the fiber supplement may have been responsible for the small
increase in Hb levels observed in the omnivores, and the high amounts of
citrate, ascorbate and fructose probably enhanced iron availability.
Zinc status, as estimated from plasma zinc, was not affected in the long-
term by the fiber supplement, although fluctuations were observed which
may be indicative of adaptive mechanisms.

1. S. J. Fairweather-Tait and A. J. A. Wright, The effect of 'fibre-
filler' (F-plan diet) on iron, zinc and calcium absorption in
rats, Br. J. Nutr. 54:585 (1985).
2. A. Eyton, "The F-plan diet", Penguin, Harmondsworth (1982).

DAILY VARIATIONS IN PLASMA ZINC IN NORMAL ADULT WOMEN

M. Jacobs Goodall, K.M. Hambidge, C. Stall,
J. Pritts, and D.E. Nelson

Department of Pediatrics
University of Colorado Health Sciences Center
Denver, Colorado 80262

The present study was designed to measure the daily variations in the plasma zinc concentrations of normal, adult women of child-bearing age with particular emphasis on the effect of meals on plasma zinc concentrations.

METHODS

The subjects were twelve healthy women, 17–42 years of age, who had been observing a normal daytime work/nighttime sleep pattern for 2 weeks prior to study. Standard meals were served immediately following the blood collections at 0700, 1200, and 1700 h. The meals provided:

	Breakfast	Lunch	Dinner
Total Energy (1750 Kcal)	21%	35%	42%
Total Protein (70 g)	15	46	38
Total CHO (209 g)	33	32	35
Total Fat (70 g)	11	34	55
Total Zinc (7.1 mg)	26	43	29

Subjects were allowed to undertake normal activities throughout the day. A plastic catheter was inserted into a superficial forearm vein at 1430 pm on day 1. Venous blood samples were collected via this catheter at 1 h intervals from 1500 h on day 1 to 1700 h on day 2. Blood was drawn at 30 min. intervals for 2 to 3 h following each meal. Plasma zinc concentrations were determined by atomic absorption spectrophotometry.

RESULTS AND DISCUSSION

Following a progressive increase overnight, the mean plasma zinc concentration peaked at 0800 h with a value of 75.8 ± 2.9 ug/dl. (Mean $\pm$ SEM) Fig 1. Levels started to decline by 0830 h and fell to 65.2 ± 2.2 ug/dl at 1000 h (13.3% below the immediate pre-breakfast value at 0700 h of 75.2 ± 3.0 ug/dl). Subsequently, the mean increased to a level of 72.4 ± 2.5 at 1200 h.

The next decline started 1 hr after lunch progressing to a level of 64.2 ± 2.3 ug/dl at 1500 h (11.3% below 1200 value and 14.6% below the 0700 value). This was followed by a slight rise before and immediately after dinner to 65.5 ± 3.8 ug/dl at 1800 h. After dinner, plasma zinc

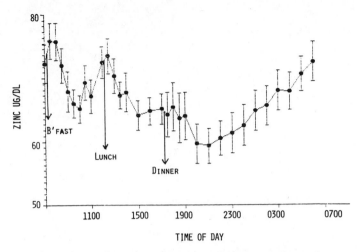

Fig 1 Daily Variations in Plasma Zn Concentrations (mean ± SEM).

levels began to decline again at 1830 h and continued to decline to a
nadir for the day of 59.3 ± 2.7 ug/dl at 2100 h (7.8% below 1700 value and
21.1% below the 0700 value). This nadir was 22% below the 0800 peak
value.

In this study, a decline in plasma zinc concentrations occurred 1-1.5
h after each of the 3 main meals of the day reaching a low point 3-4 h
after the meal. There was a trend for an overall decline throughout the
day, which may be attributable to a cumulative effect of meals. Only
during the night, when there was a 14 hr period without a major meal, did
an uninterrupted increase in plasma zinc occur. These results indicated
that meals are the primary and possibly the only factor responsible for
variations in plasma zinc during the course of any one day. The results
of earlier, less detailed studies are generally compatible with this conclusion as are those of two studies which included sampling of comparable
frequency. In one of these (1), a main meal was given at midnight and the
nadir for the day was at 0600 h. This finding provides support for the
conclusion that the uninterrupted rise in plasma zinc through the night in
the current study was attributable to the lack of meals over this period.
In another recent study (2), breakfast was given 1.5 h later than in the
current study. The peak plasma Zn level was also 1.5 h later and the
post-breakfast decline reached its nadir 1.5 h later. Thus the times of
these changes in the two studies were closely comparable with respect to
the first meal of the day rather than to the actual time of day.

In conclusion, there are variations of as much as 20% in plasma Zn
concentrations depending on the time of day at which samples are obtained.
The pattern of change in this study, together with a comparison with those
studies in which meals were consumed at different times, indicates that
the primary factor and possibly the only factor responsible for the daily
variations in plasma Zn concentration is the consumption of food.
(Supported by NIADDKD, 5R22 AM12432, and grant RR69 from NIH, GCRC).

REFERENCES

1. Lifschitz MD, Henkin RI. J Appl Physiol 31(1):88-92 (1971).
2. Markowitz ME, Kosen JF, Mizruchi M. Am J Clin Nutr 41:(4)689-96
   (1985).

URINARY EXCRETION (24 HOURS) OF ZINC AND COPPER IN NORMAL VEGETARIAN AND NON-

VEGETARIAN FEMALE SUBJECTS DURING THE DIFFERENT SEASONS: A PRELIMINARY STUDY

Vasantha Iyengar

University of Michigan School of Medicine
Department of Human Genetics
Ann Arbor, Michigan 48109

INTRODUCTION

The usefulness of a laboratory test is determined by two general classes
of variables: analytical and biological. These variations affect the spread
of values that define the normal or reference range. It is known that factors
related to climate (seasonal changes) such as temperature, humidity, baro-
metric pressure and due point affect water balance. As a consequence, urine
output, and the distribution of exchangeable water in the body are altered
(1). This may lead to great variations in the excretion of several urinary
constituents, including trace elements.Thus,climate may become a significant
factor in assessing reference values for urinary constituents. In addition to
these natural cyclic changes,dietary habits such as vegetarianism,as well as
variation in diet as a result of seasonal changes in availability of foods for
consumption, can also influence the urinary profile. In order to define the
levels of reference values (e.g. trace elements) in urine, all the above mentioned variables should be recognized. This study was initiated to determine
the urinary excretion of copper and zinc during the four seasons of the year
for women consuming vegetarian and meat-based diets.

MATERIALS AND METHODS

Three healthy female volunteers aged 50 (subject A), 26 (subject B), and
35 years (subject C) were chosen for the study:two of them, A and B, consumed
meat-based diets, and the third one, subject C, a lacto-vegetarian diet. 24-
hour urine output was collected using acid washed polyethylene.collection kit.
The collections were made for three consecutive days for each of the four
climatic seasons, thus giving a total of 38 individual samples including a
repeat of two samples. The total daily volumes of the samples were recorded
and are shown in Table 1.

Collections were made in each mid-season. No supplementation of trace
minerals, particularly of copper and zinc were taken before several weeks and
during collection periods. The participants kept a daily diary of foods eaten
during the days of collection period. The food consumption pattern can be
summarized as follows:
Subjects A and B: meat component of the diet was almost always present
during autumn, winter and spring, while vegetable and other cereal products
were in moderation. On the other hand, there was a clear difference in summer
when considerably less meat was eaten (also with respect to frequency), and

493

was substituted by vegetable and cereal-based intake; salads were consumed almost every day as part of a meal. Subject C: there was very little variation in the diet of this subject. Intake of fluids in all subjects was much higher in summer than in other seasons, as expected.

Table 1.   Urinary excretion in different seasons
(values are mean ± 1 S.D.)

Seasons	Subjects		
	Non-vegetarian		vegetarian
	A	B	C
		ml/d	
Summer	980 ± 200	957 ± 357	1144 ± 186
Autumn	1173 ± 55	1035 ± 301	1050 ± 359
Winter	1118 ± 198	773 ± -00	1440 ± 440
Spring	935 ± 363	688 ± 120	1261 ± 318
Mean	1051 ± 112	863 ± 160	1224 ± 168

The samples were analyzed after each seasonal collection. Copper and zinc were determined using flame atomic absorption spectrophotometry (Model 451, Instrumentation Laboratory). The samples were analyzed both before and after centrifugation. There were no significant variations in values before and after centrifugation. A standard reference urine issued by the United States National Bureau of Standards (SRM 2670) was used for the purpose of quality control. Aliquots of this SRM were analyzed as controls with each batch of urine samples in different seasons. This SRM is certified for copper at both low and elevated levels. Zinc was determined in this SRM, although this standard is not certified for this element.

RESULTS AND DISCUSSION

Table 1 shows the average output of urine over three days for the three subjects for different seasons.

As shown from Table 1, subject C had generally the highest urinary volume (mean output of 1224 ml). Subject A had slightly higher urinary output than subject B (mean volume of 1050 ml vs 860 ml). On a concentration basis, in general, zinc was inversely related to the volume of the daily urinary output, while copper was not (data not shown).

Table 2 shows the mean daily urinary excretion of copper and zinc for the three subjects for the four seasons of the year.

Standard Reference Material

The results obtained for a total of twelve aliquots were as follows:

Copper: 0.15 ± 0.008 and 0.37 ± 0.008 mg/l for the low and elevated levels, respectively. Certified values are 0.13 ± 0.02 and 0.37 ± 0.03 for the low and the elevated levels, respectively.

Zinc: 2.53 ± 0.25 and 0.71 ± 0.06 for the high and low levels, respectively, were observed. This element has not been certifed in this standard due to problems of contamination while packaging.

494

Table 2.   Mean daily urinary output of Cu and Zn in different seasons
(values are the mean of 3-day output ± 1 S.D.)

	Subject A		Subject B		Subject C	
	Cu	Zn	Cu	Zn	Cu	Zn
			ug/d			
Summer	60 ± 13	320 ± 35	74 ± 20	560 ± 49	50 ± 4	300 ± 41
Autumn	43 ± 2	300 ± 127	55 ± 14	680 ± 60	52 ± 2	230 ± 17
Winter	63 ± 16	360 ± 88	37 ± 10	450 ± 55	49 ± 8	240 ± 70
Spring	30 ± 5	320 ± 70	25 ± 5	470 ± 40	39 ± 8	330 ± 38
Mean	49 ± 15	330 ± 25	48 ± 21	540 ± 100	48 ± 6	270 ± 40

Zinc

There was no significant inter-seasonal variation in the excretion of
this element in any of the three subjects. However, when dietary habits were
compared, there were differences. For example, the quantity excreted per day
by subject B was twice that of the subject C (500 ± 100 ug/d vs 270 ± 40
ug/d), indicating a higher fraction of the absorbed zinc in the subject of
meat-based dietary habit. Subject (also of meat-based dietary habit) showed
a lower value for zinc excretion than subject B (330 ± 25 ug/d vs 540 ± 100
ug/d). This is probably attributable to the quantity of food intake by
subject A (approximately 60 % that of subject B). However, the zinc excretion
in subject A was also higher than in the vegetarian subject C (330 ± 25 ug/d
vs 270 ± 40 ug/d).

Copper

Some seasonal variations were observed for the subjects A and B (Table
1). The highest and the lowest quantities excreted by subject A were 60 ± 13
ug/d and 30 ± 4.2 ug/d, for summer and spring, respectively. For subject B,
the corresponding figures were 74 ± 20 ug/d for summer, and 25 ± 5 ug/d
for spring. In contrast, the changes in urinary excretion of copper observed
for subject C were minimal throughout all seasons, and varied between 50 ± 4
ug/d and 39 ± 8.5 ug/d.

The overall ranges of excretion of zinc and copper are in agreement
with the data reported in the literature (2). However, the excretion of zinc
by the vegetarian subject falls at the lower end of the range while that of
the subject of meat-based dietary habit is in the upper range.

CONCLUSION

The data indicate that variations observed for urinary zinc excretion
were not seasonal but mainly diet dependent. On the other hand, there
appears to be a seasonal variation for copper for the meat-based dietary
subjects (low in spring and high in summer). This may be related to changes
in copper intake originating from variation in food intake such as consuming
fresh vegetables in summer, and therefore, changes in the absorption of
copper for the meat-based dietary group.

REFERENCES

1. R.L. Vick (1982) Contemporary Medical Physiology, Addison Wesley Pub-
    lishing Company, London, New York.
2. G.V. Iyengar, W.E. Kollmer, H.J.M. Bowen (1978) Elemental Concentra-
    tions of Tissues and Body Fluids, Verlag Chemie, Weinheim, New York.

HEPATIC TRACE ELEMENTS IN KWASHIORKOR

Jerome Miles, Michael Golden, Dan Ramdath and Barbara Golden

Wellcome Trace Element Research Group
Tropical Metabolism Research Unit, University of the West
Indies, Mona, Kingston 7, Jamaica

It has been proposed that kwashiorkor, a common form of malnutrition,
occurs as a result of an imbalance in the level of free radicals and their
safe disposal[1]. Iron is a major catalyst of free radical generating reactions in vivo[2]. The seleno-enzyme glutathione peroxidase (GSHPx) and the
copper/zinc enzyme superoxide dismutase (SOD), catalyse the safe dissipa-
tion of free radicals. These enzymes have their essential trace elements
at their active sites. High levels of hepatic iron could propagate free
radical generating reactions. Impaired protection against free radical
generating species could result from reduced levels of the trace element
cofactors essential for their safe disposal. Waterlow[3] found stainable
iron in children with kwashiorkor, while Warren et al[4] found depressed
levels of several other trace elements. Trace element status may be reflected in hepatic levels. To investigate this, hepatic levels of iron,
copper, zinc and selenium were measured in Jamaican malnourished children.

METHODS

Seventeen liver samples were collected at postmortem from children who had
died after admission to hospital with severe malnutrition. Samples were
digested using a nitric/perchloric acid mixture followed by hydrogen peroxide. The digestion effectively broke down the organic matrix. Iron,
copper and zinc were measured by flame atomic absorption. Selenium was
measured by fluorometry using 2,3-diaminonaphthalene[5]. Moisture and fat
contents were determined for all samples. National Bureau of Standards
(NBS) bovine liver standards and duplicate samples were run to assess the
accuracy and precision of the method.

RESULTS AND DISCUSSION

Analysis of NBS standards gave results well within the acceptable limits
for all elements measured. Duplicated sample analyses were all within 1%
of each other. Figure 1. displays the results of the trace element
analysis. Control values were taken from the literature [3,4,7]. Iron
levels were approximately three times higher than controls. Selenium was
less than 50% of the control value. The levels found for the other trace
elements were essentially the same as the controls. Mean percent fat
found was 56% (dry weight basis), while that for the controls was 20%.

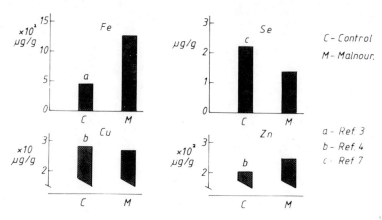

Figure 1. Trace Element Concentrations fat free dry weight basis

Of the children who died of severe malnutrition during the study, all but one were suffering from kwashiorkor. It is of great interest that these children had not only high levels of hepatic fat, but also high· levels of hepatic iron. It has been suggested that since only 0.5-2% of many chronically malnourished populations develop kwashiorkor, these patients may represent those at the upper end of a normal distribution of iron status[1]. A genetic predisposition to iron accumulation cannot, however, be ruled out[6]. If this iron is available for acting as a catalyst for free radical generating reactions, coupled with the possibility of reduced protection against free radical production, due to inadequate selenium status, the findings of this study could have far reaching implications for the treatment of kwashiorkor. A reduction in the free iron status along with selenium supplementation of these patients could improve prognosis.

REFERENCES

1. M.H.N. Golden and D. Ramdath. Free radicals in the pathogenesis of kwashiorkor. Proc. Nutr. Soc. 46: 53-68 (1987).
2. B. Halliwell and J.M.C. Gutteridge. Oxygen toxicity, oxygen radicals transition metals and disease. Biochem. J. 219: 1-14 (1984).
3. J.C. Waterlow, In: Medical Research Council Special Report. Series No. 263; London, H.M. Stationery Offices.
4. P.J. Warren, J.D.L. Hanson and B.H. Lehmann. The concentration of copper, zinc and manganese in the liver of African children with marasmus and kwashiorkor. Proc. Nutr. Soc. 26: 6A-7A (1969).
5. R.F. Byfield and L.F. Romalis. pH control in the fluorometric assay for selenium with 2,3-diaminonaphthalene. Anal. Biochem. 144: 569-576 (1985).
6. J.A. Edwards and R.M. Bannerman. Hereditary defect of intestinal iron transport of mice with sex linked anaemia. J. Clin. Invest. 49: 1869 (1970).
7. G. Norheim and Lars Ace Solberg. Hepatic selenium levels in human autopsy cases. In: "Trace elements in man and animals - TEMA 5". C.F. Mill, I. Bremner and J.K. Chesters. ed. Commonwealth Agricultural Bureau.

IMPROVED ZINC BIOAVAILABILITY FROM COLLOIDAL CALCIUM

PHOSPHATE-FREE COW'S MILK

J. Kiely[1], A. Flynn[1], H. Singh[2], and P. F. Fox[2]

Departments of [1]Nutrition and [2]Food Chemistry
University College
Cork, Ireland

INTRODUCTION

The lower bioavailability of zinc in cow's milk compared with human milk is well documented, but the basis of this difference has yet to be adequately explained. Over 90% of the Zn in cow's milk is associated with casein micelles and it has been shown (Lonnerdal et al., 1985) that the bioavailability to rats of this fraction of Zn is similar to that in cow's milk, but lower than that of human skim milk. We have recently obtained evidence that 60-70% of the Zn in bovine casein micelles is associated with colloidal calcium phosphate (CCP), while the remainder is directly bound to casein (Singh et al., 1987). The object of this study was to investigate whether the CCP fraction influences the bioavailability of Zn in cow's milk.

MATERIALS AND METHODS

Whole raw bovine milk was defatted by centrifugation at 3000g for 20 min at room temperature. Colloidal calcium phosphate-free (CCP-free) skimmed milk was prepared by the method of Pyne and McGann (1960). Zinc bioavailability was determined in rat pups by a modification of the method of Sandstrom et al. (1983). Sixteen day-old rat pups, fasted overnight, were intubated with 0.2 ml milk, extrinsically labelled with ^{65}Zn (1μCi/ml) and liver uptake of isotope determined after 6h by counting in a well gamma counter.

RESULTS AND DISCUSSION

The removal of CCP from cow's milk reduced the Ca, Pi and Zn concentrations to 45, 63 and 31%, respectively, of those in the skim milk (Table 1). Of the remaining Zn ∿15% was removable by equilibrium dialysis and the non-diffusable zinc was bound to casein aggregates, as shown by chromatography on Sephadex G-100. The bioavailability of zinc in skimmed milk was significantly improved by removal of CCP (Table 1). This was not due to the lower Zn concentration in the CCP-free milk as the bioavailability of Zn in CCP-free milk with added Zn (as $ZnNO_3$) was also significantly greater than that in skimmed milk. Readdition of Ca and Pi

Table 1. Effect of the Mineral Composition of Skim Milk
on the Bioavailability of Zn in Suckling Rats

| Milk | Mineral concentration (mg/1) | | | | Zinc bioavailability | |
	Zn	Ca	Pi	n	Liver uptake (% dose)	SEM
Skim milk	4.10	1140	630	17	18.5	0.6
CCP-free	1.26	514	400	17	21.6[a]	0.5
CCP-free + Zn	4.10	514	400	7	21.8[a]	1.0
CCP-free + Ca + Pi	1.26	1140	630	10	19.0	0.5

[a]differ significantly ($p < 0.01$) from the value for skim milk.

to the CCP-free milk reduced the bioavailability of Zn to a value similar to that in skimmed milk.

These results show that the presence of CCP in cow's milk reduces the bioavailability of Zn. This could be due to the effect of CCP on digestion of casein. Removal of CCP lowers the curd tension of renneted milk (Pyne, 1962) and may therefore improve the digestibility of casein which has been proposed to reduce the bioavailability of Zn by forming indigestible casein-Zn complexes in the gastrointestinal tract (Lonnerdal et al., 1985; Blakeborough et al., 1986). Alternatively, the effect could be due to the formation of insoluble calcium phosphate-Zn complexes in the small intestine.

The results of the present study suggest that the higher content of CCP in cow's milk compared with human milk (Singh et al., 1987) could contribute to the difference in the bioavailability of Zn in the two milks.

REFERENCES

Blakeborough, P., Gurr, M. I. and Salter, D. N., 1986, Digestion of zinc in human milk, cow's milk and a commercial babyfood: some implications for human infant nutrition, Br. J. Nutr., 55:209.

Lonnerdal, B., Keen, C. L., Bell, J. G. and Hurley, L. S., 1985, Zinc uptake and retention from chelates and milk fractions, in: "Trace Elements in Man and Animals - TEMA 5", C. F. Mills, I. Bremner and J. K. Chesters, eds., Commonwealth Agricultural Bureau, Slough, U.K.

Pyne, G. T. and McGann, T. C. A., 1960, The colloidal phosphate of Milk II. Influence of citrate, J. Dairy Res., 27:9.

Pyne, G. T., 1962, Some aspects of the physical chemistry of the salts in milk, J. Dairy Res., 29:101.

Sandstrom, B., Keen, C. L. and Lonnerdal, B., 1983, An experimental model for studies of zinc bioavailability from milk and infant formulas using extrinsic labelling, Am. J. Clin. Nutr., 38:420.

Singh, H., Flynn, A. and Fox, P. F., 1987, Binding of zinc to colloidal calcium phosphate in human and cow's milks, in: "Trace Elements in Man and Animals - TEMA 6", Plenum Publishing Co., New York.

EFFECTS OF SOYBEAN TRYPSIN INHIBITOR ON TISSUE ZINC

IN WEANLING MALE LONG-EVANS RATS

J. I. Rader, S.-H. Tao, and M. R. S. Fox

Division of Nutrition
Food and Drug Administration
Washington, DC

INTRODUCTION

Antinutritive components in raw soy flour include trypsin inhibitors (TI) and phytic acid (PA). PA increases dietary requirements for Zn and Mn when soy proteins are fed to animals. A Zn-PA-protein complex is responsible for effects on Zn (O'Dell and Savage, 1960). In diets containing PA from textured vegetable protein (from heated, defatted soy flour), supplementation with Zn overcomes adverse effects of PA on growth rate and prevents deficiency symptoms (Rader et al., 1986). TI in unheated soy flour causes pancreatic hypertrophy, hypersecretion of proteolytic enzymes, decreased growth, and reduced fat absorption when fed to rats. Reduced protein digestibility and loss of nitrogen due to excessive pancreatic secretion also contribute to the poor nutritive value of unheated soybeans. These problems are overcome by live steam treatment (toasting), which denatures TI (Rackis, 1978). Effects of TI per se on mineral bioavailability have been less extensively studied than have those of PA.

EXPERIMENTAL

Male Long-Evans rats (23 days old, body weight 43-50 g) were fed diets containing 14 or 31.5% protein as casein (Cas) or commercial soy flour (SF) (Table 1). Diets 1 and 3 served as controls for diets 2 and 4, respectively. All nutrients except Zn were included in diets 1 and 2 at near-NRC (1978) levels. Purified TI (a 300-fold concentrate from soybeans) was fed in diet 2 to approximate the endogenous TI in diet 4. Ca, P, Mg, K, Fe, and Cu in Cas diet 3 were increased to match the SF diet. PA (as sodium phytate) was added to diet 3 at 11.7 g/kg to match the endogenous PA content of diet 4. Observations included body weight, food consumption, tissue weight, and minerals determined by inductively coupled argon plasma-atomic emission spectrometry.

RESULTS

TI did not decrease growth (diets 2 vs. 1 and 3 vs. 4, Table 1). Efficiency of food utilization was decreased only in rats fed diet 4. Efficiency of protein utilization was lower in rats fed diets 3 and 4 vs. 1 and 2 and in those fed diet 4 vs. 3. Similar changes occurred in relative

501

Table 1. Responses of Weanling Rats to TI After 4 Weeks

Diet	Protein %	TI g/kg	Zn mg/kg	Fe	Mg	Ca g/kg	P	Wt. Gain g	Food Util. g gain/g food	Liver % of body wt.	Pancreas % of body wt.
1 Cas	14	<0.1	33	32	0.4	4.9	3.9	132a	0.38b	5.1c	0.59a
2 Cas	14	4.9	33	32	0.4	5.1	3.9	142b	0.39b	4.9c	0.71b
3 Cas	31.5	<0.2	161	47	1.7	15.3	17.2	117a	0.39b	4.4b	0.62a
4 SF	31.5	7.1	173	49	1.7	16.9	13.8	126a	0.29a	3.9a	0.79c

Values are means; 10 rats/group. PA/Zn molar ratio (diets 3,4) = 7.7. Means with the same letter in a column are not significantly different, p<0.05.

Table 2. Tissue Minerals in Weanling Rats After 4 Weeks

Group	Duodenum Mn μg/g	Zn	Pancreas Na μg/g	Zn	Liver Fe μg/g	Zn	Femur Ca mg/g	Fe μg/g	Zn
1 Cas	7.3b	21.8a	1028bc	25.3b	58.6b	23.2a	237b	58.4c	167bc
2 Cas	5.8b	21.1a	919a	20.7a	51.4ab	24.8ab	228b	51.3b	149a
3 Cas	2.6a	22.1a	1046c	28.5bc	38.7a	27.1b	215a	40.1a	174cd
4 SF	2.3a	21.5a	980b	31.3c	60.7b	34.7c	230b	42.8a	185d

Values are means; 10 rats/group. Means with the same letter in a column are not significantly different, p<0.05.

liver weights. Relative pancreas weight was increased in both TI-fed groups vs. their controls. TI in diets 2 and 4 caused lower Na in pancreas (Table 2). TI in diet 2 caused Zn depletion in pancreas and femur even with dietary Zn increased almost 3-fold above requirement. Increasing dietary Zn to >150 mg/kg eliminated effects of TI on pancreas and femur Zn. Duodenal Mn, liver Fe, and femur Ca were decreased in diet 3 vs. diet 1. Increased dietary protein and PA, even with increased minerals in diet 3, probably contributed to these effects. Fe and Zn in liver but not in femur were increased in SF diet 4 vs. Cas diet 3.

CONCLUSIONS

The response of weanling rats to TI varied markedly with the composition of the purified diet used. Pancreatic enlargement occurred with high dietary Zn and was separable from effects on growth. Changes in protein, PA, and mineral content of purified diets had marked effects on tissue mineral levels. Use of well-defined diets made possible the identification of effects of TI in Cas as well as in SF diets.

REFERENCES

National Research Council, 1978, Nutritional Requirements of Laboratory Animals, pp. 7-37, Natl. Acad. Sci., Washington, DC.
O'Dell, B. L., and Savage, J. I., 1960, Effect of phytic acid on zinc bioavailability, Proc. Soc. Exp. Biol. Med., 103:304.
Rackis, J. J., 1978, in: "Soybeans: Chemistry and Technology", vol. 1, pp. 185-202, Smith, A. K., and Circle, S. J., eds., AVI Publ. Co., Inc., Westport, CT.
Rader, J. I., Tao, S.-H., Gaston, C. M., Wolnik, K. A., Fricke, F. L., and Fox, M. R. S., 1986, in: "Trace Elements in Man and Animals - TEMA 5", pp. 458-460, Mills, C. F., Bremner, I., and Chesters, J. K., eds., Commonwealth Agricultural Bureau, Farnham Royal, Slough, UK.

CALCIUM AND ZINC ABSORPTION FROM HUMAN MILK,

SOY FORMULA AND DEPHYTINIZED SOY FORMULA

C. Kunz* and B. Lönnerdal

Departments of Nutrition and Internal Medicine
University of California, Davis, CA 95616

INTRODUCTION

Dietary calcium (Ca) is essential for regulation of calcium and phosphorus homeostasis and for normal growth and bone development in infants. Some clinical disturbances in the postnatal period may be due to a low bioavailability of Ca from infant diets. Therefore, an adequate supply of Ca is necessary. Little information is available on the absorption of Ca from human milk, soy formula and dephytinized soy formula. To investigate this we used 14 day old rat pups as a model to measure the absorption of Ca. As a comparison, and to verify our model, we also studied the absorption of zinc (Zn). The major objectives of this study were (1) to investigate if a certain level of dephytinization of soy formula leads to a beneficial effect on Ca and Zn absorption; and (2) to investigate the influence of various diets on the uptake of Ca by passive diffusion. At this age it is possible to study the latter process, since the vitamin D-regulated active transport of Ca does not occur until rats are about 20 days old (1).

METHODS AND MATERIAL

Human milk, soy formula and "low phytate" soy formula were labeled extrinsically with ^{47}Ca (400,000 cpm) and ^{65}Zn (100,000 cpm). Fourteen-day-old rat pups were fasted overnight, given 0.5 ml of the diet by gastric intubation, and killed after 6 h. Stomach, intestine (3 segments), cecum + colon and liver were removed and radioactivity of each tissue was measured in a gamma counter (Beckman Gamma 8500). After the decay of ^{47}Ca ($t_{1/2}$=4.5 days) tissues were recounted for ^{65}Zn activity ($t_{1/2}$=245 days).

RESULTS AND DISCUSSION

The results showed marked differences in ^{47}Ca absorption between human milk and soy formula (Table 1). Dephytinization of soy formula had a beneficial effect on ^{47}Ca absorption; however, absorption values did not reach those obtained for human milk (compare ^{47}Ca in intestine 2 + 3 and cecum + colon).

---

*Supported by the Deutsche Forschungsgemeinschaft.

TABLE 1. Tissue distribution of ^{47}Ca after intubation with different diets.
(% of total counts)

	Stomach	Intestine			Perfusate	Cecum + Colon	Liver
		1	2	3			
a) Human Milk (n = 3)							
$\bar{x}$	6.56	1.77	32.64	27.94	9.77	6.83	3.27
SEM	2.13	0.64	2.67	5.25	2.65	0.17	0.16
b) Soy Formula (n = 5)							
$\bar{x}$	5.56	1.12	10.91	14.03	15.91	55.06	0.87
SEM	0.71	0.19	1.38	0.83	3.03	5.52	0.09
c) Low Phytate Soy Formula (n = 5)							
$\bar{x}$	7.52	1.32	14.55	31.22	11.16	27.96	2.33
SEM	1.16	0.30	2.63	5.98	1.29	1.30	0.15

TABLE 2. Tissue distribution of ^{65}Zn after intubation with different diets.
(% of total counts)

	Stomach	Intestine			Perfusate	Cecum + Colon	Liver
		1	2	3			
a) Human Milk (pooled) (n = 3)							
$\bar{x}$	2.56	6.63	6.03	4.91	1.35	1.86	30.85
SEM	0.30	0.97	0.64	1.68	0.20	0.01	1.82
b) Soy Formula (n = 5)							
$\bar{x}$	1.01	3.09	2.62	5.39	9.56	63.61	8.26
SEM	0.04	0.62	0.19	0.60	2.51	5.64	0.56
c) Low Phytate Soy Formula (n = 5)							
$\bar{x}$	1.52	5.01	6.55	14.05	3.6	15.89	22.60
SEM	0.10	0.31	1.10	1.24	0.36	1.09	0.50

Results for ^{65}Zn absorption (Table 2) are similar to those obtained previously for these diets (2). The results validate the use of dual radioisotope labeling to study Ca and Zn absorption simultaneously in the same animal. The effect of dephytinization of soy formula had a more pronounced beneficial effect on Zn absorption compared to Ca absorption, suggesting that factors other than phytic acid affect Ca absorption to a greater degree than Zn absorption.

CONCLUSIONS

Calcium is well absorbed from human milk, whereas absorption is low from soy formulas. While the negative effect of soy formula on Ca and Zn absorption can in part be removed by dephytinization, it appears that most of the phytate must be removed to observe a pronounced effect on Ca absorption. This indicates that there are other factors in soy formula which limit Ca absorption. The inhibitory effect of soy formula appears to occur via inhibition of Ca uptake in the latter 2/3 of the intestine.

REFERENCES

1. B. P. Halloran and H. F. DeLuca, Appearance of intestinal cytosolic receptor for 1,25-dihydroxyvitamin$_3$ during neonatal development in the rat, J. Biol. Chem. 256:7338-7342 (1981)
2. B. Sandström, C. L. Keen, and B. Lönnerdal, An experimental model for studies of zinc bioavailability from milk and infant formulas using extrinsic labelling, Am. J. Clin. Nutr. 38:420-428 (1983).

DISTRIBUTION OF ZINC IN HUMAN MILK AND COW'S MILK

AFTER IN VITRO PROTEOLYSIS

Bo Lönnerdal and Carol Glazier

Department of Nutrition
University of California
Davis, CA 95616

INTRODUCTION

Several studies have shown that the bioavailability of zinc is higher from human milk than from cow's milk. This difference is more pronounced in infants than in adults. Blakeborough (1) suggested a difference in solubility between human and cow's milk during digestion. We hypothesized that the higher bioavailability of zinc from human milk is due to a larger proportion of low molecular weight (LMW) zinc and a smaller proportion of insoluble/undigested zinc-binding proteins. This may be accentuated in the infant, since the digestive capacity of the gastrointestinal tract is not fully developed. Gastric acid output and pancreatic enzyme secretion are known to be lower in the infant than in the adult gut, resulting in a higher stomach pH and limitations in protein digestion. Further confounding the uptake of zinc is the much faster transit time in the infant gut. Since in vivo uptake studies are limited by the fact that LMW zinc complexes are likely to show more rapid absorption than other forms of zinc, we have followed digestion of human milk (HM) and cow's milk (CM) in vitro. In these studies we have used a more physiological stomach pH for infants, i.e., pH 5, and compared this to digestion at pH 2 (more resembling adult conditions). Time periods have been kept short to mimic rapid transit time. Individual effects of pepsin digestion, low pH and pancreatic enzyme digestion as well as combinations of these have been investigated.

METHODS

Protein digestion was studied in defatted human and cow's milk which had been frozen. Defatted milk was incubated with ^{65}Zn for 1 h immediately prior to the digestion experiments.

Experiment 1: Gastric digestion was performed in vitro using milks that had been acidified to pH 2 or 5. Pepsin was added (3200 U/ml) and samples were incubated for 30 min at 37°C. The reaction was stopped by neutralization and centrifugation.

Experiment 2: To study the effect of pH on proteins, independent of pepsin digestion, samples acidified to pH 2 or 5 were incubated for 30 min at 37°C. The reaction was stopped by neutralization and centrifugation.

Experiment 3: To study intestinal digestion milks were adjusted to pH 7, pancreatin (75 ug/ml) was added and samples were incubated for 30 min at 37°C. The reaction was stopped by cooling and centrifugation.

Experiment 4: Sequential digestion was investigated to represent the physiological sequence of gastric and intestinal digestion in vivo. Pepsin was added prior to incubation for 30 min at 37°C. The reaction was stopped by cooling and neutralization. Pancreatin was added and samples were incubated for 30 min at 37°C. The reaction was stopped by cooling and centrifugation.

Supernatant and pellet fractions and were counted on a gamma-counter. The supernatants were separated into high molecular weight (protein) and low molecular weight (< 10,000) compounds on Pharmacia PD-10 columns.

RESULTS

The data obtained from the four experiments are found in Table 1.

Table 1. Effects of Acid and Digestion on Zinc Distribution in Milk.

	Expt. 1 Gastric Digestion		Expt. 2 Acid Effects		Expt. 3 Intestinal Digestion	Expt. 4 Gastric/ Intestinal Digestion	
	pH 2	pH 5	pH 2	pH 5	pH 7	pH 2/7	pH 5/7
	(%)		(%)		(%)	(%)	
Human milk							
insoluble	26	14	12	8	7	13	4
HMW	89	95	81	84	83	73	76
Cow's milk							
insoluble	3	25	7	14	23	3	11
HMW	83	93	94	95	96	85	91

DISCUSSION

Gastric (pepsin) digestion at pH 5 resulted in a larger proportion of insoluble Zn in CM as compared to HM. This effect diminished as the pH dropped. There was no significant difference in the distribution of Zn within the soluble compartment (protein Zn vs. LMW Zn). Acid alone precipitated a larger proportion of CM Zn at pH 5 and a larger proportion of soluble Zn remained bound to protein. The effect of gastric digestion on distribution of soluble Zn at pH 5 was minimal for both HM and CM. Intestinal digestion led to a large percentage of insoluble Zn in CM compared to HM. A larger proportion of soluble Zn in CM remained protein bound as compared to HM. A sequential gastric/intestinal digestion at pH 5 (infant) resulted in a larger proportion of insoluble Zn in CM than in HM. In addition, a larger proportion of the soluble Zn in CM was bound to protein than in HM. At lower pH (adult) these trends were not as consistent.

Thus, in the infant gut with comparatively high pH and low activity of proteolytic enzymes, the Zn in HM will be more soluble and present in more LMW forms than in CM. When transit time is rapid, the likelihood of absorbing Zn from soluble and LMW complexes is higher than from insoluble complexes and macromolecules (proteins) of complicated structure. This would result in a high bioavailability of Zn from HM as compared to CM.

REFERENCE

1. P. Blakeborough, M.I. Gurr, D.N. Salter, Br. J. Nutr. 55:209-217 (1986).

THE AVAILABILITY OF ZINC, CADMIUM AND IRON FROM DIFFERENT GRAINS MEASURED AS ISOTOPE ABSORPTION AND MINERAL ACCUMULATION IN RATS

Per Tidehag, Anncatherine Moberg, Bo Sunzel, Göran Hallmans, Rolf Sjöström and Kenneth Wing

Biophysics Laboratory and the Department of Nutritional Research, University of Umeå S-901 87 Umeå, Sweden

INTRODUCTION

The purpose of this study in rats was to determine if the absorption and accumulation of zinc, cadmium and iron differ amoung diets based on different grains and, if so, what differences among the grains might account for the differences in mineral absorption.

MATERIAL AND METHODS

54 male, three-week old rats were divided into six groups. Each group was fed a diet (Table 1) containing either barley, whole wheat, oats, rye, brewers' spent grain or endosperm wheat flour (the latter supplemented with iron). After 35 days on these diets the rats were placed in metabolic cages and deprived of the diets but not water for 15 hours. The rats were then given a single meal (5 g) of their original diet to which 5 μCi Zn-65 and 5 μCi Fe-59 had been added. Six hours later the rats were killed and the G-I tract and samples of serum, liver, kidney and distal femur (no marrow) were taken for measurements of zinc, cadmium and iron concentrations and Zn-65 and Fe-59 activities.

Table 1. Composition of diets

Diets	Iron (mg)	Zinc (mg)	Cadmium (μg)	Dietary fiber (g/kg)	Phytic acid (mmol/kg)
Barley	14.4	27.9	12.7	60	36
Wheat	20.9	36.5	33.6	76	71
Oats	26.9	36.5	15.1	69	51
Rye	13.8	30.5	6.1	85	66
Brew.gr.	15.6	22.2	8.2	72	15
Endo.wh.	44.3	24.2	11.8	24	22

RESULTS AND CONCLUSIONS

There was no signigicant difference in the relative absorption of zinc (% Zn-65) among the diet groups (F=3.33,p=0.02; Table 2). The zinc concentration in the different diets was the factor which dominated the absorption of zinc. The zinc concentrations in the distal femur in the oats and endosperm wheat groups were significantly higher than those in the brewer' spent grain and rye groups and the barley group was higher than the brewers' spent grain group. These differences may represent differences in the absorption of zinc in excess of the body's needs.

The cadmium content in the liver and kidneys was highest in the group given whole wheat and lowest in the group given rye. The cadmium content in the liver and kidneys was directly related to the cadmium concentration in the diets (r=0.95, p<0.001).

The relative absorption of iron (% Fe-59) was similary among the diet groups with the exception of that in the rye group which was lower than those in the barley, whole wheat and endosperm groups. The iron concentration in the diets is the sole factor determining the iron absorption with the single exception of the absorption from rye which was lower than that from the other grains. At diet iron concentrations below 24 μg Fe/g diet, the mean haemoglobin and serum iron concentrations increased in direct proportion to the diet iron content while the liver iron concentration was stable. Above 24 μg Fe/g diet, the haemoglobin and serum iron concentrations increased very little while the liver iron concentration increased dramatically.

Table 2. The relative absorption of zinc and iron from the test meals (% Zn-65, % Fe-59) and the concentrations of zinc in femur, cadmium in the liver and kidneys, hemoglobin in blood and iron in liver after 35 days on the diets presented as the means (S.E.).

Dietgroups	% Zn-65	[Zn] femur	[Cd] liver + kidney	% Fe-59	Hemo globin	[Fe] liver
Barley	17.8 (0.8)	235.9 (5.3)	89.7 (2.5)	49.8 (2.1)	99.4 (2.8)	35.8 (1.2)
Wheat	12.7 (1.1)	217.2 (9.0)	181.1 (5.7)	45.4 (2.4)	131.9 (4.4)	46.4 (6.5)
Oats	16.6 (1.8)	243.7 (6.5)	74.8 (3.0)	43.9 (2.5)	136.5 (2.1)	72.4 (13.9)
Rye	19.3 (1.3)	215.8 (5.6)	44,25 (2.5)	36.7 (1.6)	86.7 (4.8)	33.6 (1.9)
Brew. grain	18.5 (1.7)	208.4 (3.7)	65.6 (4.5)	45.8 (3.6)	107.2 (5.7)	36.3 (1.4)
Endo. wheat	19.5 (1.5)	240.6 (3.2)	62.9 (4.5)	50.8 (2.8)	108.8 (4.9)	108.8 (12.8)

BIOAVAILABILITY OF IRON FORTIFICATION COMPOUNDS ASSESSED BY A RAT MODEL

Eugene R. Morris, James T. Tanner and Catherine Adams

USDA, ARS, BHNRC, Beltsville, MD 20705
USDHHS, FDA, CFSAN, Washington, DC 20204
ILSI-NF, 1126 16th St., N.W., Washington, DC 20036, USA

INTRODUCTION

This study was part of a cooperative iron bioavailability methodology investigation under auspice of the International Nutritional Anemia Consultative Group (INACG). An overall objective was to assess in human, animal and in vitro models bioavailability of the same iron fortification compounds. This communication reports absorption by iron adequate rats of iron from intrinsically radiolabeled fortification compounds, emulating the human model for assessing iron absorption.

MATERIALS AND METHODS

Intrinsically [55]Fe radiolabeled electrolytically reduced iron powder (Fe[0]) and $FePO_4$ were prepared by commercial procedures. In the laboratory, both [55&59]Fe labeled $FeSO_4$ was prepared by adding the radiolabel to a freshly prepared water solution of $FeSO_4 \cdot 7H_2O$ and freeze drying. Recovery indicated the labeled $FeSO_4$ was predominantly the monohydrate.

Weanling, male Sprague-Dawley rats were fed ad libitum a semipurified diet containing 35 mg/kg Fe as $FeSO_4 \cdot 7H_2O$. Deionized water was provided and rats were caged in suspended stainless steel wire cages in a controlled environment. After 25 days of the ad lib. regimen, the animals were trained to meal eat and on day 29 were offered a meal of the semipurified diet that contained [59]$FeSO_4$ and one of the [55]Fe labeled compounds. Each test compound was offered to 10 rats. The test meal was removed after 1.5 hours and an initial whole body ^{59}Fe activity measurement was obtained within 2 hours after removal of the test meal. Whole body ^{59}Fe activity counts were obtained periodically after the test meal. On day 14 a final whole body count was made and blood samples were obtained for differential counting and hemoglobin determination.

Blood samples were digested in protosol and decolorized with H_2O_2 before [55&59]Fe activity measurements by liquid scintillation counting[1]. An aliquot of the test meal was ashed[2], diluted to an appropriate volume and an aliquot treated and counted in the same manner as the blood samples. Absorption of each isotope was calculated by multiplying cpm/ml blood by total blood volume (body weight x 0.075)[3] and dividing by the cpm ingested.

509

Table 1. Weight Gains, Hemoglobin Values and Iron Absorption[*]

Test Group

Parameter	Reduced Iron	Ferric Phosphate	Ferrous Sulfate
Gain (42 days), g	234 ± 22	238 ± 10	232 ± 17
Hemoglobin, g/dl	13.8 ± 0.7	14.2 ± 0.6	14.2 ± 0.5
Absorption, %			
$^{59}FeSO_4$	19.7 ± 5.2	21.2 ± 4.4	17.4 ± 1.0
^{55}Fe test cpd.	13.7 ± 3.4^b	5.5 ± 0.5^c	18.4 ± 1.1^a

[*]Mean $\pm$ SD. Values in a row not sharing common letter superscript are significantly different, $p < 0.05$.

RESULTS AND DISCUSSION

Absorption of $FeSO_4$ calculated from blood ^{59}Fe activity did not differ significantly between the test groups. This was corroborated by the whole body retention of ^{59}Fe activity. The Fe^o was absorbed to a lesser extent than the $FeSO_4$ and the $FePO_4$ absorption was even less than the Fe^o. There was no difference in absorption of $^{55\&59}FeSO_4$. Compared to $FeSO_4$ as 1.0 the relative bioavailability of Fe^o was $0.699 \pm .067$ and $FePO_4$ was $0.266 \pm .067$. Similar relative values were obtained in human Fe absorption trials at Univ. Kansas. The Fe adequate rat model can be useful in screening iron fortification compounds for relative bioavailability. However, the extrinsic tag method cannot be used if there is not complete exchange between the extrinsic tag and the compound or nonheme Fe pool being studied.

Manuscripts are in preparation for the other facets of the study, hemoglobin regeneration in the Fe depleted rat, human iron absorption measurements and an in vitro test. These will summarize the overall results from the several laboratories and compare the different methodologies.

REFERENCES

1. Eakins, J.D. and D.A. Brown. An improved method for the simultaneous determination of iron-55 and iron-59 in blood by liquid scintillation counting. Int. J. Appl. Radiat. Isot. 17: 391 (1966).
2. Hill, A.D., K.Y. Patterson, C. Veillon and E.R. Morris. Digestion of biological materials for mineral analysis using a combination of wet and dry ashing. Anal. Chem. 58: 2340 (1986).
3. Whittaker, P., A.W. Mahoney and D.G. Hendricks. Effect of iron-deficiency anemia on percent blood volume in growing rats. J. Nutr. 114: 1137 (1984).

MANGANESE ABSORPTION FROM HUMAN MILK, COW'S MILK AND INFANT FORMULAS

L. Davidsson, A. Cederblad, B. Lönnerdal, and B. Sandström

Depts. of Clinical Nutrition and Radiation Physics,
University of Gothenburg, S-413 45, Gothenburg, Sweden, and
Dept. of Nutrition, University of California, Davis, CA 95616

INTRODUCTION

The knowledge of manganese absorption and metabolism in man is limited. Consequently the requirements of manganese for humans of different ages remain largely unknown. Most diets consumed by adults are relatively high in manganese; thus a pronounced deficiency is unlikely. In early life, however, there is a greater risk for manganese deficiency as well as for toxicity. Human milk and cow's milk-based diets are low in manganese concentration, while soy formulas and cereal diets may contain high amounts of manganese. In addition, other factors in formulas, e.g., phytate, supplemental iron and calcium could interfere with manganese absorption. Since manganese is mainly excreted via bile, the risk for manganese toxicity is greater in early life before bile flow has been established.

In this study manganese absorption from human milk, cow's milk and infant formulas was measured in healthy adults using a recently developed radionuclide method (1). The method involves feeding of an extrinsically labeled diet, using the gamma-emitter ^{54}Mn, and subsequent monitoring of the whole-body retention with a sensitive whole-body counter.

MATERIALS AND METHODS

Test Meals

1. Human milk; pooled and pasteurized (from the milk bank at Östra Hospital, Gothenburg).
2. Cow's milk, 3% fat, purchased from a local vendor.
3. Cow's milk formula (BabySemp 1, Semper); whey:casein 60/40.
4. Cow's milk formula (Enfamil, Mead Johnson); whey:casein 60/40 with (a) or without (b) iron fortification.
5. Soy formula (Prosobee, Mead Johnson).

Each serving consisted of 450 ml. The milks and formulas were extrinsically labeled with 0.2 MBq ^{54}Mn (^{54}MnCl$_2$) and 1.5 MBq ^{51}Cr (^{51}CrCl$_3$) before serving. ^{51}Cr was used as a nonabsorbable marker to establish the time point when the non-absorbed fraction of ^{54}Mn had left the body and "true" whole body retention was measured. Whole body retention was monitored for approximately 20 days. Manganese absorption was calculated by extrapolation to day 0 from whole body retention measurements day 10-20.

Subjects

Twenty-six women and six men volunteered for the study. All were healthy adults, median age 25 years, with normal iron status indices and normal levels of manganese in whole blood. Fourteen of the subjects participated twice; eight were served human milk on the first occasion and cow's milk 2 months later, while 6 subjects were fed whey-adjusted cow's milk formula with/without iron fortification on two separate occasions.

RESULTS AND DISCUSSION

The fractional manganese absorption from human milk was significantly different (p < 0.01) from the absorption from cow's milk, iron-fortified cow's milk formula and soy formula. No significant difference was observed for manganese absorption from human milk and the non-iron-fortified cow's milk formulas. Due to the high manganese content the total amount of manganese absorbed was significantly higher from soy formula as compared to the other diets.

Table 1. Manganese and Iron Content and Manganese Absorption from 450 ml of diet.

Diet	Mn	Fe	Mn-absorption	
	ug	mg	n	% (x ± SD)
Human milk	7.2	0.3	8	9.1 ± 3.1
Cow's milk	44	0.1	7	2.8 ± 1.9*
Whey-adjusted cow's milk formula	23	3.0	14	6.1 ± 4.7
Whey-adjusted cow's milk formula, iron-fortified	59	5.6	6	1.6 ± 0.8*
Whey-adjusted cow's milk formula, non-iron-fortified	59	1.0	6	3.6 ± 3.2
Soy formula	154	7.2	4	1.0 ± 0.4*

* significantly different (p < 0.01) from human milk.

The lower manganese absorption from cow's milk as compared to human milk is not likely to be due to their different manganese concentration as we have found little effect of isotope dilution (2). As can be seen from the two identical formulas differing only in iron concentration, there appears to be an effect of iron level on manganese absorption. Finally, the absorption of manganese from soy formula was very low. Thus, factors such as protein composition and level, iron concentration, phytate content and ascorbic acid level need to be evaluated in a systematic fashion in order to determine their respective influence on manganese absorption.

REFERENCES

1. B. Sandström, L. Davidsson, A. Cederblad, and B. Lönnerdal, A method for studying manganese absorption in humans, Fed. Proc. 46:570 (1987).
2. B. Sandström, L. Davidsson, A. Cederblad, and B. Lönnerdal, Manganese absorption from infant formula, Am. J. Clin. Nutr. 41:842 (1985).

INFLUENCE OF CALCIUM AND PHOSPHORUS ON THE ABSORPTION OF

MANGANESE AS DETERMINED BY PLASMA UPTAKE TESTS

Jeanne Freeland-Graves and Pao-Hwa Lin

Division of Graduate Nutrition
University of Texas at Austin
Austin, TX

INTRODUCTION

An antagonistic interaction between calcium(Ca),phosphorus(P),and manganese(Mn) has been suggested in some[1-3],but not all[2-4],animal and human studies.The purpose of this study was to determine the effect that Ca and P have on the absorption of Mn in humans.

MATERIALS AND METHODS

The absorption of Mn in humans was measured via Mn plasma uptake tests in six healthy adults (18-27yrs.). The tests consist of drawing a fasting 5 ml blood sample, giving an oral load, and drawing four more 5 ml blood samples at hourly intervals. The oral loads varied as either 40mg Mn as manganese chloride, 800 mg Ca as calcium carbonate, 40 mg Mn plus 800 mg Ca, 545 ml 2% milk(contains 800 mg Ca), and 40 mg Mn plus 545 ml 2% milk. Four of the six subjects also participated in another test in which 40 mg Mn plus 800 mg P as potassium phosphate was given. The tests were separated by a minimum of 2 weeks.

Manganese concentrations in plasma were determined by graphite furnace atomic absorption spectrophotometry (AAS)[5]. Total calcium concentrations were determined by flame AAS and ionized calcium by a calcium-sensitive electrode. Mineral concentrations in blood were calculated as areas under the curve and compared for statistical differences using Sheffe's t test.

RESULTS

The response of plasma Mn to the various treatments are shown in Fig.1. Ingestion of the 40 mg dose of Mn only produced the typical rise in plasma Mn and produced an area under the curve of 1.767 *ug/L*. As expected, the 800 mg dose of Ca did not influence plasma Mn to any significant effect. However, the addition of a 800 mg dose of Ca to the 40 mg dose of Mn essentially blocked the plasma uptake of Mn completely as indicated by an area of 0.064 *ug/L* (p<0.001).

Consumption of a volume of milk equivalent in Ca content to the inorganic calcium produced similar effects. Milk alone did not affect the plasma Mn as would be expected. But the addition of milk to the 40 mg Mn blocked the uptake of the manganese as seen by the area under the curve of -0.091ug/L (p<0.001).

Plasma Ca concentrations rose only slightly in response to a load of inorganic calcium (0.384 mg/dl) or milk (0.5 mg/dl)as indicated by areas under the curves. The addition of Mn to the inorganic Ca produced a negative response of -0.923 mg/dl, but it was not significantly different from a load of Ca alone. The addition of Mn to the milk did not significantly reduce the slight plasma uptake of Ca as shown by an area of 0.313 mg/dl. Plasma levels of ionized calcium did not change during any of these treatments.

In contrast to calcium, an oral dose of phosphorus did not reduce the uptake of inorganic manganese as seen by an area of 1.523 ug/L.

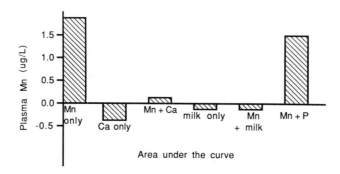

Fig.1 Response of plasma Mn to oral loads of 40 mg Mn, 800 mg Ca, 40 mg Mn +800 mg Ca, 545 ml 2% milk, milk +40 mg Mn, and 40 mg Mn+800 mg P. Data are expressed as area under the curve from time 0 to 4 hours post-load.

DISCUSSION

These data indicate that an oral load of calcium blocks the plasma uptake of the trace mineral, manganese. This inhibition occured both when the calcium was administered in the form of an inorganic salt or in a food substance such as milk. No such effects were seen with an equivalent dose of phosphorus.

REFERENCES

1. G.E.Hawkins,Jr.,G.H.Wise,G.Matrone,and R.K.Waugh. Manganese in the nutrition of young dairy cattle fed different levels of calcium and phosphorus. J.Dairy Sci. 38:536 (1955).
2. J.L.Greger and S.M.Snedeker. Effect of dietary protein and phosphorus levels on the utilization of zinc,copper,and manganese by adult males. J.Nutr.110:43 (1980).
3. H.Spencer,C.R.Asmussen,R.B.Holtzman,L.Kramer. Metabolic balances of Cd, Cu, Mn and Zn in man. Am.J.Clin.Nutr.32:1867 (1979).
4. W.G.Pond,E.F.Walker,Jr.,and D.Kirtland. Effect of dietary Ca and P level from 40 to 100 kg body weight on weight gain and bone and soft tissue mineral concentrations. J.Anim.Sci.3:686 (1978).
5. B.J.Friedman,J.H.Freeland-Graves,C.W.Bales,F.B.Behmardi,R.Shorey-Kutschke R.Willis,J.B.Crosby,P.C.Trickett and S.D.Houston. Manganese balance and clinical observations in young men fed manganese-deficient diet. J.Nutr.117:133 (1987).

TISSUE UPTAKE OF MANGANESE AS A MEASURE OF ITS BIOAVAILABILITY FOR CHICKS

C. B. Ammerman, J. R. Black, P. R. Henry and R. D. Miles

Departments of Animal Science and Poultry Science
University of Florida, Gainesville, Florida   32611 U.S.A.

Inorganic forms of Mn are used routinely as supplemental dietary sources of the element for livestock and poultry. Early research indicated that most chemical forms of manganese were of equal value to the chick, while studies by Watson et al. (1971) suggested that there were differences in the biological availability of manganese among commercially available feed grade products. The latter studies were conducted with a purified diet containing about 5 ppm manganese. Recently, studies have been conducted with manganese sources for chicks using a method described by Ammerman et al. (1985). This method is based on tissue uptake of the element when elevated levels are fed for short periods of time in a diet based on natural ingredients.

A practical-type, corn-soybean meal basal diet was used in the present studies. In Exp. 1, reagent grade sulfate, carbonate and monoxide forms of Mn were compared using a basal diet containing 116 ppm Mn and supplemental levels of 1000, 2000 and 4000 ppm Mn. In Exp. 2, reagent grade Mn sulfate and monoxide were compared with a basal diet with 15% cornstarch substituted for corn (35 ppm Mn) and supplemental levels of 40, 80 and 120 ppm Mn. Day-old chicks were fed 26 and 21 days, respectively, in the two experiments.

In both experiments, feed intake, gain, feed conversion and mortality were not influenced by treatment. There were linear increases in bone, kidney, and liver Mn with bone and kidney providing the greatest sensitivity as a measure of bioavailability. The response of bone Mn to supplemental dietary Mn at 0 to 4000 ppm is illustrated in figure 1. The relative biological availability of Mn compounds as determined with practical diets in the present studies and as determined earlier with a purified diet is shown in table 1. In the study in which a purified diet (5 ppm Mn) was used and Mn sources were added at 10 ppm Mn, differences in bioavailability were not detected among forms when evaluated on the basis of bone Mn, growth or subjective leg scores. Differences in bioavailability among the three Mn forms were evident, however, when they were evaluated on the basis of tissue Mn uptake when supplemental levels of either 40 to 120 ppm Mn or 1000 to 4000 ppm Mn were fed with practical diets.

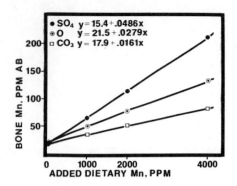

Figure 1. Effect of source and level of dietary Mn on bone Mn concentration (ash weight basis) in chicks fed 26 days.

Table 1. Relative Biological Availability of Manganese Sources When Determined by Different Techniques

Mn source	Supplemental Mn, ppm		
	10^1	$40 - 120^2$	$1000 - 4000^3$
Sulfate	100^a	100^a	100^a
Oxide	95^a	66^b	70^b
Carbonate	93^a	--	39^c

[1]Purified diet, 5 ppm Mn; Comparison based on bone Mn. (Watson al., 1971).

[2]Practical diet, 35 ppm Mn; Comparison based on bone, kidney and liver Mn. (Henry et al., 1986).

[3]Practical diet, 116 ppm Mn; Comparison based on bone and liver Mn. (Black et al., 1984).

[a,b,c]$(P < .05)$.

REFERENCES

Ammerman , C. B., P. R. Henry, J. R. Black, J. E. Margolin, M. G. Echevarria and R. D. Miles, 1985, in: Trace Elements in Man and Animals-5, C. F. Mills, I. Bremner and J. K. Chesters, ed., p. 699, Commonwealth Agricultural Bureaux, Slough, U.K.
Black, J. R., C. B. Ammerman, P. R. Henry, and R. D. Miles, 1984, Poultry Sci., 63:1999.
Henry, P. R., C. B. Ammerman, and R. D. Miles, 1986, Poultry Sci., 65:983.
Watson, L. T., C. B. Ammerman, S. M. Miller, and R. H. Harms, 1971, Poultry Sci., 50:1693.

AN EPIDEMIOLOGICAL STUDY OF A DEFINED AREA OF N. IRELAND: WHOLE BLOOD

GLUTATHIONE PEROXIDASE (EC1.11.1.9)

Dorothy McMaster, A.E. Evans, Evelyn McCrum, M. McF. Kerr,
C.C. Patterson, and A.H.G. Love

Departments of Medicine and Community Medicine, The Queen's
University of Belfast, Northern Ireland

Age standardised mortality rates for 1984 reveal that NI remains at
the top of the world mortality league for coronary heart disease. Belfast
is providing a centre for the WHO MONICA Project which is a 10 year study
of trends and determinants of cardiovascular events. Registration of
events will establish trends. Screening of independent samples of the
population on 3 occasions will monitor risk factor levels. Thus, possible
changes in incidence and case fatality may be analysed in relation to risk
factor changes. Our study area comprises Belfast, Castlereagh, North Down
and Ards Health Districts of NI with a total population of 499,111. About
65% are city dwellers, the remainder living in small country towns and
scattered farms.

Previous observations have indicated a possible link between low
Se status and heart disease but in a population of recognised low Se status
no relationship was found[1]. In the Belfast MONICA Study registration of
events began on 1 January 1983 and the first of the 3 screens took place
between October 1983 and September 1984. The overall response rate was 70%
and blood was collected from 2,327 non fasting subjects. In addition to
the assessment of serum cholesterol, blood pressure and smoking habits
we have estimated the Se status of the population by measuring whole blood
glutathione peroxidase (EC1.11.1.9) and in some selected cases serum Se.

METHODS

Glutathione peroxidase (GSHPx) was assayed at 37°C by the method of
Paglia and Valentine[2] with H_2O_2 as substrate. Serum Se was measured in a
Perkin-Elmer Zeeman 3030 atomic absorption spectrophotometer. Serum was
diluted with 3 volumes of 0.2% nitric acid containing 0.1% Triton and
injected by an AS 60 autosampler onto a pyrolysed L'vov platform. The
method of premixed standard additions was used with a matrix modifier of
copper and magnesium nitrates and maximum power heating.

RESULTS

Age had no effect on GSHPx either in males or females. For males the
mean value was 27.9±0.2(SE), n=1120, which was significantly lower than the
mean of 30.4±0.2(SE), n=1143 for females. No relationship was detected

Table 1. Serum Se levels in subjects of low and high glutathione peroxidase status.

	MALES		FEMALES	
GSHPx U/gHb	18.6 ± 0.14	38.8 ± 0.40	20.5 ± 0.15	41.3 ± 0.30
Se μmol/l	1.18 ± 0.02 (86)	1.24 ± 0.01 (109)	1.16 ± 0.02 (86)	1.25 ± 0.02 (102)
	p = 0.01		p <0.001	

between GSHPx levels and cholesterol or HDL-cholesterol. The prevalence of admitted cigarette smoking was 36% in both males and females. Cigarette smoking females had lower levels of GSHPx 29.8±0.3(SE), n=411.

Serum Se levels were measured in subjects whose GSHPx value was either equal to or less than the 10th percentile value or equal to or greater than the 90th percentile value (Table 1). No statistical difference was found between males and females but subjects with GSHPx equal to or greater than the 90th percentile had the higher levels. Se will be assayed in the remaining subjects.

COMMENT

The extent to which Se is involved in the patho-physiology of the human heart remains to be elucidated. The objective of the Belfast MONICA Project is to measure trends in death rates and morbidity from coronary heart disease in Northern Ireland and to relate these to any long term changes in living habits. Glutathione peroxidase and serum Se levels will be measured in the two remaining population screens in an attempt to define the importance of selenium as a risk factor for coronary heart disease.

ACKNOWLEDGEMENTS

We wish to thank Mr. Noel Bell BSc., AIMLS, and Mr. Paul Anderson BSc., for expert technical assistance.

REFERENCES

1. M. F. Robinson, C. D. Thomson and O. A. Levander, Low selenium status of New Zealand residents, Nutr. Res., Suppl 1: 140 (1985).

2. D. E. Paglia and W. N. Valentine, Studies on the quantitative and qualitative characterisation of erythrocyte glutathione peroxidase, J. Lab. Clin. Med, 70: 158 (1967).

HAIR CHROMIUM AS AN INDEX OF

CHROMIUM STATUS OF TANNERY WORKERS

Janis Randall and Rosalind Gibson

Applied Human Nutrition, Department of Family Studies
University of Guelph, Guelph, Ontario, Canada N1G 2W1

INTRODUCTION

Chromium (Cr) is an essential trace element involved in carbohydrate and lipid metabolism[1]. Chromium, as Cr VI from industrial compounds is readily absorbed via the lungs, skin and gastrointestinal tract and may result in adverse health effects[2]. In contrast, Cr III has been considered to be so poorly absorbed (<0.4%)[3] that routine biological monitoring has not been used for those exposed to industrial Cr III compounds. We have shown however, that Cr III compounds, used in the leather tanning industry, are indeed absorbed and retained, resulting in elevated levels of Cr in urine and serum[4]. Little is known of the health effects of absorption, arising from industrial exposure, of such Cr III compounds. The lack of routine biological monitoring to assess exposure to industrial Cr is due, in part, to the difficulties associated with sampling and analysis of Cr in biological tissues and fluids. For instance, physiological levels of Cr in serum and urine are very low, i.e. below the detection limits for many analytical methods. In contrast, levels of Cr in hair are several fold higher than in serum or urine, making Cr analysis in hair easier. Nevertheless, very little work has been carried out on the use of hair Cr as an index of industrial exposure to Cr. Hence, in this study we have investigated the use of hair as an index of industrial exposure to Cr in a group of Southern Ontario tannery workers.

METHODS

Samples were collected from 40 male tannery workers (TW) (mean age +/- SD = 38 +/- 11 years) and from 40 control subjects (CS) (mean age +/- SD = 40 +/- 13 years). A small portion of hair (~200 mg) was cut from the sub-occipital portion of the scalp. Only the proximal 1.5 cm was placed in a plastic bag for later analysis.

Hair samples were washed according to the method of Kumpulainen et al.[5] The samples were rinsed with hexane, washed twice with 1% sodium lauryl sulphate solution and subsequently rinsed six times in double distilled deionized water. The samples were dried to constant weight and then ashed for 12-16 hours in a low temperature asher (LTA 504, Waltham, Mass.). The ash was dissolved in 4 ml of 0.1 N HCL and analyzed for Cr by flameless atomic absorption spectrophotometry using a Varian Spectra 30 AA equipped with a GTA 96 furnace (Varian Canada, Georgetown, Ontario).

A reference hair material[6], certified for Cr content, was analyzed to check the accuracy of analysis. A pooled hair sample was also analyzed in each run to check the precision of the method.

The hair Cr concentrations in both groups were not normally distributed; therefore non-parametric statistics were employed, the median being used to indicate central tendency. The Kruskal-Wallis test was used to test for differences between groups. The Spearman rank correlation coefficient was used to test for correlations between hair Cr levels and levels of Cr in serum and urine.

RESULTS

As shown in Table I, the median hair Cr concentration for the tannery workers was significantly higher (p=.0001) than that for the control subjects.

Table I. Hair Chromium Levels in Tannery Workers and Control Subjects

	Tannery Workers n=40	Control Subjects n=40	p*
Hair Cr ng/g**	453 (248–783)	124 (72–210)	.0001

* Kruskal-Wallis test
**Median and 25-75 quartile

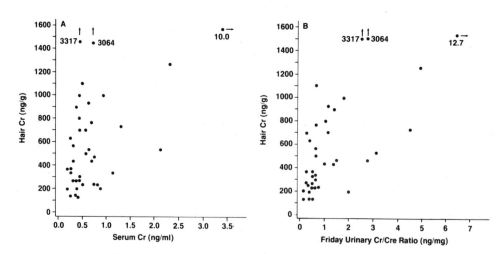

Figure 1. a) Correlation of hair Cr with serum Cr (r=.35, p=.02)
b) Correlation of hair Cr with urinary Cr/Cre ratio (r=.64, p=.0001)

For the tannery workers, hair Cr concentrations were positively correlated with serum Cr (r=.35, p=.02) and the urinary Cr/Cre ratio (r=.64, p=.0001) (Figures 1a and 1b respectively).

For the tannery workers, hair Cr levels were not correlated with length of employment in the leather tanning industry. Hair Cr concentrations were not correlated with age, height or weight for either group.

DISCUSSION

This study demonstrates that exposure to Cr III in the leather tanning industry results in elevated hair Cr concentrations in tannery workers compared with controls (Table I). Chromium in hair of the tannery workers is endogenous rather than exogenous in origin as indicated by the significant positive correlations observed between hair Cr and both serum Cr and the urinary Cr/Cre ratios (Figures 1a and 1b respectively).

Hair is a more promising biopsy tissue for monitoring industrial exposure to Cr than serum or urine. For instance, Cr concentrations in hair are two to three fold higher than in serum or urine thus facilitating analysis. Moreover, concentrations of Cr in hair are remarkably consistent compared to those for serum and urine. Trace element levels in hair are not affected by recent dietary intake, hormonal factors or circadian variation. Furthermore, collection of hair samples is a relatively non-invasive procedure and hair is easy to store. Hair trace element concentrations can provide a retrospective index of environmental exposure. Results of this study indicate that hair Cr concentrations can be used as indices of Cr exposure or status.

REFERENCES

1. J. S. Borel and R. A. Anderson, Chromium, in: "Biochemistry of the Essential Ultratrace Elements," E. Frieden, ed., Plenum Publishing Corporation, New York (1984).
2. P. L. Bidstrup, Effects of chromium compounds on the respiratory system, in: "Chromium: Metabolism and Toxicity," D. Burrows, ed., CRC Press Inc., Boca Raton (1983).
3. R. A. Anderson, M. M. Polansky, N. A. A. Bryden, K. Y. Patterson, C. Veillon and W. Glinsmann, Effects of chromium supplementation on urinary chromium excretion of human subjects and correlation of chromium excretion with selected clinical parameters, J Nutr 113:276 (1983).
4. J. A. Randall and R. S. Gibson, Serum and urine chromium as indices of chromium status in tannery workers, Proc Soc Exp Biol Med 185:16 (1987).
5. J. Kumpulainen, S. Salmela, E. Vuori and J. Lehto, Effects of various washing procedures on the chromium content of human scalp hair, Anal Chim Acta 138:361 (1982).
6. K. Okamoto, M. Morita, H. Quan, T. Vehiro and K. Fuwa, Preparation and certification of human hair powder reference material, Clin Chem 31:1592 (1985).

FAMILIAL TRENDS IN HEAVY TRACE METAL BODY BURDENS IN A SMALL RURAL COMMUNITY

D.J. Eatough, N.F. Mangelson, M.W. Hill, K.K. Nielson and L.D. Hansen, Department of Chemistry, Brigham Young University, Provo, UT 84602 USA

M.A. Skolnick, Genetic Epidemiology Group, Department of Medical Informatics, University of Utah, Salt Lake City, UT 84132 USA

INTRODUCTION

A preliminary study of relationships among exposure to heavy metals, body burdens of these metals, and cancer patterns in the exposed population has been conducted in a small agricultural community of 450 people in Utah. A suspected high incidence of cancer and the existence of some heavy metal mining in the area prompted the study. Other sources of toxic trace metals, such as smoking or industrial pollution, are minimal. Most of the present inhabitants are descendants of the original group who settled the site in 1868, so the town is a genetic isolate. Familial associations in the population are know from extensive, accurate genealogical records. Our hypothesis were (a) elevated heavy metal concentrations in the area has led to elevated concentrations of Pb, As and Cd in the population, and (b) relationships exist among the familial patterns of high concentrations of these metals in hair and familial patterns of cancer. The study group proved to be too small to establish cancer rates or familial cancer trends, but a familial trend in the response to heavy metal exposure was found.

EXPERIMENTAL

Study Site. All culinary water for the town is obtained from springs in the nearby mountains which are known to have some small Pb and Ag ore bodies. Surface water from the mountains is used to irrigate croplands and water livestock. Use of the local farm produce for food is extensive.

Collection of Environmental Samples. Long term average concentrations of trace metals in the water were evaluated by determination of trace metal concentrations in collector plants which are constantly exposed to the water supply. Samples of moss and water cress growing in the overflow water from the culinary springs and in the water storage pond for irrigation water were analyzed. Comparison samples of the same species of moss and water cress were obtained from several other sites. These included a high impact site where the plants were growing in the runoff from the tailings pond of a small lead smelter, intermediate sites where plants were growing in near proximity to active mines, and low exposure sites remote from any known mining activities. The concentrations of heavy metals in the local environment of the town were determined in soil, wheat, and alfalfa samples, and in bovine liver and kidneys from cattle grown on the surrounding croplands.

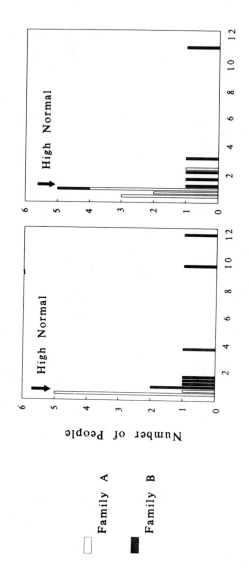

Figure 1. Frequency pattern of hair As and Cd from two study group families.

The exposure of the local population to trace metals was estimated by determining the concentrations of heavy metals in head hair samples.

Analytical Techniques. Soil samples were extracted in a pH 4.8 acetate buffer (1). Plant samples were digested in a mixture of nitric and hydrofluoric acids in Teflon containers and in Teflon-lined, sealed bombs. Hair samples were washed with trichloroethane and water, the dried hair ashed in an oxygen plasma, and the ash extracted with nitric acid. Aliquots of the resulting solutions were placed on Nuclepore filters and analyzed by proton induced X-ray emission (2).

RESULTS AND DISCUSSION

The results indicate that As, and possibly Cd and Pb, are elevated in the local environment and that the principal route of human exposure is culinary water rather than food sources. The concentrations of As, Cd, and Pb in the water cress from culinary springs in the study area and from the control locations are given in Table I. The higher concentrations of As, Cd and Pb in plants collected in the winter compared to spring are expected since the underground water supplies are diluted with snow melt and because of plant growth. Moss samples collected from the culinary springs and from the water storage pond had concentrations of As, Cd and Pb similar to the water cress samples. Concentrations of As, Cd, or Pb in alfalfa, wheat, kidney, or liver samples were not elevated. Soil samples were high in As.

An assessment of the heavy metal burden of the people was obtained by comparison of the concentrations found in hair with the results of other studies (3). The highest values observed in human hair (As 12 ppm, Cd 11 ppm, Pb 51 ppm) suggest a definite, elevated exposure. Concentrations of all three elements in hair were highly correlated for any individual. The frequency distribution of the concentrations of As and Cd found in hair samples from two family lines are given in Figure 1. Elevated concentrations of heavy metals in hair were associated only with family B.

The discovery of non-genetic, environmental factors in human cancer etiology and correlation with familial patterns of disease was an objective of this study. The study population was too small to establish a link between exposure to heavy metals and development of cancer. However, the results suggest familial differences in the uptake of toxic heavy metals.

Table I. Concentrations of As, Cd and Pb (ppm) in Water Cress from Springs at the Study and Control Sites (Number of springs sampled in parenthesis).

| Location | Samples Collected in February | | | Samples Collected in May | | |
	As	Cd	Pb	As	Cd	Pb
Study Site (3)	5.5±2.5	2.5±1.4	3.0±1.2	1.70±0.61	0.32±0.04	2.61±0.59
High Control (1)				4.66	9.3	303.
Intermediate (2)				0.91±0.02	0.45±0.22	2.25±0.18
Low Control (2)				0.38±0.23	0.31±0.06	1.45±0.30

REFERENCES

1. Black C.A., Ed. (1965) Meth. of Soil Anal., Part 2, Amer. Soc. Agronomy.
2. Mangelson N.F., et al. (1979) Anal. Chem. 51:1187-1194.
3. Underwood E.J. (1971) Trace Elem. in Human and Animal Nutr., Acad. Press.

CADMIUM, ZINC AND CALCIUM IN HUMAN KIDNEYS

R. Scott[*], E. Aughey[**], and A. Mclelland[*]
*Urology Dept. and Biochemistry Dept.,
Glasgow Royal Infirmary, Glasgow
**Dept. of Veterinary Histology
University of Glasgow

INTRODUCTION

There are various reports now available which indicate the cadmium content in kidneys from a variety of centres.(1,2,3)   Despite recent work including the importance of cadmium in renal stone formers, we are unaware of any attempt to correlate cadmium and calcium values.

MATERIAL, METHODS AND RESULTS

The lower pole of the left kidney was obtained from 1,000 post mortem samples and stored at -20°C.   All samples were subsequently analysed in the same laboratory and by the same methods as previously recorded. (4)

The cadmium and zinc contents were measured at the same time as the calcium content and for the purpose of analysis only significant differences are recorded using a Kruskal Wallis test. (table 1)   As previously recorded the cortical cadmium values along with zinc steadily rose to reach maximum value at 50 - 59 years of age.   Smoking is important with respect to renal cadmium values but is less important with respect to zinc.   Calcium deposition is not apparently affected by smoking habits.   An arbitary classification of disease as previously quoted showed no statistical significance with respect to cadmium, but significant results related to those dying from renal disease for both cadmium and zinc.(4)

DISCUSSION

Since our initial observation of a high prevalence of urinary tract stone disease in a group of coppersmiths, one of our main fields of interest has been the interplay between cadmium and calcium metabolism both in the experimental animal (5) and in the continuing clinical problem of altered calcium metabolism in cadmium exposed workers. (6)

The incidence of renal stone disease according to a variety of workers is clearly increasing in modern times. (7,8)   It is difficult to explain this increased "incidence" of a condition which is generally recognised as having a wide variety of aetological factors including metabolic diseases such as hyperparathyroidism.   There are, however, a large number of stone cases which are clearly "idiopathic".

For many years it has been recognised that calcium deposition in kidneys occurs in between 12 and 54% of individuals even in the absence of clinical stone disease.

The reason why calcium is deposited in kidneys has never been clearly understood. The present study has shown that as part of the aging process calcium deposition increases with time. Cadmium deposition however, does not follow the same pattern in that the concentration of the trace metal begins to fall after the age of 60 as does zinc. That cadmium and zinc clearly mirror each other (9) is a well recognised phenomenon and is related to the deposition of cadmium metallothionein in renal substance. There is no correlation, however, between cadmium and calcium. It is possible, nevertheless, that cadmium may induce changes in renal substance which initiate calcium deposition. It must also be borne in mind that calcium excretion with subsequent potential oestomalacia is greatly increased in chronic cadmium poisoning and tubular dysfunction clearly caused by cadmium is important in overall calcium metabolism. A recent observation shows that renal cortical cadmium is the significant factor related to microlith formation in renal tubules. (10)

Table 1   Significant Findings

	CADMIUM		ZINC		CALCIUM	
	(C)	(M)	(C)	(M)	(C)	(M)
Age	< 0.001	< 0.001	< 0.001	< 0.001	< 0.001	< 0.001
Smoking habits	< 0.001	< 0.001	< 0.05	< 0.001	NS	< 0.05
Disease	< 0.05	< 0.05	< 0.05	< 0.05	NS	NS

REFERENCES

1. Elinder C.G., Kjellstrom T., Friberg L., Lind B., Linnman L. Cadmium in kidney cortex, liver and pancreas from Swedish autopsies. Arch Environ Hlth 31: 292 (1976)
2. Tsuchyuya K., Seki Y.M., Sugita M. Cadmium concentrations in the organs and tissues of cadavers from accidental deaths. Keio Med. J. 25: 83 (1976)
3. Miller G.J., Wylie M.J., McKeown D. Cadmium exposure and renal accumulation in an Australian urban population. Med. J. Australia 1: 20 (1976)
4. Scott R., Aughey E., Reilly M., Cunningham A., McClelland A., Fell G.S. Renal Cadmium content in the West of Scotland. Urol Res. 11: 285 (1983)
5. Aughey E., Scott R. Ultra structural changes in the renal cortex of rats exposed to cadmium in trace element metabolism in man and animals. in Proc. 3rd Int. Symp. TEMA 3. Ed. Kirchgessner M Technishe Universtat Munchen Freising Weihenstaphan FRD.
6. Scott R., Cunningham C., McLelland A., Fell G.S., Fitzgerald-Finch O.P. McKellar N. The importance of cadmium as a factor in calcified upper urinary tract stone disease - a prospective 7 year study. Brit. J. Urol. 54: 584 (1982)
7. Sierakowski R., Findlayson B., Landes R.R., Findlayson C.D., Sierakowski N. The frequency of urolithiasis in hospital discharge daignosis in the United Stated. Invest. Urol. 12: 438 (1978)
8. Johnson C.M., Wilson D.M., O'Fallon W.M., Malek R.S., Kurland L.T. Renal stone epidemiology: A 25 year study in Rochester Minnesota. Kid. Int. 16: 624 (1979)
9. Elinder C-G., Piscator M. Cadmium and Zinc relationships Environ. Hlth Persp. 25: 129 (1978)
10. Hering F., Briellmann T., Luond G., Guggerhelm H., Seiler H., Rutishauser G. Stone formation in human kidneys. Urol. Res. 15: 67 (1987)

ANTE MORTEM v. POST MORTEM CHANGES IN CHRONIC
CADMIUM POISONING

R. Scott[*], and E. Aughey[**]

*Urology Dept.,
Glasgow Royal Infirmary
**Dept. of Veterinary Histology,
University of Glasgow

INTRODUCTION

The critical value of renal cortical cadmium associated with tubular
dysfunction is generally accepted as 200ug/g.(1) The geometric mean
value of cortical cadmium has been established for males as 14.7ug/g wet
weight in post mortem samples of kidneys from the U.K. (2) A group of
coppersmiths have been studied clinically over a 10 year period. Kidneys
have been obtained from six of these subjects and analysed as above. One
subject developed a renal tumour and had a kidney removed allowing analysis
of 'normal' and tumour material from the same individual.

RESULTS

Ante mortem studies of the coppersmiths showed consistently elevated
blood and urine cadmium values accompanied by variable but usually excess
excretion of Beta-2 microglobulin in urine (table 1) The concentration
of cadmium in the renal cortex of six of the coppersmiths is shown (table 2)
One subject is still alive since his kidney specimen was occasioned by the
removal of a kidney bearing a renal tumour. In subject No. 1 the cortical
cadmium exceeded 200ug/g and in this man excess Beta-2 microglobulin in
urine was found on only one occasion, In all of the other four subjects
both high and low molecular weight proteinuria occurred throughout the
period of study and all had blood and urine cadmium values in excess of
normal non-exposed individuals. In one subject, No. 6, no ante mortem
assessment was available. (table 2)

DISCUSSION

In chronic cadmium poisoning proteinuria is accepted as an outcome
of renal glomerular and tubular damage. (3,4) There are several
biochemical abnormalities including altered calcium metabolism. (5)

There has long been speculation as to the specific level of renal
cortical cadmium which produces renal tubular dysfunction. It is
generally accepted that the value of 200ug/g wet weight is critical (1)
and this figure has apparently been accepted by the W.H.O. Nomiyama
suggests a value of 300ug/g.(6) The present study would suggest that
the value should be much lower.

Table 1   ANTE MORTEM BLOOD   /   URINE CADMIUM   (Mean Values)

	BLOOD CADMIUM (ug/l)		URINE CADMIUM (ug/vol)	
	No.	Mean	No.	Mean
1975	4	16.75	4	30.5
1976	3	17.6	3	57.2
1978	4	14.2	3	15.7
1979	4	11.6	4	22.5
1980	4	20.0	5	37.8
1981	4	10.8	5	24.3
1982	3	12.3	3	12.7
1983	4	10.8	2	13.9
1984	3	11.2	3	12.6

Table 2   POST MORTEM ANALYSIS   /   TRACE METALS IN RENAL CORTEX

Subject		Cd (ug/g)	Zn (ug/g)	Ca (ug/g)
1.	Post Mortem	236	48	334
2.	Post Mortem	163	40	–
3.	Nephrectomy	147	68	139
	Nephrectomy (tumour)	3	14	97
4.	Post Mortem	55	36	0
5.	Post Mortem	26	32	0
6.	Post Mortem	149	58	156

REFERENCES

1.   Friberg L., Piscator M., Nordberg G.F., Kjellstrom T.L.   Cadmium in environment in 2nd Edition CRC Press Cleveland U.S.A. (1974)
2.   Scott R, Aughey E., Fell G.S., Quinn M.J.   Cadmium concentration in human kidneys from the U.K.   Human Toxicol 6: 111 (1987)
3.   Tsuchiya K.   Proteinuria in cadmium workers.   J. Occup. Med 18:   463   (1976)
4.   Bernard A., Butchet J.P., Roels H., Masson P., Lauwerys R.   Renal excretion of proteins and enzymes in workers exposed to cadmium. Eur. J. Clin. Invest. 9: 11 (1979)
5.   Kazantzis G.   Renal tubular dysfunction and abnormalities of calcium metabolism in cadmium workers. Environ. Hlth. Persp. 28: 199 (1979)
6.   Nomiyama K.   Does a critical concentration of cadmium in human renal cortex exist?   J. Toxicol Environ Hlth 3: 607 (1977)

ACCURATE ASSESSMENT OF THE INTAKE OF DIETARY COMPONENTS BY HUMAN SUBJECTS

Venkatesh Iyengar

USDA/NBS
Gaithersburg, MD   20899

INTRODUCTION

Planning human dietary studies pose special problems especially when representative diets are to be defined for mixed population groups. The challenge here involves development of strategies so that the investigation is scientifically comprehensive and meaningful, but does not assume unmanageable proportions in terms of complex collection procedures, too large a sample size, and the accumulation of enormous amounts of analytical data, presenting difficulties in handling and interpreting the analytical findings.

The aim of this communication is to emphasize the need for good analytical measurements, and for adopting a multidisciplinary approach to extend the scope of a dietary study to include both organic and inorganic constituents.

PLANNING DIETARY STUDIES

Elemental composition of foods represents a diverse picture. At low concentration levels, analysis of dietary material presents considerable difficulties, depending upon whether it is a simple matrix (e.g. drinking water and beverages) or a complex one (dairy products). The difficulties associated with the determination of certain elements in foods, e.g., Hg and Se, are best illustrated by a comparison of the elemental concentration profiles of total mixed diet and blood serum. The concentrations of Cu, Hg, and Se in these two materials fall in the same range, thus emphasizing the degree of analytical difficulties involved in the determination of these elements in foods. The Proximate analysis of a mixed total diet from the U.S. typically contains up to 84% total volatiles, 3% fat, 9.2% carbohydrate, 3-4% protein and an ash content of 0.7%.[1] The proportion of these components may vary significantly in different types of diets depending upon eating habits, thus creating characteristic matrix properties and unusual analytical problems.

Dietary Collections

Generally speaking, there is no single method which may be regarded as entirely satisfactory for the purpose of estimating the human dietary intake of nutrients and toxic substances in foods since each approach has its own limitations. This is because, widely differing income patterns, life-styles and heterogeneity of the food supply within a population complicate the experimental design. Therefore, a compromised approach that minimizes the limitations is of practical value. In this context, the FAO outlines 3 collection procedures. These are: (a) total diet consisting of a market basket of food reflecting a defined total diet of a consumer, (b) selective studies of individual food stuffs, and (c) duplicate portion studies. The relative merits of these methods are discussed in detail.[2] The choice of which method to adopt will depend on the objectives of the assessment and the resources available.

Multidisciplinary Approach

A recent dietary study being carried out in the U.S.A. is a good example in this context since it blends the multidisciplinary experience of the 3 participating US agencies, namely the Food and Drug Administration (FDA), the National Bureau of Standards, and the U.S. Department of Agriculture, and provides a vital link in assuring success in this kind of investigation.

In an ongoing study[1] the 201 foods from the FDA Total Diet Study (TDS)[3] are utilized to prepare a composite (USDIET-I), representative of one age-sex group. The daily intake of 3075 g (wet weight) for an adult 25-30 year male was used as the basis for calculating the contribution of each of the 201 food components to the diet. These foods were blended, freeze dried and homogenized before analysis. Three different analytical techniques were used whenever possible to cross check the accuracy.

A comparison of the results for measured intake from USDIET-I with the calculated daily intakes derived from the FDA-TDS single food analyses is shown in Table 1 for a few elements.

Table 1. Comparison of the daily intakes of various minerals through USDIET-I and FDA-TDS, based on 3075 g diet.

| | Ca | Mg | Cr | Cu | Mn | Mo | Se | Zn | As | Cd | Hg |
	mg/d					µg/d					
USDIET-I	836	305	37	1470	2740	128	128	16300	55	15.5	2.6
FDA-TDS	836	288	--	1240	2720	-	110	16150	45	15	3.9

In the TDS scheme, about 8 to 10 aliquots from each of the 201 different foods were individually analyzed and the concentrations were tabulated and stored in a computer. By using these values, the daily intake of a given element was computed by taking appropriate proportions of the 201 foods that were required to make up a day's diet. In contrast to several hundred separate foods analyzed in the TDS scheme, the results for USDIET-I were obtained by analyzing a small number of aliquots of a mixed total diet. Over 25 elements covering a concentration range of 9275 ppm for Cl to about 1 ppb for Sc were determined.

As shown in Table 1, there is good agreement between USDIET-I and FDA-TDS investigations for the daily intakes of several elements. The beneficial feature of a mixed diet is also reflected in the determination of constituents such as Cr and Mo, which appear at ultratrace concentration levels in a majority of bulk foods, yet an accurate assessment of their daily intakes presented no problems. Obtaining the same information from analysis of single foods would be analytically difficult, tedious and time consuming.

CONCLUSIONS

The results from USDIET-I demonstrate the possibility of extending the analytical coverage to over 30 elements of biological significance. The mixed diet approach is also useful for investigating organic nutrients for which presently no natural matrix SRM's are available. Currently, efforts are under way to standardize the handling steps to minimize the loss of components such as vitamin C, and to follow storage stability of thiamine, riboflavin, vitamin B6, pantothenic acid, niacin, vitamin A, vitamin B12, biotin, and total folates, over extended periods of time.

REFERENCES

1. G.V. Iyengar, J.T. Tanner, W.R. Wolf, R. Zeisler, Preparation of a Mixed Human Diet Material for the Determination of Nutrient Elements, Selected Toxic Elements and Organic Nutrients: A Preliminary Report, Sci. Total Environ. 61:235, (1987).
2. Guidlines for the study of Dietary Intakes of Chemical Contaminants, Joint FAO/WHO Food Contamination Monitoring Programme (WHO-EEP/83.53) Geneva, (1983).
3. J.A. Pennington, Revision of the Total Diet Study Food Lists and Diets, J. Am. Diet Assoc., 82:166, (1983).

REFERENCE VALUES FOR TRACE ELEMENT CONCENTRATIONS IN WHOLE BLOOD, SERUM,

HAIR, LIVER, MILK, AND URINE SPECIMENS FROM HUMAN SUBJECTS

Venkatesh Iyengar

USDA, NBS
Gaithersburg, MD 20899

INTRODUCTION

It is a common practice to use the expression "normal value" in
defining the elemental concentrations in human tissues and body fluids.
However, it appears that this terminology has been used somewhat
indiscriminately in dealing with elemental composition of biological
systems. The following example clarifies the situation. Very frequently,
this expression is used in dealing with both toxic (e.g. cadmium and lead)
and essential (e.g. copper and zinc) groups of elements. Strictly
speaking, "true" normal levels for cadmium and lead, e.g. in a specimen
such as human blood, should be close to zero. However, environmental and
other factors have contributed to the entry of these two elements into
biological systems, and therefore, one is concerned with tolerance limits
in this case. On the other hand, elements such as copper and zinc being
essential for life, need to be present in living systems and are
homeostatically regulated. Therefore, under ideal conditions, their
levels may be expected to fluctuate within narrow limits for a given
species, thereby justifying the usage of normal values. It is highly
desirable to establish such normal levels in tissues and body fluids so
that changes during disease conditions can be detected. However, in
practice it is a difficult task, requiring consideration of and
compensation for a number of possible concurrent phenomena, and
correlations may be very complex. Thus, baseline values can be deceptive
and the question as to what is normal may not be easy to answer. This
problem is particularly critical in human subjects since it is impossible
to control all the influencing factors to derive a set of values with a
statistically normal distribution.[1-4]

REFERENCE RANGE OF VALUES

In dealing with human subjects, it is more practical to consider the
usage of "reference ranges" of values based on factors such as dietary
habits and geochemical and other environmental influences. For
biologically controlled elements, such ranges are likely to be narrow for
subjects with no known health abnormalities. For nonessential elements,
such a range can be broad, depending upon the level of exposure of the
subjects in question. Taking the example of aluminum, it is not an
essential element and, therefore, it is not subject to homeostatic

Table 1. Concentrations of Selected Trace Elements in Some Clinical Samples From Adult Human Subjects

Element	Liver $\mu g\ kg^{-1}$	Whole Blood $\mu g\ l^{-1}$	Blood Serum $\mu g\ l^{-1}$	Urine $\mu g/24h$	Milk $\mu g\ l^{-1}$	Hair $\mu g\ kg^{-1}$
As	5-15	2-20?	1-10?	10-30?	0.25-3	150-300
Cd	<1000-2000	0.3-1.2[a] 1-4[b]	0.1?	1-5?	1?	400-1000
Co	30-150?	5-10	0.1-0.3?	0.5-2?	0.2-0.7	50-300?
Cr	5-50?	<1?	0.1-0.2?	0.2-2?	1.0-1.5	300-800
Cu	5000-7000	800-1100[c] 1000-1400[d]	800-1100[c] 1100-1400[d]	30-60	250-400	15000-25000
F	100-300?	200-500?	20-50?	500-1500?	10-26	-
I	100-200?	40-60	60-70?	100-200?	40-80	400-1000
Fe	150000-250000	425000-500000	800-1200	100-200?	350-600	30000-60000
Pb	350-550	90-150	<<1?	10-20?	1-5	2000-20000
Mn	1100-2100	8-12	0.5-1.0	0.6-2?	3-6	500-1500
Hg	30-150	2-20?	<1	5-20?	1-3	500-2000
Mo	500-800	1-3?	<0.5?	20-30?	1-4	50-200?
Ni	10-50?	1-5?	1-2?	2-8?	10-20	20-200?
Se	250-400	90-130	75-120	25-50	15-25	500-1000
Zn	40000-60000	6000-7000	800-1100	400-600	1500-2000	150000-250000

[a]non-smokers;
[b]smokers;
[c]males;
[d]females;
?value uncertain.

control. Its entry into the human system is highly variable with intakes fluctuating from low to very high amounts, depending upon the type of food consumed and certain other factors. Indeed, under these conditions, one should be surprised to find a normal value for this element in blood serum. This is also true for a number of other trace elements that are not essential for biological processes.

DATA BASE

In order to formulate the reference values, a literature survey of healthy, adult human subjects, was undertaken by Iyengar between 1982 and 1985. Using results, recommended by collaborators in many countries a compilation was published.[5] A summary of that compilation evaluating results for 15 trace elements in selected samples of clinical significance is presented in Table 1. The results are classified under frequent, low and high (relative to frequent values) observations. The frequent values refer to the results found in several countries. A detailed account of this approach is discussed elsewhere.[6] As seen from Table 1, for several essential trace elements included here, the range of frequently found values lies quite close (the ratio high to low being less than 1.5 in most cases). On the other hand, elements such as mercury and lead, cover a wider range, reflecting absence of homeostatic control.

REFERENCES

1. G.V. Iyengar, Presampling Factors in the Elemental Composition of Biological Systems, Anal. Chem. 54:554A, (1982).
2. G.V. Iyengar, Elemental Composition of Human and Animal Milk, Report IAEA-TECDOC-269, Vienna, (1982).
3. G.V. Iyengar, W.E. Kollmer, H.J.M. Bowen, The Elemental Composition of Human Tissues and Body Fluids, Verlag Chemie, Weinheim, (1978).
4. G.V. Iyengar, W.E. Kollmer, Some Aspects of Sample Procurement From Human Subjects for Biomedical Trace Element Research, Trace Ele. Med., 3:25, (1986).
5. G.V. Iyengar, Concentrations of 15 Trace Elements in Some Selected Adult Human Tissues and Body Fluids of Clinical Interest From Several Countries: Results From a Pilot Study for the Establishment of Reference Values, Report Juel-1974, Juelich, (1985).
6. G.V. Iyengar, Reference Values for the Concentrations of As, Cd, Co, Cr, Cu, Fe, I, Hg, Mn, Mo, Ni, Pb, Se, and Zn in Selected Human Tissues and Body Fluids, Biol. Trace Ele. Res., (1987), in press.

TRACE ELEMENTS IN HUMAN MILK: AN AUSTRALIAN STUDY

F.J. Cumming* and J.J. Fardy**

*Queensland Institute of Technology, GPO Box 2434, Brisbane
4001. Australia
**CSIRO Division of Energy Chemistry, Private Mail Bag 7
Sutherland. 2232. Australia

INTRODUCTION

The composition of breast milk substitutes is usually based on that
of breast milk. However, relatively little is known about the trace
element content of human milk from different countries.

This paper presents the results of an Australian study of the trace
elements iron, zinc, copper, manganese, selenium, cobalt, rubidium and
caesium in human milk, and some of the factors affecting their
concentration.

EXPERIMENTAL

The elemental composition of blood plasma and mature breast milk
from 14 healthy women was studied in multiple specimens collected over
16 weeks of lactation. Five of the women took progestogen-only oral
contraceptives (OC) (30 μg levonorgestrel or 250 μg norethisterone,
daily). The design of the study also allowed the effects of stage of
lactation and maternal diet on the milk composition to be observed.

Fasting plasma and fore-milk samples and dietary intake records
were collected at the same time from the 14 women at 8, 16 and 23 weeks
post-partum.

A neutron activation analysis (NAA) technique was developed,
combining radiochemical separation NAA (based on the method of Nakahara
et al (1))and instrumental NAA, to allow multi-element analysis on small
samples (200 mg) of lyophilized milk and plasma.

RESULTS AND DISCUSSION

The progestogen-only OC used by the women in this study did not
affect the milk or plasma trace element concentrations. The results
from all 14 women were therefore grouped together for further analysis.

Table 1 shows the mean milk and plasma concentrations in these
women, and the approximate mean milk: plasma elemental ratios. The
reasons for the differences in partitioning of the elements between
milk and plasma are not known, but may depend on binding characteristics,
especially to milk and plasma proteins, or on relative solubilities
in milk and plasma.

Table 1. Mean ($\pm$SD) milk and plasma elemental concentrations in 14 women between 8 and 23 weeks post-partum, and approximate mean milk: plasma ratios

Element		Weeks Post-Partum*	n	Milk	n	Plasma	Approx. Mean Milk: Plasma Ratio
Iron	mg/kg	8 + 23	25	0.48 ± 0.36	32	1.53 ± 0.39	0.3
		16	13	0.76 ± 0.54			0.5
Zinc	mg/kg	8 + 16	24	1.85 ± 0.85	42	0.94 ± 0.27	2
		23	14	1.10 ± 0.46			1
Copper	mg/kg	All	42	0.39 ± 0.13	41	1.08 ± 0.34	0.3
Manganese	μg/kg	All	37	4.5 ± 1.6	34	1.8 ± 0.5	2
Selenium	μg/kg	All	40	12 ± 2	36	77 ± 14	0.1
Cobalt	μg/kg	All	35	0.15 ± 0.06	28	0.29 ± 0.09	0.4
Rubidium	mg/kg	All	14	0.84 ± 0.12	33	0.28 ± 0.06	3
Caesium	μg/kg	All	41	4.3 ± 1.2	33	1.0 ± 0.3	4

* "All" denotes values for 8, 16, and 23 weeks post-partum.

The mean milk and plasma concentration of all elements measured did not change significantly with the progression of lactation, with the exception of iron and zinc. The mean milk iron concentration at 16 weeks was significantly higher than at 8 and 23 weeks post-partum ($p < 0.02$). This rise was, however, due to the results of only 2 of the 14 women. The reason for the rise in their milk iron concentration at this stage of lactation was not apparent.

The mean milk zinc concentration was significantly lower at 23 weeks post-partum than at 8 and 16 weeks post-partum ($p < 0.05$ and $p < 0.01$ respectively), the downward trend occurring for all but one mother.

Maternal dietary intake of iron, copper and zinc did not directly affect the milk concentrations of these nutrients (intakes of the other elements were not determined).

REFERENCES

1. Nakahara, H., Nagame, Y., Yoshizawa, Y., Oda, H., Gotoh, S. and Murakami, Y., 1979, Trace element analysis of human blood serum by neutron activation analysis, J. Radioanalyt. Chem., 54:183.
2. National Health and Medical Research Council, Recommended dietary intakes for use in Australia, Australian Governemnt Publishing Service, Canberra (1987).
3. Lonnerdal, B., 1985, Dietary factors affecting trace element bioavailability from human milk, cow's milk and infant formulas, Progress in Food and Nutrition Science, 9:35.

COPPER AND MOLYBDENUM LEVELS IN TISSUES OF HYPERCUPROTIC SHEEP DURING AND

AFTER BEING FED DIFFERENT LEVELS OF MOLYBDENUM

J.B.J. van Ryssen and P.R. Barrowman
Department of Animal Science, University of Natal
P.O. Box 375, Pietermaritzburg, South Africa

High levels of molybdenum (Mo) fed for short periods of time can reduce the level of accumulated copper (Cu) in the livers of sheep (1). This
results in elevated levels of Cu and Mo in plasma and other tissues (2,3),
including the liver (4). Such liver Cu levels will not be a reliable indication of the Cu status of the animal since some time is required after
withdrawal of Mo to resume "normal" levels. Plasma Cu was found to reach
an elevated level within 12 days of feeding Mo and return to normal within
a similar period after withdrawal of Mo (5). The effect of high levels of
dietary Mo and the withdrawal of the Mo on Cu and Mo levels in plasma and
tissues of sheep was investigated. Two sheep breeds which differ in ability
to accumulate hepatic Cu were used.

PROCEDURE

The livers of 27 Ile de France (IdF, high Cu accumulator) and 32 SA
Mutton Merinos (MM, low Cu accumulator) (2) were loaded with Cu for 64 days.
Liver biopsies were then done and 3 IdF and 8 MM sheep were slaughtered.
The remainder were allocated to 3 treatments, grouping them within breed
to have similar average levels of hepatic Cu per treatment. The treatments
were 0 (0Mo), 25 (25 Mo) and 50 (50 Mo) mg Mo per kg feed, fed individually
to each sheep. At 64 days, liver biopsies were taken and 3 sheep per treatment were slaughtered. For a further 34 days the sheep recieved a ration
with no added Mo or Cu and were then slaughtered. Copper and Mo analyses
were done on their plasma, livers and kidney cortices.

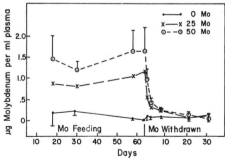

Figure 1. Changes in concentration of molybdenum
(Mo) in plasma of sheep during Mo feeding and after
withdrawal of dietary Mo. (Pooled results from
breeds).

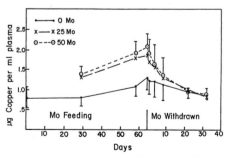

Figure 2. Changes in concentration of copper in
plasma of sheep during molybdenum (Mo) feeding and
after withdrawal of dietary Mo. (Pooled results from
breeds).

TABLE 1. Concentration of molybdenum (Mo, μg/g DM) and copper (Cu, μg/g DM) in the liver and kidney cortex of Ile de France (IdF) and SA Mutton Merino (MM) sheep fed high levels of Mo and after withdrawal of Mo.

Treatments		Molybdenum				Copper				
Mo	Breed	Liver		Kidney		Kidney		Liver		
		At days:						At days:		
		64	98*	64	98*	64	98*	0	64	98*
0	IdF	6	5	10	7	29	25	925	1012	867
	MM	7	5	12	6	29	25	750	581	478
25	IdF	27	13	210	66	297	97	942	845	832
	MM	25	16	143	44	198	72	737	499	428
50	IdF	58	37	284	85	400	122	940	681	697
	MM	34	24	200	66	259	101	743	575	474
SED		6	5	62	14	83	17	–	129	174
DF		12	24	12	24	12	24	–	42	24

* 34 days after Mo withdrawal.

RESULTS

During the Mo feeding period Cu content of the feed was 10 mg/kg and the sheep consumed 8, 23, and 45 mg Mo/day for the 0, 25, and 50 Mo treatments respectively. During the post-Mo period the concentrations of Mo and Cu in the ration were 1.5 and 12 mg/kg respectively.

Plasma Mo and Cu were elevated during Mo feeding but decreased abruptly after withdrawal of Mo (Fig. 1 & 2) to reach the 0Mo level within 8 and 16 days respectively. The reduction in liver Cu levels due to Mo feeding (Table 1) was more pronounced (P>0.05) in the IdF between the 25 and 50 Mo treatments than in the MM. Elevated levels of Cu and Mo in the kidney cortex and Mo in the livers decreased (P<0.01) after withdrawal of Mo, but were still higher (P<0.01) than the 0Mo treatment levels at day 34.

CONCLUSION

At 34 days after withdrawal of Mo, Cu and Mo levels in the tissues have not reached the levels of the 0Mo groups while hepatic Cu levels of the 0Mo group in both breeds decreased during this period. It was therefore not possible to establish the time required to reach "normal" Cu and Mo levels in tissues after withdrawal of Mo.

REFERENCES

1. van Ryssen, J.B.J., van Malsen, S. and Barrowman, P.R., 1986, S.A.J. Anim. Sci. 16:77.
2. Dick, A.T., 1956. In: Inorganic Nitrogen Metabolism, p. 445. Ed. W. D. McElroy and B. Glass, Johns Hopkins Press, Baltimore, MD.
3. Suttle, N.F., 1974. Proc. Nutr. Soc. 33:299.
4. van Ryssen, J.B.J. and Stielau, W.J., 1981, Br. J. Nutr. 45:203.
5. Smith, B.S.W. and Wright, H., 1975, J. Comp. Path. 85:299.
6. Harrison, T.J., van Ryssen, J.B.J. and Barrowman, P.R. 1987, S.A.J. Anim. Sci. 17 (in press).

INCREASED SULFIDE CONCENTRATION IN THE RUMEN FLUID OF SHEEP DOES NOT

DECREASE THE SOLUBILITY OF COPPER

H.Nederbragt, H.Kersten, M.Klomberg and A.J.Lagerwerf

Department of Pathology, Faculty of Veterinary Science
State University, Utrecht, The Netherlands

Sheep, bred under specific pathogen free (SPF) conditions died as a consequence of chronic copper (Cu) toxicity, although no source of Cu could be identified. Analysis of the rumen contents showed the absence of protozoa in the rumen and extremely low concentrations of volatile fatty acids. High concentrations of polyunsaturated fatty acids were found in adipose tissue. These symptoms point to a considerably changed rumen metabolism.

We investigated the relation between sulfide (S^{2-}) production and the solubility of Cu in the rumen of normal and SPF sheep. Cu toxicity in SPF sheep could be a consequence of increased Cu availability due to a limited synthesis of S^{2-} by the rumen flora; this is based on the same principle as that of Cu deficiency, caused by increased S^{2-} production[1].

From a normal rumen-fistulated sheep rumen samples were taken anaerobically and cultured in vitro under CO_2. S^{2-} was determined in the total sample, Cu was determined in the 35,000 g supernatant of the sample which was thought to represent soluble Cu. Rumen samples of SPF sheep were taken by stomach tube.

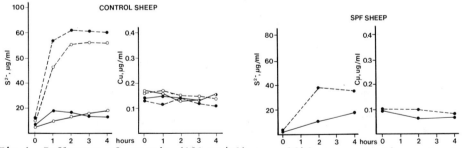

Fig.1. Influence of cysteine(180 μg/ml) on sulfide concentration and soluble Cu concentration of cultured samples of normal and SPF sheep. Open and closed circles represent different samples. ——: no cystein; --: cystein added

Addition of 180 μg cysteine /ml sample considerably increased total S^{2-} concentration in both normal and SPF sheep(Fig.1), indicating that the capacity of S^{2-} synthesis was also present in SPF sheep. Remarkably, the increase in S^{2-} did not result in a decrease in the soluble Cu concentration (Fig.1) as was expected.

This finding seemed to be confirmed by the results of an in vivo

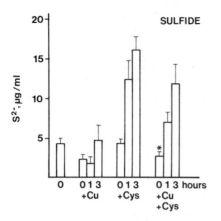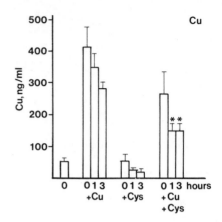

Fig. 2. Influence of addition of Cu (1 µg/ml) and cysteine (36 µg/ml) on the S^{2-} and Cu concentration of cultured rumen samples of a normal sheep. Asterisk indicates significant effect of Cu on S^{2-} (left fig.) or of crysteine on Cu (right fig.)

experiment of 24 hours in which rumen samples were taken at 90 min. intervals and analysed for S^{2-} and Cu directly. The S^{2-} concentration of the rumen fluid was increased temporarily by feeding hay ad lib. for a one hour period both in the morning and the afternoon; Cu concentrations were raised by the feeding of 500 g sheep concentrate, either in the morning or the afternoon period. Soluble Cu concentrations were influenced by the feeding of the concentrate only and not by the S^{2-} concentration of the rumen fluid.

Fig.2 shows that when ionic Cu, 1 µg/ml as $CuCl_2$, was added to rumen samples in vitro, only 30-40% of it was recovered as soluble Cu. This soluble Cu fraction was reduced significantly by addition of cysteine (36 µg/ml) to the culture which increased the S^{2-} concentration 3- to 4-fold. However, the final soluble Cu concentration always remained far above control levels. On the other hand, addition of ionic Cu also significantly reduced the S^{2-} concentration of the samples. Our results suggest that this is partly due to inhibition of S^{2-} production.

We propose that the Cu-S interaction in the rumen is not simply a consequence of S^{2-}, precipitating Cu as CuS, as was suggested recently[2]. When we correctly assume that our soluble Cu fraction represents Cu, available for absorption, than S^{2-} in the rumen seems to have little direct influence on this Cu fraction.

We conclude that, when viewing the Cu-S interaction in the rumen, compartimentalization of Cu and S has to be taken into consideration.

References

1. J.Hartmans and M.S.M.Bosman, Differences in the copper status of grazing and housed cattle and their biochemical backgrounds, in: "Trace Element Metabolism in Animals", C.F.Mills, ed., Livingstone, Edinburgh (1970) p.362.
2. M.Ivan, D.M.Veira and C.H.Kelleher, The alleviation of chronic copper toxicity in sheep by ciliate protozoa, Brit.J.Nutr.55:361 (1986)

EFFECT OF HIGH DIETARY IRON AND ASCORBIC ACID ON COPPER AND IRON

UTILIZATION DURING COPPER DEFICIENCY

Mary Ann Johnson and Cynthia Lee Murphy

Department of Foods and Nutrition
University of Georgia
Athens, GA 30602 USA

INTRODUCTION

The interactions of iron, ascorbic acid and copper may be of concern
to humans. Supplements of iron and ascorbic acid are consumed by nearly
30% and 60% of adults, respectively (1), while copper intakes may be
marginal (2). Consumption of 1500 mg ascorbic acid daily for 60 days
depressed serum ceruloplasmin and copper in adult men (3). The severity of
anemia in copper-deficient rats was increased by 1% ascorbic acid in the
diet (4). Ascorbic acid may increase the turnover or decrease the
absorption of copper (5). We observed that ascorbic acid also decreases
iron absorption in copper-deficient rats (4). It has been reported that
supplements of iron lessen the severity of copper deficiency anemia in rats
(6). However, high doses of iron do not reverse the anemia of copper
deficiency in humans (7,8) and we reported that anemia was more severe in
copper-deficient rats fed 226 rather than 54 µg Fe/g diet (9).

The effects of supplemental iron and ascorbic acid in copper-deficient
(Cu-) or copper-adequate (Cu+) rats were examined. Hematocrit, plasma
ceruloplasmin, liver Cu, Zn-superoxide dismutase (SOD), the % of absorbed
iron-59 in the red blood cells (RBC), and the apparent absorption of iron
and copper are reported in Table 1. Details of this experiment and changes
in other indices of iron and copper status are reported elsewhere (10,11).

METHODS AND MATERIALS

Male Sprague-Dawley rats (initial wt=49-61 g) were fed one of two
levels of copper (Cu-: 0.42 or Cu+: 5.74 µg Cu/g), iron (38 or 191 µg Fe/g)
and ascorbic acid (0 or 1% of the diet). Food intakes were limited to 11-
12 g/day. Iron-59 (4 µCi) was orally administered on day 15; rats were
killed on day 21. Tissue Cu, Zn-SOD was measured in ethanol/chloroform
extracts (10). Data were analyzed by two 2 X 2 ANOVA (Cu- and Cu+).

RESULTS AND DISCUSSION

Mean final body weights were not significantly different and averaged
151 g. Anemia in Cu- and ceruloplasmin in Cu+ were the most sensitive
indicators of exposure to high iron and ascorbic acid. Iron and copper
absorption were decreased by high ascorbic acid and iron, respectively, in
Cu-. Anemia in Cu- was associated with a marked reduction in RBC iron-59.

Table 1. Interactions of Dietary Copper, Iron and Ascorbic Acid in Rats

Diets		Hematocrit		Ceruloplasmin		Liver SOD		RBC Iron-59		Apparent Absorption Iron		Copper	
Ascorbic Acid	Iron	Cu-	Cu+	Cu-	Cu+	Cu-	Cu+	Cu-	Cu+	Cu-	Cu+	Cu-	Cu+
%	µg/g diet	%		IU/L		units/mg protein		% of absorbed		% of intake			
0	38	38	44	ND	91	3.8	8.2	46	53	48	44	27	22
0	191	36	46	ND	86	2.7	7.5	19	38	23	24	16	23
1	38	34	45	ND	82	3.0	7.5	38	46	39	43	20	14
1	191	29	44	ND	51	1.9	6.5	4	31	17	22	7	17
Pooled SEM:		2	1	-	6	0.4	0.4	5	4	4	3	5	3
ANOVA p-vales:													
Ascorbic Acid		0.01	NS	-	0.01	0.06	NS	0.05	NS	0.05	NS	NS	NS
Iron		0.06	NS	-	0.05	0.02	NS	0.01	0.01	0.001	0.001	0.05	NS
Interaction		NS	NS	-	NS	NS	NS	NS	NS	NS	NS	NS	NS

Means, n=5-6

It is not known if decreased RBC uptake of iron-59 or increased RBC turnover led to reduced RBC iron-59. Perhaps the impaired antioxidant status in Cu- (ie., decreased ceruloplasmin and SOD) increased the sensitivity of the RBC to the oxidative effects of iron and ascorbic acid.

REFERENCES

1. Garry, P. J., Goodwin, J. S., Hunt, W. C., Hooper, E. M. & Leonard, A. (1982) Nutritional status in a healthy elderly population: dietary and supplemental intakes. Am J. Clin. Nutr. 36:319-331.
2. Klevay, L. M., Reck, S. J. & Barcome, D. F. (1979) Evidence of dietary copper and zinc deficiencies. J. Amer. Med. Assoc. 241:1916-1918.
3. Finley, E. B. & Cerklewski, F. L. (1983) Influence of ascorbic acid supplementation on copper status in young adult men. Am. J. Clin. Nutr. 37:553-556.
4. Johnson, M. A. (1986) Interaction of dietary carbohydrate, ascorbic acid and copper with the development of copper deficiency in rats. J. Nutr. 116:802-815.
5. Van Campen, D. & Gross, E. (1968) Influence of ascorbic acid on the absorption of copper by rats. J. Nutr. 95:617-622.
6. Weisenburg, E., Halbreich, A. & Mager, J. (1980) Biochemical lesions in copper-deficient rats caused by secondary iron deficiency. Biochem J. 188:633-641.
7. Seely, J. R., Humphrey, G. B. & Matter, B. J. (1972) Copper deficiency in a premature infant fed an iron-fortified formula New Eng. J. Med. 286:109-110.
8. Karpel, J. T. & Peden, V. H. (1972) Copper deficiency in long-term parenteral nutrition. J. Pediatr. 30:32-36.
9. Johnson, M. A. & Hove, S. S. (1986) Development of anemia in copper-deficient rats fed high levels of dietary iron and sucrose. J. Nutr. 116:1225-1238.
10. Johnson, M. A. & Murphy, C. L. (1987) The adverse effects of high dietary iron and ascorbic acid on copper utilization by copper-deficient and copper-adequate rats. Am. J. Clin. Nutr. (in press).
11. Johnson, M. A. & Murphy, C. L. (1987) Anemia associated with changes in iron and iron-59 utilization in copper-deficient rats fed high levels of dietary ascorbic acid and iron. Biological Trace Elem. Res. (in press).

INTERACTION OF GOLD AND SELENIUM AFTER INJECTION TO MICE

J. Aaseth[1], J. Alexander[2] and E. Steinnes[3]

[1]Hedmark Central Hospital, Elverum, Norway
[2]National Institute of Public Health, Oslo, Norway
[3]University of Trondheim, Trondheim, Norway

INTRODUCTION

Interactions between metals and selenium have been studied extensively since the late sixties. Results of research focused on silver, copper, cadmium and mercury have indicated that in vivo, selenium is trapped in insoluble complexes by group IB or IIB metals, with or without the involvement of protein molecules (Høgberg and Alexander, 1986). However, the theoretical possibility of in vivo interactions between selenium and gold has been remarkably little studied. This is surprising since gold thiomalate (Myocrisin) is a frequently used drug in patients with rheumatoid arthritis. And such patients are reported to have lower serum selenium values than healthy controls (Aaseth et al. 1978). A possible interference of gold with the selenium metabolism might be of relevance for the therapeutic effect and/or the toxic effects observed during gold treatment. In the present study we have used mice as experimental model when searching for biological selenium-gold interactions.

MATERIALS AND METHODS

Female mice, NMRI strain, weighing 20 ± 1 g, were used. Groups of 5 mice were given either sodium selenite i.p. + gold thiomalate i.m. or one of these metal compounds + saline or saline alone. The doses used of selenite and of gold was 300 nmol per mouse. Eight days after the last injection the animals were autopsied, and the organ distribution of selenium and gold was determined by using the neutron activation method.

RESULTS

The selenite injection did not change the levels of selenium in brain, liver or kidney (table 1), and only a marginal increase ($p > 0.05$) was observed in blood selenium. The gold thiomalate injection did not influence the selenium levels in blood, brain or liver significantly, but raised the kidney selenium from 18.5 to 21.7 nmol/g in average ($p < 0.05$, Wilcoxon two-sided test).

Table 1. Organ levels of selenium (nmol/g, mean and range), in three mouse groups.

	Blood	Kidneys	Liver	Spleen	Brain
Controls (given	4.1	18.5	24.9	8.4	3.6
0.9% saline only)	(3.7-4.7)	(15.9-20.2)	(21.1-27.4)	(5.5-10.3)	(2.6-4.6)
Saline + selenite	4.5	18.5	23.4	7.5	3.4
	(3.9-5.0)	(16.1-20.0)	(20.6-27.5)	(5.1-9.7)	(2.4-4.2)
Gold + selenite	4.7	21.7	23.1	7.7	2.7
	(3.9-5.2)	(20.6-24.2)	(19.9-25.5)	(5.2-8.9)	(2.1-3.4)

The gold levels in brain, blood and liver, which in averange were 0.12, 0.6 and 10.1 nmol/g, respectively, eight days after the gold thiomalate injection, were not changed significantly by the selenite administration. However, the selenite injection raised the kidney levels of gold from 27.1 (21.1-31.1) nmol/g to 34.3 (31.3-36.7) nmol/g.

DISCUSSION

Administration of Myocrisin is known to lead to protein-bound Au^+ in the circulation (Jellum et al. 1980). Subsequently Au^+ is redistributed intracellularly, and complexed to various thiolcontaining and seleniumcontaining proteins (see Pearson 1968), possibly including glutathione peroxidase (Chaudiere and Tappel 1984). In the present study, combined injection of Myocrisin and selenite increased the kidney level of gold with about 7 nmol/g, and that of selenium with about 3 nmol/g. Formation in vivo of Au_2Se-complexes might explain this observation. Such interaction is analogues to the previously demonstrated in vivo formation of Ag_2Se (Aaseth et al. 1981) and the retention of Ag^+ in the kidneys when coadministered with selenite (Alexander and Aaseth, 1981). Whether or not protein molecules are involved in the (relatively scanty) retention of selenium and gold in the kidneys is unknown. And a possible clinical relevance of an interaction between gold and seleniumcontaining molecules remains to be studied (see Chaudiere and Tappel 1984).

REFERENCES

Aaseth, J., Munthe, E., Førre, Ø. and Steinnes, E., 1978, Trace elements in serum and urine of patients with rheumatoid arthritis. Scand J Rheumatology, 7: 237.
Aaseth, J., Olsen, A., Halse, J. and Hovig, T., 1981, Argyria - tissue deposition of silver as selenide. Scand J clin Lab Invest., 41: 247
Alexander, J. and Aaseth, J., 1981, Hepatobiliary transport and organ distribution of silver in the rat as influenced by selenite. Toxicology, 21: 179
Chaudiere, J. and Tappel, A.L., 1984, Interaction of gold(I) with the active site of selenium-glutathione peroxidase. J Inorg Biochemistry, 20: 313

Høgberg, J. and Alexander, J., 1986, Selenium, in: "Handbook on the toxicology of metals", L. Friberg, G.F. Nordberg and V.B. Vouk, eds.,
   Elsevier, Amsterdam
Jellum, E., Munthe, E., Guldahl, G. and Aaseth, J., 1980, Fate of the
   gold and the thiomalate part after intramuscular administration of
   aurothiomalate to mice. Ann Rheum Dis., 39: 155.
Pearson, R.G., 1968, Hard and soft acid and bases - HSAB. J Chem Educ.
   45: 643

EFFECTS OF ARSENIC ON THE INTESTINAL ABSORPTION OF ^{75}Se COMPOUNDS IN CHICKS

Hannu Mykkänen

Department of Nutrition
University of Helsinki
00710 Helsinki, Finland

INTRODUCTION

The protective action of arsenic against selenium toxicity has been examined by reduced soft tissue Se levels and enhanced biliary excretion of Se (Levander & Argrett, 1969; Howell & Hill, 1978; Palmer et al., 1983). Another possibility, altered intestinal absorption of Se during exposure to As, has not been explored. The purpose of the present experiments was to investigate the effects of As on the intestinal absorption of Se compounds.

METHODS

The absorption of ^{75}Se was determined in 3 wk old white Leghorn cockerels using the in vivo ligated duodenal loop procedure. The animals were raised after hatching on a commercial chick starter diet (Exp 1) or on a purified basal diet containing 0, 5 or 50 mg added As as $NaAsO_2$ per kg of diet (Exp 2). The intraduodenal dose (0.5 ml/animal) contained 0.05 µCi ^{75}Se, 0.01 mM Se and 150 mM NaCl, pH 6.5. The injected Se compounds were sodium selenite, sodium selenate and selenomethionine (Exp 1), and sodium selenite (Exp 2). The As concentration in the dose was 1 mM As as $NaAsO_2$ (Exp 1). The absorption was 30 min for ^{75}Se-selenite and -selenate, and 10 min for ^{75}Se-methionine. The following parameters of absorption were determined: 1) "total absorption" which is 100 - ^{75}Se (% dose) in the intestinal lumen; 2) "intestinal retention" which is the amount of ^{75}Se (% dose) retained by the intestinal tissue; 3) "transfer to body" which is the amount of ^{75}Se (% dose) accumulated by the liver. The effect of As on the solubility of ^{75}Se in the intestinal lumen was determined by centrifuging a sample of the luminal fluid at 5000 x g for 10 min and counting an aliquot of the supernatant fraction for ^{75}Se.

RESULTS AND DISCUSSION

The presence of 1 mM As in the dose significantly reduced the total absorption of ^{75}Se-selenite and enhanced accumulation of the label in the intestinal tissue (Table 1). None of the absorption parameters of ^{75}Se-methionine were influenced by As, while As drastically reduced the transfer of ^{75}Se-selenate to body. The latter finding was, however, not accompanied by any change in the total absorption, indicating that As alters tissue distribution of selenate rather than reduces selenate absorption. Exposure to 50 mg As/kg diet prior to determination of absorption resulted in enhanced intestinal accumulation of ^{75}Se-selenite, and consequently less ^{75}Se was transferred to body (Table 2).

The data of Exp 1 indicate that As interacts both with the mucosal uptake process of selenite and with the transfer of selenite from the epithelial cell into the circulation. Competition for mucosal binding sites due to similarity of chemical structure and/or formation of Se-As complexes in the intestinal lumen or within the intestinal cells could explain these findings. These data are in agreement with our previous results showing that intestinal interactions of Se with Pb (Mykkänen & Humaloja, 1984), Cd (Mykkänen, 1986) and Hg (Mykkänen & Metsäniitty, in press) are dependent on the chemical form of Se. The data of Exp 2 suggest that Se-As interaction can be explained, at least partly, by the reduced capacity of the intestinal tissue to transport Se. The mechanism of this interaction should be studied further.

Table 1. Effect of Arsenic on the Intestinal Absorption of ^{75}Se-selenite, ^{75}Se-selenate and ^{75}Se-methionine in Chicks (Exp 1)

Intraduodenal dose (mM)	Total absorption	Intestinal retention [75]Se (% dose)	Transfer to body	Luminal solubility (%)
Selenite	91.0±1.4[a]	24.1±1.4[a]	36.4±2.4[a]	86.4±1.8[a]
" + 1 mM As	56.3±2.5[b]	36.7±1.4[b]	4.7±0.7[b]	81.3±1.7[a]
Selenate	90.0±2.0[a]	19.1±0.7[a]	21.5±1.4[a]	87.9±1.3[a]
" + 1 mM As	82.1±3.0[a]	16.6±0.3[b]	3.8±0.2[b]	77.4±3.9[b]
Se-methionine	87.9±0.7[a]	15.8±1.0[a]	14.5±0.9[a]	89.7±0.8[a]
" + 1 mM As	85.2±0.8[a]	16.6±0.6[a]	17.3±1.0[a]	89.7±1.3[a]

Each value is the mean ± sem of 5-6 animals. Values differing significantly from the appropriate control value in each column are designated by different superscripts (P<0.05).

Table 2. Effect of Dietary Exposure to Arsenic on the Intestinal Absorption of ^{75}Se-selenite in Chicks (Exp 2)

Arsenic added to diet (mg/kg)	Total absorption	Intestinal retention	Transfer to body
0	90.5±2.5[a]	23.8±3.2[a]	31.6±4.5[a]
5	85.2±2.9[a]	26.1±5.8[ab]	25.8±4.8[ab]
50	88.7±1.3[a]	31.1±2.0[b]	23.1±2.1[b]

Each value is the mean ± sem of 5-6 animals. Values differing significantly from the control value in each column are designated by different superscripts (P<0.05).

REFERENCES

Howell, G.O. & Hill, C.H., 1978, Biological interaction of selenium with other trace elements in chicks, Environ Health Perspect, 25:147.
Levander, O.A. & Argrett, L.C., 1969, Effects of arsenic, mercury, thallium, and lead on selenium metabolism in rat, Toxicol Appl Pharmacol, 14:308.
Mykkänen, H.M., 1986, Effects of cadmium on intestinal absorption of ^{75}Se compounds in chicks. 1st Meeting of the Int'l Society for Trace Element Research in Humans (ISTERH), Annenberg Center, California, Dec 8-12, 1986.
Mykkänen, H.M. & Humaloja, T., 1984, Effect of lead on the intestinal absorption of sodium selenite and selenomethionine (^{75}Se) in chicks, Biol Trace Element Res, 6:11.
Mykkänen, H.M. & Metsäniitty, L., in press, Selenium-mercury interaction during intestinal absorption of ^{75}Se compounds in chicks, J Nutr.
Palmer, I.S., Thix, N., & Olson, O.E., 1983, Dietary selenium and arsenic effects in rats, Nutr Rep Int, 27:249.

CHROMIUM AND VANADATE SUPPLEMENTATION

OF OBESE AND LEAN MICE

B.J. Stoecker and Y.C. Li

Texas Tech University
P.O. Box 4170, Lubbock, Texas 79409

INTRODUCTION

Chromium (Cr) is essential for normal carbohydrate and lipid metabolism in both animals and humans. Severe chromium deficiency in experimental animals may contribute to insulin resistance, impaired glucose tolerance and elevated serum cholesterol. Vanadium also has been reported to be
a nutritionally important element for both chicks and rats, but its function and even its essentiality are still in question. The objectives of
the present study were to investigate the effects of Cr and/or vanadate
supplementation on Cr and vanadium retention in bone and kidney and on
serum total cholesterol, insulin, and glucose concentrations of obese and
lean mice. Possible interactions of chromium and vanadate affecting the
above parameters were also of interest.

MATERIALS AND METHODS

Animals and diets

Twenty 5-week-old male C57BL/6J obese mice and their lean controls
were studied using a split plot design with 2x2 factorial subplots. Obese
animals and lean animals were assigned at random into experimental groups:
low chromium and low vanadate (-Cr-V), low chromium and vanadate supplemented (-Cr+V), chromium supplemented and low vanadate (+Cr-V), and chromium supplemented and vanadate supplemented (+Cr+V).

All animals were housed in plastic shoebox cages and fed ad libitum
from ceramic feed cups. Distilled, deionized water in glass or plastic
bottles with glass sipper tubes was provided. After 55 days on the experimental diets, mice were fasted for 15 hours and decapitated. Serum and
tissues were stored at -70°C.

Analytical Methods

For chromium and vanadium analyses, samples of the diet, bone, and
kidney were dried and were ashed in a muffle furnace to a maximum
temperature of 480°C (1). Trace mineral concentrations were determined
using a Model 5000 atomic absorption spectrophotometer with graphite
furnace and Zeeman background correction (Perkin Elmer, Norwalk, CT).

Serum glucose was measured with a glucose analyzer (Beckman Instruments, Inc., Fullerton, CA) using the glucose oxidase method (2). Total serum cholesterol was determined enzymatically (Sigma Chemical Company, St. Louis, MO) (3). Insulin concentration was measured by a double-antibody radioimmunoassay method (Cambridge Medical Diagnostics, Inc., Billerica, MA). Analysis of variance and means separation tests at the 0.05 alpha level were performed using the Statistical Analysis System (SAS).

RESULTS AND DISCUSSION

No significant differences due to diet were found in final body weights. Likewise, serum insulin, glucose and cholesterol concentrations of obese mice were not statistically different. However, chromium appeared to have a significant effect on I/G ratios. Chromium content of the kidney of obese mice did not differ significantly between groups. Bone chromium was significantly higher ($p < 0.005$) in chromium-supplemented groups (+Cr-V and +Cr+V) vs. chromium-deprived animals (-Cr-V and -Cr+V). Vanadium concentrations of both bone and kidney were statistically greater ($p < 0.0001$) in vanadate-supplemented groups (-Cr+V and +Cr+V) than in groups fed low vanadate diets (-Cr-V and +Cr-V).

In the lean (control) mice, there were, likewise, no significant differences between groups based on mean body weights or serum insulin, glucose or cholesterol concentrations. Bone Cr concentration was significantly lower in obese than in lean mice (32 vs. 55 ng Cr/g dry wt). Mice which consumed chromium supplemented diets (+Cr-V and +Cr+V) had significantly higher ($p < 0.005$) bone Cr concentrations than the mice fed the low Cr diets (-Cr-V and -Cr+V). Chromium content of kidney of C57BL/6J mice did not differ significantly among groups. Bone and kidney vanadium contents were lower ($p < 0.0001$) in the -Cr-V and +Cr-V groups than in the -Cr+V and +Cr+V groups. Chromium addition did not affect the bone V of the -V groups, however in the +V groups, chromium-deprived animals had higher bone vanadium than did the chromium-supplemented mice. In both -Cr and +Cr groups, vanadate-supplemented groups had significantly greater bone vanadium content than those of the vanadium-deprived groups. Bone vanadium was highly correlated with renal vanadium ($p < 0.0001$) in both obese and lean mice. In this study, bone appeared to be an appropriate tissue in which to evaluate both chromium and vanadium retention.

REFERENCES

1. Li, Y.C. and Stoecker, B.J. Chromium and yogurt effects on hepatic lipid and plasma glucose and insulin of obese mice. Biol. Trace Elem. Res. 9:233-242, 1986.
2. Kadish, A.H., Litle, R.L. and Sternberg, J.C. A new and rapid method for the determination of glucose by measurement of rate of oxygen consumption. Clin. Chem. 14:116, 1969.
3. Allain, C.C., Poon, L.S., Chan, C.S.G., Richmond, W. and Fu, T.C. Enzymatic determination of total serum cholesterol. Clin. Chem. 20:4700, 1974.

THE EFFECT OF PARENTERAL IRON ADMINISTRATION UPON MANGANESE

METABOLISM

Nevenka Gruden

Institute for Medical Research and Occupational
Health, Zagreb, Yugoslavia

INTRODUCTION

The importance of iron and manganese as the essential dietary components is well established. Iron deficiency anaemia could be prevented by
iron-fortified diet - yet the inhibitory effect of such diet on manganese
absorption is well recognized (Thomson et al., 1971; Gruden, 1977; Gruden,
1982). The aim of this study was to find out whether parenteral iron administration affects manganese absorption and to what extent.

ANIMALS AND METHODS

Three-week-old female albino rats were placed in four groups of
twelve animals each according to the iron amount given intraperitoneally
daily for four days: (1) pure saline, (2) 100 µg, (3) 400 µg and (4) 1600
µg Fe/d/rat. On the fourth day manganese-54 in 0.4 ml of cow's milk was
administered to all animals by artificial feeding (Momčilović and Rabar,
1979). Three days later the animals were killed and the activity was determined in the whole body and some organs.

RESULTS AND DISCUSSION

The results for manganese-54 retention, expressed in percentage of
the administered dose, are the arithmetic means with their standard errors
(Table 1).

Manganese-54 absorption and deposition in the liver and spleen were
not significantly altered in the animals treated with 100 ug Fe daily.
The highest iron dose (1600 µg Fe/d) decreased manganese absorption by 70
per cent and its deposition in the tissues by 44-60 per cent.

In the present experiments the animals received parenterally substantially smaller doses of iron than (orally) in our previous experiments
(Gruden, 1982). Nevertheless, the dose which suppressed the absorption and
retention of manganese very efficiently - 1600 µg Fe/d, Table 1.- was 2.5
times smaller than the one given orally without any effect upon manganese
metabolism. The data presented call for additional caution when administering iron parenterally.

Table 1. Manganese-54 whole body, carcass, liver and spleen activities in three-week-old rats that received iron by intraperitoneal injections for three days before radioisotope administration[*]

Iron dose (µg/d)	Whole body	Carcass	Liver	Spleen
0	31.3 ± 1.8	15.1 ± 1.3	5.30 ± 0.39	0.038 ± 0.003
100	$27.2 \pm 2.4_s$	$13.9 \pm 1.0_s$	5.13 ± 0.38	0.037 ± 0.005
400	22.6 ± 2.3^s	10.2 ± 0.9^s	$4.55 \pm 0.43_s$	$0.032 \pm 0.004_s$
1600	8.6 ± 1.1^s	6.1 ± 0.8^s	2.99 ± 0.39^s	0.018 ± 0.003^s

[*] Percentage of the dose. Mean $(\bar{x}_{12}) \pm$ S.E.
[s] Values bearing the superscript letter are significantly different from the corresponding control values (P<0.05).

ACKNOWLEDGEMENT

The financial support of the Scientific Research Council of S.R. Croatia is gratefully acknowledged. The author thanks Mrs Mirka Buben for her valuable technical assistance, Mrs Neda Banić for correcting the language of the manuscript and Mrs Milica Horvat for preparing the manuscript.

REFERENCES

Gruden, N., 1977, Suppression of transduodenal manganese transport by milk diet supplemented with iron, Nutr. Metab., 21:305.

Gruden, N., 1982, Iron-59 and manganese-54 retention in weanling rats fed iron fortified milk, Nutr. Rep. Intern., 25:849.

Momčilović, B., and Rabar, I., 1979, Artificial feeding of infant rats, Period. Biol., 81:27.

Thomson, A.B.R., Olatunbosun, D., and Valberg, L.S., 1971, Interrelation of intestinal transport system of manganese and iron, J. Lab. Clin. Med., 78:642.

THE EFFECT OF AGE AND SEX UPON IRON-MANGANESE INTERACTION IN DIFFERENT SEGMENTS OF THE RAT'S INTESTINE

Nevenka Gruden

Institute for Medical Research and Occupational
Health, Zagreb, Yugoslavia

INTRODUCTION

Although the knowledge of iron-manganese interaction has improved a great deal, some facts are still unknown. The aim of the present study was to find out how this interaction depends upon animals' age and sex and whether it is the same in the whole small intestine.

ANIMALS AND METHODS

Experiments were performed on 5- and 25-week-old albino rats of both sexes. One half of the animals were fed pure cow's milk and the other half cow's milk fortified with iron (10 mg/100 ml), for three days. All animals were killed on the fourth day and manganese-54 transfer and intestinal retention were determined in vitro in different segments of the small intestine by the Wilson and Wiseman's method (1954). The duodenal segment was taken directly from the pyloroduodenal connection, the jejunal 20 cm distantly and the last, ileal, 20 cm above the ileocaecal connection.

RESULTS AND DISCUSSION

The inhibitory effect of iron upon manganese absorption was significant only in the proximal parts of the small intestine, where manganese transfer was by 27-50 per cent lower in iron treated animals than in those fed pure cow's milk. At the same time, the inhibition was more pronounced in the young than in old animals (manganese transfer being by 33 and 50 per cent lower in the 5-week-old and by 22 and 37 per cent lower in the 25-week-old iron treated rats) and in female than in male animals. The effect of iron on manganese retention was also highest in the duodenum, but on the whole , it was always lower than on manganese transfer. Being highest in the duodenum of young female rats, the inhibitory effect of iron upon manganese transfer is obviously correlated with body demand for iron and with the part of the intestine where iron absorption is most efficient and an active one.

557

ACKNOWLEDGEMENT

The financial support of the Scientific Research Council of S.R. Croatia is gratefully acknowledged. The author thanks Mrs Mirka Buben for her valuable technical assistance and Mrs Milica Horvat for preparing the manuscript.

REFERENCES

Wilson, T.H., and Wiseman, G., 1954, The use of sacs of everted small intestine for the study of the transference of substances from the mucosal to the serosal surface, J. Physiol., 123:116.

MINERAL-ZINC AND PROTEIN-ZINC INTERACTIONS IN MAN

Herta Spencer, S.J. Sontag and D. Osis

V.A. Hospital
Box 35
Hines, IL   60141

Interaction of calcium with zinc leads to decreased zinc absorption in animals[1] and parakeratosis[2] due to excess dietary pytic acid[3]. Phosphorus intensified this effect[4]. In man, dairy products containing calcium and phosphorus, decrease zinc absorption[5]. However, in this Research Unit, a 10-fold increase in calcium intake did not decrease zinc absorption[6,7]. In the present study the effect of calcium, phosphorus and protein on zinc metabolism was investigated in man and the effect of zinc on the absorption of calcium and iron.

MATERIALS AND METHODS

Metabolic balances of zinc were determined for several weeks in the Metabolic Research Ward in normal, fully ambulatory males. The analyzed diet[8] contained 15 mg zinc, 230 mg calcium and 850 mg phosphorus.

In the experimental studies the calcium and phosphorus intakes were increased to 2000 mg/day by adding calcium gluconate tablets and glycerophosphate. In studies of the effect of protein, the dietary protein was increased from 1 gm/kg body weight to 2 gm/kg. The effect of zinc (140 mg/day) on calcium absorption was studied with tracer doses of ^{47}Ca and the effect of 55 mg and 140 mg zinc on iron absorption with tracer doses of ^{59}Fe citrate. The diet, complete collections of urine and stool, calcium gluconate tablets, and glycerophosphate were analyzed for zinc and calcium by atomic absorption spectroscopy[9] and phosphorus according to Fiske and SubbaRow[10]. Net or apparent absorption was determined from zinc intake and fecal excretions.

RESULTS

A.   Effects of Mineral and Protein on Zinc Metabolism

1.   Effect of calcium and phosphorus. Zinc balance and ^{65}Zn studies have shown that calcium and phosphorus, in amounts up to 2000 mg/day, did not significantly affect the zinc balance or the net absorption. Phosphorus had no significant effect, regardless of the calcium intake.

Table 1 shows examples of the lack of effect of calcium and phosphorus on the zinc balance.

TABLE 1. EFFECT OF CALCIUM AND PHOSPHORUS ON THE ZINC BALANCE

DAYS	Ca INTAKE mg/day	P INTAKE mg/day	Zinc, mg/day			
			Intake	Urine	Stool	Balance
18	200	800	13.81	0.80	12.57	+ 0.44
48	2000	800	14.64	0.86	12.19	+ 1.59
48	2000	2000	15.22	0.79	12.98	+ 1.45

2. Effect of protein. Table 2 shows that the net absorption of zinc was significantly greater during a high than a normal protein intake. During a low protein-zinc intake, the zinc balances were negative (Table 3).

TABLE 2. EFFECT OF PROTEIN ON NET ABSORPTION OF ZINC

Protein Intake	Net Absorption %
Normal+	7
High*	19

+ NITROGEN Intake = 12 gm/day

* Nitrogen Intake = 24 gm/day

TABLE 3. EFFECT OF A LOW PROTEIN DIET ON THE ZINC BALANCE

Zinc Intake mg/day	Zinc Balance mg/day
6.5	-1.3
7.3	-1.8
8.1	-1.1

Protein Intake=6-8 gm NITROGEN/day

B. Effect of Zinc on Calcium and Iron Metabolism

1. Effect on calcium absorption. Large doses of zinc (140 mg) during a low calcium intake of 200 mg/day decreased intestinal calcium absorption but not during higher calcium intakes of 800 or 1100 mg/day.

2. Effect on iron absorption. ^{59}Fe absorption was the same during a dietary zinc intake of 15 mg/day and during the addition of 55 or 100 mg zinc.

DISCUSSION

In a study by others in man, calcium and phosphorus decreased zinc absorption evaluated by zinc plasma levels[5]. A review of the zinc require- ment elaborates on various factors, including calcium and phosphorus[11]. Since animal data can not be readily extrapolated to man, it is important to determine directly in man the absorption of zinc during different mineral intakes. The present controlled studies have shown that increasing the phosphorus intake by a factor of 2.5 did not significantly change the zinc balance or net absorption, regardless of the calcium intake[7].

During a high protein intake the net absorption of zinc was signifi- cantly higher than during a normal protein intake. During a low protein- low zinc intake all zinc balances were negative. Others reported that a higher protein intake increased the zinc balance slightly but not signifi- cantly[12], while other studies reported a significant increase[13,14], but the phosphorus intake played a role[14]. The effect of zinc on calcium absorption indicates competition of zinc with calcium for binding sites for absorption, leading to decreased calcium absorption during a low but not higher calcium intakes[15]. Non-heme iron decreased zinc absorption[16] as judged from zinc plasma levels. In the prsent study, in which the effect of zinc on iron absorption was examined, iron absorption remained unchanged during the high zinc intake in most cases.

REFERENCES

1. R. M. Forbes, Nutritional interactions of zinc and calcium, Fed. Proc. 19:643 (1960).
2. P. K. Lewis, W.G. Hoekstra and R.H. Grummer, Restricted calcium feeding versus zinc supplementation for the control of parakeratosis in swine, J. Animal. Sci. 16:578 (1957).
3. B. L. O'Dell, and J.E. Savage, Effect of phytic acid on zinc availability, Proc. Soc. Exp. Biol. Med. 103:304 (1960).
4. D. A. Heth, W. M. Becker, and W. G. Hoekstra, Effect of calcium, phosphorus and zinc on zinc-65 absorption and turnover in rats fed semipurified diets, J. Nutr. 88:331 (1966).
5. A. Pecoud, P. Donzel, and J. L. Schelling, Effect of foodstuffs on the absorption of zinc sulfate, Clin. Pharm. Therap. 17:469 (1975).
6. H. Spencer, V. Vankinscott, I. Lewin, and J. Samachson, Zinc-65 metabolism during low and high calcium intake in man, J. Nutr. 86:169 (1965).
7. H. Spencer, L. Kramer, C. Norris, and D. Osis, Effect of calcium and phosphorus on zinc metabolism in man, Am. J. Clin. Nutr. 40:1213 (1984).
8. D. Osis, L. Kramer, E. Wiatrowski, and H. Spencer, Dietary zinc intake in man, Am. J. Clin. Nutr. 25:582 (1972).
9. J. B. Willis, Determination of lead and other heavy metals in urine by atomic absorption spectroscopy, Analyt. Chem. 34:614 (1962).
10. C. H. Fiske, and Y. T. SubbaRow, The colorometric determination of phosphorus, J. Biol. Chem. 66:375 (1925).
11. J. A. Halsted, J. C. Smith, Jr., and M. I. Irwin, A conspectus of research on zinc requirements of man, J. Nutr. 104:345 (1974).
12. J. R. Mahalko, H. H. Sandstead, L. K. Johnson, and D. B. Milne, Effect of a moderate increase in dietary protein on the retention and excretion of Ca, Cu, Fe, Mg, P, and Zn by adult males, Am. J. Clin. Nutr. 37:8 (1983).
13. N. W. Solomons, M. Janghorbani, B. T. G. Ting, F. H. Steinke, M. Christensen, R. Bijlani, N. Istfan, V. R. Young, Bioavailability of zinc from a diet based on isolated soy protein: Application in young men of the stable isotope tracer, 70Zn, J. Nutr. 112:1809 (1982).
14. J. L. Greger, and S. M. Snedeker, Effect of dietary protein and phosphorus levels on the utilization of zinc, copper, and manganese by adult males, J. Nutr. 110:2243 (1980).
15. H. Spencer, N. Rubio, L. Kramer, C. Norris, and D. Osis, Effect of zinc supplements on the intestinal absorption of calcium, J. Am. Coll. Nutr. 6:47 (1987).
16. N. W. Solomons, and R. A. Jacob, Studies on the bioavailability of zinc in humans. Effects of heme and non-heme iron on the absorption on zinc, Am. J. Clin. Nutr. 34:475 (1981).

MINERAL BALANCE STUDY OF RATS FED MAILLARD REACTION PRODUCTS

J.M. O'Brien[*], P.A. Morrissey[+] and A. Flynn[+]

Departments of Food Chemistry[*] and Nutrition[+]
University College, Cork, Ireland

INTRODUCTION

Recent evidence suggests that certain products of the Maillard browning reaction in food may disturb mineral metabolism in humans[1] and in laboratory animals[2]. Stegink and co-workers[1] reported increases in urinary zinc, copper and iron in human subjects receiving Maillard reaction products intravenously. In this laboratory, dietary Maillard products of glucose/glutamate elevated the urinary excretion of calcium, magnesium, copper, zinc and sodium in rats[2]. The present study was conducted to examine the effects of Maillard reaction products of glucose/glutamate on mineral balance in the rat. Glutamic acid was selected as a source of amino groups because of its abundance as a free amino acid in foods.

MATERIALS AND METHODS

Maillard Reaction Product (MRP):

MRP was prepared as described previously[2] from glucose and monosodium glutamate.

Balance Study:

Groups of 8 female Wistar albino rats were fed either a diet containing 0.5%, w/w, MRP or a control diet containing unheated monosodium glutamate/ glucose for 12 days. The study was conducted as described previously[2].

RESULTS AND DISCUSSION

Twenty four hour urinary mineral output data (Table 1) for animals fed 0.5% MRP exhibited increases in values for calcium (62%), magnesium (65%), sodium (17%), zinc (158%) and copper (24%). In addition, dietary MRP appeared to have a negative effect on zinc retention compared to the control diet (Table 2). The latter finding may be explained by increased urinary and faecal zinc losses in rat fed MRP. In the case of calcium and magnesium faecal losses actually decreased slightly, which compensated for the higher urinary levels.

TABLE 1. 24-HOUR URINARY MINERAL OUTPUT VALUES FOR RATS FED CONTROL
OR 0.5% MRP DIETS

| | ---------------DIET--------------- | |
Mineral	URINARY OUTPUT / DIETARY INTAKE Control	0.5% MRP
Calcium (μg/mg)	8.12 ± 2.05[b]	13.13 ± 2.51[b]
Magnesium (mg/mg)	0.246 ± 0.018[a]	0.406± 0.037[a]
Phosphorus (mg/mg)	0.693 ± 0.012	0.694± 0.020
Sodium (mg/mg)	0.589 ± 0.020[a]	0.690± 0.022[a]
Potassium (mg/mg)	0.657 ± 0.027	0.617± 0.027
Zinc (ng/μg)	5.63 ± 0.67[c]	14.55 ± 3.81[c]
Copper (ng/μg)	14.28 ± 0.86[b]	17.77 ± 1.51[b]

Means ± SEM for 8 determinations. Values with a letter in common,
for a given parameter, differ to a statistically significant
extent (Students t-test), [a], $p < 0.001$; [b], $p < 0.02$; [c], $p < 0.05$.

TABLE 2. MINERAL RETENTION VALUES FOR RATS FED CONTROL OR 0.5%
MRP DIETS

Mineral	--------------Diet--------- % Retained	
	Control	0.5% MRP
Calcium	70.0 ± 2.1	71.8 ± 1.3
Magnesium	44.6 ± 1.4	46.2 ± 3.1
Zinc	62.2 ± 5.0*	40.6 ± 4.7*
Iron	31.2 ± 2.4	36.2 ± 2.2
Copper	51.1 ± 2.3	46.9 ± 3.9

Means ± SEM, for 8 determinations, *, differ to a statistically
significant extent (Students t-test), $p < 0.01$.

Since intravenously administered MRP also acts to increase the urinary
excretion of zinc[2], it is proposed that certain components of the MRP mixture
are capable of chelating zinc. In this respect the effects described resemble
the effect of orally and intravenously administered histidine[3]. A possible
explanation for the increases in faecal zinc in animals fed MRP may be the
presence of other zinc binding ligands in the MRP which are not absorbed.
Further work is being conducted to examine the ability of purified Maillard
reaction products to chelate metal ions *in vitro*.

REFERENCES

[1]L.D. Stegink, J.B. Freeman, L. Den Besten and L.J. Filer, Maillard reaction
products in parenteral nutrition, Prog. Fd. Nutr. Sci., 5: 265 (1981).

[2]J. O'Brien, P.A. Morrissey and A. Flynn, The effect of Maillard reaction
products on mineral homeostasis in the rat, in: "Proceedings of Euro Food
Tox II", Institute of Toxicology, Swiss Federal Institute of Technology,
Zurich (1986).

[3]R.M. Freeman and P.R. Taylor, Influence of histidine administration on zinc
metabolism in the rat, Am. J. Clin. Nutr., 30: 523 (1977).

BINDING OF ZINC TO COLLOIDAL CALCIUM PHOSPHATE

IN HUMAN AND COW'S MILKS

H. Singh[1], A. Flynn[2] and P. F. Fox[1]

Departments of [1]Food Chemistry and [2]Nutrition
University College
Cork, Ireland.

INTRODUCTION

The lower bioavailability of Zn in cow's milk compared with human milk is well documented. It is generally agreed that this is due to differences in the nature and/or concentrations of Zn-binding ligands in the two milks and in the interactions of Zn with these ligands or their digestion products in the gastrointestinal tract. Lonnerdal et al. (1985) have shown that the bioavailability to rats of the Zn associated with bovine casein micelles is lower than that in human milk. However, the nature of Zn binding in casein micelles is poorly understood. The object of this study was to investigate the distribution of Zn in human and bovine milks and to examine the nature of Zn binding in casein micelles.

MATERIALS AND METHODS

Whole raw cow's and human milks were defatted by centrifugation at 3000g x 20 min at room temperature. Colloidal calcium phosphate-free (CCP-free) skimmed milks were prepared by the method of Pyne and McGann (1960). Milks were fractionated by equilibrium dialysis, ultracentrifugation (100,000g x 1h), and gel filtration chromatography on Sephadex G-100 in 0.1M ammonium acetate buffer, pH 6.6. Zn was assayed by atomic absorption spectrophotometry.

RESULTS AND DISCUSSION

The distribution of Zn in human and bovine milks is shown in Table 1. In human milk 34% of the Zn was associated with a low molecular weight fraction, 36% with whey proteins and 30% with casein micelles, while in cow's milk these fractions contained 6%, trace amounts and 94% of the Zn, respectively. It has been suggested that there are two distinct Zn fractions in bovine casein micelles (Parkash and Jenness, 1967; McGann et al., 1983). Investigation of the nature of Zn binding in the casein micelles of cow's milk showed that 63% of the total Zn was associated with CCP and 31% was directly bound to caseins. In human skim milk only 9% of the total Zn was associated with CCP while 21% was directly bound to caseins. These differences can largely be explained by the much greater content of CCP in cow's milk (>20 times that in human milk).

Table 1.   Distribution of Zn in Human and Cow's Skim Milks

| Fraction | Zn concentration (mg/l) | | | |
| | Human Milk | | Cow's Milk | |
	Skim	CCP-free	Skim	CCP-free
Total	1.55	1.41	3.80	1.42
CCP	0.14	0	2.38	0
Casein	0.32	0.35	1.18	1.21
Whey proteins	0.56	0.57	trace	trace
Low molecular weight	0.53	0.49	0.23	0.21

In CCP-free cow's milk the Zn concentration was reduced to 37% of that of skim milk and 85% of the Zn was associated with casein aggregates, while 15% was in a low molecular weight fraction. Removal of CCP from human milk had little effect on its Zn content or distribution.

Addition of 0.2mM EDTA to cow's milk rendered ∿40% of the Zn dialysable while the remainder was gradually solublised by increasing EDTA from 0.2-50 mM. Over 80% of Zn in CCP-free cow's milk was dialysable at 0.2 mM EDTA. Thus it appears that over 60% of the Zn in cow's milk casein micelles is tightly bound to CCP while the remainder is more loosely bound to casein.

These results show that CCP is a major Zn-binding ligand in cow's milk, but not in human milk, and suggest that both CCP and casein may influence the bioavailability of Zn in cow's milk.

REFERENCES

Lonnerdal, B., Keen, C. L., Bell, J. G. and Hurley, L. S., 1985, Zinc uptake and retention from chelates and milk fractions, in: "Trace Elements in Man and Animals - TEMA 5", C. F. Mills, I. Bremner and J. K. Chesters, eds., Commonwealth Agricultural Bureau, Slough, U.K.
McGann, T. C. A., Bucheim, W., Kearney, R. D. and Richardson, T., 1983, Composition and ultrastructure of calcium phosphate-citrate complexes in bovine milk systems, Biochim. Biophys. Acta, 760:415.
Parkash, S. and Jenness, R., 1967, Status of zinc in cow's milk, J. Dairy Sci., 50:127.
Pyne, G. T. and McGann, T. C. A., 1960, The colloidal phosphate of milk II. Influence of citrate, J. Dairy Res., 29:101.

ACKNOWLEDGMENT

This work was supported by a grant from the Development Fund, University College, Cork.

IRON-ZINC INTERACTIONS IN RELATION TO INFANT WEANING FOODS

Susan J. Fairweather-Tait and Susan Southon

AFRC Institute of Food Research
Colney Lane
Norwich NR4 7UA, U.K.

INTRODUCTION

High intakes of iron have been shown to impair zinc absorption, and this may have important implications for infants fed iron-supplemented foods[1]. A series of experiments were performed in rats to investigate the interaction between iron and zinc in detail, and to determine the effects of iron fortification of weaning foods on iron and zinc status.

MATERIALS AND METHODS

Male Wistar rats were used in all experiments, caged separately and fed control semi-synthetic diet ad lib. Absorption tests were carried out by feeding ^{65}Zn-labelled $ZnCl_2$ or ^{59}Fe-labelled $FeSO_4$ in 5g cooked starch:sucrose (1:1) paste, followed by whole body counting in a small animal γ counter.

RESULTS

Experiment 1

Iron had no effect on absorption from 100µg zinc at iron:zinc molar ratios between 0 and 10.

Experiment 2

There was competitive inhibition between iron and zinc at higher doses. Non-linear regression analysis $\left(V = \dfrac{C \cdot V_{max}}{K_m + c} \right)$ of the data predicted the following:

V_{max} for zinc alone $\quad$ = 1500 ± 250µg
$\quad$ in the presence of iron = 660 ± 30µg

K_m for zinc alone $\quad$ = 3300 ± 625µg
$\quad$ in the presence of iron = 1450 ± 95µg

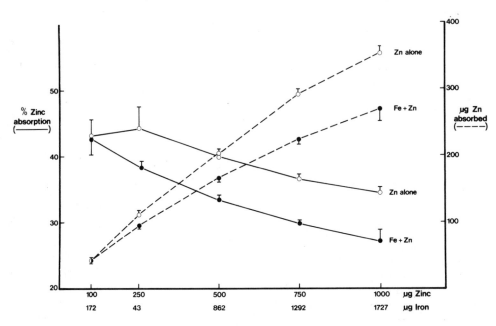

Fig. 1. Effect of Fe on Zn absorption at constant iron:zinc molar
ratio 2

Table 1. Effect of Fe fortification of infant weaning foods on iron and zinc status of rats

	Spring vegetable dinner		Oat breakfast cereal	
	-Fe	+Fe	-Fe	+Fe
Dietary iron (mg/100g dry wt)	8.0	35.9	4.2	34.5
Dietary zinc (mg/100g dry wt)	2.3	2.3	2.0	2.0
Iron:zinc molar ratio	4	18	2.4	20
Iron status				
Hb (g/100ml)	13.0	13.6	12.8*	13.5*
Liver iron (μg/g dry wt)	412*	802*	229*	685*
Zinc status				
Plasma zinc (μg/ml)	2.06	1.98	2.38*	2.10*
Femur zinc (μg/g dry wt)	209	209	224*	194*

*Significant difference between Fe-fortified and unfortified diet ($P < 0.05$).

Experiment 3

Iron reduced absorption from 500μg zinc at an iron:zinc molar ratio of 2. Zinc had no effect on iron absorption from the same test meal.

Experiment 4

Rats were fed 2 infant weaning foods ad lib, with and without additional iron, for 12 days and the effect on iron and zinc status measured.

Significant reductions in zinc status were observed in rats fed the Fe-fortified oat breakfast cereal weaning food.

CONCLUSIONS

There appears to be competitive inhibition between iron and zinc causing suppressed zinc absorption at higher doses, but no effect on iron absorption. Iron supplementation of infant weaning foods resulted in reduced zinc status in weanling rats fed an oat-based but not a vegetable based diet.

REFERENCES

1. N.W. Solomons, Competitive inhibition of iron and zinc in the diet: consequences for human nutrition, J. Nutr. 116:927 (1986).

IRON STATUS OF VERY LOW BIRTHWEIGHT PRE-TERM INFANTS RECEIVING A ZINC SUPPLEMENT

James Friel, Margaret Cox, Anna Cornel, Derek Matthew, and
Wayne Andrews

Departments of Biochemistry and Pediatrics
Memorial University
St. John's, Newfoundland, Canada   A1B 3X9

INTRODUCTION

Increased intakes of zinc may compromise iron status in humans
(1). Very low birthweight pre-term infants are known to be vulnerable
to both Zn and Fe deficiencies because of low body stores at birth (2)
and poor absorption during early infancy due to an immature gut (3).
As part of a long term study to evaluate the effect of supplemental Zn
on cognitive development in these infants, we evaluated the effect of
increased dietary Zn intakes on Fe status.

SUBJECTS AND FEEDING PROTOCOL

Forty-six infants (mean birthweight = 1125 $\pm$ 300 g, mean gestational
age = 28.9 $\pm$ 3 weeks, X $\pm$ S.D.) were recruited with informed parental
consent. All infants were eligible for this study if they were < 1500g
and < 37 weeks gestation at birth and did not suffer from severe lung,
kidney, liver or heart disease nor have any congenital malformations.
At 1850 $\pm$ 112 g, infants were randomly allocated into one of three
feeding groups. All infants were fed in hospital and then discharged
home at a mean weight of 2516$\pm$334g. Parents were then provided with
their formula in 32 ounce cans and were instructed to feed this formula
for 4 months after discharge restricting solid foods to < 10% of total
caloric intake for the first 3 months.

Infants were assigned to either: PLAC - Similac with whey and
drops (water); SUPP - Similac with whey and drops (zinc and copper);
EXPT - an experimental low birthweight formula. Mothers were given 15
mL amber dropper bottles and instructed to add 1 dropper-full to each
32 ounce can before preparing individual feedings. All formulas provided
13 mg/L Fe. The final concentrations of zinc and copper (mg/L) were:
PLAC, (6.7, 0.75); SUPP, (11, 1.2); EXPT, (9.4, 0.81).

SAMPLE COLLECTION AND ANALYSIS

All samples were collected at the start of the study (Day 0), on discharge from the nursery (Day 28) and on Day 118. Heparanized blood samples (1 mL) were obtained by heel or finger prick. Plasma was separated and frozen at -20°C until analysis. Plasma Zn was analysed by flame atomic absorption spectroscopy, while plasma ferritin (Frt) was measured by radioimmunoassay (courtesy Ross Laboratories). Dietary data from 3 day records was coded and analysed using a computerized data base. Weight was recorded on calibrated balance scales and an infantometer was used to measure length. Statistical analyses including analysis of variance and paired t-test were completed with SPSSx.

Table 1. Plasma Zinc and Ferritin From Three Different Feeding Groups

$(x \pm S.D.)$

	DAY	0		28		118
Zn	PLAC	86+38 (13)		99+44 (10)		91+26 (7)
(ug/dL)	SUPP	74+30 (14)		96+42 (11)		126+39 (8)
	EXPT	69+23 (14)		77+31 (14)		106+21 (9)
Frt	PLAC	132+141 (12)		96+45 (9)		47+41 (3)
(ng/mL)	SUPP	116+78 (12)	*	81+68 (9)	*	18+10 (3)
	EXPT	168+73 (12)	*	82+54 (13)		17+11 (6)

*Values on the same line are significantly different at $p < 0.05$.

RESULTS AND DISCUSSION

No differences in plasma Zn or Frt were found at any sampling time between the different feeding groups. This suggests that elevated intakes of Zn at dosages and Zn/Fe ratios used in the present study do not lead to decreased iron status in these infants. Yet there was a significant decline in plasma Frt in both the SUPP and EXPT groups between Days 0 and 28 and in the SUPP group between Days 28 and 118. As well, there was a trend to a decline (P = 0.09) between Days 28 and 118 in the EXPT group. Interestingly, at 118 days 2 infants with Frt values < 10 ng/mL, were from the increased Zn groups. Zn intakes were greater in the SUPP and EXPT groups compared to the PLAC group at 28 and 118 days. This data suggests that although iron status did not differ between feeding groups during the present study, increased zinc intakes over prolonged periods may lead to decreased iron status in very low birthweight preterm infants.

REFERENCES

1. P.J. Aggett, R.J. Crofton, C. Khin, S. Gvozdanovic and D. Gvozdanovic, The mutual inhibitory effects on their bioavailability of inorganic zinc and iron in: Zinc deficiency in human subjects, A. Prasad, A. Cavadar, G. Brewer and P. Aggett, eds. Alan R. Liss, Inc., New York (1983).

2. E.M. Widdowson, J. Dauncey and J.C.L. Shaw, Trace elements in early foetal and postnatal development. Proc. Nutr. Soc. 33: 275 (1974).

3. J. Dauncey and J.C.L. Shaw, The absorption and retention of magnesium, zinc and copper by low birthweight infants fed pasteurized human breast milk. Ped. Res. 2: 991 (1977).

(Supported by Health and Welfare Canada and Ross Laboratories).

PRODUCTION OF MG DEFICIENCY ANEMIA BY ZN

AND PHYTATE IN YOUNG JAPANESE QUAIL

M. R. S. Fox, S.-H. Tao, B. E. Fry, Jr., and Y. H. Lee

Division of Nutrition
Food and Drug Administration
Washington, DC

INTRODUCTION

An increase in dietary Zn to ameliorate the Zn deficiency produced by dietary phytic acid (PA) caused high mortality, decreased body weight, and elevated Fe in the liver and tibia of young Japanese quail by 14 d of age (Fox et al., 1985). Supplemental Mg prevented these adverse effects. Since marked anemia is a characteristic feature of Mg deficiency in the young quail (Tao et al., 1983), we investigated the effects of supplemental Fe in Mg-deficient birds and in those fed elevated Zn with PA.

EXPERIMENTAL PROCEDURES

Day-old Japanese quail (Coturnix coturnix japonica), 12 birds/group, were fed experimental diets for 7 d (Table 1). The adequate purified casein gelatin diet contained requirement Zn, Fe, and Mg at 20, 100, and 300 mg/kg, respectively. PA was fed to equal that contained in a diet with protein supplied by defatted soy flour. Measurements included body weight, hemoglobin (Hb), and minerals (by atomic absorption spectrometry) in liver, spleen, and tibia. Data were evaluated by Duncan's multiple range test.

RESULTS

Part of the data is presented in Table 1. Mg deficiency reduced body weight, tibia Mg, and Hb and increased relative spleen weight and Fe in the spleen and tibia. Effects of Mg deficiency on mortality and spleen size were greater with 100 than with 20 mg Zn/kg. Higher Zn slightly decreased body weight with requirement Mg.

PA with requirement Zn and Mg decreased body weight and tibia Zn and Mg, increased tibia Fe, and caused 42% mortality. PA with higher Zn and requirement Mg caused decreased Hb and 58% mortality. In these birds (vs. those fed PA with requirement Zn and Mg), Hb decreased and body weight, relative spleen weight, and tibia Zn increased. Supplemental Fe increased tissue Fe concentrations and increased mortality of Mg-deficient birds fed requirement Zn but decreased mortality of those fed PA (both dietary Zn levels). Fe did not affect Hb levels in either Mg-deficient birds or those fed PA and higher Zn.

Table 1. Effects[a] of Supplemental Fe and PA with Low and Requirement Mg and with Requirement and Higher Zn

Dietary Zn, mg/kg	Dietary Supplement[b]					
	None		Fe		PA	Fe + PA
	Dietary Mg, mg/kg					
	225	300	225	300	300	300
Mortality, %						
20	17	0	50	18	42	8
100	50	0	50	0	58	8
Body Weight, g						
20	16.2^{CD}	23.8^{A}	16.3^{CD}	23.2^{AB}	15.3^{D}	15.1^{D}
100	17.3^{C}	21.5^{B}	17.4^{C}	22.2^{AB}	17.9^{C}	17.8^{C}
Hb, g/dL						
20	4.0^{D}	7.8^{AB}	3.4^{D}	8.6^{A}	6.5^{BC}	6.4^{C}
100	3.6^{D}	6.8^{BC}	4.1^{D}	8.4^{A}	4.3^{D}	3.9^{D}
Liver Fe, µg/g						
20	121^{DEF}	68^{EF}	652^{A}	307^{C}	82^{EF}	227^{CD}
100	198^{D}	43^{F}	581^{A}	205^{CD}	167^{DE}	413^{B}
Tibia Mg, mg/g Fat-free Dry Weight						
20	1.21^{C}	1.54^{B}	1.19^{C}	1.72^{AB}	1.19^{C}	1.23^{C}
100	1.14^{C}	1.56^{B}	1.19^{C}	1.76^{A}	1.17^{C}	1.33^{C}

[a] Values are means. Means with the same superscript for each measurement are not significantly different ($P < 0.05$).
[b] Weight/kg diet: Fe, 50 mg as the sulfate; PA, 11 g as the sodium salt.

CONCLUSIONS

The combination of PA and elevated Zn with requirement Mg produced an anemia (probably due to accelerated erythrocyte destruction) and other changes characteristic of Mg deficiency. Supplemental Fe was absorbed and retained in tissues; however, it did not protect against the anemia of Mg deficiency or that produced by PA and elevated Zn. Higher Zn plus PA may reduce Mg or Fe utilization and/or erythrocyte membrane stability.

REFERENCES

Fox, M. R. S., Tao, S.-H., Fry, B. E., Jr., and Lee, Y. H., 1985, Dietary interactions among zinc, iron, magnesium, and phytate, XIII Int. Congr. Nutr. Abstr., p. 60.
Tao, S.-H., Fry, B. E., Jr., and Fox, M. R. S., 1983, Magnesium stores and anemia in young Japanese quail, J. Nutr., 113:1195

HEPATIC AND SPLENIC CONTENT OF IRON,

COPPER, ZINC, AND MANGANESE IN ANEMIC MINK

Jouko T. Työppönen and Paul O. Lindberg

College of Veterinary Medicine, Department of Biochemistry
P.O.Box 6, SF-00551 Helsinki, Finland, and Finnish Fur
Breeders Association, P.O.Box 5, SF-01601 Vantaa, Finland

INTRODUCTION

Nutritional anemia in mink is usually a consequence of feeding the animals with a high content of raw fish of the Gadidae family. Anemia during the early growth period leads to the "cotton-fur" syndrome in dark minks (Stout et al., 1960). It is characterized by the lack of pigmentation in the underfur and causes great economical losses to fur farmers in the form of poor quality pelts. Imbalance in dietary trace element composition, especially copper deficiency, may underlie anemia and "cotton-fur" syndrome (Underwood, 1977). The present paper describes dietary and tissue concentrations of some trace elements in mink.

MATERIALS AND METHODS

A control group received normal farm diet of mixed protein source consisting of herring and slaughter offal. The protein compartment in the feed of two anemiogenic groups of mink originated exclusively from raw marine fish (coalfish 36 %, capelin 19 % and blue whiting 12 %). One of the fish protein groups received a basal diet, while the other group received supplemental iron and copper.

The animals were weighed and hemoglobin was determined every two weeks after weaning. After two months on experimental diets the animals were killed at the age of 130 days, and blood and tissue samples were collected. Non-hemin iron in liver and spleen was determined according to Brückmann and Zondek (1940). Other trace elements in liver and spleen (Cu, Zn and Mn) and all four trace elements in the diets were analyzed by plasma emission spectroscopy.

RESULTS AND DISCUSSION

Both groups of mink on fish diets developed severe microcytic hypochromic anemia. The mean hemoglobin was only 55 % (P<0.001) of the values in the control group at the same age (166 ± 16 g/l). The body iron stores were totally depleted in both anemic groups (Table 1 and 2). The iron contents in control feed and basal fish diet were 94 and 53 mg/kg, respectively. Minks on fish feed were totally insensitive to supplemental

Table 1. Hepatic trace element concentrations (μg/g)· in healthy and anemic minks.

Group	n	Fe	Cu	Zn	Mn
Control	10	198 ± 88	11 ± 4	40 ± 9	2.5 ± 0.6
Anemic[a]	6	13 ± 2***	11 ± 8	47 ± 17	3.0 ± 1.0
Anemic[b]	6	18 ± 6***	47 ± 36	57 ± 9**	3.2 ± 0.7*

[a] basal diet
[b] basal diet supplemented with Fe and Cu
*P<0.05; **P<0.01; ***P<0.001 compared to control group

Table 2. Splenic trace element concentrations (μg/g) in healthy and anemic minks.

Group	n	Fe	Cu	Zn	Mn
Control	10	193 ± 59	1.3 ± 0.4	22 ± 3	0.3 ± 0.1
Anemic[a]	6	21 ± 7***	1.4 ± 0.3	22 ± 2	0.3 ± 0.1
Anemic[b]	6	24 ± 7***	2.0 ± 0.7	24 ± 5	0.3 ± 0.1

[a] basal diet
[b] basal diet supplemented with Fe and Cu
***P<0.001 compared to control group

iron (+40 mg/kg feed). The hemoglobin was normal within a wide range of non-hemin Fe in liver and spleen, but a sharp fall in hemoglobin was observed at liver and spleen Fe below 30 μg/g. The Cu levels in control and basal fish diets were 2.2 and 1.5 mg/kg, respectively. The Cu content in liver and spleen were similar in the control and basal anemic groups, but doubling the Cu level in the fish group increased the Cu level especially in liver, but there were great interindividual variations (Table 1). The Zn and Mn content in the control feed were 30 and 11.1 mg/kg, in the fish diets 21 and 8.7 mg/kg, respectively. The Zn and Mn contents in liver were similar to or higher than in the controls, especially during Fe/Cu supplementation (Table 1).

The results indicate that Cu, Zn and Mn are normally or better absorbed during low-Fe absorption, and slightly increased dietary Fe and Cu enhance absorption of Zn and Mn. According to the present study, anemia in minks on fish feed was due to strong and selective inhibition of iron absorption. Tissues seemed to respond normally to different dietary levels of the other trace elements evaluated.

REFERENCES

Brückmann, G. and Zondek, G.S., 1940, An improved method for the determination of non-hemin iron, J. Biol. Chem., 135:23.
Stout, F.M., Oldfield, J.E. and Adair, J., 1960, Nature and cause of the "cotton-fur" abnormality in mink, J. Nutr., 70:421.
Underwood, E.J., 1977, "Trace elements in human and animal nutrition," Academic Press, Inc., New York.

COPPER STATUS AND CADMIUM ADMINISTRATION: EFFECTS ON ZINC DISTRIBUTION,

METALLOTHIONEIN, AND SUPEROXIDE DISMUTASE ACTIVITY IN LIVER

Keun Chung, Donald Tinker, Nadia Romero, Kenjie Amemiya,
Carl Keen and Robert Rucker

Department of Nutrition
University of California
Davis, CA 95616

INTRODUCTION

In part, this work focuses on the extent to which cadmium intoxication
can modulate SOD and MT levels when dietary copper is limiting. Although the
ability of cadmium to stimulate SOD activity directly or cause the induction
of MT have been described (1-3), less is known about temporal and dietary
interrelationships (e.g., copper deficiency) that may influence responses
when cadmium is administered.

METHODS

Nutritional parameters were growth, mortality, tissue levels of the
metals (4), and hematocrit (5). Weanling rats were fed diets containing 0 or
25 ppm copper for 6 weeks. The rats were then injected with cadmium (10
mg/kg B.W.) and 0, 6, or 12 hours later were killed. Livers were assayed for
MT and SOD. The results that were obtained were contrasted with those for
rats fed 1000 ppm zinc or 20 ppm cadmium containing diets throughout the 6
week period. The changes in MT levels were assessed by determining the
tissue distribution of zinc by G-75 Sephadex chromatography (6), or by a
cadmium saturation method (7). SOD was determined using pyrogallol as substrate (8). Further, liver SOD was isolated and polyclonal antibodies were
prepared in chickens. An ELISA assay (9) was developed in which the antibody
was coupled to peroxidase for detection of SOD. Western blots were used to
characterize the antibody complex (10).

RESULTS

At 6 weeks, weights were 176, 271 288 or 270 g for rats fed the Cu-def.,
Cu-suppl., Zn-suppl. or Cd-suppl. diets, respectively. About 40% of the
Cu-def. rats died during the course of the experiment and all of the Cu-def.
rats had hematocrits that were 50-65% of normal values. Liver copper was
only 10% normal (0.3 vs. 3.5 µg/g liver).

Copper-deficiency did not influence the accumulation of MT. Also, at 6
or 12 h after the dose of Cd, MT was stimulated 4- or 8-fold, respectively,
independent of copper status. This was demonstrated both by the appearance
of Cd eluting in the 6000-8000 m.w. range following gel filtration and MT
assessed by a Cd saturation method (7). Only modest levels of MT were
observed in normal rats, but MT was stimulated 3X normal in rats fed zinc
(1000 ppm) or 5-fold in rats fed Cd for 6 weeks.

Enhanced zinc binding by MT was not significant 12 hours after MT was induced acutely by injection of Cd, however enhanced Zn binding to MT (assessed by gel filtration) was very apparent following chronic Cd administration, i.e., after 6 weeks.

With respect to SOD, the most important finding was that immunologically measured levels of SOD protein were not decreased by Cu-def. The values for SOD varied from 0.9 to 2.2 mg per gram of liver. The lower values were typically found in livers from control rats. Functional SOD activity, however, was depressed to less than 50% normal in Cu-def. rats. Further, liver SOD activity was increased following acute Cd from 110 U/g (control values) to 150 U/g at 6 hours or 195 U/g at 12 hours. In contrast, post-injection values for Cu-def. rats remained abnormally low (50-60 U/g liver) and there was no apparent stimulation following Cd administration.

DISCUSSION

Copper status does not appear to be a major factor in the expression of Zn- or Cd-related MT when measured by assays that assess accumulation. Although Cu-def. alters the tissue distribution and levels of some trace elements, e.g., iron, it is not clear that this happens with zinc (11). Likewise, copper status does not appear to influence the ability of liver to express and accumulate the Zn form of SOD. Using a highly specific antibody to SOD, no differences were observed in liver SOD levels, although it functional enzymatic activity was reduced to 50% of normal values. Experiments similar to some of those reported here have been carried out with Zn-suppl. and -def. rats (12). It has been demonstrated that the copper stimulated induction of MT is not influenced by Zn status, although Cu induced MT is more rapidly degraded in Zn-def. rats (12). It has been postulated that Zn binding protects MT from self-association and polymerization, perhaps a signal for MT degradation. To the extent that one can interpret from our net accumulation data, it would appear that there may be less communication between Cu and the mechanisms of control for MT than for Zn and mechanisms of control for Cu induced MT. Finally, an unique aspect of the work was the ability to assess both SOD protein levels and activity of SOD. It is of interest that the actual amount of SOD measured immunologically and functionally are not well correlated.

ACKNOWLEDGEMENTS

Funded in part by CNRU grant PHS AM-35747 and HL-15965.

REFERENCES

1. D. H. Hamer, Ann. Rev. Biochem. 55:913-951 (1986).
2. I. Bremner, J. Nutr. 117:19-24 (1987).
3. I. Fridovich, Annu. Rev. Pharmacol. Toxicol. 23:239-257 (1983).
4. M. S. Clegg, C. L. Keen, B. Lonnerdal and L. S. Hurley, Biol. Trace Element Res. 3:107-115 (1981).
5. M. Dubick, C. Keen and R. Rucker, Expl. Lung Research 8:227-241 (1984).
6. C. L. Keen, N. L. Cohen, L. S. Hurley and B. Lonnerdal, Biochem. Biophys. Res. Commun. 118:697-703 (1984).
7. S. Onosaka and M. Cherian, Toxicol. Appl. Pharmacol. 63:270-274 (1982).
8. S. Marklund and G. Marklund, Eur. J. Biochem. 47:469-474 (1974).
9. A. Voller, D. Bidwell and A. Bartlett, in: "Manual of Clinical Immunology," N. R. Rose and H. Friedman, eds., Academic Press, NY (1980).
10. W. N. Burnett, Anal. Biochem. 112:195-203 (1981).
11. C. L. Keen, N. H. Reinstein, J. Goudy-Lefevre, M. Lefevre, B. Lonnerdal, B. O. Schneeman and L. S. Hurley, Biol. Trace Element Res. 8:123-136 (1985).
12. I. Bremner, W. G. Hoekstra, N. T. Davies and B. W. Young, Biochem. J. 174:883-892 (1978).

BORON AND METHIONINE STATUS OF THE RAT AFFECTS THE PLASMA AND BONE MINERAL RESPONSE TO HIGH DIETARY ALUMINUM

T.R. Shuler and F.H. Nielsen

USDA, ARS, Human Nutrition Research Center
Grand Forks, ND 58202

INTRODUCTION

Aluminum toxicity in humans with chronic renal failure has been associated with dialysis osteomalacia syndrome. Recent findings from our laboratory suggest that boron affects calcium metabolism and bone mineralization. Two experiments were done simultaneously to ascertain whether dietary boron or magnesium affects the response of rats to high dietary aluminum, and whether the response is influenced by methionine status.

METHODS

Male weanling Sprague Dawley rats were randomly assigned to groups of six and fed their respective diets for 49 days. The diets, based on 70% acid washed ground corn-16% casein, will be described (1), but were similar to those described by Uthus, et al (2). The following supplements factorially arranged as variables were added to the diet: boron, 0 and 3 µg/g; aluminum 0 and 1.0 mg/g; and magnesium 100 and 400 µg/g. In experiment one, the diet was supplemented with 2.5 mg methionine/g (methionine adequate); in experiment two, this supplement was omitted. Blood was drawn by cardiac exsanguination from the ether anesthesized rats. The rats were decapitated; then the right leg was removed and frozen. The femur was cleaned with cheese cloth and dried at 85°C for 24 hours. The cleaned femurs were ashed with concentrated ultrapure nitric acid and 30% hydrogen peroxide in teflon tubes. The heparinized plasma was separated from the whole blood and treated with trichloroacetic acid to precipitate out the proteins. The supernatant was used for analysis. All analyses were performed on an inductively coupled argon plasma atomic emission spectrometer. This procedure will be described in detail (1).

RESULTS AND DISCUSSION

Selected findings from the analyses of plasma and bone are presented in Table 1. When dietary methionine was adequate, high dietary aluminum depressed calcium, magnesium, phosphorus and manganese in bone, and depressed the concentrations of zinc and magnesium and elevated the concentration of copper in plasma. Boron deprivation tended to enhance the depression in bone phosphorus caused by high dietary aluminum. When dietary methionine was reduced to a marginal level, none of the above

Table 1. Effects in Rats of Boron, Magnesium, and Aluminum with and without Supplemental Methionine on Mineral Content of Bone and Plasma

Treatment			Methionine Adequate (Exp 1)							Methionine Marginal (Exp 2)						
			Plasma			Femur				Plasma			Femur			
B	Mg	Al	Mg	Cu	Zn	Ca	P	Mg	Mn	Mg	Cu	Zn	Ca	P	Mg	Mn
µg/g	µg/g	mg/g	µg/ml	µg/ml	µg/ml	mg/g	mg/g	mg/g	µg/g	µg/ml	µg/ml	µg/ml	mg/g	mg/g	mg/g	µg/g
0	100	0	6.69	0.98	1.45	229	101	1.12	0.52	7.44	1.02	1.41	217	92	1.25	0.48
3	100	0	6.44	1.00	1.50	221	91	1.29	0.43	6.08	1.02	1.30	209	90	1.14	0.35
0	100	1.0	5.31	1.16	1.38	210	90	1.05	0.39	6.13	1.20	1.35	208	89	0.93	0.49
3	100	1.0	5.64	1.08	1.37	215	90	1.15	0.39	6.34	1.15	1.30	212	92	0.94	0.40
0	400	0	19.18	0.89	1.46	210	96	4.39	0.45	18.44	0.85	1.44	212	92	4.42	0.38
3	400	0	19.88	0.88	1.44	222	99	4.51	0.47	19.03	0.87	1.47	205	87	4.04	0.34
0	400	1.0	17.93	0.98	1.31	207	92	4.21	0.35	18.59	0.85	1.46	213	87	4.06	0.45
3	400	1.0	18.19	0.94	1.34	208	91	4.12	0.35	17.64	0.90	1.39	214	92	4.05	0.40
Significant Effects:																
B			NS	NS	NS	NS	NS	NS	NS	NS	NS	NS	NS	NS	NS	0.02
Al			0.01	0.02	0.005	0.007	0.0001	0.0001	0.0001	NS	0.04	NS	NS	NS	0.009	NS
Mg			0.0001	0.0002	NS	NS	NS	0.0001	NS	0.0001	0.0001	0.02	NS	NS	0.0001	NS
B x Mg			NS	NS	NS	NS	0.02	NS	NS	NS	NS	NS	NS	NS	NS	NS
Al x Mg			NS	NS	NS	NS	NS	0.03	NS	NS	NS	NS	NS	NS	NS	NS
B x Al			NS	NS	NS	NS	NS	NS	NS	NS	NS	NS	NS	0.04	NS	NS
Al x B x Mg			NS	NS	NS	NS	0.006	NS	NS	0.03	NS	NS	NS	NS	NS	NS

significant effects were found except that bone magnesium was still depressed and plasma copper was still, but less markedly, elevated by high dietary aluminum. Bone phosphorus concentrations were slightly elevated in boron-supplemented rats, and slightly depressed in boron-deprived rats, by high dietary aluminum. The findings show that methionine status markedly affects the response of the rat to oral aluminum toxicity. Boron and magnesium nutriture also occasionally modified the response to high dietary aluminum. These modifications affect mineral metabolism in bone and plasma. The findings support the hypothesis that boron is an essential nutrient involved in major mineral metabolism.

REFERENCES

1. F.H. Nielsen, T.R. Shuler, T.J. Zimmerman and E.O. Uthus, Magnesium and Methionine Deprivation Affect the Response of Rats to Boron Deprivation, submitted to Biol. Trace Elem. Res. (1987).

2. E.O. Uthus, W.E. Cornatzer and F.H. Nielsen, Consequences of Arsenic Deprivation in Laboratory Animals in: Arsenic Industrial, Biomedical, Environmental Perspectives, W.H. Lederer and R.J. Fensterheim ed., Van Nostrand Reinhold Company, New York (1983).

EFFECT OF MOLYBDENUM ON HEPATIC ENZYMES AND MINERALS OF FEMALE RATS

Meiling T. Yang and Shiang P. Yang

Food and Nutrition
Texas Tech University
Lubbock, TX 79409, USA

INTRODUCTION

Molybdenum (Mo) supplementation at a level of 10 ppm in drinking water has been demonstrated to exert inhibitory effect on the N-nitrosomethylurea-induced mammary carcinoma incidence in female Sprague-Dawley rats [1]. The total number of palpable tumors was significantly lowered by the 10 ppm Mo supplementation, whereas 0.1 and 1.0 ppm Mo showed no significant effect[2]. Twenty ppm Mo supplementation, on the other hand, exhibited a promoting effect on the carcinogenesis. In the present study, the effect of graded levels of Mo supplementation on hepatic Mo- and some Cu-enzymes was investigated in an attempt to discern possible participations of the enzymes in anticarcinogenesis.

METHODS

Seventy 3-week-old female Sprague-Dawley rats were randomly divided into 7 dietary groups of 10 animals each. They were given ad libitum for 8 weeks deionized water supplemented with 0, 0.1, 0.5, 1.0, 2.0, 5.0, or 10.0 ppm Mo from sodium molybdate and a semi-purified AIN-76A diet. The diet contained 0.026 ppm Mo and the deionized water contained no detectable Mo.

At the end of 8-week feeding the animals were sacrificed. The hepatic trace mineral contents and activities of hepatic xanthine dehydrogenase/oxidase (XDH)[3], sulfite oxidase (SOX)[4], and superoxide dismutase (SOD)[5] were determined. Mo was determined by a catalytic polarographic procedure[6], while the other minerals, by flame atomic absorption.

RESULTS AND CONCLUSION

Growth of rats was not significantly (P > .05) affected by up to 10 ppm Mo supplementation. The weights of livers of animals supplemented with 0.1 and 0.5 ppm Mo were disproportionately lower than those of other groups. The mean hepatic Mo content increased about 2-fold from non-supplemented to 0.1 ppm group, but then there was no further significant increase until the supplementation reached 5 to 10 ppm. Hepatic Cu, Fe and Zn were highest in the 0.1 ppm Mo supplemented group. As shown in Table 1, the mean hepatic XDH activity increased markedly from group 1 to group 2, and remained fairly

Table 1. Effect of graded levels of Mo supplementation on the activities of hepatic xanthine dehydrogenase/oxidase (XDH), sulfite oxidase (SOX) and superoxide dismutase (SOD)[1,2]

Group	ppm Mo in Water	XDH[3]	SOX[4]	SOD[5]
1	0	$.642 \pm .054^a$	$9.11 \pm .60^a$	5870 ± 680^a
2	0.1	$.684 \pm .064^a$	$12.25 \pm .98^b$	7430 ± 910^b
3	0.5	$.693 \pm .073^a$	$11.02 \pm .84^c$	6430 ± 660^a
4	1.0	$.703 \pm .077^a$	$10.87 \pm .91^c$	5860 ± 630^a
5	2.0	$.698 \pm .060^a$	$10.47 \pm .73^c$	5990 ± 420^a
6	5.0	$.700 \pm .071^a$	$10.51 \pm .78^c$	5810 ± 680^a
7	10.0	$.705 \pm .087^a$	10.79 ± 1.05^c	6210 ± 660^a

[1] Mean $\pm$ SD.
[2] Values with different superscripts are significantly ($P < 0.05$) different.
[3] μmoles uric acid formed/min/g liver.
[4] μmoles ferricyanide reduced/min/g liver.
[5] units SOD/g liver.

constant thereafter for the rest of the groups. But there were no significant differences in the XDH activity among all seven groups. The hepatic SOX activity increased significantly from group 1 to group 2. But it then decreased to a third significant level intermediate between the first two for the rest of the groups. Both enzymes appeared to be fully saturated by the time the Mo supplementation reached 0.1 ppm. The mean hepatic SOD activity was significantly higher in the 0.1 ppm group than in the others.

The Mo-enzymes and SOD appeared not to play roles in anticarcinogenetic activity of Mo. (Supported by PHS grant No. 1 RO1 CA 39418 awarded by National Cancer Institute, DHHS.)

REFERENCES

1. H. Wei, X. Luo, and S. P. Yang, Effects of molybdenum and tungsten on mammary carcinogenesis in SD rats, JNCI, 74:469 (1985).
2. M. T. Yang, C. D. Seaborn, and S. P. Yang, Effect of molybdenum on N-nitrosomethylurea-induced mammary carcinogenesis and hepatic molybdenum-enzymes of female rats, Fed. Proc., 46:749 (1987).
3. F. Stirpe and D. Corte, The regulation of rat xanthine oxidase. Conversion in vitro of the enzyme activity from dehydrogenase (type D) to oxidase (type O), J. Biol. Chem., 244:3855 (1969).
4. H. J. Cohen and I. Fridovich, Hepatic sulfite oxidase purification and properties, J. Biol. Chem., 246:359 (1971).
5. H. P. Misra and I. Fridovich, Superoxide dismutase: A photochemical augmentation assay, Arch. Biochem. Biophys., 181:308 (1977).
6. C. C. Deng, N. S. Wang, and C. H. Chen, Study of catalytic polarography. III. Catalytic current of molybdenum-chlorate in mandelic acid or benzilic acid medium, Acta. Sci. Natl. Univ. Fudan, 11:197 (1966).

AN INTERACTION BETWEEN DIETARY VANADIUM AND RIBOFLAVIN IN CHICKS

C. H. Hill

Dept. Poultry Science, N. C. State University
Raleigh, NC 27695-7635

We have reported that the toxic effects of vanadium in chicks can be ameliorated by the inclusion of iron, copper, mercury, and ascorbic acid in the diet, but that zinc was ineffective (1,2,3). These findings lead to the conclusion that there was a reduction-oxidation component in the counteraction of vanadium toxicity and lead to the speculation that components of the diet important in oxidation reactions would also affect vanadium toxicity. We have examined the effect of riboflavin deficiency. A corn-soybean meal basal diet was used in which the riboflavin was omitted from the vitamin mix. The basal diet contained ca. 1.7 mg/kg riboflavin, approximately half the requirement for growth.

The addition of riboflavin to this diet resulted in a faster weight gain in those chicks not receiving vanadium, but the chicks receiving vanadium were unresponsive to the riboflavin supplement (table 1). This phenomenon was observed even when the riboflavin supplement was increased to exceed the molar concentration of vanadium.

The specificity of this interaction was examined by determining whether or not the interaction could be observed with the toxicities of selenium, arsenic, or cadmium. In each of these cases chicks receiving a toxic level of the element could respond to the addition of riboflavin to the diet by increased growth indicating that the interaction of vanadium with riboflavin was relatively specific.

The well-known propensity for vanadium to interfere with phosphorylation reactions led us to examine the effect of feeding riboflavin-5-phosphate and injecting FAD on their interaction with vanadium (tables 2 and 3). The interaction was evident in both cases indicating that the interaction was not the result of failure to phosphorylate riboflavin to its coenzyme form. Furthermore, the injection experiment indicated that the interaction was not the result of interference with the absorption of riboflavin by vanadium.

Analysis of livers revealed that neither riboflavin + FMN or FAD concentrations were affected by the addition of vanadium to the diet. While it appeared that the interaction of vanadium with riboflavin was reflected in the activity of glutathione reductase, an FAD-containing enzyme, rigorous statistical analysis of the data indicated that the interaction was not significant.

Table 1. Inhibition of Riboflavin Response by Vanadium

Riboflavin mg/kg	V[1], ppm	
	0	40
	3 wk gain, g	
0	222 ± 12[2,3]	170 ± 7
3.5	339 ± 4	179 ± 9
35	344 ± 5	177 ± 7
175	361 ± 9	172 ± 14

[1]Fed as ammonium metavanadate.
[2]Each value the mean ± SEM of 6 lots of 20 chicks each.
[3]Tukey's LSD .05 = 42 g.

Table 2. Effect of Vanadium on Response to Riboflavin and Riboflavin-5-Phosphate

Treatment	Exp. 1 V[1], ppm		Exp. 2 V[1], ppm	
	0	50	0	50
	3 wk gain, g		3 wk gain, g	
Control	308±12[2,3]	232±12	317± 7[2,5]	197± 8
Riboflavin[4]	428±15	192±12	473± 6	199±12
Riboflavin-5-Phosphate[4]	448±12	196±12	467±12	234±14

[1]Fed as ammonium metavanadate.
[2]Each value the mean ± SEM of 8 lots of chicks each.
[3]Tukey's LSD .05 = 53 g.
[4]Fed at 8.8 mg/kg.
[5]Tukey's LSD .05 = 46

Table 3. Effect of Vanadium on Growth Response to Riboflavin and FAD

	V, ppm	
	0	50
	12 day gain, g	
- Riboflavin[1]	120	92
I.P. Riboflavin[2]	154	89
I.P. FAD[2]	166	95
Dietary Riboflavin[3]	194	105

[1]Means of 4 lots of 10 chicks each.
[2]10 chicks inj. I.P. every 3rd day with 0.4 mg Riboflavin or 0.8 mg FAD.
[3]8 mg/kg added to feed of 6 lots of 10 chicks each.

The evidence presented here does not reveal any reason why vanadium should prevent a response to riboflavin. The interaction can also be viewed as vanadium being less toxic to riboflavin deficient animals than to riboflavin sufficient ones. It is possible that riboflavin deficiency results in a redox state of cells which renders vanadium less able to exert its toxic effects.

REFERENCES

1. T. L. Blalock and C. H. Hill, Metabolism of Cadmium, Cobalt, Nickel and Vanadium in Iron Deficient Chicks, Fed. Proc. 43, 679 (1984).
2. C. H. Hill, Interactions of Copper and Mercury With Vanadate in the Chick, Fed. Proc. 44, 751 (1985).
3. C. H. Hill, Studies on the Ameliorating Effect of Ascorbic Acid on Mineral Toxicities in the Chick, J. Nutr. 1009, 84 (1979).

DEMONSTRATION OF RUMINAL SYNTHESIS OF THIOMOLYBDATES AND THEIR SUBSEQUENT

ABSORPTION IN SHEEP

J. Price, A. M. Will, G. Paschaleris and J. K. Chesters

Biochemistry Division
Rowett Research Institute
Bucksburn, Aberdeen AB2 9SB

Previous studies, using spectrophotometric techniques, have been limited to the detection of thiomolybdates in the liquid phase of digesta and required the addition of abnormally high concentrations of molybdenum to the diet or incubated rumen suspensions. The relevance of these studies to grazing ruminants is questionable since Mo and the factors inhibiting copper utilisation are associated mainly with the digesta solids in sheep. In the present study, [99]Mo compounds were injected into the rumen of sheep fed rations similar in Mo content to those found in field cases of Mo-induced Cu deficiency. Labelled Mo species appearing in digesta and plasma were identified after displacement from their carriers in vitro.

EXPERIMENTAL

Four cannulated sheep were fed dried grass containing (/kg DM) 6.0 mg Cu, 6.2 mg Mo and 4.3 g S. Rumen, duodenal and ileal digesta and blood samples were obtained 16 h after the rapid injection of tracer doses of [99]MoO_4Na_2 or [99]MoS_4Na_2 into the rumen. Digesta was fractionated by centrifugation at 2200 g for 10 min, the supernatant removed and centrifuged at 30,000 g for 1 h. Because virtually all of the label in the solids was found in the 2200 g pellet, this fraction was washed with Tris/HCl buffer (pH 7.6) and used in subsequent studies on bound Mo species. Plasma was chromatographed on Sephadex G25 and the fraction containing [99]Mo bound to macromolecular material retained. Using the technique applied in plasma studies by Mason[1], bound [99]Mo species in digesta solids and the plasma macromolecular fraction were displaced from their carriers in vitro by addition of unlabelled tetrathiomolybdate in high concentration. Displaced Mo species were separated and identified[2] by chromatography on Sephadex G25 columns calibrated with molybdate (MoO_4^{2-}), paramolybdate ($Mo_7O_{24}^{6-}$), mono- (MoO_3S^{2-}), di- ($MoO_2S_2^{2-}$), tri- ($MoOS_3^{2-}$) and tetrathiomolybdate (MoS_4^{2-}). Elution was with Tris/HCl buffer (10 mM, pH 7.6), oxymolybdates and monothiomolybdate eluting with distribution coefficients (K_d) <1.0, while di-, tri- and tetrathiomolybdates were retarded (K_d>1.0) by the column.

MOLYBDENUM SPECIES IN DIGESTA AND PLASMA

At 16 h after injection of [99]MoO_4^{2-} into the rumen, the label was associated mainly with the solids in rumen (74%), duodenal (99%) and ileal

Table 1. ^{99}Mo Species Displaced from Rumen, Duodenal and Ileal 2200 g Pellets 16 h after Injection of ^{99}MoO$_4{}^{2-}$ into the Rumen.

K_d	Identity of ^{99}Mo species	^{99}Mo activity (% of activity displaced)		
		Rumen[a] Mean (SE)	Duodenum[a]	Ileum[a]
0.10	Unknown	3.7 (1.0)	11.2; 5.3	0.0
0.75-0.95	Unknown	16.3 (2.8)	47.4; 37.6	14.1
1.10	Dithiomolybdate	4.2 (0.2)	3.5; 2.4	5.0
1.28	Trithiomolybdate	41.2 (9.2)	21.4; 31.6	37.7
1.71	Tetrathiomolybdate	34.1 (9.9)	16.5; 23.1	43.2
		Percent displacement of ^{99}Mo from pellet		
		50.0 (3.2)	54.9; 62.7	51.2

[a] n = 4 (rumen); n = 2 (duodenum); n = 1 (ileum)

digesta (66%), virtually all of this being in the 2200 g pellet. Tri- and tetrathiomolybdate accounted for a substantial proportion of the ^{99}Mo displaced from the solids, but dithiomolybdate was present to a minor extent (Table 1). To verify that these species had not formed in vitro, ^{99}MoO$_4{}^{2-}$ was added to unlabelled rumen solids in vitro and the displacement procedure applied; di-, tri- and tetrathiomolybdate were not detected. A substantial proportion of the label (Table 1) was not displaced from the solids by tetrathiomolybdate which would suggest that this Mo fraction either does not contain thiomolybdates or that they are bound as complexes in which the thiomolybdate core is not exchangeable. In contrast to previous studies, thiomolybdates were not detected in the liquid phase from rumen (or ileal) digesta. Di- and trithiomolybdate were present in the duodenal supernatant, but accounted for only 0.03% of the total digesta Mo.

After ^{99}MoO$_4{}^{2-}$ injection into the rumen, complete displacement of bound ^{99}Mo species in plasma was achieved. Di- and trithiomolybdate accounted for 18 and 50% of the total label, but tetrathiomolybdate was not detected. Although it accounted for 81% of the ^{99}Mo displaced from rumen solids after injection of ^{99}MoS$_4{}^{2-}$ into the rumen, only traces of tetrathiomolybdate were detected in plasma. Since thiomolybdate hydrolysis in plasma decreases di- > tri- > tetra-, our findings suggest that tetrathiomolybdate is poorly absorbed relative to di- or trithiomolybdate.

The results of the present study indicate that, in ruminants, the inhibition of Cu absorption by thiomolybdates is likely to be due to tri- or tetrathiomolybdate while systemic effects on Cu metabolism are probably due to di- and trithiomolybdate.

REFERENCES

1. J. Mason, C. A. Kelleher, and J. Letters, The demonstration of protein bound ^{99}Mo-di- and trithiomolybdate in sheep plasma after infusion of ^{99}Mo-labelled molybdate into the rumen, Br. J. Nutr. 48:391 (1982).

2. J. Price, A. M. Will, G. Paschaleris, and J. K. Chesters, Identification of thiomolybdates in digesta and plasma from sheep after administration of ^{99}Mo-labelled compounds into the rumen, Br. J. Nutr. 58, in press (1987).

HIGH RESOLUTION GAMMA-RAY SPECTROSCOPY AS AN IN VIVO TOOL FOR FOLLOWING

THE ZINC-SELENIUM INTERACTION

K. R. Zinn and J. S. Morris

University of Missouri Research Reactor
Columbia, Missouri

INTRODUCTION

The interaction between zinc and selenium has been demonstrated by the increased incidence of acute selenium deficiency diseases when animals on low selenium diets are fed supplemental zinc. Jensen[1] demonstrated this effect in chicks fed 2100 ppm or 4100 ppm zinc sulfate, added to a basal diet containing 100 ppm zinc and 0.2 ppm selenium, with adequate vitamin E. He reported an increase in mortality due to increased prevalence of muscular dystrophy and exudative diathesis. Whanger[2] also found that adding zinc carbonate at 6000 ppm to a selenium and vitamin E deficient diet resulted in the promotion of liver necrosis. This was observed in rats when the dam and young were placed on the selenium-vitamin E deficient diet at 7 days, and the young were placed on the zinc carbonate treatment at 21 days. At a much lower level of added zinc (200 ppm in drinking water) Schrauzer[3] was able to demonstrate a zinc-selenium interaction by following the incidence of spontaneously occurring mammary tumors in C_3H mice. He found the protective effect afforded by selenium, as seen by a decreased incidence of tumors, could be antagonized by zinc.

The purpose of this study was to investigate the zinc-selenium interaction by following both Se-75 and Zn-65 tracers simultaneously in the living animal. This was accomplished using high resolution gamma-ray spectroscopy.

MATERIALS AND METHODS

Male weanling rats (40 x 3 replicates) were placed on selenium adequate (0.075 ppm) or selenium deficient (0.005 ppm) torula yeast diets. The added selenium in the adequate diet was in the form of selenite. Within each dietary treatment one-half of the rats were given 200 ppm additional zinc ($ZnCl_2$) in the drinking water, for a total of 4 treatments. At the beginning of the study one-half of the rats in each of the 4 treatments were injected IP with combined Zn-65 and Se-75 (as selenite) tracers. These animals were individually counted daily for the first week after injection, and then weekly thereafter, for 7 weeks (part 1 of the study). At that point the other one-half of the rats were injected, and counted in a similar fashion (part 2 of the study). Metabolism studies were conducted during both parts.

The tracers were followed using a whole body counting system that included a solid state lithium drifted germanium detector, coupled to a multichannel analyzer. Rats were counted in a reproducible geometry and long enough to attain 10,000 counts in both the 264 KeV peak for Se-75 and the 1115 KeV peak for Zn-65. Data from whole body counting were reduced in the following manner. First, the counts per minute data were corrected for physical decay back to the injection day, and normalized for the exact amount of activity that had been injected. The log to the base 10 of the decay corrected counts was taken as the dependent variable with the day of the count as the independent variable for linear regression analysis. Linear regression analyses were done for each rat, and included 2 separate lines, namely the faster elimination phase during the first week (phase I), and the slower elimination phase from week 1 to week 7 (phase II). Analysis of variance was used to determine treatment effects for the slopes and Y-intercepts from each phase.

RESULTS AND DISCUSSION

There was no significant difference in weight gain for any of the 4 treatments. Data from the metabolism studies showed no significant difference in food consumption.

In both parts 1 and 2 of phase I there was no significant zinc effect on Se-75 retention. However, selenium adequate animals eliminated Se-75 at a faster rate (more negative slope) than selenium deficient animals. These results from whole body counting were confirmed by metabolism studies. For phase II in part 1, a significant ($p < 0.05$) zinc effect was noted for the selenium deficient animals, with additional zinc increasing the elimination of Se-75. Estimated biological half-lives for Se-75 from phase II of part 1 are as follows: selenium adequate diets=37.2 days, selenium deficient diet (additional zinc)=42.8 days, and selenium deficient diet (no additional zinc)=50.0 days. Dietary selenium was also shown to influence the retention of Zn-65. This was observed in part 2 of the study, with selenium adequate animals (no additional zinc) retaining significantly more Zn-65 that selenium deficient animal (no additional zinc).

High resolution gamma-ray spectroscopy proved to be an acceptable means of following multiple tracers in the living animal. In this study 2 different tracers were used. With the described counting system it would be possible to follow as many as 8-10 different tracers simultaneously, provided each has an appropriate gamma-ray emitted during its decay.

REFERENCES

1. L. S. Jensen, Precipitation of a Selenium Deficiency by High Dietary Levels of Copper and Zinc, Proc. Soc. Exp. Biol. Med. 149:113 (1975).
2. P. D. Whanger and P. H. Weswig, Influence of 19 Elements on Development of Liver Necrosis in Selenium and Vitamin E Deficient Rats, Nutr. Rep. Int. 18:421 (1978).
3. G. N. Schrauzer, D. A. White, and C. T. Schneider, Inhibition of the Genesis of Spontaneous Mammary Tumors in C_3H Mice: Effects of Selenium and Selenium-Antagonistic Elements and Their Possible Role in Human Breast Cancers, Bioinorg. Chem. 6:265 (1976).

EFFECT OF Fe SUPPLEMENTATION ON Zn STATUS AND THE OUTCOME OF PREGNANCY

Joan M. McKenzie-Parnell, P. Don Wilson and George F. S. Spears

Departments of Nutrition, Obstetrics and Gynecology, and
Preventive and Social Medicine, University of Otago
Dunedin, New Zealand

INTRODUCTION

A new dimension has been added to the concerns regarding the Zn status
of healthy pregnant women: the effect of Fe or Fe/folate supplements on Zn
nutrition. This recent concern is an example of the increasing importance
attached to interactions among the elements, particularly when those interactions are competitive and when they concern two or more essential elements. Such biologically important interactions were predicted[1] almost 20
years ago on the basis of chemical similarities among elements.[1] Metabolic
studies in non-pregnant human subjects have confirmed in most cases a competitive interaction, manifested as[2] a reduction in Zn absorption in the
presence of inorganic ferrous iron,[2] and there is one report of the con-
verse, i.e., a reduction in Fe absorption in the presence of excess Zn in
human subjects.[3]

Pregnancy is a physiological condition for which daily Fe and Fe/folate
supplements are often recommended. Recent studies have suggested that Fe/
folate, and folate supplements on their own, also might impair the absorption of Zn in the form of a Zn supplement in pregnant women[4]; whether such
supplements impair the absorption of usual dietary amounts of Zn has not yet
been demonstrated in pregnant women; in non-pregnant women they do.

The concern in regard to the Zn status of pregnant women is the acknowledged importance of Zn for normal fetal growth and development, and for
pregnancy outcome. In this study, therefore, we determined the effect of
Fe, and Fe/folate supplementation on maternal Zn status, and on the outcome
of the pregnancy.

SUBJECTS AND METHODS

Forty-five women from a larger study on Zn status in pregnancy were
included in this study. Group 1 (n=27) did not take any Fe or Fe/folate
supplements during their normal pregnancy. Group 2 (n=18) did take Fe or
Fe/folate supplements throughout pregnancy. The amount of Fe varied from
25-100 mg/d, and the amount of folic acid was 350 μg/day. The groups did
not differ in age, parity, height, weight at 12-16 weeks, or dietary Zn
(10.7 and 10.0 mg Zn/d) or dietary Fe intakes (12.7 and 11.4 mg Fe/d).

RESULTS

Serum Zn concentration showed the anticipated decline as pregnancy
progressed; however the women in Group 2 did have a consistently lower serum
Zn concentration throughout their pregnancy (Table 1). Cord blood serum Zn
concentration however did not differ between the two groups: means of 111

and 107 µg Zn/100 ml for Groups 1 and 2, respectively. Maternal hair Zn concentration did not differ between the two groups in any trimester.

Table 1. Maternal Serum Zn for Group 1 and Group 2 (mean ± SE)

Group	Serum Zn, µg/100 ml			
	12-16 weeks	24 weeks	36 weeks	delivery
1	82 ± 5	73 ± 4	67 ± 3	83 ± 8
2	77 ± 6	68 ± 3	57 ± 3*	69 ± 4*

*Significantly lower than for Group 1, p<0.05.

Several measured of the outcome of pregnancy were compared between the two groups. There was no significant difference for mode of labour onset, length of labour, blood loss, birth weight, birth weight centile, placental weight, neonatal length, neonatal head circumference, or for fetal maturity scores such as the Apgar score and the Dubowitz score.

DISCUSSION

In those studies which have reported thorough data analysis, maternal serum Zn concentration does not appear to be a predictor of fetal growth, but a low maternal plasma Zn concentration has been associated with fetal distress in the neonate,[5] suggesting that deficient Zn nurture might be an etiological factor in the occurrence of the complications. Other studies have reported that maternal Zn depletion, measured by leucocyte Zn content,[6] is associated with fetal Zn depletion and intrauterine growth retardation. A low bioavailability of zinc from the diet would increase the risk of maternal zinc depletion during pregnancy. The dietary Zn intake of[7] these women was below the ADI of 20 mg for pregnant women in New Zealand.[7] Presumably, the lower serum Zn status of the supplemented women in this study did reflect a reduced bioavailability of their dietary Zn intake, as the supplementary Fe and Fe/folate reportedly cause.[2] Although serum Zn concentration is used commonly as a means of assessing Zn status, it is <u>not</u> a good predictor of tissue Zn nutrition. On the other hand, the consistency in the serial serum Zn measurements in our study does add reliability to the meaning of any change in serum Zn. Fortunately the implied detrimental consequences of a lowered maternal Zn status have not been observed in this study. However, Fe and Fe/folate supplements continue to be used regularly. Further studies are warranted, to confirm the lack of consequence in pregnant women with a lowered serum Zn concentration resulting from the ingestion of Fe or Fe/folate supplements.

REFERENCES

1. C.H. Hill and G. Matrone, Chemical parameters in the study of in vivo and in vitro interactions of transition elements, <u>Fed. Proc. Am. Soc. Exp. Biol.</u> 29:1474 (1970).
2. N.W. Solomons, Competitive interaction of iron and zinc in the diet: consequences for human nutrition, <u>J. Nutr.</u> 116:927 (1986).
3. P.J. Aggett, R.W. Crofton, C. Khin, S. Gvozdanovic and D. Gvozdanovic, The mutual inhibitory effect on their bioavailability of inorganic zinc and iron, <u>in</u>: "Zinc Deficiency in Human Subjects," A.S. Prasad et al., eds., Alan R. Liss, New York (1983).
4. K. Simmer, C.A. Iles, C. James and R.P.H. Thompson, Are iron-folate supplements harmful?, <u>Am. J. Clin. Nutr.</u> 45:122 (1987).
5. M.D. Mukherjee, H.H. Sandstead, M.V. Ratuaparkhi, L.K. Johnson, D.B. Milne and H.P. Stelling, Maternal zinc, iron, folic acid and protein nutriture and outcome of human pregnancy, <u>Am. J. Clin. Nutr.</u> 40:496 (1984).
6. K. Simmer and R.P.H. Thompson, Maternal zinc and intrauterine growth retardation, <u>Clin. Sci.</u> 68:395 (1985).
7. J.M. McKenzie-Parnell and P.D. Wilson, Intake of zinc and other nutrients throughout pregnancy, <u>Proc. Nutr. Soc. NZ</u> 11:98 (1986).

HUMAN METABOLIC STUDY OF MILK SIMULTANEOUSLY FORTIFIED WITH ZINC, IRON AND COPPER

Berislav Momčilović, Malcolm J. Jackson, Joan M. Round
and Timothy B. Weir

Institute for Medical Research and Occupational Health,
Zagreb, Yugoslavia; The Rayne Institute, University College
London, Great Britain

INTRODUCTION

Milk is a rich source of calcium and[1] an attractive vehicle for zinc, iron and copper fortification as it is widely consumed[2,3]. In its pulverised form milk could help the nutritional status of large segments of the population. The aim of this balance study is to enquire into the metabolism of milk simultaneously fortified with zinc, iron and copper in man.

SUBJECT AND METHODS

The study was performed on a white male aged 44 in apparently good health. He was successively fed one of three diets for eight days each: (1) Normal balanced diet (2300 Cal/day) (Control), (2) Whole cow's milk only (2700 Cal/4 Lit/day)(Milk) and (3) Whole cow's simultaneously fortified with zinc, iron and copper to ensure a daily intake of 15,18 and 2 mg respectively[4]. Each dietary regime was divided into for-day balance periods /(Balance = Diet - (Urine + Feces)/[5].

RESULTS AND DISCUSSION

Initial body weight was 77.2 kg and final body weight 74.0 kg, where 1.2 kg being lost on the Control diet indicating that the subject was not fasting on the Milk diet[6]. Milk induces highly positive calcium and phosphorus balance especially at the beginning of the milk (high calcium) diet (Table 1). Magnesium balance was positive only on Fortified milk. Zinc balance was positive throughout the study. Copper balance was negative on Control and Milk but positive on the Fortified diet. Iron balance also tends to be positive on the Fortified diet. Iron serum was 25.7 ± 0.1 μmol/L and 19.8 ± 0.2 μmol/L on the Control and Milk diet respectively ($P < 0.01$); tending to be low in the first period (17.3 ± 0.7 μmol/L) and with a tendency to increase on the second period (18.4 μmol/L) on the

Table 1. Human metabolic balance study on milk and milk
fortified with zinc, iron and copper

	Major elements (mmol/L)			Trace elements (μmol/L)		
	Ca	P	Mg	Zn	Cu	Fe
DIET						
Control	17.9[a]	51[a]	11.6[a]	201.5[a]	29.2[a]	298[a]
Milk	120.0[b]	123[b]	15.7[b]	256.5[b]	16[b]	122[a]
Fortified	113.5[b]	118[b]	19.3[c]	301.0[b]	60.2[c]	400[b]
URINE						
Control	5.53[a]	36.5[a]	4.75[a]	12.7	0.72	7.19[a]
Milk	8.55[b]	57.5[b]	6.00[b]	12.1	1.72	5.68[a]
Fortified	9.00[b]	59.1[b]	5.50[a,b]	11.0	1.32	55.83[b]
FECES						
Control	14.1[a]	17.3[a]	7.8	151.5	38.4	306
Milk	83.2[b]	52.2[b]	10.7	199.5	20.5	83.5
Fortified	96.1[b]	49.4[b]	9.4	237.5	48.2	263.5
BALANCE						
Control	-1.7[a]	-2.7	-0.90	+37.3	-10.1[a]	-15.2
Milk	+28.3[b]	+13.4	-1.00	+44.9	-6.2[a]	+32.8
Fortified	+8.4[a,b]	+9.6	+4.42	+37.5	+40.7[b]	+109.1

[a-c] P < 0.05 for the same element

Fortified diet. Serum values of Ca, P, Mg, Zn and Cu were unchanged. Serum cholesterol and triglycerides were decreased
on Fortified milk (Control 5.7 ± 0.2 v.s. 4.7 ± 0.2 mmol/L
and 1.55 ± 0.22 v.s. 1.17 ± 0.2 mmol/L respectively). Complex
fortification of milk with physiological doses of zinc, iron
and copper increased milk nutritional density without adverse
affects[7]. Not all the changes in gastrointestinal motility
should be ascribed to glucose intolerance as they may represent the changes in intestinal flora when on milk diet[8].

REFERENCES

1. R. R. Becker, R. P. Heaney, Am.J.Clin.Nutr. 41:254
    (1985).
2. G. C. Becking, Fed.Proc.35:2480 (1976).
3. D. Carmichael, J. Christopher, J. Hegenauer, and
    P. Saltman, Am.J.Clin.Nutr. 28:487 (1975).
4. A. E. Harper, Nutr.Rev. 31:393 (1979).
5. E. Dent, Klin.Ernahrungslehre, 3:10 (1965).
6. F. G. Benedict, Carnegie Inst.Wash. Publ. No. 203:1
    (1977).
7. A. Wretlind, Nutr.Metab. 21:120 (1977).
8. N. W. Solomons, A-M Guerrero, B. Torun, Am.J.Clin.
    Nutr. 41:199 (1985).

AGE-RELATED CHANGES IN ZINC AND COPPER CONCENTRATIONS OF SERUM, KIDNEY, LIVER, AND FIVE PARTS OF THE BRAIN OF THE ADULT RAT

Ragnar Palm, Göran Wahlström, and Göran Hallmans

Biophysical Laboratory and Departments of Neurology
Pharmacology, Pathology, and Nutritional Research
University of Umeå, S-901 85 UMEÅ, Sweden

INTRODUCTION

Previous studies on zinc (Zn) and copper (Cu) concentrations and total amounts in organs from rats of different ages have mainly compared suckling and adult rats. There are only a few investigations that have[1-3] compared adult rats of different ages and the results are conflicting. In the brain, there is an uneven distribution of the trace elements[4]. In this study we have studied serum, liver, kidney, and five parts of the brain (cortex, corpus striatum, hippocampus, midbrain-medulla, and cerebellum) from adult rats aged 15 weeks and 49 weeks. Organ weights and the dry weight % of the organs have been registered as well as the concentrations of Zn and Cu. The total amounts of the trace elements have also been calculated.

MATERIAL AND METHODS

Two groups of adult male albino rats, 15 in each group, were fed ordinary pelleted food for breeding purposes ad libitum. At the age of 15 weeks and 49 weeks respectively, the rats were killed by decapitation and the liver, kidneys, and brain were removed. The fresh weights of the organs were registered. The right half of the brain and samples from liver and kidneys were used for determination of the dry weight % of the organs. The left half of the brain was dissected in five parts[5]. The organ samples were dried for 3 days at 110^0 C and then ashed at 550^0 C overnight. Zn and Cu concentrations were determined by flame AAS[6]. For the statistical analysis, differences between the two age groups were evaluated by Student's t-test.

RESULTS

The weights of body, liver, kidneys, and whole brain increased from 15 weeks to 49 weeks. In all parts of the brain, with the exception of corpus striatum, the weights also increased. The dry weight % increased in the brain and the kidneys and decreased in the liver between 15 weeks and 49 weeks. The concentrations of Zn and Cu in serum and different organ samples are given in table 1 and the calculated total amounts of the trace elements in table 2.

TABLE 1. Zn and Cu concentrations. The results are given as the mean + S.D. Significant differences between 15 weeks and 49 weeks are indicated (a=p<0.05, b=p<0.01, c=p<0.001).

	ZINC		COPPER	
	15 w	49 w	15 w	49 w
Serum (μmol/l)	18.8+1.3	20.7+2.0^b	15.0+2.4	19.4+1.8^b
Liver (μmol/kg)	528+35	467+28^c	69+6	62+4^c
Kidney (μmol/kg)	344+18	367+15^c	234+81	348+111^b
Cortex (μmol/kg)	277+4	232+5^b	27+2	36+5^c
Corpus striatum (μmol/kg)	215+11	224+9^a	42+4	67+6^c
Hippocampus (μmol/kg)	275+4	261+5^c	36+5	51+8^c
Midbrain-medulla (μmol/kg)	146+2	140+4	35+2	50+5^c
Cerebellum (μmol/kg)	181+3	172+4^c	38+3	47+5^c

TABLE 2. Calculated total amounts of Zn and Cu. For the brain the amounts are calculated for half brain. The results are given as the mean ± S.D. Significant differences are indicated as in table 1.

	ZINC		COPPER	
	15 w	49 w	15 w	49 w
Liver (μmols)	8.11+0.59	9.05+1.03^b	1.06+0.10	1.20+0.16^a
Kidneys (μmols)	1.09+0.10	1.27+0.12^c	0.75+0.27	1.19+0.33^b
Half brain (nmols)	173.2+6.3	191.4+6.8	27.1+2.0	42.2+4.1^c
Cortex (nmols)	95.0+4.1	101.7+4.8^c	11.3+1.2	15.8+1.9^c
Corpus striatum (nmols)	9.9+1.7	10.9+1.1	1.9+0.3	3.3+0.4^c
Hippocampus (nmols)	16.5+1.0	17.9+2.2^a	2.2+0.3	3.5+0.7^c
Midbrain-medulla (nmols)	31.3+3.2	36.4+3.7^c	7.6+1.0	13.1+1.4^c
Cerebellum (nmols)	19.8+2.5	24.5+3.4	4.1+0.7	6.6+1.1^c

CONCLUSIONS

Zinc and copper concentrations change in the adult rat from 15 weeks to 49 weeks. In serum, the Zn and Cu concentrations increase from 15 weeks to 49 weeks. In the liver, the Zn and Cu concentrations decrease from 15 weeks to 49 weeks and in the kidney the concentrations increase. In the brain, the Zn levels in cortex and corpus striatum increase from 15 weeks to 49 weeks while in hippocampus and cerebellum the concentrations decrease. In all brain regions investigated for Cu, the concentrations increase from 15 weeks to 49 weeks. The total amounts of Zn and Cu increase in the liver and the kidneys from 15 to 49 weeks. In the brain, the total amounts of the trace elements, with the exception of Zn in corpus striatum, also increase from 15 to 49 weeks.

The observed trace element changes in the adult rat from 15 to 49 weeks may be a normal physiological event reflecting changes in binding of the metals to different ligands or differences in the concentrations of metal binding ligands. The present investigation stresses the importance of using age-matched controls when analyzing Zn and Cu in different organs of the adult rat.

REFERENCES

1. B.Kofod, Eur J Pharmacol 13:40(1970).
2. B.Bergman, R.Sjöström, and K.R.Wing, Acta Physiol Scand 92:440(1974).
3. R.Kishi, T.Ikeda, M.Miyake, et al, Brain Reserach 251:180(1980).
4. R.Palm, G.Wahlström, and G.Hallmans, Neurochem Pathol In press (1987).
5. J.Glowinski, and L.L.Iversen, J Neurochem 13:655(1966).
6. G.Hallmans, Acta Derm Venereol 58 (Suppl 80):8-13(1978).

ZINC UPTAKE BY BRUSH BORDER MEMBRANE VESICLES (BBMV)

FROM PRE-WEANLING RAT SMALL INTESTINE

M. L. Kennedy and B. Lönnerdal

Departments of Nutrition and Internal Medicine
University of California, Davis, CA 95616 USA

INTRODUCTION

Zinc (Zn) is required for normal growth and development. To provide
for rapid growth, newborn animals absorb Zn to a higher degree than adult
animals (1). However, research on the mechanisms behind this observation
has been limited. The initial stage of Zn absorption involves transport of
Zn across luminal brush border membranes. This process can be specifically
investigated (in the absence of cellular and metabolic variables) through
isolation of intestinal brush border membrane vesicles (BEMV) (2). Results
from studies of Zn uptake into BBMV from various ages of pre-weanling rat
pups are presented in this study.

MATERIALS AND METHODS

Sprague-Dawley rat pups at 1, 7, 14 or 21 days of age were fasted 3-5
h and killed by decapitation. The upper 3/4 of intestine was removed and
rinsed with saline; the mucosa was then scraped and immediately frozen at
-70°C.

Mucosal scrapings were suspended in 15-20 volumes of ice-cold buffer
(300 mM D-mannitol, 10 mM HEPES/Tris, pH 7.3); calcium was added, then
samples were centrifuged to remove non-microvillous components. Membrane
vesicles were pelleted through high-speed centrifugation (20,000 g). Purity was assessed by lactase (D 1, 7, 14) or sucrase (D 21) enzyme activity
in homogenate and final membrane pellet. Correct vesicle orientation was
measured by comparison of enzyme activity before and after addition of
Triton X-100. Protein content was determined by a modified Lowry method
and adjusted to 2 mg/ml buffer.

Vesicles were incubated at 37°C with Zn (10 uM) radiolabeled with
^{65}Zn, and at set time points the reaction was stopped by addition of a 30x
volume of ice-cold saline. Samples were immediately filtered (0.2 um pore
size), filters rinsed twice and then counted on a gamma counter. A second
series of experiments involved addition of increasing concentrations of Zn
(20-500 uM) (Zn concentration in mature rat milk averages 20 uM (4)) at a
reaction time of 1/2 min (time point in linear range of Zn uptake). Non-
specific binding of Zn to filters and isotope dilution were accounted for
(note that isotope dilution by endogenous Zn is not a factor since BBMV Zn

599

levels are 1-3 pmoles/mg pro. (3)). In addition, binding of Zn to the membrane was differentiated from actual uptake by performing the above procedures at 0°C.

RESULTS

Measurements of lactose and sucrase specific activities revealed a 10-15 fold purification of BBMV. In addition, the majority of vesicles (90-100%) were oriented right-side-out. Uptake of 10 uM Zn was linear to 1/2 - 1 min., with a mean value of 2.3 ± 0.1 nmoles/mg pro./min. (mean ± SEM). "Uptake" approached saturation by 5 min., at a concentration of 3.5 ± 0.5 nmoles/mg pro. Results were age-independent. Measurements of Zn uptake (1/2 min.) with increasing concentrations of Zn approached saturation at 400-500 uM. Again the results were essentially independent of age of the pups. A double-reciprocal plot of uptake rates (10-50 uM Zn) revealed a maximum uptake capacity, J_{max}, of ∿70 nmole/mg pro./min. and an affinity constant, K_m, of ∿175 uM. Total Zn uptake, i.e., ug Zn/total BBMV protein, increased with age and was correlated with increasing body weight. The addition of histidine to the extravesicular medium in a 2:1 molar ratio with Zn had no effect on any of the Zn uptake parameters.

CONCLUSIONS

1) Uptake of Zn^{+2} into intestinal BBMV from pre-weanling rat pups (per mg membrane protein) was essentially age-independent. However, total uptake increased with age, and was correlated with increasing body weight.
2) Zn uptake was saturable, both over time and with increasing concentrations of Zn.
3) The percent of Zn bound to the membrane (as determined by uptake measurements at 0°) decreased from 32% to 13% with increasing Zn concentration.

The observation that Zn uptake is saturable suggests a specific uptake process. This is further verified by decreased binding of Zn to the membranes with increasing Zn concentration. It is therefore likely that one or more luminal membrane components are involved in Zn absorption. (The uptake media contained only mannitol and HEPES/Tris.)

It is interesting to note that Zn uptake values in BBMV from pre-weanling rat pups are higher than values obtained in adult rats, and considerably higher than values from control adult rats (3). Net Zn absorption is higher in Zn deficiency and at young ages. Therefore, transport of Zn through the luminal brush border membrane is one possible regulator of Zn absorption and metabolism.

REFERENCES

1. I. Bremner and C. F. Mills, Absorption, transport and tissue storage of essential trace elements, Proc. Nutr. Soc. 294:73-89 (1981).
2. U. Hopfer, Isolated membrane vesicles as tools for analysis of epithelial transport, Am. J. Physiol. 233(6):E445-E449 (1977).
3. M. P. Menard and R. J. Cousins, Zn transport by brush border membrane vesicles from rat intestine, J. Nutr. 113:1434-1442 (1983).
4. B. Lönnerdal, C. L. Keen, and L. S. Hurley, Trace elements in milk from various species, in "Trace Elements in Man and Animals - 4," J. McC. Howell, J. M. Gawthorne and C. L. White, eds., Griffin Press, Ltd., Netly, South Australia, pp. 249-252 (1981).

MATERNAL GLUCOSE HOMEOSTASIS IN RATS GIVEN MARGINAL ZN DIETS

Susan Southon, C.M. Williams, and S.J. Fairweather-Tait

AFRC Institute of Food Research
Colney Lane
Norwich NR4 7UA, U.K.

INTRODUCTION

It has been demonstrated that rats given a marginal-Zn diet toward the end of pregnancy produce large for gestational age (LGA) pups.[1] The reasons for this are unclear, but the fact that glucose is the major metabolic fuel for the developing fetus, together with evidence of altered glucose utilisation in animals and humans given suboptimal Zn diets,[3,4] suggests that the increased weight of pups may be linked to changes in carbohydrate metabolism. The present study was undertaken to investigate the effect of marginal Zn intake during the latter stage of pregnancy in rats on fetal growth, blood glucose and insulin levels, and maternal glucose tolerance and circulating insulin concentration.

METHODS

Two experiments were performed using adult female Wistar rats.

Experiment 1

Pregnant rats were given control (62mg Zn/kg:Group 1) or marginal-Zn diet (8mg Zn/kg:Group 2) throughout pregnancy, or control diet until day 16 of pregnancy followed by marginal-Zn diet (Group 3). On day 20, after an overnight fast, half the rats from each group were anaesthetised, blood taken for glucose and insulin analysis, pancreas remove for Zn analysis and fetuses removed for wet/dry weight, DNA and protein determination. The remaining animals were given 1mg glucose/g body wt by gavage and tail-blood taken 30 and 60 min later. They were then subjected to the same procedures as the other rats.

Experiment 2

Pregnant rats were given diets as described for Groups 1 and 3 in Expt. 1. Rats were fed ad lib until day 17 of pregnancy and then meal-fed for 3 days. On day 20, after a 12h fast, a 50% glucose solution (2.5mg/g body wt) was given orally and tail-blood glucose concentration measured at 0, 10, 30, 50 or 0, 20, 40, 60 min after the dose. Cardiac blood was taken 70 min post-dosing for glucose and insulin analysis. Fetuses were removed and a pooled blood sample taken for glucose and insulin analysis.

RESULTS

Maternal and fetal blood glucose and insulin concentrations, measured 70 min post-dosing, were similar in both groups.

Expt. 1. Effect of marginal-Zn intake on fetal growth and maternal blood glucose, insulin concentration and pancreatic Zn

Group	Dietary Treatment					
	1		2		3	
	mean	se(n)	mean	se(n)	mean	se(n)
Rat fetuses:						
No./litter	12.8	0.7 (22)	12.7	0.8 (21)	12.8	1.0 (17)
Mean fresh wt (g)	3.39[a]	0.08(22)	3.52[a]	0.05(21)	3.64[b]	0.11(17)
Water (g/100g)	87.3	0.1 (22)	87.2	0.1 (21)	87.3	0.1 (17)
Protein:DNA (mg:mg)	19.1[a]	0.5 (22)	19.0[a]	0.5 (21)	21.4[b]	0.6 (17)
Adult females:						
Fasting glucose (mmol/l blood)	2.5	0.1 (10)	2.6	0.2 (10)	2.2	0.1 (10)
Insulin (µunits/ml plasma)	74	14 (10)	49	7.0 (10)	47	14 (6)
Pancreas – fresh wt (g)	0.66	0.02(22)	0.60	0.03(21)	0.65	0.02(17)
Pancreatic Zn- total (g)	13.0[a]	0.6 (22)	9.6[b]	0.6 (21)	9.4[b]	0.6 (17)
Glucose 0-30 min rise post-dosing[+] (mmol/l blood)	1.01[a]	0.13(11)	1.33[a]	0.18(11)	1.61[b]	0.16(10)

[a,b]Values with different superscript letters within a line are statistically different (P<0.05).
[+]Oral dose 1mg/g body wt.

Expt. 2. Maternal glucose tolerance following an oral glucose dose of 2.5mg/g body wt

	Dietary treatment			
	Control		Marginal Zn	
	mean	se(n)	mean	se(n)
Time post-dosing (min):	(Glucose mmol/l whole-blood)			
0	3.5	0.2(19)	3.7	0.2(19)
10	5.9	0.4(10)	5.7	0.3 (9)
20	5.7	0.3 (9)	6.7*	0.3(10)
30	5.4	0.3(10)	5.5	0.3 (9)
40	5.6	0.1 (9)	5.4	0.3(10)
50	5.6	0.2(10)	6.1	0.2 (9)
60	6.0	0.2 (9)	6.2	0.3(10)

*Value significantly different (P<0.05) from value for Control group at the same time point.

CONCLUSION

Mean wt of fetuses from rats given marginal-Zn diet during late pregnancy was higher than for the Control group, as found previously,[1] with a higher protein:DNA ratio. Weight of fetuses from rats given marginal-Zn diet from the onset of pregnancy, however, was not significantly different from the Controls, suggesting adaptation to the longer period of marginal-Zn intake.

Circulating insulin levels were similar between groups but the possibility that the substantial decrease in pancreatic Zn in pregnant rats given marginal-Zn diet may be associated with diminished insulin stores should be considered.

The significant increase in the early rise in blood glucose concentration in rats given marginal-Zn diet during late pregnancy indicated a difference in glucose utilisation. Further investigation is required to determine if this change is associated with the increment in fetal growth.

REFERENCES

1. S. J. Fairweather-Tait, A. J. A. Wright, J. Cooke and J. Franklin, Studies of zinc metabolism in pregnant and lactating rats, Br. J. Nutr. 54:401 (1985).
2. P. G. Reeves and B. L. O'Dell, The effect of zinc deficiency on glucose metabolism, Br. J. Nutr. 49:411 (1983).
3. M. T. Baer, J. C. King, T. Tamura, S. Margen, R. B. Bradfield, W. L. Weston and N. A. Daugherty, Nitrogen utilisation, enzyme activity, glucose intolerance and leukocyle chemotaxis in human experimental zinc depletion, Am. J. Clin. Nutr. 41:1220 (1985).

PREGNANCY COMPLICATIONS, LABOR ABNORMALITIES AND ZINC STATUS

N. Lazebnik, B. R. Kuhnert, P. M. Kuhnert, and K. Thompson

Dept. Ob/Gyn and Perinatal Clinical Research Center, Cleveland
Metropolitan General Hospital, Case Western Reserve University
3395 Scranton Road, Cleveland, Ohio 44109

INTRODUCTION

Several authors have shown that lowered zinc is related to risk factors
during pregnancy and labor complications.[1] The best index of zinc status is
unclear. The purpose of the present study was to test the hypothesis that
certain complications during the course of pregnancy and delivery would be
related to one or more lowered indices of zinc status. Plasma zinc, erythrocyte zinc and serum alkaline phosphatase activity were used as indices of
zinc status and were compared to the incidence of complications and major
dysfunctional labor patterns.

MATERIAL AND METHODS

Two hundred seventy nine women were studied at delivery. Antepartum
risk factor, labor progress, intrapartum events and immediate neonatal outcome were recorded as part of a large computerized perinatal data base in
use in the hospital. Friedman's criteria[2] were used for diagnosis of the 6
major dysfunctional labor patterns. Venous blood (20 ml) was obtained using
a plastic syringe and a stainless steel needle. Samples were analyzed for
plasma zinc, RBC zinc and heat labile alkaline phosphatase. The median values for plasma zinc, RBC zinc and alkaline phosphatase were used as cut
points to subdivide the 279 patients into "low group" and "high group."

RESULTS

Plasma zinc was associated with more complications during the antepartum and intrapartum periods than RBC zinc or alkaline phosphatase. The
median values, means, standard deviations and ranges for each group for each
variable are shown in Table 1. Table 2 summarizes the significant finding
for plasma zinc in the antenatal period. Table 3 summarizes the significant
findings for plasma zinc in the intrapartum period.

DISCUSSION

The findings of this study are supported by previous research in several disciplines. For example, we found relationships between low plasma

Table 1. Ranges and Mean Values of Zinc Indices in Low and High Groups

Index (units)	Number of Subjects	Mean ± S.D.	Range
Plasma Zn (ug%)			
Low	144	49.4 ± 6.0	29 – 57
High	135	65.2 ± 7.5	> 57 – 100
Alkaline Phosphatase (units/100 ml)			
Low	135	4.3 ± 0.9	1.4 – 5.6
High	141	8.0 ± 2.1	> 5.6 – 15.5
RBC Zinc (ug%)			
Low	143	1110 ± 118	666 – 1261
High	133	1376 ± 92	>1261 – 1710

Table 2. Relationship Between Plasma Zinc and Antenatal Risk Factors

| | Plasma Zn | | |
	Low (N=144) percent	High (N=135) percent	P value
Mild toxemia	5.6	0.7	0.02
Vaginitis	12.6	4.4	0.01
Post term > 42 wks	4.2	0	0.01

Table 3. Relationship Between Plasma Zinc and Intrapartum Complications

| | Plasma Zn | | |
	Low (N=144) percent	High (N=135) percent	P value
Prolonged latent phase	2.8	0	0.05
Protracted active phase	28.7	18.2	0.04
Labor > 20 hours	6.3	1.5	0.03
Second stage > 2.5 hours	6.3	0.7	0.01
Lacerations > 3rd degree	7.0	1.5	0.02

zinc levels and pre-eclamptic toxemia (hypertension). Others have suggested that zinc deficiency might confer an increased sensitivity to cadmium toxicity which may, in turn, result in vasoconstriction, sodium retention, altered catecholamine metabolism, increased renin activity and resultant hypertension.[3] We reported a tripled incidence of vaginitis in patients with low plasma zinc levels. Others have reported a relationship between lowered zinc levels and a reduced immunity to <u>Candida</u> <u>albicans</u> infection in both humans and mice.[4] We report lower zinc levels associated with gestations longer than 42 weeks, i.e., a failure to initiate labor. Others have suggested that lowered zinc may be related to an inadequate number of gap junctions in the myometrium at term since these steroid receptors are zinc dependent metalloproteins.[5] While labor abnormalities have been reported in the past in patients with low zinc levels, this appears to be the first study to use Friedman's criteria[2] to look at the 6 major dysfunctional labor patterns in regard to lowered plasma zinc levels. Our major finding was that patients with lowered plasma zinc levels have a significantly more prolonged latent phase and a more protracted active phase of dilatation. Again, poor uterine activity might be explained as a result of possible association between lowered zinc levels and the number of gap junctions in the myometrium.[6]

SUMMARY

The results of this study suggest that low plasma zinc is a valid indicator of pregnancy complications and abnormal labors compared to indices of zinc status such as alkaline phosphatases and RBC zinc. Plasma zinc screening as part of the patient's work up early in gestation and possible zinc supplementation need to be evaluated.

REFERENCES

1. J. Apgar, Zinc and reproduction, <u>Ann. Rev. Nutr.</u> 5:43 (1985).
2. E. A. Friedman, "Labor: Clinical Evaluation and Management," Appleton-Century-Crofts, New York (1978).
3. J. C. Chisolm and C. R. Handorf, Zinc, cadmium, metallothionein and progesterone; do they participate in the etiology of pregnancy induced hypertension?, <u>Med. Hypoth.</u> 17:231 (1985).
4. J. Edman, J. D. Sobel and M. L. Taylor, Zinc status in with recurrent vulvo vaginal candidiasis, <u>Am. J. Obstet. Gynecol.</u> 155:1082 (1986).
5. D. P. Dylewski, F. D. C. Lytton and G. E. Bunce, Dietary zinc and parturition in the rat. II. Myometrial gap junctions, <u>Biol. Trace Element Res.</u> 9:165 (1986).
6. F. D. C. Lytton and G. E. Bunce, Dietary zinc and parturition in the rat. I. Uterine pressure cycles, <u>Biol. Trace Element Res.</u> 9:151 (1986).

IS SPINA BIFIDA IN MAN LINKED TO A GENETIC DEFECT OF

CELLULAR ZINC UPTAKE IN CHILDREN ?

Alain Favier, Brigitte Dardelet, Marie-Jeanne Richard,
and Josiane Arnaud

Laboratoire de Biochimie C - Hôpital A. Michallon
38043 Grenoble Cédex - France

The etiology of spina bifida and other neural tube defects (NTD) is not yet fully known. It seems to be linked to genetic and nutritional factors [1,2]. Different hypothesis may be found in the literature: vitamin deficiency (more often folic acid) and zinc deficiency[3,4].

In a first study, we try to corroborate these different connections by realising a large biological check up in mothers and in spina bifida babies at each abnormal delivery (spina bifida and other NTD). Then, we measure [65]zinc uptake in growing fibroblasts of all spina (newborn and children).

MATERIAL AND METHODS

First study : 23 mothers and their spina bifida newborns, 41 controls mothers and 14 normal neonates were included in this study. We collected whole blood on trace element free tube and we took hair sample. We measured nutrition markors, vitamins and trace element.

Second and third study : Skin fibroblasts were taken from 8 spina bifida (6 neonates and 2 oldest children) and 5 controls (3 neonates and 2 infants). These fibroblasts (subculture limited to 5) were grown in RPMI 1640 medium containing 15 % fetal calf serum, penicillin, streptomycin and fungizone. Then, they were transfered in a similar medium with 0.1 uCi/ ml of [65]Zn, during 1 to 48 hours incubation. After washing twice with Puck saline, and trypsination, radioactivity of whole cells was measured after 2, 4, 12, 24 and 48 hours. Proteins were determined by a modified Lowry method.

RESULTS

First study : Results of maternal and newborns's status are presented in table 1. As we see, among all nutritional parameters only serum zinc, alkalin phosphatase and transferrin are significatively lower and serum amylase higher in spina bifida's mothers. So, we think that zinc deficiency contribut to human neural tube defects. These results are in agreement with previous works[3,4,5].

Table 1 : Results. Degrees of significance ** $p < 0.001$, * $p < 0.05$.

	MOTHERS OF SPINA n = 23	MOTHERS OF NORMAL n = 41	SPINA BIFIDA NEONATES n = 23	NORMAL NEONATES n = 14
Zn serum ,umol/l	** 10.4 ± 2.2	12.3 ± 1.5	* 15.4 ± 3.4	15.2 ± 2.6
Zn hair ,ug/g	160 ± 41	170 ± 28	* 193 ± 29	235 ± 40
Cu serum ,umol	39.7 ± 9.1	47.3 ± 7.8	* 12.2 ± 5.7	16.1 ± 3.2
Cu hair ,ug/g	** 20.6 ± 7.5	16.1 ± 6.0	14.5 ± 3.9	24.3 ± 17.3
Plasma folic ac. nmo/l	* 10.4 ± 3.9	6.0 ± 2.6	62.0 ± 53.0	–
Serum Albumin g/l	* 42.4 ± 4.6	38.4 ± 10.0	–	–
Serum Transferrin g/l	* 3.6 ± 0.6	4.1 ± 0.6	–	–
Alkaline phosphatase U.I	76.0 ± 26.0	98.2 ± 23.6	–	–
Plasma Amylase nKat/l	** 5.5 ± 1.0	4.1 ± 1.5	–	–

Second and third studies : As ZIMMERMAN [6] , we observed that ^{65}Zn uptake from newborn with spina bifida did not differe from control at 2 hours of incubation. At 24 and 48 h, ^{65}Zn incorporation is significatively depressed in spina bifida's neonates (Fig. 1). We found the same abnormal ^{65}Zn kinetic with fibroblasts of oldest spina bifida (Fig. 2).

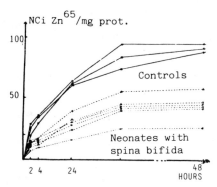

Fig.1 Kinetics uptake of ^{65}Zn in cultured fibroblasts

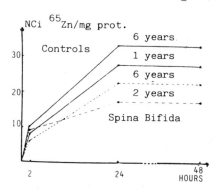

Fig. 2 Kinetics uptake of ^{65}Zn in fibroblasts of oldest children

DISCUSSION

During pregnancy, zinc is decreased in mothers with spina bifida newborn, but numerous mothers with very low zinc level delivered normal babies. An abnormal kinetic of zinc is found in growing fibroblasts of all spina (newborn and children). So, we emit the hypothesis that spina bifida may result from the simultaneous co-existence of a maternal nutritional zinc deficiency and a genetic disorder metabolism in children. Each of this two disorders being enable, separatly to create the defect.

REFERENCES

1. L.S. Hurley, P.B. Mutch, J. Nutr., 103 : 649 (1973)
2. A.O. Cavdar, E. Babacan, S. Asik, et al., Nutr. Res. supp. I : 331, (1985)
3. K.E. Bergmann, M.D.G. Nakosch, K.H. Tews, Am. J. Clin. Nutr., 33 : 2145 (1980)
4. M.H. Soltan, O.H. Jenkins, Br. J. Obstet. Gynecol. 89 : 56 (1982)
5. P.K. Buamah, M. Russel, M. Bates, A.M. Ward, A.W. Skillen, Br. J. Obstet. Gynecol., 91 : 788 (1984)
6. A.W. Zimmerman, D.W. Rowe, Z. Kinderchir, supp. II, 38 : 64 (1983)

EFFECTS OF VARYING DIETARY ZINC IN MICE DURING RECOVERY FROM UNDERNUTRITION

P. N. Morgan, C. L. Keen, G. H. Cardinet III and B. Lönnerdal

Tropical Metabolism Research Unit, UWI, Kingston, JA, and
Depts. of Nutrition and Anatomy, University of California
Davis, CA 95616

INTRODUCTION

Children suffering from severe malnutrition may also have low plasma and liver concentrations of Zn.[1] Currently, it is unclear how reduced tissue Zn may affect pathogenesis or therapy of severe malnutrition. From Zn supplementation studies in children during recovery, Golden and Golden[2,3] concluded that rapid weight gain induced a relative Zn deficiency and lean body mass (LBM) gain was particularly limited. These observations are important because composition of gain during "catch-up growth" is often compromised compared to anthropometric standards.[4] Thus, the purpose of this study was to investigate whether dietary Zn was a limiting factor in LBM gain following early undernutrition using direct analysis, which is feasible only in animal models.

MATERIALS AND METHODS

To induce undernutrition in mice, dams with 12 pups/litter were fed low protein diets. From d10-21 of lactation, dams were fed ad libitum complete, isoenergetic diets containing 4% (LP) or 25% (C) casein and 40 µg Zn/g. During the recovery phase from d21-40, in Expt. 1, pups were fed C diets modified to contain 5, 10 or 40 µg Zn/g. Pair fed (pf) groups were included for pups fed $\leq$ 5 µg Zn/g. The control or reference group was C40, male pups nursed by dams given C diets and then fed the same diet with 40 µg Zn/g during recovery. Carcass composition was determined by proximate analysis and regression equations were derived from direct measures to predict composition of gain. Liver Zn levels were measured by atomic absorption spectrophotometry. Statistical analysis was performed by ANCOVA, with food intake as covariate.

To investigate whether reduced muscle mass in Zn deficient pups could be related to loss or atrophy of skeletal muscle fiber types reported to contain different Zn concentrations, histochemical analyses were done. In Expt. 2, from d21-40, C, LP and pf pups were fed 40 or 1.5 µg Zn/g diets. Gastrocnemius muscles were removed, slow and fast oxidative (high Zn) and fast glycolytic (low Zn) fibers were identified by staining with myofibrillar ATPase and NADH tetrazolium reductase,[5] and the cross sectional area of each fiber type was determined by planimetry.

RESULTS AND DISCUSSION

At d21, LP pups had 30% lower liver Zn concentration and 50% less body weight (BW) than C pups. By d40, all LP pups had recovered except those fed ≤ 5 µg Zn/g diets. These had particularly low protein gains compared to C40 and pf controls. LP pups fed 10 µg Zn/g diets recovered but had low protein gains and high fat gains and therefore had altered final body composition compared to C40 pups (Fig. 1).

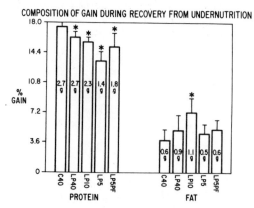

FIG. 1. Percent (and absolute, number within bars) protein and fat gain by C (previously well-nourished) and LP (previously undernourished) male mouse pups fed 25% casein diets containing 40, 10 or 5 µg Zn/g, and pair fed controls, from d21-40. Each value represents x̄ ± SD for ≥ 6 animals. Asterisks indicate significant differences from C40 pups at p < 0.05.

Gastrocnemius muscle weight (Mwt) was severely stunted in LP1.5 and LP1.5pf pups and was 19% and 40%, respectively, of C40 Mwt (0.23 ± 0.05 g, sd). In LP1.5 pups, muscle to body weight ratios were significantly reduced. Histochemical analysis of gastrocnemius muscle demonstrated the presence of slow oxidative, fast oxidative, and fast glycolytic fibers. Mean fiber size was reduced by more than 20% in LP compared to C40 pups.

Early undernutrition in mice stunted growth and reduced liver Zn concentration. During recovery, dietary Zn deficiency limited growth and protein gain, and marginal Zn intake limited protein gain. Gastrocnemius Mwt was particularly stunted in previously malnourished, severely Zn deficient pups, and overall muscle fiber size was reduced. Data from these studies support observations that dietary Zn intake can be a limiting factor in LBM gain as well as growth during recovery from malnutrition.

REFERENCES

1. B. H. Lehmann, J. D. L. Hansen, and P. J. Warren, Br. J. Nutr. 26:197-202 (1971).
2. M. H. N. Golden and B. E. Golden, Am. J. Clin. Nutr. 34:900-908 (1981).
3. B. E. Golden and M. H. N. Golden, Am. J. Clin. Nutr. 34:892-899 (1981).
4. A. Ashworth, Br. J. Nutr. 23:835-845.
5. G. H. Cardinet III and T. A. Holliday, Ann. N.Y. Acad. Sci. 317:290-313.

ALTERED MINERAL METABOLISM AS A MECHANISM UNDERLYING

THE EXPRESSION OF FETAL ALCOHOL SYNDROME IN RATS

S. Zidenberg-Cherr, J. Rosenbaum and C. L. Keen

Department of Nutrition, University of California
Davis, CA 95616

INTRODUCTION

Ethanol (EtOH) intake during pregnancy can cause birth defects in humans and is referred to as Fetal Alcohol Syndrome (FAS). The exact mechanism(s) underlying the teratogenic effects of EtOH are unknown. One hypothesis is that EtOH functions as a teratogen by inducing a state of maternal and/or fetal Zn deficiency. Zn has been shown to be required for DNA and protein synthesis, microtubule polymerization, membrane stabilization and protection from free radical damage (1). Because of the similar characteristics of FAS and Zn deficiency, we have investigated the interaction of Zn and EtOH with regard to fetal outcome in Sprague-Dawley rats.

METHODS

Female rats were fed liquid diets in which EtOH contributed 0% (control; C) or 36% (Et) of kcals. Rats in the Et group were gradually adapted to EtOH by increasing the percentage of EtOH from 10-36% of total kcals. Following the adaptation period, females were bred and continued to be fed their respective diets. A restricted fed group (RF) was included as the consumption of EtOH has been shown to result in reduced food intake.

On d 14 of gestation one-half of the rats in each group were given access to 10 ml of diet spiked with 10 uCi ^{65}Zn. Following its consumption, additional diet without isotope was fed. Rats were killed on d 15 of gestation and the uterus examined intact. The uterus was cut open and fetuses and placentas removed and cleaned of adhering tissues. Tissues were analyzed for Zn concentration and ^{65}Zn activity, and expressed on a specific activity (SA) basis.

On d 21 of gestation the remaining dams were anesthetized and each uterus was examined as described above. Two fetuses and four placentas from each litter were frozen whole and analyzed for mineral concentration. Livers from the remaining fetuses in each litter were removed, pooled and assayed for superoxide dismutase (SOD) activity.

RESULTS

In d 15 fetuses, % resorptions were higher in Et litters compared to C (15% vs 5%, respectively). In addition, fetuses weighed less from Et (0.25 g) vs C litters (0.32 g).

Whole body fetal and placental trace element concentrations were unaffected by EtOH treatment. However, EtOH consumption as well as reduced food intake resulted in low fetal ^{65}Zn specific activity (SA) (Table 1).

Table 1. Effect of EtOH Consumption on Fetal and Placental Zn Retention

	N	Fetal ^{65}Zn SA ($\times 10^3$)	Placental ^{65}Zn SA ($\times 10^3$)	Combined fetal and placental ^{65}Zn retention ($\times 10^2$)
Control	5	6.04 ± 1	8.14 ± 1	1.2 ± 0.5
EtOH	6	4.64 ± 0.8	5.15 ± 0.6	0.24 ± 0.05
Restricted-fed	5	4.84 ± 1	9.45 ± 1	0.51 ± 0.2

Values are mean ± SEM.

Placental ^{65}Zn SA was low in the EtOH relative to control and RF groups. EtOH consumption also resulted in lower ^{65}Zn retention in the fetal placental unit. These results support the hypothesis that the availability of Zn to the fetus may be limited by maternal EtOH consumption. Although fetal Zn concentrations at d 15 of gestation were similar among the EtOH, RF and control groups, we suggest that this reflects the fact that at this stage of development Zn availability is a determinant of fetal growth. Consistent with results from d 15 litters, d 21 fetal liver Zn was similar among the groups.

Consumption of EtOH resulted in higher than normal fetal liver CuZnSOD activity. Values from control and RF litters were 160 ± 10 units/g liver while values from EtOH fed litters were 210 ± 8 units/g liver.

That EtOH exposure resulted in higher than normal activity of fetal liver CuZnSOD is consistent with the results of Dreosti and Record (2) and suggests that the fetus is responding to an increased superoxide ($O_2^{\cdot-}$) load resulting from EtOH metabolism. Since excessive free radicals can result in tissue damage, O_2^- produced during EtOH metabolism may contribute to the teratogenicity of EtOH.

REFERENCES

1. C. L. Keen and L. S. Hurley, Neurotoxicology (in press).
2. I. E. Dreosti and I. R. Record, Br. J. Nutr. 41:399-402 (1979).

ACTION OF HIGH DIETARY COPPER IN PROMOTING GROWTH OF PIGS

G.C. Shurson, P.K. Ku, G.L. Waxler,
M.T. Yokoyama and E.R. Miller

Animal Science Department
Michigan State University
East Lansing, MI 48824 USA

High levels of dietary copper (125 to 250 ppm) are used routinely in Europe (Braude, 1967) and the USA (Wallace, 1967) to promote weight gain of growing pigs. The growth response of pigs to high copper feeding is similar to that of subtherapeutic levels of antibiotics and is a consistent and inexpensive positive response.

This study was designed to determine the importance of the microbial population in the gastrointestinal tract of the pig in the high dietary copper response. To do this, several response criteria were measured in germ-free and conventionally reared pigs receiving either a control or high copper diet.

EXPERIMENTAL PROCEDURE

Germ-free pigs were obtained on the 112th day of gestation by the procedure of Waxler and Drees (1972). Conventional pigs were farrowed normally and nursed their dam for one day. Germ-free pigs were reared in germ-free isolators (Waxler and Drees, 1972) and fed sterilized evaporated cow's milk from a nipple bottle 4 times daily. Conventional pigs were reared in stainless steel cages on the same diet and feeding procedure as that of the germ-free pigs also. At four weeks of age, all pigs were individually weighed and all pigs were offered an autoclaved dry starter diet and autoclaved water ad libitum for the next three weeks. At the end of the trial, all pigs were weighed again individually, and peripheral and portal blood samples were obtained from all pigs 4 hours after feed was withdrawn. Pigs were then euthanized and liver samples were obtained. Liver samples were analyzed for concentration of Cu, Zn and Fe by atomic absorption spectrophotometry. Peripheral and portal plasma were analyzed for concentrations of ammonia and glucose.

RESULTS AND DISCUSSION

The results of the growth trial, the liver element analyses and the plasma ammonia and glucose analyses are summarized in Table 1. Germ-free pigs gained weight at a faster rate (P<.001) than conventionally reared pigs. There was a diet x environment interaction (P<.07) in that high Cu feeding increased growth rate of conventional pigs (23%) but depressed growth rate of germ-free pigs (13%).

In this study, germ-free pigs had higher liver Cu and Zn concentrations than conventional pigs (P<.001) while liver Fe levels were not different. High Cu feeding of germ-free pigs resulted in greatly elevated levels of liver, Cu and Zn, levels which would be indicative of Cu and Zn toxicity and could well account for the reduced growth rate of germ-free pigs fed the high Cu diet. The liver Cu concentrations of the conventional pigs fed the high Cu diet are high but not above that found in growth promotion.

Portal plasma ammonia values were much higher (P<.001) for conventional pigs than for germ-free pigs. Furthermore, portal plasma glucose concentrations were higher (P<.01) in germ-free pigs than in conventional pigs. There was a tendency (P<.07) for high Cu feeding to elevate portal plasma glucose concentration.

CONCLUSION

The data from this study tend to verify that the main effect of high Cu feeding of pigs is antimicrobial in the gut since the response of high Cu feeding in these measures tended to be like that of the germ-free pigs.

Table 1. Comparison of measures of growth, liver element concentrations and plasma ammonia and glucose concentrations of germ-free vs. conventional pigs fed either a control or high Cu diet

Environment	Germ-free		Conventional			P values		
Diet	Control	High Cu	Control	High Cu	MSE[a]	E[b]	D[c]	ExD[d]
Growth								
Initial wt., kg	5.27	5.51	5.97	6.03				
Daily gain, g	223	195	102	126	102	.001	NS	.07
Daily feed, g	520	500	320	340				
Gain/feed	.43	.39	.32	.37				
Liver element concentration (ppm on dry basis)								
Cu	63	1911	15	537	7436	.001	.001	.001
Zn	436	969	105	104	13124	.001	.001	.001
Fe	155	93	126	112	773	NS	.01	.01
Plasma concentrations								
Ammonia, ug/dl								
Peripheral	126	101	139	114	8	NS	.07	NS
Portal	348	327	869	729	420	.001	NS	NS
Glucose, mg/dl								
Peripheral	92	88	70	86	199	.08	NS	NS
Portal	148	226	85	109	33	.01	.07	NS

[a]Mean square error.
[b]Environment effect.
[c]Diet effect
[d]Environment x diet interaction.

REFERENCES

Braude R., 1967, Copper as a stimulant in pig feeding, World Rev. Anim. Prod., 3:69.
Wallace, H.D., 1967, High level copper in swine feeding, Internat. Copper Res. Assoc., Inc., New York.
Waxler, G.L. and Drees, D.T., 1972, Comparison of body weights, organ weights and histological features of selected organs of gnotobiotic, conventional and isolator-reared contaminated pigs, Can. J. Comp. Med., 36:265.

THE ACCUMULATION OF COPPER, ZINC, MANGANESE AND IRON IN THE FOETUS OF DEER

R. Hill, Margaret Leighton, Vivienne Heys and D.M. Jones

The Royal Veterinary College (University of London)
Boltons Park, Potters Bar
Hertfordshire, U.K.

The objects were to determine variations in the accumulation of Cu, Zn, Mn and Fe during the major phase of foetal growth and their distribution in the foetus. The Cu contents of foetal liver of numerous species, particularly of deer, were known to be high (Ashton et al, 1979). In this study Chinese water deer (Hydropotes inermis) were used because fecundity is high, and at Whipsnade Park culling was necessary, providing suitable material for these analyses.

Most calves are born from mid-May to mid-June and pregnant females were culled in mid-March, April and May, providing foetuses from mid to late pregnancy. Within each monthly group there was, inevitably, a fairly wide range of foetal ages. However, mean values gave a helpful picture of the major features of the accumulation of these elements during the period of rapid growth of the foetus.

The foetuses, six for each month, were weighed and dissected into five parts, carcase (C), heart and lungs (H), digestive tract plus contents and reproductive tissue (D), kidney (K) and liver (L). These were weighed and analysed for Cu, Zn, Mn and Fe. From these data, values for whole foetuses were calculated; these are given in Table 1.

Table 1. Weights of foetuses and concentrations of the elements (mean-se)

	Fresh wt. g	Dry wt. g	Concentration mg/kg DM			
			Cu	Zn	Mn	Fe
March	103-19	11.9- 2.7	64.6-9.1	633-30	15.2-1.6	663-80
April	372-54	74.8-11.8	36.9-3.3	382-51	4.2-0.4	74- 8
May	916-88	183.0-27.0	25.3-1.6	393-44	8.1-0.9	83- 6

The concentration and quantity of Zn in the foetus was greatest, followed by values for Fe, Cu and Mn.

The concentrations of the elements in different parts of the foetus are given in Table 2 for March and May; those for April did not alter the overall picture.

617

Table 2. Concentration of the elements in different parts of the foetus in March and May - mg/kg DM

Part of foetus		Cu		Zn		Mn		Fe	
		March	May	March	May	March	May	March	May
C	x̄	15.9	4.88	554	360	15.9	4.40	623	71
	se	3.9	0.59	29	48	2.3	0.55	83	7
H	x̄	23.5	4.95	228	176	10.4	1.53	1194	189
	se	6.1	0.45	27	48	2.8	0.22	211	23
D	x̄	23.2	10.8	362	194	18.7	80.1	569	107
	se	3.8	2.4	55	24	3.8	13.3	180	13
K	x̄	24.5	10.9	662	204	1.69	2.60	257	130
	se	1.8	0.8	40	24	0.44	0.34	42	32
L	x̄	490	1036	1548	1603	7.92	9.70	824	514
	se	27	86	69	56	0.65	0.49	89	46

Concentrations of Cu were much greater, more than 10x, in the liver than in any other part of the foetus, and although the overall Cu concentration in the foetus decreased from March to May (Table 1) the concentration in the liver increased. Concentrations of Zn were also greater in the liver than other parts of the foetus, but differences between values for the liver and other parts were not nearly as great as for Cu.

The proportions of total weights of the elements in different parts of the foetus varied markedly among elements. For Cu 0.80 of the total was in the liver and only 0.20 in the carcase, while for the other elements 0.02 to 0.20 was in the liver and 0.60 to 0.90 was in the carcase.

Among values for Mn, that of D was highest for both March and May foetuses, but in May very much greater than any other part. Analyses of further May foetuses for the three items separately in D, digestive tract tissue, the contents of the digestive tract and reproductive tissue, showed this very high concentration was given almost entirely by Mn in the contents of the digestive tract; the concentration was 433 mg/kg DM, suggesting excretion of Mn from the liver as bile. Some Zn was evidently excreted also in this way, but Cu was not.

These results directed further studies to the, apparently, very special place of Cu, and to a lesser degree of Zn, in the metabolism of the foetal liver

REFERENCES

Ashton, D.G., Jones, D.M., Lewis, G. and Cindirey, R.H., 1979, Preliminary studies on blood and liver copper levels in ungulates at Whipsnade.
Verdandlungersbericht Des XXI Int. Symp. Uber die Erkrangkungen der Zootiere, Mulhouse, p.135-144.

EFFECTS OF MOLYBDENUM ON REPRODUCTION AND MOLYBDENUM/COPPER

ENZYME ACTIVITY IN THE FEMALE RAT

T.V. Fungwe, F. Buddingh, M.T. Yang and S.P. Yang

Dept. of Pathology, Texas Tech Univ. Health Sci. Ctr.
Food and Nutrition, Texas Tech Univ.
Lubbock Texas 79409 USA

INTRODUCTION

According to Wei et al[1] molybdenum (Mo) inhibits mammary tumors. Analogies have been drawn between carcinogenesis and embryogenesis. The effect of Mo on non-ruminant reproduction has not been extensively inves- tigated. The main objective of this study was to investigate the effect of dietary levels of Mo on the estrus cycle, fertility, and reproduction and some molybdenum and Copper (Cu) enzyme activity in the female rat.

MATERIALS AND METHODS

Weanling female SD rats (21/trt.) randomly assigned to 5 dietary treatments were fed AIN-76A diet (0.025 ppm Mo, 6.3 ppm Cu) ad libitum. Sodium molybdate was supplemented in DI drinking water to contain 0, 5, 10, 50, or 100 ppm Mo. The estrus cycle was determined by vaginal smears. Six weeks into the study 6 animals per treatment (non-mated) were sacrificed while in estrus. The remaining 15 animals were mated in pro-estrus with fertile males. Gestation was monitored to day 21, when the dams were weighed and then sacrificed. Tissue, serum, and fetuses were collected and frozen for biochemical analysis. Serum ceruloplasmin (Cp)[2] , hepatic xanthine dehydrogenase/oxidase (XDH) [3,4] , and sulfite oxidase activities (SOX)[5] were determined. All data collected were subjected to analysis of variance (SAS).

RESULTS AND DISCUSSION

No differences were observed in food and water consumption during the first 6 weeks prior to mating, consequently growth and weight gain did not differ. However during gestation treatments fed 10-100 ppm Mo gained less weight than either the controls or the 5 ppm group. This was significantly reflected in the total fetal weights per litter. Litter sizes did not differ significantly. Fetuses from dams on 10, 50, and 100 ppm Mo were significantly ($P > 0.05$) smaller than controls and the 5 ppm group. Mo supplements above 10 ppm significantly prolonged the estrus cycle. This interference was also reported by Wei et al. However this did not affect fertility at the treatment levels in this study. This conclusion is based on the fact that the conception rate exceeded 80% in all treatment groups.

Embryo resorption was more prevalent when Mo was supplemented at 10 ppm and above. The high resorption rates and fewer or no uterine deaths leads us to speculate that Mo may be directly or indirectly affecting fetal development at the embryonic cell and tissue differentiation stage.
Dietary Mo supplementation increased hepatic XDH activity in both mated and non-mated animals. A comparison of both groups showed a difference only at the 100 ppm treatment levels. XDH activity in the mated dams was significantly higher than in the non-mated animals. Although XDH activity is known to increase with Mo supplementation[6], it's role in reproduction is not clear. Hepatic SOX activity also increased significantly with Mo supplementation. In the gestating animals the increase in SOX activity was followed by a significant decline above 5 ppm. It has been demonstrated that a high sensitivity to sulfide[7] due to Cu restriction resulted when high levels of Mo were fed. Whether the decrease in SOX activity in the gestating dams reflects an increase demand for sulfur remains to be investigated. Serum Cp activity did not differ in the non-mated animals. There was a significant increase in the mated dams at 10, 50, and 100 ppm. Nederbragt[8] reported that plasma Cp increased with Cu deficiency in male rats. Coughlan[9] noted that individuals in Mo rich regions of Armenia and India exhibited significant increases of serum Cp. The question to be posed is whether the latter is indicative of a prevention of cellular uptake of copper or mobilization of tissue copper or both.

In summary, Mo supplemented at 10-100 ppm does not affect fertility but could prolong the estrus cycle as well as modulate fetal growth during gestation. XDH and SOX activity increased in normal animals, but SOX activity eventually declined during gestation. (This study was supported in part with funds from the AMAX Inc., Texas Tech University Institute for Nutritional Sciences and College of Home Economics Research Institute.)

REFERENCES

1. H. J. Wei, X. M. Lou, and S. P. Yang, Effect of molybdenum and tungsten on mammary carcinogenesis in SD rats. J. Natl Cancer Instit 74:469 (1985).
2. K. H. Schosinsky, H. P. Lehmann and M. F. Beeler, Measurement of ceruloplasmin from it's oxidase activity in serum by use of O-dianisidine dehydrochloride. Clin. Chem. 20 (12): 1556 (1974).
3. J. L. Johnson, K. V. Rajagopalan and H. J. Cohen, Molecular basis of the biological function of molybdenum. Effects of Tungsten in Xanthine Oxidase in the rat. J. Biol. Chem. 249:859 (1974).
4. F. Stirpe and E. D. Corte, The regulation of rat liver xanthine oxidase conversion in vitro of enzyme activity from dehydrogenase (type D) to oxidase (type O). J. Biol. Chem. 244:3855 (1969).
5. H. J. Cohen and I. Fridovich, Hepatic sulfite oxidase purification and properties. J. Biol. Chem., 246:359 (1971).
6. E. J. Underwood, "Trace Elements in Human and Animal Nutrition". 4th ed. Academic Press NY. (1977).
7. B. Dutt and C. F. Mills, Reproductive failure in rats due to Cu deficiency. J. Comp. Path. 70:120 (1960).
8. H. Nederbragt, The influence of molybdenum on copper metabolism of the rat, different Cu levels of the diet. Br. J. Nutr. 43:329 (1980).
9. M. D. Coughlan, The role of molybdenum in human biology. J. Inher. Metab. Dis. 6 Suppl. 1:70 (1983).

Defective Hepatic Copper Storage in the Brindled Mouse

A. Garnica, M. Perry, J. Bates, and O. Rennert

Department of Pediatrics, The University of Oklahoma
Health Sciences Center, Oklahoma Children's Hospital

The mouse mutant brindled is a model for the human Menkes Kinky Hair syndrome, an X-linked inborn error of copper metabolism (Hunt 1974). The primary metabolic defect in the brindled mouse has not been defined, but the data available are suggestive of a generalized defect of copper metabolism, which is demonstrable in the liver of mutant animals during fetal life (Darwish et al. 1983, Prohaska 1983). Hepatic copper uptake and efflux, cytosol copper binding, and metallothionein metabolism are apparently normal in mutant animals, while intracellular copper retention is impaired. A comparison of copper binding in the subcellular fractions of normal and brindled mice suggests that copper binding in the particulate fractions may be functionally deficient in the mutant mice (Evans and Wiederanders, 1967).

Materials and Methods

Animals: Female C3H and brindled mice were obtained from Jackson Laboratories (Bar Harbor, ME) and are cared for according to guide-lines of the National Research Council. 64Cu was obtained from New England Nuclear (Boston,MA) dissolved in 1 N HCl, diluted, and adjusted to pH 7.0. Stable copper was prepared as cupric chloride solution, pH 7.0, [Cu] 400 mcg/ml. Liver homogenates were subjected to differential ultracentrifugation by an adaptation of the method of Fleischer and Kervina (1974).

Results

Copper concentrations of hepatic subcellular fractions of 5 week-old brindled heterozygotes were not significantly different from those of age-matched controls (Table I). Copper concentrations of hepatic subcellular fractions of brindled hemizygotes of age 10 days, however, were consistently lower than those of control animals ($p < 0.005$) [Table II]. Copper treatment of 5 week-old control mice, 100 mcg/kg injected intraperitoneally, 24 and 12 hours before sacrifice resulted in increases in the copper concentration of all subcellular fractions, but only the increase in the nuclear fraction was significant ($p < 0.005$) [Table I]; the same treatment in 5 week-old brindled heterozygotes produced little change in fraction copper concentrations. Copper treatment of 10 day-old controls over a period of 4 days, 100 mcg/kg/d injected subcutaneously, produced a significant increase in the copper concentration of all fractions ($p < 0.005$); the same treatment in brindled hemizygotes resulted in little change in the copper concentration of any fraction.

Table I. Copper treatment over a 24 hour period in 5 week-old normal and heterozygous brindled mice [total dose, 200 mcg/kg].

fraction	1N	2NI	*3Br	**4BrI
270 x g	7.0+3.0	25 (21-28)	6 (4-10)	6 (5-7)
2500 x g	26.3+8.4	33 (30-35)	24 (17-31)	30 (26-33)
8500 x g	44.8+9.5	55 (29-44)	38 (14-33)	24 (14-33)
165,000 x g	20.1+6.6	35 (30-40)	14 (11-17)	14 (13-15)
Supernatant	41.9+10.2	59 (50-67)	35 (27-42)	46 (39-53)

*Brindled heterozygote
1 N: normal, uninjected (n=8)   3 Br UI: brindled, uninjected (n=3)
2 NI: normal, injected (n=2)   4 Br I: brindled, injected (n=2)

Table II. Copper treatment over a 4 day period in 10 day-old control and hemizygous brindled mice [total dose 400 mcg/kg].

fraction	1 N	2 NI	3 ** Br	4 ** BrI
270 x g	136.3+10.1	545	29 (18-41)	35 (33-37)
2500 x g	162.6+22.4	646	25 (22-29)	64 (56-71)
8500 x g	48.8+8.3	610	14 (10-20)	57 (53-61)
165,000 x g	24.8+4.9	109	9 (6-10)	18 (17-19)
Supernatant	61.2+9.0	378	21 (17-26)	128 (111-140)

**Brindled hemizygote
1 N: normal, uninjected (n=8)   3 Br UI: brindled, uninjected (n=3)
2 NI: normal, injected (n=1)   4 BrI: brindled, injected (n=3)

The incorporation of 64Cu into all subcellular hepatic fractions 8 hours after intraperitoneal injection of 1.0 mcCi/g of isotope was somewhat greater in the normal than the brindled heterozygote (Table III). The greater uptake of copper into the carcass of the brindled mouse suggests that the isotope might not be available for incorporation into liver because of its sequestration in extra-hepatic tissues. Despite pretreatment with parenteral copper, 100 mcg/kg/d for 4 days to reduce the sequestration of label in extrahepatic tissues, the 64Cu uptake into the hepatic subcellular fractions of the normal mouse remained greater than in the brindled mouse, despite similar carcass 64Cu uptakes (Table III). Copper pretreatment increased the uptake of 64Cu in all subcellular fractions in normal and brindled mice. 64Cu uptakes into all subcellular fractions of an 8 day old normal mouse were less than in a normal 4 day old mouse, even after copper pretreatment (Table I).

Table III. Distribution of 64Cu among the subcellular hepatic fractions of normal and brindled mice, cpm x 10^3/mg protein.

fraction	*Nl UI	Br UI	Nl PI	Br PI	NY UI	NY PI
270 x g	--	9.7	--	--	101.2	51.3
2500 x g	17.0	13.9	76.4	44.5	263.8	72.8
8500 x g	21.3	9.9	84.0	52.8	160.9	42.2
165,000 x g	10.0	7.3	39.7	17.3	77.2	51.3
Supernatant	38.7	20.6	76.8	46.7	149.6	117.0
Carcass, cpm/g	3075	4626	3470	3856	5961	6149

*Nl UI, normal uninjected; Br UI, brindled uninjected; Nl PI, normal pre-injected; Br PI, brindled pre-injected; NY UI, normal 4 day-old, uninjected; NY PI, normal 8 day-old, pre-injected.

Conclusion

The critical period(s) of copper requirement during development are undefined, although copper has been suggested to be essential for myelin synthesis and development. The major sites of copper storage in developing mammals are brain, liver, muscle, and bone.

Studies in normal newborn animals indicate that the greater proportion of storage copper is bound in the particulate subcellular fractions (Evans et al. 1970). The data presented here suggest the following: 1) the binding of copper to the particulate hepatic subcellular fractions in brindled mice is reduced, particularly the nuclear (270 x g) and large mito-chondrial (2500 x g) fractions; 2) the capacity of the particulate subcel-lular fractions to bind copper decreases with age but may be induced to a limited degree by copper loading; 3) the inducibility of copper binding to the hepatic subcellular fractions in brindled heterozygotes, and to a greater degree hemizygotes, is impaired.

REFERENCES

Darwish HM, Hoke JE, Ettinger MJ: Kinetics of Cu(II) transport and accumu-lation by hepatocytes from copper-deficient mice and the brindled mouse model of Menkes disease. J. Biol. Chem. 258(22):13621-13626, 1983.

Evans GW, Myron DR, Cornatzer NF, Cornatzer WE: Age-dependent alterations in hepatic subcellular copper distribution and ceruloplasmin. Am. J. Physiol. 218(1):298-302, 1970.

Fleischer S, Kervina M: Subcellular fractionation of rat liver. Methods in Enzymology. 31:6-40, 1974.

Hunt DM: Primary defect in copper transport underlies mottled mutants in the mouse. Nature. 249:852-853, 1974.

Prohaska JR: Comparison of copper metabolism between brindled mice and dietary copper-deficient mice using 67Cu. J. Nutr. 113(6):1212-1220, 1983.

ZINC, COPPER AND IRON METABOLISM BY TURKEY EMBRYO HEPATOCYTES

Mark P. Richards and Norman C. Steele

U. S. Department of Agriculture
Agricultural Research Service
Beltsville, MD 20705 U.S.A.

INTRODUCTION

Growth and development of the avian embryo are dependent on adequate trace element nutriture (1,2). The primary source of trace elements for the avian embryo is the egg yolk which is mobilized through the digestive and absorptive actions of the yolk sac membrane (2). The absorbed metals are then transported to tissue storage sites such as the liver. We have demonstrated developmental changes in hepatic trace element levels during incubation of the turkey embryo (3). However, it is difficult to determine if these changes are the result of developmentally regulated events within the cells of the liver or if they reflect availability of metals derived from the actions of the yolk sac membrane on the yolk metal stores. In order to study hepatic trace element metabolism during avian embryonic development in a defined system, we have devised techniques for the isolation and primary monolayer culture of turkey embryo hepatocytes (4). The purpose of these studies was to investigate the uptake, intracellular partitioning and efflux of Zn, Cu and Fe by embryonic hepatocytes maintained in serum-free culture.

MATERIALS AND METHODS

Primary monolayer cultures of hepatocytes were prepared from the livers of 16-day-old turkey embryos by digestion with collagenase (0.05%). The isolated cells were plated in medium 199 (M199) containing 10% fetal bovine serum (FBS) and 0.5 µg/ml bovine insulin. After 16 hr and at subsequent 24 hr intervals, the medium was replaced with M199 without serum (M199-FBS). For metal uptake and efflux experiments, the medium was replaced with M199-FBS containing 0 to 50 µM Zn, Cu or Fe for periods up to 24 hr. In addition, metal-binding compounds, serum supplements, isotopes and hormones were added at this time. At the end of the experiment, each plate was washed twice with phosphate-buffered saline and the monolayer harvested by scraping with a rubber policeman into 2 ml of deionized water. The cell suspensions were sonicated and analyzed for metals using atomic absorption spectrophotometry (AAS) and for protein using a dye-binding technique (Bio Rad protein assay). A soluble fraction was obtained by centrifugation (166,000 xg for 30 min at $4^{O}C$) of the sonicates and analyzed for metals using AAS. The soluble proteins were fractionated using Sephadex G-75 column chromatography or, for the determination of metallothionein (MT)-bound Zn (MT-Zn), using reversed-phase high-performance liquid chromatography coupled to AAS (5). The data were expressed as µg metal per mg cellular protein.

Table 1. Uptake (mean ± SEM) of Zn, Cu and Fe Over 24 hr by Hepatocytes

Metal Concentration (μM)	Zn	Cu	Fe
	---------------- (μg/mg) -----------------		
0	0.25 ± 0.01	0.03 ± 0.002	0.56 ± 0.03
10	0.84 ± 0.04	0.88 ± 0.01	1.42 ± 0.05
20	1.26 ± 0.04	1.70 ± 0.07	2.52 ± 0.14
40	1.51 ± 0.03	1.83 ± 0.12	3.80 ± 0.41

RESULTS AND DISCUSSION

The uptake of Zn, Cu and Fe by hepatocytes cultured under serum-free conditions was dependent on the medium metal concentration (Table 1) and on the time of exposure. Within 24 hr after removing metal supplementation, 60% of the Zn was lost from the cells, whereas, only 10% of the Cu and 14% of the Fe were lost. Similarly, MT-Zn declined rapidly (half-life=12.3 hr). Since as much as 60% of the newly accumulated Zn was bound to MT, this may account for the rapid turnover of the Zn by hepatocytes in culture. Dexamethasone (10^{-7} M) together with Zn (10 μM) increased the incorporation of ^{65}Zn (3-fold) and ^{35}S-cysteine (9-fold) into MT and the uptake of Zn (3-fold) more so than Zn alone, demonstrating a glucocorticoid responsiveness of the embryonic hepatocytes. Proportionally less of the accumulated Cu was bound to MT (22%) and more (41%) to a high-molecular-weight protein fraction. This is similar to what occurs during hepatic Cu accretion by the turkey embryo in ovo (6). A Zn-Cu interaction was observed in culture. Increasing the medium Zn concentration from 25 to 50 μM resulted in a 50% reduction in Cu uptake (1.46 vs. 0.73 μg/mg). Conversely, increasing the Cu concentration to 20 μM increased Zn uptake by as much as 1.6-fold (0.96 vs. 1.51 μg/mg).

The uptake of Fe was influenced by the form of the metal provided to the cells. Both $FeCl_3$ and ferric nitrilotriacetate (5 to 50 μM Fe) increased cellular Fe levels up to 10-fold. Conalbumin (Fe-saturated) at equimolar Fe levels, on the other hand, was only capable of elevating cellular Fe by a maximum of 1.4-fold, indicative of a receptor-mediated mechanism for Fe uptake. Conalbumin (Fe-free), added to the medium at 1 mg/ml reduced the efflux of Fe from hepatocytes, possibly by promoting a recycling of Fe back into the cells. Serum (FBS) added to the medium (10%) reduced the uptake of Fe by 45% as well as the uptake of Zn (74%) and Cu (47%). The simultaneous addition of all three metals to the medium (10 μM) reduced the uptake of Fe by 25%, Zn by 29% and Cu by 32% compared to singular additions of each metal.

In conclusion, embryonic turkey hepatocytes cultured under serum-free conditions are capable of actively accumulating and metabolizing Zn, Cu and Fe. These findings contribute to an understanding of hepatic metabolism of trace metals during avian embryonic development. Further studies with hepatocytes from embryos of different incubational ages will be useful in determining developmental changes in hepatic trace metal metabolism.

REFERENCES

1. J.E. Savage, Fed. Proc. 27:927-931 (1968).
2. M.P. Richards, and N.C. Steele, J. Exp. Zool. Suppl. 1:39-51 (1987).
3. M.P. Richards, and B.C. Weinland, Biol. Trace Elem. Res. 7:269-283 (1985).
4. S.E. Darcey, M.P. Richards, K.C. Klasing, and N.C. Steele, Fed. Proc. 45:1084 (1986).
5. M.P. Richards, S.E. Darcey, and N.C. Steele, Fed. Proc. 45:1084 (1986).
6. M.P. Richards, R.W. Rosebrough, and N.C. Steele, Comp. Biochem. Physiol. 78A:525-531 (1984).

PERINATAL DEVELOPMENT OF SUPEROXIDE DISMUTASE

IN RAT LIVER AND KIDNEY

Theodor Günther and Jürgen Vormann

Institute of Molecular Biology and Biochemistry
Free University of Berlin
Arnimallee 22, D-1000 Berlin 33, FRG

INTRODUCTION

Superoxide dismutase (SOD) (EC 1.15.1.1.) catalyses the dismutation of the superoxide radical anion (O_2^-) according to the equation:
$2O_2^- + 2H^+ \rightarrow H_2O_2 + O_2$.
There are various isoenzymes of SOD. Dimeric Cu,Zn-SOD and the tetrameric Mn-SOD and EC-SOD (EC = extracellular) have been found in animals[1,2]. The activities of Cu,Zn-SOD and Mn-SOD are highest in liver and kidney[3], where Cu,Zn-SOD amounts to 96-98 % of total SOD activity[3].

O_2^- radicals which are generated by various spontaneous and enzymatic oxidations[1], produce deleterious effects, e.g. lipid peroxidation[1,4]. Lipid peroxidation, performed by O_2^-, was increased by Zn deficiency[5] and reduced by addition of Zn[4]. Zn deficiency[6] also induced fetal malformations[6] and enhanced drug-induced malformations[6] as well as toxicity of drugs[7] whereas Zn supplementation reduced hepatotoxicity of drugs[8,9].

In order to get some insight into the possible mechanism for the protective effect of Zn, we measured perinatal development of SOD activity in liver and kidney of rats fed a normal or Zn-deficient diet.

MATERIALS AND METHODS

Pregnant Wistar rats, weighing 200 g, were fed a control or Zn-deficient diet (Zn content: 1.5 $\pm$ 0.2 ppm) and distilled water ad libitum from day 0 of gestation. At days 17, 19 and 21 of gestation, the uteri were dissected under nembutal anesthesia (50 mg/kg s.c.), and fetal and maternal liver and kidneys were removed. For experimental details see Vormann and Günther[10]. Similarly, liver and kidneys were taken from newborn rats and from normal and Zn-deficient rats, with an age up to 3 months. 700 xg supernatants from liver and kidney homogenates in 50 mM potassium phosphate buffer, pH 7.4, were taken for measurement of SOD activity according to Heikkila and Cabbat[11].

RESULTS

In fetal liver and kidney, SOD activity at pH 7.4 is almost completely

627

inhibited by 5 mM KCN, indicating that in fetal tissues almost all SOD consists of Cu, Zn-SOD as in tissues of adult rats. SOD activity in fetal liver to 13 % of maternal SOD activity. Immediately after birth, SOD activity increased continuously. After 3 months, maternal values were reached. Maternal values were the same as in adult male and non-pregnant female rats. SOD activity in liver was twice the activity in kidney. In Zn-deficient fetal rats, SOD was not changed, although the concentration of metallothionein (MT) in liver was reduced.[6,9]. Probably, the remaining Zn-MT is sufficient to support normal SOD activity.

DISCUSSION

The low SOD activity in fetal tissues and the slow increase to adult values may have phylogenetic reasons. Only with the occurrence of aerobic conditions did this enzyme become essential. The rather late increase of SOD activity may be of clinical significance. Treatment of newborns with high oxygen concentrations may cause a high risk of tissue damage by O_2^-. The low activity of SOD in fetal tissues may partly be compensated by MT because MT can in part protect cells against superoxide and hydroxyl radicals by liberation of Zn after oxidation of thiolates[4] and competition of the released Zn with membrane-bound Fe^{4}. So, Zn-MT by its high concentration in fetal and newborn tissues, besides other actions, has a protective effect.

REFERENCES

1. I. Fridovich, Superoxide dismutases, Advances Enzymol. 41:35 (1974).
2. I. Fridovich, Superoxide dismutases, Advances Enzymol. 58:61 (1986).
3. S. L. Marklund, Extracellular superoxide dismutase and other superoxide dismutase isoenzymes in tissues from nine mammalian species, Biochem.J. 222:649 (1984).
4. J. P. Thomas, G. J. Bachowski and A. W. Girotti, Inhibition of cell membrane lipid peroxidation by cadmium- and zinc-metallothioneins, Biochim. Biophys. Acta, 884:448 (1986).
5. J. P. Burke and M. R. Fenton, Effect of a zinc-deficient diet on lipid peroxidation in liver and tumor subcellular membranes, Proc. Soc. Exp. Biol. Med. 179:187 (1985).
6. R. M. Hackman and L. S. Hurley, Interactions of salicylate, dietary zinc, and genetic strain in teratogenesis in rats, Teratology, 30:225 (1984).
7. J. Vormann, V. Höllriegl, H. J. Merker and T. Günther, Effect of salicylate on zinc metabolism in fetal and maternal rats fed normal and zinc-deficient diets, Biol. Trace Elem. Res. 9:55 (1986).
8. A. S.Prasad, Clinical, biochemical and pharmacological role of zinc, Ann. Rev. Pharmacol. Toxicol. 20:393 (1979).
9. C. P. Chengelis, D. C. Dodd, J. R. Means and F. N. Kotsonis, Protection by zinc against acetaminophen induced hepatotoxicity in mice, Fund. Appl. Toxicol. 6:278 (1986).
10. J. Vormann and T. Günther, Development of fetal mineral and trace element metabolism in rats with normal as well as magnesium- and zinc-deficient diets, Biol. Trace Elem. Res. 9:37 (1986).
11. R. E. Heikkila and F. Cabbat, A sensitive assay for superoxide dismutase based on the autoxidation of 6-hydroxydopamine, Analyt. Biochem. 75:356 (1976).

THE AMENITY AND ANNOYANCE: CIRCADIAN RHYTHM, FUNCTION OF THE LIMBIC
SYSTEM AND METAL METABOLISM IN MICE

K. Hoshishima
2, Watari-Yamanoshita, Fukushima-shi, 960   Japan

INTRODUCTION

At TEMA-5,[1] I reported that the prenatal administration of trace
amounts of 13 metals and fluoride was effective in changing circadian
rhythms in mice. In a subsequent study,[2] treatments (bilateral lesions of
the central nucleus of amygdala, partial removal of the olfactory bulb, and
chemical sympathectomy) and the function of the limbic system were also
studied.

In the present study, the combined effects of metal administration
(chromium) and the three treatments upon circadian rhythms were analysed to
obtain information on relationships regarding biological rhythms, the
emotional state, and metal metabolism in mice.

MATERIALS AND METHODS

Male mice of the CFW strain were subjected to a mild change of light-
dark cycle, and circadian rhythm was recorded and analysed. Details of the
experimental procedure applied in this experiment are given in reference 2.

RESULTS

Changes in drinking patterns indicated the effects of the treatments.
The sympathectomized mice had more precise 24 hr rhythms than did the other
groups. A more pronounced effect was seen in mice exposed to Cr. However,
the normalization to proper rhythm during the recovery phase was observed to
occur sooner in bulbectomized and amygdala-lesioned mice than in sympathectomized mice. The prenatal administration of Cr (see figure) masked the
effects of these two treatments.

The figure demonstrates the clear effects of Cr exposure and the three
treatments upon the measured value. Other metals and F also caused specific
changes in drinking patterns with a combination of the three treatments.

The endorphin contents in brain did not show any significant difference
among the groups of mice tested.

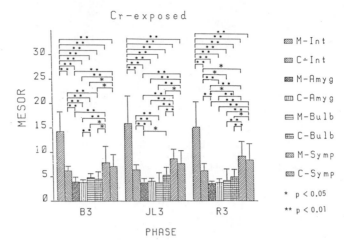

Fig. 1 An example of the rhythmicity analysis at the third day of each phase.
C: control, M: prenatally metal administered, Int: intact, Amyg: amygdala lesioned, Bulb: olfactory bulb removed, Symp: sympathectomized, B: basic phase, JL: Jet Lag phase, R: recovery phase.

COMMENTS

The changes in circadian rhythm due to the three different treatments for the function of the limbic system indicate that the emotional state has a bearing upon the drinking pattern rhythm in mice, and that metal administration affects this "emotion-related" rhythmicity.

It is a well-known fact that the disturbance of proper circadian rhythm causes "annoyance." The so-called Jet-Lag phenomenon is a common example of this disturbance.

The present results reinforce the author's proposal at TEMA-5 that research projects concerning metal metabolism and circadian rhythm pave the road for attractive and newer research concerning "Amenity-Annoyance" problems.

REFERENCES

1. K. Hoshishima and S. Shimai, Trace amounts of metal(s) prenatally
   administered and the circadian drinking rhythm in mice, in TEMA-5, D.F.
   Mills, I. Bremner and J.F. Chesters (eds.), Commonwealth Agricultural
   Bureaux, Farnam Royal, UK, pp. 292-294 (1985).
2. K. Hoshishima and S. Shimai, Trace amounts of metal(s) prenatally
   administered and the circadian drinking rhythm in mice, in Trace
   Substances in Environmental Health, XIX, D.D. Hemphill (ed.), Univ. of
   Missouri Press, pp. 417-424 (1986).

MARGINAL Zn DEFICIENCY AFFECTS PUP BRAIN MICROTUBULE ASSEMBLY IN RATS

P. I. Oteiza, C. L. Keen, B. Lonnerdal and L. S. Hurley

Dept. of Nutrition, University of California, Davis, CA 95616

INTRODUCTION

Zinc has been shown to bind to tubulin (Hesketh, 1983; Eagle et al., 1983) and to have a stimulatory effect on tubulin assembly in vitro (Haskins et al., 1980; Hesketh, 1982). There is evidence which supports a physiological role of Zn on microtubule (Mct) formation. High concentrations of Zn have been found to be associated with Mct obtained from sea urchin spermatozoa and in the mitotic apparatus from sea urchin eggs (Morisawa & Mohri, 1972). Hesketh (1981) has reported that tubulin polymerization in brain supernatants from severely Zn deficient (< 1 ug Zn/g diet) pigs and rats was slower than in controls.

In contrast to a severe Zn deficiency syndrome, marginal Zn deficiency is more likely to occur in human populations. We have therefore evaluated the effect of a maternal marginal Zn diet (10 ug Zn/g diet) fed during pregnancy and lactation on in vitro brain tubulin reassembly in the 20 day old pups, and compared these results to those from control pups from dams fed adequate Zn (50 ug Zn/g diet). Virgin female Sprague-Dawley rats weighing 200-225 g were fed either 10 or 50 ug Zn/g diets from day 0 of pregnancy and pups were killed at 20 days after birth. Experiments were carried out in pairs comprised of both experimental groups, and the size of the litters was adjusted to be the same for each pair.

Mct polymerization in vitro was measured in brain supernatants (SN) obtained after centrifuging the homogenate (1 g brain in 3 ml of 0.1 M Pipes, pH 7.0) at 100,000 g for 30 min at 4°C. The SN was decanted, and aliquots were taken to determine protein and Zn concentrations. Tubulin assembly was started by placing 1 ml aliquots of SN at 37°C and the increase in absorbance at 350 nm was recorded over a 60-80 min period.

Body and brain weights and protein concentration in the brain SN were not significantly different between the groups. Plasma zinc concentration was lower in the 10 Zn group (1.3 ± 0.1 ug/g) compared to the 50 Zn group (2.0 ± 0.1 ug/g). There were no significant differences in the brain SN Zn concentration (1.43 ± 0.03 (50 Zn); 1.35 ± 0.05 (10 Zn) ug/g).

The characteristics of the polymerization kinetics were different between the groups (Table 1). In the 10 Zn SN, the lag period was longer and the initial velocity of the reaction was slower compared to controls.

631

Table 1. Kinetics of Tubulin Polymerization in Brain Supernatants of 20 Day Old Pups from Dams fed 50 or 10 ug/g Zinc Diets.

Dietary Group	Lag Period (min)	Initial Velocity	Elongation Velocity
50 Zn	1.1 ± 0.3	0.0030 ± 0.0001	0.0021 ± 0.0001
10 Zn	2.9 ± 0.8[a]	0.0025 ± 0.0001[b]	0.0019 ± 0.0001

The initial velocity was calculated between 5 to 20 min and the x intercept of this initial curve was defined as the lag period. The elongation velocity was calculated between 30 to 60 min. Both velocities express the increase in absorbance at 350 nm per min. Data are expressed as mean ± S.E.M., and are the average of 4 litters in the 50 Zn group and 6 litters in the 10 Zn group. Data analysis by paired t-test showed differences to be significantly significant: [a] $p < 0.001$; [b] $p < 0.01$.

Table 2. Effect of Added Zinc on Tubulin Assembly Kinetics.

Zinc added (uM)	Lag period (%)	Initial velocity (%)	Elongation velocity (%)
0	100	100	100
37	59 ± 2	146 ± 6	113 ± 5

Results are expressed as percentage of the data when no zinc was added.

The elongation velocity was not significantly different between the groups. The lag period and the initial velocity reactions are parameters of the nucleation phase of Mct polymerization. The effect of added Zn ($ZnCl_2$) in concentrations (11-38 uM) known to produce normal microtubules in vitro, was characterized in SN of 20 day old control pups. Table 2 shows the effect of adding 37 uM Zn on Mct polymerization parameters. The additional Zn decreased the lag period, increased the initial velocity and had little effect on the elongation velocity of Mct polymerization.

As previously suggested by Hesketh (1984), the effect of Zn seems to be mainly associated with the nucleation phase. Changes in the kinetic parameters of tubulin assembly in the presence of added Zn agree with changes found in the marginal Zn animals, suggesting that a decrease in a pool of Zn available for tubulin polymerization is responsible for the alterations observed in the 10 Zn animals. Consistent with these data, we have previously found that brain tubulin assembly is also altered in dams fed marginal Zn diets during pregnancy and lactation (Oteiza et al., 1987).

Since Mct participate in a multitude of cellular functions, an impairment in tubulin polymerization could explain some of the structural and neurological alterations observed with zinc deficiency during prenatal and early postnatal development.

REFERENCES

Eagle, G., Zombola, R., and Himes, R., 1983, Biochemistry, 22:221-228.
Haskins, K. M., Zombola, R. R., Boling, J. M., Lee, Y. C., and Himes, R. H., 1980, B.B.R.C. 95:1703-1709.
Hesketh, J., 1981, Int. J. Biochem. 13:921-926.
Hesketh, J., 1983, Int. J. Biochem. 15:743-746.
Hesketh, J., 1984, Int. J. Biochem. 16:1331-1339.
Morisawa, M., and Mohri, H., 1972, Expl. Cell. Res. 70:311-316.
Oteiza, P. I., Keen, C. L., Lonnerdal, B., and Hurley, L. S., 1987, Fed. Proc. 45:596.

EFFECTS OF MATERNAL DIETARY ZINC ON THE EXPRESSION OF DIABETES-INDUCED TERATOGENICITY IN THE RAT

Janet Y. Uriu-Hare, Judith S. Stern and Carl L. Keen

Department of Nutrition and the Food Intake Laboratory
University of California, Davis, California 95616

INTRODUCTION

Although prenatal care of pregnant diabetic women has improved fetal outcome, there is still a two to four times higher incidence of congenital anomalies in the offspring of diabetic mothers compared to the general population (1). Fetuses from diabetic rats have low liver Zn levels and a high malformation frequency (2,3) compared to controls. The malformations are similar to those seen in fetuses from Zn-deficient dams. These data suggest that the teratogenic effect of diabetes is due in part to an induced fetal Zn deficiency. In this study, we tested the hypothesis that varying dietary intakes of Zn would affect the expression of diabetes-induced teratogenicity. Rats were injected with STZ (40 mg/kg) or buffer (controls) 14 days prior to mating, and fed diets containing 4.5, 25 or 500 ug Zn/g diet on day 0 of gestation. Fetuses were taken by caesarian section on day 20.

RESULTS AND DISCUSSION

Increasing dietary Zn increased maternal Zn and metallothionein (MT) levels in liver and kidney. Diabetic rats consistently had higher MT levels than controls, suggesting both a hormonal and a dietary influence on its synthesis and/or degradation. The function of the Zn accumulation in the form of MT and whether this Zn is available to the animal is unknown. Plasma Zn levels within each dietary treatment were similar between control and diabetic dams. As dietary Zn increased, plasma Zn levels also increased. It has been suggested that due to glycosylation of proteins in diabetes, there may be shifts in the distribution of Zn bound to albumin (exchangeable Zn pool) and Zn bound to alpha-2-macroglobulin (non-exchangeable Zn pool). However, there were no significant differences in the percentages of albumin-bound and alpha-2-macroglobulin-bound Zn between control and diabetic dams.

Fetuses from control dams fed the marginal Zn diet had lower liver Zn and MT levels compared to controls with adequate or supplemental Zn intakes. In contrast, there were no differences among liver Zn or MT levels of fetuses from diabetic dams, regardless of maternal dietary Zn intake. The liver Zn and MT levels of fetuses from diabetic dams were similar to the low levels seen in fetuses from control dams fed the marginal Zn diet. Thus, there is a marked difference in liver Zn uptake and/or retention between fetuses from diabetic and control dams. This is in contrast to what is seen in the diabetic dams, where Zn and MT levels were higher in the diabetic dams compared to controls.

Although fetal liver Zn levels in pups from diabetic dams did not reflect changes in dietary Zn levels, fetal outcome was improved in the diabetic animals by increasing maternal dietary Zn intake. In the controls, fetal length was shortest in the low Zn group. With adequate Zn, linear growth increased but did not improve further with high Zn diets. In contrast, length of fetuses from diabetic dams continued to improve with increasing dietary Zn such that the length of fetuses from the high Zn group approached that of the low Zn controls. The same positive effect of supplemental Zn diets in the diabetic dam was seen with fetal weight. The development of the fetal skeletal system was evaluated using a differential staining technique. Fetuses from marginal Zn diets had the fewest calcified sternal sites. Calcification of sternal sites was improved with adequate Zn diets, with an 11% increase in the control fetuses and a 29% increase in the diabetic fetuses. The data indicate that low Zn diets in the diabetic rat had a more deleterious effect on fetal calcification than in controls. There was no further increase with high Zn diets in either the control or diabetic groups. The number of ossified caudal vertebrae, metacarpals and anterior phalanges followed similar patterns. It is possible that even with supplemental Zn in the diet, fetuses from diabetic rats may still not have adequate Zn for saturation of Zn pools and bone development. Alternatively, bone calcification may be affected by the high glucose environment seen in the fetuses from diabetic dams.

In addition to lower bone calcification in fetuses from diabetic dams, there were skeletal and other gross anomalies. Overall malformation frequency was higher in diabetic animals vs. controls but there was a marked effect of dietary Zn in the diabetic rat. An adequate Zn diet lowered the incidence of malformations dramatically. Again, it is evident that a low Zn diet coupled with diabetes has a potent teratogenic effect. However, the high Zn diets did not further reduce the malformation frequency of fetuses from diabetic dams. As liver Zn levels were low in fetuses from diabetic dams regardless of dietary Zn intake, there may be a problem of Zn transfer across the placenta, or of Zn handling in the fetuses.

There were marked alterations in trace element metabolism in both the diabetic dams and their fetuses. While diabetic dams had an accumulation of Zn in liver and kidney compared to controls, fetuses from diabetic dams had low levels of liver Zn vs. controls, regardless of dietary Zn intake. Fetal outcome was improved in diabetic dams fed adequate and supplemental Zn diets; fetal length and weight increased, ossification of skeletal sites improved, and most importantly, fetal abnormalities decreased. These results support our hypothesis that altered Zn metabolism plays in important part in the teratogenicity of the diabetic state. Marginal Zn intakes during pregnancy may be a problem of many women. Based on our findings, we suggest that pregnant diabetic women who are already at high risk for poor fetal outcome, should be closely monitored with regard to their Zn status.

REFERENCES

1. J. L. Simpson, E. Sherman, A. O. Martin, M. S. Palmer, E. S. Ogata,
      and R. A. Radvany, Am. J. Obstet. Gynecol. 146:263, (1983).
2. J. Y. Uriu-Hare, J. S. Stern, G. M. Reaven, and C. L. Keen, Diabetes
      34:1031 (1985).
3. U. J. Eriksson, J. Nutr. 114:477 (1984).

THE EFFECT OF MONENSIN SODIUM ON TRACE ELEMENT STATUS OF CATTLE AT PASTURE

Nick D. Costa

School of Veterinary Studies
Murdoch University
Murdoch, W A Australia 6150

INTRODUCTION

When fed to sheep and cattle the carboxylic ionophore, monensin sodium, can produce a wide variety of effects on ruminal function with the main effect being a decrease in the acetate:propionate ratio[1]. Monensin has been used to increase feed conversion efficiency in cattle and sheep fed high concentrate diets and also to improve average daily gain in cattle at pasture[2]. Although the improvement in performance has been attributed mainly to shifts in ruminal fermentation towards an increase in propionate production, this increase alone can not account for all the improvement in productivity[1]. More recently monensin has been shown to increase retention of Se and Zn in steers[3]. Monensin supplementation in sheep increases Mg, P and Zn retention and Zn absorption[4] and also increases blood glutathione peroxidase activity[5]. The increase in retention of selenium in steers[3] and of zinc in lambs[4] observed during monensin supplementation is of sufficient magnitude to suggest that monensin may help alleviate problems of dietary deficiency of selenium and zinc. Ruminal delivery devices (RDD) designed to release 150 mgs of monensin per day for 150 days provided a means for assessing the hypothesis that monensin could significantly alter trace element status in cattle at pasture.

MATERIALS AND METHODS

Five Angus cross heifers of mean weight 176±3.4 kg received a single RDD and five heifers of mean weight 173.2±4.3 were left untreated. All ten heifers were grazed on unirrigated oat and clover pasture at the School of Veterinary Studies farm from 15th August (beginning of Spring; the main growing season) to 15th March (end of Summer). All animals were fed hay, oats and lupins from the end of January to mid-March, i.e. the period after exhaustion of the monensin from the RDDs. Each heifer was weighed fortnightly at which time blood samples were collected from the jugular vein for assay of glutathione peroxidase, selenium, haemoglobin and plasma copper, iron and zinc.

RESULTS

During the first 90 days of the trial both treatment groups gained 1.35 kg per day. For the next 60 days when pastures had dried off, the monensin treated heifers gained 0.49±0.28 kg per day while the untreated heifers gained 0.35±0.38 kg per day. The higher apparent rate of gain in the monensin group could be attributed to the fact that these animals lost less weight during early December when there was a sharp decline in the productivity of the pasture. In the period from mid-January to mid-March the RDDs were no longer releasing momensin and the untreated group showed compensatory growth under supplementation with hay, oats and lupins such that there was no difference in weight between the treatment groups at the end of the trial. Selenium status in all heifers was low to marginal (i.e. <40 ng/ml blood) with blood selenium concentrations in the momensin treated heifers of 20.5±1.2 ng/ml which were consistently and significantly greater (P<0.001) than in untreated heifers (15.1±1.2 ng/ml). This was not associated with significantly higher blood glutathione peoxidase (GSH.Px) activity in monensin treated heifers (23.4±1.8 Eu/g Hb) than in untreated heifers (21.5±1.5 Eu/g Hb). Mean plasma copper concentrations were 0.60±0.018 µg/ml in the untreated heifers and 0.595±0.024 µg/ml in the monensin treated heifers. Plasma zinc concentrations in the monensin treated heifers (0.953±0.017 µg/ml) were significantly higher (P<0.05) than in the untreated heifers (0.882±0.021 µg/ml). Plasma iron and blood haemoglobin concentrations were respectively 1.586±0.102 µg/ml and 12.3±1.3 g/100 ml in untreated heifers and 1.627±0.114 µg/ml and 12.3±1.2 g/100 ml in monensin treated heifers.

DISCUSSION

   Monensin did not increase the efficiency of production in heifers during the period that both groups were growing at over 1.0 kg per day. In fact the only improvement in productivity came during the period of lowest average daily gain (ADG) of less than 0.5 kg per day. This is consistent with the findings of Potter et al[2] who reported an overall 14.4% increase in ADG in monensin treated cattle at pasture but substantially lower percentage increase in groups growing at approximately 1.0 kg per day. Although monensin significantly increases blood selenium in cattle at pasture, this did not substantially alter the selenium status from deficient (<20 ng/ml) to adequate (>40 ng/ml). The blood glutathione peroxidase activity was not significantly affected by monensin in direct contrast to the findings of Anderson et al[6] in sheep. Thus monensin released from this type of RDD in heifers will not compensate a deficient selenium intake nor the need to supplement selenium. Plasma copper, iron and zinc concentrations were in the normal range for cattle indicating that the pastures were not limiting for these nutrients. Nothwithstanding this, the monensin treated heifers had significantly higher plasma zinc concentrations. For this reason, judgement must be reserved as to whether monensin will compensate a deficient zinc intake.

REFERENCES

1.   G.T. Schelling, Monensin Mode of Action in the Rumen, J. Anim. Sci., 58:1518 (1984).
2.   E.L. Potter, R.D. Muller, M.I. Wray, L.H. Carroll and R.M. Meyer, Effect of Monensin on the Performance of Cattle on Pasture or Fed Harvested Forages in Confinement, J. Anim. Sci., 62:583 (1986).
3.   N.D. Costa, P.T. Gleed, B.F. Sansom, H.W. Symonds, and W.M. Allen, Monensin and Narasin Increase Selenium and Zinc Absorption in Steers in "Trace Elements in Man and Animals (TEMA-5)", pp. 472, CAB.
4.   D.J. Kirke, L.W. Greene, G.T. Schelling and F.M. Byers, Effects of Monensin on Mg, Ca, P and Zn Metabolism and Tissue Concentration in Lambs, J. Anim. Sci., 60:1485 (1985).
5.   P.H. Anderson, S. Berrett, J. Catchpole, M.W. Gregory and D.C. Brown, An Observation on the Effect of Monensin on the Selenium Status of Sheep, Vet. Record, 113:498 (1983).

THE EFFECTS OF DIFFERING COMPOSITIONS OF SOLUBLE PHOSPHATE GLASS BOLUSES ON THE COPPER, COBALT AND SELENIUM STATUS OF SWALEDALE EWES

P. M. Driver, C. Eames and S. B. Telfer

Department of Animal Physiology & Nutrition
The University of Leeds
Leeds   LS2 9JT   England

INTRODUCTION

Several studies have shown that the copper, cobalt and selenium status of ewes can be improved by the administration of soluble glass boluses (Zervas, 1983; Driver et al., 1985). To study the relationship between the utilisation of these elements and their supply from glass boluses, we have examined the effects of boluses of differing solubilities and containing two levels of P_2O_5. The effects of 3 compositions of soluble phosphate glass boluses on ewes are reported.

MATERIALS AND METHODS

The predicted solubilities ($mg/cm^2/day$) of the boluses were: Bolus A = 1, low P_2O_5; Bolus B = 1, high P_2O_5; Bolus C = 4, high P_2O_5.

Each of 45 breeding ewes in 3 groups of 15, received one bolus (either A or B or C) in November (day 0). A further 15 ewes were kept as untreated controls. The ewes grazed a known Cu/Co deficient pasture except for days 73-129 when they were housed for lambing and fed hay. Blood samples were taken at intervals. Plasma copper was determined by atomic absorption, serum vitamin B12 by RIA (Becton-Dickinson), and erythrocyte glutathione peroxidase (GSHPx) by an automated modification of the method described by Al-Tekrity (1986).

RESULTS

In tables, values with differing letters are significantly different.

Table 1. Mean Plasma Cu (μ moles/litre) ± sem (n).

Day	Control	Bolus A	Bolus B	Bolus C
0	14.1±0.5 (15)	14.2±0.4 (15)	15.2±1.0 (15)	13.1±1.1 (15)
73	10.9±0.7 (15)	12.4±0.5 (12)	12.7±0.5 (15)	12.4±0.5 (13)
129	13.5±0.7 (14)	13.7±0.7 (15)	14.1±0.4 (15)	15.1±0.6 (14)
233	9.8±0.8 a(14)	14.8±0.8 b(15)	14.7±0.4 b(15)	15.8±0.9 b(14)
315	9.8±0.7 a(10)	15.0±1.3 b (5)	14.1±0.7 b(11)	14.6±0.9 b (8)
365	14.1±1.1 a(14)	17.8±1.0bc(13)	14.8±0.7ab(12)	18.3±1.1 c (9)

Table 2. Mean Erythrocyte GSHPx (units/cm^3 pcv.) ± sem (n).

Day	Control			Bolus A			Bolus B			Bolus C		
0	45±5	a	(15)	40±3	a	(15)	79±9	b	(15)	94±10	b	(15)
73	80±7	a	(15)	62±5	a	(14)	134±15	b	(15)	84±7	a	(13)
129	75±4	a	(14)	120±11	b	(13)	103±12	b	(15)	167±8	c	(14)
233	43±4	a	(14)	72±7	b	(15)	98±10	bc	(15)	116±10	c	(14)
315	17±2	a	(10)	37±5	b	(5)	40±4	b	(11)	66±9	c	(8)
365	22±2	a	(14)	50±5	b	(13)	50±4	b	(12)	74±6	c	(9)

Table 3. Mean Serum Vitamin B12 (pg/cm^3) ± sem (n).

Day	Control			Bolus A			Bolus B			Bolus C		
0	460±32		(14)	427±37		(14)	512±78		(15)	519±67		(15)
73	1302±74	a	(13)	1759±258	a	(15)	1743±192	a	(15)	2585±318	b	(13)
129	2801±248	a	(14)	3017±285	a	(15)	3955±333	b	(15)	4176±397	b	(14)
233	341±52	a	(14)	1531±179	c	(14)	997±181	b	(13)	1472±212	c	(12)
315	272±45	a	(10)	1455±192	b	(5)	1518±142	b	(11)	2170±291	c	(8)
365	370±32	a	(14)	1144±135	b	(12)	1183±91	b	(12)	1230±133	b	(9)

Plasma Cu and vitamin B12 status improved in all groups upon housing. The mean Cu, Se and B12 status of the untreated ewes became marginal over the summer with some individuals becoming deficient. By day 129, ewes with Bolus C had significantly higher Cu, GSHPx and B12 levels compared with controls. Significant improvements in the trace element status in all the bolused groups were seen from day 233 onwards, and were maintained until the end of the trial. The differences in GSHPx on day 0 are probably a result of some ewes having initially grazed a different pasture.

DISCUSSION

Housing with extra feeding may have prevented severe deficiencies in the untreated ewes during the winter. However all 3 boluses resulted in an improvement of the trace element status of the treated ewes (cf. controls) by day 129 (i.e., during housing) and subsequently in preventing deficiencies during the summer. Furthermore these data indicate that the boluses remained effective for one year. Boluses recovered from cast ewes gave estimated solubilities of A=1.0, B=1.4 and C=3.1 mg/cm^2/day. Bolus C had the better overall performance, probably attributable to its higher solubility. There were no apparent differences between the effects of boluses A and B, although B may have been more effective during the first few months.

In conclusion we feel that these results confirm that soluble glass boluses are an effective means of providing supplementary trace elements to breeding ewes (even at a solubility of 1.0 mg/cm^2/day) but that glass of composition C is the most successful bolus extensively tested.

REFERENCES

Al-Tekrity, S.S.A., 1986, Selenium Metabolism and Deficiency in Ruminant Animals, Ph.D. Thesis, University of Leeds.
Driver, P.M., Kidger, M., Anderson, P.J., Carlos, G., Illingworth, D.A.V., and Telfer, S.B., 1985, An investigation into the copper, vitamin B12 and selenium status of winter housed ewes following treatment with soluble glass boluses, Anim. Prod. 40:566.
Zervas, G., 1983, Prevention of Copper Deficiency in Ruminants by Means of Soluble Glass Rumen Bullets, Ph.D. Thesis, University of Leeds.

THE EFFECT SOLUBLE GLASS BOLUSES UPON THE COPPER AND SELENIUM STATUS OF GRAZING CATTLE

P. M. Driver, C. Eames and S. B. Telfer

Department of Animal Physiology & Nutrition
The University of Leeds
Leeds   LS2 9JT   England

INTRODUCTION

Previous trials have shown that soluble glass boluses are an effective means of improving the copper, cobalt and selenium status of cattle (Telfer et al., 1984; Driver et al., 1986). This study was carried out to confirm the results of previous years and to compare the efficacy of the bolus with an injection of copper-calcium edetate.

MATERIALS AND METHODS

Twenty-four steers were divided into 3 groups. On day 0 (April) 12 steers each received 2 soluble glass boluses (containing Cu, Co & Se), 6 steers were injected with 100 mg Cu-Ca edetate and 6 animals were left as untreated controls. The cattle grazed a Cu/Se deficient pasture over the summer. Blood samples were taken at mid-summer (day 93) and in autumn (day 156). The blood samples were analysed for plasma Cu by atomic absorption, for plasma caeruloplasmin (CP) by a modification of the method of Henry et al. (1974); and for erythrocyte Glutathione Peroxidase (GSHPx) by the method described by Al-Tekrity (1986).

RESULTS AND DISCUSSION

Table 1 shows that although the injected cattle had lower mean GSHPx levels throughout the trial, they followed a similar trend to the untreated group. Both these groups became selenium deficient over the summer. The bolused animals had significantly elevated GSHPx levels by day 93 and remained Se sufficient over the 5 months.

In tables, values with differing letters are significantly different.

Table 1. Mean Erythrocyte GSHPx (units/cm^3/pcv.) ± sem (n).

Day	Control	Injected	Bolused
0	28.8±3.3 (6)	19.7±3.4 (6)	27.1±3.1 (12)
93	30.5±2.2 a (5)	21.5±2.5 b (6)	41.1±1.7 c (12)
156	19.8±2.9 a (6)	13.6±0.9 a (5)	45.5±2.7 b (10)

Table 2. Mean Plasma Copper (μ moles/litre) ± sem (n).

Day	Control		Injected		Bolused	
0	17.8±2.4	(6)	15.5±1.1	(6)	18.1±1.0	(12)
93	7.4±1.8 a	(5)	12.0±0.8 b	(6)	14.1±0.5 b	(12)
156	8.7±2.4	(6)	8.4±2.4	(5)	13.3±1.6	(10)

Table 3. Mean Plasma Caeruloplasmin (mg /cm^3) ± sem (n).

Day	Control		Injected		Bolused	
0	24.7±4.6	(6)	22.8±2.0	(6)	24.4±2.2	(12)
93	11.2±3.9 a	(5)	19.7±2.6 b	(6)	19.9±1.3 b	(12)
156	5.2±1.6	(6)	7.0±1.3	(5)	9.9±1.4	(10)

By mid-summer the untreated animals were Cu deficient whilst both bolused and injected groups had significantly higher plasma Cu levels. After 156 days the injected steers had also become Cu deficient whereas the bolused group appeared still (but non-significantly) sufficient. These results suggest that, on this pasture at least, treatment with soluble glass boluses was more effective than the injection of 100 mg Cu-Ca edetate in maintaining adequate plasma copper levels.

Mean values obtained from analyses of the pasture herbage were: Copper = 7.15 mg/kg; Molybdenum = 3.70 mg/kg; Sulphur = 0.38%; (calculated available Copper = 2.3%); Selenium 0.05 mg/kg. These results indicate a severe copper and selenium challenge to the animals.

Boluses recovered from 4 steers at slaughter (212 days) had a range of dissolution rates 2.6 - 3.9, mean = 3.13, mg/cm^2/day.

We conclude that these soluble glass boluses will improve and maintain the copper and selenium status of cattle, at least over the summer grazing period.

ACKNOWLEDGEMENTS

We acknowledge the assistance of Mr. G.C. Pritchard (V.I.C.) during this study.

REFERENCES

Al-Tekrity, S.S.A., 1986, Selenium Metabolism and Deficiency in Ruminant Animals, Ph.D. Thesis, University of Leeds.

Driver, P.M., Carlos, G.M., Eames, C. and Telfer, S.B., 1986, The efficacy of soluble glass boluses (Cosecure) in grazing cattle, Anim. Prod., 42:472.

Henry, R.J., Cannon, D.C. and Winkelman, J.W., 1974, Clinical Chemistry, 2nd Edition, pub. Harper and Row.

Telfer, S.B., Zervas, G. and Carlos, G., 1984, Curing and preventing deficiencies in copper, cobalt and selenium in cattle and sheep using Tracerglass, Can. J. Anim. Sci., 64(suppl):234.

THE EFFECT OF DIFFERENT COMPOSITIONS OF SOLUBLE PHOSPHATE GLASS BOLUSES ON THE COPPER AND VITAMIN B12 STATUS OF GROWING LAMBS

P. M. Driver, C. Eames and S. B. Telfer

Department of Animal Physiology & Nutrition
The University of Leeds
Leeds   LS2 9JT   England

INTRODUCTION

Soluble glass boluses have been found to be effective in improving the copper, cobalt and selenium status of young lambs (Care et al., 1985). We report the results of a trial, carried out under commercial conditions, to study the effect of boluses of differing solubilities in lambs.

MATERIALS AND METHODS

The predicted solubility of each bolus type was $(mg/cm^3/day)$: Bolus X = 1.0, Bolus Y = 2.3 and Bolus Z = 3.7.

Eighty Suffolk-cross lambs (mean weight = 20.6 kg) were divided into 4 groups (3 treated and 1 untreated control). Each treated lamb received one (17 g, lamb size) soluble glass bolus (containing Cu, Co & Se) in May (day 0). The lambs grazed a known copper deficient pasture throughout the trial. Boluses were recovered at slaughter on days 69 and 272. Blood samples were taken at intervals and analysed for plasma copper by atomic absorption; for plasma caeruloplasmin by an adaptation of the method of Henry et al. (1974); and serum vitamin B12 by RIA (Becton-Dickinson).

RESULTS AND DISCUSSION

Table 1 shows that the overall solubilities of boluses Y and Z (i.e., on day 272) were lower than either the predicted, or the in vivo, solubilities on day 69. However, there appeared to be a direct relationship between the in vivo dissolution rates and the predicted solubility of the glass.

In tables, values with differing letters are significantly different.

Table 1. Mean Bolus Dissolution Rate $(mg/cm^2/day)$ ± sem (n).

Day	Bolus X	Bolus Y	Bolus Z
(predicted)	(1.0)	(2.3)	(3.7)
69	2.0±0.1 a (9)	2.5±0.1 b (12)	3.3±0.2 c (10)
272	1.0±0.1 a (6)	1.9±0.2 b (4)	2.2±0.2 b (6)

Table 2. Mean Plasma Cu (μ moles/litre) $\pm$ sem (n).

Day	Control	Bolus X	Bolus Y	Bolus Z
0	3.8±0.5 a (20)	4.2±0.5 a (20)	6.1±0.4 b (20)	6.4±0.4 b (20)
56	5.6±0.5 a (17)	10.4±0.7 b (19)	10.3±0.6 b (20)	12.6±0.9 c (18)
140	5.3±0.8 a (7)	8.8±0.5 b (9)	9.6±0.8 bc (7)	11.5±0.8 c (11)
175	5.7±0.3 a (6)	9.1±1.0 b (7)	9.3±0.9 b (7)	11.0±0.9 b (8)

Table 3. Mean Plasma Caeruloplasmin (mg/100 cm^3) $\pm$ sem (n).

Day	Control	Bolus X	Bolus Y	Bolus Z
0	4.6±1.1 (20)	5.1±0.8 (20)	3.9±0.5 (20)	4.7±0.9 (20)
56	7.6±1.0 a (17)	16.6±1.5 b (19)	15.3±1.4 b (20)	18.1±1.6 b (18)
140	4.1±1.0 a (7)	9.0±1.1 b (9)	9.0±1.6 b (7)	12.6±1.1 c (11)

Table 4. Mean Serum Vitamin B12 (pg/cm^3) $\pm$ sem (n).

Day	Control	Bolus X	Bolus Y	Bolus Z
0	793±70 (20)	760±58 (19)	883±82 (19)	849±96 (20)
56	637±102 a (14)	1620±188 b (16)	1534±102 bc (13)	1942±106 c (17)
140	571±79 a (7)	836±68 ab (9)	1099±111 bc (5)	1249±97 c (11)
175	768±91 a (6)	1653±231 b (7)	1417±190 b (7)	1831±130 b (8)

The differences in plasma Cu seen on day 0 between the control and bolus X groups, and groups Y and Z were probably the result of some of the lambs grazing different pastures prior to the trial.

All the treatments significantly elevated the Cu and B12 status of the lambs (compared with controls) in line with the solubilities of the glasses, the more soluble bolus Z being the most effective at increasing and maintaining the Cu status of these growing lambs. The untreated lambs remained Cu deficient throughout the trial indicating a severe copper challenge on this pasture.

As this trial was conducted under commercial conditions the earlier maturing lambs were removed for slaughter first and therefore meaningful liveweight data were not available. However, Livesey et al. (1985) reported increased growth rate and higher B12 levels in lambs given glass boluses.

In conclusion, these results confirm that soluble glass boluses are an effective means of providing supplementary trace elements to lambs. The most soluble bolus (Z), with a dissolution rate of 2.2-3.2 mg/cm^2/day, would appear to be the most beneficial under these grazing conditions.

REFERENCES

Care, A.D., Anderson, P.J.B., Illingworth, D.V., Zervas, G. and Telfer, S.B., 1985, The effect of soluble-glass on copper, cobalt and selenium status of Suffolk cross lambs, T.E.M.A.-5, 717.

Henry, R.J., Cannon, D.C. and Winkelman, J.W., 1974, Clinical Chemistry, 2nd Edition, pub. Harper and Row.

Livesey, C.T., Peers, D.G. and Stebbings, R. st J., 1985, Cobalt pine and selenium deficiency in lambs grazing herbage of apparently adequate cobalt status in the spring, in "Efficiency in Grassland Production and Utilisation," pub. A.D.A.S.

A PRELIMINARY ASSESSMENT OF CONTROLLED RELEASE DEVICES

FOR SUPPLEMENTING GRAZING SHEEP WITH TRACE ELEMENTS

Duncan W. Peter* and Keith J. Ellis**

CSIRO Division of Animal Production, *Private Bag, PO Wembley,
WA 6014 and **Private Bag, PO Armidale, NSW 2050, Australia

INTRODUCTION

A major problem associated with studies of trace element nutrition and
metabolism in grazing livestock is the ability to supplement animals with
different chemical forms of elements in a continuous and controlled manner.
Intraruminal controlled release devices (CRDs)[1] were therefore tested for
this purpose using different forms of selenium (Se).

MATERIALS AND METHODS

CRDs designed to release either no Se (Control-CRD), 50 or 100 µg Se/d
as sodium selenite (Sel.-50; Sel.-100) or selenomethionine (Se-met-50;
Se-met-100) or 100 µg Se/d as elemental Se (Ele.Se-100) or iron selenide
(FeS_2-100) for a period of 90d were administered in late October, at the
time of pasture senescence, to two year old Merino wethers (5 sheep/
treatment) at pasture at Bakers Hill, Western Australia. The effectiveness
of the different forms was monitored by regular measurements of plasma and
whole blood Se concentrations[2] over the following 84d. On d84 liver and
semitendinosus muscle samples were collected by biopsy for Se analysis[2].
Liver samples were also collected on d0 as well as on d84 from a further
group of nine sheep which received no CRDs. CRD reliability was confirmed
by measuring the rate of release from groups of identical CRDs suspended in
the rumen of fistulated sheep at pasture at Armidale, NSW.

RESULTS AND DISCUSSION

The mean release rates (µg Se/d) and standard deviations for each
formulation were as follows: Elem.Se-100, 85±4.0; FeS_2-100, 91±15.4;
Sel.-50, 43±1.4; Sel.-100, 88±3.5; Se met.-50, 42±1.3; Se met.-100,
94±9.3.

Table 1. Plasma, Whole Blood and Tissue Se Concentrations (µg/L and µg/kg DM; Mean ± SEM)

Treatment		Time after CRD administration (d)[a]				Liver	Muscle
		0	21	42	84	84	84
Cont.-CRD	Plasma	29±1	28±1	25±1	24±1	303±15	65±8
	Whole blood	43±5	47±5	44±3	44±3		
Ele.Se-100	Plasma	28±2	28±3	24±2	24±2	274±15	66±15
	Whole blood	42±5	46±3	41±3	43±3		
FeS$_2$-100	Plasma	29±3	26±2	23±2	23±1	268±11	63±5
	Whole blood	47±6	50±5	46±4	45±3		
Sel.-50[b]	Plasma	26±2(27)	38±7(31)	34±4(30)	32±1(32)	370±30	75±7
	Whole blood	39±3(39)	58±10(48)	54±9(45)	57±7(51)	(349±27)	(67±1)
Sel.-100[c]	Plasma	30±3(28)	60±12(41)	56±9(42)	51±4(50)	554±41	112±19
	Whole blood	41±6(41)	79±16(58)	80±15(59)	99±14(82)	(533±28)	(90±4)
Se-met-50	Plasma	26±1	42±3	47±3	50±4	579±44	113±6
	Whole blood	40±3	55±3	60±2	79±3		
Se-met-100	Plasma	28±1	54±4	72±5	86±6	887±55	192±13
	Whole blood	39±2	66±5	80±6	129±11		

a. Data from intermediate sampling times not shown
b. Values in parenthesis are the mean with values for sheep with a probable CRD malfunction excluded; 1 and 2 sheep excluded in Sel-50 and Sel-100 groups respectively.

Both elemental Se and iron selenide appeared to be totally ineffective forms of Se supplement while "50 µg Se/d" as selenite also produced only small, statistically non significant increases in the levels of Se in blood and muscle. CRDs delivering "100 µg Se/d" as selenite did however result in elevations in both blood and tissue concentrations, although CRDs delivering "50 µg Se/d" as Se-met. were equally as effective. The higher dose rate of Se met. resulted in yet further increases in Se concentrations with values 3.0 to 3.5 times those found in control sheep. Differences in response between selenite and Se-met., although of greater magnitude, were in accord with previous findings with sheep fed indoors[3,4]. Since these differences appear to emanate, in part, from differences in rumen metabolism[4], the low availability of Se from selenite observed in the current study may have been a reflection of the lower digestibility and protein content of dry pasture compared with diets used indoors.

The results clearly demonstrate the need to compare different forms of an element when studying trace element nutrition and metabolism in grazing animals. They likewise indicate that CRDs, through their ability to deliver different dose levels of a given chemical form, are ideally suited for this purpose. CRDs would also appear to have considerable commercial potential for trace element supplementation.

REFERENCES

1. R.M. Laby, Aust. Patent Application No. 35908/78 (1978).
2. J.H. Watkinson, Semi-automated fluorimetric determination of nanogram quantities of selenium in biological material, Anal. Chim. Acta. 105:319 (1979).
3. D.W. Peter, P.D. Whanger, J.R. Lindsay and D.J. Buscall, Excretion of selenium, zinc and copper by sheep receiving continuous intraruminal infusions of selenite or selenomethionine. Proc. Nutr. Soc. Aust. 7:178 (1982).
4. D.W. Peter, D.J. Buscall and P. Young, Excretion, apparent absorption and retention of selenium from selenite and selenomethionine by sheep, Proc. Nutr. Soc. Aust. 11:180 (1986).

EFFICACY OF SELENIUM PELLETS IN BEEF COWS

Walter H. Johnson, Ben B. Norman, John R. Dunbar,and
Michael N. Oliver

University of California Cooperative Extension
3179 Bechelli Lane, Suite 206, Redding, CA 96002 U.S.A.

Four trials testing efficacy of Se pellets (10% Se, 90% Fe) for raising and maintaining whole blood Se in beef cows were established in Shasta
County, CA in 1980. Each pelleted cow received two 30 gram pellets by ball-
ing gun. Two trials had groups that received occasional injections of Se-
vitamin E. All available cows were bled from two to four times a year with
whole blood analyzed for total Se. Statistical analyses for trials with two
groups were by T-tests. The trials with three groups were analyzed by one
way analysis of variance using Tukey's Studentized Range Test and P values
shown for these trials are from T-tests comparing pelleted to control cows.

TRIAL A - TABLE 1 - DURATION 1312 DAYS (43 MONTHS)

Blood Se was significantly higher in the pelleted cows at all dates
after treatment except the final date when cow numbers were low. Blood Se
varied seasonally in the pelleted cows but not in the control cows. No pelleted cows declined to the control mean for 36 months; at 43 months 25% of
the cows were below the control mean.

TRIAL R - TABLE 2 - DURATION 1257 DAYS (41 MONTHS)

Pelleted cows had significantly higher blood Se levels at all dates
after treatment. Seasonal variation was similar with the control and pelleted groups. 4.6% of the pelleted cows were below the control mean at 12
months, Ø at 25 and 36 months, and 12.5% at 41 months.

TRIAL T - TABLE 3 - DURATION 1227 DAYS (40 MONTHS)

Pelleted cow blood Se was significantly higher than control and inject
groups at all dates after treatment. Both control and inject groups received
injections during the latter half of the trial because of concern for their
health. There was seasonal variation in the pelleted group, but not in the
other groups. Only one cow (4.8%) was below the control mean at any time
during the trial; this occurred at 23 months.

TRIAL G - TABLE 4 - DURATION 1826 DAYS (60 MONTHS)

Pelleted cows had significantly higher blood Se than control cows for 60
months based on T-tests. Tukey's analysis showed differences between pelleted, inject, and control groups at 249 and 368 days; and showed all treatments similar at 1740 days. The inject group received no injections after
616 days. There was seasonal variation in all three groups. 3 to 6% of the
pelleted cows were below the control mean for 48 months. At 60 months 30.8%
of the pelleted cows were below the control mean.

Table 1. BLOOD SE, ppm - TRIAL A

DAYS	CONTROL MEAN	PELLET MEAN	P - VALUE
0	.028 (24)	.026 (25)	.3412
216	.015 (23)	.061 (25)	.0001
288	.023 (22)	.073 (25)	.0001
365	.026 (20)	.116 (24)	.0001
583	.025 (20)	.082 (24)	.0001
659	.024 (19)	.083 (24)	.0001
731	.025 (18)	.101 (24)	.0001
953	.026 (11)	.086 (14)	.0001
1031	.019 (10)	.083 (9)	.0011
1098	.015 (10)	.086 (7)	.0025
1312	.015 (8)	.062 (4)	.0789

Table 2. BLOOD SE, ppm - TRIAL R

DAYS	CONTROL MEAN	PELLET MEAN	P - VALUE
0	.015 (25)	.017 (25)	.4199
158	.014 (25)	.092 (25)	.0001
271	.012 (17)	.100 (11)	.0001
357	.009 (22)	.064 (22)	.0001
648	.013 (15)	.073 (12)	.0001
774	.026 (12)	.067 (8)	.0001
903	.029 (12)	.073 (11)	.0007
1023	.019 (12)	.072 (7)	.0082
1091	.018 (12)	.062 (9)	.0012
1257	.021 (7)	.051 (6)	.0486

Table 3. BLOOD SE, ppm - TRIAL T

DAYS	CONTROL	INJECT	PELLET	P - VALUE
0	.010 (30) A	.010 (30) A	.011 (30) A	.3430
134	.010 (28) B	.019 (28) B	.100 (30) A	.0001
329	.010 (28) B	.014 (26) B	.071 (26) A	.0001
513	.019 (23) B	.024 (26) B	.100 (21) A	.0001
702	.024 (20) B	.023 (23) B	.074 (21) A	.0001
871	.037 (18) B	.037 (20) B	.088 (16) A	.0001
1120	.045 (14) B	.041 (13) B	.088 (14) A	.0001
1227	.029 (13) B	.029 (13) B	.085 (11) A	.0001

Table 4. BLOOD SE, ppm - TRIAL G

DAYS	CONTROL	INJECT	PELLET	P - VALUE
0	.093 (30) A	.099 (30) A	.104 (28) A	.1338
123	.083 (28) B	.086 (29) B	.152 (29) A	.0001
249	.029 (29) C	.049 (30) B	.117 (30) A	.0001
368	.090 (26) C	.109 (30) B	.163 (30) A	.0001
482	.135 (26) B	.152 (21) B	.219 (26) A	.0001
616	.037 (26) B	.046 (22) B	.110 (26) A	.0001
728	.053 (17) B	.066 (17) B	.125 (17) A	.0001
983	.032 (19) B	.033 (6) B	.093 (17) A	.0001
1111	.079 (21) B	.088 (16) B	.131 (21) A	.0001
1379	.020 (19) B	.021 (17) B	.066 (17) A	.0001
1466	.060 (19) B	.054 (16) B	.088 (16) A	.0001
1740	.016 (16) A	.017 (14) A	.029 (14) A	.0200
1826	.042 (14) B	.040 (14) B	.055 (13) A	.0130

*Acknowledgement to Suzanne Strasser, Senior Programmer, U.C. Cooperative Extension, for the statistical analyses.
In the tables numbers in parentheses are the number of cow bloods analyzed.

THE DEVELOPMENT OF A LONG ACTING INJECTABLE PREPARATION FOR THE TREATMENT AND PREVENTION OF SELENIUM DEFICIENCY IN CATTLE

I. M. McPhee and G.D. Cawley

Rycovet Limited, Talkin, Brampton
Cumbria CA8 1LE, England

INTRODUCTION

Selenium deficiency in cattle is associated with a range of clinical conditions including white muscle disease, muscle weakness in neonatal calves, immuno-suppression, ill thrift, reduced weight gain, scour, infertility, abortion and retained placenta. This paper describes a long acting selenium therapy for the treatment and prevention of hyposeleosis.

METHODS AND MATERIALS

(Deposel, Rycovet Ltd) is a sparingly soluble formulation of micronised barium selenate containing 50mg Se/ml in an inert viscous petroleum base, given by subcutaneous injection at a rate of 1mg Se/kg bodyweight to cattle and sheep.

To assess efficacy, groups of 6 selenium deficient Friesian-cross calves weighing 70-200kg were treated with Deposel at 1mg Se/kg bodyweight, or with potassium selenate and vitamin E (60ug Se and 2.7 i.u. vitamin E per kg bodyweight), or remained untreated. Blood samples taken before treatment and at approximately monthly intervals for 11 months after treatment were assessed for erythrocyte glutathione peroxidase (EGSH-Px) activity (Paglia & Valantine, 1967). This assay is a sensitive indicator of selenium status in cattle and sheep (Anderson et al 1979).

To demonstrate maternal transfer two groups of 6 selenium deficient beef cattle were compared. One group was treated 3-4 months before calving with 500mg selenium as Deposel; the other group was not treated. EGSH-Px activity was measured in blood samples taken from cows before treatment and from cows and their calves at periods after treatment.

The measurement of selenium transfer into milk was carried out in two groups of 5 selenium sufficient Friesian cattle. In mid-lactation cows were treated with 500mg selenium as Deposel or remained untreated as controls. Milk samples taken every 2 weeks for 14 weeks after treatment were measured for their selenium concentration by atomic absorption spectrophotometry after hydride generation.

RESULTS AND DISCUSSION

The treatment of calves led to a highly significant rise in EGSH-Px activity (P<0.001) which continued for the 11 months of the trial. The control group remained deficient throughout the trial.

Calves from Deposel treated cows were born with significantly higher EGSH-Px activities (P<0.01) than calves born to control cows. The EGSH-Px activity of treated cows and their calves was maintained throughout lactation (Fig.1.)

The selenium concentration in milk from Deposel treated cows was maintained for the 14 weeks of the trial but did not rise above normal values. That from control cows fell significantly (P<0.01)

Fig.1. Maternal selenium transfer from Deposel treated and control cows to their calves at birth and throughout lactation

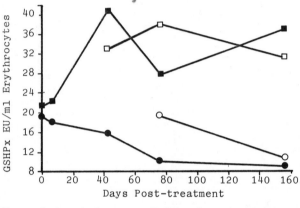

■Deposel treated cows □Calves from treated cows
●Control cows   OCalves from control cows

CONCLUSION

The treatment of selenium deficient cattle by subcutaneous injection of barium selenate as Deposel provides a steady rise in EGSH-Px activity over a period of 1-3 months. Thereafter the selenium status is maintained for at least 11 months. The improved selenium status of treated cows is transferred to calves during pregnancy and the selenium status is maintained in suckler calves by the selenium in the milk of their dams.

The use of this long acting selenium preparation offers considerable advantages to the stockman especially in extensive rearing systems, by the flexibility offered in the timing of treatments, the reduction in animal handling and the ability to prevent selenium responsive conditions in areas of known selenium deficiency.

REFERENCES

Anderson, P.H., Berrett, S., Patterson, D.S.P., (1979), Veterinary Record, 104, 235-238

Paglia, D.E., Valantine, W.N., (1967), J. Lab. Clin. Med., 70, 158-169

EFFECT OF ORAL COPPER NEEDLES AND PARENTERAL COPPER ON HYPOCUPRAEMIA,

BODY WEIGHT GAIN AND FERTILITY IN CATTLE

B.E. Ruksan*, M. Correa Luna**, and F. Lagos**

*Department Pat. Anim. CICV, INTA
**Fundación J.M. Aragon
 cc 77, Moron, (1708), Buenos Aires, Argentina

INTRODUCTION

In previous studies in the lowland areas of Santa Fe Province, Argentina, it was demonstrated that beef cattle showing leg stiffness, walking difficulties and spontaneous bone fractures were affected with severe conditioned copper deficiency. This was corrected by parenteral copper suplementation (Gonzalez Pondal et al, 1975).

Because of the frequent required dosages oral therapy has provided an inferior alternative to parenteral copper treatment when cupric sulphate was administered to heifers. Subcutaneous therapy should be repeated every 60-70 days in order to maintain the blood copper concentration at the desired value of 0.50 mcg/ml (Ruksan et al, 1981). However, site of injections of reactions were observed which is a frequent occurrence with this type of treatment.

Dewey (1977) indicated that the slow release of copper from cupric oxide needles (CuO_N) in the acid medium of the abomasum would allow CuO_N to be used as an effective oral treatment of copper deficiency. Studies of Whitelaw et al, (1980), Suttle (1981), and Judson et al (1981), support the view that oral dosing of oxidised copper needles are a useful and safe alternative to a parenteral product for cattle hypocuprosis prophylaxis.

The present study compares copper oxide needles given orally with that of sucutaneous injections of copper complex (Vascunin-Induvet) as a means of raising blood copper status, liveweight gain and fertility in cattle.

MATERIAL AND METHODS

Seventy three crossbred Zebu suckling female calves were included in the trial, which lasted two years (November 1983-October 1985). Animals were subdivided into three groups: Group I, control; Group II animals received 32 mg once, 53 mg five times and 85 mg twice of Cu-Complex (Vascunin-Induvet) and Group III, animals received 12,5 g initially and 25 g eleven months later of CuO_N needles given orally.
Blood copper was determined by the ceruloplasmin oxidase method

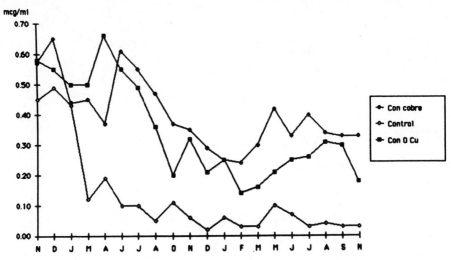

Figure 1. Effect of Oral CuO and Subcutaneous Cu Administration on Cu
Serum Values.

(Bingley et al, 1978) and live weight gain recorded monthly. Native pastures grazed by animals were analyzed periodically for copper (Eden and
Green, 1940) and molybdenum content (Bingley, 1959).

RESULTS

Differences in ovarian activity and percentage of pregnancies were
observed in Table 1.

TABLE 1. Mean values of ovarian activity and percentage of pregnancy

	Ovarian activity	Early pregnancy	Total pregnancy
Control	17.4	52.4	81.0
Oral CuO$_N$	29.2	70.8	79.1
Subcut. Cu	32.0	56.0	80.0

The differences due to the effect of oral and subcutaneous copper
administration on blood copper values of calves during the experimental
period were higher (P<0.05) than control animals. There were no differences among cupraemias of alternative treatments (Figure 1).

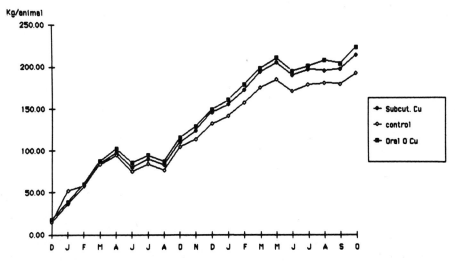

Figure 2. Effect of Oral CuO and Subcutaneous Cu Administration on Calf Weight Gains.

However the copper oxide needles treatment resulted in sustained higher total weight gain 225.0 ± 28.8 versus 215.8 ± 17.3 for parenteral treatment and 196.0 ± 24.8 Kg/animal for the control group respectively see Figure 2.

DISCUSSION

The copper treatments were sufficient to maintain the blood copper concentration of calves above 0.50 mcg/ml during the first nine months. Afterwards, neither the second treatment with 25 g of CuO_N nor the increased dosages of subcutaneous copper injections were sufficient to increase the blood copper concentration of animals above the desired concentration of 0.50 mcg/ml. However, treated animals had higher blood concentrations (P<0.05) than controls. This suggests that the dosages should be increased for hypocuprosis treatment in that area.

Differences in live weight gains began to appear after 16 weeks similar to the observations done by Phillippo et al., (1985) and they increased with time, indicating that long experiments must be carried out in order that the live weight gains reflect differences in blood copper concentration.

The experimental animals were grazing on native pasture with a copper concentration of 5.5 ± 2.2 ppm, while the molybendum was 3.1 ± 2:2 ppm.

In this study a higher ovarian activity and percentage of early pregnancy for the animals treated with CuO_N was found. This suggests that the treatment with CuO_N could overcome the inhibitory effect of molybdenum upon luteinizing hormone secretion (Shyamala and Leonard, 1980) and the delay in time of onset of first estrus in cattle observed with a diet containing 5 ppm Mo/Dm of at least 6 weeks (Phillippo et al., 1985).

It has now been demonstrated that the dosage rate of 25 g CuO_N for yearling heifers was not sufficient to increase the blood copper concentrations above 0.50 mcg/ml. However, the treatment was sufficient to overcome the inhibitory effect of molybdenum on the delay in time of onset of first estrus in cattle as well as in the increased live weight gain, both of which are very important in order to increase overall productivity in the lowland areas of Santa Fe Province.

REFERENCES

Bingley, J.B., (1959), J. Ag. Fd. Chem., 7,269.

Bingley, J.B., Ruksan, B.E., and Carrillo, B.J., (1978), Rev. Med. Vet., 59, 3-8.

Dewey, D.W., (1977), Search 8, 326-327.

Eden, A. and Green, H.H., (1940), Biochem. J. 34, 1202.

Gonzalez Pondal, D., Casaro, A.E.P., and Carrillo, B.J., (1975), Fundación J.M. Aragon Publ. No. 3.

Judson, C.J., Dewey, D.W., McFarlene, J.D., and Riley, W.J., (1981), In: Trace Element Metabolism in Man and Animals-TEMA 4 (Eds. J. McC. Howell, J.M. Gawthorne, and C.I. White), pp. 187-190, Canberra: Austral. Acad. Sci.

Phillippo, M., Humphries, W.R., Brenner, J., Atkinson, T., and Henderson, O., (1985), In: Trace Elements in Man and Animals-TEMA 5 (Eds. C.F. Mills, I. Bremner, and J.K. Chesters), pp. 176-180, Aberdeen: Commonwealth Agricultural Bureaux.

Ruksan, B.E., Casaro, A.E.P., Jaschke, J., Lagos, F., and Gonzalez Pondal, D., (1981), In: Trace Element Metabolism in Man and Animals-TEMA 4, (Eds. J. McC. Howell, J.M. Gawthorne, and C.I. White), pp. 179-182, Canberra: Austral. Acad. Sci.

Shyamala, G. and Leonard, L., (1980), J. Biol. Chem. 255, 6028-6031.

Suttle, N.P., (1981), Vet. Rec. 108, 4170420.

Whitelaw, A., Armstrong, R.H., Evans, C.C., Fawcett, S.R., and Russel, A.J.P., (1981), Vet. Rec. 107, 87-88.

OXIDIZED COPPER WIRE PARTICLES AS AN ORAL COPPER SUPPLEMENT FOR CATTLE

J. R. Dunbar, J. G. Morris, B. B. Norman, A. J. Jenkins
C. B. Wilson, N. L. Martin, and J. M. Connor

University of California
Davis, CA 95616

INTRODUCTION

Areas of endemic copper deficiency occur in California and animals show a response in production to supplementation.

Methods of supplementation which have been used included mineral blocks, concentrate mixes or liquid supplements containing copper are offered on a free choice basis. However, copper intake varies widely between animals. Injectable forms of copper are readily administered and cause tissue reactions while producing lesions, which result in carcass damage.

Studies by MacPherson (1983) indicate that oral dosing with oxidized copper wire particles is an effective method of raising the copper status of cattle.

MATERIALS AND METHODS

The University of California Sierra Foothill Range Field Station was chosen as the site for the experiment because of the control that could be exercised over the trial.

One hundred twenty Hereford heifers aged 6-9 months, averaging 210 kg, were randomly assigned to 3 treatment groups. One group of 40 heifers were left untreated, a further 40 heifers were given 10 g of orally-administered oxidized copper wire particles in a soluble capsule, while the remaining 40 received a subcutaneous injection of Molycu[R] (120 mg of copper as cupric glycinate).

Heifers grazed as one herd during the duration of the study. Liver biopsies and blood samples for copper and selenium assay were collected from a randomized sample of 13 heifers in each treatment group at day 0 and at 28, 91, and 158 days.

RESULTS

Copper injection produced a significant ($P > .05$) increase in liver copper levels over the other groups at 28 days (Table 1). Oxidized copper

wire particles produced a significantly (P>.05) higher level of liver
copper at 28 days compared to controls and at 91 days compared to controls
and copper injection heifers. Although liver concentrations in control
heifers were generally below those in treated heifers, they were indicative
of adequate copper status. No significant difference in liver copper
status existed among groups at 158 days. Mean plasma copper values were
not altered by treatment (Table 2). Copper injection caused severe tissue
reaction in 30 of 40 heifers. Blood selenium values were not altered by
copper treatment (Table 3).

Table 1. Liver Copper, mg/kg DM.

	\multicolumn{4}{c}{Days After Treatment}			
	0	28	91	158
Control	123[a]	138[a]	184[a]	233[a]
Copper Injection	152[a]	258[b]	228[a]	263[a]
Copper Pellet	142[a]	194[c]	296[b]	264[a]

a,b,c Means in the same column with the same letters
are not significantly different.

Table 2. Plasma Copper, mg/l.

	Days After Treatment			
	0	28	91	158
Control	1.03	.89	.93	.98
Copper Injection	1.02	.94	.89	.94
Copper Pellet	1.02	1.01	.96	.99

None of the values are significantly different.

Table 3. Whole Blood Selenium, mg/l.

	Days After Treatment			
	0	28	91	158
Control	.03	.04	.06	.07
Control Injection	.03	.03	.05	.06
Copper Pellet	.03	.03	.05	.08

None of the values are significantly different.

REFERENCES

MacPherson, A., 1983, Copper dioxide wire for the Bovine. Occasional
    Publications No. 7. British Society of Animal Production.
    pp 140-141.

INVESTIGATIONS ON VANADIUM DEFICIENCY IN RUMINANTS

M. Anke, B. Groppel, T. Košla, and K. Gruhn

Karl-Marx-Universität Leipzig
Wissenschaftsbereich Tierernährungschemie
6900 Jena, Dornburger Straße 24, GDR

INTRODUCTION

The essentiality of V for the fauna is controversial since
it was not possible to induce specific V deficiency symptoms.
Up to now, there have been no long-term V deficiency experiments in which the influence of V deficiency on feed-intake,
growth, reproduction and life expectancy was investigated and
which resulted in V deficiency symptoms.

MATERIAL AND METHODS

The investigations with female animals and their offspring
have been repeated 7 times since 1980. On the average of all
rations used, the V content of the V deficiency ration varied
between 1 and 9 µg/kg dry matter. The control ration contained
2 mg V/kg dry matter. The goats were kept in polystyrol sties.
Their drinking water was distilled.

RESULTS

Surprisingly, the young growing, pregnant and lactating V
deficiency goats took in 8 % more feedstuff than corresponding
control animals in the first year of life (table 1). Adult V
deficiency goats consumed the same feed amount as control animals which were older than 1 year.

Therefore, no influence of V deficiency on growth was to
be expected. V-poor rations did not influence growth, neither
at birth nor at the end of the lactation period (91st day of
life) and on the subsequent 168 experimental days up to the
end of the 2nd month of gravidity. Only the proportion of kids
with low weight (< 1.6 kg birth weight), which were not
capable of living, was significantly higher in V deficiency
goats (10 %) than in control goats (2 %).

Only 57 % of V deficiency goats became pregnant after the
first mating. Repeated matings, however, led to the same conception rate in V deficiency animals as in control goats.
25 % of V deficiency goats aborted their kids, mostly at the
end of gravidity.

Table 1. The influence of V deficiency on feed-intake, growth, reproduction, milk production and life expectancy during 7 experiments (53 control and 38 V deficiency goats)

parameter		+V	−V	p
feed-intake	1st year, (g/d)	679	731	< 0.001
	adult, (g/d)	647	656	> 0.05
growth	1st day of life, kg	2.86	2.90	> 0.05
	91st day of life, kg	17.2	16.6	> 0.05
	g/day for 1st 168 days	92	92	> 0.05
reproduction performance	success of first insemination, %	77	57	< 0.001
	conception rate, %	89	85	> 0.05
	services per gravidity	1.4	2.0	< 0.01
	abortion rate, %	0	27	< 0.001
	kids per gravid goat	1.5	1.1	< 0.001
	91-old-day kids per test goat	0.73	0.46	< 0.01
milk performance	milk, l/day, first 56 days	1.185	1.073	< 0.01
	fat, g/day, first 56 days	41.6	39.4	> 0.05
	protein, g/day, first 56 days	32.5	32.8	> 0.05
life expectancy	dead kids, %	9	41	< 0.001
	dead mothers, %	25	58	< 0.001

V deficiency goats gave birth to significantly fewer twins than control goats. Apart from one living kid, mumified foetuses were registered in several cases.

V deficiency only reduced milk production by 9 %. Its influence on milk fat production was only 5 % and, thus, insignificant. Both groups had the same milk protein production. Hence it follows that the influence of V deficiency on milk production is unimportant.

More than 40 % of V deficiency kids died between the 7th and 91st day of life. In several cases, death was preceded by convulsions. The losses during the first week of life were not taken into consideration since, during this period, it is complicated to clarify the causes of death. Thus, the loss of kids with low birth weight is not included.

More than 50 % of adult goats died in the last trimester of the suckling period. The offspring of V deficiency goats often suffered from skeleton changes at the forefoot tarsal joints. The animals had apparently pains when standing and always relieved one frontleg. The alternating relieving of the frontlegs was already obvious before the thickening of the forefoot tarsal joints. Apart from this thickening, deformations slowly developed as well. This particular deficiency symptom was neither registered in control goats nor in kids of other deficiency groups (Se, F, Ni, Li, Br, Al, Cd). Therefore, this is a V-specific deficiency symptom. Skeleton damage was not the reason for the exitus. Nongravid goats can live with it for years.

In accordance with previous investigations, the long-term V depletion did not induce hematological changes (hemoglobin, mean red blood cellular volume) in goats with visible V deficiency symptoms.

POSSIBILITIES OF DIAGNOSING IODINE DEFICIENCY IN RUMINANTS

Bernd Groppel, Manfred Anke, and Arno Hennig

Karl-Marx-Universität Leipzig
Wissenschaftsbereich T, rernährungschemie
6900 Jena, Dornburger Str. 24, GDR

INTRODUCTION

We investigated the influence of a deficient, marginal or sufficient I supply (0.04 to 0.73 mg I/kg dry ration) on the I (TI, PBI, BEI) and TT_4 and TT_3 concentration of blood serum, milk, hair (or wool) and other organs in goats, sheep, cows, calves and fattening bulls.

MATERIAL AND METHODS

After alkaline dry ashing (BEI) or after perchloric acid-sulfuric acid disintegration (TI, PBI) I analysis was carried out colorimetrically according to the Sandell-Kolthoff-procedure. TT_3 and TT_4 were determined by means of radioimmunassay.

RESULTS

Though it was possible to demonstrate that the formation of goitres points to I deficiency, only the histological investigation of thyroid tissue and the degree of morphological activation of the gland allow reliable statements on the I supply of an animal. All experimental animals with $\leq$ 0.11 mg I/kg dry diet, either their own or the maternal ration, suffered from struma.
When the I supply is sufficient, the TI, PBI, BEI and TT_4 values in the serum as well as the I content in the hair increase in the last trimester of gravidity. The I and TT_4 values decrease again in the perinatal period. The I and I hormone level of calves and lambs is influenced by the pre- and postnatal I supply. The highest values occurred on the first day of life. After 14-28 days, they had decreased by 50%.
We recommend the following limit values of a sufficient I status (nmol/1) in the blood serum:

		TI	PBI	BEI	TT_4
cows	- last trimester of gravidity	550	400	300	–
	- first month of lactation	320	300	240	–
goats	- last trimester of gravidity	450	330	300	–
	- first month of lactation	350	300	280	–
calves	- 3rd-4th week of life	500	400	360	80
	- 7th-8th week of life	450	380	330	70
fattening bulls		340	290	210	–

TT_3 is not very suitable for the diagnosis of I deficiency since, due to the peripheral monodeiodization of TT_4, the organism forms more TT_3 in the case of insufficient I.

The I content of milk also reflected the I intake. Colostrum is richer in iodine than mature milk. Goat's milk accumulates more I than cow's milk. In mature cow's milk, <189 nmol I/1, <500 nmol I/1 suggest insufficient I status.

Hair and wool belong to the I pool which is slowly exchangeable. More I is incorporated into black than into white hair. The I content of hair and wool is suited for the assessment of the long term I intake. The following I values (mg/kg dry substance) might be considered as the limit values of a sufficient I status in hair and wool:

goats (white)	0.30	fattening bulls (black)	0.25
sheep (white)	0.20	goat kids (1st–56th day of life, white)	0.60
cows (black)	1.00	calves (1st–56th day of life, black)	1.80

Therefore, the I content of organs reacts faster to changes in I intake than hair. When different I amounts are taken in, significant differences are found both in the intrathyroidal I content and the I content of extrathyroidal tissues. Summing up, the following limit values for a sufficient I status are proposed:

Organ, tissue	Adult goats and sheep	Fattening bulls	Kids/lambs/calves 1st day of life	Kids/lambs/calves 75th–110th day of life
thyroid gland (ug I/g fresh substance)	170	170	300	300
cerebrum (ug I/kg ds)	70	–	150	100
lungs (" ")	120	100	250	130
liver (" ")	120	120	220	130
spleen (" ")	120	–	250	100
heart (" ")	100	100	200	130
skeleton muscle (" ")	150	–	200	130
kidneys (" ")	150	120	220	130

NORMAL MANGANESE, ZINC, COPPER, IRON, IODINE, MOLYBDENUM,
NICKEL, ARSENIC, LITHIUM AND CADMIUM SUPPLY DEPENDENT ON THE
GEOLOGICAL ORIGIN OF THE SITE AND ITS EFFECTS ON THE STATUS OF
THESE ELEMENTS IN WILD AND DOMESTIC RUMINANTS

M. Anke, B. Groppel, Ute Krause, L. Angelow,
W. Arnhold, T. Masaoka, S. Barhoum, and G. Zervas

Karl-Marx-Universität Leipzig
Wissenschaftsbereich Tierernährungschemie
6900 Jena, Dornburger Straße 24, GDR

INTRODUCTION

The influence of the geological origin of the site of the
flora on its trace element content was systematically investigated in Central Europe (GDR, Hungary, Czechoslovakia) by means
of indicator plants. Lucerne and acre red clover in buds,
meadow red clover and rye in blossom and wheat in stalk shoo-
ting were used for that purpose. The time of collecting the
samples was fixed on the basis of phenological data so that
the stage of development was the same in all countries. 7185
plant samples were at our disposal. Depending on their biotope,
the trace element status of the ruminant species cattle, sheep,
goats, roe deer, fallow deer, red deer and moufflons was investigated in 5570 samples of liver, cerebrum, kidneys, ribs,
blood serum and hair.

Results of plant analysis

The geological origin of the material for soil formation
varies the trace element content of the indicator plants of
all tested trace elements significantly (table 1). For cla-
rity's sake, the trace element content was relativized and
the mean richest origin was equated with 100.
The Cu, Ni and Fe content of the flora on the weathering
soils of the lower strata of new red sandstone (table 1) is
significantly higher than that of the moor and peat areas.
Boulder clay and diluvial sands have a similar Cu- and Ni-
poor vegetation. Compared to the lower strata of new red
sandstone, the sediments of the triassic produce Ni-
poorer feedstuffs.
On the other hand, the flora of alluvial riverside soils
proved to be particularly J-rich and that of triassic and, particularly, of muschelkalk very J-poor. Concentrated J deficiency in ruminants occurs on this soil.

The Mo-richest feedstuff were found on the weathering soils of gneiss and granite, whereas those of the trias formations were particularly Mo-poor. Thus Mo-deficiency usually occurred in cauliflower and leguminous plants on muschelkalk and keuper weathering soils.

Table 1. The influence of the geological origin of the site on the copper, nickel, iron, iodine and molybdenum content of indicator plants in Central Europe

origin	Cu	Ni	Fe	Mn	Zn	Cd	As	Li	I	Mo
gneiss	93	70	79	52	75	78	100	98	80	100
granite	82	69	81	74	85	100	58	79	54	100
syenite	86	89	84	100	100	100	58	72	87	63
phyllite	93	83	81	73	92	81	38	69	92	73
slate	94	84	72	80	82	80	44	100	73	88
lower strata of new red sandstone	100	100	100	62	84	89	87	83	71	96
new red sandstone	80	62	69	69	61	74	49	89	71	67
muschelkalk	93	65	73	44	59	68	50	89	65	54
keuper	85	61	72	52	52	66	46	100	61	51
boulder clay	70	67	68	69	61	85	56	58	95	61
diluvial sand	70	67	90	98	79	69	59	77	95	62
loess	86	63	82	58	62	60	66	76	83	67
moor, peat	52	56	79	69	82	51	-	60	-	69
alluvium	74	68	83	69	70	70	70	54	100	79

The Mn-, Zn- and Cd-richest vegetation was found on the acid syenite sites of Central Europe, whereas muschelkalk or keuper weathering soils as well as loess produced the Mn-poorest feedstuffs. Therefore, Mn deficiency symptoms regularly occurred in cattle on these soils. The Cd intake of ruminants is lowest on moor and peat sites.

Apart from the geological origin of the site, emissions, plant species, part and age of plants influence the trace element content of the flora significantly.

The trace element status of ruminants

The relations between the site-specific offer, the influence of antagonists and the species-specificity of the trace element status are demonstrated by the example of Cu. Cu deficiency is most reliably reflected by the Cu content of cerebrum and liver. The Cu offer to cattle is worst on moor, diluvial sand and boulder clay. Primary Cu deficiency regularly occurs on these soils. Their Cu status and that of cows exposed to S, Cd and Mo is represented in table 2.

Table 2. The copper status of dairy cows with primary and
secondary copper deficiency (mg/kg dry matter)

part of the body	normal (n 177)	primary (50)	Cu - d e f i c i e n c y S (115)	secondary Cd (130)	Mo (15)	Fp
cerebrum	11	6.0	5.2	8.4	6.4	< 0.001
liver	137	14	15	64	7.1	< 0.001

The trace element status of the organs is species-specific.
The "normal Cu" content of the cerebrum of roe deer differs
significantly from that of sheep. In sheep, this Cu concentration would indicate a considerable Cu deficiency.

LITERATURE

Anke,M., Szentmihályi,S., Groppel,B., Regius,A., Lokay,D.
(1986) Mengen- u.Spurenelemente 6:108-120

PATHOLOGY OF EXPERIMENTAL NUTRITIONAL

DEGENERATIVE MYOPATHY IN RUMINANT CATTLE

Seamus Kennedy and Desmond A. Rice

Veterinary Research Laboratories
Stormont
Belfast, N. Ireland

Since 1970 there have been many reports of nutritional degenerative myopathy (NDM) in yearling cattle typically following turnout to pasture in the spring. The disease was associated with inwintering on low selenium and vitamin E (Se-E) diets and could be prevented by Se-E treatment prior to turnout.[1] However, the factors triggering the onset of clinical signs after turnout were unknown. An experimental model was devised to investigate whether polyunsaturated fatty acids (PUFA) absorbed from ingested grass could induce clinical NDM in calves already depleted of Se-E.[2] This paper outlines the clinicopathologic features of this model.

EXPERIMENTAL MODEL

Our model consists of 1.) feeding a low Se-E diet to calves for approximately 4-5 months (simulating inwintering on a low Se-E diet) followed by 2.) feeding protected PUFA for approximately 6-11 days (simulating absorption of unhydrogenated PUFA from grass). The basal diet (0.03 mg kg^{-1} Se; 1.0 mg kg^{-1} alpha tocopherol) is based on NaOH-treated, naturally Se-deficient barley while protected linseed oil (54% linolenic acid) is utilized as a source of PUFA.

CLINICAL PATHOLOGY

Depletion of Se-E is associated with a rapid fall in plasma alpha tocopherol, a gradual decline in erythrocyte glutathione peroxidase and a gradual rise in plasma creatine kinase (CK).

Feeding protected PUFA results in rapid increases in plasma linolenic acid (from 2% to 25% of total plasma fatty acids after 10 days) and plasma CK activity (to levels in excess of 100,000 UL^{-1} after 8 days).

CLINICAL SIGNS AND ECG CHANGES

No clinical signs are seen until approximately 130 days of depletion. Subsequently, there is intermittent teeth-grinding and ECG changes including ST segment elevation and increased T wave amplitude.

Feeding protected PUFA rapidly induces a variety of cardiopulmonary and locomotory signs, myoglobinuria and further ECG abnormalities. Some calves develop dyspnea while others have abnormalities of gait ranging from staggering to ataxia and complete recumbency.

SKELETAL MYOPATHOLOGY

Calves maintained on the low Se-E diet for 4-5 months may have pale skeletal muscles but usually no other gross abnormalities. However, they have histological lesions of coagulative necrosis and regeneration of small numbers of skeletal myocytes.

Feeding protected PUFA produces widespread necrosis of skeletal muscles. The lesions appear as white plaques resembling fish meat. Histologically, there is coagulative necrosis of skeletal myocytes. Regeneration is rapid even in the absence of Se-E therapy.

We have compared the histochemical fiber type profiles of muscles of Se-E depleted calves fed PUFA to those of age-, sex- and breed-matched Se-E adequate control calves. The type I (oxidative):type II (glycolytic) fiber ratio of M. triceps brachii was approximately 30:70 in controls and 10:90 in calves with experimental NDM indicating preferential necrosis of type I myocytes.

CARDIAC PATHOLOGY

The main lesion seen during Se-E depletion is accumulation of autofluorescent granules in Purkinje cardiocytes of the conduction system of the heart. These granules have histological and histochemical properties consistent with lipopigment. Ultrastructurally they appear as cell debris undergoing cytolysosomal degradation and lipopigment granules.

Addition of protected PUFA to the basal diet produces myocardial necrosis with preferential involvement of the left ventricle. There is necrosis of both Purkinje and contractile cardiocytes. Large amounts of membranous debris and lipopigment granules accumulate in both types of cell in association with mitochondrial and plasmalemmal damage and myofibrillolysis.

DISCUSSION

Evaluation of our model indicates that it reproduces both clinical and subclinical myopathy. Clinical signs produced are similar to those of spontaneous NDM.[3] Preferential necrosis of the left ventricle and type I skeletal myocytes, and Purkinje lipopigmentation have also been reported in field outbreaks of NDM.[4,5] Damage to mitochondrial and plasmalemmal membranes, and accumulation of cardiac lipopigment support a pathogenetic role for lipoperoxidation in development of the lesions.

In conclusion, this model accurately reproduces the clinicopathological features of spontaneous NDM. The Purkinje cell lesions appear to constitute a unique example of selective damage to cells of the cardiac conduction system.

REFERENCES

1. McMurray, C.H. & McEldowney, P.K. (1977) Br. Vet. J. 133, 535-542.
2. Rice, D.A., Blanchflower, W.J. & McMurray, C.H. (1981) Vet. Rec. 109, 161-162.
3. Allen, W.M., Bradley, R., Berrett, S., Parr, W.H., Swannack, K., Barton, C.R.Q. & MacPhee, A. (1975) Br. Vet. J. 131, 292-308.
4. Bradley, R., Anderson, P.H. & Wilesmith, J.W. (1986) Proc. 6th Int. Conf. on Production Disease in Farm Animals, p. 248-251.
5. Van Vleet, J.F., Crawley, R.R. & Amstutz, H.E. (1977) JAVMA 171, 443-445.

TREATMENT OF DAIRY SHEEP WITH SOLUBLE GLASS BOLUSES

George Zervas

Department of Animal Nutrition
Agricultural College of Athens
Iera Odos 75, GR-118 55, Greece

INTRODUCTION

Work previously reported (Zervas et al., 1984,1985; Telfer and Zervas, 1984; Telfer et al., 1984a,b; Zervas, 1986) has shown that soluble glass boluses, containing Cu, Co and Se, were effective in raising the status of the mentioned elements of calves, cows, lambs, ewes and goats to normal and in maintaining them at an adequate level for periods up to a year.

This work describes a field trial designed to test the efficacy of these soluble glass boluses on grazing dairy ewes in Greece.

MATERIALS AND METHODS

Fifty-two karagounico dairy ewes, aged 2 to 5 years, were randomized into two groups. Twenty-six of them were given a glass bolus at the end of their second month of pregnancy, while the rest were left as untreated controls.

The ewes were run as a single flock at Elasona area of Greece having access to hill pastures from March to October. Over the rest of the year they were housed and fed alfalfa hay and a mixture from corn and cotton seed cake without any mineral supplement.

Blood and wool samples were taken prior to treatment (day 0) and subsequently at days 103, 224 and 338 post treatment. The results were analysed by Student's t-test at each sampling time.

RESULTS AND DISCUSSION

Table 1 presents the mean concentrations of Cu, vitamin B12, Se in plasma, serum and whole blood respectively and the mean concentrations of Cu in wool of treated and control ewes.

Although the concentrations of plasma Cu of treated and control ewes were assessed as adequate, those of treated ewes were significantly ($P<0.001$) elevated and remained above that of controls during the whole period, covering at least 11 months after the administration of the glass boluses.

Table 1. Effect of Glass Bolus Supplementation on Blood and Wool Parameters

Days Post Bolus	Group	n	Plasma Cu μ moles/ litre	Vit. B12 μg/litre serum	Blood Se μ moles/ litre	Wool Cu μg/g DM.
0	control	26	11.3±0.5	953±38	0.82±0.04	5.7±0.3
	treated	26	10.3±0.3	890±42	0.84±0.02	5.5±0.4
103	control	25	11.4±0.3	907±35	0.71±0.02	5.7±0.3
	treated	25	12.8±0.2**	1132±52**	1.37±0.04***	7.5±0.3**
224	control	20	11.6±0.2	795±37	0.81±0.02	5.9±0.4
	treated	24	13.2±0.3***	1194±56***	1.15±0.02***	7.9±0.3***
338	control	21	10.8±0.3	880±42	0.82±0.02	5.7±0.3
	treated	22	12.3±0.2***	1008±43*	1.01±0.02***	7.9±0.3***

All values are means ± sem.
Significant differences: * $P<0.05$; ** $P<0.01$; *** $P<0.001$.

The glass boluses were releasing sufficient Co to raise serum vitamin B12 concentrations in treated ewes. Highly significant ($P<0.001$) responses in blood Se concentrations were also evident and maintained up to the end of the trial in treated ewes, while the controls were and stayed, during the experimental period, at low to marginal Se level. The treatment of ewes with a glass bolus increased also the Cu content of wool significantly ($P<0.001$) above that of controls.

In conclusion, the use of Cu-Co-Se containing soluble glass boluses appears to be an efficient and practical method of providing long term oral supplementation of these three elements, involves the minimum amount of collection and handling of livestock and causes no systemic toxicity and tissue damage to the animal.

REFERENCES

Telfer, S.B. and Zervas, G., 1984, The effect of Tracer glass-L on the Cu, Co and Se status of lambs, Anim. Prod., 38:561.
Telfer, S.B., Illingworth, D.V., Anderson, P.J., Zervas, G., and Carlos, G., 1984a, Effect of soluble glass boluses on Cu, Co and Se status of sheep. 609th Meeting of Biochemical Society, Leeds, 1984, Leeds, UK, pp. 529.
Telfer, S.B., Zervas, G., and Carlos, G., 1984b, Curing or preventing deficiencies in Cu, Co and Se in cattle and sheep using Tracer glass, Canad. J. Anim. Sci. 64(Suppl):234.
Zervas, G., Telfer, S.B., Carlos, G., and Anderson, P.H., 1984, The effect of soluble glass boluses on the Cu, Co and Se blood status of ewes. 35th EAAP Annual Meeting, the Hague, Netherlands, 1984, pp. S6a.6.
Zervas, G., Telfer, S.B., and Carlos, G., 1985, The use of Cu, Co and Se containing soluble glass boluses in the prevention of trace element deficiencies in calves. 36th EAAP Annual Meeting, Halkidiki, Greece, 1985, pp. C5B.32.
Zervas, G., 1986, Treatment of goats with soluble glass boluses containing Cu, Co and Se. 5th International Symposium on Iodine and Other Trace Elements, Jena, German Democratic Republic, 1986.

HORMONAL MODULATION OF RAT HEPATIC ZINC METALLOTHIONEIN LEVELS

Scott H. Garrett, Koji Arizono, and Frank O. Brady

Department of Biochemistry, University of South Dakota
School of Medicine, Vermillion, South Dakota 57069, USA

The hormonal induction of hepatic zinc thionein (ZnMT) is a complex and multifaceted process. Zinc and glucocorticoids have direct effects on MT-mRNA transcription. Other known modulators communicate with MT genes via their plasma membrane receptors and adenylate cyclase or phospholipase C, resulting in the activation of protein kinases A and C and calmodulin-dependent protein kinase (See Figure 1). Our group has demonstrated the ability of catecholamines, angiotensin II, phorbol esters, and adenosine and its analogues to induce ZnMT (1-6). Cousins' group has done the same for zinc, glucocorticoids, glucagon, cAMP analogues, and interleukin-1 (7,8). Our major hypothesis is that ZnMT induction is a key response by the liver to stresses on the animal. High levels of ZnMT may be necessary for the hepatocyte to make the large quantities of new macromolecules (DNA, RNA, and proteins) it needs, when changing its metabolism in response to stresses. Many of the enzymes involved in nucleic acid and protein biosynthesis are zinc-requiring enzymes.

Protein kinase C can be directly activated by certain tumor promoters, called phorbol esters. We have tested TPA (12-0-tetradecanoylphorbol 13-acetate) as an inducer of ZnMT in vivo and in vitro. It is the most potent inducer yet described with an in vivo $ED_{0.5}$ of 26.5 nmoles/kg b.w. and in vitro of ~ 1-10 nM. It is intermediate in its effectiveness as an inducer of MT, yielding maximal ZnMT levels in between that of catecholamines and glucocorticoids. The induction of ZnMT by phorbol esters requires de novo protein synthesis, as illustrated by [^{35}S]-cysteine incorporation/cycloheximide experiments.

Purinergic receptors (9) are also linked to adenylate cyclase or phospholipase C. P_1 (adenosine) receptors activate hepatic adenylate cyclase via the A_2 subclass. P_2 (ATP,ADP) receptors activate phospholipase C via the P_{2Y} subclass. We have tested the ability of adenosine, ATP, and various adenosine analogues to induce hepatic ZnMT in vivo. ATP was ineffective (1x or 3x, 11h). Adenosine (3x, 11h), 2-chloroadenosine (1x, 11h), NECA (1x, 11h), and 5'-chloro-5'-deoxyadenosine (3x, 11h) were all good inducers of ZnMT (4.1, 7.3, 5.5, and 3.9-fold inductions, respectively). The induction by 2-chloroadenosine produces a maximal level of ZnMT, comparable to that attained with zinc or with catecholamine inductions.

Induction of Hepatocyte Metallothionein

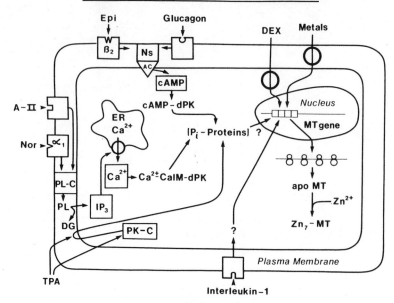

Figure 1

As can be seen from this brief description, the hormonal modulation of hepatic ZnMT levels is a complex process. The mechanisms by which the hormones which activate the protein kinases can activate MT-mRNA transcription are currently completely unknown. There is tremendous interest in these types of processes. The MT system appears to us to be an excellent one in which to explore the possibilities.

References

1. Brady, F.O., and Bunger, P. (1979) Biochem. Biophys. Res. Commun. 91, 911-918.
2. Brady, F.O. (1980) Life Sci. 28, 1647-1654.
3. Brady, F.O. (1983) Trends Biochem. Sci. 7, 143-145.
4. Brady, F.O., and Helvig, B.S. (1984) Am. J. Physiol. 247, E318-E322.
5. Helvig, B.S., and Brady, F.O. (1984) Life Sci. 35, 2513-2518.
6. Brady, F.O., Helvig, B.S., Funk, A.E., and Garrett, S.H. (1987) in Kägi, J.H.R. (ed.) Metallothionein II, Birkhäuser Verlag, Basel, in press.
7. Cousins, R.J. (1985) Physiol. Rev. 65, 238-309.
8. Dunn, M.A., Blalock, T.L., and Cousins, R.J. (1987) Proc. Soc. Exp. Biol. Med. 185, 107-119.
9. Williams, M. (1987) Ann. Rev. Pharmacol. Toxicol. 27, 315-345.

EFFECTS OF ZN-DEFICIENCY AND VALPROATE ON

ISOMETALLOTHIONEINS IN FETAL RAT LIVER

Jürgen Vormann and Theodor Günther

Institute of Molecular Biology and Biochemistry
Free University of Berlin
Arnimallee 22, D-1000 Berlin 33, FRG

INTRODUCTION

During fetal development, the content of Zn-metallothionein (Zn-MT) in liver increases, reaching a maximum at birth[1]. By either feeding pregnant rats a Zn-deficient diet or treating them with valproate, Zn-MT content in fetal liver is drastically reduced[2]. In rat liver cells, at least two iso-forms of Zn-MT are present. Therefore, we investigated whether these iso-forms are identically affected by changing the total Zn-MT content, or whether they are regulated separately.

METHODS

Livers from day 21 fetuses were used. Pregnant Wistar rats were fed a control (100 ppm Zn) or Zn-deficient (1.5 ppm Zn) diet from day 0 to 21 of gestation. In an additional experiment, normally fed dams received a daily oral dose of 300 mg/kg sodium valproate from day 16 to 20 of gestation. Zn-MT from heat-treated (2 min 96 °C) liver cytoplasm was separated by con-ventional gel chromatography on Sephadex G-75 (Pharmacia) or by FPLC on Superose 12 (Pharmacia). Zn-containing fractions with a molecular weight between 5000 and 15000 were pooled and further separated on reversed phase (RP)-HPLC or ion exchange chromatography.

Reversed Phase HPLC: RP-HPLC was performed at a flow rate of 1 ml/min on Nucleosil 10 C8 columns, 10 μM particle size (Knauer, Berlin, FRG). 20 μl of pooled MT-fractions from gel chromatography were injected and separation was achieved by a gradient (straight line) formed between buffer A (50 mM K-phosphate-buffer, pH 7.4) and buffer B (50 mM K-phosphate-buffer, pH 7.4 with 60 % acetonitrile). MT-peaks were identified by their ^{65}Zn content. For that purpose, MT-containing fractions were equilibrated with ^{65}Zn, and after separation on RP-HPLC, radioactivity of the peaks was measured. Only two peaks had a significant ^{65}Zn radioactivity.

Ion exchange chromatography: Ion exchange chromatography with 2 ml of pooled MT-fractions from gel chromatography was performed on a DE 52 (Whatman) anion exchange column equilibrated with 10 mM Tris/Cl, pH 7.4. The column was eluted with a gradient (straight line) formed between buffer A (10 mM Tris/Cl, pH 7.4) and buffer B (1 M Tris/Cl, pH 7.4). 2 ml fractions were collected at a flow rate of 1 ml/min and the Zn content of the fractions was measured by atomic absorptions spectrophotometry (Pye Unicam, SP 9).

RESULTS AND DISCUSSION

By means of RP-HPLC two Zn-MT peaks were separated. The ratio of
MT 1/MT 2 amounted to 1.3. After feeding the Zn-deficient diet or after
application of valproate, Zn-MT was drastically reduced (Zn-def. -79 %, valproate -58 %). However, the ratio of MT 1/MT 2 did not change significantly.

With ion exchange chromatography, two additional Zn-MT containing peaks
could be separated. Probably the two additional Zn-MT peaks (I,II) are
splitted from the MT 1 peak, obtained by RP-HPLC. Neither Zn deficiency nor
valproate treatment had an effect on the relative amounts of the four isometallothioneins.

So, these treatments affect all isometallothioneins to the same degree,
probably by influencing the regulation of the MT-genes.

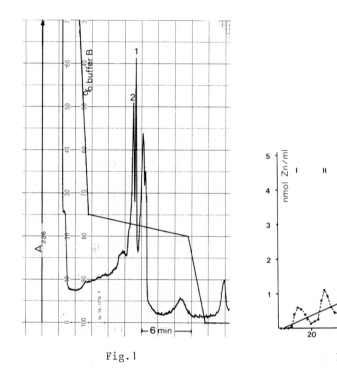

Fig.1   Fig.2

Reversed phase HPLC (Fig.1) and ion exchange chromatography (Fig.2) of
pooled Zn-MT containing fractions from gel chromatography; liver cytoplasm
from day 21 control fetuses.

REFERENCES

1. J. Vormann and T. Günther, Development of Fetal Mineral and Trace Element Metabolism in Rats with Normal as well as Magnesium- and Zinc-
   Deficient Diets, Biol. Trace Elem. Res. 9:37 (1986)

2. J. Vormann, V. Höllriegl, H. J. Merker, T. Günther, Effect of Valproate
   on Zinc Metabolism in Fetal and Maternal Rats Fed Normal and Zinc-
   Deficient Diets, Biol. Trace Elem. Res. 10:25 (1986)

UNINDUCED CU-METALLOTHIONEIN AND OTHER CU-BINDING PROTEINS IN MAMMALS

AND FISH

D. H. Petering, S. Krezoski, T. Hartmann, Pu Chen, P. Onana, and C. F. Shaw III

Department of Chemistry,
University of Wisconsin-Milwaukee
Milwaukee, WI 53201

INTRODUCTION

Among species of metallothionein (Mt), the least understood are those containing copper. Relatively few reports on native Cu-Mt, isolated from organisms not exposed to Cu. Furthermore, there has been marked variability in reported Cu to protein stoichiometry and in the coordination number for sulfhydryl groups bound to Cu^{1-3}. Hence, the experiments reported here focus on basic inorganic properties of a Cu,Zn-Mt isolated from bovine calf liver and survey fresh water fishes for the presence and characteristics of Cu,Zn-binding proteins (BP), analogous to Mt.

RESULTS AND DISCUSSION

Cu,Zn-Mt and Cu,Zn-BP were isolated from liver cytosols as pools of metal running with 10,000 dalton protein over Sephadex G-75. They were further purified on an HPLC-DEAE column.

The protein from bovine calf has a variable Cu to Zn ratio, probably reflecting the age of the animal. The properties of one with a Cu/Zn ratio of 8 was examined in detail. It had a (Cu+Zn)/protein ratio of 7.7, a sulfhydryl to protein ratio of 17.7, and a SH/(Cu+Zn) ratio of 2.3. It also has a ratio of rapidly reacting $Fe(CN)_6^{3+}$/(Cu+SH) of 0.9, indicating that all of the copper was Cu(I) and that all of the redox-active groups in the protein react with this oxidant. As Cu is displaced anaerobically from the protein below pH 1, the pH dependence show that 3 H^+ are bound to sulfhydryls per Cu released. All of these results suggest that Cu binds to Mt in the same manner as Zn or Cd and not with a larger Cu/protein stoichiometry of 11-12 to 1 as recently suggested[4]. However, the Cu content of the bovine liver protein can be raised to 10-12 by incubating the protein with Cu^{2+} and 2-mercaptoethanol.

Both bathocuproine disulfonate (BCS) and penicillamine compete with Mt for Cu(I) bound to Cu,Zn-Mt. Using BCS the reaction is triphasic, with 40% of the metal removed within the time of mixing. These results show that Cu like Zn is kinetically labile when bound to Mt and suggests that Cu-Mt could be a reactive site when penicillamine is used to extract Cu from copper-loaded mammals.

Cu,Zn-BPs are found in high concentration in livers of eleven freshwater fishes: lake trout fingerling and adult, burbot, goldfish, bullhead, brook trout, catfish, large mouth bass, carp, yellow perch, and rainbow trout. Most of these animals have only one liver isoprotein and most have Cu/Zn ratios of 0.6-1.2. They behave like Cu, Zn-Mt in several ways. Carp protein has a SH/(Cu+Zn) stoichiometry of 2.3-2.6 and resembles Mt in its reactivity with BCS and 5,5'-dithio-bis(2-nitrobenzoate). Only Zn in Cu,Zn-BP from berbot is displaced by Cd under conditions of stoichiometric competition. The finding of high levels of Cu,Zn-BP in adult freshwater fishes contrasts with the relative scarcity of Mt in adult mammalian liver of rat or mouse.

ACKNOWLEDGMENT

Supported by the Wisconsin Sea Grant Program and NIH grant ES-01985.

REFERENCES

1. K. Lerch, Chemistry and Biology of Copper Metallothioneins, in Metal Ions in Biological Systems, 13, H. Siegel, ed., Marcel Dekker, New York, 1981.
2. I. Bremner, Involvement of metallothionein in the hepatic metabolism of copper. J. Nutr., 117:19 (1986).
3. J. H. Freedman, L. Powers, and J. Peisach, Structure of the copper cluster in canine hepatic metallothionein using x-ray absorption spectroscopy. Biochem., 25:2342 (1986).

INSTABILITY AT LOW pH OF COPPER-THIONEIN-PROTEINS FROM LIVERS OF

BEDLINGTON TERRIERS WITH COPPER-TOXICOSIS

H.Nederbragt, A.J.Lagerwerf and T.S.G.A.M. van den Ingh

Department of Pathology, Faculty of Veterinary Science
State University, Utrecht, The Netherlands

Copper (Cu) toxicosis in Bedlington terriers is an autosomal recess-ively inherited disorder, leading to excessive accumulation of Cu in the liver. In a final stage this may lead to sudden death due to hemolysis and liver damage. Most of the Cu in the liver has been found as metallothionein (MT)-bound Cu in the lysosomes[1]. It was the purpose of our study to inves-tigate whether Cu-thionein of affected Bedlington terriers behaves diffe-rently in a lysosomal environment.

Livers from 3 Bedlington terriers were used. The control dog, BT1, had a liver Cu concentration of 45 µg/g; two dogs with Cu toxicosis, BT2 and BT3, had liver Cu concentrations of 1790 and 1974 µg/g resp. Different methods for treating the livers were used. They were homogenised in 20 mM phosphate buffer, pH 7.4, with or without 10 mM mercaptoethanol (ME). Both homogenates were centrifuged, either during 60 min. at 75,000 g or, after heating at 60°C during 10 min., at 6,000 g for 10 min. Supernatants were fractionated using gelfiltration on Sephadex G-75 at 22°C with a Tris-Ac buffer, pH 7.4, with or without ME as in the homogenizing buffer. Combi-nation of the four different homogenizing methods yielded four different elution patterns of Cu proteins. These patterns, as obtained for the liver

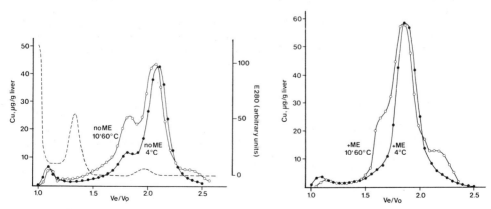

Fig.1. Gelfiltration pattern of liver supernatants of dog BT2, obtained by different methods. o—•: Cu; ---: E 280.

Table 1. Distribution of Cu after rechromatography of Cu-thionein-like protein at pH 7.4 or pH 4.0.

method	elution pH	Cu peak[a]	Cu concentration, µg/g liver		
			BT1	BT2	BT3
no ME	7.4	A	5.42	7.77	2.25
		B	–	–	–
	4.0	A	2.93	–	–
		B	–	14.60	15.43
with ME	7.4	A	5.53	10.46	39.85
		B	–	0.21	–
	4.0	A	1.45	8.48	31.74
		B	–	–	–

a: peak A: Ve/Vo = 1.82–2.20; peak B: Ve/Vo = 2.33–2.80

of dog BT2, are shown in fig.1. The Cu-peak eluting between Ve/Vo = 1.75 and Ve/Vo = 2.20 is assumed to represent a Cu-thionein-like protein (CuTLP). Addition of ME to the buffers seemed to cause an increase in the apparent molecular weight of the CuTLP and the heat treatment resulted in the occurrence of Cu containing peaks of higher molecular weight. Only the homogenate obtained in the presence of ME without heating gave one single peak at Ve/Vo = 1.8 – 2.2.

The fractions containing the CuTLP were collected and rechromatographed, one part in a buffer with the same composition as before, pH 7.4, the other part in that same buffer but with pH 4.0. Cu was eluted either at Ve/Vo = 1.82 – 2.20 or at Ve/Vo = 2.33 – 2.80 which represents free Cu. The summary of the results of this rechromatography experiment is shown in table 1. As can be seen CuTLP loses all of its Cu at pH 4.0 in affected Bedlington terriers when the protein is not protected by ME. The results of CuTLP obtained after heat treatment were identical (not shown).

We assume that this loss of Cu at lower pH is not a consequence of excessive accumulation of Cu and its binding to non-specific binding-sites on the MT molecule. This is presently investigated by subjecting livers of Cu-loaded rats and sheep to the same treatments.

When this assumption is correct, than these findings may contribute to an explanation of the Cu accumulation in the liver of affected Bedlington terriers. Thus, Cu taken up by the lysosomes as Cu-thionein may become free from the protein at the low lysosomal pH. Then it may diffuse out of the lysosome to the cytoplasma where it may induce new MT synthesis; this MT, to which Cu is bound, may in its turn be taken up by lysosomes, and so on, until a toxic Cu or Cu-thionein concentration has been reached.

Reference

1. G.F.Johnson, A.G.Morell, R.J.Stockert and I.Sternlieb, Hepatic lysosomal copper protein in dogs with an inherited copper toxicosis, Hepatology, 1:243 (1981).

ZINC, COPPER AND METALLOTHIONEIN mRNA IN SHEEP LIVER DURING DEVELOPMENT

J.F.B.Mercer, J.Smith, A.Grimes, J.McC.Howell*, P.Gill*, and D.M.Danks

Murdoch Institute, Royal Children's Hospital, Melbourne 3052, Australia
*School of Veterinary Studies, Murdoch University, Perth 6150, Australia

INTRODUCTION

During foetal and early neonatal life mammals have elevated concentrations of copper and zinc in their livers (1). Much of this copper and zinc is associated with the low molecular weight cysteine-rich proteins, metallothioneins (MTs). Our previous work with foetal and neonatal rats showed that metallothionein mRNA levels in the liver were elevated in the late foetal and early neonatal period (2). There was a reasonable correlation between hepatic zinc concentration and mRNA levels, but our data suggested that zinc was not the primary inducer of foetal liver MT mRNA.

To further investigate the regulation of MT genes in developing liver, we have examined the copper, zinc and MT mRNA levels in foetal and neonatal sheep livers. Sheep are of interest since they have an unusual copper metabolism, characterized by reduced biliary excretion and marked hepatic accumulation if dietary levels are too high. Our results are surprising in that we found very high levels of zinc and metallothionein mRNA in the foetal sheep liver.

METHODS

Merino sheep were used and foetal ages were estimated from crown rump length measurements. Copper and zinc levels were analyzed using a Perkin Elmer model 5000 atomic absorption spectrophotometer. RNA was isolated from the livers using a guanidine hydrochloride procedure (3). MT 1a mRNA was measured using dot blot hybridization and a specific probe as described by Peterson and Mercer (4). To determine the absolute levels of MT mRNA, an RNA standard was prepared using an SP 6 transcription system. An RNA/DNA ratio of 2 was assumed for sheep liver.

RESULTS

Fig 1a shows that the levels of MT 1a mRNA are very high early in foetal liver (about 10,000 molecules/cell at 30 days gestation). A progressive reduction in levels during gestation occurs, with adult levels being attained soon after birth. To put these values in perspective, the foetal rat liver contains a maximum of 400 molecules/cell of MT1 mRNA. Hepatic zinc concentrations (Fig 1b) also showed the levels of the metal

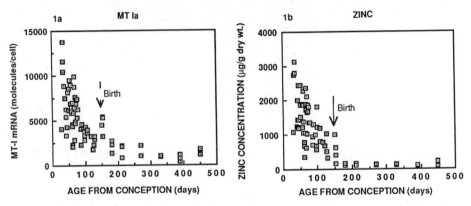

Fig.1. Variation of MT1a mRNA (a) and zinc concentration in sheep liver during development.

were strikingly elevated in the early foetal period (up to 3000 µg/g). The MT mRNA and zinc levels were well correlated (correlation coefficient 0.89). In contrast there is no correlation between MT mRNA levels and copper.

In adult sheep, MT mRNA levels are much lower, but still higher than in the uninduced adult rat liver. A plot of zinc concentration versus MT mRNA (not shown) suggests that in both species zinc and message levels are correlated, but five-fold more mRNA is present in the sheep liver per unit of zinc than in the rat liver.

DISCUSSION

The most likely interpretation of the close correlation between zinc and MT mRNA is that zinc is the primary inducer of the message in the foetal sheep. High levels of both metal and mRNA are found at times when cortisol, the other most probable inducer, is only present at a low level in the foetus. The comparison of the response of sheep and rat MT mRNA to a given concentration of zinc suggests the former species produces 5-fold more mRNA. We have yet to establish whether all this mRNA actually translates into high levels of protein.

This leads us to propose that the high levels of zinc in foetal liver, and the tendency of the adult liver to accumulate copper may be both due to the hyper-response of the metallothionein genes in sheep liver to zinc.

REFERENCES

1. Bremner, I., Williams, R.B. and Young, B.W. (1977) Br.J.Nutr. 38, 87-92.
2. Mercer, J.F.B. and Grimes, A. (1986) Biochem.J. 238, 23-27.
3. Wake, S.A. and Mercer, J.F.B. (1985) Biochem.J. 228, 425-432.
4. Peterson, M.G. and Mercer, J.F.B. (1986) Eur.J.Biochem. 160, 579-585.

METALLOTHIONEIN CONCENTRATIONS IN THE BLOOD

AND URINE OF STREPTOZOTOCIN TREATED RATS

Ian Bremner, James N. Morrison and Anne M. Wood

Rowett Research Institute
Bucksburn
Aberdeen, AB2 9SB, U.K.

INTRODUCTION

The accumulation of metallothionein (MT) in tissues is influenced by changes in Zn and Cu status and also by the occurrence of various types of stress and infection. The effects of these 'stress' factors are mediated by changes in hormonal status; glucocorticoids, glucagon and catecholamines can, for example, induce hepatic MT synthesis. In studies of the effects of adrenal and pancreatic hormones on trace metal metabolism, Failla et al. (1985) have shown that hepatic and renal concentrations of Zn, Cu and of MT are greatly increased in rats made diabetic by treatment with streptozotocin (STZ). We have recently suggested that assay of MT in blood and urine can be used in the diagnosis of Zn deficiency (Bremner and Morrison, 1987) but before this technique can be applied in practice more information is needed on the effects of other patho-physiological states on the occurrence of MT in these fluids. It is particularly important that studies be extended to conditions where tissue MT levels are altered. It was therefore of interest to determine the effects of STZ-induced diabetes on MT levels in blood plasma and cells and in urine of rats.

MATERIALS AND METHODS

Forty male Hooded Lister rats (Rowett strain) aged 5 weeks and weighing about 140 g, were allocated at random to 8 groups and given stock colony diet ad libitum. Four of these groups were injected i.p. with STZ in 10 mM citrate buffer, pH 4.5 (100 mg/kg body weight). Control rats were injected with citrate buffer alone. Groups of STZ-treated and control rats were killed after 1, 2, 4 and 7 days by exsanguination under ether anaesthesia. Blood plasma and lysed cells and samples of liver and kidneys were assayed for MT-I by radioimmunoassay. In parallel experiments urine samples were collected on ice from rats kept in metabolism cages.

RESULTS AND DISCUSSION

As expected injection of STZ seriously affected the growth rate of the rats and after 7 days the weights of the control and STZ-treated rats were 192 ± 5 and 138 ± 8 g respectively. The diabetic state of the latter rats was confirmed by their elevated blood glucose levels, 3.5 ± 0.1 mg/ml after 7

Table 1. Effect of STZ on MT-I Concentrations in Rat Tissues

Time (Days)	Liver (µg/g)		Kidneys (µg/g)		Plasma (ng/ml)		Erythrocytes (ng/ml blood)	
	Con	STZ	Con	STZ	Con	STZ	Con	STZ
1	23	17	28	25	2.1	1.5	53	36
2	18	123	39	46	4.1	6.9	50	9
4	9	152	38	133	2.2	14.9	43	8
7	7	84	42	180	4.1	13.3	33	11

days compared with only 1.5 ± 0.1 mg/ml in the controls. Injection of STZ caused large increases in both hepatic and renal concentrations of MT-I after a few days (Table 1). These were accompanied by increases in liver and kidney concentrations of Zn and to a lesser extent Cu. Plasma MT-I concentrations also increased in the diabetic rats after 4 days and were correlated with MT-I concentrations in the liver. Urinary excretion of MT-I in STZ-treated rats was also increased. Thus concentrations after 7 days (ng/mg creatinine) were 521 ± 154 in the diabetic animals and 72 ± 12 in the controls. Total excretion of MT-I over a 24 hour period was 2.00 ± 0.54 and 0.30 ± 0.04 µg in diabetic and control rats respectively.

Contrary to these findings of increased MT-I levels in tissues, plasma and urine, induction of a diabetic state caused a rapid decrease in MT-I concentrations in lysed blood cells. This was evident within 1 day and within 2 days concentrations were only 18% of control values. The proportional reduction was less after 7 days but this was a reflection of the age-dependent decrease in blood cell MT-I levels in the controls.

The increases in liver and kidney MT-I levels in the diabetic rats probably result from the changes in hormonal status of the rats, as has been suggested by Failla et al. (1985). The increased plasma and urine MT-I concentrations are in turn likely to be a consequence of the changes in tissue MT-I, since correlations between liver and plasma MT-I levels have consistently been observed by us (Bremner & Morrison, 1987). However it is difficult to explain the reduction in blood cell MT-I levels in these animals. Similar rapid decreases in cell MT-I levels occur in Zn-deficient rats. The results could reflect increased degradation of MT-I, its leakage from the cell or simply loss of bound metal. No fractionation of blood cells was carried out but it is known that MT-I is associated mainly with the erythrocytes and particularly reticulocyte-rich fractions. It will be of interest to establish whether similar changes in blood and urine MT levels occur in diabetic humans.

REFERENCES

Bremner, I., and Morrison, J., 1987, Metallothionein as an indicator of zinc status, in: "Trace Element Research in Humans", A. S. Prasad, ed., Alan R. Liss Inc., New York.
Failla, M., Caperna, T. J., and Dougherty, J. M. 1985, Influence of pancreatic and adrenal hormones on altered trace metal metabolism in the ST2-diabetic rat, in: "Trace Elements in Man and Animals-TEMA5", C.F. Mills, I. Bremner, and J.K. Chesters, eds., pp 330-333, CAB; Slough.

EFFECTS OF METALLOTHIONEIN BINDING CAPACITY FOR VARIOUS METALS ON HEME OXYGENASE ACTIVITY

Koji Arizono, Eiji Okanari, Kiyoshi Ueno, and Toshihiko Ariyoshi

Faculty of Pharmaceutical Sciences
Nagasaki University
Nagasaki 852, Japan

It is well known that metallothionein (MT), which is involved in protection from metal toxicity and homeostasis of essential metals, is induced by various metals (1). In general, these metals are divided into two groups: those that bind to MT in vivo or those that do not (2).

Our previous study indicated that a single treatment with cadmium (Cd), a typical metal that binds to MT, increased the activity of heme oxygenase (HO), the rate-limiting enzyme of heme degradation, 11-fold as compared with the control. Daily administration of Cd for 3 days caused an increase of only 4-fold in that activity. From this experiment, we suggested that there may be a relation between MT content and HO activity (3). Therefore, it seemed interesting to investigate if MT binding capacity for various metals regulates the induction of HO. Cd, Zinc (Zn), Copper (Cu) and Silver (Ag) are used as MT binding metals, whereas Cobalt (Co), Nickel (Ni) and Manganese (Mn) are used as nonbinding metals.

Adult male rats of the Wistar strain, 250 g, were used for these experiments. Metals used and doses were as follows:
$CdCl_2$ $2H_2O$ (3 mg/kg), $Zn(CH_3COO)_2$ $2H_2O$ (33 mg/kg), $CuSO_4$ $5H_2O$ (15 or 10 mg/kg), $AgNO_3$ (10 mg/kg), $CoCl_2$ $6H_2O$ (20 mg/kg), $NiCl_2$ $6H_2O$ (40 mg/kg) and $Mn(CH_3COO)_2$ $4H_2O$ (45 mg/kg).

MT concentration and HO activity were significantly increased, namely MT was 55.6-, 33.9-, 35.3- and 34.4-fold, and HO was 11.0-, 2.6-, 4.4-, and 5.2-fold at 24 h after the single injection of Cd, Zn, Cu and Ag, respectively, as compared with the control. At 72 h after the injection, both MT concentration and HO activity were almost restored to the control level.

After repeated treatment with metals daily for 3 days, MT concentration was markedly increased by 95.4-, 53.8-, 31.0- and 61.3-fold at 24 h after the last injection of Cd, Zn, Cu and Ag, respectively, as compared to the control.

In the case of a single injection of Co, Ni and Mn, MT concentration was increased 4.9-, 9.1- and 2.7-fold at 24 h after respective treatment, while HO activity was also enhanced 2.4-, 5.7- and 3.6-fold, respectively. However, at 72 h after the injection MT concentration, HO activity was restored nearly to the control levels.

683

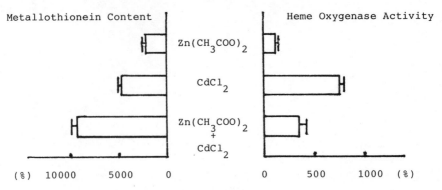

Fig. 1. Effects of CdCl₂ on Hepatic Metallothionein Content and Heme
Oxygenase Activity in Zn(CH₃COO)₂-pretreated Rats. Each
column represents % of control and the mean ± S.E. of 3-6 rats.

The repeated injection of Co, Ni and Mn daily for 3 days showed marked
increases in the MT concentration, 12.3-, 24.7-, and 23.0-fold at 24 h after
the last injection, respectively. In addition, HO activity was also
enhanced 4.8-, 9.9-, and 4.6-fold as compared to the control.

By treatment with Cd of Zn, Cu or Ag pretreated rats, MT concentration
was markedly increased, when compared with Cd injection alone. However, HO
activity was at a very low level, as compared with that of Cd treatment
alone.

At 24 h after the injection of Ni to Zn pretreated rats, both MT concentration and HO activity were more increased than with Ni treatment alone.

It was reported that induction of HO by metals is due to metals acting
directly at cellular HO regulatory sites (4). Administration of Cd to MT
preinduced animals allowed the capture of Cd by this preinduced MT. Ni
which does not bind to MT in vivo was not captured by preinduced MT. As a
result, the decreased ratio of metals which can modulate HO regulatory sites
caused less HO activity enhancement effect.

It is interesting that both repeated Cd treatment and Zn pretreatment
showed good preventive effects on enhanced HO levels.

Judging from these experiments, we conclude that MT binding capacity
for various metals affects HO activity and homeostasis of essential metals.

References

(1) Brady, F.O. (1982) Trends Biochem. Sci. 7, 143-145.
(2) Suzuki, Y., and Yoshikawa, H. (1976) Ind. Health 14, 25-31.
(3) Ueno, K., Matsumoto, H., Arizono, K., and Ariyoshi, T. (1985) Eisei
     Kagaku 30, P-32.
(4) Maines, M.D., and Kappas, A. (1977) Pro. Natl. Acad. Sci. 74,
     1875-1878.

THE EFFECT OF DIETARY ZN BEFORE AND AFTER 65-ZN ADMINISTRATION ON

ABSORPTION AND TURNOVER OF 65-ZN

Janet R. Hunt, Phyllis E. Johnson and Patricia B. Swan

USDA-ARS Grand Forks Human Nutrition Research Center
Grand Forks, North Dakota, and University of Minnesota
St. Paul, Minnesota

Zn absorption may adapt rapidly to alterations in Zn intake and may depend on the diet fed after Zn consumption. The present study tested the hypotheses that 65-Zn absorption was affected by a) altering Zn intake 4 d before 65-Zn administration, and b) altering Zn intake after 65-Zn administration.

METHODS

Male Long-Evans rats 32 d old and weighing 126 ± 10 g (SD) were fed an AIN-76A diet modified in Zn content. For 6 dietary treatments, diets contained 85.0 mg Zn/kg for all but the 4 d before 65-Zn administration, when diets contained 7.5, 11.5, 18.1, 22.6, 39.5, and 85.0 mg Zn/kg. A 7th dietary treatment continued the 11.5 mg Zn/kg diet used for 4 d prior to 65-Zn administration, rather than returning to the diet containing 85.0 mg Zn/kg. For 20 d, food was made available only between 8:00 and 12:00 hours. On day 21 between 8:00 and 9:00 hours, 6 rats from each dietary treatment received 1 μCi carrier-free 65-ZnCl_2 in 0.2 ml deionized water by gavage and 6 received the isotope in 0.1 ml of saline containing 5 mM glycine intramuscularly. Diets were fed ad libitum beginning at 16:00 hours that day. Absorption was determined by whole body counting, comparing retention of 65-Zn after injection or gavage (1). Percent retention was corrected to zero time by extrapolating from the linear portion of the semilogarithmic retention plot. Biological half-lives were calculated from the linear portion of the plot (between 5 and 26 d).

RESULTS AND DISCUSSION

Altering Zn intakes before 65-Zn administration affected retention of both injected and gavaged 65-Zn (Figure 1), but did not affect biological half-life of 65-Zn. The effect on absorption was not significant, but differences in retention of gavaged 65-Zn were greater than those using injected 65-Zn, suggesting that absorption was affected, and to a greater degree than short-term excretion (the latter being indicated by retention of injected 65-Zn). Differences in Zn retention after 4 d of altered Zn intakes were not as great as when Zn intakes were varied for 3 weeks before and 4 weeks after 65-Zn administration (2).

Feeding Zn concentrations of 11.5 rather than 85 mg/kg after 65-Zn administration increased retention (whether 65-Zn was gavaged or injected),

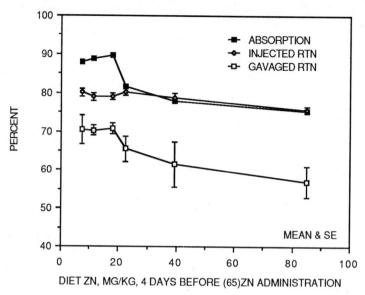

PERCENT

- ■ ABSORPTION
- ◆ INJECTED RTN
- □ GAVAGED RTN

MEAN & SE

DIET ZN, MG/KG, 4 DAYS BEFORE (65)ZN ADMINISTRATION

Figure 1. Effect of altering Zn intakes 4 d before 65-Zn administration.

absorption, and the biological half-life of 65-Zn (Table 1). Thus, 65-Zn absorption was affected by Zn consumed several hours after 65-Zn administration, when presumably the colon was the major site of 65-Zn absorption. Differences in short-term excretion (indicated by retention of injected 65-Zn) were slightly greater than differences in absorption.

In conclusion, homeostatic adjustments in Zn absorption and excretion occurred within a few days of alterations in Zn intakes. Previous Zn intakes primarily affected absorption, while subsequent intakes had a somewhat greater effect on excretion.

Table 1. Effect of dietary zinc consumed after 65-Zn administration on retention, absorption, and biological half-life of 65-Zn (Values are mean ± SD)

Diet Zn after 65-Zn Admin (mg/kg)	Retention of Gavaged 65-Zn (%)	Retention of Injected 65-Zn (%)	Absorption of Gavaged 65-Zn (%)	Biological Half-life of 65-Zn (d)
11.5	81.8 ± 3.4	86.8 ± 4.1	94.3 ± 3.9	76.0 ± 9.7
85.0	70.2 ± 3.3	78.9 ± 2.6	88.9 ± 4.1	35.0 ± 2.2
Probability	$p < 0.001$	$p < 0.003$	$p < 0.05$	$p < 0.001$

REFERENCES

1. D.A. Heth & W.G. Hoekstra, Zinc-65 absorption and turnover in rats I. A procedure to determine zinc-65 absorption and the antagonistic effect of calcium in a practical diet, J. Nutr. 85:367-374 (1965).
2. J.R. Hunt, P.E. Johnson and P.B. Swan, The influence of usual zinc intake and zinc in a meal on 65Zn retention and turnover in the rat, J. Nutr. (in press).

EFFECT OF LONG-TERM TRACE ELEMENT SUPPLEMENTATION ON BLOOD TRACE ELEMENT LEVELS AND ABSORPTION OF ^{75}Se, ^{54}Mn and ^{65}Zn

Brittmarie Sandström, Lena Davidsson, Robert Eriksson, and Magne Alpsten

Departments of Clinical Nutrition and Radiation Physics
University of Gothenburg, S-413 45 Gothenburg, Sweden

INTRODUCTION

The trace element content of a "refined western" diet is often low and as Sweden is also a low selenium area it can be difficult to reach recommended or the so called "safe and adequate levels" of selenium as well as of other trace elements on a normal diet. This has led to widespread use of trace element supplements in various combinations. To increase our knowledge of the metabolic handling of trace elements in humans, we have measured the effect of daily intake of a trace element supplement over a long period on commonly used indices of trace element status and on the absorption of zinc, selenium and manganese from the supplement in healthy subjects.

MATERIAL AND METHODS

Ten subjects (8 women, 3 men; 23-31 years) volunteered for the studies. They were all apparently healthy and a brief dietary history revealed normal Swedish dietary habits. Each subject took a vitamin and trace element supplement containing RDA or "safe and adequate" levels of nutrients, including 18 mg of iron, 50 µg of selenium, 15 mg of zinc and 2.5 mg of manganese (Vitamineral-ACO, ACO Läkemedel, Solna, Sweden) daily for a minimum of 12 weeks. Seven of the subjects continued the supplementation for 35 weeks. Twenty-four hour urine collections were made and a blood sample was drawn every other week for the first 16 weeks and thereafter once a month up to 44 weeks. In seven of the subjects, the absorption of zinc, selenium and manganese was measured after 30-31 weeks of supplementation.

Zinc in serum and urine was measured by flame atomic absorption spectroscopy (Perkin Elmer 360). Manganese in blood and selenium in plasma and urine were measured by flameless atomic absorption spectroscopy (Perkin Elmer Zeeman 3030)[1,2,3]. Activity of glutathionperoxidase (GSHPx) was measured in platelets and plasma according to Levander et al[4]. Iron status indices were determined by routine clinical laboratory methods.

The absorption of zinc, selenium and manganese from the supplement was measured by administration of an aqueous solution of the supplement labelled with ^{65}Zn, ^{54}Mn, and ^{75}Se together with a small meal and measurement of the whole body retention of the radionuclides on day 5-30. Absorption was ^{75}Se calculated from the retention measurements with the addition of urinary excretion during the first 24 h[5].

RESULTS

The initial blood values were all within the normal range for our laboratory: P-Se 1.01+0.14 µmol/l, S-Zn 12.8+2.2 µmol/l, S-Fe 19.2+6.3 µmol/l, S-TIBC 59.2+11.1 µmol/l, B-Hb 139+8.6 g/l, B-Mn 263+48.3 nmol/l. Urinary excretion of zinc was 6.7+3.3 µmol/24 h and of selenium 0.34+0.11 µmol/24 h. Activity of GSHPx in plasma before supplementation was 4.8+0.6 munits/mg protein. Due to technical problems, GSHPx-activity in platelets was not possible to determine before the study. After 7 weeks of supplementation, the activity was 119+58 munits/mg protein. After two weeks of supplementation, plasma selenium values had increased to 1.08+0.16 (p<0.05, paired t-test). GSHPx-activity in platelets after 16 weeks of supplementation was 169+42 munits/mg protein (p<0.05 cf week 7). However, the increase in both indices was transient and at the end of the supplementation no difference from initial values was observed. A small increase of serum zinc values, to 14.0+2.6 µmol/l, was observed after 30 weeks of supplementation. No significant changes were observed in the other blood and urine parameters studied.

The absorption of zinc, selenium and manganese was 7.2+3.5%, 39+5% and 1.0+0.8%, respectively.

DISCUSSION

Doubling of normal dietary intake of trace elements for up to 35 weeks obviously had only marginal effects on commonly used indices of trace element status. To some extent, this lack of effect could have been due to an impaired absorption. We have earlier observed in a study of nonsupplemented subjects an absorption of 10+2%, 87+4%, and 2.0+1.0% for zinc, selenium and manganese, respectively[5]. However, even with the degree of absorption observed in the present study, substantial amounts of trace elements were provided from the supplement over the supplementation period. It is possible that the absorption of trace elements from the diet was also affected and that the excretion via other routes than the kidney was increased. However, the study clearly shows that the indices of trace element status used in this study are not sensitive enough to distinguish individuals with approximately twice the normal trace element intake from those on ordinary intake levels.

REFERENCES

1. M. S. Clegg, B. Lönnerdal, L. S. Hurley, C. L. Keen. Analysis of whole blood manganese by flameless atomic absorption spectrophotometry and its use as an indicator of manganese status in animals. Anal Biochem 157:12-18 (1986).
2. W. Slavin, G. R. Carnich, D. C. Manning, E. Pruszkowska. Recent experiences with the stabilized temperature platform furnace and Zeeman background correction. Atomic Spectroscopy 4:69-86 (1983).
3. G. Schlemmer, B. Welz. Palladium and magnesium nitrates, a more universal modifier for graphite furnace atomic absorption spectroscopy. Spectrochim Acta 41B:1157-1165 (1986).
4. O. A. Levander, D. F. DeLoach, V. C. Morris, P. B. Moser. Platelet glutathion peroxidase activity as an index of selenium status in rats. J Nutr 113:55-63 (1983).
5. B. Sandström, L. Davidsson, R. Eriksson, M. Alpsten, C. Bogentoft. Retention of selenium (^{75}Se), zinc ([65]Zn) and manganese ([54]Mn) in humans after intake of a labelled vitamin and mineral supplement. J Trace Elements and Electrolytes in Health and Disease. In press.

RETENTION OF LEAD IN GROWING AND ADULT RATS DEPENDENT ON VARYING LEAD SUPPLY

Manfred Kirchgessner, Anna M. Reichlmayr-Lais and
Norbert K. Stöckl

Institut für Ernährungsphysiologie
Technische Universität München
D-8050 Freising-Weihenstephan

INTRODUCTION

The metabolism of essential trace elements is characterized by homeostatic regulation. This regulation guarantees, over a relatively wide and
element-specific supply range, a physiologically sufficient and tolerable
concentration of each trace element in the cells and their compartments.
The objective of the study presented here was to test whether or not the
metabolism of lead is subject to such regulation because in several earlier
experiments, we were able to prove the essentiality of this element
(Reichlmayr-Lais and Kirchgessner, 1981a,b). Whole body retention was
chosen as the study criterion, and was determined in correlation to a rising
lead supply in both growing rats (Experiment 1) and in adult rats (Experiment 2).

MATERIALS AND METHODS

Experiment 1

One hundred and forty-four female rats with a beginning weight of 35 ± 3 g
were divided into 18 groups of 8 animals each. The rats were provided ad
libitum a half-synthetic diet which contained an increasing lead concentration of 0.025 to 600 mg/kg. The experiment lasted 23 days.

Experiment 2

One hundred thirty-two female adult rats with a mean body weight of 235 ±
12 g were fed ad libitum a half-synthetic diet containing 0.4 mg Pb++/kg for
17 days. Subsequently, 120 of the animals were divided into 15 groups. The
remaining rats served as reference animals for the lead-retention determination. The animals were provided a half-synthetic diet in an amount which
just met the energy maintenance requirement (10.2 g per animal and day).
The diet varied in lead supply from 0.02 to 800 mg Pb++/kg. The experiment
lasted 29 days.

At the end of both experiments, the animals were painlessly killed with
an overdose of ether and each stomach-intestinal tract was removed. Carcasses were then frozen at -20°C until needed for analysis. Lead retention
was determined by measuring the lead concentration in the whole body; here,

the mean concentration in the reference animals sacrificed at the beginning of the experiment was subtracted from the lead value for each of the experimental animals. In order to perform an inverse voltametric detection of lead, the carcasses were dry-ashed at 600°C, 6 N HCl was repeatedly added and allowed to evaporate and then the remaining material was dissolved in an 0.6 N HCl solution.

The composition of the diet, the care of the animals as well as the results are described in detail in Kirchgessner et al. (1987) and Reichlmayr-Lais et al. (1987).

RESULTS

In the growing rats the selected lead supplements led to neither clinical symptoms nor to an effect on the live mass development. Feed intake was also not affected. An elevated absolute lead retention first occurred when the lead supplementation in the diet was 5 ppm. Below this level, the lead retention value initially climbed to a lead concentration of 0.225 ppm, and subsequently declined. With reference to lead intake, the retention in the lower lead supply range was very high, but declined rapidly and approached a constant value with increasing lead supply.

In the adult rats an elevated absolute retention was first ascertained at 10 mg Pb++/kg diet. In the lower range there was again at first a slight increase of lead retention followed by a decrease which in turn was followed by a clear elevation. In the lower supply range, the relative retention in adult rats was elevated, too.

DISCUSSION

These experiments clearly demonstrate that the metabolism of the element lead is subject to homeostatic regulation. On the one hand, this regulation causes an increase in retention only after a relatively high lead supplementation is achieved. On the other hand, homeostatic regulation induces, in comparison to intake, a high retention in the very low supply ranges. In this way a deficiency is guarded against. Accordingly, the lowest supply ranges in both experiments could be rated as suboptimal, for both growing as well as adult rats. In both experiments, a conspicuous result of the lower lead supply range is the initial slight increase of lead retention followed by a decrease. The phenomenon has been observed before for other essential trace elements as well (e.g. Weigand and Kirchgessner, 1978; Kirchgessner et al., 1984) and can be interpreted as an overcompensation of a counter-regulation.

The experiments presented here also show that growing rats react more sensitively to higher lead concentrations than do adult rats. This is evident from the earlier increase of lead retention in growing rats as compared to adult rats.

REFERENCES

Kirchgessner, M., Reichlmayr-Lais, A.M. and Maier, R., 1984, Z. Tierphysiol., Tierernährg. u. Futtermittelkde. 46:1.
Kirchgessner, M., Reichlmayr-Lais, A.M. and Stöckl, N.K., 1987; J. Trace Elem. Electrol. Health Dis., in preparation.
Reichlmayr-Lais, A.M, and Kirchgessner, M., 1981a, Z. Tierphysiol., Tierernährg. u. Futtermittelkde. 46:1.
Reichlmayr-Lais, A.M, and Kirchgessner, M., 1981b, Ann. Nutr. Metab. 25:281.
Reichlmayr-Lais, A.M., Stöckl, N.K., and Kirchgessner, M., 1987, J. Trace Elem. Electrol. Health Dis., in preparation.
Weigand, E., and Kirchgessner, M., 1978, Nutr. Metab. 22:101.

ZINC LOSSES DURING PROLONGED COLD WATER IMMERSION

P.A. Deuster[1], D.J. Smith[2], A. Singh[1], L.L Bernier[1],
U.H. Trostmann[1], B.L. Smoak[1] and T.J. Doubt[2]

[1]Uniformed Services University of the Health Sciences
and [2]Naval Medical Research Institute, Bethesda, MD

INTRODUCTION

Fluid and ion shifts have been reported during head-out immersion, cold exposure and under hyperbaric conditions, with diuresis, natriuresis, and kaliuresis consistent observations (1). The finding that the time courses of water, Na, K, Ca, Mg and Zn excretion coincide after volume expansion in man (3) suggests that similar changes may occur for Zn during immersion. However, this has not been studied. Zn serves vital roles in many cellular responses, including energy metabolism and muscle function, and depletion could impair performance during prolonged immersion in cold water. Thus, characterizing changes in Zn was of interest.

MATERIALS AND METHODS

After giving informed consent 16 male divers each participated in two five-day air saturation dives (ASD) in a hyperbaric chamber maintained at 6.1 meters of sea water. During each ASD series divers participated in two whole body immersions in water maintained at $5.0 \pm 0.1°C$; one began at 1000 hrs (AM) and the other at 2200 hrs (PM); 54 hrs separated the AM and PM immersions. Divers were equipped with full face masks and dry suits for passive thermal protection. While in the chamber physical activity was restricted and sleep hrs of 2200 - 0600 were enforced prior to AM immersions; no sleep periods were instituted prior to the PM immersions. Control diets, providing approximately 3,000 calories and 15 mg of zinc per day, were served to the divers while in the chamber.

Blood samples (25 ml) were obtained before and after each immersion and urine was collected for 12 hours prior to, during and after immersion for a total of 24 hours. Blood samples were analyzed for plasma Zn, creatinine, hematocrit (Hct), hemoglobin (Hb), and the Zn-binding proteins, α_2-macroglobulin (α_2-MG) and albumin (ALB). Urine was analyzed for Zn and creatinine. Clearance of creatinine and Zn was calculated.

RESULTS

Plasma volume, calculated from changes in Hct and Hb, decreased significantly by $17.3 \pm 1.1\%$ and $16.9 \pm 1.3\%$ for AM and PM immersions, respectively. The increases in plasma Zn and α_2-MG noted during both immersions (Table 1) could be accounted for by the reduction in plasma volume, whereas the increase in ALB was less than would be expected during PM immersions. A significant diuresis and increase in Zn excretion were noted during immersion (Figure 1), with the magnitude of the increase less during PM immersions. The increase in the rate of Zn excretion was $264.8 \pm 37.7\%$. No association was noted between the magnitude of the increase in

Table 1. Concentrations of plasma constituents before and after cold water immersions lasting from three to six hours in the AM and PM (Mean ± SEM).

		AM	PM
Plasma Zn (μg/dl)	Before	74.0 ± 1.7	68.1 ± 1.6
	After	84.3 ± 2.1	80.3 ± 1.8
Serum ALB (g/dl)	Before	4.9 ± 0.1	4.9 ± 0.2
	After	5.5 ± 0.2	5.3 ± 0.2
Plasma α_2-MG (mg/dl)	Before	217.4 ± 8.8	223.9 ± 8.5
	After	256.3 ± 9.4	257.7 ± 9.4

Zn excretion and drop in core temperature during immersion. Clearance of Zn increased significantly and to the same extent during both immersions (AM: 0.60 ± 0.07 to 0.97 ± 0.13 ml/min vs PM: 0.50 ± 0.07 to 1.22 ± 0.18 ml/min). Fractional excretion of Zn increased significantly only during the PM (AM: 0.7 ± 0.2 to 0.8 ± 0.2% vs PM: 0.5 ± 0.1 to 1.1 ± 0.3%).

DISCUSSION

While there are no reports describing immersion-induced changes in plasma or urinary Zn, changes in tissue distribution of Zn have been reported in response to cold stress (2). Further, there is evidence for an increased excretion of Zn during volume expansion and diuretic therapy (3). The results of the present study provide evidence that when men are immersed in cold water for prolonged periods, there is a significant increase in plasma Zn which parallels the reduction in plasma volume. In addition, a significant increase in the rate of urinary Zn excretion ensues. That there was no net loss of Zn from plasma suggests that Zn must be coming from other tissues, such as muscle or kidney.

The primary stimulus and mechanisms responsible for the increased urinary Zn excretion in response to cold water immersion cannot be determined from the present results. However, immersion appears to be a more powerful stimulus than cold because none of the responses were related to drop in core temperature. It is likely that alterations in renal function resulting from the immersion-induced central hypervolemia are responsible. Further work will be required to identify specific mechanisms and ascertain implications of the observed responses.

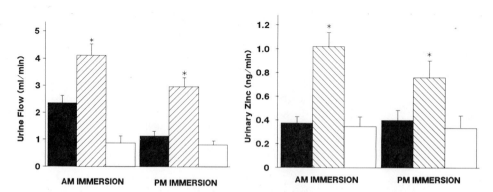

Figure 1. Urinary flow rates and excretion of Zn before (solid), during (hatched) and after (open) AM and PM immersions.

REFERENCES

1. Epstein, M. Renal effects of head-out immersion in man: Implications for an understanding of volume homeostasis. Physiol Rev 58:529-581, 1978.
2. Ohno, HT, K Yahata, R Yamashita, R Doi, N Taniguchi and A Kuroshima. Zinc metabolism in human blood during acute exposure to cold. Res Comm Chem Path Pharm 52:251-256, 1986.
3. Reyes, AJ, WP Leary, CJ Lockett and L Alcocer. Diuretics and zinc. S Afr Med J 62:373-375, 1982.

METABOLIC ADAPTATION OF GILTS ON TRACE-MINERALS

IN RESPONSE TO PREGNANCY AND LOW DIETARY ZINC

Eduardo R. Chavez and Juan Kalinowski

Department of Animal Science
Macdonald College of McGill University
Montreal, Quebec, Canada

Adaptive mechanisms of a different nature exist in animal metabolic systems. They represent the flexibility of the biological unit to cope with continuous environmental and physiological changes and maintain homeostatic conditions compatible with normal growth, maintenance and reproduction during the life-cycle of the animal.

Although maternal adjustments on the endocrine system are known to occur just after conception and continuously through fetal development, neither the true metabolic needs nor the exact pathways by which improvement in nutrient utilization occur as pregnancy progresses are well understood (Davies & Williams, 1976; Swanson & King, 1982). Balance and "tracer" studies were conducted to assess the adaptive mechanisms that take place in the female pig in response to pregnancy and low dietary zinc intake.

I. MATERIALS AND METHODS

Paired littermate Landrace first-pregnancy gilts were fed either a control (50 ppm) or a low-Zn (10 ppm) semi-purified diet during the entire gestation. Non-pregnant gilts of similar age and size were also fed the control diet for comparative balance studies. Animals were maintained in stainless steel metabolic cages and one-week balance studies were conducted in the middle (60 d) and late (100 d) gestation; a "tracer" (Zn-65) study was superimposed during the late balance. All the animals fed were restricted to 2.0 kg/d during the gestation period. Drinking water containing 2.7 ug Zn/L was provided ad libitum. Urine was collected continuously through a urethral catheter using a Bordex Foley Teflon catheter (22 fr) held by a 30 cc balloon. Components of the balance output included feces, urine, sloughed hair and skin desquamation by the animals. The tracer dose was injected through the ear vein cannulated in advance and collection of feces continued until radioactive count was very low.

II. RESULTS AND DISCUSSION

Restricted feeding during gestation did not allow differences in feed consumption between the control and low-Zn groups, although feed consumption was slightly lower for the latter group during late-gestation. However, apparent dry-matter digestibility was not affected by pregnancy or low dietary Zn intake.

Skin desquamation was quantitatively a more important route of integumental trace-mineral losses than was hair. In the non-pregnant and mid-pregnancy gilts skin losses ranged from 4.7 to 5.4 g/d while during late gestation the rate ranged from 4.1 to 4.7 g/d. Hair losses accounted for only 1.6 to 2.9 and 0.3 to 0.5 g/d, respectively.

Non-pregnant gilts excreted slightly but non-significantly greater urine volume (38±7dL/d) compared to their pregnant counterparts (28±2dL/d).

Non-pregnant gilts showed an apparent retention of the total intake of 13% for Cu, 12.8% for Mn and 3.8% for Zn. In contrast, gilts in mid-gestation showed an almost zero balance for Cu, but a negative balance for Mn and Zn equivalent to 5.0 and 7.6% of the total intakes, respectively. During late gestation all these trace-mineral balances were greatly improved. Thus, Mn was slightly positive, Cu balance was shifted to levels similar to or higher than non-pregnant gilts and Zn balance became positive, equivalent to 1.1% of the total intake.

Low dietary Zn intake during gestation of the gilt had a profound effect on the trace-mineral balance gilt. Iron balance, apparently unaffected by pregnancy (20% retention of intake) when compared to non-pregnant gilts (18.5%) suffered a significant reduction on apparent retention (12%) of the total Fe intake. Copper appeared more effectively handled under low dietary Zn intake since apparent retention values were much higher than control counterparts (11 and 22% vs -0.1 and 15% for mid- and late gestation, respectively). Manganese was in a similar negative balance in the two groups at mid-gestation and both improved toward the end of gestation with the control becoming slightly positive and the low-Zn group still slightly negative. Zinc balance under limited supply showed the wide metabolic flexibility of the pregnant gilt to modify the intermediate transactions that result in a highly positive nutrient economy under a low dietary intake. Thus, while the control pregnant gilt showed a negative balance equivalent to almost 8% of an abundant intake that improved late in gestation to almost balance, the low-Zn group kept a highly positive balance during mid- and late gestation, being about 26% of the reduced intake. Tracer studies showed that the "true absorption" of dietary Zn was increased from 38% in the control to 95% in the low-Zn group. Endogenous Zn fecal excretion was reduced from 34 in the control to 10 mg/d in the low-Zn group. Consequently, the saving and prolonged retention of endogenous Zn was corroborated by the tremendous increase in the "biological half-life" of Zn in the pregnant animal from 46 d in the control to 96 d in those with limited dietary supply.

Except for Fe, at similar trace-mineral intake levels, pregnant gilts were less efficient than non-pregnant gilts in handling the metabolic balance of these nutrients, and more so at the mid- than later in pregnancy. They reduced absorption and/or increased endogenous fecal and urinary excretion. Thus, a negative balance was observed for Zn, Cu and Mn during the middle of pregnancy of gilts fed the same diet which then resulted in positive balance in all cases when fed to the non-pregnant gilts. The evaluation of pregnancy on trace-mineral utilization through balance studies is limited. Apparent Zn retention in women was not affected by pregnancy (Swanson et al. 1982;83). A low dietary Zn fed during pregnancy appeared to set stronger adaptive mechanisms in motion. Metabolic adjustments were so severe that they appeared to override those previously referred to in response to pregnancy status. Under low dietary Zn intake, pregnant gilts improved the normal overall economy of trace-mineral nutrition; increased absorption, reduced endogenous fecal and urine losses and prolongated the metabolic active life of Zn.

REFERENCES

Davies, N.T. & Williams, R.B., 1976. The effects of pregnancy of uptake and distribution of Cu in the rat. Proc. Nutr. Soc., 35:4A-5A.
Swanson, C.A. & King. J.C. 1982. Zinc utilization in pregnant and non-pregnant women fed controlled diets providing the Zn RDA, J. Nutr. 112:697
Swanson, C.A., Turnlund, J.R. & King, J.C. 1983. Effect of dietary Zn source and pregnancy on Zn utilization in adult women fed controlled diet. J. Nutr. 113:2557.

AN EPISODE OF ZINC TOXICOSIS IN MILK-FED HOLSTEIN BULL CALVES:

PATHOLOGIC AND TOXICOLOGIC CONSIDERATIONS

T. W. Graham, C. L. Keen, C. A. Holmberg, M. C. Thurmond and
M. S. Clegg

Departments of Medicine and Pathology, School of Veterinary
Medicine, and Department of Nutrition, University of
California, Davis 95616

INTRODUCTION

The potentially severe effects of Zn deficiency in mammals are documented; however, amounts of Zn required to induce toxicosis and signs of
toxicosis are poorly understood.[1-6] Though reports indicate that there may
be a wide margin of safety in Zn supplementation of feeds for ruminants,[5,6]
other investigators have found that dietary Zn supplementation used to increase rates of gain and feed efficiency, to prevent footrot in sheep and
infectious pododermatitis in cattle can result in death, fetal loss or
reduced performance. The purposes of this investigation were to describe
clinical signs associated with a natural outbreak of Zn toxicosis as well as
tissue mineral concentrations, gross and histopathology.

MATERIALS AND METHODS

In 1984, a San Joaquin Valley, California, veal calf producer fed
calves a specially prepared feed containing a zinc supplement. Ninety-five
calves were fed a milk replacer that contained 706 ug of Zn sulfate/g of the
replacer for a 35 day period. Management practices and clinical observations for these animals have been previously reported.[2-4] Briefly, animals
were from two rooms containing 48 and 47 calves. All calves had received a
commercially prepared milk replacer containing 706 ug Zn/g milk replacer for
28 days before the first death occurred. Of the original 95 calves present,
1 died before the outbreak, 16 died during the outbreak, 12 were euthanized,
1 was culled for poor performance, 1 was lost and 64 were slaughtered.[2]
Surviving calves were sent to slaughter on May 19, 1984.

Fourteen animals (Group A) were examined after spontaneous death or
were euthanized in an agonal state. Twelve of the surviving calves were
euthanized 23 days after being placed on a diet containing lower concentrations of Zn (150 ug/g) and examined postmortem (Group B). Complete gross
examinations were done on all dead calves. Representative tissues were
fixed in 10% neutral buffered formalin, paraffin embedded, sectioned at 6 um
and stained with hematoxylin and eosin (HE). Samples of fat, liver, kidney,
brain and skeletal muscle were selected for chemical analysis (Table 1) and
frozen at -40°C. Trace element concentrations in diets and tissues were
determined following wet ashing with nitric acid.

Table 1. Zinc, Copper, Iron and Manganese Concentrations in Tissues* from 12 Animals in Group A, 12 Animals from Group B.

Tissue	Trace Elements			
	Zn	Cu	Fe	Mn
Group A				
Liver	346 ± 5	52 ± 7	40 ± 5	0.6 ± 0.0
Kidney	219 ± 31	9 ± 1	15 ± 1	0.4 ± 0.0
Group B				
Liver	345 ± 20	43 ± 5	42 ± 10	1.2 ± 0.1
Kidney	252 ± 41	14 ± 4	30 ± 4	0.7 ± 0.0

*ug of metal/g tissue ± S.E. on wet weight basis.

RESULTS

Levels of Zn, Cu, Mn and Fe from Groups A and B are reported in Table 1. Tissue manganese, copper, and iron levels were within normal limits. Tissue fractionization was done and almost all Zn was found associated with metallothionein. Notable gross necropsy changes included pneumonia, fluid digestive content, petechiae and infarcts in liver, kidney and heart, reflecting various complicating bacterial infections. Histological changes that were directly attributed to high zinc intake were in the pancreas, adrenal gland, liver and kidney. Pancreatic changes in Group A calves included marked atrophy of acinar tissue with individual cell to complete acinar unit necrosis. Group B calves had multifocal fibrosis of pancreatic acinar tissue. Renal changes observed in Group A and B calves were multifocal cortical fibrosis with necrosis in both the cortical convoluted tubules and loop of Henle with intratubular mineralization. The most notable hepatic change was mid-zonal mineralization in Group B calves. Increased fibrosis of the zona glomerulosa was observed in the adrenal glands of Group B calves.

In summary, we present evidence for a large-scale outbreak of accidental dietary zinc toxicity in a population of veal calves. The data show that effects of excess dietary Zn can occur quickly and can be severe, potentially resulting in the death of the animal within four weeks. In our opinion the currently suggested National Research Council's maximum "safe" dietary level of Zn for cattle of 500 ug/g is too high for young calves being fed a diet consisting solely of milk replacer. Since there are no data to support the need to feed a diet containing more than 100 ug/g Zn, we suggest that 100 ug/g Zn may be a safer maximum level of dietary Zn for calves, at least until toxic levels of Zn can be better defined or evidence for improved performance at higher levels is documented.

REFERENCES

1. J. G. Allen, H. G. Masters, R. L. Peet et al., J. Comp. Pathol. 93:363 (1983).
2. T. W. Graham, M. C. Thurmond, M. S. Clegg, C. L. Keen, C. A. Holmberg, M. R. Slanker and W. J. Goodger, J. Am. Vet. Med. Assoc. 190:1296 (1987).
3. T. W. Graham, W. J. Goodger, V. Christiansen and M. C. Thurmond, J. Am. Vet. Med. Assoc. 190:668 (1987).
4. T. W. Graham, C. A. Holmberg, C. L. Keen et al., Vet. Path. (in press).
5. E. A. Ott, W. H. Smith, R. B. Harrington et al., J. An. Sci. 25:419 (1966).
6. E. A. Ott, W. H. Smith, R. B. Harrington et al., J. An. Sci. 25:432 (1966).

COPPER TOLERANCE IN RATS - A HISTOCHEMICAL AND IMMUNO-CYTOCHEMICAL STUDY

*M. Elmes. S. Haywood. *J.P. Clarkson,
I.C. Fuentealba, and *B. Jasani

*Department of Pathology
The University of Wales College of Medicine
Cardiff, CF4 4XN

Department of Veterinary Pathology
University of Liverpool
P.O. Box 147. Liverpool. L69 3BX

INTRODUCTION

Metallothionein (MT) has been associated with the detoxification of heavy metals (1) and it is possible that the acquired tolerance of rats (2) is related to the induction of MT synthesis.

The relationship of MT to histochemically demonstrable copper (Cu) was explored in the liver and kidneys of Cu-loaded rats.

MATERIALS AND METHODS

Male rats fed a high Cu diet (1500ppm) for 16 weeks until Cu-tolerant were killed at intervals. Copper was measured by AA spectrophotometry in liver and kidneys and demonstrated in these tissues with Rubeanic acid and Rhodanine stains. MT was demonstrated with the DNP hapten-antihapten sandwich immuno-peroxidase method and antibody to MT-1 (3).

RESULTS (a) LIVER

Liver copper concentrations rose to 2840 ± 45ug (M+SEM dry weight) (control 22.0 ± 0.61ug/g) at 4 weeks, but fell subsequently to 1928 ± 90ug/g at 16 weeks. Histochemically demonstrable copper increased uniformly from 2 - 4 weeks in the periportal and mid zones of the liver lobules in which it occurred in particulate form in the cytoplasm of hepatocytes. MT positive hepatocytes first appeared in the periportal and midzones at 2 weeks and at 3 - 4 weeks there was a marked increase in numbers and intensity of staining of MT positive cells in these outer zones These hepatocytes exhibited both cytoplasmic and nuclear staining or mainly nuclear staining.

From 5 - 16 weeks MT positive staining of periportal hepato-cytes persisted although the nuclei appeared to lose the stain.

(b) KIDNEY

Kidney copper concentrations rose to 1142 ± 88ug/g (M+SEM dry weight) at 4 weeks stabilising thereafter. Copper stains demonstrated granules and droplets within the proximal convoluted tubules (PCT) related to overall Cu content. MT positive staining of both cytoplasm and nuclei of proximal renal tubular cells became marked as the renal copper concentration rose.

CONCLUSIONS

Copper tolerance is associated with an increase in, and altered distribution of immunoreactive MT-1 in liver and kidney which supports the concept that MT has a protective function in copper overload in the rat and tolerance is associated with a fundamental change in copper metabolism.

MT-1 reaction product occurs in different intracellular localisations in both liver and kidney from that of the histochemically demonstrable metal which suggests that copper occurs in different molecular complexes and / or ionic forms within the cell.

REFERENCES

1.   Bremner I (1981)   The Nature and Function of Metallo-thionein in Trace Element Metabolism in Man and Animals. TEMA-4. Ed. J. McC Howell, J.M. Gawthorne & C.L. White. Aus. Acad. Science, Canberra, 1981.

2.   Haywood S (1985) Copper Toxicosis and Tolerance in the Rat I Changes in Copper Content of Liver and Kidney. J. Path. 145 : 149-158.

3.   Clarkson J P, Elmes M E, Jasani B, Webb M (1985) Histological Demonstration of Immunoreactive Zn Metallothionein in Liver, Ileum of Rat and Man. Histochem. J. 17 : 343-352.

TOXICOSIS OF DEVELOPING PIGS FED SELENIUM FROM VARIOUS SOURCES:

CLINICAL PATHOLOGY

D. Baker, L. James, K. Panter,
H. Mayland, and *; J. Pfister

USDA-ARS Poisonous Plant Research Laboratory, 1150 E. 1400 N.,
Logan, UT 84321, and * USDA-ARS Snake River Conservation
Research Center, Rt. 1, Box 186, Kimberly, ID 83341

INTRODUCTION

Selenium is an essential trace element for many animal species and is a
cofactor for some forms of the enzyme glutathione peroxidase. Glutathione
peroxidase in concert with vitamin E protects tissues from endogenous oxidative damage. Clinical deficiency syndromes are well described in cattle,
horses, chickens, pigs and sheep. The gross and microscopic lesions in selenium deficient animals are variable between species, but include skeletal
and myocardial degeneration and necrosis, hepatic degeneration and necrosis,
steatitis, encephalomalacia and exudative diathesis. Fibrinoid necrosis of
arterioles, thrombi and endothelial loss are evident histologically.

Animals intoxicated by selenium have similar lesions as selenium deficient animals. Experimental selenium toxicosis has been produced in animal
species by feeding sodium selenite, selenomethionine, corn grown on high
selenium soil, and selenous acid. Elemental selenium and selenide are not
biologically active and do not produce intoxication.

Serum biochemical alterations during intoxication by selenium in pigs
has not been reported previously. The objective of this study was to feed
pigs selenium in organic, inorganic and protein incorporated sources and
compare the gross, microscopic and serum biochemical changes between groups.

MATERIALS AND METHODS

Experimental groups - Forty, male and female, 8-14 week old, hampshire
and cross-breed pigs were divided into 7 groups. Pigs were fed a complete
commercial feed, or a commercial feed with selenium added. Selenium was
added to the diet by mixing the feed with either dried and finely chopped
Astragalus praelongus, A. bisulcatus, or sodium selenite, sodium selenate or
D,L-selenocysteine. Seleniferous wheat was substituted for chopped barley
in the commercial feed of the group fed high selenium containing wheat.
Pigs were fed twice daily and feed intake recorded. The daily average selenium consumption was calculated for each group. Pigs were weighed before
feeding and at necropsy.

Group	Selenium in Feed	Mean Selenium Intake
Control	trace	trace
A. praelongus	31.6 ppm Se	21.5 mg Se/pig/day
A. bisulcatus	31.7 ppm Se	18.7 mg Se/pig/day

Na$_2$SeO$_3$	29.6 ppm Se	32.3 mg Se/pig/day
Na$_2$SeO$_4$	26.6 ppm Se	12.5 mg Se/pig/day
Selenocysteine	24.8 ppm Se	11.2 mg Se/pig/day
Se-wheat	15.9 ppm Se	15.9 mg Se/pig/day

Blood was drawn from the external jugular vein prior to initiation of feeding and at approximately 7 day intervals for 9 weeks. A portion of the collected blood was added to tubes containing EDTA and the remaining blood was allowed to clot and serum was harvested from the clotted blood then frozen at -70°C until analysis. Blood added to EDTA-containing tubes was used to determine the packed cell volume. Serum was analyzed for calcium, phosphorus, total protein, albumin, glucose, creatinine and urea nitrogen concentrations; and alkaline phosphatase (AP), aspartate transaminase (AST), gamma-glutamyltransferase (GGT) and creatine kinase (CK) activities using an automated batch analyzer and prepared reagents.

RESULTS

All groups of pigs fed selenium had decreased feed consumption, reduced weight gain, loss of hair and cracked hoof walls. All groups of pigs except those fed selenocysteine and Se-wheat had pigs which developed rear limb paresis. Pigs fed selenocysteine and Se-wheat did not develop poliomyelomalacia or polioencephalomalacia. Hepatocellular vacuolar degeneration and portal fibrosis occurred in most groups, with occasional myocardial and skeletal muscular degeneration.

Serum calcium concentration was significantly decreased in all groups fed selenium except those pigs fed Se-wheat mixed into the ration. Phosphorus was significantly decreased at some time in all groups fed selenium except those pigs with A. bisulcatus or Se-wheat mixed into the ration. Those pigs fed Se-wheat mixed in the ration had significantly elevated phosphorus following initiation of feeding compared with pretreatment concentration, however the concentrations were not different from control pig phosphorus concentration or pretreatment phosphorus concentrations of all pigs. All groups of pigs had significantly reduced serum total protein concentrations which was associated with a reduced serum albumin concentration in those groups fed either A. praelongus, A. bisulcatus, Na$_2$SeO$_4$ or selenocysteine. Serum urea nitrogen or creatinine significantly but variably increased or decreased from pretreatment concentration for all groups administered selenium, however neither serum urea nitrogen nor creatinine concentrations deviated outside the reference range for pigs. Serum GGT and AP activities significantly increased in all groups administered selenium and serum AST activity increased in all groups fed selenium except those fed Na$_2$SeO$_3$. While serum CK activity was increased in all groups of pigs fed selenium, the activity was significantly increased only in selenocysteine, Na$_2$SeO$_4$ and A. bisulcatus fed groups. Serum CK activity was significantly decreased on day 21 compared with all other sampling dates. Serum amylase activity was significantly increased in all groups fed selenium. The PCV of the control group of pigs decreased from days 21 through 50 and was significantly decreased on days 27 and 50. All groups of pigs fed selenium had either no change or had significantly increased PCVs.

DISCUSSION

All forms of selenium induced clinical symptoms of weight loss, loss of hair, cracked hooves and inflamed coronary bands to varying degrees. Clinical chemistry alterations were not pathognomic for selenium intoxication, and were likely due to hepatic dysfunction, muscular degeneration or at least in part to reduced feed intake.

702

ALUMINUM INDUCED ALTERATION IN HEPATIC GLUCURONYL TRANSFERASE ACTIVITY

Mark S. Meskin, Robert C. Brown and Wayne R. Bidlack

Department of Pharmacology and Nutrition
USC School of Medicine
Los Angeles, California   90033

INTRODUCTION

Total parenteral nutrition (TPN) and intravenous solutions used in the United States, England and New Zealand have been found to be inadvertently contaminated with aluminum (Al). Among the solution constituents currently identified as contaminated with Al are casein hydrolysates used as the protein/amino acid source, calcium and phosphate salts, heparin and albumin. In addition, a number of investigators have reported Al contamination of products given to young infants including a variety of intravenous fluids, infant formulas, vaccines and toxoids. Aluminum contamination is of concern because it has been associated with encephalopathy, bone disease, kidney damage and cholestatic liver disease.

The data reported here are part of a larger study attempting to determine the mechanism(s) of Al induced cholestatic liver disease. In a pilot experiment, infusion of Al into pigs and rats has resulted in cholestasis as indicated by a fourfold increase in serum bile acids (Klein et al., 1985). The effect of Al on the hepatic mixed function oxidase and drug metabolism showed that Al treated animals had hepatic glucuronyl transferase (GT) activity 4 times greater than the GT activity in control animals, and this activity was not increased further by detergent activation (Bidlack et al., 1987). The experiments reported here were designed to evaluate two possible mechanisms for the observed GT stimulation: a cation regulation or bile acid stimulation of GT activity. Specifically our experiments evaluated the in vitro effects of magnesium (Mg^{++}), a known cation regulator of GT, with Al^{+++} and the detergent effect of sodium (Na) cholate, the most abundant bile acid in humans.

METHODS

Livers from male Sprague-Dawley rats were homogenized (10% w/v) in Tris (3 mM)-sucrose (0.25 M) buffer, pH 7.4. The homogenate was centrifuged at 28,000 x g for 20 min. The resulting supernatant was then centrifuged at 105,000 x g for 30 min. to obtain the microsomal fraction. The microsomal membrane was rinsed and resuspended in Tris (50 mM)-KCl (150 mM) buffer, pH 7.4, containing 0.1 mM EDTA. Following recentrifugation the membrane was resuspended in the same buffer without EDTA.

Glucuronyl transferase was quantitated using p-nitrophenol (pNP; 200

uM) and UDPGA (2.5 mM). Each reaction tube was incubated 10 minutes with all reagents except UDPGA. Reaction tubes contained 1.0 mg of microsomal protein per ml and a final volume of 1.0 ml. Reactions were initiated with the addition of UDPGA and terminated after 20 min. with 0.5 ml of 20% TCA. The amount of non-conjugated pNP was determined spectrophotometrically at 401 nm. The difference between the final pNP and the initial pNP concentration is the amount of pNP conjugated.

RESULTS AND CONCLUSIONS

The effects of Mg and Al on GT activity are indicated in Table 1. Magnesium provided a 19.2% increase in GT activity at 1.0 mM and a 41.6% increase at 5.0 mM. Aluminum stimulated GT activity from 10 to 56% at concentrations of 1.0 to 2.0 mM. (Aluminum does not stay in solution above 2.0 mM.) Both Mg and Al exert positive cation effects on the native enzymatic activity of GT. When the GT activity was measured in the presence of Na cholate, between 1.0 and 2.0 mM, the activity was stimulated 50 to 200% (Table 2). The Na cholate effect on GT activity is more than two times greater than the cation effects produced by either Mg or Al. These results suggest that the activation of GT observed in the Al treated animals resulted from the cholestasis effect on the membrane rather than a direct effect of Al on the enzyme. Liver samples from these animals have been frozen and will be analyzed for bile acid content in the near future.

Table 1. Effect of Magnesium and Aluminum On Glucuronyl Transferase Activity

Cation (mM)	Glucuronyl Transferase Activity[1]					
	Aluminum			Magnesium		
	(n)	Activity	% Stimulation	(n)	Activity	% Stimulation
0	(3)	100.0 + 5.5	---	(2)	100.0 + 3.9	---
0.1	(3)	106.1 + 8.1	6.1			
1.0	(3)	119.2 + 3.3	19.2	(2)	110.2 + 5.5	10.2
2.0				(2)	156.8 + 9.9	56.8
5.0	(3)	141.6 +14.1	41.6			

[1]Native rate standardized to 100. Mean native rate = 48.9 + 14.9 nMoles pNP Conjugated/mg Protein/20 min.

Table 2. Effect of Sodium Cholate On Glucuronyl Transferase Activity

Concentration (mM)	(n)	Glucuronyl Transferase Activity[1]	% Stimulation
0	(3)	100.0 + 2.2	---
0.5	(3)	103.3 + 12.3	3.3
1.0	(3)	146.9 + 16.7	46.9
1.25	(3)	185.7 + 28.5	85.7
1.5	(3)	234.4 + 30.5	134.4
2.0	(3)	307.7 + 62.7	207.7

[1]Native rate standardized to 100. Mean native rate = 48.9 + 14.9 nMoles pNP Conjugated/mg Protein/20 min. Stimulation due to sodium cholate was in the presence of 4.0-5.0 mM magnesium.

DECREASE OF SERUM TRIGLYCERIDE AND ESSENTIAL METAL CHANGES

IN NORMAL RATS FED WITH 2000 ppm ALUMINUM DIETS

C. Sugawara and N. Sugawara

Department of Public health
Sapporo Medical College
S-1, W-17, Central Ward
Sapporo, 060 Japan

INTRODUCTION

In evaluating the effect of oral Al on normal rats, we found a decrease of serum triglyceride (TG) concentrations:

1. With the addition of $AlCl_3$, $Al(OH)_3$, or $AlK(SO_4)_2$,
2. In feeding with sucrose, lactose, whole milk, casein and soy-protein diet,
3. In young and retired rats.

No Al effect was observed on serum cholesterol and P-lipid concentrations (reported in the VI UOEH International Symposium and the III COMTOX on Bio-and Toxicokinetics of Metals, 1986 JAPAN). A significant correlation between serum TG and NEFA (nonesterified fatty acid) concentrations suggested the effect of AL on the TG cycle under starvation. To clarify the mechanism of serum TG decreases, metal shift were surveyed in rats fed with a sucrose diet.

METHODS

An experimental sucrose diet was prepared using a commercial diet (oriental Yeast, Tokyo, 55g), potato starch (15g) and sucrose (30g). Male weanling and retired (30 weeks of age on the average) Wistar rats were fed for 67 days on the sucrose diet with the addition of 2000 ppm AL ($AL(OH)_3$ or $ALK(SO_4)_2$). After a half-day starvation, rats were sacrificed. To determine metals, serum was diluted directly and the intestine, liver, epiditymal adipose tissue and femur were ashed and diluted appropriately with saturated EDTA or dilute acid solutions. Al concentrations were determined by a polarized Zeeman flameless AAS (Hitachi 180-80). Overall group differences wer compared by a two-way analysis of variance test with a significance level of $P=0.05$. When differences were identified, Scheffé's test was used to make pairwise comparisons.

RESULTS AND DISCUSSION

Al compounds had no effect on body weight gain.

Al: Serum Al concentrations were near the detection limit and did not exceed 20 ppb. Al concentrations were increased in intestine and femur but did not increase in other organs. The decrease of serum TG could not be explained directly by Al accumulation in liver or adipose tissue.

Fe: It is well known that Al interferes with Fe metabolism and that osteomalasia in patients with renal deficiencies can be ameliorated by Fe-chelator DFO. Hypertiriglyceridemia observed in rats under Fe deficiencies can be explained in part by a decrease of lipoprotein lipase activity and changes in lipogenesis from flucose in liver and intestine (1). In our experiment, hepatic fe concentration tended to decrease. Al had no effect of Fe concentrations in serum and other organs and had no anemic effect.

An and Cu: It was reported that the rate of TG absorption markedly decreased in Zn-deficient rats as a result of block to the movement of lipid droplets out of the mucosa (2). In this experiment, dietary Zn was sufficient but its absorption could be interrupted by Al. However, an increase of intestinal

Table 1

Constituents of diet			Serum	
			Tri-glyceride (mg/dl)	NEFA (mEq/l)
Al	12.9 ppm			
Fe	94.3 "			
Zn	51.3 "	Young	Control	188 ± 51	1.41 ± 0.59
Cu	9.4 "		$Al(OH)_3$	158 ± 49	1.14 ± 0.26
Ca	849 mg/100g		$AlK(SO_4)_2$	140 ± 36	1.15 ± 0.35
Mg	430 "			
P	425 "	Retired	Control	227 ± 46	1.35 ± 0.28
Protein	15.1 g/100g		$Al(OH)_3$	167 ± 52	1.08 ± 0.24
TG	1.3 "		$AlK(SO_4)_2$	176 ± 40	1.23 ± 0.32
P-lipid	0.2 "			
Cholesterol	53 mg/100g			

These data represent M±SD.

		Liver		Intestine	
		Fe (ppm)	Zn (ppm)	Fe (ppm)	Zn (ppm)
Young	Control	54.5 ± 7.3	44.4 ± 3.0	16.4 ± 1.7	19.4 ± 1.2
	$Al(OH)_3$	51.3 ± 12.1	41.3 ± 2.0	17.2 ± 1.2	22.5 ± 0.9
	$AlK(SO_4)_2$	40.6 ± 10.6	43.7 ± 2.8	14.8 ± 1.8	22.0 ± 1.8
Retired	Control	99.2 ± 21.5	44.0 ± 8.0	20.1 ± 2.5	18.4 ± 1.8
	$Al(OH)_3$	74.0 ± 20.2	40.8 ± 4.7	19.6 ± 1.7	21.5 ± 0.9
	$AlK(SO_4)_2$	79.7 ± 26.1	38.8 ± 4.8	18.9 ± 1.3	21.7 ± 1.8

Zn concentration was observed. This may have been the result of a stimulative effect of Al on mucosa cells but a morphological study revealed no such Al effect. No Al effect on Cu concentrations was observed.

Ca, Mg and P: Ca and Mg concentrations were not decreased by Al. Serum P concentration was increased significantly by Al. No decrease of P concentrations was observed in any organ

This survey did not clarify the mechanism for the serum TG decrease induced by ingested Al.

REFERENCES

1. E.K. Amine, E.J. Desilets and D.M. Hegsted, Effects of dietary fats on lipogenesis in iron deficiency anemic chicks and rats. J. Nutr. 106: 405-411 (1976).

2. S.I. Koo and D.E. Turk, Effect of zinc deficiency on intestinal transport of triglyceride in the rat. J. Nutr. 107: 909-919 (1977).

TISSUE CONCENTRATIONS OF MOLYBDENUM AFTER CHRONIC DOSAGE

IN RATS DO NOT FOLLOW SIZE OF DOSE

P. W. Winston and L. J. Kosarek

Dept. of E.P.O. Biology
University of Colorado
Boulder, Colorado, USA, 80309

The concentration of a potentially toxic element in tissues has been widely used to estimate exposure to excess amounts, especially of a substance that accumulates in cells with time. Molybdenum, like other essential trace elements, is toxic at high intake levels and an increase of two or more orders above "normal" is necessary to produce clinical symptoms. Though it does not accumulate with time, the level of intake is reflected in the amount retained (Winston 1981). We assumed that concentrations of Mo in tissues and its deleterious effects would vary more or less directly with intake. As will be shown below, this was only partly upheld.

Female Sprague-Dawley rats were reared from weaning on solutions of 0 ppm, 10 ppm, 100 ppm, and 1000 ppm Mo as Na_2MoO_4 as their sole source of drinking water, and their offspring were continued on the same regime for 6-12 months before sacrifice. Rats were killed by ether overdose and tissues excised immediately. Samples of liver, kidney, adrenal, gonads, heart, bone, skeletal muscle, and brain were taken for analysis by X-ray fluorescence (Alfrey et al 1976). Frequent checks on accuracy were made using duplicate samples and "wet-chemistry" analysis. The terms parts-per-million (ppm) and ug/g will denote treatment groups and concentration of Mo in tissues, respectively, Data for the effects on growth were taken from records of a group of male rats which were actually used in other experiments, to insure an homogenous sample. Conditions were identical to those for the rats taken for tissue samples.

Results

The concentration of Mo in tissues ranged from 0.151 ug/g in control muscle to 109 ug/g in kidney of rats on 1000 ppm. All tissues were 60-70 times higher than controls at maximum intake except for liver which increased only about 10 times. Average concentrations in all tissues adjusted for the different mass of organs, are shown in Table 1. It can be seen that, though excess intake resulted in elevated tissue levels, it does not show a direct relationship to intake concentrations. Instead, those at 10 ppm retained significantly more Mo than those at 100 ppm. This difference from the expected held in all 8 tissues; thus retention does not bear a direct relationship to intake. There was no effect of sex discernable and only a slight, insignificant indication of accumulation with age between 6 and 12 months.

Table 1. Pooled Average Concentrations of Mo (dw) in 8 Tissues Adjusted For Mass of organs of Rats on 4 Levels of Mo in Drinking Water. All Values were Different (P 0.05). N for each group in parentheses. Data are mean $\pm$ S.E.

PPM Mo	0 (32)	10 (25)	100 (26)	1000 (19)
ug/g Tissue	0.608 $\pm$0.09	17.72 $\pm$ 1.7	6.25 $\pm$0.56	29.7 $\pm$ 1.8

Elevated intake of Mo reduced growth as was expected. Rats in each group were significantly smaller than those on the next lower intake, a relationship which also held for over 200 rats of both sexes raised in the same way.

Retention of Mo in controls and at higher intake agrees surprisingly well with most published results on rats, e.g. Higgins et al (1956), even though most tests covered less than lifetime and none were from conception. Almost no one exposed rats to levels as low as 10 ppm and none of these showed higher retention than at 100 ppm. The question then arises as to why there is this apparently anomalous retention at 10 ppm, a level at which toxicity is very low. An explanation is possible using the different thiomolybdates which can be formed in the reducing segments of the gut (Mills et al 1981). Molybdenum is toxic primarily because of the reduction of $(MoO_4)^{-2}$ to $(MoS_4)^{2-}$, tetrathiomolybdate (TTM). This compound combines with proteins and copper to fix them in insoluble, unavailable forms. The proteins are large molecules which appear to be induced in favor of smaller ones such as metalothionen, which could explain the accumulation of Mo in the 100 and 1000 ppm groups.

In the formation of TTM, the oxythiomolybdates, dithio- and trithio-molybdate, are intermediates. These have been shown to be much less toxic than TTM; and, though they are bound to proteins and Cu, the binding is weak, the proteins are small, and they are soluble (Bremner et al 1982). We suggest, therefore that, under the conditions of the treatment used here, oxythiomolybdates are the primary forms retained at 10 ppm. These could accumulate to high levels before an equilibrium were reached without causing severely toxic symptoms. Why this effect might be dependent on the rats' being exposed from conception is unclear.

REFERENCES

Alfrey, A.C., Nunnelley, L. L., Rudolf, H., and Smyth, W. R., 1976, Medical application of a small-sample X-ray fluorescence system, Adv. in X-Ray Anal., 19:497.

Bremner, I., Mills, C.F., and Young, B.W. 1982, Copper metabolism in rats given di- or trithiomolybdates, J. Inorg, Biochem., 16:109.

Higgins, E. S., Richert, D. A., and Westerfeld, W. W., 1956, Molybdenum deficiency and tungstate inhibition studies, J. Nutr., 59;539.

Mills, C.F., El-Gallad, T. T., and Bremner, I., 1981, Effects of molybdate, sulfide, and tetrathiomolybdate on copper metabolism in rats, J. Inorg. Biochem., 14:189.

Winston, P.W., 1981, Molybdenum, in: "Disorders of Mineral Metabolism", F. Bronner and J. W. Coburn, eds., Academic Press, New York

METALLOTHIONEIN AND COPPER IN HUMAN LIVER - A HISTOPATHOLOGICAL STUDY

M.E. Elmes, J.P. Clarkson, B. Jasani, and N.J. Mahy

Department of Pathology
University of Wales College of Medicine
Cardiff, CF4 4XN, Wales, U.K.

Levels of copper above 55 ug/G dry weight in adult human liver are considered to be abnormal and may be associated with liver cell damage. In primary biliary cirrhosis (PBC) Epstein et al (1981)[1] failed to find any correlation between liver copper concentration and the degree of liver damage. In Wilson's disease (WD) no histochemically detectable copper was found in some livers with high levels of copper measured by neutron activation analysis[2]. Conventional histochemical stains for copper such as rhodanine and rubeanic acid demonstrate CuII[3]. Orcein demonstrates copper associated protein (CAP) which is probably polymerised metallothionein (MT)[4]. The molecular state of copper appears to determine both its hepatotoxicity and our ability to demonstrate it histochemically and we suggest that 'invisible' copper is non-toxic CuMT[5]. This study examines the relationship between Cu, CAP and MT in 132 human liver biopsies.

SUBJECTS AND METHODS

Diagnostic liver biopsies from 132 patients with conditions associated with copper retention[6] were stained with rubeanic acid for Cu[7], orcein for CAP[8] and MT using a specific immunoperoxidase technique[9] and antibody to MTI[10]:

25 biopsies were histologically normal; 47 non specific cirrhosis and cholestasis; 14 primary biliary cirrhosis (PBC); 25 chronic active hepatitis (CAH); 17 Indian childhood cirrhosis (ICC); 4 Wilson's disease (WD)

RESULTS

In histologically normal liver biopsies no Cu or CAP was found and the MT staining was seen in both nucleus and cytoplasm of liver cells. MT positive cells were orientated round the central hepatic veins of the lobules, Cu and CAP when present were seen in hepatocytes in the periportal areas. When cirrhosis was present and the normal architecture lost Cu and CAP were found in hepatocytes bordering fibrous septa in large intracytoplasmic granules. The normal centrilobular orientation of MT was absent in livers containing Cu and CAP. In PBC, ICC and WD large amounts of cytoplasmic MT was seen. Necrotic hepatocytes were intensely MT positive and were morphologically similar to cells containing Cu and CAP. When cholestasis was present some bile plugs were MT positive. The most intense staining of non-necrotic hepatocytes was seen in Wilson's disease including one asymptomatic case and was considerably decreased in a post-treatment biopsy after penicillamine therapy.

DISCUSSION AND CONCLUSION

We suggest that when copper accumulates in the liver either due to excess intake (ICC), hepatocyte damage and/or chronic cholestasis (PBC) or congenital hepatocyte malfunction (WD) it is in the form of copper metallothionein in which copper is in the CuI form and invisible to routine histochemical stains. As the copper load increases polymerised MT becomes visible as CAP in lysosomal granules. If breakdown of this detoxifying mechanism occurs histochemically detectable CuII appears associated with hepatocyte necrosis and previously tightly packed MT becomes available to antibody rendering it visible.

ACKNOWLEDGEMENTS

J.P. Clarkson was supported by the Welsh Scheme for the Development of Health and Social Research and this study received funds from the International Copper Research Association Inc.

REFERENCES

1. Epstein O, Arborgh B, Sagiv M, Wroblewski R, Scheuer PJ, Sherlock S. Is copper hepatoxic in primary biliary cirrhosis? J Clin Pathol 1981, 34, 1071.
2. Jain S, Scheuer PJ, Archer B, Newman SP, Sherlock S. Histological demonstration of copper and copper associated protein in chronic liver disease. J Clin Pathol 1978, 31, 784.
3. Pearse AGE. Histochemistry, Theoretical and Applied, 4th Edition Vol 2, Churchill Livingstone 1985, p 991-992.
4. Vaux DJT, Watt F, Grime GW, Takacs J. Hepatic copper distribution in primary biliary cirrhosis shown by the scanning proton microprobe. J Clin Pathol, 1985, 38, 653.
5. Sato M, Bremner J. Biliary excretion of metallothionein and a possible degradation product in rats injected with copper and zinc. Biochem J. 1984, 223, 475.
6. Smallwood RA. In : Metals and the Liver, Ed. LW Powell, Marcel Dekker Inc. 1978, p 313.
7. Uzman LL. Histochemical localisation of copper with rubeanic acid. Lab Invest 1956, 5, 299.
8. Shikata T, Uzama T, Yoshiwara N, Akatsuka T, Yamazaki S. Staining methods of Australia antigen in paraffin section. Japan J Exp Med 1974, 44, 25.
9. Clarkson JP, Elmes ME, Jasani B, Webb M. Histological demonstration of immunoreactive zinc metallothionein in liver and ileum of rat and man. Histochem J 1985, 17, 343-352.
10. Vander Mallie RS, Garvey JS. Radioimmunoassay of Metallothionein. J Biol Chem 1979, 254, 8416-8421.

THE ELEMENTAL CONTENT OF SOME TRACE ELEMENT MEDICAMENTS

BY PROTON-INDUCED X-RAY AND GAMMA-RAY EMISSION ANALYSES

Päivi Nikkinen[1], Mervi Hyvönen-Dabek[2] and Tuomas Westermarck[3]

[1]Helsinki University Central Hospital, [2]Department of Physics, University of Helsinki and [3]the Helsinki Central Institution for the Mentally Retarded, Kirkkonummi, Finland

INTRODUCTION

A number of metals (e.g., Fe, Ti, Cu, Pb, Zn, Mg, Ca, Al and Ni) have been shown[1] to have inhibitory or stimulatory effects on the rate of lipid oxidation. The purpose of this paper is to study different kinds of trace element pills, particularly those which have been argued to have an anti-oxidative response in humans; especially the concentrations and pureness as far as trace elements are concerned.

Proton-induced x-ray emission (PIXE) and proton-induced gamma-ray emission (PIGE) analyses are especially suited for detecting both major, minor and trace elements in various materials because of the sensitivity, multielementarity and the small mass of sample needed.[2]

METHOD

When a sample is bombarded by a beam of protons both x-rays and gamma-rays are emitted by the target atoms and nuclei. These phenomena make possible the use of proton-induced x-ray emission (PIXE)[3] and proton-induced gamma-ray emission (PIGE)[4] methods as analytical tools.

Sample Preparation. Several kinds of pills or powders aimed at pro-moting health or curing disease, containing major, minor and trace elements were purchased either from the chemists, at so-called nature shops or ob-tained from research people abroad. Hard pills were crushed by a pestle and a mortar and pressed into the centre of circular collars cut from filter paper to form pellets. The masses of the samples were ∿30 mg, their thick-ness ∿0.5 mm and surface area 130 mm².

Experiments. The measurements were performed at the 2.5 MV Van de Graaff accelerator at the University of Helsinki. For the PIXE measurements the proton beam (E_p ∿ 2.2 MeV) was brought into the air using a system described earlier[5] and a Kapton foil, 7.5 μm thick as the exit foil. The samples were positioned at 45° to the incident beam and x-rays detected with a Seforad Ge detector positioned at an angle of 80° to the beam, 12 mm from the sample surface. Absorbers made of 5 layers of Kapton, 125 μm thick, were used to reduce the proportions of low energy x-rays due to Cl, Ar, K and Ca.

The PIGE measurements were performed in vacuo by bombarding the samples with 2.4 MeV protons. The samples were positioned in the target holder at

45° to the incident beam and the gamma-rays detected with a Ge(Li) detector (Princeton PGT) at 55° to the beam.

RESULTS

The elemental concentrations were calculated by comparison with spectra of the NBS reference material Orchard leaves and for elements not present in Orchard leaves by using cross-sections for the production of x-rays.[3]

The ranges of the elemental masses detected by PIXE per pill or capsule were as follows (detection limits are given in parentheses): Ti 0.3-25 μg (0.4 ppm), Cr 0.4-60 μg (0.4 ppm), Mn 0.5-2100 μg (1.8 ppm), Fe 13-9800 μg (1.2 ppm), Co 0.8-140 μg (3 ppm), Cu 0.2-420 μg (0.7 ppm), Zn 1.0-1500 μg (2.9 ppm), Pb 0.1-13 μg (0.5 ppm), Se 32-1130 μg (2.1 ppm), Br 0.5-480 μg (1.0 ppm), Rb 4.2-84 μg (4.0 ppm), Sr 0.4-420 μg (2.0 ppm), and in one pill As 50 μg (4.0 ppm).

The ranges of the elemental masses detected by PIGE per pill or capsule were as follows (detection limits are given in parentheses): F 1-85 μg (2 ppm), Ca 0.02-60 mg (1%), Mg 130-17000 μg (300 ppm), B 3-110 μg (4 ppm), Na 7-4500 μg (2 ppm), Li 0.1-2 μg (0.3 ppm), Al 14-780 μg (24 ppm), P 0.5-27 mg (0.03%) and K 0.8-5 mg (0.5%).

In Fig. 1 an x-ray spectrum of a medicament is shown. The spectrum was obtained using a beam current of 0.25 μA and collecting a charge of 50 μC.

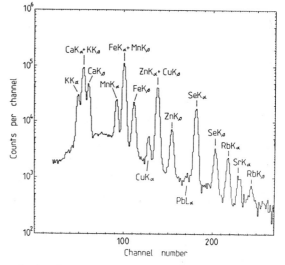

Fig. 1. X-ray spectrum of a Seleros pill. The peaks are identified by atomic symbols.

DISCUSSION

The content of Co^{2+}, Cu^{1+}, Fe^{2+}, Mn^{2+} and Ti^{3+} ions in pills may decompose hydrogen peroxide in a Fenton-type reaction with formation of toxic hydroxyl (OH·) radicals.[1] The pureness of the trace element medicaments as concerning these metals was not good in every case. The concentration of toxic elements in certain pills, such as Pb and As was neither expected. Thus, in the manufacture of trace element preparations quality assurance regarding the purity and safety of products should receive more attention.

REFERENCES

1. J.M.C. Gutteridge, T. Westermarck & B. Halliwell: Oxygen radical damage in biological systems. In: Modern Aging Research, Eds. R.C. Adelman et al., Vol. Free Radicals, Aging and Degenerative Diseases, Ed. J.E. Johnson, Publ. Alan R. Liss, 1985, pp. 99-139.
2. M. Hyvönen-Dabek, P. Nikkinen & J.T. Dabek, Clin.Chem. 30:529 (1984).
3. S.A.E. Johansson & T.B. Johansson, Nucl.Instr.Meth. 137:473 (1976).
4. A. Anttila, R. Hänninen & J. Räisänen, J.Radioanal.Chem. 62:293 (1981).
5. J. Räisänen, M. Hyvönen-Dabek & J.T. Dabek, Int.J.Appl.Radiat.Isot. 32: 165 (1981).

THE APPLICATION OF A SYNCHROTRON RADIATION MICROPROBE

TO TRACE ELEMENT ANALYSIS

B. M. Gordon, A. L. Hanson, K. W. Jones, W. M. Kwiatek,
G. J. Long, J. G. Pounds, G. Schidlovsky, P. Spanne,
M. L. Rivers[1] and S. R. Sutton[1]

Dept. of Applied Science, Brookhaven National Laboratory
Upton, NY 11973 and [1]Dept. of the Geophysical Sciences,
University of Chicago, Chicago, IL 60637

INTRODUCTION

Synchrotron radiation is light emitted by electrons when accelerated in
a circular orbit. Some properties of synchrotron radiation important to
trace element analysis by x-ray fluorescence analysis are: 1) a broad, continuous and tunable energy spectrum for K- and L-shell excitation of all
elements; 2) a linearly polarized source reducing the scattered radiation
backgrounds; 3) low energy deposition in the target; and 4) an appreciable
flux in narrow energy bandwidths for chemical speciation.

EXPERIMENTAL

Experiments to date have generally used "white" continuous spectra with
a low energy absorber and no focussing. Future runs will use focussing mir-
rors which increase intensities by a factor of more than 1000. Monochroma-
tors will be used to select the energy and bandwidths appropriate to the
experiment.

Detection Limits

Detection limits for thin biomedical samples using a solid-state detector, a 0.5 mm beam and a 5 min counting interval were in the range of 30 ppb
for calcium to 50 ppb for zinc. Future developments will include a focussed
microprobe beam line producing a 30 μm beam spot and wavelength-dispersive
detectors.[1] Detection limits will range down to a few ppb in 5 sec counts.

Trace Element Analysis of Living Cells

A prototype wet cell was designed, contructed and tested using cat cardiac myocytes with the result that major trace elements such as iron could
be quantitated in single myocytes. The cell compartment is formed by two
thin-film windows separated by a thin-film spacer with a cut-out forming the
cell. Nutrient solution can be passed through the cell during the run.

Support: DOE DE-AC02-76CH00016, NIH P41RR01838, NIH F05-TWOP3735.

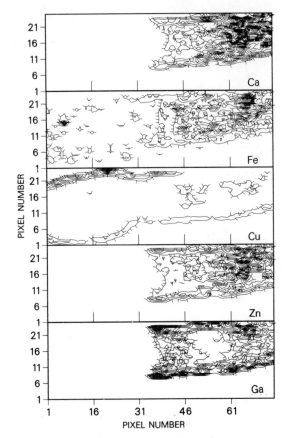

Fig. 1. Contour plots for localization of gallium and other elements
in fetal rat bone explants. The area is 1 × 3 mm^2. The tick
marks point in the direction of lesser counts/pixel.

Scanning Techniques

The x-ray microprobe was used to localize gallium in fetal rat bone explants after being cultured in BGJ media containing 25 μM Ga(NO$_3$)$_3$. The 1 ×
3 mm^2 area was scanned with a 40 μm square pixel size for 2 sec/pixel. Contour plots for a number of elemental concentrations are shown in Fig. 1.[2]

Microtomography

The high brightness of x rays from a synchrotron source makes possible
the development of computerized tomography on a micrometer scale. A tomo-
gram of a freeze-dried caterpillar head was produced in a 50 min scan. The
pixel size was 30 μm using a 20-μm beam. Maps of trace element concentrations in cross section can also be made by detection of fluorescence x rays
using a point source, or in thin samples with a line source.

REFERENCES

1. B. M. Gordon and K. W. Jones, Design criteria and sensitivity calcula-
     tions for multielemental trace analysis at the NSLS x-ray microprobe,
     Nucl. Instr. Methods B10/11:293 (1985).
2. R. Bockman, J. G. Pounds and G. J. Long, unpublished results.

AUTOMATED DETERMINATION OF BLOOD GLUTATHIONE PEROXIDASE AND THE

EFFECT OF STORAGE, CYANIDE AND DRABKIN'S REAGENT

W.J. Blanchflower, D.A. Rice, and W.B. Davidson

Veterinary Research Laboratories
Stormont
Northern Ireland, BT4 3SD

INTRODUCTION

Since glutathione peroxidase (GSH-Px) activity in whole blood has been shown to be significantly correlated to selenium intake in domestic animals, it is frequently used as an indicator of selenium status.

Most methods for assaying enzyme activity are based on the kinetic assay of Paglia and Valentine[1]. While assessing the suitability of these methods for use in our laboratory, a number of problems became apparent. These included decreases in enzyme activity in blood samples stored at -20°C for several months and in haemolysates which were diluted in Drabkin's reagent or cyanide. A method, for clinical use[2], using a Hitachi 705 analyser, has therefore been developed to circumvent these probelms.

MATERIALS AND METHODS

Reagent 1 for the Hitachi analyser contained 40 mM phosphate pH7, 8 mm EDTA (disodium salt) pH7, 2 U/ml glutathione reductase (Sigma), 3.25 mM GSH, and 0.58 mM NADPH. Reagent 2 contained 0.38 mM cumene hydro-peroxide and 1 mM phosphate pH7. For the assay, whole blood samples were diluted 20X using 2 mM GSH in 10 mM phosphate pH7. The Hitachi was programmed as a rate assay (rate - 19-23), the volume of reagents 1 and 2 were 200 µl, wavelengths 1 and 2 were 376 and 340 nm respectively, the factor was 13,354 and the temperature was 37°C. The sample volume was 7 µl. Enzyme activity could also be measured manually on a spectrophotometer by increasing the volumes 5X and recording the kinetic reaction for 3 minutes at 340 nm. Results were expressed as IU/g Hb.

RESULTS AND DISCUSSION

When blood samples had been stored for several months at -20°C, there was an apparent loss in GSH-Px activity and the kinetic assays were non-linear (fig 1). Enzyme activity, however, was full restored by pre-incubating the diluted blood samples for 10 minutes in the presence of 2 mM GSH and the assay became linear. A possible explanation for this is that the enzyme is normally maintained in the reduced state in blood

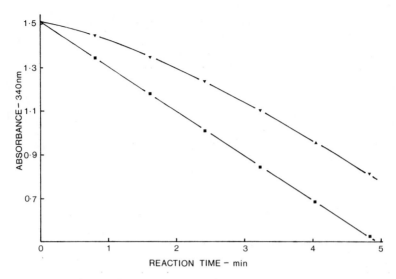

Fig 1   Reactivation and measurement of GSH-Px in a sheep blood sample
which had become inactivated after storage at -20°C for 3 months.
Haemolysates were prepared in either phosphate buffer alone (▼)
or phosphate buffer + GSH (■) and pre-incubated for 10 minutes
before assay.

by the presence of red cell GSH, but as this becomes depleted on storage,
the enzyme is converted to the oxidised form, which is inactive. Pig,
human, sheep and bovine samples stored at -20°C for several months had
their original GSH-Px activities completely restored by this procedure.

Several methods for GSH-Px have recommended the use of Drabkin's
reagent or cyanide in the assay to prevent non GSH-Px peroxidative
interference. Using cumene as a substrate, however, we have no evidence
of any such interference. Furthermore, we have found that the presence
of cyanide or Drabkin's reagent can cause an irreversible loss in enzyme
activity. This is particularly noticeable in stored blood samples where
the use of 3 mM KCN causes a complete loss of enzyme activity. This loss
could be prevented by first diluting the blood sample with GSH before the
addition of KCN, but in any case the use of cyanide or Drabkin's reagent
is not recommended. This is in agreement with Prohaska et al[3], who
showed that a purified form of GSH-Px was irreversibly inactivated by
cyanide.

Using the Hitachi analyser, the recommended method was found to
yield good accuracy and precision. CV's for within run precision using
high and low GSH-Px sheep blood samples ranged from 0.38 to 0.78% and day
to day precision from 2.76 to 6.66 respectively.

REFERENCES

[1]Paglia, D.E. and Valentine, W.N. (1967) J. Lab. Clin. Med. 70, 158-169.
[2]Blanchflower, W.J., Rice, D.A., Davidson, W.B. (1986) Bio. Trace Element
    Res. (in press).
[3]Prokaska, J.R., Oh, S.H., Hoekstra, W.G., Ganther, H.E. (1977) Biochem.
    Biophys. Res. Comm. 74, 64-71.

CHEMICAL SPECIATION OF METAL IONS IN THE RUMINAL AND ABOMASAL FLUID OF SHEEP

Julian Lee

Biotechnology Division, D.S.I.R.
Private Bag, Palmerston North, New Zealand

The chemical form of a metal or ligand is more relevant to bioactivity than the total amount present and therefore the quantitation and characterization of specific metal ion species will increase our understanding of factors controlling availability, transport and absorption of elements.

METHODS

Chemical equilibrium models (GEOCHEM, PHREEQE), chemical-physical fraction (diafiltration, ion-exchange) and chromatographic techniques (Bio-Gel P6, HPLC) were used to study the nature and distribution of metal ion species in a 30,000G supernatant fraction from the rumen and abomasum fluid of sheep fed fresh pasture. Quantitative data on total cations and anions were obtained by plasma emission spectrometry and HPLC techniques. This data, along with pH, Eh, temperature and partial gas pressures were incorporated into an ion-association model to calculate ion activities and identify dominant ion pairs.

RESULTS AND DISCUSSION

Major ion pairs for several metals are shown in Table 1. These are first approximations only however. Comparisons with data from diafiltration, ion-exchange (Table 2) and chromatographical separations show that, although Ca, Mg and Mn ion activities may be reasonably well predicted, limitations in accounting for the solubility of Cu, Fe and Zn are clearly evident. Calcium in the rumen is oversaturated with respect to phosphate system and not surprisingly, sulphide ion controls the activity of Cu, Fe and Zn in the rumen and, to a lesser extent, in the abomasum (apart from Zn, which is more soluble at the lower pH). However the activity of these metal ions in the soluble phase is substantially greater than that shown by the equilibrium model which fails to adequately predict the metal complexation of the soluble organic component. Strong binding is necessary to stabilize Cu (II) in particular, and prevent reduction to Cu (I) and subsequent formation of solid Cu_2S. Such moieties are undoubtedly present as seen from Table 2.

Fractionation of rumen supernatant on Biogel P-6 and P-60 indicated the presence of both high (>60,000 daltons) and low M.W. (2000-5000 daltons) components containing Cu, Fe and Zn (similar to that reported by Bremner[1]). These fractions also contained S and on acid hydrolysis a number of amino acids including cysteine, threonine, glycine, glutamate and aspartate. In one experiment using [65]Zn infused into the rumen until steady state reached, fractions were pooled, concentrated on uM-2 Amicon membranes and further examined by HPLC on reverse phase and ion exchange resins. At least 3 further components containing ^{65}Zn were obtained from the low MW fraction.

Table 1. Major metal-ion species in the 30,000 G supernatant in the rumen and abomasum identified by ion-association model.

Species	Rumen[a]	Abomasum[b]	% composition Species	Rumen	Abomasum
Ca^{2+}	55.7	88.0	K^+,Na^+	96	99
$CaHCO_3^+$	4.6	-			
$CaHPO_4^0,CaH_2PO_4^+$	19.2	9.4	Mn^{2+}	24.3	92
Ca (Acetate)$^+$	16.6	-	$MnHCO_3^+$	16.1	-
Ca (Citrate)$^-$	2.5	-	$MnHPO_4^0,MnH_2PO_4^+$	39.5	7.4
			Mn (Acetate)$^+$	13.9	-
Mg^{2+}	50	88	Mn (Citrate)$^-$	5.3	-
$MgHPO_4^0,MgH_2PO_4^+$	24.4	9.4			
Mg (Acetate)$^+$	18.8	-	Zn^{2+}	-	87.6
Mg Glycine	2.4	-	$ZnHPO_4^0,ZnH_2PO_4^+$	-	11.8
			ZnS (s)	100	-
Fe^{II} S (s)	100	63			
Fe^{III} (Citrate)	-	37	Cu_2^IS (s)	100	100

[a] pH 6.3, pe -3.5, I 01.6, pCO_2 0.185   [b] pH 2.4, pe - 0.1, I 0.10, pCO_2 0.003

Table 2. Comparison of chemico-physical methods to fractionate several metal ions from rumen 30,000 G supernatant.

	Ca	Mg	Cu	Zn	Fe
Total 30,000G (μM)	2794	2564	2.68	3.05	20.6
Diafiltration: [a]		% of total			
MW <10,000	86	90	59	57	13
MW < 5,000	84	90	26	57	16
Ion-exchange[b]					
Ag-Mp-50	95	84	nd	4	13
Chelex-100	73	63	15	54	24

[a] Dialysed against 12mM KH_2PO_4, 0.1M NaCl, 10mM NaN_3, pH6.5: cell volumes.
[b] Batch equilibration, 4 hrs - labile and moderately labile complexes exchanged. Distribution coefficient calculated using standards of metal ion as their chlorides - buffer as in diafiltration experiment. 20ml sample, 1g resin.

Because of gaps in our understanding of the relevant thermodynamic data for some metals in the digesta of rumen and abomasum, there are limitations in computer equilibria simulations. Nevertheless they provide information on important interactions. Current and future work will focus on further characterization of those fractions of Zn, Cu and Fe which are associated with organic moieties in both rumen and abomasum supernatant.

REFERENCES

1. Bremner, I. (1970) Zinc, copper and manganese in the alimentary tract of sheep. Br. J. Nutr. 24: 769.

Copper (continued)
  Metalloproteins, 675
  Metallothionein, 37, 39, 103,
        281, 287, 293, 295, 297,
        315, 579, 675, 677, 679,
        699, 711
  Milk, 445, 539, 595
  Molybdenum interaction, 303, 309,
        311, 313, 315, 317, 321,
        473, 541, 543, 619
  Plasma localization, 141
  Pancreas, 113, 381
  Reproduction, 185, 653
  Selenium, 657
  Sexual dimorphism, 117
  Status, 161, 163, 171, 185, 189,
        197, 383, 391, 435, 439,
        451, 461, 463, 473, 485,
        493, 539, 577, 663, 711
  Subcellular localization, 179,
        297
  Sulfur interaction, 543
  Superoxide dismutase, 103, 135,
        381, 435, 451, 473, 477,
        579, 613, 627
  Supplementation methods, 595,
        637, 639, 641, 653, 657
  Tissue levels, 311, 523, 535
  Toxicity, 179, 319, 405, 677, 699
  Transcuprein, 141
  Transport, 129, 141, 145, 405
  Zinc interactions, 173, 181, 235,
        281, 303

Development, 425, 433, 673, 679
Diabetes, 381, 389, 481, 633, 681
Dietary fiber, 229, 237

E. coli, 239, 301, 401, 405
Electron transfer, 23, 29
Endocrine function, 53, 55, 59, 103
Ephemeral fever, 403
Epidemiology, 1, 123, 149, 155,
        161, 163, 167, 171, 197,
        517, 519, 527, 697
Epistemology, 173
Exercise, 383, 391
Exocrine function, 43, 103

Fetal Alcohol Syndrome, 613
Fiber, 229, 237, 431, 487, 493
Fluoride, 273
Free radicals, 197, 359, 563

Gamma-ray emission analysis, 591,
        713
Gene-nutrient interactions, 203,
        239, 281, 287, 293, 299,
        301, 313, 323, 405, 435,
        523, 553, 557, 579, 679
Glucoronyl transferase, 471, 703

Glucose metabolism, 49, 113, 601
Glutathione peroxidase, 13, 19, 21,
        97, 189, 197, 245, 249,
        257, 345, 361, 461, 517,
        637, 639, 673, 717
Gold, 547

Heart disease, 357, 375, 453, 667
Hemachromatosis, 67, 439
Heme oxygenase, 683
Hepatitis, 711
Hormones, 43, 49, 53, 55, 59, 103,
        109, 113, 115, 117, 119,
        121, 123, 281, 385, 629,
        661, 671

Immunity, 85, 91, 101, 395, 397,
        401
Infant development, 189, 197, 211,
        215, 219
Infant foods, 189, 219, 511, 567,
        571
Inflammation, 97, 403
Iodine, 123, 661, 663
Iron
  Absorption, 189, 233, 407, 507,
        509, 577
  Bioavailability, 189, 413, 447,
        595
  Calcium interactions, 221
  Cellular uptake, 433, 625
  Chelation, 61, 67, 95
  Copper, 181, 231, 303, 309, 545
  Deficiency, mink, 577
  Developmental changes, 617
  Fiber interaction, 237
  Fortification, 509, 595
  Hepatocytes, 625
  Immunity, 85
  Infection, 95
  Interactions, 173, 181, 231, 233
  Manganese interaction 233, 555,
        557
  Metabolism, 433
  Milk, 189, 423, 445, 595
  Status, 163, 185, 391, 439, 463,
        487, 497, 571, 577, 663
  Toxicity, 67, 197
  Zinc interaction, 173, 181, 231,
        233, 303, 507, 567, 571,
        593

Kashin-Beck disease, 149
Kidney, 527, 529
Kwashiorkor, 197, 497

Lead, 73, 121, 365, 485, 529, 535,
        689
Lipid metabolism, 187, 235, 253,
        363, 365, 367, 371, 375,
        379, 497, 553, 627, 705,
        713

722